Lecture Notes in Electrical Engineering

Volume 133

Dehuai Yang (Ed.)

Informatics in Control, Automation and Robotics

Volume 2

Springer

Dehuai Yang
Huazhong Normal Universiy,
Hubei,
China, People's Republic
E-mail: dehuai2020@126.com

ISBN 978-3-642-25991-3 e-ISBN 978-3-642-25992-0

DOI 10.1007/978-3-642-25992-0

Lecture Notes in Electrical Engineering ISSN 1876-1100

Library of Congress Control Number: 2011923552

Typeset by Scientific Publishing Services Pvt. Ltd., Chennai, India.

Printed on acid-free paper

9 8 7 6 5 4 3 2 1

springer.com

Foreword

2011 3rd International Asia Conference on Informatics in Control, Automation and Robotics (CAR 2011) is held on December 24–25, 2011, Shenzhen, China.

The purpose of the 2011 3rd International Asia Conference on Informatics in Control, Automation and Robotics (CAR 2011) is to bring together researchers, engineers and practitioners interested in the application of informatics to Control, Automation and Robotics.

The conference is organized in three simultaneous tracks:"Intelligent Control Systems and Optimization", "Robotics and Automation" and "Systems Modeling, Signal Processing and Control". We welcome papers from all these areas.

The book is based on the same structure, and topics of particular interest include, but not limited to: Intelligent Control Systems and Optimization, Robotics and Automation, Signal Processing, Systems Modeling and Control, Communication Theory and Technology, Sensor and sensor network, Computer Theory and Technology, Electrical and Electronic Engineering, Multimedia and Network Information, Photonic and Optoelectronic Circuits, Communication Systems and Information Technology, Electrical and Electronic Engineering, Signal Processing

Although CAR 2011 receives 850 paper submissions, only 219 where accepted as regular papers, based on the classifications provided by the Conference Committee and the reviewers. The selected papers also reflect the interdisciplinary nature of the conference. The diversity of topics is an important feature of this conference, enabling an overall perception of several important scientific and technological trends. These high quality standards will be maintained and reinforced at CAR 2011, and in future editions of this conference.

Furthermore, CAR 2011 includes 2 plenary keynote lectures given by internationally recognized researchers. They are Prof. Patrick S.P. Wang, Ph.D., Fellow, IAPR, ISIBM and WASE Northeastern University Boston, and Prof. Jun Wang, Department of Mechanical & Automation Engineering, The Chinese University of Hong Kong, Shatin, New Territories, Hong Kong. Their presentations presented an important contribution to increasing the overall quality of the conference, and are partially included in the first section of the book. We would like to express our appreciation to all the invited keynote speakers who took the time to contribute with a paper to this book.

On behalf of the conference organizing committee, we would like to thank all participants. First of all to the authors, whose quality work is the essence of the conference and to the members of the program committee, who help us with their expertise and time.

As we all know, producing a conference requires the effort of many individuals. We wish to thank all the people from IERI, whose work and commitment are invaluable.

Committee

Honorary Chair

ChinChen Chang	National Chung Hsing University,Taiwan
Patrick S.P. Wang	Fellow, IAPR, ISIBM and WASE Northeastern University Boston, USA

Program Co-chairs

Yuntao Wu	Wuhan University of Technology, China
Weitao Zheng	Wuhan University of Technology, China

Publication Co-chairs

Honghua Tan	Wuhan Insitute of Technology, China
Dehuai Yang	Huazhong Normal Universiy, China

International Committee

Wei Li	Asia Pacific Human-Computer Interaction Research Center, Hong Kong
Xiaoming Liu	Nankai University, China
Xiaoxiao Yu	Wuhan University, China
Chi Zhang	Nanchang University, China
Bo Zhang	Beijing University, China
Lei Zhang	Tianjin Institute of Urban Construction, China
Ping He	Liaoning Police Academy, China
Alireza Yazdizadeh	International Science and Engineering Center, Hong Kong
Wenjin Hu	Jiangxi Computer Study Institute, China
Qun Zeng	Nanchang University, China

Contents

Robotics and Automation

Simulation of Smog Diffusing Based on Particle Swarm Optimization

Jianfeng Xu[1], Lan Liu[3], and Qun Liu[2]

[1] College of Software, Nanchang University, Nanchang 330047
jianfeng_x@ncu.edu.cn
[2] Chongqing University of Posts and Telecommunications, Chongqing 400065
[3] Department of Computer Science & Application Technology,
Nanchang University, Nanchang 330031

Abstract. In the field of Computer Graphics the simulation of irregular fuzzy object still is a difficult challenge. Based on particle swarm optimization, the method of combining smog particle density diffusion and inter particles collision each other was studied in this research, which simulates the diffusing and real-time performance of smog. Diffuse force field was established by the diffusion equation, which could ensure the accuracy of the trajectories of diffusion. In order to reduce the time on collision between particles each other, a collision technology based on spatial hashtable is introduced in the research. Experimental results via simulation experiment show that the proposed method in the paper not only can improve the simulation speed of real-time requirements, but also demonstrate the authenticity of the smog diffusion.

Keywords: Particle swarm, Diffusion equation, Spatial hashtable, collision detection, Real time.

The simulation of irregular fuzzy object such as smog is one of the most difficult challenge in the field of Computer Graphics. It can't be accurately described in classic geometry, because of irregular shape and complicated transformation. However, with the development of the requirement increaseing of fuzzy object emulation in the filed of Moves, Games and Computer Cartoon,a lot of mew simulation methods come forth. The method based on particle swarm (a technology of simulate fuzzy object motion, which was proposed by Reeves [1] in1983) is used in many simulation of irregular fuzzy object . In the theory, object can be defined as particle set which is consist of thousands of irregular and random distribution particles. Every particle has individual attributions such as color, position, size and speed .And at every moment it consecutive moves and changes forms .In addition, it has a certain lifecycle which includes the process of birth, activity and death. So Particle Swarm can represent the modality of irregular fuzzy object just as motility, randomicity and the Visual authenticity.

The simulation of irregular fuzzy object based on Fluid function is another important method, which can display the authenticity of moving object by solving fluid equation.

Stam [2] combined semi Lagrangian method to implicitly resolved NS function, this method improved the development of the Fluid third dimension simulation. At the same time, Fedkiw [3]and Selle [4] are the outstanding representative of using

D. Yang (Ed.): Informatics in Control, Automation and Robotics, Volume 2, LNEE 133, pp. 1–8.
springerlink.com

hydrodynamics to simulate the movement of smog atoms. their job of simulant smog fuid has strong third dimension. Traditional method technology based on Particle Swarm Optimization can not solves the occasion of accuracy request [5,7]. Similarly, the method based on fluid function need calculate a great deal of function. Although, the method such as gridding and interpolation can improve the compute speed, the real-time reflection still is the bottleneck of the method using currently hardware. So it is important that combining the method of particle swarm and other physics models for the simulation of irregular fuzzy floating object.

After that, this paper imports a method based on space hash table [6], which is used to collision detection inside a deformable objects such as smog particles, to improve the efficiency of real-time reflection. While this simple and effective method be executed, the accuracy of collision simulation can be adjusted to accord the system's need and the time of collision detection can be greatly reduced.

Furthermore, the effect of density diffusion force is added to this paper to simulating the diffusion of smog accurately .The model ,which combined with the technology of particle swarm and other physics models, can balances the physical authenticity and the real-time ability of simulation . Finally this research gets better remarkable effect of simulation.

1 Density Diffusion

Suppose there is no effection of other factors such as earth and wing, the smog's diffusion in the endless space obeies the rule of heat exchange. Because the density of smog changes along with the diffusion, then the function of the density change can be expressed as Equation (1):

$$\begin{cases} \frac{\partial C}{\partial t} = k\left(\frac{\partial^2 C}{\partial x^2} + \frac{\partial^2 C}{\partial y^2} + \frac{\partial^2 C}{\partial z^2}\right), (-\infty < x, y, z < \infty, t > t_0) \\ C(x, y, z, t_0) = C_0(x, y, z) \end{cases} \tag{1}$$

In Equation (1): C as the density scalar quantity of density field, k as the diffusion coefficient, t0 as the time when the particle is created. Since function (1) just simulate the change of the density when the particle is created, C_0 should be zero as Equation(2).

$$C_0(x, y, z) = 0 \tag{2}$$

Then Equation (3) is the solution of equation (1) based on Equation (2), which was resolved by Jos Stam [5] and his partners.

$$C(x, y, z, t) = \frac{1}{(2k\sqrt{\pi(t - t_0)})^3} e^{-\frac{(x-x_0)^2+(y-y_0)^2+(z-z_0)^2}{4k^2(t-t_0)}} \tag{3}$$

In this Equation (3) (x_0, y_0, z_0) as the position of the smog source.

If $\begin{cases} l = t - t_0 \\ r = \sqrt{(x - x_0)^2 + (y - y_0)^2 + (z - z_0)^2} \end{cases}$

And l represents the currently particle alive time (life value currently), r as the distance from the particle to the source. So the equaption (3) can change to equation (4)

$$C(r,l)=\frac{1}{(2k\sqrt{\pi l})^3}e^{-\frac{r^2}{4k^2 l}} \quad (4)$$

It is a solution of equation (4), that the isosurface of density is a sphere which has equal radius, and the density continues degression with the enhance of the radius. As shown in Figure 1, the smog is diffusing from high density to low density.

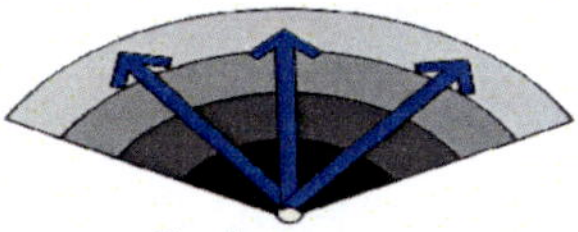

Fig. 1. Diffusion of smog

Suppose there is a smog particle in air, the distance from current position to the source is $r_{1,}$ the current life-value is l_1, the distance from the position of prevframe to the source is r2, the life-value is l2, Next the density difference equation between the two frames should be Equation (5).

$$\Delta C = C(r_2, l_2) - C(r_1, l_1) \quad (5)$$

Based on ideal gas-state Function: $PV = nRT$, the pressure and the density are in the direct ratio, on the contrary the density difference will bring pressure and stress. For convenience, suppose the diffusing stress brought from density difference is Equation (6).

$$f_{diffusion} = \Delta C \times s \quad (6)$$

S is the constant which is proposed by experience value.

Since the smog diffuses along with the every direction of the sphere radius, the direction of $f_{diffusion}$ points from the source to the currently position. Though the power of density diffusion is more small than other powers, the effect to the track of smog movement is still important. So, if the power of the density diffusion be added, the stimulation of smog diffuse should be more true.

2 Inside Collision Detection

In the movement of smog particles, the collision between the particles can't be negatived. And the powers is the main reasons of smog diffusion which are produced from the collision. Obviously, the collisions between every two particles will spend great deal of times. Moreover the real-time requirement is difficulty to get, especially the number of the particles is large. So disposing collision detection and collision response is a very difficult question in computer graphics. Morever, it still is the focus of computer graphics nowadays that how to abbreviate the time of collision detection and how to move of particles after collision .Thus, in the paper, Space Hash-Table and the Impulse theory are introduced to detect collision, and to accurately work out the particles moving states happened after the particles collision.

2.1 Space Hash-Table

Just like simple hash-table, space hash-table sets up a mapping relation between the data-entity and the data-address. using space hash-table, three-dimensional space can be divided into many little three-dimensional gridding. Moreover, the smog information of the system should be stored in the little three-dimensional gridding, and the index code of the three-dimensional gridding just as the smog particle memory address. For example, as shown in figure2, the two-dimensional space be divided into, 0、1、2、3, four little grid. the index codes of particles, A、B、C、D、E、F、G、H, which stay in the two-dimensional spaces are reflected by hash-table.

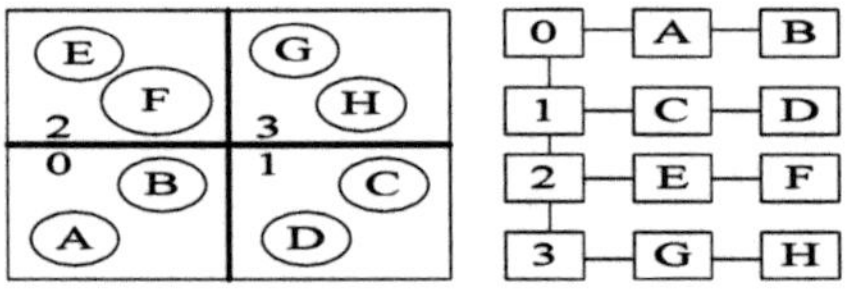

Fig. 2. Two-dimensional space and hash-table

2.2 Diffusion of Particle Space

In three-dimensional space, the hash-table key data h is the reflection of a three-dimensional particle(x,y,z): $h = hash(x, y, z)$. Then, this paper sets a special hash function according to the shape of smog. which can improve the average degree of diffusion. Since the direction of smog diffuses is from down to up, the Y-shaft has more long span than that of X-shaft or Z-shaft. Therefore, Y-shaft can be divided into several layers and each layer can be divided into four sections.

$$h(x, y, z) = D(y) \times 4 + D(x, z)$$

Thereinto,

$$D(y) = \begin{cases} 0 & y \le L \\ \lfloor y / L \rfloor & L < y < L \times (n-1) \\ n-1 & y \ge L \times (n-1) \end{cases}$$

$$D(x, z) = \begin{cases} 0 & x \ge 0, z \ge 0 \\ 1 & x < 0, z \ge 0 \\ 2 & x < 0, z < 0 \\ 3 & x \ge 0, z < 0 \end{cases}$$

In the Equation given above, L as the altitude of each layers, n as the number of layers. While this equation executing, particles can be equably distributed into hash-table and the hash collision can be reduced.

If the data structure of the above equation be defined, $c_i (0 <= i < n)$ should be the three-dimensional space whose key-data is i, S as the particles numbers in the space, and $p_{i,j}(0 <= j < m)$ as the first j particle in the first i gridding.

Similarly, every particles need some definite space, so particle border should be taked into account when particles distributing. Since particles can be regarded as a 3D square, the eight vertex of the 3D square should be reflected into hash-table which be considered as entire particle. So a particle maybe holds more than one three-dimensional gridding, and reflects more than one hash key data. Figure 3 is the example of the theory with two-dimensional space. Particle s holds four space of gridding such as 1、3、4、6, so the four gridding be reflected to hash-table as the entire particle S.

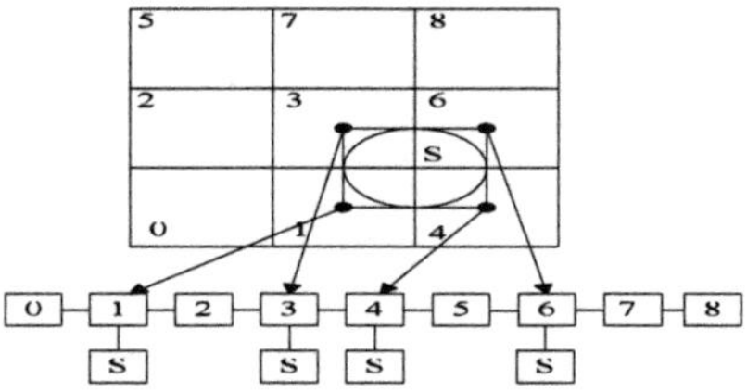

Fig. 3. A particle in several space gridding

2.3 Collision of Each Particles

In order to reduce the complexity of measuring the collision of each particles, this paper just detections this particles which in same gridding, namely, have same hash-table key.

If Particle p both in space C_i and C_j, the measuring to the collision of b should be separate into two sections include the collisions happens in C_i and it happens in C_j. So the pseudocode of colliding each Particle should be expressed as:

```
for (int i =0 to n-1)                    // Traversing from C_0 to C_{n-1}
{ for (int j =0 to sizeof( C_i )-1 )      // Traversing each griddings
    for (int k =0 to sizeof( C_i )-1)    // Measuring each two Particles in one gridding.
{   if ( p_{i,j} != p_{i,k} )
        Collision detection ( p_{i,j} , p_{i,k} );    //Measuring
    }
}
```

As figure4 showing, particle S1 collide with particle S2, the movement speeds of the two particles S1 and S2 are v_{1-} and v_{2-} before collision, then the speeds change to v_{1+} and v_{2+} after collision. And parameter n is the unit vector of collision action line.

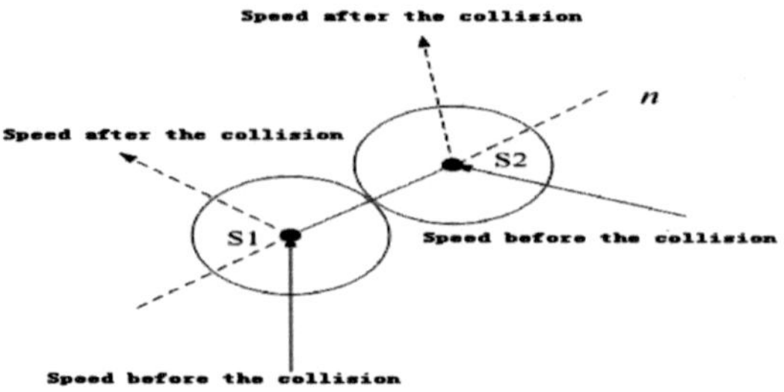

Fig. 4. Collision of two particles

For example there is a collision in one-dimensional space, the speeds of particles after collision can be resolved by momentum conservation theorem. According to condition, there are three equations:

$$J = m_1(v_{1+} - v_{1-}) \tag{7}$$

$$-J = m_2(v_{2+} - v_{2-}) \tag{8}$$

$$e = -(v_{1+} - v_{2+})/(v_{1-} - v_{2-}) \tag{9}$$

Equation (7) and (8) are impulse theorem, Equation (9) is function of resuming parameter. In above equation, m1 and m2 separately express the quality of S1 and S2, J as the impulse data by collision, e is the resuming parameter of collision which is a constant depended on material and structure of object, while the impulse direction of S1 is contrary for S2. Equation (7) and (8) can be changed to Equation (10) and (11).

$$v_{1+} = J/m_1 + v_{1-} \tag{10}$$

$$v_{2+} = -J/m_2 + v_{2-} \tag{11}$$

Then, v_{1+} and v_{2+} be substituted to Equation (9). Equation (12) should be educed.

$$J = -(v_{1-} - v_{2-})(e+1)/(1/m_1 + 1/m_2) \tag{12}$$

And the speeds of two particles after collision should be resolved when equation (12) be substituted to equation (10) and (11). But, in three-dimensional space, speed should be changed to vector, only the speed changing should produce impulse of which the direction is the same as the normal line, and only the component of the impulse can influence the speed. So the equation (12) can be changed to equation (13).

$$J = \frac{-(v_{1-} - v_{2-})(e+1) \bullet n}{1/m_1 + 1/m_2} \tag{13}$$

And equation (10) and (11) can educe the equation as follows.

$$\begin{cases} v_{1+} = J \bullet n / m_1 + v_{1-} \\ v_{2+} = -J \bullet n / m_2 + v_{2-} \end{cases}$$

Finally, when above equation be substituted equation (13), the speeds after collision, v_{1+}, and v_{2+}, can be educed.

$$v_{1+} = v_{1-} - \frac{(v_{1-} - v_{2-})(e+1)}{1 + m_1 / m_2}$$

$$v_{2+} = v_{2-} + \frac{(v_{1-} - v_{2-})(e+1)}{1 + m_2 / m_1}$$

In addition, after collision, the two particles consecutive movement with the speeds of v_{1+}, and v_{2+}, then the two particles collide with other particles in the same one frame , and the collision should be detectiond sequentially .

3 Simulation Experiment

In the phase of simulation, smog particles be Rendered based on particles swarm optimization method union texture mapping and billboard technologys. In this way, a texture piece can instead of many particles, so computing resource be saved and the rendering speed be improved. In this paper, the texture with size of 40×40 be used to particle map, and simulating big particles be adopted.

This algorithm had been realized by use Personal computer. Hardware environment include CPU : INTEL E2180, graphics card: Gforce 7300GT 128M, EMS memory : 1GB RAM. R&d platform is VS2005 C++. Figure 5 is the diffusing phenomena of smog which be affected by random wind. Figure 6 is the exhaling phenomena of smog, which be produced by the collision of smog and sphere in space.

Fig. 5. Smog freely diffuse in the wind

Fig. 6. Smog diffuse while the particles collide in space

Table 1. The comparison results of the algorithms

Emission rate of particle（A/SEC）	Frames per second of this algorithm	Frames per second of this traditional Particle System algorithm
10	72	46
20	60	32
30	51	22
40	43	17
50	38	14

Finally, this algorithm be compared with particle system method which be proposed in reference [1]. In the experiment, 32 grids be adopted in hash-table, the height of 3D grid is $L = 100$, the number of layers is n =8, the recovery coefficient of collision is $e = 0.5$ .The proportional constant of consistenceis diffusive force is $s = 0.1$. The life cycle of particle between 6 sec to 8 sec. The comparison results be showed in Table 1.

As Table1 shown, the algorithm of this paper is faster than traditional particle system algorithm .Similarly the more numbers of particles in the system, frame rate more than 20, the less real-time could be achievedwhile traditional particle system algorithm be used, however the algorithm of this paper can get better performance.

4 Conclusion

This study of simulating the smog particle diffusion is based on the particle swarm technology, Kinetics and the method of collision detection based on space hash-table. And the effect of density diffusion force is added to the scene.

The result of the above method has better performance such as Real-time, physical authenticity and visual authenticity than traditional method. The main research of the next step includes two main aspects. firstly, there is a interesting study of polymerization and deformation effect after the collision of particles . Secondly, the photosensitive photos of smog particles should be simulated, so that the emulation effect may be improved. Above research can be widely used to environment detection, air pollution index measure, etc.

References

1. Reeves, W.T.: Particle Systems–A Technique for Modeling a Class of Fuzzy Objects. Computer Graphics 17(3), 359–376 (1983)
2. Stam, J.: Stable fluids. In: Proceedings of SIGGRAPH, Los Angeles, California, USA, pp. 121–128 (1999)
3. Fedkiw, R., Stam, J., Jensen, H.: Visual simulation of smoke. In: Proceedings of SIGGRAPH, Los Angeles, California, USA, pp. 15–22 (2001)
4. Selle, A., Mohr, A., Chenney, S.: Cartoon rendering of smoke animations. In: Proceedings of the 3rd International Symposium on Non-Photorealistic Animation and Rendering, Annecy, France, pp. 57–60 (2004)
5. Zhuo, N.-W., Rao, Y.-B.: Real Time Dense Smoke Simulation Based Particle System. In: IITAW 2008, pp. 809–813 (2008)
6. Teschner, M., Heidelberger, B., Muller, M., Pomeranets, D., Gross, M.: Optimized spatial hashing for collision detection of deformable objects. In: Proceedings of Vision, Modeling, Visualization, VMV 2003, pp. 47–54 (2003)
7. Zhang, Q., Xie, J., Wu, H.: Of such of Research review of Modeling method for irregular objects: Fire, smoke and clouds China. Image and Graphics Academic Journal 5(3), 186–190 (2000)
8. Zhan, Y., She, M., et al.: Real time cartoon smog simitation algorithm based on physics model, China. Image and Graphics Academic Journal 12(2), 261–265 (2007)

Self-compensating Technology for Disturbing Torque of Space Swing Camera

Wang Xiaoyong

Northwestern Polytechnical University, Xi'an, 710072
w8320@126.com

Abstract. The disturbing torque generated by space swing camera during imaging affects the attitude stabilization of satellite. To solve this problem, self-compensation technology for reducing disturbing torque of camera was proposed. By using self-compensating equipment and synchronization control technology, 90% of the disturbing torque generated by camera can be reduced, hence the significant reduction in its influence on the satellite stability. Through simulation and testing, the technology was proved to be able to reduce the disturbing torque by camera significantly and meets the requirement of satellite attitude control.

Keywords: swing camera, disturbing torque, self-compensation, satellite attitude.

1 Introduction

Satellite-carried swing camera uses lateral scanning and imaging technology to fulfill the extra-wide land coverage. The optical system of the camera adopts the all reflecting optical system with little angle of field and small relative aperture, with built-in scanning reflection lens in front of the camera's optical system. The camera provides enlarged scanning coverage through lateral scanning along the path vertical to the satellite flight direction. In order to compensate the velocity of image motion produced by the satellite forward movement during imaging process, the camera's optical axis needs to process image compensation along the opposite direction of the satellite flight trajectory, as illustrated in Fig. 1:

When a satellite flies along its trajectory, a complete action cycle is indicated from point A to point D. From A to C, the camera scanning reflection lens rotates with constant speed along x-axis (satellite flight direction) to scan from left to right during imaging process; From B to C, the camera produces image of the right area; From C to D, the camera reflection lens rotate in opposite direction along x-axis until the scanning reflection lens backs to the original position, which concludes a complete working cycle.

A satellite uses three axis stable attitude control mechanism. Its attitude stability requires to be 0.005°/s (3σ). Due to the disturbing torque generated during the scanning process by scanning reflection lens that hampers the attitude stability as well as image quality, it is necessary to compensate the disturbing torque generated in scanning reflection lens movement.

D. Yang (Ed.): Informatics in Control, Automation and Robotics, Volume 2, LNEE 133, pp. 9–17.
springerlink.com

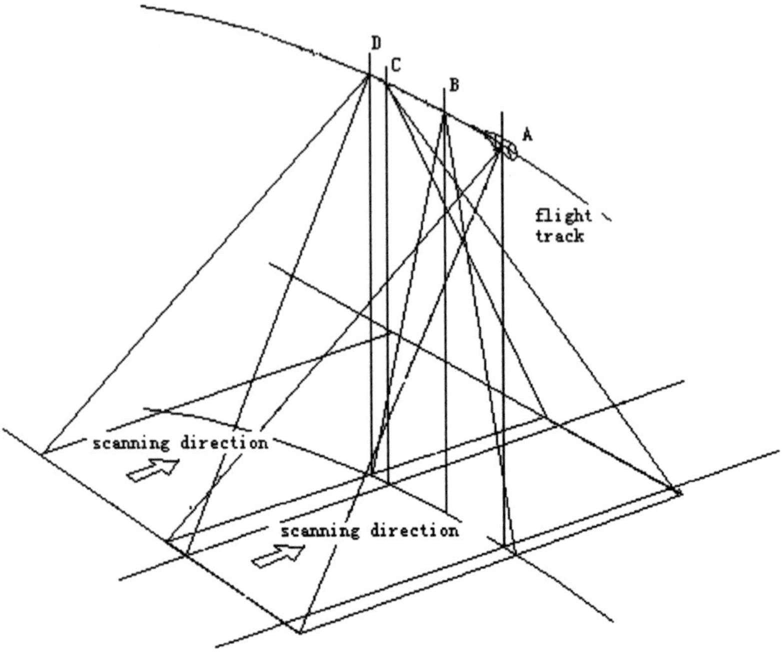

Fig. 1. Scan imaging mechanism illustration

2 The Disturbing Torque Self-compensation Proposal

The printing area is 122 mm × 193 mm. The text should be justified to occupy the full line width, so that the right margin is not ragged, with words hyphenated as appropriate. Please fill pages so that the length of the text is no less than 180 mm, if possible.

Swing camera is tilt positioned on the satellite to fulfill image compensation. The scanning axis of the scanning reflection lens was positioned at the angel of α=3.6340 in plane X_bOY_b, inertia of the scanning lens along the rotation axis J=0.035kgm2, the damped coefficient is 0.0188kgm2/s. The camera scans along with single axis cyclically, with cycle time of 6.28s. The scanning movement equation is as below:

$$\omega_x=10.396t \qquad 0\leqslant t\leqslant 0.205s \tag{1}$$

$$\omega_x=2.151 \qquad 0.205s\leqslant t\leqslant 4.205s \tag{2}$$

$$\omega_x=2.151-10.296(t-4.205) \qquad 4.205s\leqslant t\leqslant 5.345s \tag{3}$$

$$\omega_x=-9.693+6.997(t-5.345) \qquad 5.345s\leqslant t\leqslant 6.28s \tag{4}$$

The disturbing torque generated by camera scanning lens movement is illustrated by Fig.2:

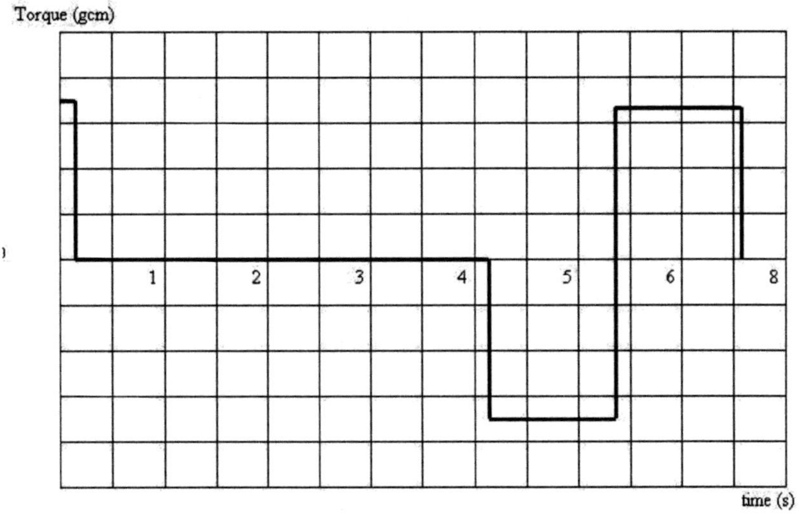

Fig. 2. Camera scanning disturb torque curve

With the anti-disturbing torque equipment installed on the camera, when scanning lens are moving, the anti-disturbing torque equipment generates the disturbing torque opposite to the scanning lens movement direction. By doing so the self-compensation is achieved.

Camera disturbing torque self-compensation mechanism uses electrical synchronous mechanical torque compensation technology through the shafting of the torque compensation momentum wheel installed on the camera. The rotation inertia of scanning mirror equals to that of momentum wheel carried on torque self-compensation equipment. The torque compensation equipment leverages the same kind of motor that has the similar torque feature curve observed in the scanning driving motor, is driven by the same source of power, and has the same damping torque. The control power drives the opposite direction movement of scanning mirror and momentum wheel. When the motor driven camera scanning mirror makes the swing movement of±4.5°, the momentum wheel of the torque compensation system does the same pendular movement of±4.5° in the opposite direction, hence the torque compensation is achieved via torque balancing, resulting in the elimination of most of disturbance generated by the pendular movement of camera, and rendering less than 10% compensation accuracy variation.

In simulation of total output torque of the camera with torque compensation, assume compensation accuracy variation is less than 10% and response time no greater than 10ms, the output disturbing torque after compensation is charted in Fig. 3.

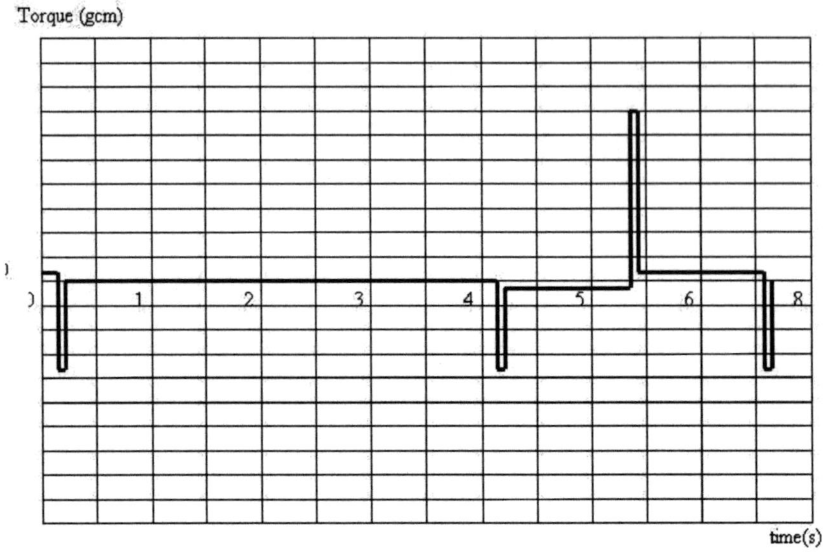

Fig. 3. Residual Torque Curve After Camera Compensation

3 Satellite Attitude Mathematical Simulation Analysis

The following two steps are used to analyze the mathematical simulation of the satellite attitude with the existence of disturbing torque:

Firstly, without the consideration of disturbing torque, analyze the impact on satellite attitude based on the established model parameters of camera scanning equipment.

Secondly, with the consideration of disturbing torque, analyze the impact on satellite attitude based on the established model parameters of camera scanning equipment and torque compensation equipment.

Based on the general satellite feature, assume the following satellite quality parameters:

Weight M=600kg, rotation inertia Ixx=193 kgm2, Iyy=70 kgm2, Izz=204 kgm2 Based on the camera structure feature, assume the camera scanning reflection lens are ellipse, long axis =220mm, short axix=140mm, weight=0.98kg, rotation inertia=0.04 kgm2.

3.1 Simulation Analysis without Self-compensation

Based on the simulation of the scanning movement according to the given ideal movement equation without camera torque self-compensation, satellite attitude stability during camera imaging process is charted in Fig. 4. It is shown that rotation angel speed has been impacted significantly. rotation angel speed on steady state was larger than 2×10-3o/s. The fluctuation was obviously observable.

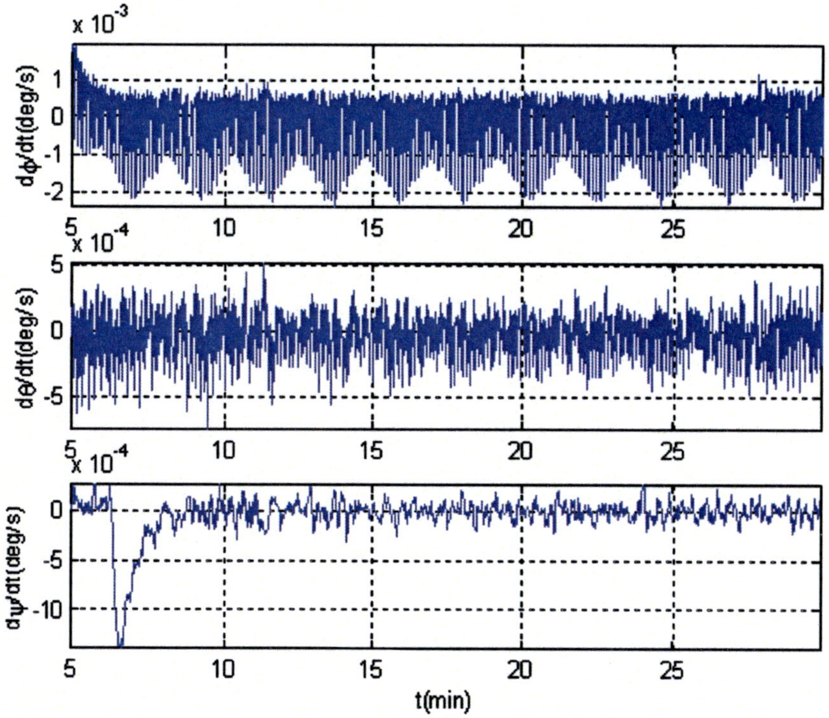

Fig. 4. Satellite Attitude Angel Speed Simulation Curve (without scanning torque self-compensation)

3.2 Simulation Analysis with Self-compensation

With torque self-compensation, due to the synchronization phase difference existent in scanning and compensation, the torque on x-axis of the satellite in a short period of time was 0.0086×2=0.0172Nm. In the simulation, the phase difference of camera pendular movement and compensation driving movement was set up as 10ms. Fig. 5 shows the satellite attitude simulation curve. Compared with the result without self-compensation, the satellite attitude angel speed has improved significantly. The rotation angel speed on steady state was less than 5×10-4o/s.

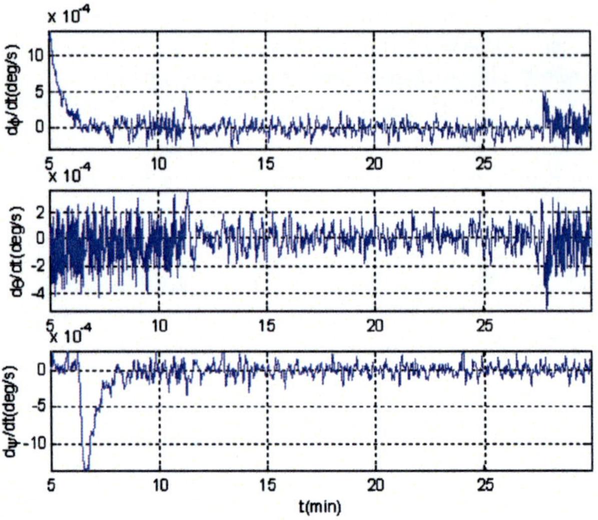

Fig. 5. Satellite Attitude Angel Speed Simulation Curve (with scanning torque compensation)

4 Disturbing Torque and Self-compensation Effect Test

To prove the effectiveness of anti-torque compensation equipment technology, camera torque compensation test was conducted. To precisely capture the impact of scanning movement on satellite attitude, the scanning torque disturbance was measured on a single-axis flotation physical simulation platform. The measurement was charted in the Fig.6. The chart shows that the disturbing torque generated in the camera movement without self-compensation equipment installed is no greater than 8.5 gcm.

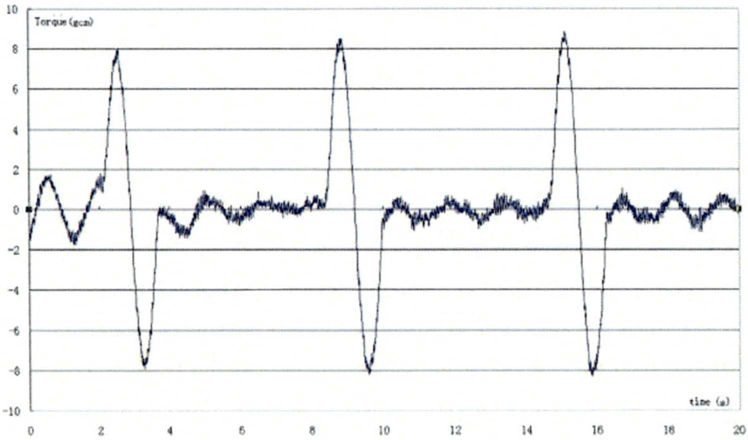

Fig. 6. Camera Disturbing Torque Measurement (without self-compensation)

Further measurement on disturbing torque was conducted when self-compensation was adopted. The measurement result is shown in Fig.7. The chart shows that the disturbing torque generated in the camera movement with self-compensation equipment installed is no greater than 2 gcm.

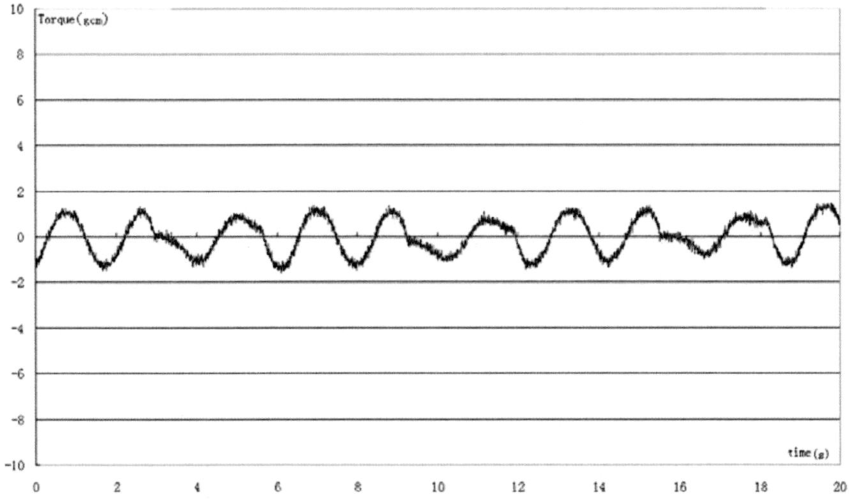

Fig. 7. Camera Disturbing Torque Measurement (with self-compensation)

After the camera scanning torque disturbance test was completed and the relevant data was captured, a mechanics coupling physical simulation test of swing camera and satellite control system was implemented. The test system is composed of wide-coverage camera system, satellite attitude control system, momentum wheel, satellite rotation inertia analogy module, etc.

During the test, after the camera's scanning movement, the attitude control systems controls the satellite according to the given parameters. Fig.8 shows the test result. In the diagram, horizontal axis means time, the three graphs on the right hand side show 1) offset of satellite attitude angel, 2) offset of satellite attitude angel speed, and 3) offset of rotation speed of the satellite control system execution element -- momentum wheel, respectively.

It is observed from the measurement curve that with camera self-compensation, under the impact of disturbing torque generated in camera imaging process, the satellite attitude angel accuracy variation is less than 0.01 °; the attitude angel speed control accuracy variation is less than 0.001 °/s, which has met the requirements that 1) satellite control attitude stability should be no greater than 0.005°/s, 2) has no impact on imaging process, and 3) momentum wheel is within the saturation control range.

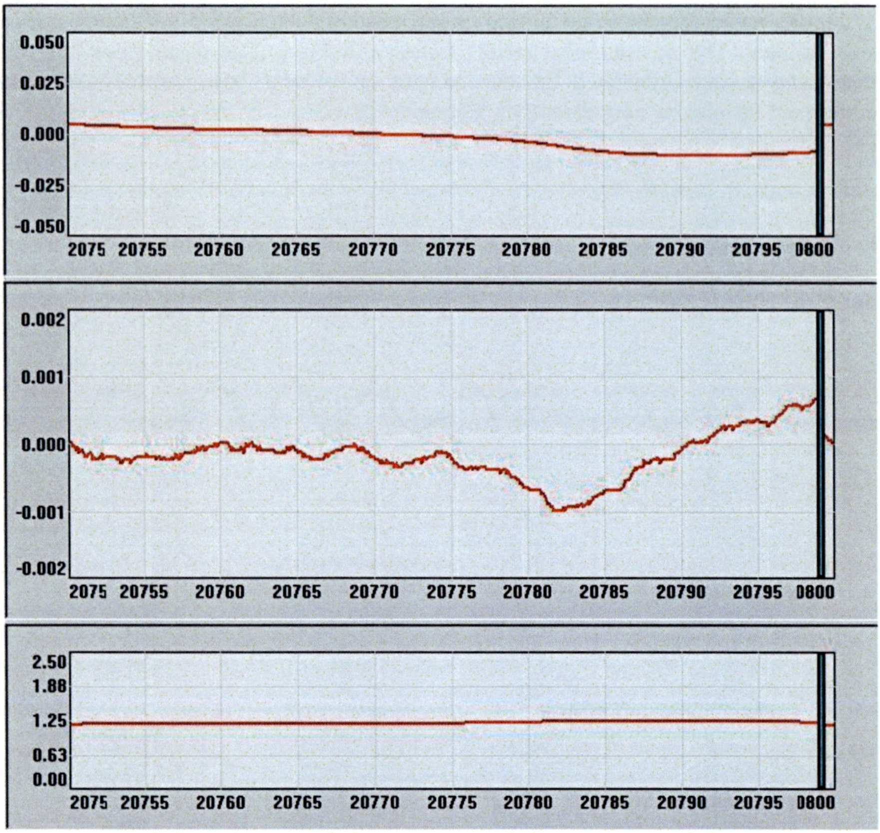

Fig. 8. Mechanics Coupling Physical Simulation of Camera and Satellite Control System Test Result

5 Conclusion

Satellite carried space swing camera influences the satellite flight attitude during the imaging process, which in turn influences the image quality. In response to this problem, a self-compensation solution by installing anti-torque compensation equipment on the camera was proposed. To prove the effectiveness of the torque compensation, mathematical simulation and physical simulation testing were implemented to measure the fluctuation generated by camera scanning movement before and after the compensation equipment was installed. The test result shows the compensation effectiveness of greater than 90%, and the residual torque disturbance reduction to less than 10% of its original level. The technology has met the stability requirement. With proven feasibility, it can be applied to all kinds of satellite space swing cameras.

References

1. Liu, W., Yao, D.-Z., Hua, Y.-Y., Wu, J.-W.: The Design of TDI CCD Camera System. Infrared Technology 32(9), 505–508 (2010)
2. Fan, C., Li, Y.-C., Yi, H.-W.: Influence Analysis of Buffeting on Image quality of TDICCD Camera. Acta Photonica Sinica 36(9), 1714–1717 (2007)
3. Fan, C., Li, Y.-C., Yi, H.-W.: Influence analysis of drift angle on image quality of TDICCD camera. Opto-Electronic Engineering 34(9), 70–73 (2007)
4. Zhang, Y.-X., Liu, Y., Ge, W.-Q.: Development and prospect of image motion compensation technology. Chinese Journal of Optics and Applied Optics 3(2), 112–118 (2010)
5. Liu, M., Wu, H.-S., Kuang, K.-P., Xiu, J.-H.: Image motion compensation in aerospace camera. Optics and Precision Engineering 12(z2), 30–32 (2004)

Matrix Multiplication Parallelization on a Many-Core Platform

Pollawat Thanarungroj and Chen Liu

Department of Electrical and Computer Engineering
Florida International University Miami, Florida, USA
{pthan001,cliu}@fiu.edu

Abstract. This paper introduces an approach to analyze the power and energy consumption of a many-core system. The investigation has been done by using the Intel SCC system as an experimental platform. The approach is to collect the time and power profiling of an executing application on the Intel SCC system. And then, we find the total energy consumed for the entire execution. We studied the effects of power and energy consumption in many-core systems by varying different hardware configuration parameters such as number of cores, clock frequency and voltage level. Thus, the many-core system can be explored for its scalability, fitness in operational cost and performance.

1 Introduction

Many hardware techniques have been proposed to increase the parallelization of these transistors. Lots of hardware implementations are already existed for parallelizing sequential program execution such as Superscalar, Superpipelining, Simultaneous Multithreading (SMT)[1-2], Chip Multiprocessors (CMP) [3] (also known as multi-core processors) and most recently, many-core processors.

There are advantages that multi-core and many-core gain while single core processor does not [3-5, 16-18]. Both multi-core and many-core processors gain the benefit from shorter wiring which minimized the delay among cores instead of going off-chip. Similarly, in multi-core and many-core processors power and energy consumption increase linearly when the number of cores increases; while increasing the complexity in single processor design, it consumes a quadratic increase or even more in power and energy consumption. Moreover, the use of identical processing elements in homogeneous architecture reduces the complexity of hardware design and verification hence the entire development cycle.

The Single Chip Cloud Computer (SCC) experimental processor [6-10] is a 48 core 'concept vehicle' created by Intel Labs as a platform for many-core software research. This SCC system is a second generation processor design that has been successfully developed by Tera-Scale Computing Research Program. SCC is the microprocessor system that has the highest number of cores integrated onto a single chip which intended to encourage more researches on many-core processor research and parallel programming research. The many-core processor researches [11] are high-performance power-efficient fabric, fine-grain power management and message-based programming support.

D. Yang (Ed.): Informatics in Control, Automation and Robotics, Volume 2, LNEE 133, pp. 19–26.
springerlink.com

SCC is a research platform that allows voltage and frequency scaling. Thus, it is possible to derive the power and energy consumption characteristics of the SCC by applying different frequency and voltage configuration to the different cores. We observe the power and energy scalability of the SCC system by executing a program with different numbers of cores. By performing such experiment, we can also verify that the message passing interface can hold the power and energy consumption scalability.

The rest of the paper is organized as follows. In Section 2 we discuss on our matrix multiplication test program. In the later section, we present our results from varying the number of cores and varying two different frequency and voltage level of all the cores. Finally, a conclusion is presented in Section 4.

2 Experimental Setup

In our experiment, we implemented our work on SCC system with SCCKit version 1.4.0. SCC system setup consists of two components, one is SCC itself and the other one is called Management Console Personal Computer (MCPC). MCPC is installed with a 64-bit Linux operating system, Ubuntu 10.04 in this case. MCPC system supports basic C programming language compilers such as gcc, g++, icc and icpc. Moreover, MCPC also supports the need of FORTRAN compiler ifort and MKL compilers mkl. The executables are generated by MCPC and run on the SCC cores.

We implemented a mixed sequential and parallel workload program that performs a multiplication of two matrices (C=A*B). We tried to imitate the program behavior according to the original program in [12, 13]. This program is written based on C programming language. To calculate the value of C=A*B in parallel the program executes as follows. First, we assign one of the cores as root core and other cores as computing core. At the beginning of the program, the root core creates matrix A and matrix B. The value of the elements in matrix A are calculated by summing up the x-y coordinates of the matrix. The value of the elements in matrix B are calculated by multiplying the x-y coordinates of the matrix. Then, the root core divides matrix A into different number of rows and distributes the fragments of matrix A to other executing cores. Root core also sends the information of the offset of the fragmented matrix A, number of rows of matrix A and the entire matrix B to be multiplied with. This centralized communication within the root core is the sequential part of the program execution. To plot the experimental data, the longer program execution yields a clearer graphical plot. Therefore, in this experiment, we assigned 3,000 rows in matrix A and 150 columns in both matrix A and matrix B. The matrix multiplication program has been implemented on icc compiler version 8.1.038 with the optimization flag O3. Only the matrix multiplication calculation itself can be computed in parallel but there are overheads that cannot be parallelized such as the overheads from data communication, conditional loop, timer, measurement and results collection. An execution flow of our matrix multiplication program is illustrated in Fig. 1.

RCCE [14-15] is a specific API library for programming the message passing functions in SCC system. RCCE library has been inserted into the matrix multiplication program to adjust the frequency and voltage level on the SCC cores. There are two sets of setups in our experiment. First, we set the frequency and voltage of every core to 800 MHz and 1V respectively. Second, we set the frequency and voltage of every core to 400 MHz and 0.7V respectively. These settings apply to all the computing cores and also the root core. By varying the number of cores, we are able to observe time, power and energy characteristics. Because not all cores are involved in the computation, there will be idle cores which are needed to be taken into consideration. Due to the fact that we cannot completely turn off any core even it is in an idle state, we put the idle cores into the lowest stable operational frequency of 200 MHz with a supplying voltage of 0.6V. The operating frequency was fixed at the beginning of the execution. Thus, the program was executed at the same frequency and voltage until the program finishes its execution.

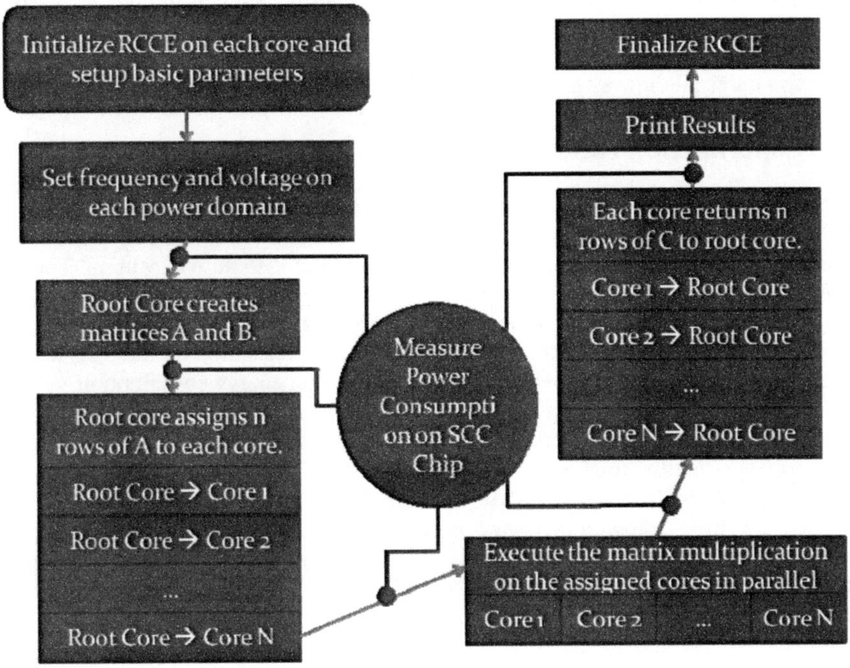

Fig. 1. Matrix multiplication program execution flow

In order to observe the power and energy consumption characteristics, we implemented a software design that can fetch the reading of the instantaneous power consumption of the SCC chip. The code was merged into the matrix multiplication program. To counter the time spent to stop the execution of the matrix multiplication program in order to get the power reading, we modified the matrix multiplication program so that it is now containing a RCCE timer which only counts when the program is running and stops counting when the program is fetching the instantaneous power measurement from the SCC chip. However, to start and stop timer is not enough to totally eliminate the overheads from such context switching. Other timing overheads include frequency switching at the beginning of the execution and power measurement all along the execution which cannot be easily removed. We try to run our program long enough so that those overhead has the least effect on our results. As we execute the matrix multiplication program, we performed the power measurement 5 times in total. The power measurement cannot be done during the multiplication on each core is being computed because the power measurement will interrupt the sequential communication between root core and executing cores. An executing core returns its own computed result only when the result is completed. Hence, the completeness sequence from each core cannot be predicted. The power measurement is needed to be done after the all the executing cores have finished their computation to avoid the interruption. The power measurement involves the product of the total voltage dropped on the SCC chip and the total current flowed through the SCC chip. The measurement is similar to the one done by the GUI performance meter provided by SCC. At the end of the matrix multiplication program, the overall average power consumption was calculated from the mean of all power measurements done within the execution. We execute the matrix multiplication program starting from two cores all the way to entire 48 cores.

3 Experimental Results

The experimental results are separated into two sets, the setup with frequency of 800MHz and the setup with frequency of 400MHz. In our experiment, we recorded total execution time and average power consumption of the entire execution. Then we calculate total energy consumed by the SCC chip during the execution by multiplying the recorded total execution time and the recorded average power consumption. Therefore, we have three different parameters for each execution on the SCC system: total elapsed time, average power consumption and total energy consumption.

Firstly, Fig. 2 presents the relationship between the number of cores utilized for the matrix multiplication execution and the time it takes to finish the execution. We can obviously see that the total execution time needed to finish the matrix multiplication program in 800MHz frequency case is much less than the frequency of 400MHz. Despite the results showed that we can reduce the execution time by increasing the number of cores within the 48 cores execution, the time decays does not convince us to massively increase the number of cores because it is an exponential decay that slowly converges to a time instant. This limitation cannot be resolved due to the on-chip communication overhead among the cores. The total execution time gap between the 800MHz and 400 MHz frequencies are close to each other as the number of cores increases.

Secondly, Fig. 3 presents the relationship between the number of cores utilized for the matrix multiplication execution and the average power consumption of the entire execution. There is a dramatic increase in power consumption for the 800MHz frequency setup. The power consumption increases almost linearly when the number of cores increases for both operational frequencies. The average power consumption slope of the 800MHz frequency execution is much higher than the one from 400MHz frequency execution. Hence, the number of cores has less contribution to the power consumption for low frequency execution.

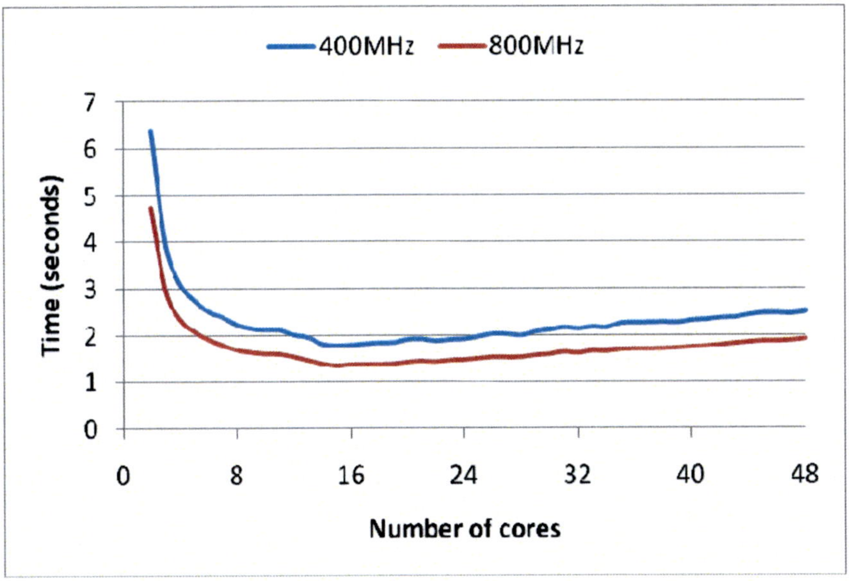

Fig. 2. Execution time comparison between 400MHz and 800MHz

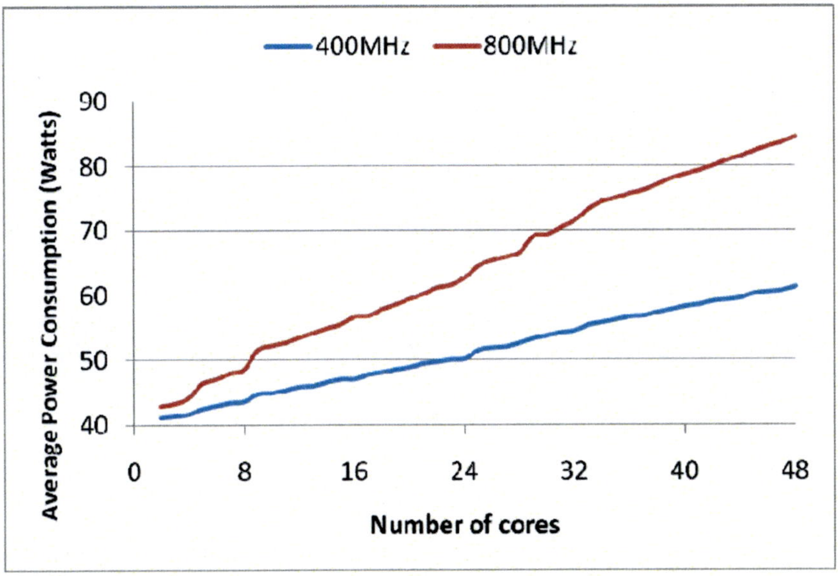

Fig. 3. Average power consumption comparison between 400MHz and 800MHz

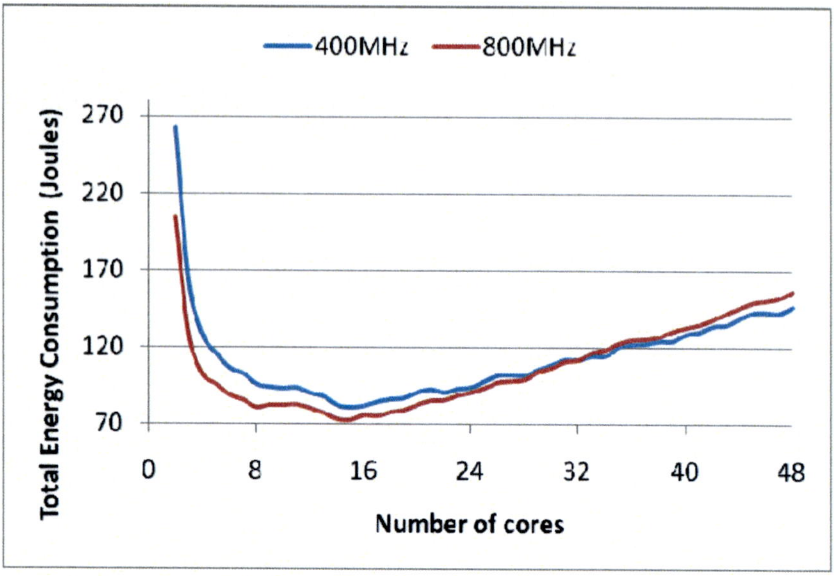

Fig. 4. Overall energy consumption comparison between 400MHz and 800MHz

Lastly, Fig. 4 presents the relationship between the number of cores utilized for the matrix multiplication execution and overall energy consumption of the entire execution. This plot represents the multiplication product of Fig. 2 and Fig. 3. On the left of the intersection with the number of cores lowers than 34, the frequency operation of 800 MHz needs less energy consumption than the 400MHz frequency. At 34 cores the total energy consumption of 400MHz is first less that the total energy consumption of 800MHz. Therefore, in term of energy optimization, we better run the program at high frequency for lower number of cores and run the program at low frequency for higher number of cores. Moreover, the results from 800MHz frequency shows that after 34 cores if we keep increasing the number of cores, instead of saving the energy, it will consume more energy to keep all the cores busy. The lowest energy consumption of 800MHz frequency operation is at 34 cores which mean we should not increase the number of core more than 34 cores on this configuration. Otherwise, it will consume more energy in order to execute the same amount of workload.

4 Conclusion

In this paper, we have investigated the time, power and energy relationship in a many-core system, SCC. Our results show that there is a boundary on the number of cores in many-core system and we realized that power consumption of a many-core system has an almost linear characteristic increment with the number of cores for a mixed sequential and parallel program execution. Our experimental results suggest us to run higher number of cores at low frequency operation and run lower number of cores at high frequency operation to optimize energy consumption of SCC system.

Acknowledgment. We would like to thank Intel Labs for providing us with SCC system in material transfer to conduct this research. This work is partly supported by the National Science Foundation under grant number ECCS-1125762. Any opinions, findings, and conclusions or recommendations expressed in this material are those of the authors and do not necessarily reflect the views of the National Science Foundation.

References

1. Tullsen, D.M., Eggers, S.J., Levy, H.M.: Simultaneous multithreading: Maximizing on-chip parallelism. In: Proceedings of 22nd Annual International Symposium on Computer Architecture, June 22-24, pp. 392–403 (1995)
2. Eggers, S.J., Emer, J.S., Leby, H.M., Lo, J.L., Stamm, R.L., Tullsen, D.M.: Simultaneous multithreading: a platform for next-generation processors. IEEE Micro. 17(5), 12–19 (1997)
3. Lance, H., Basem, N.A., Kunle, O.: A Single-Chip Multiprocessor. IEEE Computer 30(9), 79–85 (1997)
4. Flynn, M.J., Hung, P., Rudd, K.W.: Deep submicron microprocessor design issues. IEEE Micro. 19(4), 11–22 (1999)
5. Flynn, M.J., Hung, P.: Microprocessor design issues: thoughts on the road ahead. IEEE Micro. 25(3), 16–31 (2005)
6. Mattson, T.G., Van der Wijngaart, R.F., Riepen, M., Lehnig, T., Brett, P., Haas, W., Kennedy, P., Howard, J., Vangal, S., Borkar, N., Ruhl, G., Dighe, S.: The 48-core SCC Processor: the Programmer's View. In: 2010 International Conference for High Performance Computing, Networking, Storage and Analysis (SC), pp. 1–11, 13–19 (November 2010)
7. Howard, J., Dighe, S., Hoskote, Y., Vangal, S., Finan, D., Ruhl, G., Jenkins, D., Wilson, H., Borkar, N., Schrom, G., Pailet, F., Jain, S., Jacob, T., Yada, S., Marella, S., Salihundam, P., Erraguntla, V., Konow, M., Riepen, M., Droege, G., Lindemann, J., Gries, M., Apel, T., Henriss, K., Lund-Larsen, T., Steibl, S., Borkar, S., De, V., Van Der Wijngaart, R., Mattson, T.: A 48-Core IA-32 message-passing processor with DVFS in 45nm CMOS. In: 2010 IEEE International Solid-State Circuits Conference Digest of Technical Papers (ISSCC), pp. 108–109, 7–11 (February 2010)
8. Intel Labs, SCC Platform Overview. Intel Many-core Applications Research Community, Revision 0.75 (September 2010)
9. Jim, H.: Single-Chip Cloud Computer. In: Intel Labs Single-chip Cloud Computer Symposium (February 2010)
10. Salihundam, P., et al.: A 2 Tb/s 6 4 Mesh Network for a Single-Chip Cloud Computer With DVFS in 45 nm CMOS. IEEE Journal of Solid-State Circuits 46(4), 757 (2011)
11. Intel-Tera Scale Computing Research, `http://techresearch.intel.com/ResearchAreaDetails.aspx?Id=27`
12. National Center for Computational Science, `http://www.nccs.gov/user-support/training-education/hpcparallel-computing-links/mpi-examples/`
13. Snir, M., Otto, S., Huss-Lederman, S., Walker, D., Dongarra, J.: MPI: The Complete Reference. MIT Press (1996)

14. Van der Wijngaart, R.F., Mattson, T.G., Haas, W.: Light-weight communications on Intel's single-chip cloud computer processor. SIGOPS Oper. Syst. Rev. 45(1), 73–83 (2011)
15. Tim, M., van der Rob, W.: RCCE: a Small Library for Many-Core Communication. Intel Many-core Applications Research Community, document version 0.75 (September 2010)
16. Naffziger, S.: High-Performance Processors in a Power-Limited World. In: VLSI Circuits, 2006 Symposium on Digest of Technical Papers, pp. 93–97 (2006)
17. Krste, A., et al.: The Lanscape of Parallel Computing Research: A View from Berkeley, EECS techinical report (December 2006)
18. Brey, B.B.: The Intel Microprocessors 8086/8088, 80186/80188, 80286, 80386, 80486, Pentium and Pentuim Pro Processor: Architecture, Programming and Interfacing, 4th edn., ISBN: 81-203-1220-1

Solar Power Station Monitoring System with Five Remote Functions and Double Intelligent Control Loop

Dingjin Huang[1], Yunqiang Zheng[1], and Hadi Malek[2]

[1] School of Optoelectronic Engineering, Xi'an Technological University
Xi'an, 710032, China
[2] Department of Electrical & Computer Engineering, Utah State University
4120 Old Main Hill, Logan UT 84322-4120, USA
zhengyunqiang585@126.com, hdjing18@yahoo.com.cn,
hadi.malek@nceis.com

Abstract. In this paper, by using ARM Cortex-M3 embedded high-performance processor, we implement a remote monitoring system for a solar power station with five essential functions which are telemetry, remote control, remote signal, remote dispatch and remote vision. In this system, Data Acquisition Unit, Data Transmit Unit and Monitoring Centre Unit are combined to form this remote monitoring system. This remote intelligent monitoring system enables power station's operators to check the station status anytime and anywhere through the internet or cell phone, and they can send pre-defined command in the special conditions or doing supervision tasks. This system improves the monitoring system's integrity, reliability, flexibility and intellectuality.

Keywords: Five remotes, Double intelligent control loop, Remote monitoring system, Data fusion, ARM Cortex-M3.

1 Introduction

The era of fossil fuels, As a Main part of our energy supplies, is coming to the end by next some years. Then using new forms of energy instead of traditional forms of energy is necessary. Remarkable daily amount of energy from the sun, and the fact that using solar energy is free of environmental pollution and harmful gases, makes solar energy as a proper source to replace fossil Fuels. In areas where there is sufficient amount of solar energy, a solar power plant is the best way of generating electricity. Optimum use of a great permanent, endless and always available source of energy is another advantage of this energy. Nowadays, main problem of using solar power stations is high costs of establishing and operating of these stations. Among all methods suggested for solar station monitoring, those methods which contain lower costs, shorter construction time and higher reliability are more desirable.

Most application of solar power station is in the desert, grassland, basin, shallow water and other regions, where are not covered by national power grid, but these areas contain a huge resources of solar energy. Nowadays, as a part of dispatch automation system, most of solar power station monitoring systems contain data acquisition and

D. Yang (Ed.): Informatics in Control, Automation and Robotics, Volume 2, LNEE 133, pp. 27–34.
springerlink.com

monitoring function, which are isolated from other systems and they form an island of information effect in the power systems.

Moreover, current monitoring systems have only four remote functions and also the common used data processing methods add the collected information together simply or using the ballot choices, which these methods reduces the reliability and intellectuality.

In this paper, Solar Power Station Monitoring System with five remote functions and double intelligent control loop (SPSMS) is presented. SPSMS enables operator to check the station's status anytime and anywhere and increases power station safety level.

In the following, in section 2 we present the main function and proposed system configuration. Hardware and software design are described in section 3 and section 4 and finally, section 5 is conclusion.

2 System Architecture and Functions

SPSMS overall configuration is shown in Figure1.

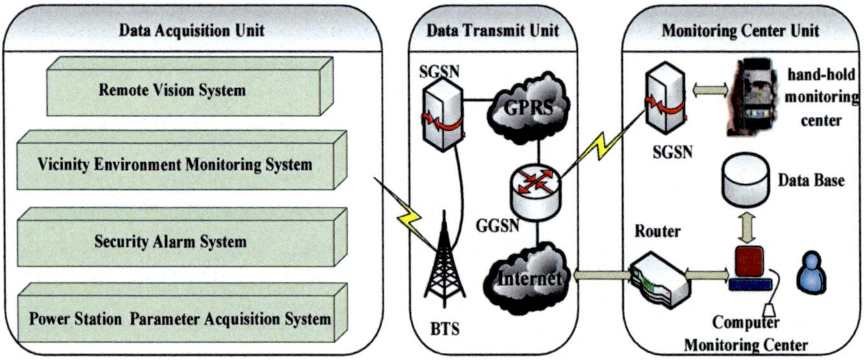

Fig. 1. SPSMS overall configuration

This system is divided into three units. First unit is data acquisition unit which is constituted by remote terminal unit (RTU) and local instruments. This unit has functions of vicinity environment monitoring, remote vision, security alarm and power station operation parameters acquisition. In this unit, Field data are collected by sensors and will be preprocessed comprehensively and intelligently. It can also communicate with the local equipments based on MODBUS communication protocol to debug or maintain them.

Second unit is data transmit unit which contains optical fiber, GSM/GPRS wireless network and Ethernet cable to realize long distance communication.

Third unit is monitoring center unit with double intelligent control loop. The primary control loop is a centralized monitoring center consisting of industrial personal computer and Web Server. It works through operating serial port or network port to realize Data analyzing, data sorting, data displaying, control instructions sending, and so on. All of processed information will be released through TCP/IP protocol and therefore will be accessible through a designed website according to the requirements eventually. The secondary control loop is the hand-held mobile terminals. Control software is based on Wince. It can check the station's status anytime and anywhere through cell phone.

SPSMS has five remote duties including telemetry, remote control, remote signal, remote dispatch and remote vision.

3 Hardware Design

Data acquisition unit circuit configuration is shown in Figure2:

The embedded controller of SPSMS is STM32F107VC based on the ARM Cortex-M3 produced by STMicroelectronics CO. Hardware circuit can be divided into five parts which is sensor data collecting circuit, external devices communication circuit, input/output controlling circuit, security alarm circuit and network communication circuit.

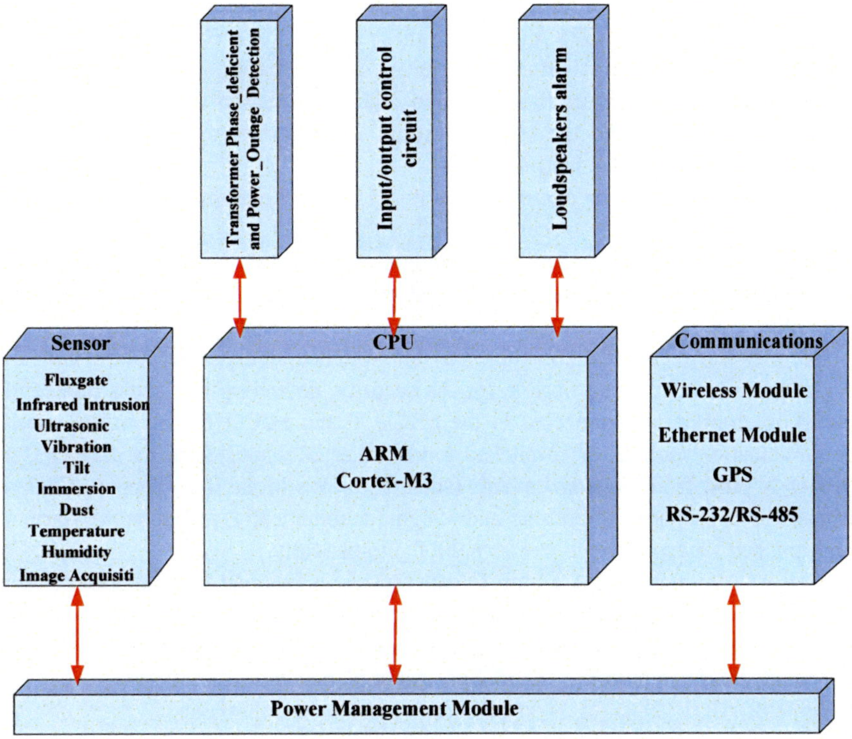

Fig. 2. Data acquisition unit circuit configuration

Sensor data collecting circuit processes sensors signals including humidity and temperature sensors, dust content sensors, water immersion sensors, vibration sensors, tilt sensors, windows state sensors, infrared intrusion sensors, ultrasonic sensors, transformer phase lack and power failure sensors. Humidity and temperature sensor communicate with the CPU via single-wire bus; other sensors can be directly connected to GPIO. Transformer phase lack and power failure sensor will produce a low voltage if there is any lack of phase in transmission line. All sensors connected to CPU are isolated by the photon coupled isolators.

External communications circuit communicates with the local equipments based on MODBUS communication protocol. This circuit is responsible for collecting local equipments data including image data, GPS geographic data, solar controller working state, current data and voltage data. Also, SPSMS enables operator to debug the equipments via the external communication circuit. Image acquisition circuit is constituted by the CMOS sensor, image acquisition module and mass storage. This circuit is designed with infrared illuminator in order to collect images in the dark condition. Image acquisition module is responsible for CMOS sensor exposure setting, image resolution setting, formats and contrast settings .The image is compressed into JPEG file format and transmits via RS485 serial port. The image will be stored in SD card. GPS module connect to CPU via RS232 serial port and produces data containing latitude, longitude, speed, direction Angle, year, month, hours, minutes, seconds, milliseconds and other information. The circuit reads the voltage, current data and the solar controller information, running state and other information from the local instruments.

Input/output controlling circuit is constituted by relay controlling unit. The circuits can control power supply, security and alarm equipments to change switch's state. The relay controlling unit adopts independent power supply mode, add the backflow inhibit circuit and electric spark suppression circuit to increase the stability and reliability.

Security alarm circuit is constituted by high-pitched loudspeakers, alarm and audio surveillance unit. It is responsible for the security, defending and audio intercepting etc. A microphone is connected to the MICO_P and MICO_N pins in the wireless communication module. High-pitched loudspeaker is connected to the SPKO_P and SPKO_N pins. If any unusual events (such as thieves broke into) happen, CPU will trigger the microphone to pick-up audio signal automatically, and open the alarm and high-pitched loudspeakers to overawe thief automatically.

Network communication circuit is constituted by Ethernet communication circuit and GSM/GPRS wireless communication circuit. It is responsible for sending the data via Ethernet and GSM/GPRS wireless network to the monitoring center and receiving the data from the monitoring centre. Specific implementation as follows:

(a) The data collected is sent to the monitoring centre by CPU through Ethernet.

(b) CPU calls AT instructions to send the data to computer monitoring centre and the hand-held monitoring center in forms of GPRS transparent transmission or the SMS/MMS mode.

4 Software Design

4.1 Communication Protocols and Data Fusion Algorithm

Communication protocol between the RTU and the computer monitoring center should be established. Data frame format can be divided into command message and response message. A detailed command message protocol format is as table1:

Table 1. Command message

SeqNr	Addr0	Addr1	Com	Len	Data[0...n]	CRC	End

SeqNr: the message head, command message serial number identification (1 byte) Addr0 Addr1: RTU address, divided into provincial address and regional address (2 bytes)

Com: the command sender command type identification (1 byte)

Len: length of data (2 bytes)

Data [0...n]: Data bytes

CRC: CRC checking (2 bytes)

End: end identification (1 byte)

A detailed response message protocol format is as table2:

Table 2. Rresponse message

SeqNr	Addr0	Addr1	Type	Len	Data[0...n]	CRC	End

SeqNr: the message head, command message serial number identification (1 byte) Addr0 Addr1: RTU address, divided into provincial address and regional address (2 bytes)

Type: the data sender data Type identification (1 byte)

Len: length of data (2 bytes)

Data [0...n]: Data bytes

CRC: CRC checking (2 bytes)

End: end identification (1 byte)

Following industrial communication standard, Modbus_RTU communication protocol is adopted between RTU and other instruments in the power station.

Using data fusion technology to improve reliability and intellectuality, prevent error and omission alarms. SPSMS determines the weights and membership functions of each sensor based on the estimation method and classical statistical method. For example, there are N sensors to measure physical quantities and the collected data from the first sensor is X_i, in which i=l, 2, …, n. SPSMS averages each sensor's data with its weighting factor W_i ($\sum_{i=0}^{n} W_i = 1$), the weighted average of the results is:

$$\overline{X} = \sum_{i=0}^{n} W_i X_i = WX \tag{1}$$

Because of different quality of each sensor and other interference factors, X_i has randomness characteristics. Suppose X_i are independent and distributed as N (μ_i,σ_i). Sensor precision can be calculated by:

$$\sigma = \frac{\sigma_i}{\sqrt{n}} \tag{2}$$

Obviously by doing this averaging process, the result data is more reliable than using individual sensor signal.

4.2 Remote Terminal Unit Software Design

The structured top-down programming method is used in the RTU software design. According to the function, system software can be divided into initialization program module, networking and data sending program, image acquisition program, SD card reading and writing program, GPS geographic data reading and writing program, sensor data collection and fusing program, short message and MMS sending program, external device communications and debugging program, input/output controlling program, terminal services program and the main program.

The main program sets the configuration of the I/O pins, timer, registers, and USART after power on. Then calls instruction to complete GPS module, camera module, Peripheral modules to initialize, and defines ring buffer array FIFO and the frame format of the upload data.

The main program coordinates and controls the subprogram through the timer interruption, external interruption and USART interruption. The image acquisition program uses USART DMA mode in the image data acquisition. Reading and writing data will be controlled by the DMA controller instead of interruption. Thus the efficiency can be greatly improved. The main program flow diagram is shown in Figure3:

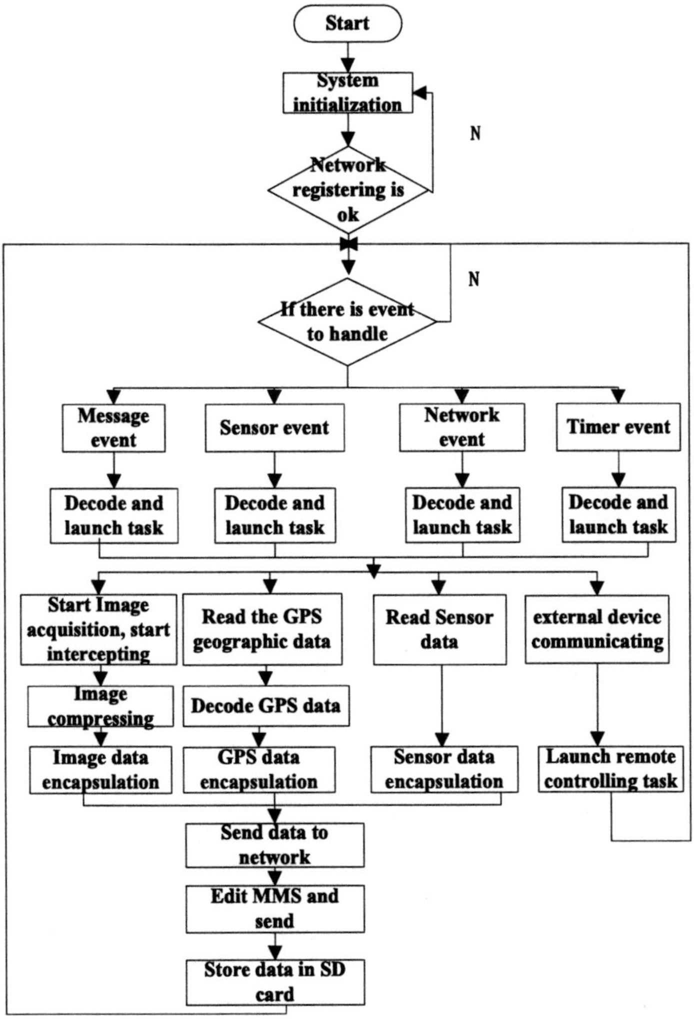

Fig. 3. The main program flow diagram

4.3 Monitoring Centre Software Design

According to the function, the monitoring software can be divided into three subprograms: Data transmission and analyzing section, database section, and data display section. The whole software is C/S structure. Socket operation receives and decodes the data, then saves the effective data SQL. The data will be released as a WEB form, which enables the user to search conveniently. TCP/IP protocol is used in the monitoring centre. SQL database servers accept client request, return processed results or middle data to the client. Also, SQL database serve as the centre of the system controlling, maintenance, analyzing, and data statistics. The real-time state of the power station will be displayed on the monitoring centre. Any emergency will be displayed immediately. The alarm information includes alarm level, location, time and other information.

5 Conclusion

By using high-performance embedded ARM Cortex-M3 processor, this paper design a remote monitoring system constituted by data acquisition unit, data transmit unit and monitoring center unit to form, and realize unattended solar power station monitoring with five remotes and double intelligent control loop.

There are two innovations in this system:

(a) SPSMS overcomes the islanding effect in the solar power station monitoring system, forms measurement and control system with organic relationship. When emergency happens, system has a nice dynamic linkage. Functions of remote dispatch and remote vision will improve the modernization of the solar power station monitoring system.

(b) SPSMS is designed with double intelligent control loop that enable operator to check the station status anytime and anywhere. Data fusion technology is used to reduce the error and omission alarms. This system improves the monitoring system's integrity, reliability, flexibility and intellectuality.

References

1. Flores, A., Sáez, D.: Fuzzy Predictive Control of a Solar Power Plant, pp. 58–66. IEEE Press (2005)
2. Kang, D.-J., Kim, H.-M.: Development of test-bed and security devices for SCADA communication in electric power system. In: 2009 Telecommunications Energy Conference (2009)
3. Yang, J., Jia, X.: Study Of Image Recognition Used For Unattended Substation. Master's Thesis of North China Electric Power University, D.BeiJing (2004)
4. Varshney, P.K.: Multisensor data fusion. Electronics & Communication Engineering Journal 12, 245–253 (1997)
5. Foresti, G.L.: A real-time system for video surveillance of unattended outdoor environments Circuits and Systems for Video Technology. IEEE Transactions 8(6), 697–704 (1998)
6. Itschner, R., Pommerell, Rutishauser, M.: Remote monitoring of embedded systems in power Engineering. IEEE Internet Computing (1998)
7. Andrea, A., Casini, A.L.: Embedded Java in a web-based teleradiology system. IEEE Internet Computing 2(3), 60–68 (1998)

Design of Portable Solar Energy System with Fast-Stationed Feature

Dingjin Huang and Yanwu Ma

School of Optoelectronic Engineering, Xi'an Technological University
Xi'an, China, 710032
jiao_251406174@qq.com

Abstract. He article had designed a portable solar energy system with fast-stationed feature. The system uses CAN-Bus and the parallel and sharing current technology that enables the system to achieve a distributed redundant expansion and rapidly set up power stations. The system is able to offer power to equipment on the environment without electricity environments, such as the outdoor travel, scientific investigation and field working ones. By the way, power station organized fast can meet the long-term, high-power power supply needs on field condition.

Keywords: Fast-stationed, Portable system, CAN-Bus, MPPT.

1 Introduction

In travel, scientific exploration or field work condition, people often use batteries, portable solar power, generators and other equipments to offer power to electrical equipment. Portable solar power device, due to its technical conditions, it is difficult to achieve high-capacity, high power output requirements, unable to meet the complex application environment and load demand.

Development of power system is a distributed power system instead of centralized power supply system. Compared the collective power system, distributed power system has more advantages: increased flexibility of the system; higher the module's power density lead to small size and height; improved system reliability because the electrical stress of the power semiconductor device in each module has been reduced; easier redundant expansion due to a distribution system; easier standardization due to small range.

The portable solar energy system with fast-stationed feature adopts distribution design, can enlarge the capacity and output power of system. The portable solar energy system had achieved the management of power in the station with CAN-Bus, and adjusted the parallel and sharing current with parallel and sharing unit in order to decrease the system damage and ensure the system stable outputting.

D. Yang (Ed.): Informatics in Control, Automation and Robotics, Volume 2, LNEE 133, pp. 35–40.
springerlink.com

2 The Overall Design of the System

2.1 Design of System Perform

The portable solar energy system with fast organized feature begins with aims that being portable, being fast charging, multi-output, and redundant expansion .In order to make full use of electricity power transited from solar energy , the MPPT technology and the charging method of approximation of ideal charge curve is used in the charging unit of this system. The power output unit of this system consists of the isolated DC / DC conversion circuit and DC / AC inverter circuit to increase the output power. Besides, this system has expansion feature, which allows a single portable system of solar cells and energy storage unit to expand. At same time, this system adopts CAN-Bus technology and the parallel and sharing current technology, enabling the system to achieve the expansion of distributed redundancy which makes N +1 portable system form into a series-parallel connected solar power station.

2.2 System Composition

PV module changes solar energy into electrical energy. In under the control of the controller, MPPT charging unit automatic tracks the PV maximum power point, adjusts the voltage of the PV module output, charges up energy storage unit. The voltage of Energy storage unit output goes through DC/DC adjusting circuit or DC/AC inverter circuit to supply DC or AC load. Operation parameters of each unit were collected by the controller, and displayed in the LCD monitor. Then, the controller intelligent controls the operation status of each unit. A number of systems can organize a solar power station through the CAN-Bus network. The system frame shown in figure 1.

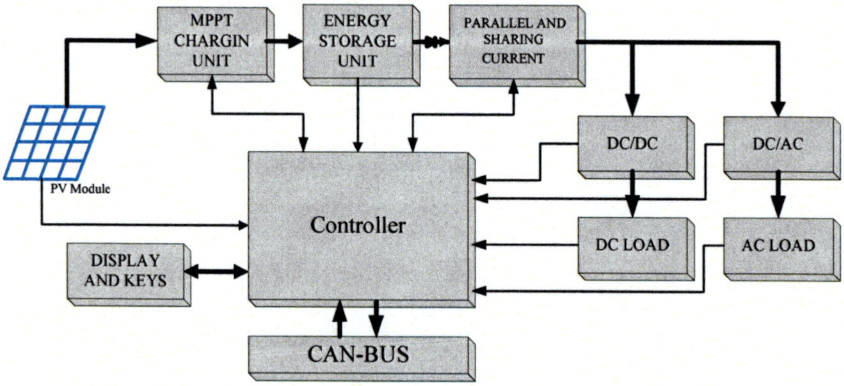

Fig. 1. The frame of portable solar energy system with fast organized feature

The microprocessor with ARM Cortex-M3 core was used in the system. The system has the functions of data acquisition, data storage, intelligent control, over voltage, over charge, over discharge, reverse polarity protection, alarm and display.

MPPT Charging Unit: It can track maximum power point of the solar cell, and complete the charging function.

Energy Storage Unit: It uses a modular design, the battery is made into the energy storage module, which can be easily demolished and replaced. It can be used a backup power.

DC / DC: It adjusts the direct voltage of the battery output to supply voltage to DC loads.

DC / AC: It transforms the direct voltage into alternating current voltage, then supplies AC load.

Parallel and Sharing Unit: it adjusts the current of the system output, reduce the difference of each portable system output current when they are used to from a power station.

CAN-Bus: In the formation of power station, we can establish local area communications network with CAN-Bus, manage electric quantity of each portable systems in the power station with distributed. CAN-Bus is the premise to achieve system redundancy, and rapid formation of power station.

3 CAN-Bus

3.1 The Application of CAN-Bus in Portable System

It is told that CAN is a short call for Controller Area Network. CAN-Bus nodes on the network can actually reach 112 to meet the need of organizing distribution and redundant expansion station.

CAN-Bus is a connection of managing the power distribution between several portable systems. CAN-Bus makes a whole power web, using several portable systems. Each system makes a communication with others with CAN-Bus. When one portable system is lack of power, CAN-Bus cut off the one or the branch of this system so that the power station can work normally. Not the system does get access to the station until the offering power condition of this system is conformed.

3.2 CAN-Bus Communication Circuit

With fast combine-station function, the CAN-Bus communication circuit of solar portable system shown in figure 2, MCU's serial data input and output connect CAN-Bus transceiver MCP2551 with high-speed optical isolated devices TLP113. In order to enhance anti-jamming capability of CAN node, using both ends of the TLP113 isolated power supply, produce a isolated CAN 5V signal by DC-DC devices, while the power supply of input +5V and STM32 is in common ground. In order to reduce the device's radio frequency interference, MCP2551'RS side grounded through resistance, turn the MCP2551 into the slope-control mode, reduce the output current of the RS side, thereby limiting the CANH and CANL rise and fall times, and further reduce EMI. In order to filter out high frequency interference on the bus, to prevent electromagnetic radiation, in the MCP2551 CANH and CANL side grounded through 30pF capacitors. In order to protect MCP2551 against over-current impact, series connect a 5Ω resistor between each side of CANH and CANL with bus to limit current. In order to impedance matching with cable and ensure the data signals would not reflected at both ends of bus, connect a 120Ω resistor between CANH and CANL, which is defined in ISO11898 linear topology rating resistance at both ends of bus.

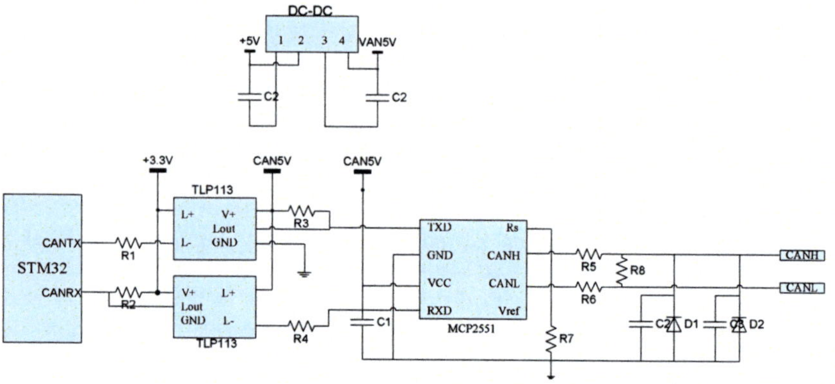

Fig. 2. CAN-Bus communication circuit

4 Parallel and Sharing Unit

4.1 Parallel and Sharing Current Technology

In the development of the switching power supply the parallel and sharing current technology is proposed. The bias existed in parameters and changes caused by operating environment will lead to the current distributed not evenly. Some of the output current is large and some is small while the others does not have the output. The switch in the power module which shares more current will have thermal stress augment. The system reliability therefore is reduced and the working life of the power module is shorter or the module may be damaged that can not maintain the normal work.

Parallel and sharing current technology enables the power modules which are working in parallel share the load current evenly. The current in this case is called load current sharing. The role that the load current sharing plays is to make sure that each module in the system can output power effectively and in the best working condition ensuring that the power system is stable, reliable, and efficient. Parallel current sharing technology makes the power systems flexible to expand its capacity. The power system has a very high serviceability once a single power supply module fails it can be easily hot-swap replaced or repaired.

4.2 Method OF Parallel AND Sharing Current

The common used methods of Parallel and sharing current are: the output impedance, master-slave setting, the average current automatic sharing current, the maximum current automatic sharing current and so on.

Though the parallel DC module has many solutions to carry out the current sharing, while as for the distributed solar electric power, there will be more or less defectives that lead to the output impedance of current-sharing accuracy is not very high. Master set method and the average current laws can not achieve redundancy, the maximum current automatic current sharing law is better. However, all the three methods are the hardware current sharing program which has many problems in current sharing such as the high cost, hard to change and to expand. Aiming at that, this paper proposed a software current sharing plan. Software current sharing method possesses a convenient, low cost, easy to change and upgrade features.

Software current sharing method is to collect the output current of each portable system and then calculate the total output current of the power station. The average output current of each portable system is calculated according to the number of the portable system. Then, the system according to current sharing requirements, compared to the unit's output current, adjustable output current of the unit, ensure that the current offset within the scope and conduct a continuous comparison of the current continued to promote the approximation of each module, each module to achieve balance between the output current, achieve parallel and sharing current.

5 Conclusion and Outlook

Distributed power systems instead of centralized power supply system is one of the power system development, and distributed solar power generation systems have enabled the solar energy toward the distributed development , The portable solar energy system with fast-stationed feature will be one of solar power systems' design development direction. The application of CAN-Bus technology and Parallel and sharing current technology enable portable system flexibly realize redundant expansion, improve the system limitations of portable solar power current application, which make portable solar energy systems have a wider range of applications.

References

1. Zhang, L., Huang, Z., Liu, C., et al.: Experimental Study on Parallel Connecting of MOSFETs. Telecom Power Technologies 6, 5–7 (2007)
2. Jiang, J.-A., Huang, T.-L., Hsiao, Y.-T., Chen, C.-H.: Maximum Power Tracking for Photovoltaic Power Systems. Tamkang Journal of Science and Engineering 8(2), 147–153 (2005)
3. Sun, Q., Wang, X., Gao, Y.: Study of the Current Balance of IGBTs in Paralleling. Power ElectronicTechnology 38(1), 4–6 (2002)
4. Zhang, Z., Shen, H., Chen, W., Shu, J.: The Design and Performance of Portable Integrated PV System. In: The 8th PV Conference Proceedings, pp. 612–616 (2004)

5. Zhang, S., Xiang, L., Yao, G.: Controlling Means of Parallel Current Sharing in Blocking DC Power Supply. World Power Supply, 12–15 (2007)
6. Shi, J., Shan, C., Tong, H., Fang, P.: Solar Cell MPPT Controller based on MCU. Power Electronics 42(11), 45–46 (2008)
7. Lu, L., Zhang, S., Shu, G.: Research on Max Power Point Tracking for Solar-Electric Charging System based on Micro-Controller. Power Electronics 41(2), 96–98 (2007)
8. Xiao, W., Dunford, W.G.: A Modified Adaptive Hillclimbing MPPT Method for Photovoltaic Power Systems. In: IEEE PESC 2004, vol. (3), pp. 195–196 (2004)
9. Ai, B., Yang, H., Shen, H., Liao, X.: Computer-aided design of PV/Wind hybrid system. Renewable Energy 28, 1491–1512 (2003)
10. Lin, X.: Power Electronic Technology. Tsinghua University Press (2009)

The Fuzzy Binary Relations on n- Frustum Pyramid Fuzzy Number Space and Its Application

Jiaxi Zhang, Guixiang Wang, and Jie Du

Institute of Operational Research and Cybernetics
Hangzhou Dianzi University, Hangzhou, 310018, China
mathdogged@126.com

Abstract. In this paper, we give a definition of $n-$ frustum pyramid fuzzy numbers (a kind of $n-$ dimensional fuzzy numbers which is easy to be used). And we define two fuzzy binary relations on $n-$ frustum pyramid fuzzy number space according to the characteristics of the special fuzzy numbers, study its properties. And then, we infer the computational formula which is easy to be programmed. At last we give a practical example to show the application in classification which is based on fuzzy approximation relation on $n-$ frustum pyramid fuzzy numbers space.

Keywords: fuzzy numbers, $n-$ frustum pyramid fuzzy numbers, fuzzy approximation relation, classification.

1 Introduction

The concept of fuzzy set was first put forward by Zadeh in 1965 [1], and Chang and Zadeh proposed the concept of fuzzy numbers in 1972 [2]. With the development of mathematics, more and more researchers studied the properties and applications of fuzzy numbers to meet the engineering research. In 2002, Wang [3] introduced the concept of fuzzy $n-$ cell numbers. In 2007, Wang proved that fuzzy $n-$ cell numbers and $n-$ dimensional fuzzy vectors can represent each other [4]. Afterward, Wang studied the ranking, pattern recognition and classification in an imprecise or uncertain environment based on fuzzy $n-$ cell numbers [5].

Compared with general fuzzy numbers, fuzzy $n-$ cell number has possessed of satisfying properties, which provide some convenience in both theory and application. However, its calculation in application is somewhat complicated. Therefore, we propose a special kind of $n-$ dimensional fuzzy numbers which is easy to be calculated, namely $n-$ frustum pyramid fuzzy number. We define two fuzzy binary relations on $n-$ frustum pyramid fuzzy number space according to the characteristics of the special fuzzy numbers, study its properties. And then, we infer the computational formula which is easy to be programmed. At last we give a practical example to show the application in classification which is based on fuzzy approximation relation on $n-$ frustum pyramid fuzzy numbers space.

D. Yang (Ed.): Informatics in Control, Automation and Robotics, Volume 2, LNEE 133, pp. 41–48.
springerlink.com

2 Basic Definitions and Notations

Fuzzy sets defined in R^n ($n-$ dimensional Euclidean Space) is a function. $u: R^n \to [0,1]$. For each such fuzzy set u, we denote by $[u]^r = \{x \in R^n : u(x) \geq r\}$ for any $r \in (0,1]$, we name it the $r-$level of fuzzy set u. The definition of the support of fuzzy set u is $\{x \in R^n : u(x) > 0\}$. Then, $[u]^0 = \overline{\{x \in R^n : u(x) > 0\}}$. Fuzzy number is a special fuzzy set which should meet the following four conditions: u is normal; u is fuzzy convex; $u(x)$ is upper semi-continuous; $[u]^0$ is a compact set. Let E^n denote the collection of all fuzzy sets of R^n. If $u \in E^n$, and $[u]^r$ can be represented by $\prod_{i=1}^n [\underline{u}_i(r), \overline{u}_i(r)]$, $\forall r \in [0,1]$, where $\underline{u}_i(r) \leq \overline{u}_i(r)$, ($i = 1,2,\cdots,n$), then we call u a fuzzy $n-$cell number. And, we denote the collection of all fuzzy $n-$cell numbers by $L(E^n)$. If $u \in E$, and there exist $a,b,c,d \in R$ with $a \leq b \leq c \leq d$ such that

$$u(x) = \begin{cases} \dfrac{x-a}{b-a}, & if \quad x \in [a,b] \\ 1, & if \quad x \in [b,c] \\ \dfrac{x-d}{c-d}, & if \quad x \in [c,d] \\ 0, & if \quad x \notin [a,d] \end{cases}$$

Then u is called a trapezoidal fuzzy number and denoted as $u = (a,b,c,d)$. If $b = c$, u is a triangle fuzzy number, $u = (a,b,d)$.

Let $u_i \in E$, ($i = 1,2,\cdots,n$), $n-$dimensional fuzzy vector is refer to ordered group with n elements as $u = (u_1, u_2, \cdots, u_n)$. In [4], it is proved that fuzzy $n-$cell numbers and $n-$dimensional fuzzy vectors can represent each other. So, for any $u \in L(E^n)$, we can represent it as a vector.

Let $u = (u_1, u_2, \cdots, u_n) \in L(E^n)$, $v = (v_1, v_2, \cdots, v_n) \in L(E^n)$, $\alpha = (\alpha_1, \alpha_2, \cdots, \alpha_n) \in R^n$, and α satisfies $\sum_{i=1}^n \alpha_i = 1, \alpha_i \geq 0$, ($i = 1,2,\cdots,n$). Let $M_\alpha(u) = \sum_{i=1}^n \alpha_i \int_0^1 r[\underline{u}_i(r) + \overline{u}_i(r)]dr$,

$D_\alpha(u) = \sum_{i=1}^n \alpha_i [\overline{u}_i(r) - \underline{u}_i(r)]dr$, then we call $M_\alpha(u)$ and $D_\alpha(u)$ the weighted mean and the weighted fuzzy degree respectively. $\rho_\alpha(u,v)$ is a metric between u and v, and $\rho_\alpha(u,v) = \sum_{i=1}^n \alpha_i \int_0^1 r[|\underline{u}_i(r) - \underline{v}_i(r)| + |\overline{u}_i(r) - \overline{v}_i(r)|]$. Define $u \leq v$ if and only if $\underline{u}_i(r) \leq \underline{v}_i(r)$ and $\overline{u}_i(r) \leq \overline{v}_i(r)$, ($i = 1,2,\cdots,n$), $\forall r \in [0,1]$.

3 Fuzzy Binary Relations

Definition 1. Let $u = (u_1, u_2, \cdots, u_n) \in L(E^n)$. If $u_1, u_2, \cdots, u_n$ are all trapezoidal fuzzy numbers, then we call u a $n-$frustum pyramid fuzzy number. And if $u_i = (a_i, b_i, c_i, d_i)$, ($i = 1,2,\cdots,n$), then we denote u by

$$u=\begin{pmatrix} a_1 & a_2 & \cdots & a_n \\ b_1 & b_2 & \cdots & b_n \\ c_1 & c_2 & \cdots & c_n \\ d_1 & d_2 & \cdots & d_n \end{pmatrix}$$

And we denote the $n-$ frustum pyramid fuzzy number space by $FP(E^n)$. If $b_i = c_i$, ($i=1,2,\cdots,n$), u is a $n-$ pyramid fuzzy number.

Lemma 1. Let $u \in FP(E^n)$, $v \in FP(E^n)$, $u_i = (a_i, b_i, c_i, d_i)$, $v_i = (a_i', b_i', c_i', d_i')$, ($i=1,2,\cdots,n$).

(1). $u \le v \Leftrightarrow a_i \le a_i'$, $b_i \le b_i'$, $c_i \le c_i'$, $d_i \le d_i'$, $(i=1,2,\cdots,n)$, $\forall r \in [0,1]$.

(2). $u \subset v \Leftrightarrow a_i' \le a_i$, $b_i' \le b_i$, $c_i \le c_i'$, $d_i \le d_i'$, $(i=1,2,\cdots,n)$, $\forall r \in [0,1]$.

Theorem 1. $\forall u \in FP(E^n)$, $\alpha = (\alpha_1, \alpha_2, \cdots, \alpha_n) \in R^n$, and $\sum_{i=1}^n \alpha_i = 1, \alpha_i \ge 0$ $(i=1,2,\cdots,n)$.

$$M_\alpha(u) = \frac{1}{6}\sum_{i=1}^n \alpha_i (a_i + d_i) + \frac{1}{3}\sum_{i=1}^n \alpha_i (b_i + c_i),\quad D_\alpha(u) = \frac{1}{6}\sum_{i=1}^n \alpha_i (d_i - a_i) + \frac{1}{3}\sum_{i=1}^n \alpha_i (c_i - b_i)$$

Proof. It is easy to know $\underline{u_i}(r) = (b_i - a_i)r + a_i$, $\overline{u_i}(r) = (c_i - d_i)r + d_i$. So, we have

$$\begin{aligned}
M_\alpha(u) &= \sum_{i=1}^n \alpha_i \int_0^1 r[\underline{u_i}(r) + \overline{u_i}(r)]dr \\
&= \sum_{i=1}^n \alpha_i \int_0^1 r[(b_i - a_i)r + a_i + (c_i - d_i)r + d_i]dr \\
&= \sum_{i=1}^n \alpha_i [(a_i + d_i)\int_0^1 r dr + (b_i + c_i - a_i - d_i)\int_0^1 r^2 dr] \\
&= \sum_{i=1}^n \alpha_i [\frac{a_i + d_i}{2} + \frac{b_i + c_i - a_i - d_i}{3}] \\
&= \frac{1}{6}\sum_{i=1}^n \alpha_i (a_i + d_i) + \frac{1}{3}\sum_{i=1}^n \alpha_i (b_i + c_i)
\end{aligned}$$

$$\begin{aligned}
D_\alpha(u) &= \sum_{i=1}^n \alpha_i \int_0^1 r[\overline{u_i}(r) - \underline{u_i}(r)]dr \\
&= \sum_{i=1}^n \alpha_i \int_0^1 r[((c_i - d_i)r + d_i) - ((b_i - a_i)r + a_i)]dr \\
&= \sum_{i=1}^n \alpha_i [(d_i - a_i)\int_0^1 r dr + (a_i + c_i - b_i - d_i)\int_0^1 r^2 dr] \\
&= \sum_{i=1}^n \alpha_i [\frac{d_i - a_i}{2} + \frac{a_i + c_i - b_i - d_i}{3}] \\
&= \frac{1}{6}\sum_{i=1}^n \alpha_i (d_i - a_i) + \frac{1}{3}\sum_{i=1}^n \alpha_i (c_i - b_i)
\end{aligned}$$

Theorem 2. $u, v \in FP(E^n)$, then

$$\rho_\alpha(u,v) = \sum_{i=1}^n \alpha_i \int_0^1 r\{|[(b_i - a_i) - (b_i' - a_i')]r + a_i - a_i'| + |[(c_i - d_i) - (c_i' - d_i')]r + d_i - d_i'|\}dr$$

Definition 2. Let $u, v \in FP(E^n)$, $p \in [0,1]$, $\alpha = (\alpha_1, \alpha_2, \cdots, \alpha_n) \in R^n$, and $\sum_{i=1}^n \alpha_i = 1$, $\alpha_i \ge 0$, $(i=1,2,\cdots,n)$. Let $d_{\alpha,p}(u,v) = p(M_\alpha(u) - M_\alpha(v))^2 + (1-p)(D_\alpha(u) - D_\alpha(v))^2$.

Definition 3. $p \in [0,1]$, we define two fuzzy binary relations on $FP(E^n)$ as follows:

$$G_{\alpha,p}(u,v) = \frac{1}{1+\sqrt{d_{\alpha,p}(u,v)}}, \quad H_{\alpha,p} = \frac{1}{1+\sqrt{p\rho_\alpha{}^2(u,v)+(1-p)[\sum_{i=1}^{n}\alpha_i|D(u_i)-D(v_i)|]^2}}$$

And we call the fuzzy binary relations $G_{\alpha,p}$ and $H_{\alpha,p}$ fuzzy approximation relations.

Theorem 3. $\forall u,v \in FP(E^n)$, $\alpha = (\alpha_1,\alpha_2,\cdots,\alpha_n) \in R^n$, and α satisfies $\sum_{i=1}^{n}\alpha_i = 1$, $\alpha_i \geq 0$, $i = 1,2,\cdots,n$.

(1). $d_{\alpha,p}(u,u) = 0$.

(2). $d_{\alpha,p}(u,v) = d_{\alpha,p}(v,u)$.

(3). $d_{\alpha,p}(ku,kv) = k^2 d_{\alpha,p}(u,v)$.

Proof. Obviously, (1) and (2) are right. So we only show (3). According to the definition of $M_\alpha(u)$ and $D_\alpha(u)$, we know $M_\alpha(ku) = kM_\alpha(u), D_\alpha(ku) = |k|D_\alpha(u)$. So we have $d_{\alpha,p}(ku,kv) = p(kM_\alpha(u) - kM_\alpha(v))^2 + (1-p)(|k|D_\alpha(u) - |k|D_\alpha(v))^2 = k^2 d_{\alpha,p}(u,v)$.

Remark 1. $d_{\alpha,p}(u,v) + d_{\alpha,p}(v,w) \geq d_{\alpha,p}(u,w)$ is not necessarily true, so $d_{\alpha,p}$ is not a metric.

Theorem 4. $\forall u,v \in FP(E^n)$, $\alpha = (\alpha_1,\alpha_2,\cdots,\alpha_n) \in R^n$ and $\sum_{i=1}^{n}\alpha_i = 1, \alpha_i \geq 0$.

$$d_{\alpha,p}(u,v) = \frac{p}{36}\{\sum_{i=1}^{n}[(a_i - a_i') + 2(b_i - b_i') + 2(c_i - c_i') + (d_i - d_i')]\}^2$$
$$+\frac{1-p}{36}\{\sum_{i=1}^{n}[(a_i' - a_i) + 2(b_i' - b_i) + 2(c_i - c_i') + (d_i - d_i')]\}^2$$

Proof. According to Theorem 1, we can get the formula easily.

Deduction 1. If u,v are $n-$pyramid fuzzy numbers, i.e. $b_i = c_i, b_i' = c_i'$ then

$$d_{\alpha,p}(u,v) = \frac{p}{36}\sum_{i=1}^{n}[(a_i - a_i') + 4(b_i - b_i') + (d_i - d_i')]^2 + \frac{1-p}{36}\sum_{i=1}^{n}[(a_i' - a_i) + (d_i - d_i')]^2$$

Theorem 5. $\forall u,v \in FP(E^n)$, $\alpha = (\alpha_1,\alpha_2,\cdots,\alpha_n) \in R^n$, and $\sum_{i=1}^{n}\alpha_i = 1, \alpha_i \geq 0$.

(1). $v \leq u \Rightarrow \rho_\alpha(u,v) = M_\alpha(u) - M_\alpha(v)$.

(2). $v \subset u \Rightarrow \rho_\alpha(u,v) = \frac{1}{6}\sum_{i=1}^{n}\alpha_i(a_i' - a_i + d_i - d_i') + \frac{1}{3}\sum_{i=1}^{n}\alpha_i(b_i' - b_i + c_i - c_i')$.

Proof. According to the definition of $v \leq u$, we know $a_i' - a_i \leq 0$, $b_i' - b_i \leq 0$, $c_i' - c_i \leq 0$, $d_i' - d_i \leq 0$, $(i = 1,2,\cdots,n)$, $\forall r \in [0,1]$.

Let $f(r) = [(b_i - b_i') - (a_i - a_i')]r + a_i - a_i'$, it is easy to know $f(0) \geq 0$, $f(1) \geq 0$, and the function $f(r)$ is either monotonically nonincreasing function, or monotonically nondecreasing function. Therefore, we know $f(r) \geq 0$, $\forall r \in [0,1]$. Simultaneously, Let $g(r) = [(c_i - c_i') - (d_i - d_i')]r + d_i - d_i'$, we get $g(r) \geq 0$, $\forall r \in [0,1]$. So, we have

$$\rho_\alpha(u,v)=\sum_{i=1}^n\alpha_i\int_0^1 r\{|[(b_i-a_i)-(b_i'-a_i')]r+a_i-a_i'|+|[(c_i-d_i)-(c_i'-d_i')]r+d_i-d_i'|\}dr$$
$$=\sum_{i=1}^n\alpha_i\int_0^1 r\{[(b_i-a_i)-(b_i'-a_i')]r+a_i-a_i'+[(c_i-d_i)-(c_i'-d_i')]r+d_i-d_i'\}dr$$
$$=\sum_{i=1}^n\alpha_i\int_0^1 r\{[(b_i-a_i)r+a_i+(c_i-d_i)r+d_i]-[(b_i'-a_i')r+a_i'+(c_i'-d_i')r+d_i']\}dr$$
$$=\sum_{i=1}^n\alpha_i\int_0^1 r[\underline{u_i}(r)+\overline{u_i}(r)]dr-\sum_{i=1}^n\alpha_i\int_0^1 r[\underline{v_i}(r)+\overline{v_i}(r)]dr$$
$$=M_\alpha(u)-M_\alpha(v)$$

(2). According to the definition of $v\subset u$, we know $a_i'-a_i\ge 0$, $b_i'-b_i\ge 0$, $c_i'-c_i\le 0$, $d_i'-d_i\le 0$, ($i=1,2,\cdots,n$), $\forall r\in[0,1]$. Let $h(r)=[(b_i-b_i')-(a_i-a_i')]r+a_i-a_i'$. From (1), we know $h(r)\le 0$, $\forall r\in[0,1]$. Therefore, we have

$$\rho_\alpha(u,v)=\sum_{i=1}^n\alpha_i\int_0^1 r\{|[(b_i-a_i)-(b_i'-a_i')]r+a_i-a_i'|+|[(c_i-d_i)-(c_i'-d_i')]r+d_i-d_i'|\}dr$$
$$=\sum_{i=1}^n\alpha_i\int_0^1 r\{[-(b_i-a_i)+(b_i'-a_i')]r-a_i+a_i'+[(c_i-d_i)-(c_i'-d_i')]r+d_i-d_i'\}dr$$
$$=\sum_{i=1}^n\alpha_i[\frac{(a_i'-a_i)+(d_i-d_i')}{2}+\frac{(a_i-a_i')+(b_i'-b_i)+(c_i-c_i')+(d_i'-d_i)}{3}]$$
$$=\frac{1}{6}\sum_{i=1}^n\alpha_i(a_i'-a_i+d_i-d_i')+\frac{1}{3}\sum_{i=1}^n\alpha_i(b_i'-b_i+c_i-c_i')$$

Theorem 6. $\forall u,v\in FP(E^n)$, $p=1$, $\alpha=(\alpha_1,\alpha_2,\cdots,\alpha_n)\in R^n$, and $\sum_{i=1}^n\alpha_i=1,\alpha_i\ge 0$, $(i=1,2,\cdots,n)$. If $v\le u$ or $u\le v$, we can get $G_{\alpha,p}(u,v)=H_{\alpha,p}(u,v)$.

Proof. According to Theorem 5, we know if $v\le u$, then $\rho_\alpha(u,v)=M_\alpha(u)-M_\alpha(v)$. In the same way, if $u\le v$, then $\rho_\alpha(u,v)=M_\alpha(v)-M_\alpha(u)$. Therefore, we have

$$G_{\alpha,p}(u,v)=\frac{1}{1+\sqrt{p}|M_\alpha(u)-M_\alpha(v)|}=\frac{1}{1+\sqrt{p\rho_\alpha(u,v)}}=H_{\alpha,p}(u,v).$$

Theorem 7. $\forall u,v,w\in FP(E^n)$, $\alpha=(\alpha_1,\alpha_2,\cdots,\alpha_n)\in R^n$, and α satisfies $\sum_{i=1}^n\alpha_i=1$, $\alpha_i\ge 0$, $(i=1,2,\cdots,n)$. $w=(w_1,w_2,\cdots,w_n)$, $w_i=(a_i'',b_i'',c_i'',d_i'')$.

(1). $G_{\alpha,p}(u,u)=1$.

(2). $G_{\alpha,p}(u,v)=G_{\alpha,p}(v,u)$.

(3). $G_{\alpha,p}(u+w,v+w)=G_{\alpha,p}(u,v)$.

(4). $\dfrac{1}{G_{\alpha,p}(ku,kv)}=\dfrac{|k|}{G_{\alpha,p}(u,v)}+1-|k|$.

Proof. It is obviously that (1) and (2) hold.
(3). According to Theorem 1, we know

$$M_\alpha(u+w)=\frac{1}{6}\sum_{i=1}^n\alpha_i(a_i+d_i+a_i''+d_i'')+\frac{1}{3}\sum_{i=1}^n\alpha_i(b_i+c_i+b_i''+c_i'')$$
$$D_\alpha(u+w)=\frac{1}{6}\sum_{i=1}^n\alpha_i(d_i-a_i+d_i''-a_i'')+\frac{1}{3}\sum_{i=1}^n\alpha_i(c_i-b_i+c_i''-b_i'')$$

$$M_\alpha(v+w)=\frac{1}{6}\sum_{i=1}^{n}\alpha_i(a_i'+d_i'+a_i''+d_i'')+\frac{1}{3}\sum_{i=1}^{n}\alpha_i(b_i'+c_i'+b_i''+c_i'')$$

$$D_\alpha(v+w)=\frac{1}{6}\sum_{i=1}^{n}\alpha_i(d_i'-a_i'+d_i''-a_i'')+\frac{1}{3}\sum_{i=1}^{n}\alpha_i(c_i'-b_i'+c_i''-b_i'')$$

So we have

$$M_\alpha(u+w)-M_\alpha(v+w)=M_\alpha(u)-M_\alpha(v)\,,\quad D_\alpha(u+w)-D_\alpha(v+w)=D_\alpha(u)-D_\alpha(v)\,.$$

And then $d_{\alpha,p}(u+w,v+w)=d_{\alpha,p}(u,v)$. Combined with Definition 3, (3) is proved.

(4).By $d_{\alpha,p}(ku,kv)=k^2 d_{\alpha,p}(u,v)$, then,

$$G_{\alpha,p}(ku,kv)=\frac{1}{1+\sqrt{d_{\alpha,p}(ku,kv)}}=\frac{1}{1+|k|\sqrt{d_{\alpha,p}(u,v)}}\,,$$

Then, $\frac{1}{G_{\alpha,p}(ku,kv)}-1=|k|(\frac{1}{G_{\alpha,p}(u,v)}-1)$, i.e. $\frac{1}{G_{\alpha,p}(ku,kv)}=\frac{|k|}{G_{\alpha,p}(u,v)}+1-|k|$.

4 The Application of Fuzzy Approximation Relation

In the following, we give a practical example to show the application in classification which is based on fuzzy approximation relation on $n-$ frustum pyramid fuzzy numbers.

Suppose there be four kinds of flowers, A, B, C, D. There are four characteristic, length of the calyx, width of the calyx, length of the petal, width of the petal. We want to classify A, B, C, D into two classes based on the four characteristic. There are twenty samples for each kind of flower. The data is in the following matrixes. The row and the array of each matrix represent the characteristic of the flowers and the samples, respectively.

$$A:\begin{pmatrix} 5.02 & 4.75 & 4.73 & 4.65 & 5.08 & 5.35 & 4.65 & 5.07 & 4.49 & 4.95 & 5.45 & 4.84 & 4.85 & 4.35 & 5.15 & 5.25 & 5.28 & 5.13 & 5.72 & 5.13 \\ 3.55 & 3.15 & 3.25 & 3.29 & 3.52 & 3.85 & 3.34 & 3.45 & 2.94 & 3.35 & 3.45 & 3.44 & 3.02 & 3.05 & 4.13 & 4.42 & 3.99 & 3.56 & 3.85 & 3.83 \\ 1.35 & 1.45 & 1.35 & 1.63 & 1.44 & 1.75 & 1.35 & 1.49 & 1.53 & 1.54 & 1.55 & 1.65 & 1.42 & 1.15 & 1.27 & 1.55 & 1.36 & 1.49 & 1.75 & 1.55 \\ 0.21 & 0.18 & 0.23 & 0.31 & 0.21 & 0.29 & 0.33 & 0.20 & 0.21 & 0.19 & 0.30 & 0.20 & 0.17 & 0.14 & 0.20 & 0.24 & 0.28 & 0.31 & 0.25 & 0.30 \end{pmatrix}$$

$$B:\begin{pmatrix} 7.18 & 6.43 & 6.56 & 5.78 & 6.85 & 5.58 & 6.46 & 4.97 & 6.79 & 5.90 & 5.30 & 5.99 & 6.18 & 6.23 & 5.76 & 6.76 & 5.61 & 5.82 & 6.39 & 5.99 \\ 3.26 & 3.28 & 3.11 & 2.39 & 2.80 & 2.83 & 3.33 & 2.45 & 2.92 & 2.71 & 2.00 & 3.01 & 2.24 & 2.91 & 2.95 & 3.14 & 3.02 & 2.74 & 2.25 & 2.54 \\ 4.75 & 4.58 & 4.91 & 4.01 & 4.63 & 4.55 & 4.77 & 3.31 & 4.68 & 3.94 & 3.58 & 4.24 & 4.09 & 4.71 & 3.61 & 4.45 & 4.51 & 4.10 & 4.55 & 3.91 \\ 1.49 & 1.54 & 1.55 & 1.36 & 1.52 & 1.32 & 1.67 & 1.02 & 1.35 & 1.41 & 1.08 & 1.53 & 1.02 & 1.49 & 1.39 & 1.44 & 1.52 & 1.09 & 1.53 & 1.18 \end{pmatrix}$$

$$C:\begin{pmatrix} 5.28 & 5.03 & 4.96 & 4.98 & 5.25 & 5.68 & 4.86 & 5.37 & 4.69 & 5.30 & 5.70 & 5.19 & 5.08 & 4.63 & 5.36 & 5.56 & 5.41 & 5.42 & 5.99 & 5.49 \\ 3.86 & 3.38 & 3.51 & 3.49 & 3.80 & 4.03 & 3.63 & 3.65 & 3.22 & 3.51 & 3.70 & 3.61 & 3.34 & 3.21 & 4.45 & 4.64 & 4.22 & 3.74 & 4.15 & 4.04 \\ 1.55 & 1.58 & 1.51 & 1.71 & 1.63 & 1.85 & 1.57 & 1.51 & 1.78 & 1.64 & 1.78 & 1.74 & 1.69 & 1.21 & 1.41 & 1.65 & 1.51 & 1.50 & 1.95 & 1.61 \\ 0.39 & 0.24 & 0.35 & 0.46 & 0.32 & 0.52 & 0.47 & 0.32 & 0.35 & 0.21 & 0.48 & 0.33 & 0.22 & 0.29 & 0.39 & 0.54 & 0.52 & 0.49 & 0.33 & 0.48 \end{pmatrix}$$

$$D:\begin{pmatrix} 7.38 & 6.73 & 6.76 & 6.08 & 7.05 & 5.88 & 6.66 & 5.27 & 6.99 & 6.20 & 5.50 & 6.29 & 6.38 & 6.53 & 5.96 & 7.06 & 5.81 & 6.12 & 6.59 & 6.29 \\ 3.56 & 3.48 & 3.41 & 2.59 & 3.10 & 3.03 & 3.63 & 2.65 & 3.22 & 2.91 & 2.30 & 3.21 & 2.54 & 3.11 & 3.25 & 3.34 & 3.32 & 2.94 & 2.55 & 2.74 \\ 4.95 & 4.68 & 5.11 & 4.11 & 4.83 & 4.65 & 4.97 & 3.41 & 4.88 & 4.04 & 3.78 & 4.34 & 4.29 & 4.81 & 3.81 & 4.55 & 4.71 & 4.20 & 4.75 & 4.01 \\ 1.59 & 1.64 & 1.65 & 1.46 & 1.62 & 1.42 & 1.77 & 1.12 & 1.45 & 1.51 & 1.18 & 1.63 & 1.12 & 1.59 & 1.49 & 1.54 & 1.62 & 1.19 & 1.63 & 1.28 \end{pmatrix}$$

For A, B, C, D, we get the mean value and the standard deviation of each characteristic, respectively. The mean value of A is $o_i = [4.9945; 3.5215; 1.4810; 0.2375]$. And the standard deviation of A is $\delta_i = [0.3376; 0.3884; 0.1526; 0.0551]$, $(i = 1,2,3,4)$. We will make the construction of trapezoidal fuzzy numbers. Let $u = (u_1, u_2, u_3, u_4)$.

$$u_i(x_i) = \begin{cases} \dfrac{x_i - (o_i - \beta_1\delta_i)}{(\beta_1 - \beta_2)\delta_i} & x_i \in [o_i - \beta_1\delta_i, o_i - \beta_2\delta_i] \\ 1 & x_i \in [o_i - \beta_2\delta_i, o_i + \beta_3\delta_i] \\ \dfrac{x_i - (o_i + \beta_4\delta_i)}{(\beta_3 - \beta_4)\delta_i} & x_i \in [o_i + \beta_3\delta_i, o_i + \beta_4\delta_i] \\ 0, & \text{else} \end{cases}, \quad i = 1,2,3,4$$

Then $u_i = (o_i - \beta_1\delta_i, o_i - \beta_2\delta_i,\ o_i + \beta_3\delta_i,\ o_i + \beta_4\delta_i)^{\mathrm{T}}$, $\beta_i \geq 0$ are parameters. So, we get $n-$ frustum pyramid fuzzy number u which represents flower A. In the same way, we can get $n-$ frustum pyramid fuzzy number v, w, q represent flower B, C, D, respectively. Let $\beta_1 = \beta_4 = 3, \beta_2 = \beta_3 = 1$, we have

$$u = \begin{pmatrix} 3.9817 & 2.3562 & 1.0232 & 0.0721 \\ 4.6596 & 3.1331 & 1.3248 & 0.1824 \\ 5.3321 & 3.9099 & 1.6336 & 0.2926 \\ 6.0073 & 4.6868 & 1.9388 & 0.4029 \end{pmatrix}, v = \begin{pmatrix} 4.4446 & 1.6626 & 2.9323 & 0.7872 \\ 5.5659 & 2.4169 & 3.8401 & 1.1791 \\ 6.6871 & 3.1711 & 4.7479 & 1.5709 \\ 7.8084 & 3.9254 & 5.6557 & 1.9628 \end{pmatrix}$$

$$w = \begin{pmatrix} 4.2272 & 2.5778 & 1.1257 & 0.0716 \\ 4.9167 & 3.3653 & 1.4546 & 0.2805 \\ 5.6063 & 4.1527 & 1.7834 & 0.4895 \\ 6.2958 & 4.9402 & 2.1123 & 0.6884 \end{pmatrix}, q = \begin{pmatrix} 4.7392 & 1.9047 & 3.0348 & 0.8872 \\ 5.8307 & 2.6642 & 3.9743 & 1.2791 \\ 6.9223 & 3.4238 & 4.9137 & 1.6709 \\ 8.0138 & 4.1833 & 5.8532 & 2.0628 \end{pmatrix}$$

Let $\alpha = (0.25, 0.25, 0.25, 0.25)$, and $p = 0.9$. We have

$$G_{\alpha,p}(u,v) = 0.5437; G_{\alpha,p}(u,w) = 0.8669; G_{\alpha,p}(u,q) = 0.5033;$$
$$G_{\alpha,p}(v,w) = 0.5931; G_{\alpha,p}(v,q) = 0.8710; G_{\alpha,p}(w,q) = 0.5455.$$

Obviously, we can put A, C in a group and B, C in another group.

Acknowledgments. This work is supported by Natural Science Foundation of China (No.60934009).

References

1. Zadeh, L.A.: Fuzzy Sets. Information Control, 338–353 (1965)
2. Chang, S.S.L., Zadeh, L.A.: On fuzzy mappings and control. IEEE Trans. Syst. Man Cybernet. 2, 30–34 (1972)
3. Wang, G.X., Wu, C.X.: Fuzzy n-cell Numbers and the Differential of Fuzzy n-cell Value Mappings. Fuzzy Sets and System 130, 367–381 (2003)
4. Wang, G.X., Li, Y.M., Wen, C.L.: On Fuzzy n-cell Numbers and n-dimension Fuzzy Vectors. Fuzzy Sets and system 158, 71–84 (2007)
5. Wang, G.X., Shi, P., Messenger, P.: Representation of uncertain Multi-Channel Digital Signal Spaces and Study of Pattern Recognition Based on Metrics and Difference Values on Fuzzy n-cell Number Spaces. IEEE Trans. on Fuzzy systems 17(2), 421–439 (2009)
6. Wang, G.X., Shi, P., Wen, C.L.: Fuzzy approximation relations on fuzzy n-cell number space and their applications in classification. Information Sciences 181, 3846–3860 (2011)

Similarity Relation of Two-Dimensional Frustum Pyramid Fuzzy Numbers Based on the Geometry and Its Application

Jie Du, Guixiang Wang, and Jiaxi Zhang

Institute of Operational Research and Cybernetics,
Hangzhou Dianzi University, Hangzhou, 310018, China
dujie_41@163.com

Abstract. In this article, we define a similarity relation of two-dimensional frustum pyramid fuzzy numbers, and discuss its properties, and use the similarity relation to rank some uncertain attributes (e.g. the observing of Eigen values of target is uncertain or imprecise observation). The similarity relation is given out according to the definition of the virtual center of gravity and volume of two-dimensional frustum pyramid fuzzy numbers. In this article, we give an example to illustrate the similarity relation is reasonable in the actual application.

Keywords: Virtual center of gravity, volume, similarity relation of fuzzy numbers, ranking.

1 Introduction

In recent years, many scholars have studied on the similarity relation of fuzzy numbers. Lee [1], Hsieh-and-Chen [2], Chen-and-Chen [3], and Chen-and-Lin [4], discuss in the similarity relation, but these methods have some shortcomings. Then, based on the traditional center of gravity method, Chen improved the past insufficient. But in the life, there're many targets with some uncertain attributes can't be solved by the one-dimensional trapezoidal fuzzy numbers because they have multidimensional attributes. Therefore, seeking a method for similarity relation of multidimensional fuzzy numbers is a useful job. This article mainly gives the definition of similarity relation of two-dimensional frustum pyramid fuzzy numbers, discusses its properties, and ranks the targets with some uncertain attributes by the similarity relation. This method combines geometry distance with the virtual center of gravity and volume of two-dimensional frustum pyramid fuzzy numbers together to depict the similarity relation of two-dimensional frustum pyramid fuzzy numbers. The method has some characteristics, such as simple calculation, convenient application, and the results obtained are more rationality, and so on.

D. Yang (Ed.): Informatics in Control, Automation and Robotics, Volume 2, LNEE 133, pp. 49–57.
springerlink.com

2 Preliminaries

In the literature [5], [6], the fuzzy numbers that author discussed is regular, i.e. the maximum value of membership function is 1. In order to apply more widely, the fuzzy numbers this paper relates to does not necessarily require that they are normal, the value of membership function is not 1 can be allowed, i.e. the fuzzy numbers whose maximum value of membership function is ω $(0<\omega\leq 1)$.

2.1 Trapezoidal Fuzzy Numbers

Denote the R as the set of real numbers, $F(R)$ as all the fuzzy subset of R, i.e. $F(R)=\{u \mid u:R\rightarrow[0,1]\}$. $u\in F(R)$, $a,b,c,d\in R$ and $a\leq b\leq c\leq d$, $\omega\in(0,1]$. If the membership function of u satisfies

$$u(x)=\begin{cases}\dfrac{\omega(x-a)}{b-a}, & a\leq x<b\\ \omega, & b\leq x<c\\ \dfrac{\omega(x-d)}{c-d}, & c\leq x<d\\ 0, & x\notin[a,d]\end{cases},$$

then u is called trapezoid fuzzy numbers, denotes as $u=(a,b,c,d;\omega)$.

2.2 Two-Dimensional Frustum Pyramid Fuzzy Numbers

Definition 1: Assume $a_i,b_i,c_i,d_i\in R$ and $a_i\leq b_i\leq c_i\leq d_i$, $\omega\in(0,1]$. $u\in F(R^2)$ $F(R^2)=\{u \mid u:R^2\rightarrow[0,1]\}$, which R^2 is the Cartesian product of R and R. If u satisfies: when $(x_1,x_2)\notin[a_1,d_1]\times[a_2,d_2]$, then $u(x_1,x_2)=0$; when $(x_1,x_2)\in[a_1,d_1]\times[a_2,d_2]$, then the value of $u(x_1,x_2)$ is formed by the side and up bottom of 4 pyramid whose up bottom and down bottom are respectively composed by $\{(x_1,x_2,x_3):(x_1,x_2)\in[b_1,c_1]\times[b_2,c_2],x_3=\omega\}$ and $\{(x_1,x_2,x_3):(x_1,x_2)\in[a_1,d_1]\times[a_2,d_2],x_3=0\}$, i.e. the point $(x_1,x_2,u(x_1,x_2))$ is on the side and up bottom of 4 pyramid, then the u is called two-dimensional frustum pyramid fuzzy numbers, denotes as $\begin{bmatrix}a_1 & b_1 & c_1 & d_1 & \\ a_2 & b_2 & c_2 & d_2 & \end{bmatrix}\omega$.

Assume $a_i,b_i,c_i,d_i\in R$ and $a_i\leq b_i\leq c_i\leq d_i$, $\omega_i\in(0,1]$. $u_i=(a_i,b_i,c_i,d_i;\omega_i)$, $(i=1,2)$ are trapezoidal fuzzy numbers, then $u=\begin{bmatrix}a_1 & b_1 & c_1 & d_1 & \\ a_2 & b_2 & c_2 & d_2 & \end{bmatrix}\omega$ is called two-dimensional frustum pyramid fuzzy numbers which is induced by u_1 and u_2, denotes as $u=[u_1,u_2]^T$ or $u=\begin{bmatrix}a_1 & b_1 & c_1 & d_1 & \omega_1\\ a_2 & b_2 & c_2 & d_2 & \omega_2\end{bmatrix}$, i.e. $\begin{bmatrix}a_1 & b_1 & c_1 & d_1 & \omega_1\\ a_2 & b_2 & c_2 & d_2 & \omega_2\end{bmatrix}=\begin{bmatrix}a_1 & b_1 & c_1 & d_1 & \\ a_2 & b_2 & c_2 & d_2 & \end{bmatrix}\omega$, in which $\omega=\min(\omega_1,\omega_2)$. The Fig.1 is shown as follows.

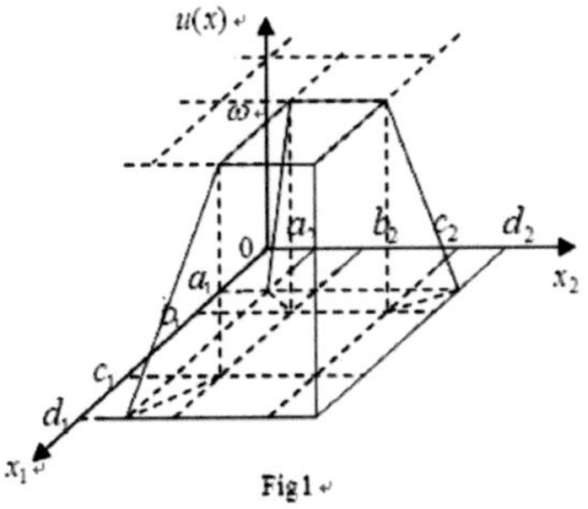

Fig1

2.3 Several Special Two-Dimensional Frustum Pyramid Fuzzy Numbers

Assume $u=\begin{bmatrix} a_1 & b_1 & c_1 & d_1 & \omega \\ a_2 & b_2 & c_2 & d_2 & \end{bmatrix}$ is a two-dimensional frustum pyramid fuzzy numbers, in which $a_i, b_i, c_i, d_i \in R$ and $a_i \le b_i \le c_i \le d_i$, $\omega \in (0,1]$. Then

i) If $\omega = 1$, then u is called a normal two-dimensional frustum pyramid fuzzy numbers;

ii) If $a_1 = a_2, b_1 = b_2, c_1 = c_2, d_1 = d_2$, then u is called a isosceles two-dimensional frustum pyramid fuzzy numbers;

iii) If $b_1 = c_1, b_2 = c_2$, then u is called a pyramidal fuzzy numbers;

iv) If $a_1 = b_1, c_1 = d_1, a_2 = b_2, c_2 = d_2$, then u is called a prismatic fuzzy numbers;

v) If $a_1 = b_1 = c_1 = d_1, a_2 = b_2 = c_2 = d_2$, then u is called a fuzzy point.

2.4 The Fuzzy Operations between Two-Dimensional Frustum Pyramid Fuzzy Numbers

In the literature [7], the author has given out the operations between trapezoidal fuzzy numbers. It is shown as follows:

Assume $u = (a_u, b_u, c_u, d_u; \omega_u)$ and $v = (a_v, b_v, c_v, d_v; \omega_v)$ are two trapezoidal fuzzy numbers, and k is a constant, then

a) $u + v = (a_u + a_v, b_u + b_v, c_u + c_v, d_u + d_v; \min\{\omega_u, \omega_v\})$;

b) $ku = (ka_u, kb_u, kc_u, kd_u; \omega_u)$;

c) $u \cdot v = (a_u \times a_v, b_u \times b_v, c_u \times c_v, d_u \times d_v; \min\{\omega_u, \omega_v\})$.

Then we give out the definition of fuzzy operations between two-dimensional frustum pyramid fuzzy numbers:

Definition 2: Assume $u=\begin{bmatrix} a_{u1} & b_{u1} & c_{u1} & d_{u1} & \omega_u \\ a_{u2} & b_{u2} & c_{u2} & d_{u2} & \end{bmatrix}, v=\begin{bmatrix} a_{v1} & b_{v1} & c_{v1} & d_{v1} & \omega_v \\ a_{v2} & b_{v2} & c_{v2} & d_{v2} & \end{bmatrix}$ are two-dimensional frustum pyramid fuzzy numbers, and k is a constant, then the fuzzy operations are shown as follows:

$$1)\ u+v=\begin{bmatrix} a_{u1}+a_{v1} & b_{u1}+b_{v1} & c_{u1}+c_{v1} & d_{u1}+d_{v1} & \min\{\omega_u,\omega_v\} \\ a_{u2}+a_{v2} & b_{u2}+b_{v2} & c_{u2}+c_{v2} & d_{u2}+d_{v2} & \end{bmatrix};$$

$$2)\ ku=\begin{bmatrix} ka_{u1} & kb_{u1} & kc_{u1} & kd_{u1} & \omega_u \\ ka_{u2} & kb_{u2} & kc_{u2} & kd_{u2} & \end{bmatrix};$$

$$3)\ u\cdot v=\begin{bmatrix} a_{u1}\times a_{v1} & b_{u1}\times b_{v1} & c_{u1}\times c_{v1} & d_{u1}\times d_{v1} & \min\{\omega_u,\omega_v\} \\ a_{u2}\times a_{v2} & b_{u2}\times b_{v2} & c_{u2}\times c_{v2} & d_{u2}\times d_{v2} & \end{bmatrix}.$$

Proposition 1: Assume u_1,u_2,v_1,v_2 are trapezoidal fuzzy numbers, $u=[u_1,u_2]^T$, and $v=[v_1,v_2]^T$ are two-dimensional frustum pyramid fuzzy numbers which are induced by u_1,u_2 and v_1,v_2 respectively, and k is a constant. Then

1) $u+v=[u_1+v_1,u_2+v_2]^T$;

2) $ku=[ku_1,ku_2]^T$;

3) $u\cdot v=[u_1\cdot v_1,u_2\cdot v_2]^T$.

Proof: Assume $u_i=(a_{ui},b_{ui},c_{ui},d_{ui};\omega_{ui})$ and $v_i=(a_{vi},b_{vi},c_{vi},d_{vi};\omega_{vi})$ $(i=1,2)$.

1) On one hand, according to the definition of two-dimensional frustum pyramid fuzzy numbers which is induced by two trapezoidal fuzzy numbers, we can see that

$$u=[u_1,u_2]^T=\begin{bmatrix} a_{u1} & b_{u1} & c_{u1} & d_{u1} & \omega_{u1} \\ a_{u2} & b_{u2} & c_{u2} & d_{u2} & \omega_{u2}\end{bmatrix}=\begin{bmatrix} a_{u1} & b_{u1} & c_{u1} & d_{u1} & \min\{\omega_{u1},\omega_{u2}\} \\ a_{u2} & b_{u2} & c_{u2} & d_{u2} & \end{bmatrix},$$

$$v=[v_1,v_2]^T=\begin{bmatrix} a_{v1} & b_{v1} & c_{v1} & d_{v1} & \omega_{v1} \\ a_{v2} & b_{v2} & c_{v2} & d_{v2} & \omega_{v2}\end{bmatrix}=\begin{bmatrix} a_{v1} & b_{v1} & c_{v1} & d_{v1} & \min\{\omega_{v1},\omega_{v2}\} \\ a_{v2} & b_{v2} & c_{v2} & d_{v2} & \end{bmatrix}.$$

Also according to the definition of two-dimensional frustum pyramid fuzzy numbers addition, we can see that

$$u+v=\begin{bmatrix} a_{u1}+a_{v1} & b_{u1}+b_{v1} & c_{u1}+c_{v1} & d_{u1}+d_{v1} & \min\{\omega_{u1},\omega_{u2},\omega_{v1},\omega_{v2}\} \\ a_{u2}+a_{v2} & b_{u2}+b_{v2} & c_{u2}+c_{v2} & d_{u2}+d_{v2} & \end{bmatrix}.$$

On the other hand, according to the definition of the trapezoidal fuzzy numbers addition and the definition of two-dimensional frustum pyramid fuzzy numbers which is induced by two trapezoidal fuzzy numbers, we can see that

$$[u_1+v_1,u_2+v_2]^T=\begin{bmatrix} a_{u1}+a_{v1} & b_{u1}+b_{v1} & c_{u1}+c_{v1} & d_{u1}+d_{v1} & \min\{\omega_{u1},\omega_{v1}\} \\ a_{u2}+a_{v2} & b_{u2}+b_{v2} & c_{u2}+c_{v2} & d_{u2}+d_{v2} & \min\{\omega_{u2},\omega_{v2}\}\end{bmatrix}$$
$$=\begin{bmatrix} a_{u1}+a_{v1} & b_{u1}+b_{v1} & c_{u1}+c_{v1} & d_{u1}+d_{v1} & \min\{\omega_{u1},\omega_{v1},\omega_{u2},\omega_{v2}\} \\ a_{u2}+a_{v2} & b_{u2}+b_{v2} & c_{u2}+c_{v2} & d_{u2}+d_{v2} & \end{bmatrix}.$$

Therefore, $u+v=[u_1+v_1,u_2+v_2]^T$ is established.

2) On one hand, according to the definition of two-dimensional frustum pyramid fuzzy numbers which is induced by two trapezoidal fuzzy numbers, we can see that

$$u=[u_1,u_2]^T=\begin{bmatrix} a_{u1} & b_{u1} & c_{u1} & d_{u1} & \omega_{u1} \\ a_{u2} & b_{u2} & c_{u2} & d_{u2} & \omega_{u2} \end{bmatrix}=\begin{bmatrix} a_{u1} & b_{u1} & c_{u1} & d_{u1} & \min\{\omega_{u1},\omega_{u2}\} \\ a_{u2} & b_{u2} & c_{u2} & d_{u2} & \end{bmatrix},$$

Also according to the definition of two-dimensional frustum pyramid fuzzy numbers scalar multiplication, we can see that

$$ku=\begin{bmatrix} a_{u1} & b_{u1} & c_{u1} & d_{u1} & \min\{\omega_{u1},\omega_{u2}\} \\ a_{u2} & b_{u2} & c_{u2} & d_{u2} & \end{bmatrix}.$$

On the other hand, according to the definition of the trapezoidal fuzzy numbers scalar multiplication and the definition of two-dimensional frustum pyramid fuzzy numbers which is induced by two trapezoidal fuzzy numbers, we can see that

$$[ku_1,ku_2]^T=\begin{bmatrix} ka_{u1} & kb_{u1} & kc_{u1} & kd_{u1} & \omega_{u1} \\ ka_{u2} & kb_{u2} & kc_{u2} & kd_{u2} & \omega_{u2} \end{bmatrix}=\begin{bmatrix} ka_{u1} & kb_{u1} & kc_{u1} & kd_{u1} & \min\{\omega_{u1},\omega_{u2}\} \\ ka_{u2} & kb_{u2} & kc_{u2} & kd_{u2} & \end{bmatrix}.$$

Therefore, $ku=[ku_1,ku_2]^T$ is established.

3) The proof method is similar with the two former methods.

3 Geometrical Quantity of Two-Dimensional Frustum Pyramid Fuzzy Numbers

We know that calculating the center of gravity of a three-dimensional graphics is used the method of triple integral. The method is shown as follows: Let V be an object in space whose density function is $\rho(x,y,z)$, and $\rho(x,y,z)$ is continuous in the V. When the density of the object V is uniform, i.e. $\rho(x,y,z)$ is a constant, then the center of gravity coordinate of the space object is

$$\bar{x}=\frac{1}{V}\iiint xdV \qquad \bar{y}=\frac{1}{V}\iiint ydV \qquad \bar{z}=\frac{1}{V}\iiint zdV \tag{1}$$

Then the points which two-dimensional frustum pyramid fuzzy numbers corresponds to are on the side and up bottom of 4 pyramid, and the graphics which is composed by two-dimensional frustum pyramid fuzzy numbers can be seen as an object with uniform density in space. Therefore, according to (1), we can get the center of gravity of two-dimensional frustum pyramid fuzzy numbers. It is complex to compute the center of gravity of two-dimensional frustum pyramid fuzzy numbers with the method of triple integral and the amount of calculation is large. And this method can not calculate the center of gravity of fuzzy point. So we put forward to a new method to calculate the center of gravity of two-dimensional frustum pyramid fuzzy numbers.

Assume $u=\begin{bmatrix} a_1 & b_1 & c_1 & d_1 & \omega \\ a_2 & b_2 & c_2 & d_2 & \end{bmatrix}$ is a two-dimensional frustum pyramid fuzzy numbers, $a_i,b_i,c_i,d_i\in R$ and $a_i\le b_i\le c_i\le d_i, i=1,2. \omega\in(0,1]$. Then the definition of virtual center of gravity is shown as follows:

$$y=\begin{cases}\dfrac{\omega\left(\dfrac{(c_1-b_1)(c_2-b_2)}{(d_1-a_1)(d_2-a_2)}+2\right)}{6}, & d_1\neq a_1, d_2\neq a_2,\\ \dfrac{\omega}{2}, & \text{others}\end{cases}$$

$$\begin{cases}x_1=\dfrac{c_2+b_2-(d_2+a_2)}{2\omega}y+\dfrac{d_2+a_2}{2}\\ x_2=\dfrac{c_1+b_1-(d_1+a_1)}{2\omega}y+\dfrac{d_1+a_1}{2}\end{cases}, d_1\neq a_1, d_2\neq a_2, \begin{cases}x_1=\dfrac{d_2+a_2}{2}\\ x_2=\dfrac{d_1+a_1}{2}\end{cases}, \text{others}. \quad (2)$$

The definition of volume is shown as follows:

$$V=\begin{cases}\dfrac{\omega}{2}[(d_1-a_1)+(c_1-b_1)+(d_2-a_2)+(c_2-b_2)] & a_1=d_1 \text{ or } a_2=d_2\\ \dfrac{\omega}{2}[(d_1-a_1)(d_2-a_2)+3(c_1-b_1)(c_2-b_2)] & \text{others}\end{cases} \quad (3)$$

4 Similarity Relation between Two-Dimensional Frustum Pyramid Fuzzy Numbers

In discussing the problem of ranking two-dimensional frustum pyramid fuzzy numbers whose two component are both trapezoidal fuzzy numbers, in order to unify dimension, we restrict that the parameters a,b,c,d satisfies $0\le a\le b\le c\le d\le 1$.

Definition 3: Assume $u=\begin{bmatrix}a_1 & b_1 & c_1 & d_1 & \\ a_2 & b_2 & c_2 & d_2 & \end{bmatrix}\omega$ and $v=\begin{bmatrix}a_1' & b_1' & c_1' & d_1' & \\ a_2' & b_2' & c_2' & d_2' & \end{bmatrix}\omega'$ are two two-dimensional frustum pyramid fuzzy numbers, in which $0\le a_i\le b_i\le c_i\le d_i\le 1$, $0\le a_i'\le b_i'\le c_i'\le d_i'\le 1, i=1,2$, and $\omega,\omega'\in(0,1]$. Then the similarity between u and v is shown as follows:

$$S(u,v)=\left(1-\frac{\sum_{i=1}^{2}\left(|a_i-a_i'|+|b_i-b_i'|+|c_i-c_i'|+|d_i-d_i'|\right)}{2\sum_{i=1}^{2}\dfrac{|d_i+d_i'|}{B(a_i,a_i')}+C(a_i,a_i')}\right),$$

$$\times\frac{1}{1+|x_1-x_1'|+|x_2-x_2'|}\times[1-(V_u-V_v)]\times\frac{\min\{\omega,\omega'\}}{\max\{\omega,\omega'\}}$$

in which $B(a_i,a_i')=\begin{cases}1, & a_i=0, a_i'=0\\ |a_i+a_i'|, & \text{others}\end{cases}$, $C(a_i,a_i')=\begin{cases}4, & a_i=0, a_i'=0\\ 0, & \text{others}\end{cases}$.

4.1 The Properties of Similarity Relation

Property 1: Two-dimensional frustum pyramid fuzzy numbers u, v are identical if and only if $S(u,v)=1$.

Proof: ($\Rightarrow$) If u and v are identical, we can see that $a_i = a_i^{'}, b_i = b_i^{'}, c_i = c_i^{'}$, $d_i = d_i^{'}$, and . Thus, according to (2), (3) we can get the virtual center of gravity and the value of volume. We can see that $x_1 = x_1^{'}, x_2 = x_2^{'}, V_u = V_v$. If $a_i = 0, a_i^{'} = 0$, then $B(a_i, a_i^{'}) = 1$, So $S(u,v) = 1$. If a_i and $a_i^{'}$ are others, then $B(a_i, a_i^{'}) = \left| a_i + a_i^{'} \right|$, $C(a_i, a_i^{'}) = 0$. So $S(u,v) = 1$. Therefore we can see that $S(u,v) = 1$.

($\Leftarrow$) If $S(u,v) = 1$, then $\sum_{i=1}^{2}\left(\left|a_i - a_i^{'}\right| + \left|b_i - b_i^{'}\right| + \left|c_i - c_i^{'}\right| + \left|d_i - d_i^{'}\right|\right) = 0$, it implies that $a_i = a_i^{'}, b_i = b_i^{'}, c_i = c_i^{'}, d_i = d_i^{'}, i = 1,2$. According to the definition of similarity relation, we can see that $\frac{\min\left\{\omega, \omega^{'}\right\}}{\max\left\{\omega, \omega^{'}\right\}} = 1$, so $\omega = \omega^{'}$. From (2) and (3), we can see that $x_1 = x_1^{'}$, $x_2 = x_2^{'}, V_u = V_v$. Therefore, we can get $u = v$.

Property 2: $S(u,v) = S(v,u)$.

Proof: According to the definition of similarity relation of two-dimensional frustum pyramid fuzzy numbers, we can see that it is right obviously.

Property 3: If $u = \begin{bmatrix} a & a & a & a & \\ b & b & b & b & 1 \end{bmatrix}$ and $v = \begin{bmatrix} a^{'} & a^{'} & a^{'} & a^{'} & \\ b^{'} & b^{'} & b^{'} & b^{'} & 1 \end{bmatrix}$ are two-dimensional frustum pyramid fuzzy numbers. Then $S(u,v) = 1 - \left|a - a^{'}\right| - \left|b - b^{'}\right|$.

Proof: From (2) and (3), we can see that $x_1 = x_1^{'}, x_2 = x_2^{'}, V_u = V_v$. If $a_i = 0, a_i^{'} = 0$, $a_i^{'} = 0$, then $B(a_i, a_i^{'}) = 1, C(a_i, a_i^{'}) = 4$. Thus $S(u,v) = 1 - \left|a - a^{'}\right| - \left|b - b^{'}\right|$. If a_i and $a_i^{'}$ are other values, then $B(a_i, a_i^{'}) = \left|a_i + a_i^{'}\right|, C(a_i, a_i^{'}) = 0$, thus $S(u,v) = 1 - \left|a - a^{'}\right| - \left|b - b^{'}\right|$.

Assume $u = (u_1, u_2)$ and $v = (v_1, v_2)$ are two-dimensional fuzzy vectors, u_1, u_2 and v_1, v_2 are both trapezoid fuzzy numbers, then the similarity between u and v is defined as the similarity between the two-dimensional frustum pyramid fuzzy numbers which are induced respectively by u_1, u_2 and v_1, v_2 , and denotes as $S(u,v)$, too.

5 The Example for Practical Application

Example: To a producer, the good or bad working state of the producer is related to the product quantity and quality in one working day. The following data is obtained from a factory with 10 producers' work performance in 10 days.

	P_{11}	P_{21}	P_{31}	P_{41}	P_{51}	P_{61}	P_{71}	P_{81}	P_{91}	P_{101}	P_{12}	P_{22}	P_{32}	P_{42}	P_{52}	P_{62}	P_{72}	P_{82}	P_{92}	P_{102}
D_1	1200	1173	1107	994	1092	1092	1153	1200	1051	1093	98.3	98.4	94.3	87.8	89.4	89.3	93.7	96.3	98.4	95.6
D_2	1000	1042	1128	1200	1038	1086	1129	1198	1033	1120	99.2	97.1	97.5	89.9	94.5	94.5	91.4	98.6	94.6	97.1
D_3	1101	997	928	1036	1259	1102	1084	1145	998	1139	9801	97.9	89.9	92.6	93.4	88.8	91.8	99.1	95.4	98.3
D_4	980	935	1148	1029	994	1117	1078	1138	1173	1140	98.9	99.4	90.3	90.6	96.6	93.5	89.9	98.3	97.5	95.3
D_5	993	1102	974	1147	1100	1185	992	983	1039	1203	94.8	98.1	91.5	89.7	96.3	93.8	97.7	91.3	96.1	96.2
D_6	968	1098	1250	1083	1048	950	994	995	1167	1240	99.3	98.0	96.7	86.8	97.6	84.9	98.9	91.2	94.6	94.8
D_7	1150	1163	985	981	1176	1130	1041	998	1193	993	98.1	96.9	88.7	90.5	95.4	89.5	96.4	93.3	96.8	93.5
D_8	1012	1058	990	937	1093	989	1021	1001	1129	989	98.5	97.1	92.9	93.4	97.1	92.7	96.5	94.9	99.5	96.4
D_9	1121	954	1046	1102	1094	994	1232	1038	1102	993	99.1	98.5	94.2	91.8	98.8	94.4	95.3	94.7	95.4	97.2
D_{10}	1034	1203	1092	1096	1101	1019	1192	1129	1029	1067	98.7	99.3	94.6	93.8	94.4	91.5	94.9	97.5	96.4	95.9

We denote P_{i1} where $P_{i1} \in [0,+\infty], i=1,2$, as the number of products what the first i person (denotes as P_i) produce (unit: part) in one day. And we denote P_{i2} where $P_{i2} \in [0,100]$, $i=1,2,\cdots 10$ as the production of the qualified rate of products what the first i individual produce (unit :%) in one day. Denote D_i as the first i day. We need to sort the 10 producers.

For each of their "ten day's working state", we can construct a two-dimensional fuzzy vectors $u_{P_i} = \left(u_{P_i}^j, u_{P_i}^j\right), i=1,2,\cdots 10, j=1,2$. According to (2) and (3), we can obtain respectively the volumes and virtual centers of gravity of ten two-dimensional frustum pyramid fuzzy numbers. To everyday, take the maximum value of the number of products and the highest qualified rate of products what the 10 individuals product, and then we can construct an "ideal state standard" P' . To P' , we can also construct a two-dimensional fuzzy vector $u_{P'} = \left(u_{P'}^1, u_{P'}^2\right)$. According to the definition of similarity relation between two-dimensional frustum pyramid fuzzy numbers, we can obtain the similarity between $u_{P'}$ and $u_{P_i} (i=1,2,\cdots 10)$,i.e. $S(u_{P_3}, u_{P'})=0.5502$, $S(u_{P_4}, u_{P'})=0.5479$, $S(u_{P_5}, u_{P'})=0.5152$, $S(u_{P_6}, u_{P'})=0.5886$, $S(u_{P_7}, u_{P'})=0.6274$, $S(u_{P_8}, u_{P'})=0.6019$, $S(u_{P_9}, u_{P'})=0.7057$, $S(u_{P_{10}}, u_{P'})=0.5695$. Then in order to know which producer is better performance, we can make a comparison between the "ten day's working state" of ten produces and the "ideal state standard". Obviously, $S(u_{P_1}, u_{P'})=0.7233$ is the biggest. Therefore, we can believe the producer P_1 is better performance than others.

Acknowledgments. This work is supported by Natural Science Foundations of China (No. 60934009).

References

1. Lee, H.S.: An Optimal Aggregation Method for Fuzzy Opinions of Group Decision. In: Proc. 1999 IEEE Int. Conf. Systems, Man, Cybernetics, vol. 3, pp. 314–319 (1999)
2. Hsieh, C.H., Chen, S.H.: Similarity of Generalized Fuzzy Numbers with Graded Mean Integration Representation. In: Proc. 8th Int. Fuzzy Systems Association World Congr., Taipei, Taiwan, Republic of China, vol. 2, pp. 551–555 (1999)

3. Chen, S.J., Chen, S.M.: A New Simple Center-of-gravity Method for Handling the Fuzzy Ranking and the Defuzzification Problems. In: 8th National Conf. Fuzzy Theory Applications, Taipei, Taiwan, Republic of China (2000)
4. Chen, S.M.: New Methods for Subjective Mental Workload Assessment and Fuzzy Risk Analysis. Cybern. Syst.: Int. J. 27(5), 449–472 (1996)
5. Wang, G.X., Wu, C.X.: Fuzzy n-cell Numbers and the Differential of Fuzzy n-cell Value Mappings. Fuzzy Sets and System 130, 367–381 (2003)
6. Wang, G.X., Li, Y.M., Wen, C.L.: On Fuzzy n-cell Numbers and n-dimension Fuzzy Vectors. Fuzzy Sets and System 158, 71–84 (2007)
7. Chen, S.J.: Fuzzy Risk Analysis Based on Similarity Measures of Generalized Fuzzy Numbers. IEEE Transactions on Fuzzy Systems 11, 45–56 (2003)

Robust Controller Design for One Arm Space Manipulator with Uncertainties Compensated

Guangyan Xu, Meng Zhang, and Huanqiang Wang

School of Automation, Shenyang Aerospace University
110136, Shenyang, China
Guangyan_xu@163.com

Abstract. According to trajectory tracking control of space robot with parameters uncertainties, the disturbance on space robot trajectory tracking is taken into consideration. Firstly, robust controller is applied to compensate the coriolis force, centrifugal force and other disturbance. Robust control law with items uncertainties compensated is improved. The desired time-varying trajectory can be real-time tracked by manipulators while the attitude of space robot can be controlled. The compensation matrix is proposed. Uncertainty item range can be solved based on PSO and it can be applied to control manipulators. The simulation results show that the robust controller has a good performance.

Keywords: space manipulator, trajectory tracking, robust control, disturbance compensated.

1 Introduction

Free flying space robot will be widely applied to accomplish space tasks. Trajectory tracking is not only helpful to complete space missions but also reduce the disturbance to attitude. Therefore, the attitude can be controlled while space tasks are accomplished.

Many domestic and foreign experts have a thorough research in the fields of dynamics analysis, attitude control and trajectory tracking. An adaptive robust controller for one-arm space manipulator is designed to accomplish trajectory tracking and the stability analysis is based on Lyapunov theory[1]. RBF neutral network is applied to approach the nonlinear dynamics model of space robot, and then tracking control based on robust controller in space task is researched[2]. In paper[3], an adaptive controller is designed by GU YL. The results show that it has a good performance in trajectory tracking. However, dynamics equations are needed to be linearization and complex pre-calculation is needed to derive regression matrix. In fact, the regression matrix is difficult to obtain for the complex of space robot. The dynamics model of the space robot is derived based on Generalized Jacobin Matrix by Chen L and the corporative motion of joint and attitude is completed by the robust adaptive controller [4]. An optimization of minimum fuel consumption is derived by WF XU. Joint motion is planned to control the attitude and joint angles [5]. A.Green and J.Z.Sasiadek respectively apply inverse flexible dynamics control (IFDC) and fuzzy logic system adaptive control (FLSAC) to control the flexible space

D. Yang (Ed.): Informatics in Control, Automation and Robotics, Volume 2, LNEE 133, pp. 59–66.
springerlink.com

manipulators. Simulation results show that joint motion has a greatly disturbance to the orbits of space robot when IFDC is the control strategy. In contrast, FLSAC control strategy has a better performance[6]. However, there are too much computing. A robust adaptive controller is designed to control trajectory based on E.Platform[7]. In paper[8], on the base of PD control strategy, a decentralized nonlinear robust control is proposed. The tracking error is global convergence. In paper [9], the optimization of trajectory tracking can be derived by Karol when the space craft is docking. Experiments prove that the controller has a good effect, but the attitude of space manipulator is no taken into consideration.

In the paper, space robot with parameters uncertainties compensated is controlled based on robust control theory of uncertainty-system. Firstly, centrifugal, coriolis forces and attitude disturbance are compensated and the control law is improved to avoid the singular. Stability analysis is proved by Lyapunov theory. In order to solve the ranges of uncertainty-parameters and improve the performance, Practical Swarm Optimization is applied to obtain the compensated matrix. Global exponential stability of Robust controller is the difference.

Compared with paper[10], control law is improved and Robust controller is global exponential stability. Especially, control torque is more slightly. Simulation results show that it has a better performance than the previous controller.

2 Dynamics Equations of Space Robot

Dynamics equations of space robot with n rigid body are derived based on Lagrange Equations. Geopotential energy and Perturbing force are ignored. Dynamics equations of multi-rigid body space robot are given by:

$$M_{11}(\mathbf{q})\ddot{\boldsymbol{\varphi}}+M_{12}(\mathbf{q})\ddot{\boldsymbol{\theta}}+C_1(\mathbf{q},\dot{\mathbf{q}})\dot{\mathbf{q}}=G\boldsymbol{\tau}_{\varphi} \tag{1}$$

$$M_{21}(\mathbf{q})\ddot{\boldsymbol{\varphi}}+M_{22}(\mathbf{q})\ddot{\boldsymbol{\theta}}+C_2(\mathbf{q},\dot{\mathbf{q}})\dot{\mathbf{q}}=\boldsymbol{\tau}_{\theta} \tag{2}$$

$$C_2(\mathbf{q},\dot{\mathbf{q}})=\sum_{j=1}^{3+n}\dot{q}_j C_{2j}(\mathbf{q}) \tag{3}$$

Where $\varphi \in R^{3\times 1}$ is attitude vector, $\boldsymbol{\theta} \in R^{n\times 1}$ is the vector of space manipulator joint angles. n stands for the number of kinematic pairs, $q=[\boldsymbol{\varphi}^T \ \boldsymbol{\theta}^T]^T \in R^{3+n}$ is the vector of generalized coordinates of the manipulator, the control torque of attitude $\boldsymbol{\tau}_c \in R^{3\times 1}$ is relative to the center of mass coordinate. $\boldsymbol{\tau}_{\theta} \in R^{n\times 1}$ is the joint of generalized control torque, $M_{11}(\mathbf{q}) \in R^{3\times 3}$, $M_{12}(\mathbf{q}) \in R^{3\times n}$, $M_{21}(\mathbf{q}) \in R^{n\times 3}$, $M_{22}(\mathbf{q}) \in R^{n\times n}$ are inertial matrix, $C_1(\mathbf{q},\dot{\mathbf{q}}) \in R^{3\times n}$, $C_2(\mathbf{q},\dot{\mathbf{q}}) \in R^{n\times n}$ are the n dimensional vector representing centrifugal and coriolis forces, $G(\mathbf{q}) \in R^{3\times 3}$ is the transform matrix.

In this paper, attitude control is the key to research so that there are no parameters relative to the orbit position of space robot.

3 Control System Design of Space Robot

In attitude control flying, space robot has some implies that existing:

1) Body's inertia is much greater than joint
2) Attitude, velocity of attitude, joint angle and speed of joint angle can be obtained.

Joint control torque can be obtained based on (2)

$$\boldsymbol{\tau}_{\boldsymbol{\theta}} = M_{22}(\mathbf{q})\mathbf{u} \tag{4}$$

So the acceleration of joint can be given by

$$\ddot{\boldsymbol{\theta}} = \mathbf{u} - \boldsymbol{\eta} \tag{5}$$

Where $\mathbf{u}$ is the control, the disturbance $\boldsymbol{\eta}$ is due to centrifugal and coriolis forces and it is given by

$$\boldsymbol{\eta} = M_{22}^{-1}\left(M_{21}\ddot{\boldsymbol{\varphi}} + C_2(\mathbf{q},\dot{\mathbf{q}})\dot{\mathbf{q}}\right) \tag{6}$$

Feedback control law is

$$\ddot{\boldsymbol{\theta}}^d - \ddot{\boldsymbol{\theta}} = -K_{p1}\left(\boldsymbol{\theta}^d - \boldsymbol{\theta}\right) - K_{d1}\left(\dot{\boldsymbol{\theta}}^d - \dot{\boldsymbol{\theta}}\right) - \left(\Delta\mathbf{u} - \boldsymbol{\eta}\right) \tag{7}$$

Where θ^d is the vector of desired joint angles, K_{p1} and K_{d1} are feedback gains, $\Delta\mathbf{u}$ is the compensate term of disturbance $\boldsymbol{\eta}$.

Based on equation (6), the closed-loop system can be written as

$$\dot{\mathbf{x}} = A\mathbf{x} + B(\boldsymbol{\eta} - \Delta\mathbf{u}) \tag{8}$$

Where

$$\mathbf{x} = \begin{bmatrix} \mathbf{x}_1 \\ \mathbf{x}_2 \end{bmatrix} = \begin{bmatrix} \boldsymbol{\theta}^d - \boldsymbol{\theta} \\ \dot{\boldsymbol{\theta}}^d - \dot{\boldsymbol{\theta}} \end{bmatrix} \tag{9}$$

$$A = \begin{bmatrix} 0 & I \\ -K_{p1} & -K_{d1} \end{bmatrix} \quad B = \begin{bmatrix} 0 \\ -I \end{bmatrix} \tag{10}$$

Several important properties of space robot may be derived

1) Generalized coordinates are bounded
2) With the fuel exhaust, inertia improved, and many other uncertainty prosperities as follows

$$\|M_{21}(\mathbf{q})\| \le \alpha_1 \tag{11}$$

$$\lambda_m \le \|M_{22}(\mathbf{q})\| \le \lambda_M \tag{12}$$

$$\|\boldsymbol{\eta}\| \le \boldsymbol{\rho} \tag{13}$$

α_1, λ_m and λ_M are the given positive parameters.

Define

$$\boldsymbol{\eta} = \sum_{j=1}^{3+n} \dot{q}_j D_{1j} \dot{\mathbf{q}} + D_2 \ddot{\boldsymbol{\varphi}} \tag{14}$$

Where

$$D_{1j} = M_{22}^{-1} C_{2j} \qquad D_2 = M_{22}^{-1} M_{21} \tag{15}$$

So the results can be obtained based on equation (14), (15)

$$\rho(t) = \eta^{\max}(i) = \sum_{j=1}^{3+n} \sum_{k=1}^{3+n} |d_{ijk}| |\dot{q}_j \dot{q}_k| + \sum_{k=1}^{3} |d_{ik}| |\ddot{\varphi}_k| \tag{16}$$

d_{ijk} and d_{ik} are respectively the k elements of the i column in D_{1j} and D_2.That $d_{ijk}=D_{1j}(i,k)$ $d_{ik}=D_2(i,k)$. Solution of $\eta^{\max}(i)$ can be referenced paper[10].

Lmma1 (Global exponential stability) taken into the system

$$\dot{\mathbf{x}} = \mathbf{f}(\mathbf{x}, t) \qquad \mathbf{x}(t_0) = \mathbf{x}_0 \tag{17}$$

Assumption that there exist $V(\mathbf{x}, t)$, λ_1 and λ_2, yields to

$$\lambda_1 \|\mathbf{x}\|^2 \le V(\mathbf{x}) \le \lambda_2 \|\mathbf{x}\|^2 \qquad \dot{V}(\mathbf{x}) \le -\lambda_2 \|\mathbf{x}\|^2 + \gamma \exp(-\beta t) \tag{18}$$

If $\lim_{t \to \infty} \gamma \exp(-\beta t) = 0$ then (17) is global exponential stability. Because state matrix A is positive definite, there is positive definite matrix $P > 0$, yields to

$$A^T P + PA = -Q \tag{19}$$

Where Q is positive definite matrix. The control law is designed to compensate the disturbance $\boldsymbol{\eta}$.

$$\Delta \mathbf{u} = -\frac{\rho(t) \cdot PB\mathbf{x}}{\|PB\mathbf{x}\| + \gamma \exp(-\beta t)} \tag{20}$$

Stability of the robust controller can be proved based on (21)

$$V(\mathbf{x}) = \frac{1}{2} \mathbf{x}^{\mathrm{T}} P \mathbf{x} \tag{21}$$

Then derivation (21)

$$\begin{aligned}\dot{V}(\mathbf{x}) &= \frac{1}{2}\dot{\mathbf{x}}^{\mathrm{T}}P\mathbf{x}+\frac{1}{2}\mathbf{x}^{\mathrm{T}}P\dot{\mathbf{x}} \\ &= -\frac{1}{2}\mathbf{x}^{T}Q\mathbf{x}+\mathbf{x}^{\mathrm{T}}PB(\Delta\mathbf{u}-\boldsymbol{\eta}) \\ &\le -\frac{1}{2}\mathbf{x}^{T}Q\mathbf{x}+\mathbf{x}^{\mathrm{T}}PB\Delta\mathbf{u}+\left\|\mathbf{x}^{T}PB\right\|\left\|\boldsymbol{\eta}\right\| \\ &\le -\frac{1}{2}\mathbf{x}^{T}Q\mathbf{x}+F_1+F_2 \\ &\le -\frac{1}{2}\mathbf{x}^{T}Q\mathbf{x}+\frac{\left\|\mathbf{x}^{T}PB\right\|\rho(t)}{\left\|PB\mathbf{x}\right\|+\gamma\exp(-\beta t)}\gamma\exp(-\beta t) \\ &\le -\frac{1}{2}\mathbf{x}^{T}Q\mathbf{x}+\rho(t)\gamma\exp(-\beta t)\end{aligned} \tag{22}$$

Where

$$F_1=\frac{-\mathbf{x}^{\mathrm{T}}PB(PB\mathbf{x})\rho(t)+\left\|\mathbf{x}^{T}PB\right\|\left\|PB\mathbf{x}\right\|\rho(t)}{\left\|PB\mathbf{x}\right\|+\gamma\exp(-\beta t)} \tag{23}$$

$$F_2=\left\|\mathbf{x}^{T}PB\right\|\rho(t)\gamma\exp(-\beta t) \tag{24}$$

Therefore, the robust controller is taken as

$$\tau_u=M_{22}(\theta)\Delta\mathbf{u} \qquad \tau_L=M_{22}(\theta)(\mathbf{u}-\Delta\mathbf{u}) \tag{25}$$

$$\boldsymbol{\tau}_{\boldsymbol{\theta}}=\boldsymbol{\tau}_L+\boldsymbol{\tau}_u \tag{26}$$

4 Simulation and Analysis

In order to verify the effect of the designed robust controller, the corporative simulation of trajectory tracking and attitude control is accomplished. Space robot with 5 degree of freedom and its mathematical model can be obtained in equations (1) and (2).

Attitude controller is by PD controller and the control parameters

$$K_{p2}=[800 \quad 2900 \quad 2000] \; K_{d2}=[460 \quad 1970 \quad 1270]$$

besides, mass of the system center is the inertial mass coordinates origin and the original parameters in inertial mass coordinates as follows

Initial Attitude (rad): $\varphi(0)=[0.0698 \quad 0.0349 \quad 0.0349]^T$

Desired trajectory of attitude: $\varphi^d=[0 \quad 0.0010663t \quad 0]^T$

Initial joint angels: $\theta=\mathbf{0}$

Table 1. Parameters of space robot

Parameters	Base	L1	L2	L3	units
Link length		0.6	1	1	m
mass	400	40	30	30	kg
Ixx	80	1	1	1.3	
Iyy	96	5	3	3	
Izz	40	0.8	2.8	0.5	(Kg*m2)
Ixy	35	-0.08	-0.28	0.01	
Ixz	-38	-0.8	4	3	
Iyz	-35	0.17	0.08	-0.3	

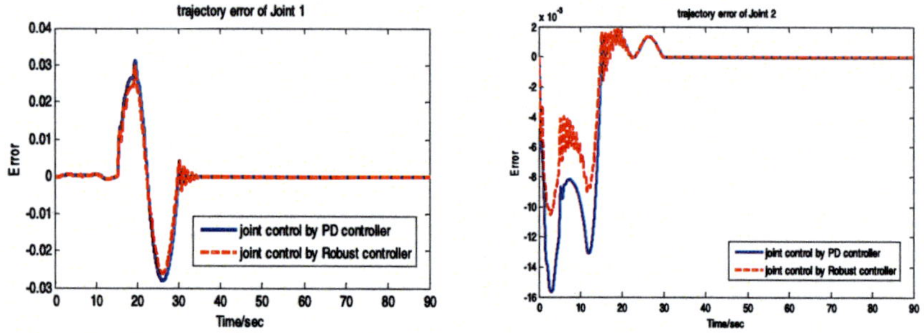

Fig. 1. Trajectory tracking error (end effecter no load)

Fig.1 shows the result of joint trajectory tracking. There is still minimal oscillation and tracking error when PD is the control strategy. However, space manipulator can be controlled in accordance with the desired position. In contrast, robust controller has better performance than PD controller.

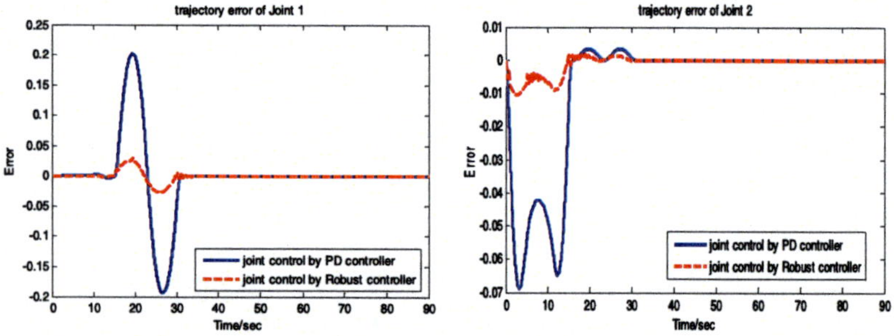

Fig. 2. Trajectory tracking error (end effecter with load)

Fig.2 shows the result of trajectory tracking error and the target is in end effecter. In task space, there is serious tracking error for vary of inertia when joint motion is controlled by PD controller. Conversely, robust controller has a better performance.

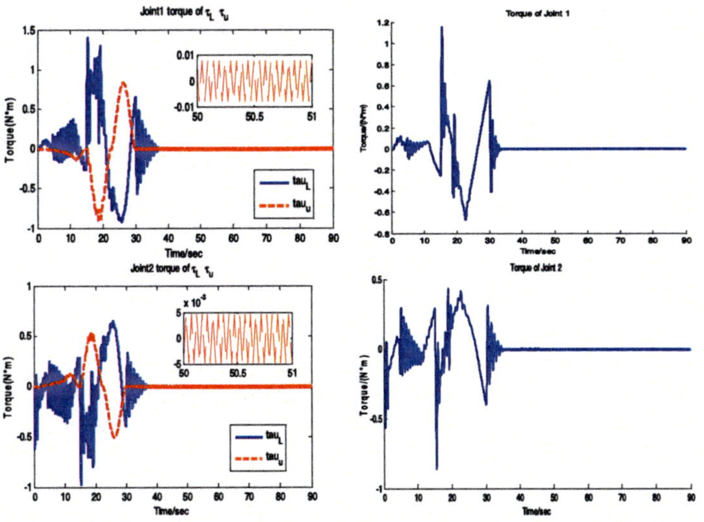

Fig. 3. Torque of Joint 1 and Joint 2 (end effecter no load)

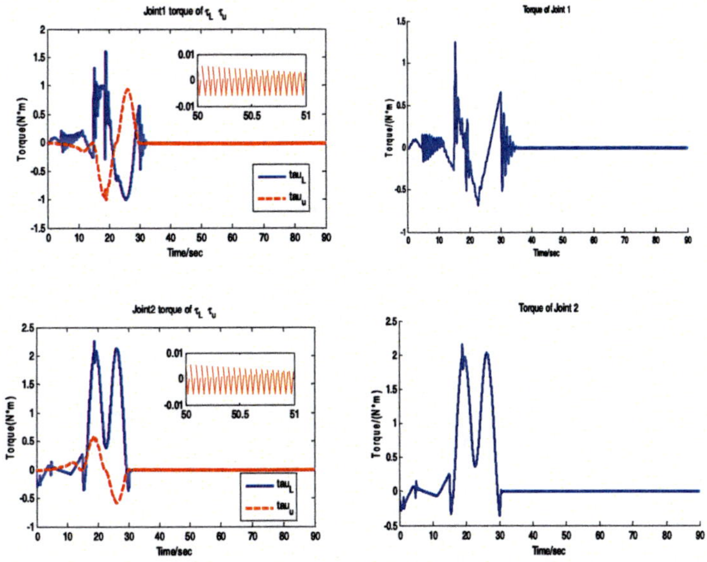

Fig. 4. Torque of Joint 1 and Joint 2 (end effecter with load)

From Fig.3-Fig.4, it is that the torque is different between end effecter no load and end effecter with load; τ_L and τ_u are a complementary relation so that the control torque can be decreased; The torque of Joint 2 is higher when space manipulator in docking mode.

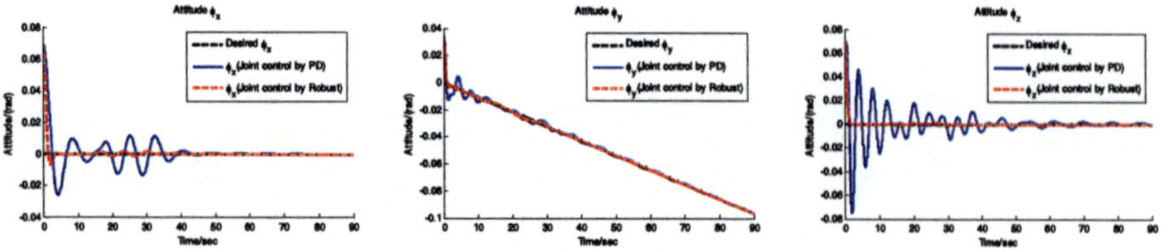

Fig. 5. Attitude trajectory tracking (end effecter with load)

Conclusion

In this paper, robust controller is improved. Trajectory tracking is much better than paper[11]; Torque of Joint can be analyzed and there are much more fuel will be saved. Besides, attitude control is more easily.

Acknowledgements. The authors would like to thank the editors and referees of this proceeding. This work is supported by East red Satellite Company.

References

1. Matsuda, K., Kanemitsui, Y.H., Kijimoto, S.: Attitude control of a space robot using the nonholonomic structure. In: AIAA (1998)
2. Feng, B., Ma, G., Wen, Q., Wang, C.: Design of Robust Intelligent Controller for Space Robot in Task Space. Journal of Astronautics (April 2007)
3. Gu, Y.L., Xu, Y.S.: A normal form augmentation approach to adaptive control of space robot systems. In: Conf. on Robotics and Automation, Atlanta, USA, pp. 731–737 (1993)
4. Chen, L.: Adaptive and robust composite control of coordinated motion of space robot system with prismatic joint. In: Intelligent Control and Automation, China (2002)
5. Xu, W.: Path Planning for Base Attitude Adjustment of a Free-floating Space Robot System. Robotics 28(3), 291–296 (2006)
6. Green, A., Sasiadek, J.Z.: Intelligent Tracking Control of a Free-Flying Flexible Space Robot Manipulator. In: Navigation and Control Conference, AIAA-2007-6320 (2007)
7. Yime, E., Saltaren, R., Diaz, J.: Robust adaptive control of the Stewart-Gough robot in the task space. In: American Control Conference (ACC), pp. 5248–5253 (2010)
8. Liu, M.: Decentralized PD and robust nonlinear control for robot manipulators. Journal of Intelligent and Robotic Systems 20(2), 319–332 (1997)
9. Seweryn, K., Banaszkiewicz, M.: Optimization of the Trajectory of a General Free – Flying Manipulator during the Rendezvous Maneuver. In: AIAA Guidance, Honolulu (2008)
10. Xu, G., Zhang, M.: Robust control of One Arm Space Manipulator. In: IPTC (2011)

Design of Quasi-periodic Satellite Formation at Critical Inclination

Guangyan Xu, Huanqiang Wang, and Meng Zhang

School of Automation, Shenyang Aerospace University
110136, Shenyang, China
Guangyan_xu@163.com

Abstract. Periodic or quasi-periodic satellite formations can minimize the active control of long-term satellite formation flying. In this paper, a design of quasi-periodic satellite formation is proposed. The quasi-periodic constrains and mean relative motion of the member satellite under J_2 perturbation are used. The reference orbit must be at the critical inclination with a small eccentricity. The quasi-periodic constrains can release more DOFS to design the geometry formation than strict J_2 invariant trajectory conditions. The osculating orbital elements are used in numerical simulation, the design is proved to be correct through the simulation.

Keywords: satellite formation; J_2 perturbation; quasi-periodic conditions; relative orbit; osculating orbital elements.

1 Introduction

With the development of the space technology, the small satellite formation technology has become an issue research. There are many important properties in the frozen orbit, so designing satellite formations on the orbit has important significance. It is very necessary to design the satellite formations, analyses the effect of all kinds of perturbations and research the formation maintenance and control at critical inclination.

Many scholars have already done a lot of work on satellite formation research. Chris Sabol, Hanspter and Schaub etc have got important achievements by using the Hill or improved Hill equation in [1][2]. The design has been researched by Zhang Yu-kun through the kinematics and dynamics and the perturbation is analyzed in [3], however, the small eccentricity elliptical orbit of reference satellite is not analyzed; Guoqiang Zeng has put forward a general design of the satellite formation and the effect of various perturbations is analyzed through numerical method in [4]. We do a lots work on the periodic or quasi-periodic satellite formation design and perturbation analysis at critical inclination and make some achievements in [5][6].

In this paper, a design of quasi-periodic satellite formation is proposed. The reference orbit must be at the critical inclination with a small eccentricity. The osculating orbital elements are used in numerical simulation, it show the satellite formation can accommodate J_2 perturbation effectively, so the design is proved to be correct.

D. Yang (Ed.): Informatics in Control, Automation and Robotics, Volume 2, LNEE 133, pp. 67–74.
springerlink.com

2 Design of Quasi-periodic Satellite Formation

A passive and periodic satellite formation is critical to sustain a long-term formation flying mission. It is proved in [7][8] that the passive and periodic satellite formations in classical Keplerian motion can be achieved as long as every satellite in the formation has the same orbital period *T*, which is equivalent to having the same mean motion $\dot{M}$, or same semi-major axis *a*.

$$\delta T = 0 \Leftrightarrow \delta \dot{M} = 0 \Leftrightarrow \delta a = 0 \tag{1}$$

A passive and periodic satellite formation accommodating the J_2 perturbation will ensure long-term flying formation practically possible. Including the J_2 gravity term into the classical Keplerian motion causes the secular growths of three mean orbit elements $(\bar{\Omega}, \bar{\omega}, \bar{M})$ i.e., the right ascension of the ascending node $\bar{\Omega}$, the argument of perigee $\bar{\omega}$ and the mean anomaly $\bar{M}$. The J_2 perturbation lead the rates of differential orbit elements in [6] as

$$\delta\dot{\bar{\Omega}} = \frac{3J_2R_e^2\sqrt{\mu}}{4\bar{a}^{9/2}\bar{\eta}^5}\left(7c_{\bar{i}}\bar{\eta}\delta\bar{a} + 2\bar{a}\left(4c_{\bar{i}}\delta\bar{\eta} + s_{\bar{i}}\bar{\eta}\delta\bar{i}\right)\right) \tag{2}$$

$$\delta\dot{\bar{\omega}} = -\frac{3J_2R_e^2\sqrt{\mu}}{8\bar{a}^{9/2}\bar{\eta}^5}\cdot 7\left(5c_{\bar{i}}^2 - 1\right)\bar{\eta}\delta\bar{a} - \frac{3J_2R_e^2\sqrt{\mu}}{8\bar{a}^{9/2}\bar{\eta}^5}\cdot 2\bar{a}\left(4\left(5c_{\bar{i}}^2 - 1\right)\delta\bar{\eta} + 5\bar{\eta}s_{2\bar{i}}\delta\bar{i}\right) \tag{3}$$

$$\delta\dot{\bar{M}} = \frac{3\sqrt{\mu}}{2\bar{a}^{5/2}}\left(\frac{7J_2R_e^2}{4\bar{a}^2\bar{\eta}^3}\left(1 - 3c_{\bar{i}}^2\right) - 1\right)\delta\bar{a} + \frac{9J_2R_e^2\sqrt{u}}{4\bar{a}^{7/2}\bar{\eta}^4}\left(\left(1 - 3c_{\bar{i}}^2\right)\delta\bar{\eta} - \bar{\eta}s_{2\bar{i}}\delta\bar{i}\right) \tag{4}$$

Where $s_{(\cdot)} = \sin(\cdot), s_{(\cdot)} = \cos(\cdot)$ .Over bars denotes the secular part of a variable. $\bar{\eta} = \sqrt{1 - \bar{e}^2}$ is another expression of eccentricity *e*. *i* denotes orbit inclination. J_2 , R_e and$\mu$are J_2 coefficient, equatorial radius and gravitational constant of the Earth respectively.

Some researchers [9] suggested confining a formation by only one constraint, i.e., matching the nodal orbital period only

$$\delta T = 0 \Leftrightarrow \frac{\delta\dot{\bar{M}}}{\eta} + \delta\dot{\bar{\omega}} + \delta\dot{\bar{\Omega}}\cos\bar{i} = 0 \tag{5}$$

This condition is known as the no-drift condition. On one side, this condition can be taken as the general case of the period matching condition (1) of the unperturbed Keplerian motion. On another side, since both the strict J_2 invariant relative orbit conditions in [10] and the J_2 invariant relative orbit conditions in [11] ensure the no-drift condition (5), the no-drift condition can also be viewed as a general case of another two J_2 invariant relative orbit conditions.

However, although these conditions cannot guarantee a satellite under J_2 perturbation tracking a passive and periodic relative trajectory and a satellite may still tumble in its relative trajectory and requires compensation to maintain the periodic relative trajectory. The satellite orbit is a frozen orbit when the orbit inclination is critical inclination i_c, it can inhibit the satellite tumbling in its relative trajectory effectively.

The critical inclination i_c may be either 63.435° or 116.565° .In order to get quasi-periodic relative orbit conditions, we can set three conditions as follows:

$$\delta\dot{\bar{M}} = 0 \tag{6}$$

$$\delta\dot{\bar{\lambda}} = \delta\dot{\bar{\omega}} + \delta\dot{\bar{\Omega}}\cos\bar{i} = 0 \tag{7}$$

$$\bar{i} = i_c \tag{8}$$

In the study of the satellite relative motion, without loss of generality, a 2-satellite system is considered i.e. a passive or a virtual reference satellite S_0(without control force) and a member satellite S_j (with or without control force). Generally, the relative position of member satellite S_j is presented in a LVLH (Local Vertical, Local Horizontal) coordinate frame centered at the reference satellite as a function of time or a monotonic increasing parameter. Where x is a vector pointing in the direction of the reference satellite position from Earth-centered position, z is a vector normal to the orbital plane pointing in the direction of the orbit angular momentum, and y completes the right-hand frame.

The mean relative motion of the member satellite S_j under J_2 perturbation is derived in [6] as

$$\begin{aligned}
x_j\left(\bar{v}^t, \delta\bar{M}^t\right) &= \frac{\bar{\eta}^2\delta\bar{a}}{1+\bar{e}c_{\bar{v}^t}} - \bar{a}\delta\bar{e}\cdot c_{\bar{v}^t} + \frac{\bar{a}\cdot\bar{e}}{\bar{\eta}} s_{\bar{v}^t}\delta\bar{M}^t \\
y_j\left(\bar{v}^t, \delta\bar{M}^t, \delta\bar{\Omega}^t, \delta\bar{\omega}^t\right) &= \frac{2+\bar{e}c_{\bar{v}^t}}{1+\bar{e}c_{\bar{v}^t}}\bar{a}\delta\bar{e}\cdot s_{\bar{v}^t} + \frac{\bar{a}}{\bar{\eta}}\left(1+\bar{e}c_{\bar{v}^t}\right)\delta\bar{M}^t \\
&\quad + \frac{\bar{a}\bar{\eta}^2}{1+\bar{e}c_{\bar{v}^t}}\left(c_{\bar{i}}\delta\bar{\Omega}^t + \delta\bar{\omega}^t\right) \\
z_j\left(\bar{v}^t, \delta\bar{\Omega}^t, \bar{\omega}^t\right) &= \frac{\bar{a}\bar{\eta}^2}{1+\bar{e}c_{\bar{v}^t}}\left(\delta\bar{i}\cdot s_{\left(\bar{v}^t+\bar{\omega}^t\right)} - s_{\bar{i}}c_{\left(\bar{v}^t+\bar{\omega}^t\right)}\delta\bar{\Omega}^t\right)
\end{aligned} \tag{9}$$

The $\delta\bar{\lambda}^t = \delta\bar{\lambda} = \delta\bar{\omega} + \delta\bar{\Omega}\cos\bar{i}, \delta\bar{M}^t = \delta\bar{M}$ are constant because the conditions (6), (7). Superscript t denotes time varying variables. The v denotes true anomaly . Substituting (6), (7) into (9), the secular relative motion of the member satellite S_j under J_2 perturbation can be expressed as

$$\begin{aligned}
x_j\left(\bar{v}'\right) &= \frac{\bar{\eta}^2\delta\bar{a}}{1+\bar{e}c_{\bar{v}'}} - \bar{a}\delta\bar{e}\cdot c_{\bar{v}'} + \frac{\bar{a}\bar{e}\delta\bar{M}}{\bar{\eta}}s_{\bar{v}'} \\
y_j\left(\bar{v}'\right) &= \frac{2+\bar{e}c_{\bar{v}'}}{1+\bar{e}c_{\bar{v}'}}\cdot\bar{a}\delta\bar{e}\cdot s_{\bar{v}'} + \frac{\bar{a}\delta\bar{M}}{\bar{\eta}}\left(1+\bar{e}c_{\bar{v}'}\right) + \frac{\bar{a}\bar{\eta}^2\delta\bar{\lambda}}{1+\bar{e}c_{\bar{v}'}} \\
z_j\left(\bar{v}',\delta\bar{\Omega}',\bar{\omega}'\right) &= \frac{\bar{a}\bar{\eta}^2}{1+\bar{e}c_{\bar{v}'}}\left(\delta\bar{i}\cdot s_{\left(\bar{v}'+\bar{\omega}'\right)} - s_{\bar{i}}c_{\left(\bar{v}'+\bar{\omega}'\right)}\delta\bar{\Omega}'\right)
\end{aligned} \tag{10}$$

Applying (2)-(4), (6), (7) is derived to (11), (12)

$$7\left(1-3c_{\bar{i}}^2\right)\bar{\eta}\delta\bar{a} = 8\bar{a}\left(s_{2\bar{i}}\bar{\eta}\delta\bar{i} - \left(1-3c_{\bar{i}}^2\right)\delta\bar{\eta}\right) \tag{11}$$

$$\left(\frac{7\bar{\eta}}{4\bar{a}}\left(1-3c_{\bar{i}}^2\right) - \frac{\bar{a}\bar{\eta}^4}{J_2R_e^2}\right)\delta\bar{a} = \frac{3}{2}\left(\bar{\eta}s_{2\bar{i}}\delta\bar{i} - \left(1-3c_{\bar{i}}^2\right)\delta\bar{\eta}\right) \tag{12}$$

Considering the constraints (11) and (12), we can get as follow:

$$\left(\frac{7\bar{\eta}}{4\bar{a}}\left(1-3c_{\bar{i}}^2\right) - \frac{\bar{a}\bar{\eta}^4}{J_2R_e^2}\right)\delta\bar{a} = \frac{3}{2}\frac{7\left(1-3c_{\bar{i}}^2\right)\bar{\eta}}{8\bar{a}}\delta\bar{a} \tag{13}$$

If $\delta\bar{a} \neq 0$, the (13) can be derived $e \approx 0.97$ when the orbit inclination at critical inclination, it's impossible for satellite formation, so $\delta\bar{a} = 0$ is necessary for (13). Then (13) can be expressed as

$$\delta\bar{a} = 0,\quad \bar{\eta}s_{2\bar{i}}\delta\bar{i} = \left(1-3c_{\bar{i}}^2\right)\delta\bar{\eta} \tag{14}$$

When $c_{\bar{i}} \neq 0$ and under the condition (14), the (2)-(4) are simplified as

$$\delta\dot{\bar{\Omega}} = \frac{3n_{J_2}}{4\bar{\eta}^5}\cdot\frac{5c_{\bar{i}}^2+1}{c_{\bar{i}}}\delta\bar{\eta},\ \delta\dot{\bar{\omega}} = -\frac{3n_{J_2}}{4\bar{\eta}^5}\left(5c_{\bar{i}}^2+1\right)\delta\bar{\eta},\ \delta\dot{\bar{M}} = 0 \tag{15}$$

Where $n_{J_2} = \frac{J_2R_e^2\sqrt{\mu}}{\bar{a}^{7/2}}$. When $\bar{i} = i_c$, the (15) can be simplified further as

$$\delta\dot{\bar{\Omega}} = \pm\frac{3\sqrt{5}n_{J_2}}{2\bar{\eta}^5}\cdot\delta\bar{\eta}\ ,\ \delta\dot{\bar{\omega}} = -\frac{3n_{J_2}}{2\bar{\eta}^5}\cdot\delta\bar{\eta}\ ,\ \delta\dot{\bar{M}} = 0 \tag{16}$$

We can get the $\delta\eta$ relation as follow because of $e^2+\eta^2=1$.

$$\delta\eta = -\frac{e\delta e}{\eta} \tag{17}$$

So (14) is equivalent to (18).

$$\delta\bar{a}=0\ ,\ \bar{\eta}^2 s_{2i}\delta\bar{i}=\bar{e}\left(3c_i^2-1\right)\delta\bar{e} \tag{18}$$

Substituting (18) into (10), the secular relative motion of the member satellite S_j can be expressed as

$$\begin{aligned} x_j\left(\bar{v}^t\right) &= -\bar{a}\delta\bar{e}\cdot c_{\bar{v}^t}+\frac{\bar{a}\bar{e}\delta\bar{M}}{\bar{\eta}}\cdot s_{\bar{v}^t} \\ y_j\left(\bar{v}^t\right) &= \frac{2+\bar{e}c_{\bar{v}^t}}{1+\bar{e}c_{\bar{v}^t}}\cdot\bar{a}\delta\bar{e}\cdot s_{\bar{v}^t}+\frac{\bar{a}\delta\bar{M}}{\bar{\eta}}\left(1+\bar{e}c_{\bar{v}^t}\right)+\frac{\bar{a}\bar{\eta}^2\delta\bar{\lambda}}{1+\bar{e}c_{\bar{v}^t}} \\ z_j\left(\bar{v}^t,\delta\bar{\Omega}^t,\bar{\omega}^t\right) &= \frac{\bar{a}}{1+\bar{e}c_{\bar{v}^t}}\cdot\left(\frac{\bar{e}\left(3c_{\bar{i}}^2-1\right)\delta\bar{e}}{s_{2\bar{i}}}\cdot s_{\left(\bar{v}^t+\bar{\omega}^t\right)}-\bar{\eta}^2 s_i\delta\bar{\Omega}^t c_{\left(\bar{v}^t+\bar{\omega}^t\right)}\right) \end{aligned} \tag{19}$$

Given the orbital elements $\left(a,e,i,\omega^t\right)$ of reference satellite S_0, the formation is decided by the four parameters $\left(\delta e,\delta M,\delta\lambda,\delta\Omega^t\right)$ in the equation (19). That is to say, there are four DOFS can be used for relative orbit design. The equation (19) can also expressed as

$$\begin{aligned} x_j &= -\bar{a}\delta\bar{e}\cdot c_{\bar{v}^t}+\frac{\bar{a}\bar{e}\delta\bar{M}}{\bar{\eta}}\cdot s_{\bar{v}^t} \\ y_j &= \left(s_{\bar{v}^t}+\frac{s_{\bar{v}^t}}{1+\bar{e}c_{\bar{v}^t}}\right)\cdot\bar{a}\delta\bar{e}\cdot+\frac{\bar{a}\delta\bar{M}}{\bar{\eta}}\left(1+\bar{e}c_{\bar{v}^t}\right)+\bar{a}\delta\bar{\lambda}\cdot\frac{\bar{\eta}^2}{1+\bar{e}c_{\bar{v}^t}} \\ z_j &= \bar{a}\left(\frac{\bar{e}c_{\bar{\omega}^t}\left(3c_{\bar{i}}^2-1\right)\delta\bar{e}}{s_{2\bar{i}}}+\bar{\eta}^2 s_{\bar{i}}\delta\bar{\Omega}^t s_{\bar{\omega}^t}\right)\frac{s_{\bar{v}^t}}{1+\bar{e}c_{\bar{v}^t}} \\ &\quad +\bar{a}\left(\frac{\bar{e}s_{\bar{\omega}^t}\left(3c_{\bar{i}}^2-1\right)\delta\bar{e}}{s_{2\bar{i}}}-\bar{\eta}^2 s_{\bar{i}}\delta\bar{\Omega}^t c_{\bar{\omega}^t}\right)\frac{c_{\bar{v}^t}}{1+\bar{e}c_{\bar{v}^t}} \end{aligned} \tag{20}$$

There are some relations between the true anomaly v and eccentric anomaly E as follows:

$$\frac{s_v}{1+ec_v}=\frac{s_E}{\eta},\ \frac{c_v}{1+ec_v}=\frac{c_E-e}{\eta^2},\ \frac{1}{1+ec_v}=\frac{1-ec_E}{\eta^2} \tag{21}$$

Substituting (28) into (27), the secular relative motion of the member satellite S_j turns to

$$
\begin{aligned}
u_j &= -\delta\bar{e}\cdot c_{\bar{v}} + \frac{\bar{e}\delta\bar{M}}{\bar{\eta}}\cdot s_{\bar{v}} \\
v_j &= \left(\delta\bar{e}\cdot s_{\bar{v}} + \frac{\bar{e}\delta\bar{M}}{\bar{\eta}}\cdot c_{\bar{v}}\right) + \left(\frac{\delta\bar{e}}{\bar{\eta}}\cdot s_{\bar{E}} - \bar{e}\delta\lambda\cdot c_{\bar{E}}\right) + \left(\delta\lambda + \frac{\delta\bar{M}}{\bar{\eta}}\right) \\
w_j &= -\frac{s_{\bar{E}}}{\bar{\eta}}\left(\frac{\bar{e}c_{\bar{\varpi}}\delta\bar{e}}{2} - \frac{2\bar{\eta}^2 s_{\bar{\varpi}}\delta\bar{\Omega}_t}{\sqrt{5}}\right) \\
&\quad - \frac{c_{\bar{E}}}{\bar{\eta}^2}\left(\frac{2\bar{\eta}^2 c_{\bar{\varpi}}\delta\bar{\Omega}_t}{\sqrt{5}} + \frac{\bar{e}s_{\bar{\varpi}}\delta\bar{e}}{2}\right) + \frac{\bar{e}}{\bar{\eta}^2}\left(\frac{2\bar{\eta}^2 c_{\bar{\varpi}}\delta\bar{\Omega}_t}{\sqrt{5}} + \frac{\bar{e}s_{\bar{\varpi}}\delta\bar{e}}{2}\right)
\end{aligned}
\tag{22}
$$

Firstly, define a transformation,

$$
\begin{aligned}
A_{j1} &= \sqrt{\left(\delta e_j\right)^2 + \left(\frac{e\delta M_j}{\eta}\right)^2},\ s_{\phi_{j1}} = \frac{-\delta e_j}{A_{j1}},\ c_{\phi_{j1}} = \frac{e\delta M_j}{\eta A_{j1}}, B_j = \delta\lambda_j + \frac{\delta M_j}{\eta} \\
A_{j2}^t &= \sqrt{\left(s_i\delta\Omega_j^t\right)^2 + \left(\frac{e\left(3c_i^2-1\right)\delta e_j}{\eta^2 s_{2i}}\right)^2},\ s_{\phi_{j2}^t} = \frac{e\left(3c_i^2-1\right)\delta e_j}{\eta^2 s_{2i} A_{j2}^t},\quad c_{\phi_{j2}^t} = \frac{s_i\delta\Omega_j^t}{A_{j2}^t}
\end{aligned}
\tag{23}
$$

Substituting (23) into (22) and considering the math relation (24),

$$
\begin{aligned}
-s_x s_y + \eta c_x c_y &= \frac{\eta+1}{2}c_{(y+x)} + \frac{\eta-1}{2}c_{(y-x)} \\
\eta s_x s_y - c_x c_y &= -\frac{\eta+1}{2}c_{(y+x)} + \frac{\eta-1}{2}c_{(y-x)}
\end{aligned}
\tag{24}
$$

So the relative motion (22) is converted to

$$
\begin{aligned}
u_j &= A_{j1}s_{\left(v^t+\phi_{j1}\right)} \\
v_j &= A_{j1}\left(c_{\left(v^t+\phi_{j1}\right)} + \frac{1+\eta}{2\eta}c_{\left(E^t+\phi_{j1}\right)} - \frac{1-\eta}{2\eta}c_{\left(E^t-\phi_{j1}\right)}\right) - eB_j\cdot c_{E^t} + B_j \\
w_j &= -A_{j2}^t\left(\frac{1+\eta}{2}c_{\left(E^t+\omega^t+\phi_{j2}^t\right)} + \frac{1-\eta}{2}c_{\left(E^t-\left(\omega^t+\phi_{j2}^t\right)\right)}\right) + eA_{j2}^t c_{\left(\omega^t+\phi_{j2}^t\right)}
\end{aligned}
\tag{25}
$$

The five parameters $\left(A_{j1}, A_{j2}^t, B_j, \phi_{j1}, \phi_{j2}^t\right)$ are the functions of $\left(\delta e, \delta M, \delta\lambda, \delta\Omega^t\right)$, so there are four DOFS.

We apply $\left(A_{j1}, A_{j2}^t, B_j, \phi_{j1}\right)$ on the geometry design of satellite relative orbit, then the ϕ_{j2}^t can be decided by $\phi_{j2}^t = f\left(A_{j1}, A_{j2}^t, B_j, \phi_{j1}\right)$.

3 Numerical Simulations

Numerical simulations are performed to verify the proposed quasi-periodic relative orbit conditions. In the simulations, a reference satellite flies in small elliptic orbit at the critical inclination, while a member satellite flies around the reference satellite.

Suppose $A_{j1} = A_{j2}^{t_0}$ =0.0001, B_j=0, $\phi_{j1} = pi/3$ and represented the mean orbit elements of the reference satellite as follows

$$\bar{a} = 8000\,km, \bar{e} = 0.005, \bar{i} = i_c = 63.435^\circ$$
$$\bar{\omega} = pi/2, \bar{\Omega} = 0, \bar{M} = 0$$

Use (18), (23) can get the mean orbit elements of member satellite. Convert the mean orbit elements to osculating orbital elements. The osculating orbital elements for the two satellites in the formation are given in Table 1.

Table 1. The Osculating Orbital Elements for the Two Satellites

Satellite No.	The osculating orbital elements					
	a/km	*e*	*i/deg*	*ω/deg*	*Ω/deg*	*v/deg*
S_0	7993.2836	0.00410	63.423	90	0	0
S_1	7993.2856	0.00401	63.423	88.2	0.02	1.77

Substituting the orbital elements of S_0 and S_1 in Table 1 to exact equation including J_2 perturbation and getting the relative orbit.

The simulation result Fig. 1

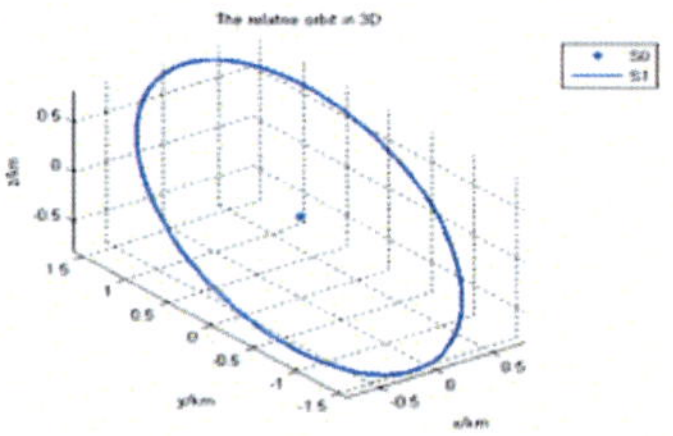

The Fig.1 show the relative orbit in 3D over 2 days. The simulation results show that the satellite formation has accommodated J_2 perturbation. The design of quasi-periodic satellite formation at critical inclination is proved to be correct.

4 Conclusion

A design of quasi-periodic satellite formation is proposed. The quasi-periodic constrains and mean relative motion of the member satellite under J_2 perturbation are used. The reference orbit must be at the critical inclination with a small eccentricity. The quasi-periodic constrains can release more DOFS to design the geometry formation than strict J_2 invariant trajectory conditions. The osculating orbital elements are used in numerical simulation, the design is proved to be correct through the simulation.

Acknowledgment. The authors would like to thank the editors and referees of this proceeding. This work is supported by State Natural Sciences Fund(61074159) and State Department of Human Resources and Social Security funding students who preferred(LNY2010001).

References

1. Sabol, C., Burns, R., McLaughlin, C.A.: Satellite "Formation Flying Design and Evolution". Journal of Spacecraft and Rockets 38(2), 270–278 (2001)
2. Hanspter, S.: Incorporating Secular Drifts Into the Orbit Element Difference description of Relative Orbits. In: 13th AAS/AIAA Space Flight Mechanics Meeting, Puerto Rico, NM, Paper No. AAS03-115 (February 2003)
3. Zhang, Y.-K.: The Research on Theory and Experiment about Steel-Deck SMA Pavement Technology. National University of Defense Technology, Changsha (2002)
4. Zeng, G.-Q.: Formation Flight Spacecraft Collaborative Control Theories and Methods. National University of Defense Technology, Changsha (2002)
5. Xu, G., Wang, D., Wu, B., Poh, E.K.: Passive and Periodic Circular-Like Formations at Critical Inclination around an Oblate Earth. In: 2010 Chinese Control and Decision Conference (2010)
6. Xu, G., Wang, D., Wu, B., Poh, E.K.: Periodic and Quasi-Periodic Satellite Relative Orbits at Critical Inclination. In: Proceeding of 2009 (2009)
7. Gurfil, P.: Relative Motion Between Elliptic Orbits: Generalized Boundedness Conditions and Optimal Formationkeeping. Journal of Guidance, Control, and Dynamics 28(4), 761–767 (2005)
8. Xing, M.J., Tang, G., Xi, X., Li, H.: Satellite Formation Design and Optimal Stationkeeping Considering Nonlinearity and Eccentricity. Journal of Guidance, Control, and Dynamics 30(5), 1523–1527 (2007)
9. Tragesser, S.G., Skrehart, B.: Relative Trajectory Design to Minimize Stationkeeping Effort. In: Proceeding of 2008 AAS/AIAA Space Flight Mechanics Conference, AAS 08-210, Galveston, TX, U.S. (2008)
10. Schaub, H., Alfriend, K.T.: J2 Invariant Reference Orbits for Spacecraft Formations. Celestial Mechanics and Dynamical Astronomy 79(2), 77–95 (2001)
11. Schaub, H., Junkins, J.L.: Analytical Mechanics of Space Systems. American Institute of Aeronautics and Astronautics, Inc., Reston (2003)

Deadbeat Control Technique Applied to Active Power Filter

Lan Li, Hai Chen, and Xiaoxia Cao

College of Electrical and Power Engineering, Taiyuan University of Technology, 030024 Taiyuan, China
lilan030@sina.com, 13934135668@163.com, caoxiaoxia610@163.com

Abstract. In order to improve current compensation effect of shunt active power filter based on three-phase three-wire system, deadbeat current tracking control technique based on repetitive predictor was adopted. The working principle of shunt active power filter and the method of the deadbeat current control were introduced, and the formula of deadbeat current control algorithm was derived. In addition, predictive instruction current was obtained by constructing harmonic current repetitive predictor. Finally, active power filter system under the platform of Matlab/Simulink was established and simulated. The results of simulation indicate that not only the active power filter can restrain harmonic current effectively, but also its compensation current can track instruction current precisely in both amplitude and phase, which has achieved the purpose of deadbeat control.

Keywords: Deadbeat control, Shunt active power filter, Repetitive predictor.

1 Introduction

As plenty of nonlinear loads are widely used in modern power system, the harmonic pollution of power grid has become increasingly serious. The application of active power filter (APF) is an effective method to improve the modern power quality. In the technology of active power filter control, traditional methods are triangular wave comparison control strategy [1] and hysteresis current tracking control strategy [2]. However, both of two control methods have hysteretic nature. Even though increasing the sampling points and computing speed in one power frequency period, the compensation current is always later than the instruction current, which inevitably occur errors. Deadbeat control belongs to the control methods of predictable nature [3], [4] , which makes the output of system at every sampling time in full accord with the instruction by forecasting the instruction, to fulfill the purpose of precise control and error-free.

The research object of this paper is shunt active power filter based on three-phase three-wire system. Its basic principle was analyzed, and the basic structure of active power filter control system was introduced. Meanwhile, deadbeat current tracking control technique was adopted, and the formula of deadbeat control algorithm was derived. Besides, the predictive instruction current can be obtained by constructing

D. Yang (Ed.): Informatics in Control, Automation and Robotics, Volume 2, LNEE 133, pp. 75–83.
springerlink.com

the harmonic current repetitive predictor. Under the platform of Matlab/Simulink, the active power filter system was established and simulated. And the results show that: the compensation current is exactly identical to the instructions current, without any amplitude and phase deviation, and the harmonic current from the line side is restrained effectively.

2 The Working Principle of Active Power Filter

The main circuit basic structure of three-phase three-wire shunt active power filter (APF) is shown as the Fig. 1. Inside, u_{sa}, u_{sb}, u_{sc} are the three-phase system voltage; i_{sa}, i_{sb}, i_{sc} are the three-phase system current; i_{La}, i_{Lb}, i_{Lc} are the three-phase load current; i_{ca}, i_{cb}, i_{cc} are the three-phase compensation current. Three-phase electric network supplies three-phase working current for the nonlinear load, which is harmonic source. The load current contains the fundamental component (i_{fa}, i_{fb}, i_{fc}) and harmonic component (i_{ha}, i_{hb}, i_{hc}), namely:

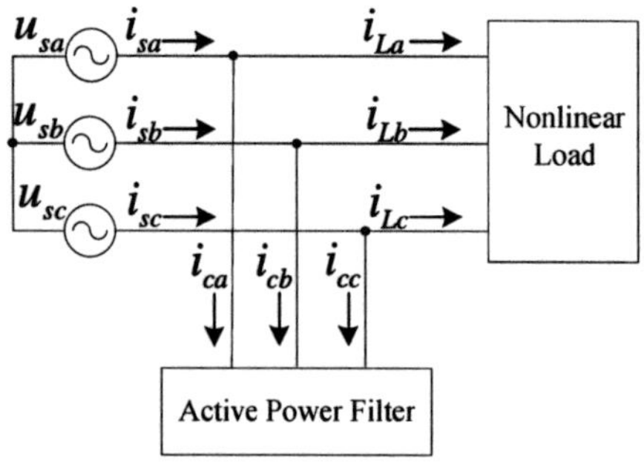

Fig. 1. Main circuit of three-phase three-wire shunt APF

$$\begin{cases} i_{La} = i_{fa} + i_{ha} \\ i_{Lb} = i_{fb} + i_{hb} \\ i_{Lc} = i_{fc} + i_{hc} \end{cases} . \tag{1}$$

At first, the active power filter detects the harmonic current instantaneous value of the compensated object (load current) and calculates its harmonic component. Then it can generate the compensation current through PWM modulation and the technology of power switching control, which is equal with the quantity and opposite to the direction of the harmonic current, and meet:

$$\begin{cases} i_{ca} = -i_{ha} \\ i_{cb} = -i_{hb} \\ i_{cc} = -i_{hc} \end{cases} . \tag{2}$$

Superimpose the compensation current on the power grid, because:

$$\begin{cases} i_{sa} = i_{La} + i_{ca} \\ i_{sb} = i_{Lb} + i_{cb} \\ i_{sc} = i_{Lc} + i_{cc} \end{cases} . \tag{3}$$

And it makes the system current only contain the fundamental component.

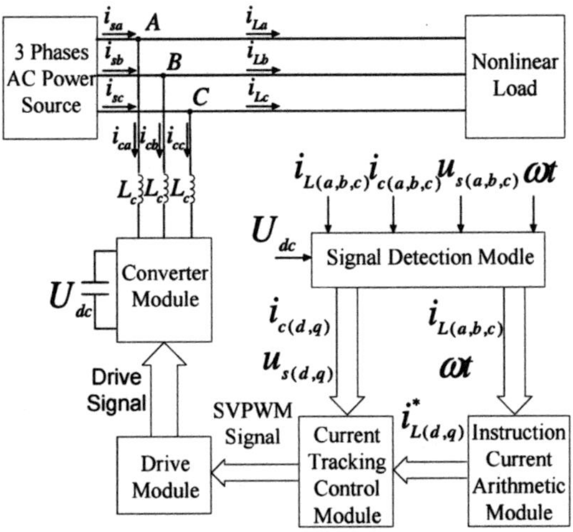

Fig. 2. Control system structure of APF

According to the working principle of active power filter, the control system was designed, as shown in Fig. 2 [5]. The keys of the whole control system are the detection of load harmonic current and the tracking control of instruction current.

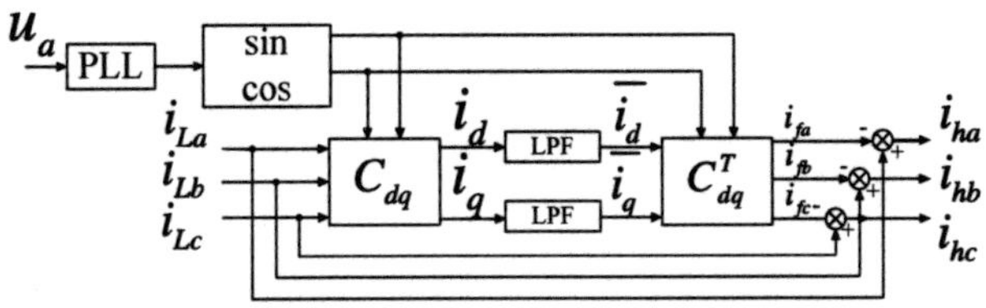

Fig. 3. Detection method of i_d-i_q harmonic current

The current detection method based on Instantaneous Reactive Power Theory is adopted in the harmonic current detection [6], [7], as shown in Fig. 3. Through the transformation of three-phase static coordinate to two-phase rotational coordinate, the three-phase load current i_{La}, i_{Lb}, i_{Lc} convert into d and q axes components i_d, i_q. After passing the low pass filter, the DC components of d and q axes $\overline{i_d}$ and $\overline{i_q}$ can be got respectively. Convert the DC components located on d, q axes from two-phase rotational coordinate to three-phase static coordinate, current fundamental components i_{fa}, i_{fb}, i_{fc} are got. At last, load current subtracts its fundamental component, and three-phase harmonic components i_{ha}, i_{hb}, i_{hc} can be obtained.

3 Deadbeat Control Algorithm

The equivalent circuit of active power filter is shown as the Fig. 4, inside, Rc, Lc present AC reactor equivalent resistance and filter inductance of active power filter respectively.

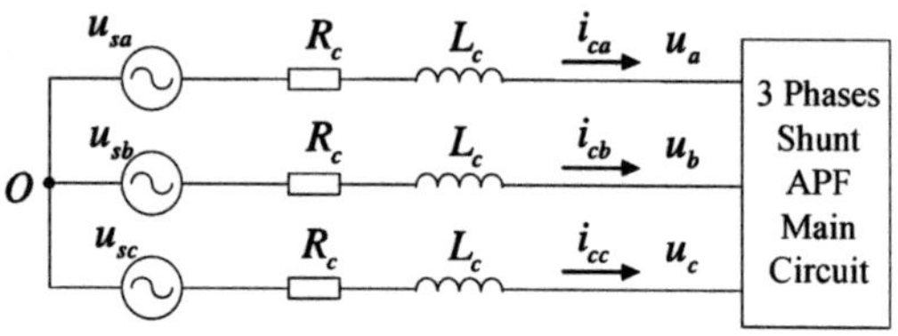

Fig. 4. Equivalent circuit of APF

We can get the following formula according to Fig. 4:

$$\begin{bmatrix} \frac{di_{ca}}{dt} \\ \frac{di_{cb}}{dt} \\ \frac{di_{cc}}{dt} \end{bmatrix} = \begin{bmatrix} -\frac{R_c}{L_c} & 0 & 0 \\ 0 & -\frac{R_c}{L_c} & 0 \\ 0 & 0 & -\frac{R_c}{L_c} \end{bmatrix} \begin{bmatrix} i_{ca} \\ i_{cb} \\ i_{cc} \end{bmatrix} + \frac{1}{L_c} \begin{bmatrix} u_{sa}-u_a \\ u_{sb}-u_b \\ u_{sc}-u_c \end{bmatrix} . \quad (4)$$

After transforming it into d-q coordinate system, we can use the state space form to express it as the following:

$$\dot{\mathbf{X}} = \mathbf{A} \cdot \mathbf{X} + \mathbf{B} \cdot \mathbf{U} . \quad (5)$$

Where,

$$\mathbf{X} = \begin{bmatrix} i_{cd} & i_{cq} \end{bmatrix}^T . \quad (6)$$

$$\mathbf{U} = \begin{bmatrix} u_{sd} - u_d & u_{sq} - u_q \end{bmatrix}^T . \quad (7)$$

$$\mathbf{A} = \begin{bmatrix} -\frac{R_c}{L_c} & \omega \\ -\omega & -\frac{R_c}{L_c} \end{bmatrix} . \tag{8}$$

$$\mathbf{B} = \begin{bmatrix} \frac{1}{L_c} & 0 \\ 0 & \frac{1}{L_c} \end{bmatrix} . \tag{9}$$

Let T be the sampling interval, the equivalent discrete model of equation (5) is expressed as the following:

$$\mathbf{X}(k+1) = \mathbf{G} \cdot \mathbf{X}(k) + \mathbf{H} \cdot \mathbf{U}(k) \ . \tag{10}$$

Inside the formula, k stands for the kth sampling point (k = 1,2,3 ... N) in every working frequency sample period, N stands for the sampling number per period.

$$\mathbf{G} = e^{\mathbf{A}T} \ . \tag{11}$$

$$\mathbf{X}(k) = \left[i_{cd}(k) \quad i_{cq}(k) \right]^T \ . \tag{12}$$

$$\mathbf{H} = (e^{\mathbf{A}T} - \mathbf{I}) \cdot \mathbf{A}^{-1} \cdot \mathbf{B} \ . \tag{13}$$

$$\mathbf{U}(k) = \left[u_{sd}(k) - u_d(k) \quad u_{sq}(k) - u_q(k) \right]^T \ . \tag{14}$$

From the formula (10), the current instruction voltage can be derived.

$$\begin{bmatrix} u_d(k) \\ u_q(k) \end{bmatrix} = \mathbf{H}^{-1} \cdot \mathbf{G} \cdot \begin{bmatrix} i_{cd}(k) \\ i_{cq}(k) \end{bmatrix} - \mathbf{H}^{-1} \cdot \begin{bmatrix} i_{cd}(k+1) \\ i_{cq}(k+1) \end{bmatrix} + \begin{bmatrix} u_{sd}(k) \\ u_{sq}(k) \end{bmatrix} . \tag{15}$$

The formula (15) is the deadbeat algorithm used for active power filter based on three-phase three-wire system. To get the instruction voltage, nevertheless, the predictive instruction current $i_{cd}(k+1)$ and $i_{cq}(k+1)$ at the next time need to be calculated in advance. d and q axes predictive instruction current can be generated by constructing harmonic current repetitive predictor, which is shown in Fig. 5.

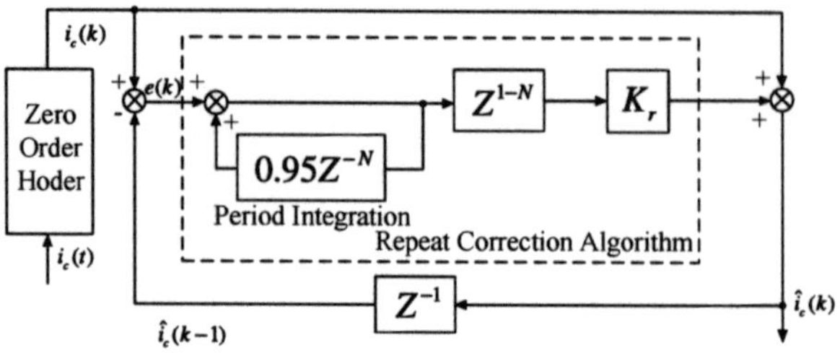

Fig. 5. Harmonic current repetitive predictor

Harmonic current basically meets the characteristics of periodic repetition. On the assumption, the error between the prediction value and the actual value will also appear repeatedly. This error can be eliminated by using the repeated correction algorithm. Passing throng the zero-order holder at t time, harmonic current $i_c(t)$ becomes discrete value of the kth sampling point $i_c(k)$ according to predetermined sampling frequency. $\hat{i}_c(k)$ denotes predictive value of harmonic current at the next time. The correcting value of error $e(k)$, which is calculated by repeated correction algorithm at k time, plus $i_c(k)$ is $\hat{i}_c(k)$. Kr is the proportion coefficient of repetitive predictor, which could regulate the size of compensation, and the effect of repeated correction. After one beat delay, $\hat{i}_c(k-1)$, which denotes harmonic current predictive value of the (k-1)th sampling point, becomes $\hat{i}_c(k)$. If harmonic current met the demand of periodic repetition characteristics, the predictor could make the predictive value of instruction current completely equal to the instruction current value at the next beat, namely, give the instruction current one beat in advance. And finally, the instruction value is identical to the actual output both in phase and amplitude, which can achieve the effect of deadbeat control.

4 The Results of System Simulation

The system simulation model is built under the platform of Matlab/Simulink, which is shown in Fig. 6.

The Simulation parameters are expressed as follows: three-phase three-wire system line voltage is 380V, power frequency is 50Hz, AC reactor equivalent resistance of active power filter is 0.5Ω, the equivalent inductance is 7mH, DC side voltage is 900V, switching frequency in the control system is 10KHz. Taking three-phase uncontrolled rectifier bridge with Resistance-inductance Load as the harmonic sources, inside, load resistance is 10 Ω, inductance is 48 mH.

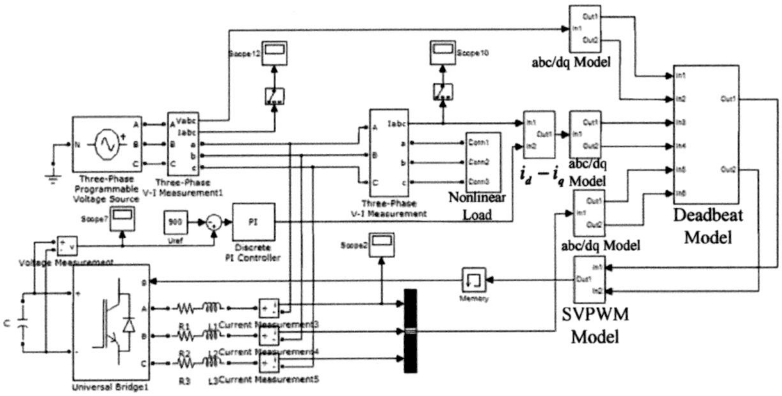

Fig. 6. Simulation model of APF control system

System simulation results are shown in Fig. 7, where (a) ~ (e) are A-phase current and DC side voltage; (f) and (g) are the results of Fast Fourier Transform analysis. Because of the impact of harmonic source, the load current total harmonic distortion (THD) reached 20.78%. While system current appeared ideal sine-wave on account of APF, and its THD is only 1.03%. (h) and (i) are the input and output current waveforms of d-axis repetitive predictor respectively. We can see that, output of repetitive predictor is earlier one beat than input. It indicates that repetitive predictor has played the role of predicting the instruction current of next time, namely, it gives the instruction current one beat in advance. Moreover, through the observation of the DC-side voltage waveform, it can be known that the result of PI control is ideal, and the voltage could keep at the reference value 900V stably, with fluctuations within ± 1V.

When the harmonic source load change suddenly, the compensation current based on deadbeat control algorithm is shown in Fig. 8. It can be seen from the figure, increasing load suddenly can cause the significant change of load current, nearly 2 times in amplitude, but the deadbeat control algorithm can make adjustments quickly within one cycle, and the response speed is relatively fast. THD value of the load current changed from 20.78% (before the load increase) to 16.68% (after the load increase), and THD values of the compensated system current are both 1.03%. DC-side voltage of PI control only has small fluctuation. In addition, the voltage can regulate to the given value quickly.

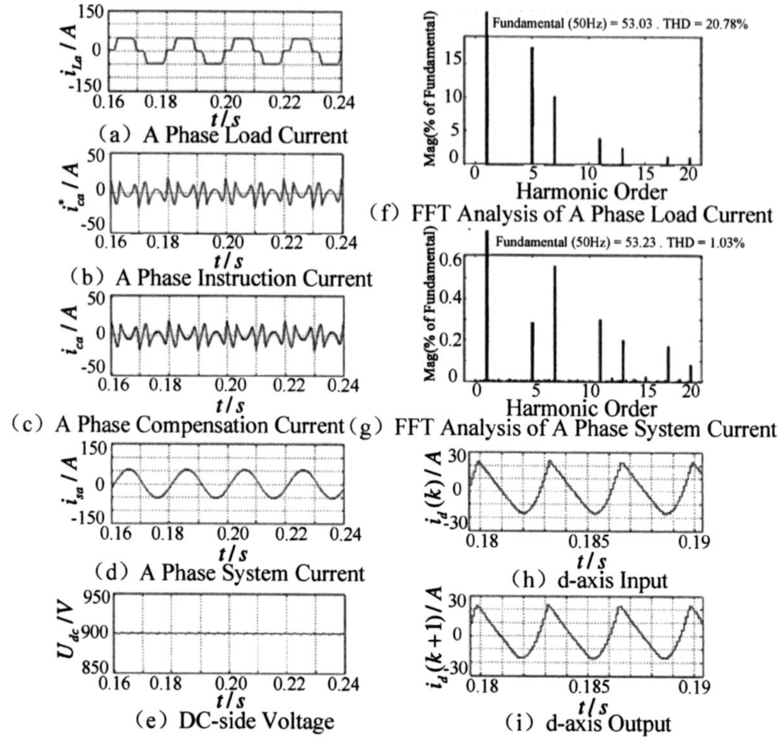

Fig. 7. System simulation results of deadbeat control

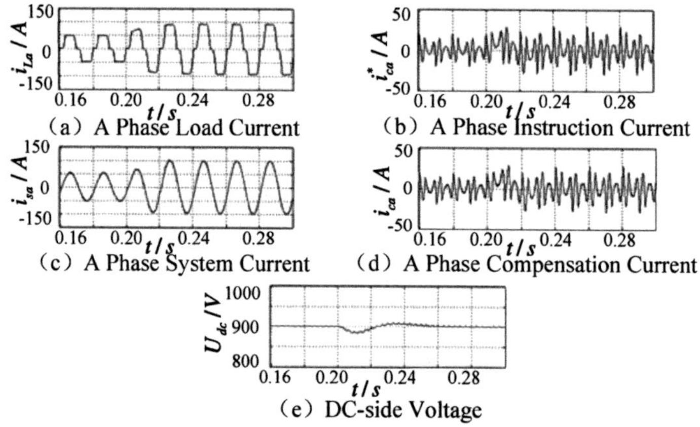

Fig. 8. Simulation results under abrupt change of load

5 Conclusion

Introducing deadbeat control into the application of active power filter, it can conquer the disadvantages of traditional hysteresis control methods—compensation current always delays one beat compared with instruction current. System with deadbeat control algorithm can track the change of instruction current one beat in advance, so that the following output current is identical to instruction current in both amplitude and phase. The Simulation results show that the deadbeat control algorithm has achieved the expected effect. Although calculation burden of this method is heavy, with the development of hardware level, the shortcomings will become more and more unconspicuous. Especially, the algorithm is suitable for the achievement of digital control, so the prospect of its applications is bright.

References

1. Zhang, G.: Control Method and Simulation Research of Active Power Filter. Tianjin University (2004)
2. Fan, X., Zhang, D., Sun, Q., et al.: Hysteresis Current Control Strategy for Three-phase Three-wire Active Power Filter. Automation of Electric Power System 31(18), 57–60 (2007)
3. Paolo, M., Pinhabel, M.F.: Repetitive-based control for selective harmonic compensation in active power filters. IEEE Transactions on Industrial Electronics 51(5), 1018–1024 (2004)
4. Jeong, S., Woo, M.: DSP-based active power filter with predictive current control. IEEE Transactions on Industrial Electronics 44(3), 329–336 (1997)
5. Wang, K.: The research on active power filter based on TMS320F2812. Taiyuan University of Technology (2010)
6. Xing, X.: Study on controller of the active power filter based on DSP. Taiyuan University of Technology (2011)
7. Herrera Reyes, S., Patricio, S., Kim, H.: Instantaneous reactive power theory applied to active power filter compensation: Different approaches, assessment, and experimental results. IEEE Transaction on Industrial Electronics 55(1), 184–196 (2008)

The Development of Spectral Responsivity Testing System

Jin Ren-cheng, Dong Huan-huan, He Cheng-qun, and Chu Jin-kui

School of Mechanical Engineering, Dalian University of Technology,
Dalian 116024, China
rcjin@dlut.edu.cn

Abstract. The spectral responsivity testing system based on optical fiber transporting light has been designed and developed. The system includes five parts: optical source, monochromator system, darkroom, picoammeter and stepping motor driver. Using the standard photodetector substitution to calibrate the photodetector under testing, with the optical fiber translating monochromatic light, the main technical problem emphasizes on solving the limitation of photodetector's active area involving in the test of spectral responsivity. The testing result shows that the error is less than 0.46%.

Keywords: photodetector, optical fiber, spectral responsivity, VC++6.0, MSComm.

1 Introduction

Spectral responsivity, which represents ability of detector transforming the input optical signal to electrical signal, is a parameter describing the sensibility of photoelectric pickoff. The accurate measurement of spectral responsivity is very important to the development of fabrication process of detectors.

As it is more difficult to measure the optical power directly, spectral responsivity is always acquired indirectly by the help of the detector whose spectral responsivity is already known. The means usually adopted is using radiation field to irradiate the effective photosensitive area of detectors [1]. The irradiance of the radiation field is uniform. As the exit spot of irradiate field was not able to focus to be small enough because of the bad quality of imaging, the radiation field whose irradiance was uniform was adopted to make the output area greater than the effective photosensitive area of the detector. But most monochromator have the problem that their emergent light is not uniform [2]. So the spectral responsivity testing system based on optical path consisting of optical fiber was designed.

2 Testing Principles

Definition of spectral responsivity is the output current of detector which is under per unit radiant power of monochromatic light whose wavelength is λ [3]. Where

$$R(\lambda) = I(\lambda)/P(\lambda) \tag{1}$$

D. Yang (Ed.): Informatics in Control, Automation and Robotics, Volume 2, LNEE 133, pp. 85–91.
springerlink.com

$$I(\lambda) = R(\lambda) \times H(\lambda) \times S \tag{2}$$

Where $R(\lambda)$ is the spectral responsivity of detector which is under light. Wavelength of this light is λ. $I(\lambda)$ is the photocurrent under light. $H(\lambda)$ is the irradiance of monochromatic light λ. S is the actual photosensitive area of detector that receives light.

During the testing, the standard detector was placed at the end of the optical fiber. $I_s(\lambda)$, photocurrent of the standard detector, was measured under the monochromatic light. Then the detector was removed. $I_d(\lambda)$, photocurrent of the testing detector, was measured under the same light.

From the formula (2), two formulas below could be got:

$$I_s(\lambda) = R_s(\lambda) \times H_s(\lambda) \times S_s \tag{3}$$

$$I_d(\lambda) = R_d(\lambda) \times H_d(\lambda) \times S_d \tag{4}$$

Using formula (3) to divide formula (4), formula below could be got:

$$\frac{I_s(\lambda)}{I_d(\lambda)} = \frac{R_s(\lambda) \times H_s(\lambda) \times S_s}{R_d(\lambda) \times H_d(\lambda) \times S_d} \tag{5}$$

During the testing, the testing detector should be kept in the same place with the standard detector. So, S_s, which is photosensitive area of the standard detector under light, is equal to S_d , photosensitive area of the testing detector. They are both areas of emergent light of the optical fiber. As both the standard detector and the testing detector were placed in the same testing environment, irradiance of them were the same, that is $H_s = H_d$. So formula (5) could be transformed to:

$$R_d(\lambda) = \frac{I_d(\lambda)}{I_s(\lambda)} \times R_s(\lambda) \tag{6}$$

Where $R_s(\lambda)$, spectral responsivity of the standard detector was already known. $I_s(\lambda)$ and $I_d(\lambda)$ could be measured by picoammeter. So, spectral responsivity of the tested detector had been got from formula (6).

3 Composition of Testing System

As shown in Fig.1, this testing system included five parts:

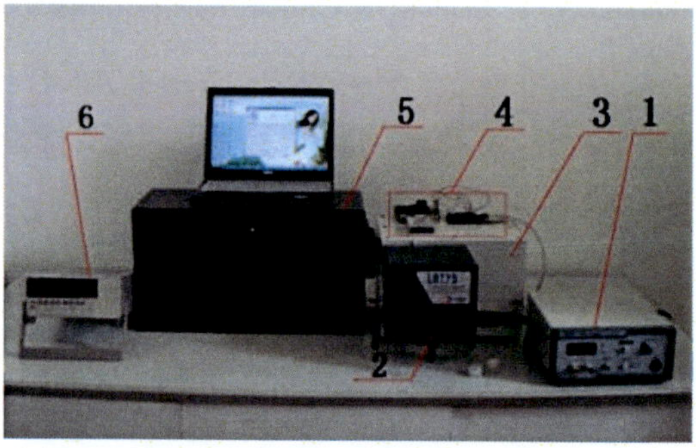

Fig. 1. The testing system of spectral responsibility

1) Light source (including the regulated power supply)

This testing system got monochromatic light from monochromator whose wavelength had error in accuracy and multiplicity. Supposing this error was $\Delta\lambda$. Actual monochrome radiant energy was got from wavelength $\lambda+\Delta\lambda$ when computing the responsivity at wavelength λ. If there was a large gradient in the energy wavelength, though $\Delta\lambda$ was very small, ΔE, the monochrome radiant energy, got to be very big. This had obvious influence on measuring accuracy. If the spectral curve of light source was flat, though there were errors in accuracy and multiplicity, it did not influence the measuring accuracy too much.

Because of this, the 75W tungsten light (Serial number 2) was used. The tungsten light had characteristics of high light efficiency, high color temperature, and stably luminous flux. Its spectral range was 350-2500nm. To increase the stability of irradiance of the output light, the voltage source (Serial number 1) which was matched with the tungsten light was used. It ensured wave range of the luminous flux to be 0.12% - 0.14%.

2) Spectroscope system

Uniform and steady monochromatic light should be chosen to measure the spectral responsivity of the testing detector. SSM301 monochromator (Serial number 3) which was the production of Zolix was chosen. SSM301 provided 330nm-900nm continuous spectrum. Its resolution was 0.1nm. Its accuracy was $\pm$0.2nm.

3) Darkroom(The light path adjusting and transmission unit)

Photoelectric detector could transform the incident light to electrical signal. If there was disturbance from the outside light when the detector was being tested, even the weakest stray light could affect the result a lot. That made the result distort. To avoid the effect from different period of time, and to remove the disturbance from outside light, this darkroom was designed (Serial number 5). It ensured the link between the darkroom and the monochromator when outside light was isolated by it. The exit slit of SSM301 monochromator was a 3mm $\times$ 10mm rectangle. So the emergent light was rectangle too. SSM301 used optical lens to collect emergent light to improve the uniformity, and to increase the strength of the monochromatic light.

4) Data acquisition unit(DAU)

When the power of light source was decided, light current varied with wavelength. The minimum rank of the light current was nA. Number 6485 picoammeter (Serial number 6) was chosen to measure such a tiny light current. 20fA-20mA current could be measured by 6485 picoammeter. It can make the measurement at the maximum speed of 1000 reading per second. Its resolution was 10fA.

5) Automatic control unit

SSM301 monochromator was only manual operated. To simplify the experiment process and operation, a stepper motor controller (Serial number 4) which was based on L297 and L298 was designed. This controller was driven by SCM which was being communicated with PC in VC6.0 environment. The aim of automatic control of this testing system was achieved through it.

4 Problem That the Testing System Has Solved Mainly

4.1 Testing of Emergent Area Affecting Spectral Responsivity

Effective photosensitive area of photoelectric detector received incident light. It reacted differently when receiving monochromatic light with different wavelength. When wavelength and intensity of the incident light were certain, the output current or voltage was proportion to the photosensitive area. Normally, irradiated area of the monochromatic light which went through the spectroscope was fixed. But the photosensitive area of standard detector was different from the tested detector. Supposing photosensitive area of the standard detector and the tested detector to be S_1 and S_2. S_{max} and S_{min} was the bigger one and smaller one between S_1 and S_2. To accomplish the measurement of spectral responsivity of the testing detector, it must be guaranteed that irradiated area of the monochromatic light was greater than S_{max} or smaller than S_{min}. Because it was difficult to measure the irradiated area of the monochromatic light at a certain location, when the area of emergent light was between S_{max} and S_{min}, the photosensitive area of the detector which had a bigger photosensitive area was the irradiated area of the emergent light. So, there was an unknown photosensitive area in formula 6. It could not be computed.

Fig. 2. Ultimate sink of optical fiber

The collected monochromatic light was received by ultimate sink of optical fiber as shown in Fig.2. Monochromatic light exited from the end of the optical fiber after transmitted by it. Irradiate area of the monochromatic light was about $0.28mm^2$ when core parameter of the optical fiber was $600\mu_m$. Detector was placed at the end of the optical fiber, so the emergent monochromatic light was received both by the standard detector and the testing one completely. When the standard detector and the testing one were placed at the same position, photosensitive area of the standard detector was equal to the testing one. Because of this, testing of the spectral responsivity had not been limited by the actual photosensitive area of the detector. The micro movement platform was used to position the detector accurately and to ensure the accuracy was 0.1mm after the testing detector changed to another one.

4.2 Automatic Control of Monochromator

Spectral function of monochromator was realized by rotating the grating. Wave length scanning mechanism of SSM301 monochromator was sine mechanism. Fig.3 was the functional scheme of sine mechanism.

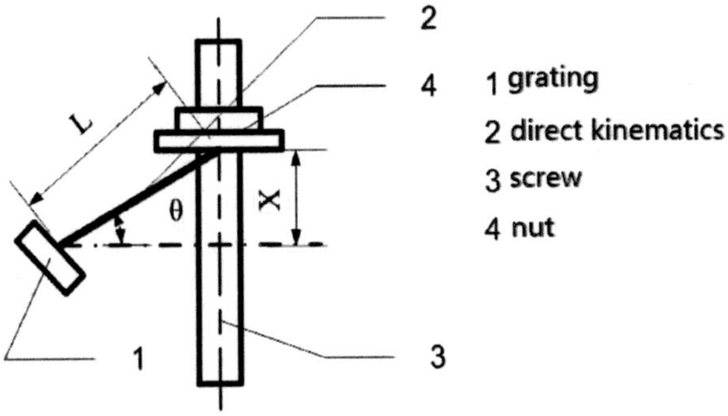

Fig. 3. Schematic diagram of Sine organization chart

Formula $\sin\theta = \frac{X}{L}$ could be got from Fig.3. Because the grating was fixed together with lever arm, rotary angles of the lever arm and the grating resulted from the change of x was the same [4]. So:

$$\frac{\lambda}{2d\cos\frac{\alpha+\beta}{2}} = \frac{x}{L} \tag{7}$$

$$\lambda = 2d\cos\frac{\alpha+\beta}{2}\frac{x}{L} \tag{8}$$

Where α is diffraction angle. β is incident angle of light irradiating the grating.

Therefore, linear sweep of the wavelength of the grating could be realized by keeping the uniform motion of the screw. To realize the automatic control of monochromator, a stepper motor was installed on the wavelength scanning mechanism. Screw was drove by this motor, making the grating rolling.

STC89C52RC SCM and L297, L298 driver chip were used to realize controlling and driving of the stepper motor. By using L297 and 1298 union IC chip, not only the disadvantage of control plan that use discrete component could be avoid [5], but also the program could be simplified a lot. It only need clock, direction and analog input signal. Phase was produced by itself. So, it lightened the burden of computer (or SCM) and programming. It had the advantages of using less component, low module dissipation, high reliability and easy software development. For visualization operation, the software was wrote with MFC in VC++6.0, by using MSComm ActiveX. It realized the communication with STC89C52RC SCM. The control interface was shown in Fig.4.

Fig. 4. Interface of the testing system

5 Testing of the Comparison between Results of Irradiating the Detector Directly and Indirectly

This test was held every 10nm. S1087 photoelectric detector which was produced by HAMAMATSU was used as the standard detector. The responsivity of S1133 was measured in two different conditions which were irradiated by emergent light of monochromator directly and irradiated by optical fiber. Results were shown in Fig.5. S1087 and S2233, whose spectral responsivity were the same from each other, were in the same series. Their responsivity curves overlapped completely [6]. As shown in Fig.6, these two curves were in better agreement when being irradiated by optical fiber, which showed that better results could be got when optical fiber was used for transforming. The test of spectral responsivity was carried out discretely at wavelength 400nm-730nm every 10nm. Error was accord in every test. The maximum error was 0.46% at wavelength 730nm.

There were two reasons why so big error occurred when being irradiated directly by emergent light of monochromator. One is the difference of photosensitive area between S1087 and S1133. The effective photosensitive areas of S1087 and S1133 are 1.6×1.6mm^2 and 2.4×2.8mm^2; another is lacking of the image registration mechanism causes non-uniform irradiance of the emergent light of monochromator. But the irradiance of emergent light of the whole radiation surface was supposed to be non-uniform when formula 6 was used.

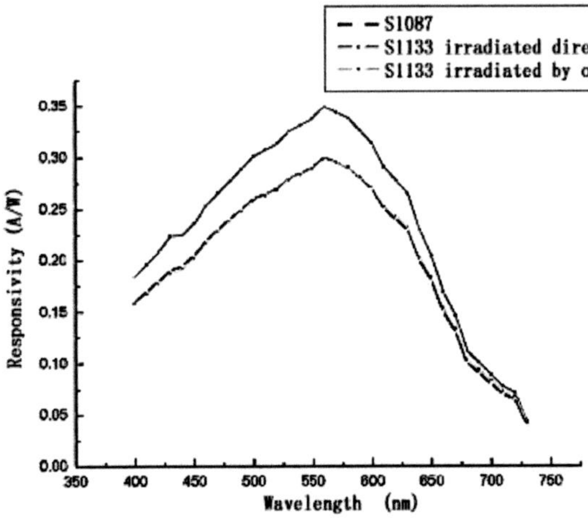

Fig. 5. The spectral responsivity comparison between direct radiation and optical radiation

6 Conclusion

The spectral responsivity testing system, which was designed, using optical fiber transmitting light path, takes advantage of substitution method of standard detectors. Compared with the traditional methods, it has higher accuracy. Moreover, it breaks the limitation of detector's photosensitive area for the measurement of spectral responsivity.

Acknowledgments. This research is supported by National Basic Research Program of China (973 Program) under project No. 2011CB302105.

References

1. Thomas, L.C., Sally, B.S., Albert, P.C.: NIST Measurement Service: Spectroradiometric Detector Measurements. NIST Special Publication (1998)
2. Xian, T.D.: Semiconductor photoelectric device. China Machine Press, Beijing (1982)
3. Song, J.Y., Ping, Y., Yu, L.Z.: Photoelectric technology and experiment. Beijing Institute of Technology Press, Beijing (2000)
4. Hu, D.S.: Development Of Intelligentized Multi-grating Monochromator. Journal of Infrared and Millimeter Waves, 133–137 (2002)
5. L297 Stepper Motor Controllers, STMicroelectronics (2000)
6. HAMAMATS, Si photodiode S1087 and S1133 series

A Robust and Fast Eyelash Detection Basted on Expectation Maximization and Gaussian Mixture Model

Ting Wang[1], Min Han[1], Honglin Wan[2], and Yong Yin[3]

[1] School of Information Science and Engineering, Shandong University, 27 Shanda Nanlu, 250100, Jinan, P.R. China
wangting198765@mail.sdu.edu.cn, hanmin@sdu.edu.cn
[2] School of Physics and Electronic Science, Shandong Normal University, East Wenhua Road 88, 250014, Jinan, China
visage1979@sdu.edu.cn
[3] Department of Ratiation Oncology, Shandong Cancer Hospital and Institute, 250117, Jinan, China
yinyongsd@yahoo.com.cn

Abstract. During recent years, iris recognition has been accepted by public as a high-discriminative personal identification technology duo to its unique, stable and noninvasiveness properties. However, eyelash occlusion degrades iris equality and leads low recognition rate of nonideal iris. In this paper we propose a novel eyelash detection method based on expectation maximization (EM) algorithm and Gaussian Mixture Model (GMM). This method is robust, and does not rely on the fixed threshold and the detection of eyelid. Experimental results depict this method can detect eyelashes accurately and effectively.

Keywords: iris recognition, eyelash detection, the EM algorithm, GMM; iris segmentation.

1 Introduction

With a rapid increasing demand for social security, personal identification is becoming more and more important. As a biometric technology, iris recognition has been a hot issue [1,2].Unfortunately, in nonideal cases acquired images always contain some undesirable components, such as eyelashes, eyelid and so on. Especially eyelashes, which are considered to be a part of iris texture and to be encoded, will be mismatched if they can not be detected and removed. It will deteriorate the accuracy of an iris recognition system. Therefore, a robust method to reduce the influence of eyelashes is demanded.

Some eyelash detection methods have been reported. Kong and Zhang[3] used a 5×5 window to locate multiple eyelashes and a 1D Gabor filter to detect separable eyelashes. But it needs a fixed threshold and a high time consuming. J. Huang[4] detected eyelashes based on phase congruency in which a decision has to be made that whether a spot belongs to eyelash. In Kong and Park's algorithm[5], eyelashes are classified into

D. Yang (Ed.): Informatics in Control, Automation and Robotics, Volume 2, LNEE 133, pp. 93–97.
springerlink.com

separable and multiple eyelashes. But when the separable eyelashes are detected, horizontal and vertical direction convolution kernels are applied from six different directions, which leads to a heavy computational load.

In this paper we design an effective eyelash location method based on expectation maximization (EM) algorithm[6] and Gaussian Mixture model (GMM). Because the effect of upper eyelash is larger than the lower; we only attempt to remove the upper eyelash in our experiments.

The remainder of this paper is constructed as follows. Firstly, we introduce the EM algorithm in Section 2, and its application in GMM is presented in Section 3. The detection of eyelashes will be described in Section 4. Experimental results will be showed in Section 5 and Section 6 will give the conclusion.

2 The EM Algorithm and the EM Algorithm for GMM

The data obtained is often incomplete or has missing value. In this case, the EM algorithm is always used to find the maximum-likelihood estimate of the model parameters. The EM algorithm includes two steps: E-step calculates the expectation of the log-Likelihood based on the initialized parameter model set, and M-step uses a certain optimization method to maximize the expectation [7].

GMM is regarded as a general tool for density function estimation, object modeling and pattern classification. Under many circumstances, one characteristic of an object, such as probability density function, can be seen as a mixture of Gaussian function.

In unsupervised learning, only the data sample x is observable but the information of the class and state are unknown. θ is the parameter model set and $p(x|\theta)$ is the probability density functions over θ . Here, we give two-parameter estimation approaches for a general K component GMM that[8]

$$p(x|\theta) = \sum_{i=1}^{K} \alpha_i p_i(x|\theta_i) \tag{1}$$

where

$$p_i(x|\theta_i) \sim N(\mu_i, \Sigma_i) \tag{2}$$

$$\sum_{i=1}^{K} \alpha_i = 1 \tag{3}$$

α_i is the mixing coefficients and $p_i(x|\theta_i)$ is a Gaussian probability density function.

Hence, the parameter set is $\Theta = \{\alpha_1 \cdots \alpha_K, \mu_1 \cdots \mu_K, \Sigma_1 \cdots \Sigma_K\}$.

The EM algorithm is always used to estimate the parameter of GMM. Based on EM algorithm, the parameter estimation for the GMM is obtained [9].

$$\alpha_i^{t+1} = \frac{1}{S} \sum_{j=1}^{S} p(i|x_j, \Theta) \tag{4}$$

$$\mu_i^{t+1} = \sum_{j=1}^{S} x_j p(i|x_j,\Theta) / \sum_{j=1}^{S} p(i|x_j,\Theta) \tag{5}$$

$$\Sigma_i^{t+1} = \sum_{j=1}^{S} (x_j - \mu_i^{t+1})(x_j - \mu_i^{t+1})^T p(i|x_j,\Theta) / \sum_{j=1}^{S} p(i|x_j,\Theta) \tag{6}$$

$$p(i|x_j,\Theta) = p_i(x_j|\Theta)\alpha_i^t / \sum_{l=1}^{K} p_t(x_j|\Theta)\alpha_l^t \tag{7}$$

S is the total number of observations. $p(i|x_j,\Theta)$ is the posterior probability and the observation x_j is generated by ith Gaussian component. To obtain the final parameters estimation, we iterate through above equations until they convergence.

3 Eyelash Detection

It is supposed that an eye image's intensity distribution can be represented by a mixture of Gaussian function. The eyelash region or the part composed of eyelashes and some other components in the image yielding to a Gaussian distribution. We need to find the parameter estimation of the Gaussian function of the eyelashes and then detect the eyelash region.

With regard to the selection of K, there are some papers to introduce the algorithm. However, if we choose a value that can have a better performance, it is not necessary to select a value of K by an algorithm since that need to consume time. In our paper, the value of K is 5 which is selected through a great amount of experiments.

In this section, we present eyelash detection as follows (we use a sample of our Database to display the result of every step in Fig.1~Fig.3):

Step1: Use the EM algorithm to estimate the parameter of iris with GMM. Five parts (shown from b to f in Fig.1) will be described by mixing coefficients, means and variances.

Step2: Localize Pupil based on the information of part1. We take the centroid of part1 as pupil's center, and the mean of the distances from edge points of the lower half of the part1 to center as radius since the other sometimes contains some eyelash points. Fig.2 gives the result of pupil location.

Step3: Exclude the points that are not eyelashes on the basis of step2 and the mean of part2, and then obtain eyelashes as shown in Fig. 3.

4 Experiment Results and Discussion

The tested images are from CASIA1 Iris Image Database provided by the Institute of Automation & Chinese Academy of Sciences [10]. The Database is composed of 756 iris images from 108 different eyes of 80 persons. As shown in Fig.3 and Fig.4, almost all eyelashes of the pictures can be fined. Only a few of eyelash details can not be detected, however, that has little effect on the recognition. We define the accuracy of eyelash detection as follows. If most eyelashes in the iris region are located, we determine it as a correct detection. In our experiments, the accuracy of eyelash location with our scheme is 98.42%.

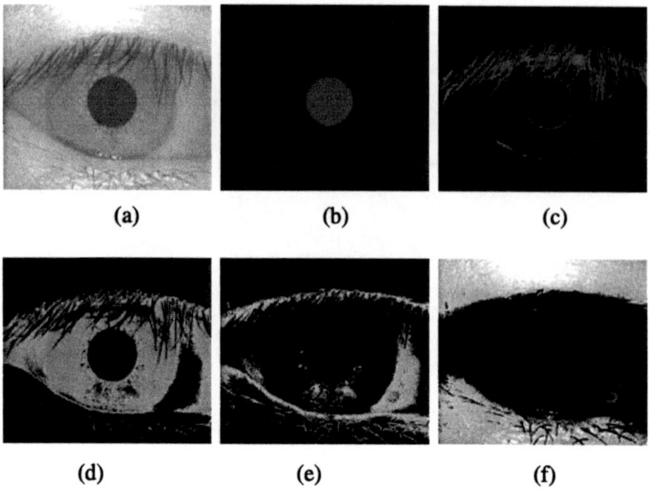

Fig. 1. (a) Original image (b) ~ (f)part1~part5 obtained using EM algorithm and GMM

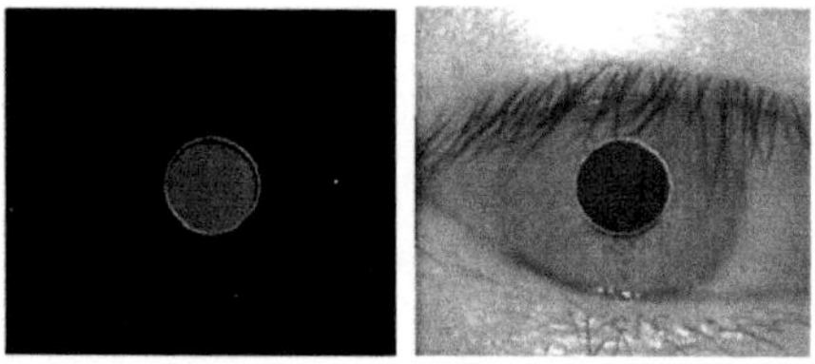

Fig. 2. Inner boundary location based on part1

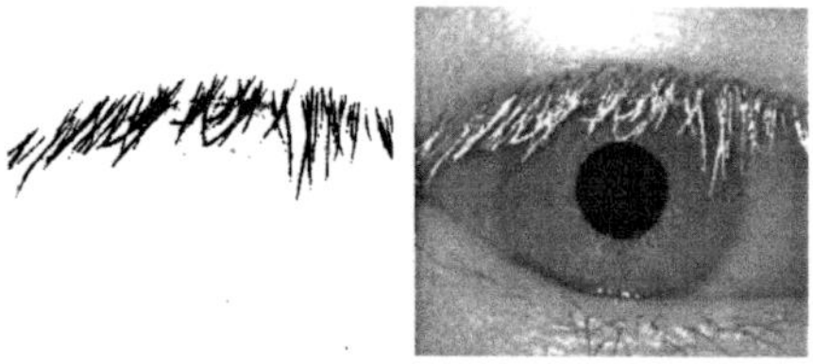

Fig. 3. Eyelashes detected

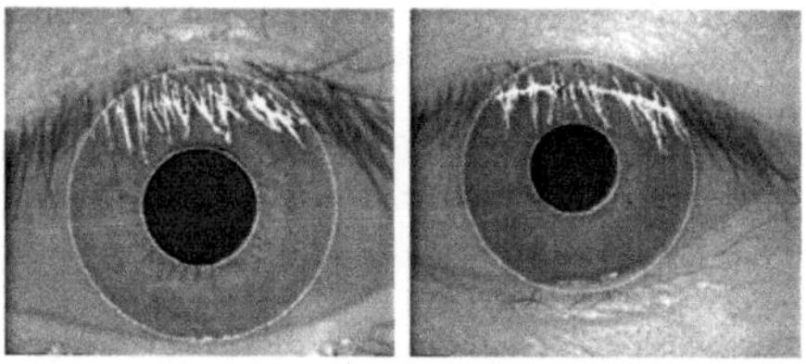

Fig. 4. Examples of iris segmentation results

In addition, when the interior boundary of iris is located, the method mentioned in step2 is a good choice because it can locate the boundary fast and accurately. Then we use the other methods to locate the exterior boundary so that we can obtain accurate iris segmentation. There are some examples of iris segmentation results in Fig.4, where interior and exterior boundary location by our proposal and integral differential operator are displayed respectively.

5 Conclusion

This paper has presented a novel eyelash detection method based on EM algorithm and GMM. A number of eye images are selected to verify the accuracy of the method and the results are excellent. This approach does not need to search eyelid first and classify the eyelashes, so it is better compared to other method. After excluding the eyelashes, the detected eyelash regions are not used for code generation, and iris recognition accuracy can be greatly improved.

References

1. Daugman, J.: How iris recognition works. IEEE Transactions on Circuits and Systems for Video Technology 14(1), 21–30 (2004)
2. Wildes, R.P.: Iris recognition: an emerging biometric technology. Proceedings of the IEEE 85(9), 1348–1363 (1997)
3. Kong, W.K., Zhang, D.: Accurate iris segmentation based on novel reflection and eyelash detection model. In: Proceedings of 2001 International Symposium on Intelligent Multimedia, Video and Speech Processing (2001)
4. Huang, J.: A New Iris Segmentation Method for Recognition. In: 17th International Conference on Pattern Recognition (ICPR 2004), vol. 3, pp. 554–557 (2004)
5. Kang, B.J., Park, K.R.: A robust eyelash detection based on iris focus assessment. Pattern Recognition Letters 28(13), 1630–1639 (2007)
6. Dempster, A.P., Laird, N.M., Rubin, D.B.: Maximum Likelihood from Incomplete Data via the EM Algorithm. Journal of the Royal Statistical Society. Series B (Methodological) 39(1), 1–38 (1977)
7. Tran, D., Wagner, H.: Fuzzy expectation-maximisation algorithm for speech and speaker recognition. In: 18th International Conference of the North American Fuzzy Information Processing Society, NAFIPS (1999)
8. Zhu, Y., Fujimura, K.: Driver face tracking using Gaussian mixture model(GMM). In: Proceedings of Intelligent Vehicles Symposium. IEEE (2003)
9. Bilmes, J.A.: A Gentle Tutorial of the EM Algorithm and its Application to Parameter Estimation for Gaussian Mixture and Hidden Markov Models, pp. 2–7 (1998)
10. Institute of Automation, Chinese Academy of Science, `http://www.sinobiometrics.com/english/IrisDatabase.asp.2006`

Application Study on Internet of Things in Environment Protection Field

Shixing Li[1,2], Hong Wang[1], Tao Xu[3], and Guiping Zhou[1,2]

[1] Shenyang Institute of Automation, Chinese Academy of Sciences, Shenyang 110016
[2] Graduate University of Chinese Academy of Sciences, Beijing 100039
[3] Institute of Industry Technology Guangzhou & Chinese Academy of Sciences, Guangzhou, 511458, China
lisx@sia.cn

Abstract. The conception of internet of things (IoT) and current situation are introduced and the disadvantages of traditional environment on-line monitor systems are discussed. According to the demands of environment protection and management, a basic framework of the environment internet of things is presented. The structure of the environment IoT contains three layers, namely data collection and process layer, network layer and application layer. Different layers play different parts in the environment IoT. Further, the cloud computing application is mentioned. In general, IoT is feasible and has a favorable prospect in the environment protection field in future.

Keywords: Environment, internet of things, on-line monitor.

1 Introduction

In recent years, so many environment pollution events have come up, so that the health of people has been threaten heavily. The twelfth Five-Year Plan Period is crucial time for the target of total construction of a relatively rich society in China, so environment protection is an arduous task. In environment technology, the gap between China and developed countries is large. On June 9th, 2011, Ministry of Environmental Protection printed and distributed the document of National Environment Protection Technological Development Plan for the twelfth Five-Year Plan. The plan requires that a technological innovation and environment technology management system based on environment protection and ecological degradation control will have been built up by the end of the twelfth Five-Year Plan and have achieved a bath of controlling pollution sources and emission reduction technologies with own intelligent property right under the constrains of environment norms.

At present, an outstanding problem in environment protection in China is too low information level to fit the need of environment protection development. Aiming at the present environment protect situation and environment monitoring circumstance, establishing a efficient, real time, and intellectualized environment mentoring system is imperative. IoT integrates sensors, RFID, intellectual chips and embedded systems technology, making information exchanged and shared among any things and

D. Yang (Ed.): Informatics in Control, Automation and Robotics, Volume 2, LNEE 133, pp. 99–106.
springerlink.com

providing technique support for the construction of environment monitor system. It is worthy to point out that the construction of IoT was written into the premier Wen's government work report of the year of 2010, marking that the concern and support for IoT have been lifted to the level of national strategy. Meanwhile, the characteristics of IoT determine its bright prospect in environment protection field. Thus grasping characteristics of IoT, combining them with the construction of environment monitoring system, expressing its huge application value in environment protection field and constructing the environment IoT are key points of improving environment protecting ability in the twelfth Five-Year Plan Period.

2 Current Situation of IoT

2.1 The Concept and Development of IoT

In 2005, ITU put forward the concept of IoT officially, pointing out that using RFID, sensors, new nanometer materials, embedded system and other technologies, all the things in the world can exchange information through interconnection networks[1]. Obviously, IoT is mainly related to RFID and sensor technologies. RFID's aim is to mark things, giving items different identifications. In a general way, in order to trace goods and anti-counterfeit, the global only identification for every item is necessary in production and circulation field. In this case, RFID is a leading role, the IoT based on RFID can meet the application needs. From sensor technology perspective, on the basis of sensor networks, the IoT based on sensor technology can be built up, that is sensor networks mainly consisted of sensors, communication networks and information processing system with real time data acquisition, monitoring and control, information share, information storage administration and other functions. It develops functions of the present network technology greatly, making it possible to monitor environment in real time and facilities and their inner operation mechanism. In other words, previously things with a far cry from networks such as home electronics, traffic management, agricultural production and safety of buildings and environment protection can be monitored by networks effectively, and even can be controlled remotely via networks.

At present, IoT is in the early stage. China, as a dominant country of making international standards, has research ability of top level in the whole world. As early as in 1999, Chinese Academy of Sciences started to develop sensor networks and made great progress in sensor network communication, sensor terminal devices and other aspects, forming a complete industry chain. On December 8, 2009, Wuxi Institute of Internet of Things Industry and Chongqing University of posts and telecommunications signed agreement officially on the first construction in western China of research center on IoT in Chongqing. Now, the China's standard system of IoT has formed a preliminary framework and applied partly. For example, sensor products of sensor network center in Wuxi has put into use successfully in Shanghai Pudong International Airport and the Exhibition Hall of EXPO2010 in Shanghai. In foreign countries, CERP-IoT formulated the research strategy route chart, the models of RFID and IoT and other position papers financed by European Commission. In

August 2009, Japan put forward newer vision of national information strategy called "i-Japan strategy 2015", with the main point of developing electronic government and electronic local governing groups, to promote the electronic of medical treatment, health and education.

2.2 Architecture and Characteristics of IoT

The architecture of IoT contains three layers: data collection and process layer, network transmission layer, application layer, as shown in Figure 1.

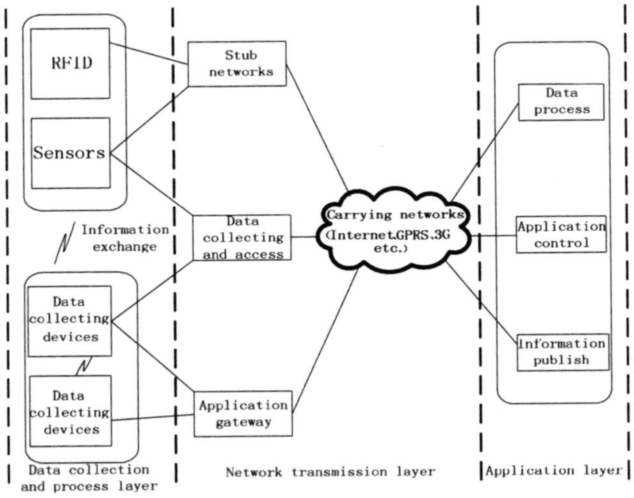

Fig. 1. The structure of IoT

The data collection and process layer uses RFID, sensors and others information collection devices to sense and collect environment information together, thus different sensors can communicate, realizing interconnection among things. The network transmission layer puts data from the data collection and process layer in uniform infrastructure transit network by different access methods. Due to the difference and large quantity of different sense devices, this layer needs heterogeneous network access technologies and base core network technologies including Internet, 3G, Bluetooth, UWB and other access technologies. The application layer faces masses of data sources and data. On the one hand, the application layer analyzes and makes use of data in specific flow and rules according to the definite demands, on the other hand, the application layer needs have the strong ability of data processing and data distribution. Consequently, this layer will adopt new data calculation and information distribution platform like cloud computing.

The following features are obvious through the architecture of IoT. First, IoT is an application of all kinds of awareness technologies. Masses of multi-type sensors are contained in IoT, and every sensor is an information source. In addition, the content and format of information get by different kinds of sensors are different. Data from sensors must be real time and sensors collect and refresh environment information data according to a certain period.

Second, it is a ubiquitous network established on communication. The base and core of IoT is communication. It transmits the information of tings accurately in real time by kinds of wired and wireless communication networks. In IoT, data from sensors is transmitted by networks and the amount of information is tremendous, forming mass data. The transmission process must adapt to various kinds of heterogeneous networks and protocols in order to ensure the data correct and timely.

Third, IoT doesn't only provide the connection among sensors but it has the ability of information processing, and it can give things intelligent control. Using cloud computing, pattern recognition and other intelligent technologies, IoT combines sensors and intelligent processing, expanding its application fields. Meaningful data which is analyzed and processed is to adapt different users' demands and developing new application fields and application modes.

2.3 The Great Significance of Developing IoT

The application of IoT is abroad, including intelligent traffic, environment protection, government operation, public security, intelligent home furnishing, intelligent fire control, industry monitor, agricultural management, the aged nursing, individual health and other application fields. The widespread and development of IoT accompany the gradual mature and application of related technologies, not when all technologies are of full maturity. Some fields will apply IoT firstly, and at the same time, with the progress of technologies, IoT will go into all aspects of our life.

In times of IoT, items will be clever. Sensors can read information automatically and information can be transmitted by Internet and communication networks. The connection of the whole process of information obtaining, transmitting and processing is a great liberation for human productivity once again.

3 Traditional Environment On-Line Monitoring

The previous environment on-line monitoring devices are mannered reading instruments. Thus, lots of time is wasted and the transmitting function of devices becomes ornaments. With the technologies progressing especially networks, the auto transmission systems of on-line devices get applied gradually, but this kind of application is limited to single pollution source monitoring device's data uploading to monitoring center such as real time data, history data and state data as well as the monitoring center give extremely control to the field monitoring devices and the devices carry out the control instructions in the pollution fields [2]. Figure 2 is a chart of typical traditional environment on-line monitoring system. In the system, isolated monitoring stations were built up through GPRS and other communication manners with severs in the back stage [3]. From figure 2, we can see that traditional system adopted star-shaped networks to exchange data and interact. Its disadvantages are that pollution source nodes can't sense each other, have no linkage and no coordination. In the system, things can't connect through networks and the source monitoring networks have no self-adaption, have no active self tune mechanism and function. The traditional systems were not able to ensure pollution sources monitored on-line and controlled with intelligence.

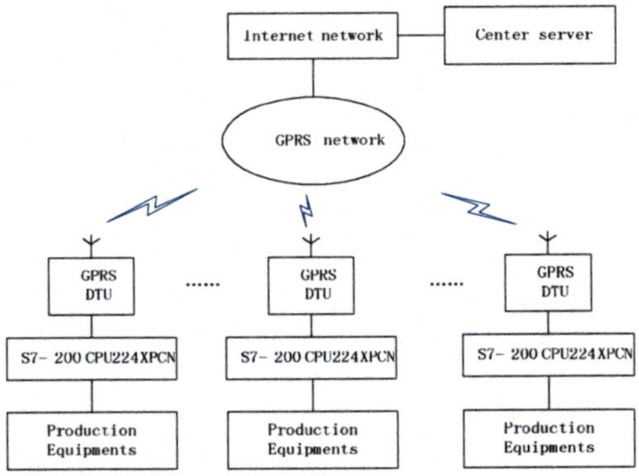

Fig. 2. The traditional environment on-line monitoring system

In addition, in the process of old environment monitoring system construction, different equipment suppliers and system integrators established their own systems. The monitoring software platforms' communication protocols were independent respectively and didn't open to each other, making that there was no information sharing and the barriers of system upgrade and network connection. Thus, auto-monitoring systems' construction was enslaved to equipment suppliers or system integrators, blocking the pace of monitoring automation and information construct [4].

From the analysis above, there are lots of defects in traditional environment on-line monitoring system. If we want to solve the problems and improve the ability of environment on-line monitoring, we need to change the traditional node-centre network mode into the IoT structure of both node-node and node-centre mode.

4 Environment System Based on IoT

The IoT's essential features of networked, things connected, automatic, perceptive intelligent and spatial determine its wide application prospect in environment application field. In the aspect of environment automatic monitoring, the environment IoT can apply in many cases. In particular, it is suit to the current pollution source on-line monitoring, motor emission management, early warning for environment quality, environment risk source monitoring and management and other application demands supported by information business, resolving the information problems of proceduring, high density, fining and rapid acquisition for environment [5]. Now, IoT for environment protection in the world is at the primary stage, but with preferable foundation. In foreign countries, there are some application cases of the environment IoT, playing an important part in pollution prevention and ecology protection such as

CitySense system in America, Australian toad distribution monitoring system and the Tai Lake monitoring system in China. With the maturity and mass application gradually, the environment IoT is developing on a good base [2]. The basic framework of the environment of things is shown as figure 3, including data collection and process layer, network layer and application layer with the functions vividly called sense and control and transmission and convey. The data collection and process layer contains sensors, actuators and instruments with communication function that can sense the pollution factor. The network layer contains all kinds of wired and wireless networks. The application layer is consisted of all kinds of services and business based on cloud computing platform.

4.1 The Data Collection and Process Layer

The data collecting is mainly used for collecting real time data of monitoring points and environment monitoring index, including physical, chemical, biological indicators and identification of environment management targets, sound and visual information, satellite positioning and satellite remote sensing. The data collection of the environment IoT is related to sensors, detectors, RFID, multi-media information acquisition, two-dimension code, real time location and all the other monitoring, detection and sensor techniques. In geography, the scale and range contain rivers, sea, lands, sky and space, etc. The data collection and process layer makes masses of devices connected, environment business evident, proving to be the core and base of the environment IoT. New and professional sensor technologies need to be developed to satisfy rising monitoring demands and improve sensors' ability to perception and data process. At the same time, this layer calls for sensor network organizing and collaborative information processing and control technologies, realizing the process of sensors and RFID's data transmission, network self-organization, multi-sensors' data collaborative processing, control of instruments, valves and other actuators. In this layer, all the automatic monitoring nodes sense, exchange information, cooperate, to realize information exchanging among things, business cooperation, constant self-adjusting and optimizing the process of pollution emission according to a certain goal. Thus, setting up efficient smart environment monitoring and control mechanism and forming a dynamic information chain for ensuring environment quality steady are the best application scenarios of IoT and have powerful application potential.

4.2 The Network Layer

The network layer realizes functions of environment information transmission and systems interconnection, transmitting the available environment information in a barriers free, high reliability and high safety manner. The mobile communication networks and private networks for environment information techniques are already very mature, satisfying the data transmission demands of the environment IoT well. IoT needs supporting networks with strong fault self-recovery, reliable routing transmission and other network service abilities. Considering the features of outdoor multi-sites and large range, the environment IoT merges 3G/4G mobile networks, Internet, satellite communication networks together, providing transparent network address allocation, network management, monitoring, safety, service quality assurance and other network service functions. The network layer provides network support for things' connection between environment auto-monitoring nodes.

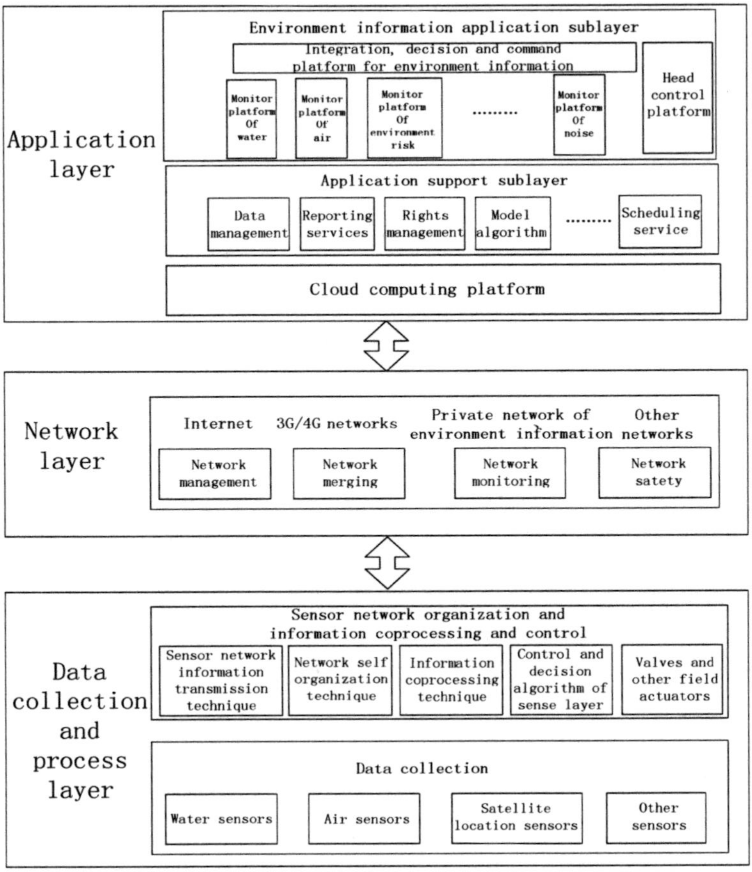

Fig. 3. The framework of the environment IoT

4.3 The Application Layer

The application layer of the environment IoT is mainly consisted of application support platform, information application platform and cloud computing platform. The application layer provides the functions of organizing, storing and processing the environment data acquired by bottom sensors as well as the functions of identity authentication, permission management, the workflow of GIS environment model calculating, environment management accounting and report form services. These services help to show pollution emission information and environment quality information in real time and to show the comprehensive statistics of environment elements from the information. The application layer is an information door of the environment IoT for environment decision management and planning service. In addition, the sewage draining exits' data of the serious pollution enterprises can be monitored in real time, even can be shut up the sewage draining exits by remote, preventing abrupt environment contamination accidents.

With the monitoring scale becoming larger and more refined, data from bottom layer devices is more and more and need stronger calculation ability. In the application layer of the environment IoT, the technology of efficient processing mass data and scientific model will come out and apply. In recent years, cloud computing was presented and developed fast, which provides a completely new train of thought for data processing and calculation in the environment IoT. The cloud computing is based on virtualization technique, using networks as the carrier, taking basic structure platform software service as the form, integrating large scale extensible data calculation and storage and other distribute resources to work together and calculate in supercomputing mode. The superiorities of high reliability, extensity and high flexibility are obvious in distributed application based on cloud computing platform. In a word, the large scale application of the environment IoT is an inevitable trend.

5 Epilogue

The combination of environment monitoring and IoT is feasible and there are natural relations between the two. The merging of environment monitoring and IoT is a development tendency for the environment protection in future. In particular, when the infrastructure of IoT is built widely and the industrialization of IoT comes into being gradually, environment monitoring based on IoT will certainly have more application fields and greater market value. Environment protection is beneficial to the modern time and the generations to come. It is a crying need to take the lead in constructing IoT in the environment protection field in contemporary China.

References

1. Liu, Z., Wei, F., Chai, Y., Shen, X.: The think about the development of IoT in China. Comprehensive Transportation 2, 37–40 (2010) (in Chinese)
2. Li, G., Li, X., Wen, X.: Influence of Internet of Things Technology on the Development of Automatic Environmental Monitoring System. Environmental Monitoring in China 27(1), 5–10 (2011) (in Chinese)
3. Chen, L., Feng, J., Xu, B.: Design of Remote Environmental Monitoring System. Journal of Tongji University (Natural Science) 38(11), 1659–1663 (2010) (in Chinese)
4. Lou, H.: Multielement Integration Online Monitoring System and Its Application. Sichuan Environment 25(2), 51–53 (2006) (in Chinese)
5. Tian, T., Cheng, G., Mao, S., Tan, H.: Discussions on the Development of Internet of Things for Environmental Protection. Information and Communications Technologies 5, 31–36 (2010) (in Chinese)

Multiresolution Based Segmentation for Nonideal Iris with Nonlinear Diffusion*

Honglin Wan[1], Min Han[2], and Ting Wang[2]

[1] Shandong Normal University, East Wenhua Road 88, 250014, Jinan, China
[2] Shandong University, South Shanda Road 27, 250100, Jinan, China
{visage1979,hanmin}@sdu.edu.cn, wangting19870605@yahoo.com.cn

Abstract. Nonideal iris segmentation is a challenge for iris recognition. In this paper, we propose a multiresolution frame based segmentation algorithm for nonideal iris with nonlinear diffusion. Our proposal includes two stages: interior boundary and exterior boundary detection. In process of exterior boundary localization, a simplified region-based curve evolution is invented to search weak boundary. Experiments reveal validity of our proposal and a high correct ratio is obtained.

Keywords: Nonideal Iris Segmentation, Nonlinear Diffusion, Biometrics, Partial Differential Equations, Multiresolution Analysis.

1 Introduction

Among biometric techniques, people prefer iris recognition because of its highly discriminative and stable texture patterns [1]. Many researchers have stressed this issue, including Daugman [2], Wildes [3] and their extensions. However if iris images are captured in nonideal cases caused by light, uncooperativeness of persons, unconscious body motions, etc, they can not be segmented correctly. Some people have investigated this problem and gained meaningful results [4, 5]. Although literatures above, nonideal iris segmentation still remains difficult. To response to such a challenge, we design an iris segmentation algorithm employing multiresolution analysis and nonlinear diffusion which are incorporated into both interior and exterior boundary detections.

This rest of this paper is organized as follows. In Section 2 of this paper, we illustrate the proposed segmentation algorithm for nonideal iris data. Nonlinear diffusion is employed to smear small-scaled reflection while maintain large-scaled iris boundary [6]. Experiment results are given in Section 3. Finally, Section 4 concludes the paper and discusses prospective future work.

* This paper is supported by national major fund of China, Scientific and Technological Developing Project of Shan Dong Province (No. 2010GGX10136), and Excellent Youth Scientist Award Foundation of Shandong Province (No. BS2011DX036, No. BS2009DX008).

D. Yang (Ed.): Informatics in Control, Automation and Robotics, Volume 2, LNEE 133, pp. 107–111.
springerlink.com

2 Nonideal Iris Segmentation

In this section we focus on two issues. One is interior boundary localization where the key is searching iris center. In this circumstance, the scale-dependent evolution property of anisotropic diffusion provides a practical utility for the nonideal iris segmentation. The other is exterior boundary detection where with Laplacian pyramid (LP), nonlinear order statistical filter and a simplified region-based curve evolution are utilized. Explicitly iris data should be projected into polar coordinate and only the segment without eyelid occlusion is available to search exterior boundary. Moreover, for bettering active contour model (ACM) segmentation, we take advantage of nonlinear order statistical filter (OSF) regionally to maintain boundary of iris and sclera while flatten their interior variations respectively. The details of the two-step method are described as follows. Flowcharts have been shown in Fig.1 and Fig.2.

2.1 Interior Boundary Localization

(S1-1) LP decomposition This substep can reduce iris data. In the following substeps only coarse level obtained from LP.

(S1-2) Pupil centers searching It includes:

a) Nonlinear diffusion Here we use P-M method to smooth iris while preserve edges according to scale to eliminate small-scaled reflections.

b) Threshold A simple Gaussian Mixture Model (GMM) can separate pupil and other components of iris image.

c) Binarization and close operation This substep aims at reflection removal. Thresholding region is set to bright while others set to dark, and such binarization is followed by morphologic close operation which eliminates isolated reflections.

(S1-3) Mapping pupil region from coarse level into original image We use nearest interpolation to map this small mask to original level. Naturally, mapped mask will cover real region of pupil that we compute the pupil's centroid as the iris center.

(S1-4) Interior boundary detection We use Sobel operator to find maximum of each column and view them as interior boundary of iris in polar-transformed image resulted from previous substeps. Then we record distance of each maximum to pupil' center, and form its histogram in which the distance with highest probability is the radius.

Explicitly, searching pupil's center as in*(S1-2)*, founds the algorithm of interior boundary location. As pupil has been modeled as a darkness disk, evidently its center is the centroid. But eyelash forms the darkness similar to pupil thus it is difficult to separated them. Eyelash removal becomes feasible by employing P-M formulation-included PDE evolution, because small-scaled eyelash will be eliminated and on the contrary, large-scaled pupil will be preserved when iris data has been anisotropically diffused. Specifically, by gradient descent method the P-M formulation can be calculated by

$$I_t = I_{t-1} - a \cdot div\left(g(|\nabla I_{t-1}|)\nabla I_{t-1}\right) \quad (1)$$

where the parameter a, as K in (2) or (3), controls the evolution speed and can be assigned experientially. Fig.1 shows the diffusion where $a = 0.25$ and $g(\cdot)$ is specified by (2). Notably, unnecessary time consuming in diffusion for iris data can be avoided by selecting a large K. In fact, if variation of K or iteration of the formulation (4) is not very large, little will be changed to smoothing effect when carrying out anisotropic diffusion and edge-preserving still can be guaranteed.

2.2 Exterior Boundary Localization

(S2-1) LP decomposition of reflection-removed iris image

(S2-2) Region Selection By observation we find that eyelid occlusion will lead to false exterior boundary detection. We choose a segment within an angular range without eyelid occlusion to detect the boundary circle. Conveniently an angular range selecting of top or bottom of an iris image should be avoided.

(S2-3) Order statistical filtering of selected region Reliable exterior boundary detection depends on not only grey level of iris and background, but also a distinct edge between them. In our algorithm, selected iris region in polar domain is split uniformly into three parts: each pixel value of the part nearest to iris is replaced by local minimum iteratively, and similarly, the part farthest to iris is by local maxima. To protect exterior boundary the middle part is processed by a median filter.

(S2-4) Region-based segmentation Region-based segmentation in this substep is a simplified version of active contour model. Exterior boundary is initialized as a line near to iris interior boundary, and mean values of both separated regions are computed. Then the line is pushed far away from interior boundary until the difference of the two mean values is smaller than the new one.

3 Experiments

We adopt CASIA-IrisV3-Interval database [7] to test our approach. CASIA-IrisV3-Interval database is released by Institution of Automation, Chinese Academy of Science, which consists of 2903 images with 256 grey levels and 320×280 size. All experiments are carried out using Matlab 7.0 software on a Pentium 2.4 GHz processor and 2Gbyte RAM. Some experiment results have been shown in Fig.3.

The correct segmentation ratio (CSR) and averaged processing time of different methods are evaluated. Since the methods of Daugman and Wildes only deal with interior boundary, we attach our exterior boundary detection to them to evaluate these to our method equivalently. The methods are listed as below.

- Method-1: our algorithm proposed in Section 2.
- Method-2: Daugman's integro/differential operator with LP framework for interior boundary, and our algorithm for exterior boundary.
- Method-3: Wildes' method consisting of Hough transform and Canny detector for interior boundary with LP framework, and our algorithm for exterior boundary.

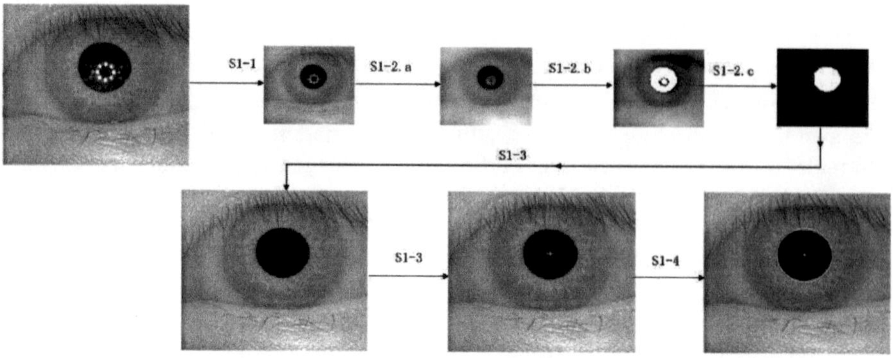

Fig. 1. Flowchart of interior boundary detection

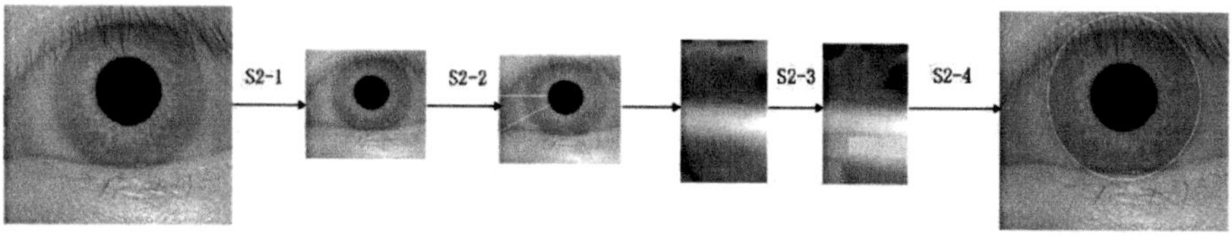

Fig. 2. Flowchart of Exterior Boundary Detection

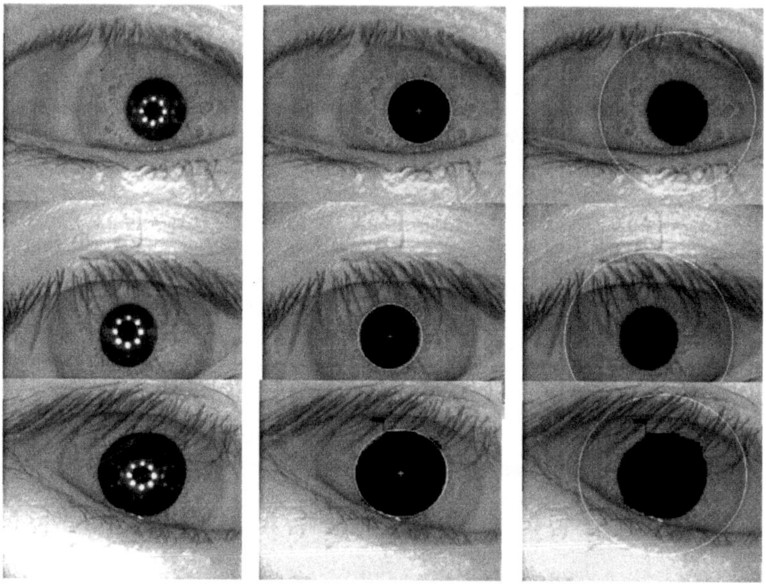

Fig. 3. Some results of our segmentation experiment

In this paper, the one aspect of segmentation quality is assessed by Equality Ration (ER) that calculates the ration of pupil area over interior circle area. A good ration in experiments is 95%~105%. Our method achieves 97.29%, higher than 96.84% of the Method-2 and 93.32% of the Method-3.

The other aspect of segmentation quality is the efficiency which is evaluated by averaged processing time. Not only interior but exterior boundary is considered Our method achieves a better performance, 1.67 seconds, faster than 2.17 and 4.90 of the Method-2 and Method-3 respectively.

4 Conclusion

In this paper, a novel nonideal iris segmentation method has been proposed. In the first stage, nonlinear diffusion is utilized to smear small-scaled reflection while maintain large-scaled iris boundary. During the second stage, we devise a simplified active contour model to detect exterior boundary. Utilizing order statistical filters regionally facilitates a correct convergence of this model. We compare our method to other classical ones of Daugman and Wildes. Experimental results show that our proposal researches to a high correct segmentation ratio and lower averaged time consuming.

References

1. Jain, A.K., Ross, A., Pankanti, S.: Biometrics: a tool for information security. IEEE Trans. Information Forensics and Security 1(2), 125–143 (2006)
2. Daugman, J.: How iris recognition works. IEEE Trans. Circuits and Systems for Video Technology 14(1), 21–30 (2004)
3. Wildes, P.R.: Iris recognition: an emerging biometrics technology. Proc. IEEE 85(9), 1348–1363 (1997)
4. He, Z., Tan, T., Sun, Z., Qiu, X.: Toward accurate and fast iris segmentation for iris biometrics. IEEE Trans. PAMI 31(9), 1670–1684 (2009)
5. Zuo, A., Schmid, N.A.: On a methodology for robust segmentation of nonideal iris images. IEEE Trans. Systems, Man, and Cybernetics, Part B: Cybernetics - Special Issue on Game Theory 40(3), 703–718 (2010)
6. Strong, D., Chan, T.: Edge-preserving and scale-dependent properties of total variation regularization. Inverse Problem 19(6), 165–187 (2003)
7. `http://www.cbsr.ia.ac.cn/irisdatabase.htm`

Gait Programming and Data Acquisition User Interfaces, for Modular Snake Robots

Kamilo Melo, Laura Paez, Andrea Polo, and Carlos Parra

SIRP Research Group Department of Electronics, Pontificia Universidad Javeriana, Bogota D.C., Colombia
{melok,laura.paez,poloa,carlos.parra}@javeriana.edu.co

Abstract. This paper presents three graphic user interfaces, that have been developed as a first functional versions. Two of them, are currently used to fulfill requirements of gait programming for scripted and parametrized gaits. The third one, relies on the need to acquire variables to measure gait performance. Gait designing procedures can be improved by understanding the gait analyzing the data acquired. Examples of the use of these interfaces are presented. Finally, a projection is made according to the integration and future use of these tools.

Keywords: Gait design, graphic interface, snake robots, locomotion.

1 Introduction

There are several ways to control the locomotion of a modular snake robot. Some of them depend on the type of gait that the robot is executing which can be preprogrammed. Others depend on an environmental feedback of the interactions with the place where the robot is performing the locomotion. Relevant work on feedback control for modular snake robots can be found in recent developments [1], regarding obstacle aided locomotion feedback control. On the other hand, central pattern generators have been widely used for more than a decade as a method to govern the snake robot movements. This can be specially seen in the contributions of professor Shugen Ma [2], among others.

Turning a planar (2D) gait into a spatial one (3D), increases considerably the complexity of the feedback locomotion control system. The simplest way to program snake robot movements is a step by step assignation of a complete set of generalized coordinate values. Independent joint control of actuators ensures that the robot will change its configuration to the desired shape programmed. Assigning joint variables references in a defined time scheme will lead the robot to perform locomotion. This seems to be the worst way to program the robot, but in some difficult terrains the use of these scripted gaits yields to a success locomotion. Professor Choset at CMU[1] and his team, have demonstrated that scripted gaits will be the solution for a complicated gaits like stair climbing and gap crossing, as reported on [3]. The scripted gait programming can be more

[1] ModSnake, Biorobotics Lab, The Robotics Institute, Carnegie Mellon University.

D. Yang (Ed.): Informatics in Control, Automation and Robotics, Volume 2, LNEE 133, pp. 113–117.
springerlink.com

effective, if a indexed table is used. This table contains the information of joint variable values and timing for each actuator. The indexing processes can be repeated, in order to generate cyclic gaits with a simple recurrent table.

On the other hand, an open loop gait generator based on a parametric equation [3], also addressed by professor Choset, have demonstrated that several cyclic gaits for a 16-DOF robot, can be programmed by assigning a reduced set of sinusoidal equation parameters that commands the movements of robot actuators. These parameters include joint offset, joint amplitude, phase shift between modules reference signal, time and spatial frequency, allowing timing and spatial distribution of the joint reference signal value. The last method for generate gaits only work properly with a robot as seen on the figure 1. This robot presents a series of 16 actuators whose joints are twisted 90° and with ±90° of amplitude for each joint. It can be called an active body, not wheeled, bending type, modular snake robot [4].

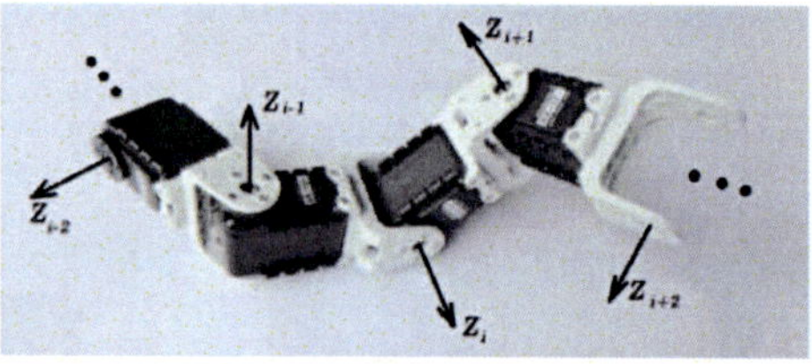

Fig. 1. Active body not wheeled bending type modular snake robot. Coordinate frame assignments. Four of the 16 modules implemented are shown.

Once a gait can be programmed by any of both methods mentioned before (scripted or parametrized), there is a good experimental practice to acquire information from the actuator's built in sensors (i.e. position, speed, torque, temperature) and external sensors (i.e. accelerometers and gyroscopes). Not only to determine the dynamics of locomotion, but to evaluate energy efficiency metrics of the performed movement [4], for a better understanding of the gait [3].

2 Scripted and Parametrized Gaits Interface

Our laboratory[2] currently has three robots which are constructed using Dynamixel AX-12+ servomotors as actuators (Figure 1). For experimental purposes the mechanical structure of the three robots differs. This allow us to use different length robots with the same control interfaces. An external computer host the control interface and by means of a serial communication, send position instructions to the servomotors and receive status of several sensors way back. A brief description of each interface and some examples of its utilization will be described as follows.

[2] www.gruposirp.org/sirp/sirp-research-CP-msr.html

2.1 Scripted Gait

Using the scripted gait interface on figure 2, programming a controlled movement, is an easy task. The interface allows the user to load text files with joint angle values indexed by a time component. Once the file is loaded, there is a execution button that starts communication with the robot to send the proper information to the actuators. If a gait or movement wanted to be programed and saved on a file, the interface has a palette to assign the respective desired joint angle values to each actuator. The execution button moves the robot. If the movement was performed in a right manner, the configuration are saved by pushing the "good" button. If the movement was wrong, the "bad" button will return the robot to the last configuration without save.

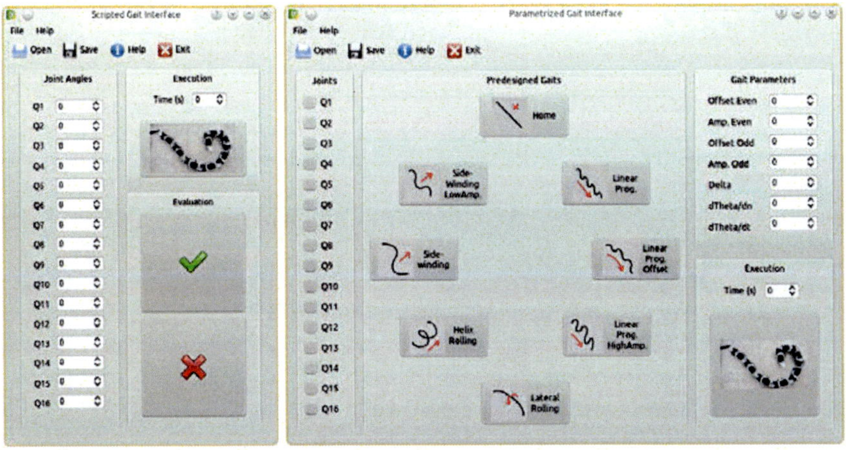

Fig. 2. Scripted and Parametrized Gait interfaces

Examples of the use of this interface are currently done by our lab. A proposed robot configuration for recognizance purposes [4], can be programed with the interface. Recognizance configuration are intended to be executed for a robot with end module's cameras and is stable to guarantee image acquisition. Other movements programed are the swing locomotion of a hanged modular snake robot [5], and a ellipsoidal shape rolling modular robot [6].

2.2 Parametrized Gait

The parametrized gait interface as seen on figure 2, allows the user to input different parameters that will generate cyclic motions of the robot for locomotion. An equation is evaluated with these parameters, generating a set of functions that describe each of the joint angle values that the robot must perform in time.

Once these values are computed and stored on a data structure, a communication protocol will send information packages through a serial wire to deliver the proper instructions to the low level control of the actuators.

Some gaits addressed by [3], can be programed by the interface and executed by our robots [4]. These gaits include: linear progression, side-winding and lateral rolling. A special rolling gait intended to be performed on the outside of pipes, was also studied in order to establish metrics of energy efficiency and mechanical stability [7].

3 Data Acquisition Interface

An interface dedicated to acquire data from the robot was implemented. It increases experimental rigor and ensures better understanding of the movements programed by the other interfaces. In figure 3 the layout of the interface is shown.

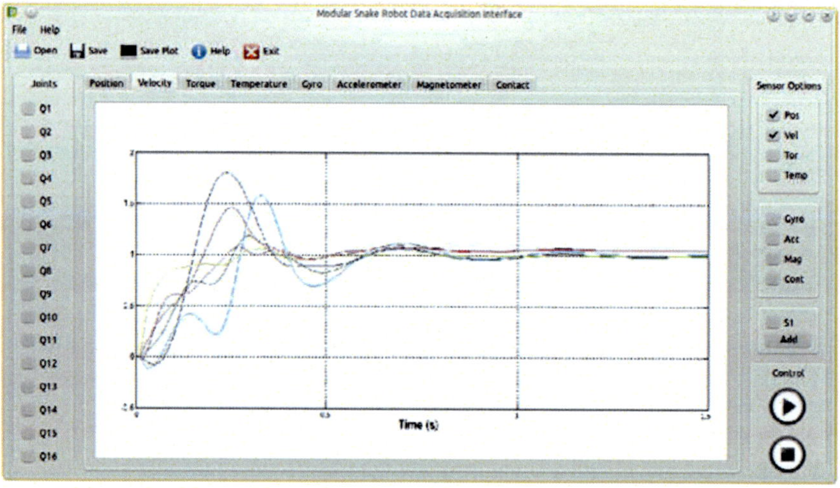

Fig. 3. Modular snake robot data acquisition interface

This interface operates receiving information from 2 different ports. First one, the robot actuators serial communication itself. The same that send locomotion instructions to the robot. It sends back information of the actuator built in sensors like position, speed, torque and temperature. On the other hand, a second communication channel dedicated for external sensors added to the robot will retrieve data from gyroscopes and accelerometers.

The interface allows users to define the start and end of the acquisition time. There is a tab for each sensor. A selection palette allows the user to select the desired actuator to being monitored. Once the acquisition ended, a plot of the data acquired is generated and saved. The second channel allows in a future the addition of new sensors.

There is a interest to know the robot status to evaluate metrics that describe quantitatively the performance of the robot [4]. Contributions on predictive control has been developed. A distributed model predictive control [8], evidence the need of acquire experimental data from the robot during a gait execution.

4 Concluding Remarks

Our lab seeks improvement of the interfaces and proper integration of them. To ensure a better and flexible robot control, and a proper data analysis regarding locomotion. The flexibility of the three interfaces is a key topic on the future. Allowing timing and smoothness for scripted gaits, the real time manipulation of equation parameters for parametrized gaits and the flexibility on the addition of new sensors for data acquisition, are projections to be done by our lab. A database for off-line data analysis can be used to store and analyze data from experiments. In this way ensure a better understanding of gaits and at the same time, allowing the creation of models to control and improve the locomotion capabilities of this robots. Real time execution-acquisition-analysis will be one of the challenges in the future. This will allow the implementation of a gait planner that measures the environment and dynamics of the robot in order to self adapt gaits to increase energy efficiency and performance in locomotion.

References

1. Liljebäck, P.: Modelling, Development, and Control of Snake Robots. PhD thesis, NTNU, Norwegian University of Science and Technology (2010)
2. Inoue, K., Sumi, T., Ma, S.: Cpg-based control of a simulated snake-like robot adaptable to changing ground friction. In: IEEE/RSJ International Conference on Intelligent Robots and Systems, IROS 2007, pp. 1957–1962 (2007)
3. Tesch, M., Lipkin, K., Brown, I., Hatton, R.L., Peck, A., Rembisz, J., Choset, H.: Parameterized and scripted gaits for modular snake robots. Advanced Robotics 23, 1131–1158 (2009)
4. Melo, K., Paez, L., Hernandez, M., Velasco, A., Calderon, F., Parra, C.: Preliminary studies on modular snake robots applied on de-mining tasks. In: Proc. IEEE IX Latin American Robotics Symposium and IEEE Colombian Conference on Automatic Control (2011)
5. Trujillo, A., Igua, J., Melo, K., Parra, C.: Passive dynamic swing locomotion for a 6dof coplanar articuled chain. In: Proc. IEEE IX Latin American Robotics Symposium and IEEE Colombian Conference on Automatic Control (2011)
6. Melo, K., Velasco, A., Parra, C.: Motion analysis of an ellipsoidal kinematic closed chain. In: Proc. IEEE IX Latin American Robotics Symposium and IEEE Colombian Conference on Automatic Control (2011)
7. Paez, L., Melo, K., Parra, C.: Center of mass displacements using rolling gaits for modular robots on the outside of pipes. In: Proc. IEEE IX Latin American Robotics Symposium and IEEE Colombian Conference on Automatic Control (2011)
8. Cortes, F., Linares, D., Patino, D., Melo, K.: A distributed model predicitve control (d-mpc) for modular robots in chain configuration. In: Proc. IEEE IX Latin American Robotics Symposium and IEEE Colombian Conference on Automatic Control (2011)

Design of a Mobile Robotic Arm for People with Severe Disabilities

Hiroki Higa, Hideyuki Uehara, and Takashi Soken

University of the Ryukyus,
1 Senbaru, Nishihara, Okinawa 903-0213, Japan
hrhiga@eee.u-ryukyu.ac.jp

Abstract. This paper presents a design of a mobile robotic arm for people with severe disabilities. The main body of the robotic arm can be contained in a laptop computer's briefcase. Its weight is only 5 kg, including two 12-V lead acid rechargeable batteries. A drinking tea task was experimentally performed by a person with muscular dystrophy. An eating task was also conducted by two able-bodied subjects. The experimental results showed that they could carry out drinking and eating tasks.

Keywords: Robotic arm, feeding assistance, muscular dystrophy, mobile system.

1 Introduction

There are many people who cannot perform voluntary movements of their extremities due to their diseases and who need someone's help. For example, a muscular dystrophy (MD) is characterized by progressive weakness and degeneration of the skeletal muscles that control movement [1]. On the other hand, we live longer than we used to. Many of caregivers are getting older, and it is difficult for them to take good care of people with disabilities because of their advanced ages. It is very important for us to support their lives. In step with the advances in assistive technology, some robots, such as Handy 1 [2] and My Spoon [3], have fortunately been developed. These systems enable people with disabilities to eat foods by themselves to some extent. However, the Handy 1 is too large and heavy to use in a restricted housing environment in Japan. The My Spoon requires users to prepare foods to be cut and to use the dedicated tray for meals. In addition, they can hardly eat out with these robotic systems.

A mobile robotic arm for people with severe disabilities, such as stroke, spinal cord injury, and muscular dystrophy (MD), has been developed to explore the optimum solution for the above-mentioned problems. Our final goal is to provide a reasonable and useful mobile robotic arm system to them. To evaluate performance of the robotic arm system, a drinking tea task and eating task with it were performed experimentally by a person with MD and two able-bodied subjects, respectively.

D. Yang (Ed.): Informatics in Control, Automation and Robotics, Volume 2, LNEE 133, pp. 119–122.
springerlink.com

2 System Configuration

Fig. 1 shows the prototype robotic arm. As can be seen in Fig. 1 (a), the robotic arm's main body is small enough to put in a laptop computer's briefcase (36 cm × 25 cm × 9 cm). This system is composed of a main body of robotic arm, a microcontroller, and a controller. The microcontroller, AT91SAM7S256, has a 32-bit ARM7TDMI RISC processor (Atmel Corporation), which is low-power, small, cost-effective, and excellent real-time interrupt response. The robotic arm also can be mounted on a wheelchair. One of the fundamental concepts for the robotic arm system is portability. This feature allows users to enjoy not only having dinner with their companions in their houses but also eating out with them. The robotic arm system can be utilized when the users go on a trip, which enables them to try a variety of foods at restaurant as much as they wish while traveling.

The robotic arm has seven servos. These servos are controlled by the microcontroller. The gripper, the end effector, is also controlled by a servo, which is attached to the wrist portion. Some materials, such as carbon plate, balsa wood, and low foamable vinyl chloride, were used to reduce the weight of the end effector of the robotic arm. We also made a controller for the person with MD, as shown in Fig. 1 (c). An emergency stop switch was installed for emergency purposes. Turning this switch on allows users to shut down the robotic arm system. Furthermore, a control program written in the form of C language was developed with the gcc compiler. The program was then written into a flash memory of the microprocessor which was connected to a PC using an RS-232C cable.

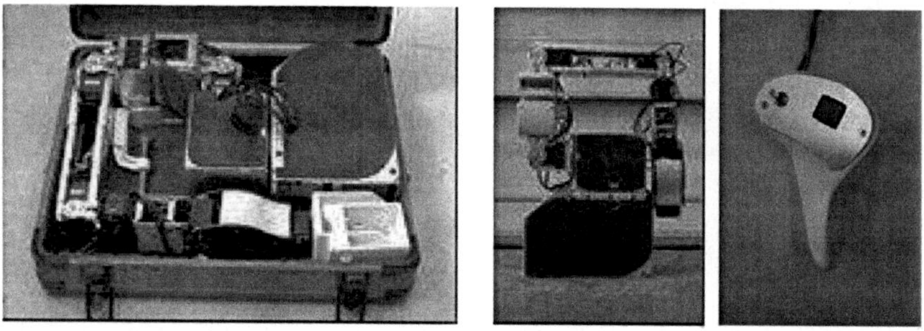

(a) (b) (c)

Fig. 1. Mobile robotic arm (a) when folded in a briefcase. (b) its side view. (c) controller for the person with MD. The left toggle switch is an emergency stop switch. The right is a button for giving the robotic arm control commands.

3 Methods

The main body of the robotic arm was placed at a subject's right side. Each subject was asked to push the button to control the mobile robotic arm. First, a person with MD participated in a drinking tea task. He was seated on his wheelchair in front of an over bed table. The subject was required to drink tea with the mobile robotic arm. The tea was in a 500-ml plastic bottle, which was commercially available.

Then, two able-bodied subjects participated in an eating task. Each subject was seated on a chair in front of a table. To issue a command to the robotic arm, a notebook PC, Lenovo R61e powered by the Celeron processor of 1.86 GHz with RAM of 2 GB, was used in this task. The subject was asked to eat some foods with the mobile robotic arm. The foods were rice gruel, porridge, and Chinese dumpling, shaomai. In this task, the subject was asked to use a plastic spoon when eating rice porridge and to use a plastic fork when eating shaomai. These utensils were commercially available. The spoon and fork were set in the stands to facilitate grasping and releasing them by the gripper. All tasks were recorded by a digital video camera.

4 Results and Discussion

An experimental result of the drinking tea task is shown in Fig. 2. The experimental result was captured from the video data. The elapsed time for drinking tea was 30 s. The subject made no error in operation during this task. It was clear from the experimental result that the subject could appropriately drink tea with the mobile robotic arm system.

The eating task result is shown in Fig. 3. The result was also captured from the video data. The elapsed time was indicated in each image at the upper left corner. In this task, the subject B could eat a spoonful of rice porridge at 36 s, and have a shaomai by the plastic fork at 79 s. The time to have each food can be adjustable and determined by each subject. The exchange of the spoon for fork was smoothly performed by the subject with our system [from 54 s to 66 s in Fig. 3]. The subject C also could conduct the same task handily. Hence, it was found from the experimental result that our system allowed users to have their meals at their paces and to use the commercial spoon and fork when eating.

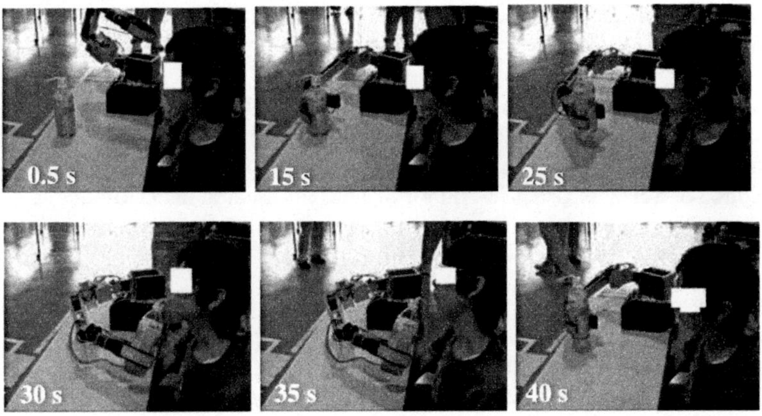

Fig. 2. Experimental result of drinking tea task (subject A who is suffering from MD). The elapsed time was indicated in each image at the lower left corner.

Fig. 3. Experimental result of eating task (able-bodied subject B). The foods were rice porridge in the right dish and shaomais in the center dish.

A feeding assistance with assistive robotics is the most desired assistance for people with disabilities [4]. We hope people with severe disabilities will be more comfortable to manipulate our system, which leads to a better option for them. An optimum controller for a manually control mode in the mobile robotic arm system is under ongoing consideration.

5 Conclusion

This paper described the demonstration of the mobile robotic arm. We have built the robotic arm which could firmly grasp the 500-ml plastic bottle as well as the plastic spoon or folk. The experimental results showed that we could drink tea in the commercial plastic bottle and eat rice porridge and shaomais with commercially available utensils. Manually control mode for our robotic arm will be considered in our future work.

Acknowledgments. The authors would like to thank the subject with MD and staff in National Hospital Organization Okinawa Hospital for their valuable cooperation to the progress of our work. This work was partly supported by Japan Society for the Promotion of Science, Grant-in-Aid for Scientific Research under Grant C-22500509.

References

1. NINDS Muscular Dystrophy Information, `http://www.ninds.nih.gov/disorders/md/md.htm` (accessed January 2010)
2. Topping, M.J., Smith, J.K.: The Development of Handy 1. A Robotic System to Assist the Severely Disabled. Journal of Tech. and Disability 10(2), 95–105 (1999)
3. My Spoon, `http://www.secom.co.jp/personal/medical/myspoon.html` (accessed December 2008)
4. Miller, D.P.: Assistive robotics: an overview. Assist. Tech. and AI, 126–136 (1998)

An Emergency Call Device for Wheelchair Users with Severe Disabilities

Bintao Wang, Hiroki Higa, Hideyuki Uehara, and Takashi Soken

University of the Ryukyus,
1 Senbaru, Nishihara, Okinawa 903-0213, Japan
k118555@eve.u-ryukyu.ac.jp, hrhiga@eee.u-ryukyu.ac.jp

Abstract. In this paper, an emergency call device for wheelchair users with severe disabilities is proposed. The system is composed of a touch sensor circuit, microcontroller, circuit which is sending an emergency e-mail, and mobile phone. To verify the performance, three able-bodied subjects participated in the experiment. From experimental results, it was clear that the system could detect whether the subject's hand was touched the control lever or not.

Keywords: Assistive device, electric powered wheelchair, emergency call system, microcontroller, people with disabilities.

1 Introduction

There are people who have difficulty in walking from various causes. Some are disabled due to their diseases, others by some accidents. As a mobility assistive device, a wheelchair plays an important role for them. For example, muscular dystrophies (MD) are characterized by progressive weakness and degeneration of the skeletal muscles that control movement [1]. Most of the individuals with MD cannot move their arms freely due to their deteriorated muscles. When they drive their electric powered wheelchairs, they make their maximum efforts not to detach their hands from the control joystick of the wheelchair. If this occurs accidentally, they cannot control their wheelchair again until caregivers help them. Wheelchair technology has been developed. We can have a power-assisted wheelchairs and multifunctional wheelchairs [2], [3]. Wheelchair users always go outside along with caregivers. Some of persons with disabilities, however, have strong desire to spend private time alone.

In order to improve the quality of lives (QOLs) for the wheelchair users with severe disabilities, we have developed an emergency call device which can detect whether users' hands hold the joysticks of the wheelchairs. When their hands are not touched them for a couple of seconds, the device sends the message that the hands of the wheelchair users are gotten away from the joysticks and that they cannot move from the place where they are to the caregivers. We tested the device to evaluate the performance.

D. Yang (Ed.): Informatics in Control, Automation and Robotics, Volume 2, LNEE 133, pp. 123–126.
springerlink.com

2 System Configuration

A block diagram of the emergency call device is shown in Fig. 1. The system is composed of a touch sensor circuit, microcontroller, circuit sending e-mail, and mobile phone. In addition, the LED and electronic buzzer are utilized to inform people nearby the user about the emergency situation. Using the touch sensor circuit, we first monitor whether the user holds the joystick of the wheelchair. If the user's hand is accidentally detached from the joystick for a few seconds, the circuit outputs an emergency signal to the microcontroller. We used a PIC12F683, which is an 8-pin PIC Flash microcontroller, Microchip, for the microcontroller. The microcontroller then activates the emergency call device's outputs: the LED is flashing, the electronic buzzer is sounding, the circuit sends an emergency e-mail to the pre-registered e-mail addresses (e.g., a personal doctor and/or caregiver), and the device automatically dials the pre-registered phone number (e.g., the personal doctor, caregiver, or rescue life guard) to get help through the mobile phone.

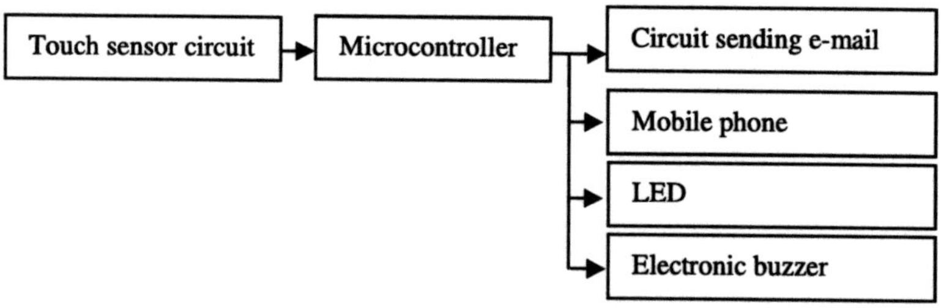

Fig. 1. Block diagram of emergency call device

3 Method

In the experiment, we covered the control lever of an electric powered wheelchair with meshed copper, as an input portion of the touch sensor circuit. The emergency call device was fixed to the wheelchair. Three able-bodied subjects (two males and a female, average age: 24.0-year old) participated in the experiment.

To send the emergency e-mail wirelessly, we used the wireless router, AirStation WHR-G301N, Buffalo Technology, Inc., and the Ethernet converter, WLI-TX4-AG300N, Buffalo Technology, Inc. [4]. The wireless router was set to the third floor of the University building. The Ethernet converter was also fixed to the wheelchair. The emergency call device with the Ethernet converter was powered by a rechargeable 12-V lead-acid battery. A caregiver (one of able-bodied subjects) waited ready for an emergency call in the laboratory at the third floor of the same building. The phone number for receiving the emergency call was set to that on the caregiver's mobile phone. The e-mail address for receiving the emergency e-mail was fixed to the University e-mail account.

Each subject was asked to sit the electric powered wheelchair and to go outside of the building by using the wheelchair by her/himself. Then, she/he intentionally got away her/his hand from the control lever during driving the wheelchair. When the caregiver received the emergency call, he went out to help the wheelchair user. All the outputs of the emergency call device were experimentally checked.

4 Results and Discussion

Examples of experimental results are shown in Figs. 2-4. Figs. 2 (a) and (b) show the cases when the subject A holds the control lever and when he does not hold it, respectively. In the latter case, it was confirmed that the LED was flashing and the electronic buzzer was sounding. Figs. 3 (a) and (b) show the transmission of the emergency call from the mobile phone that is fixed to the user's wheelchair and the incoming message on the caregiver's mobile phone during the emergency situation, respectively. A received emergency e-mail is shown in Fig. 4. We also obtained the same results in the cases when two other subjects tested.

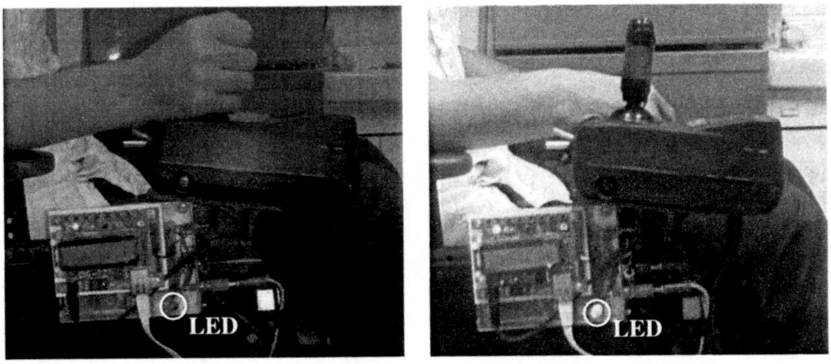

(a) When the subject A holds the joystick (b) When he does not hold it

Fig. 2. Experimental results of emergency call device's outputs: LED and buzzer

(a) wheelchair's mobile phone (b) caregiver's mobile phone

Fig. 3. Experimental results of emergency call device's outputs: mobile phones

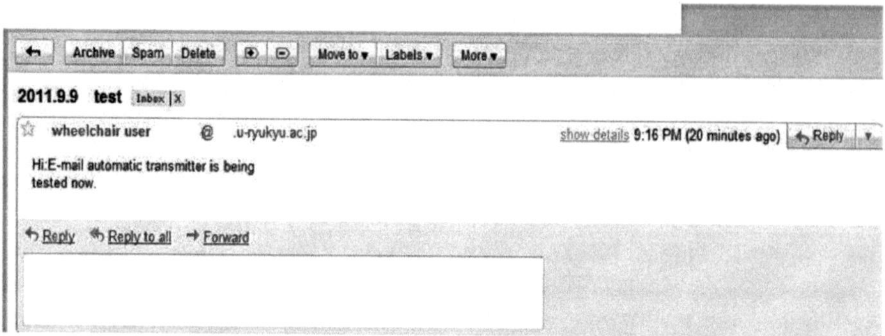

Fig. 4. Experimental results of emergency call device's output: emergency e-mail. The message is also pre-registered in the circuit which sends the e-mail.

When the caregiver received the emergency call from the wheelchair user, he can ask the wheelchair user where she/he is through the mobile phone. However, some people with disabilities, for example the persons who are inserted through a tracheotomy, cannot speak in a loud voice. Some mobile phones are already equipped with GPS locator. To specify the location of these people, applying the GPS locator function to our emergency call device will be more useful.

5 Conclusion

We proposed the emergency call device for people with severe disabilities to improve their QOLs. From the experimental results, it was found that the emergency call device was performed appropriately. An experiment with people with disabilities will be needed in our future work.

Acknowledgments. The authors would like to thank staff in National Hospital Organization Okinawa Hospital for their valuable discussions. This work was partly supported by Japan Society for the Promotion of Science, Grant-in-Aid for Scientific Research under Grant C-22500509.

References

1. NINDS Muscular Dystrophy Information Page, `http://www.ninds.nih.gov/disorders/md/md.htm` (accessed February 2011)
2. Cooper, R.A., Ohnabe, H., Hobson, D.A.: An Introduction to Rehabilitation Engineering, pp. 129–155. Taylor & Francis (2007)
3. Enderle, J.D., Blanchard, S.M., Bronzino, J.D.: Introduction to Biomedical Engineering, pp. 211–254. Elsevier Academic Press (2005)
4. Wang, B.: A Study on Improvement of Wheelchair Users' QOL. University of the Ryukyus graduation thesis (2011)

Toward Human-Robot Interaction Design through Human-Human Interaction Experiment

Yutaka Hiroi[1] and Akinori Ito[2]

[1] Faculty of Engineering, Osaka Institute of Technology, 5-16-1 Omiya, Asahi-ku, Osaka, Japan
[2] Graduate School of Engineering, Tohoku University, 6-6-05 Aramaki aza Aoba, Aoba-ku, Sendai, Japan
hiroi@med.oit.ac.jp, aito@spcom.ecei.tohoku.ac.jp

Abstract. In this paper, we examined a possibility of assessing a human's impression of a behavior of a robot through a human-human interaction (HHI), that enables us to design a robot's behavior without actually implementing that behavior to the robot. Experiments were conducted for comparing users' impressions of human-robot interaction (HRI) and HHI. We employed paper-scissors-rock (janken) game as a task of interaction. Virtual reality technology was used for realizing the robot's behavior. From the experimental result, we observed significant difference for the impression of friendliness, but we could observe no significant difference for other items. The result suggests that the behavior of a robot can be assessed through HHI.

Keywords: Human-robot interaction, human-machine interaction, interaction design, virtual reality.

1 Introduction

Designing how a robot interacts to a human is an important part of robotic design, especially for a robot that has a social role [1]. Not only designs of physical property [2,3] or appearance [4] of a robot but also design of its behavior [5] is a very important issue. On designing a robot including its interaction, iterative improvement of its design through a field test is indispensable. However, it is costly to construct a number of prototype robots for evaluating robot design. Therefore, we have proposed a method for evaluating an appearance of a robot using augmented reality technology [6].

In this paper, we investigate how an impression of a human-robot interaction (HRI) can be assessed through human-human interaction (HHI) experiment and virtual reality. Our final goal is to obtain a design strategy of HRI through an observation of HHI augmented by a computer graphics. If a user's impressions from HRI and HHI are similar, it suggests that an HRI could be designed by observing an HHI of the same task, which means that we could avoid building many robots with different behavior. If a human-robot interaction experiment can be substituted by a human-human experiment, we can save huge amount of time and money when developing an interaction robot.

D. Yang (Ed.): Informatics in Control, Automation and Robotics, Volume 2, LNEE 133, pp. 127–130.
springerlink.com

To this end, we investigated similarity between a user's impressions by an HRI and an HHI of the same task.

2 Experimental Paradigm

In this work, we chose a "paper-scissors-rock" (*janken*) game as a task of interaction [7]. A subject (*a player*) is asked to play *janken* game with a robot or a human (*a partner*), and answered a questionnaire for surveying impressions of the interactions to the robot and human.

While playing, we displayed the partner's next hand around the partner's figure using an augmented reality. Figure 1 shows an example of a partner robot playing with a player. The next hand ("paper" in this case) is displayed in a balloon.

We conducted two sessions for each player. In the first session, the balloon shows the user the partner's next hand; in the second session, the partner throws a different hand from that previously shown in the balloon as a "betrayal" behavior. This betrayal is introduced for investigating whether the betrayal behavior affects the impression on the partner.

Fig. 1. An example of a partner (robot)

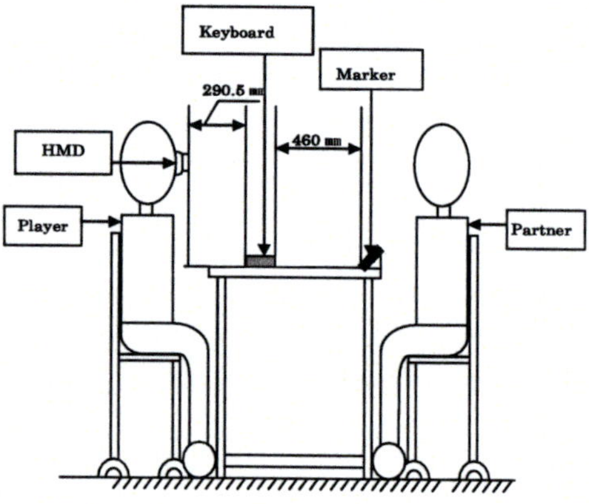

Fig. 2. Arrangement of the experiments.

3 Experimental Setup

The experiments were conducted in a room. Figure 2 shows the arrangement of the experiment. A player wore a head-mounted display (VUZIX iWare VX920) and sat on a chair. The player used a keyboard with a paper/scissors/rock mark for throwing a hand. The marker is used to generate the balloon using the AR. The ARToolkit [8] was used for overlaying the generated image onto the captured image.

First, an experiment was carried out employing a human partner. The subjects were 8 male students who are from 20 to 22 years old. These subjects were employed as players. We explained them the experiment and made agreement with the subjects. After that, the above-mentioned two sessions were conducted for each subject in this order. The *janken* plays were performed according to calls made by the partner.

Next, an experiment for robot partner was conducted. In this experiment, the subjects were 8 male students who are from 21 to 22 years old, who were different from those of the previous experiment. The robot had only body and arms; its hands and head were generated using virtual reality [5] to realize hands for *janken* and facial expression. Same as the previous experiment, the *janken* play were performed according to the partner robot's call, as which a synthesized voice was used.

The questionnaire sent to the subjects had the following four evaluations, in which a subject scored on a one-to-seven scale, as well as its reason.

- How did you feel the partner *trustworthy*?
- How did you feel the partner *serious*?
- How did you feel the partner *friendly*?
- How did you feel the partner *kind*?

4 Results and Discussion

Figure 3 shows the experimental result. In this result, scores of a question are averaged. We conducted Student's t-test for comparison of scores for the robot and human partner, and we observed statistically significant difference between human and robot partner for only "*friendliness*" question ($p<0.05$).

The score for *kindness* was lower than 4, which seems to be caused by the betrayal behavior. The result for *friendliness* could be explained by the robot's exaggerated appearance. This result could be change if we use a robot with more human-like apperance.

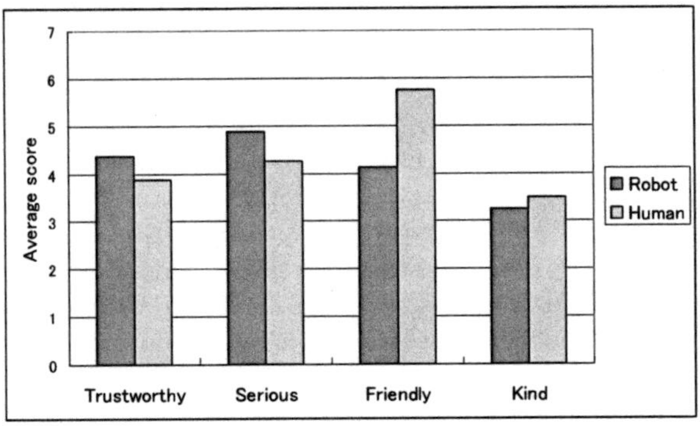

Fig. 3. Result of the questionnaire

From these results, we observed no big difference for evaluation of interaction with robot and human, which suggests a possibility for using HHI instead of HRI for designing behavior of a robot except the robot's appearance. If we can evaluate interaction using HHI, we can reduce development cost by implementing a robot an interaction function according to results of HHI evaluations.

5 Summary

In this paper, we investigated a possibility for designing HRI by observing evaluation of HHI. We conducted experiments for comparing subjects' impression for human and robot, and the result was that there were no big difference between the impressions for HRI and HHI. From this work, we could find a new methodology for interaction design of a robot.

As a future work, we will conduct further experiments using robots with different appearances and more subjects.

References

1. Berazeal, C.: Social Interactions in HRI: A Robot View. IEEE Trans. Systems, Man & Cybernetics —Part C: Applications and Reviews 34, 181–186 (2004)
2. Tonietti, G., Schiavi, R., Bicchi, A.: Design and Control of a Variable Stiffness Actuator for Safe and Fast Physical Human/Robot Interaction. In: Proc. IEEE Int. Conf. on Robotics and Automation (ICRA 2005), pp. 526–531 (2005)
3. Hiroi, Y., Ito, A.: Effect of the Size Factor on Psychological Threat of a Mobile Robot Moving Toward a Human. KANSEI Engineering International 8, 51–58 (2009)
4. DiSalvo, C.F., Gemperle, F., Forlizzi, J., Kiesler, S.: All robots are not created equal: the design and perception of humanoid robot heads. In: Proc. 4th Conf. on Designing Interactive Systems: Processes, Practices, Methods, and Techniques (DIS 2002), pp. 321–326 (2002)
5. Gockley, R., Bruce, A., Forlizzi, J., Michalowski, M., Mundell, A., Rosenthal, S., Sellner, B., Simmons, R., Snipes, K., Schultz, A.C., Wang, J.: Designing Robots for Long-Term Social Interaction. In: Proc. Int. Conf. on Intelligent Robots and Systems (IROS 2005), pp. 1338–1343 (2005)
6. Hiroi, Y., Hisano, S., Ito, A.: Evaluation of Head Size of an Interactive Robot Using an Augmented Reality. In: Proc. Int. Symp. on Robotics and Applications (2010)
7. Kanda, T., Ishiguro, H., Imai, M., Ono, T.: Development and evaluation of interactive humanoid robots. Proceedings of IEEE 92, 1839–1850 (2004)
8. ARtoolkit Home Page (retrieved December 20, 2010), `http://www.hitl.washington.edu/artoolkit/`

Research Survey of Technique about Wire-Driven Parallel Suspension Systems Used in Forced Oscillation Experiments in Low-Speed Wind Tunnels for Obtaining Dynamic Derivatives

Yaqing Zheng and Shuaihe Zhao

College of Mechanical Engineering and Automation, Huaqiao University, 361021, Xiamen, China
yq_zheng@hqu.edu.cn, feitianbao_@163.com

Abstract. It is pointed out that there are two key issues that should be tackled deeply if the wire-driven parallel suspension system is introduced to be used in forced oscillation experiment in low-speed wind tunnels for obtaining dynamic derivatives, i.e., the implementation of the single-degree-of-freedom oscillation of the model of aircrafts as well as the accurate identification of the dynamic derivatives. In order to solve the two key issues, the related scientific and technological problems are summarized, as following: (1) In the premise of satisfaction of the desired requirements of forced oscillation experiments, the mechanism design of the wire-driven parallel suspension system is given.(2) The strength, stiffness, dynamic characteristics, measuring precision and control precision should be satisfied as well.(3) Also the aerodynamic interference of the system should be eliminated as much as possible. The coupled fluid-solid interaction between the wires and the streamline flow should be tackled well and the suitable 6-element force-measuring system to measure the force and moment of the streamline flow should be used.

Keywords: low-speed wind tunnel, forced oscillation experiment, wire-driven, parallel suspension system.

1 Introduction

In order that the aircrafts can have good flying qualities, more demands of dynamic behavior of the aircrafts should be put forward. The dynamic derivatives should be obtained to predict the dynamic behavior, to design the flying control system, and to get an analysis of stall and spin [1].

Wind tunnel experiment is an important way to determine dynamic derivatives [2]. Forced oscillation used in the experiment of small angle of attack and small disturbance is the traditional method. Model suspension system of the wind tunnel experiment plays an important role in the whole process. Traditional suspension systems with strong interference between strut and streamline flow can easily cause vibration between the model, the balance and the suspension system, also one excitation device of the system

D. Yang (Ed.): Informatics in Control, Automation and Robotics, Volume 2, LNEE 133, pp. 131–138.
springerlink.com

can just achieve one kind of oscillation in the experiment. Therefore it is necessary to develop a new kind of the suspension system [3].

As a new suspension system, the cable-mounted suspension system, which can solve the contradiction between the stiffness of model support and aerodynamic interference, has been successfully used in obtaining the static derivatives as well as the stability parameters of the aircrafts [4]. But from the viewpoint of mechanism, the cable-mounted suspension system, which is different from the wire-driven parallel suspension system in posture control and the principle of force measuring, is not robotic to achieve automatic control [5, 6].

Wire-driven parallel manipulators, which are very suitable for measuring dynamic derivatives, being used in model suspension systems of wind tunnel tests, is a new concept with the development of parallel manipulators and force control technique. Like cable-mounted suspension systems, this kind of manipulators can solve the contradiction between the stiffness of model support and aerodynamic interference. Owing to the adoption of advanced robot control technology, wire-driven parallel suspension systems can implement the experiment of attitude control with the combination of two or three attitude angles, as well as dynamic experiments. Compared with cable-mounted suspension systems, wire-driven parallel suspension systems have a broad prospect especially in dynamic experiments [7-9].

Investigations has been made about the application of wire-driven parallel manipulators in the wind tunnel tests and some important research results have been obtained in the currently well-known project SACSO supported by ONERA. In the project, the model is suspended by nine wires motorized to realize six-degree-of-freedom (DOF) motion in a vertical wind tunnel. The wire-driven parallel suspension systems used in measuring dynamic derivatives is feasible, which has been enlightened by the progress of SACSO project [10, 11]. A series of studies on wire-driven parallel suspension systems with 8 wires (WDPSS-8 for short) used as the suspension system of static force measurement in low-speed wind tunnels have been done by the authors and the cooperators, and a series of achievements have been made. Meanwhile, that WDPSS-8 can be used in forced oscillation experiment to extract the dynamic derivatives of the model has also been pointed out.

But there are two key issues should be tackled deeply if WDPSS-8 is introduced to be used in forced oscillation experiment, i.e., the implementation of the single-degree-of-freedom oscillatory of the model of aircraft as well as the accurate identification of the dynamic derivatives. In order to realize the research aims, the study, focusing on problems of the mechanism design of WDPSS-8 and some related scientific and technological problems, is in the following sections.

2 Design and Investigation of WDPSS

2.1 The Implementation of the Single-Degree-of-Freedom Motion of WDPSS

2.1.1 The Principle of the Implementation of the Single-Degree-of-Freedom Motion of Aircraft Model

In accordance with the requirement of the experiment, WDPSS which can realize the 6-DOF motions is suitable. So WDPSS-8 shown in Fig. 1, proposed by the authors, which can control 6-DOF motions of the aircraft model, can be used. The scale model of aircraft of the wire-driven parallel suspension system is suspended in the streamline flow by eight wires, shown in Fig. 1. The wires are motorized to realize the required five single-degree-of-freedom forcedly oscillatory motions of the scale model [12]. And the parameters of the mechanism are determined by the requirement of the forced oscillation experiment.

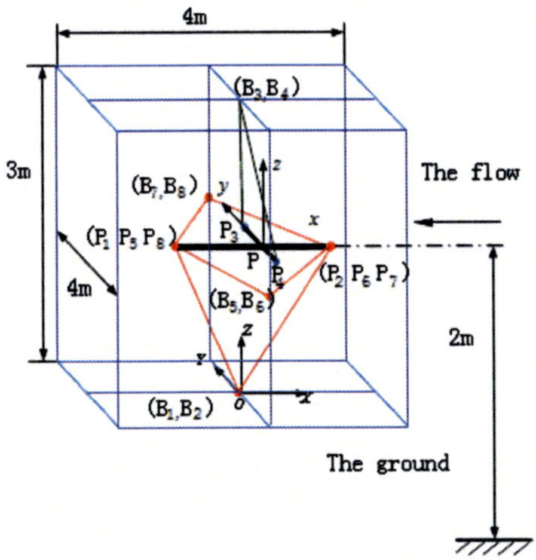

Fig. 1. Configuration schematic diagram of WDPSS-8

2.1.2 Problems of Science and Technique Involved

(1) The requirements of the forced oscillation experiment are as follows:

1) The angle of attack (α) and yaw (β) of the aircraft model should meet the required values; 2) The linear displacement A_1 and angular displacement A_2 of amplitude of oscillatory motions should meet the required values too, and the frequency is generally between 0.3 and 4 HZ ; 3) The low-speed wind tunnel should be medium-sized (the diameter of the test section is generally greater than or equals to 3 meters), and the speed of the flow approximately equals to 35 meters per second; 4) The model of aircraft should be standard model, the size and material of which are required and the mass should be less than or equal to 20Kg; meanwhile the 6-element force-measuring system is assembled within the model, and natural frequency of the system composed of the two parts mentioned above should be greater than or equal to 10 times of the frequency of the corresponding oscillatory motion; 5) The resolution and sampling rate of the data acquisition system are available; 6) The method of experiment and data processing should be appropriate [13].

(2)Problems of science and technique involved

The system of WDPSS should realize forced oscillatory motions with five single-degree-of-freedoms (pitch, roll, yaw, heave and translation). From the viewpoint of mechanism design, WDPSS-8 can achieve a large range of α, $\beta$and A in theory, but actually it can not, because of the limit of structure size of mechanism component. So the parameters of α, β and A should be chosen relatively small temporarily, that 6-DOF WDPSS-8 can meet the requirement.

In the premise of satisfaction of the range of oscillatory motions, WDPSS-8 should meet the requirements as follows: 1) The brace stiffness of the system should be enough, 2) The system must be not failure (the strength should be satisfied as well); 3) Vibration of the wires should be eliminated, 4) Measuring precision and control precision of the pose of model should meet the requirements.

2.1.3 Specific Study on Key Issues

(1) Brace stiffness and strength should be enough.

Theoretical analysis holds that the matrix of brace stiffness, named static stiffness too, of the wire-driven parallel suspension system can be decomposed into two parts, one is resulted from the changes of position and orientation of the movable platform, while the other one is caused by the deformation and deflection of the wires [14]. Although in previous studies system failure rarely occurs, this is very important, possible failure modes must be estimated.

(2) Vibration of the wires should be eliminated.

In order that the WDPSS-8 system can be used in forced oscillation experiment, first of all, the stiffness of the system should be enough and resonance of the system can not occur. Ref. [4] indicates that the factors of influence of resonant frequency of the support system are as the follows: 1) Longitudinal and transverse vibration of the wires; 2) Vortex shedding of the wires; 3) Vortex shedding of the end-effector. So it's necessary to investigate the vibration of the wires to design suitable structure of the wires, to make that the aerodynamic interference of the model is as little as possible in the forced oscillation experiment and the force can be measured accurately. Specific solution is described in Refs [4, 15].

(3) Measuring precision and control precision of the pose of the model should meet the requirements.

In the wind tunnel experiment, the pose of the model is needed to be controlled and measured accurately. The measurement accuracy of the pose is in the first place, for the pose of the model is measured accurately first in order to correspond with the result of the force-measuring balance. During the design process of the system, the choice of a suitable measurement system is important, such as using an optical position measuring system or using an accelerometer and an angular displacement sensor fixed in the model, etc. Optical position measuring system mainly refers to position measuring device based on computer vision [16].

The control precision of the pose should satisfy the desired requirements of the experiment. This supporting system not only is a strongly coupled and non-linear cable parallel robot but also introduces over-constraint in its redundant drive. So when designing the control scheme, dynamic model should be linearized and decoupled; the aerodynamic interference needs to be eliminated by compensation method and optimization algorithm can be used to solve the uncertainty of the tension caused by over-constraint. Vibration of the wires and end-effector can be inhibited effectively by a control scheme, which can make sure that the tension of the eight wires can satisfy the requirement of the forced oscillation experiment and the fluid-solid coupled interaction of the wires can be decreased. The design of the control scheme can reference Ref. [17], while the constraint condition of the tension of wires should be investigated particularly on the basis of this supporting system.

2.2 Accurate Identification of the Dynamic Derivatives

2.2.1 The Calculation Principle and Identification Method of the Dynamic Derivatives

(1) The calculation principle of the dynamic derivatives

The force condition, which is measured before and after airflow is exerted on the scale model, of the model and the balance is shown in Fig. 2.

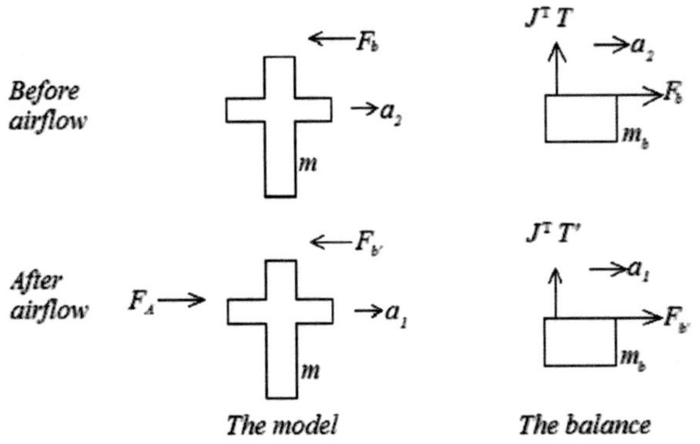

Fig. 2. The force condition of the model and the balance

Before blowing:

$$ma_2 = F_b \tag{1}$$

After blowing:

$$ma_1 = F_A + F_{b'} \tag{2}$$

$$\Rightarrow F_A = m(a_1 - a_2) + (F_b - F_{b'}) \tag{3}$$

Dynamic derivatives can be solved on the basis of F_A by appropriate identification method [18].

(2) The identification method

Based on the calculation principle mentioned above, parameter identification method such as maximum likelihood identification method etc. can be used in extracting the dynamic derivatives from F_A.

2.2.2 Problems of Science and Technique Involved

(1) The aerodynamic interference should be eliminated as much as possible.

In order that this supporting system can be used in the wind tunnel experiment successfully and the interference of the system should be eliminated as much as possible, it's necessary to consider the aerodynamic interference of each mechanism component when designing a suitable model-balance-support system. To avoid breaking the aircraft model when it is directly fixed by the wires, struts which can be fixed directly by the wires can be assembled between the model and the balance. And the balance is set in the model to make the model, the balance and the struts be one part. There is no exact theory for this design of the structure, so the structure can refer to the existing one with little interference that it also can be designed, CFD analyzed and verified by experiments [19].

(2) Fluid-solid coupled interaction of the wires should be investigated.

In practical application of the wire-driven parallel suspension mechanism, there is always wind vibration of the wires which can cause even breaking of the end-effector, so it's necessary to study the fluid-solid coupled interaction of the wires and then a suitable structure of the wires is designed or chosen. Three different analysis modules for WDPSS-8 based on the function of fluid-solid coupled interaction provided by ANSYS should be designed during the investigation of the wire fluid-solid coupled interaction under the laminar flow with a speed of 35 meters per second, as followings: 1)The discretization of fluid domain; 2) The discretization of solid domain; 3) Fluid-solid interface conditions. Also arbitrary Lagrange-Euler method, which is more mature analysis method at present, should be used to analyse the problem. WDPSS-8 is restricted to the low frequency oscillation; meanwhile the high frequency oscillation caused by the wind excitation can be solved by adding dampers or using other motion control method [20].

(3) The measurement of the aerodynamic forces is feasible.

During the solution to the dynamic derivatives the parameters of the aerodynamic forces are necessary. So the measuring precision of the aerodynamic forces directly influences the precision of the dynamic derivatives, that choosing a reasonable device of aerodynamic forces measuring is very important. Meanwhile the inner-placed 6-element force-measuring balance can be used in measuring the aerodynamic forces.

3 Conclusions

There are two key issues should be tackled deeply if the wire-driven parallel suspension system is used in forced oscillation experiment in low-speed wind tunnel s for obtaining dynamic derivatives, i.e., the implementation of the single-degree-of-freedom oscillation of the model of aircrafts as well as the accurate identification of the dynamic derivatives.

The problems of science and technique involved can be described as the follows: 1) In the premise of satisfaction of the desired requirements of forced oscillation experiments, the mechanism design of the wire-driven parallel suspension system should be reasonable. 2) The strength, stiffness, dynamic characteristics, measuring precision and control precision should be satisfied as well. 3) The aerodynamic interference of the system should be eliminated as much as possible. 4) Also the coupled fluid-solid interaction between the wires and the streamline flow should be tackled well and the suitable 6-element force-measuring system to measure the force and moment of the streamline flow should be used.

References

1. Sun, H.: The Development of 96-Test System for Measuring Dynamic Derivatives at High Angle of Attack. Experiments and Measurements in Fluid Mechanics 3, 31–37 (1999) (in Chinese)
2. Bergmann, A.: Modern Wind Tunnel Techniques for Unsteady Testing – Development of Dynamic Test Rigs. In: Radespiel, R., Rossow, C.-C., Brinkmann, B.W. (eds.) Hermann Schlichting – 100 Years. Notes on Numerical Fluid Mechanics and Multidisciplinary Design, vol. 102, pp. 59–77. Springer, Heidelberg (2009)
3. Du, X., Wang, X., Zhou, Y.: Investigation of Support Interference in Low Speed Wind Tunnel Oscillation Tests. Experiments and Measurements in Fluid Mechanics 6(2), 37–40 (2003) (in Chinese)
4. Magill, J.C., Wehe, S.D.: Initial Test of a Wire Suspension Mount for Missile Virtual Flight Testing. In: 40th AIAA Aerospace Sciences Meeting and Exhibit, 2002-0169, Reno, NV (2002)
5. Shen, L., Shen, Z., Huang, Y.: A Wire Type Support System for High Angle of Attack Test in Low Speed Wind Tunnel. Experiments and Measurements in Fluid Mechanics 12(4), 15–21 (1998) (in Chinese)
6. Bian, Y., Li, X., Li, L., et al.: Development of Model Cable Mount and Strain Gauge Balance Measuring System in Low Speed Closed Wind Tunnel. Experiments and Measurements in Fluid Mechanics 13(3), 85–90 (1999) (in Chinese)
7. Zheng, Y.: Research on Key Theoretical Issues of Wire-Driven Parallel Kinematic Manipulators and the Application to Wind Tunnel Support Systems. PhD Dissertation, pp. 4–104. Huaqiao University, Quanzhou (2004) (in Chinese)
8. Liu, X., Zheng, Y., Lin, Q.: Overview of Wire-driven Parallel Kinematic Manipulators for Aircraft Wind Tunnels. Acta Aeronautica Et Astronautica Sinica 25(4), 393–400 (2004) (in Chinese)
9. Zheng, Y., Lin, Q., Liu, X., Mitrouchev, P.: On Wire-driven Parallel Suspension Systems for Static and Dynamic Derivatives of the Aircraft in Low-Speed Wind Tunnels. Acta Aeronautica ET Astronautica Sinica 30(8), 1549–1554 (2009) (in Chinese)
10. Lafourcade, P., Llibre, M., Reboulet, C.: Design of a Parallel Wire-Driven Manipulator for Wind Tunnels. In: Proceedings of the Workshop on Fundamental Issues and Future Directions for Parallel Mechanisms and Manipulators, Quebec City, pp. 187–194 (2002)
11. Farcy, D., Llibre, M., Carton, P., Lambert, C.: SACSO: Wire-Driven Parallel Set-Up for Dynamic Tests in Wind Tunnel – Review of Principles and Advantages for Identification of Aerodynamic Models for Flight Mechanics. In: 8th ONERA-DLR Aerospace Symposium, Göttingen (2007)
12. Zheng, Y., Lin, Q., Liu, X., Mitrouchev, P.: Preliminary Step Towards Wire-Driven Parallel Suspension Systems for Static and Dynamic Derivatives of the Aircraft in Low-Speed Wind Tunnels. The Journal of Engineering Research 7(1), 22–30 (2010) (in Chinese)

13. Li, Y., Chen, H.: Wind Tunnel Techniques for Studying Spin, pp. 116–124. National Defence Industrial Press (2002) (in Chinese)
14. Li, H., Zhu, W.: Static Stiffness Analysis of Flexible-cable-driven Parallel Mechanism. Journal of Mechanical Engineering 46(3), 8–16 (2010) (in Chinese)
15. Diao, X.M.: Vibration Analysis of Cable-Driven Parallel Manipulators. Multibody System Dynamics 21, 347–360 (2009)
16. Liu, X., Qiu, Y., Sheng, Y.: Dynamic Modeling and Disturbance-rejection Control of 6-DOF Cable Driven Parallel Manipulator. Journal of System Simulation 21(18), 5801–5806 (2009) (in Chinese)
17. Liu, X., Qiu, Y., Sheng, Y.: Enhancement and Motion Control of a 6-DOF Wire-driven Parallel Manipulator with Redundant Actuations for Wind Tunnels. Acta Aeronautica Et Astronautica Sinica 30(6), 1156–1164 (2009) (in Chinese)
18. Zhang, Z.: Data Processing of Dynamic Derivatives Experiment. Aerodynamic Research and Experiment 5, 39–46 (1994) (in Chinese)
19. Chen, Z., Liu, M., Liu, Z.: Experiment Study on Aerodynamic Interference of Tandem Cables Based on Aeroelastic Model. Journal of Vibration and Shock 27(8), 7–12 (2008) (in Chinese)
20. Detter, W.G., Peric, D.: A fully Implicit Computational Strategy for Strongly Coupled Fluid-Solid Interaction. Arch. Comput. Methods Eng. 14, 205–247 (2007)

A Novel Miniature Microstrip Antenna for GPS Applications

Hong-Gang Hao[1], Hua-Xiao Lu[1], Wei Chen[1], and Chao An[2]

[1] School of Optical and Electronic Engineering, Chong Qing University of Posts and Telecommunications, Chongqing 400065, China
[2] School of Software Engineering, Chong Qing University of Posts and Telecommunications, Chongqing 400065, China

Abstract. The design of a microstrip antenna which operates at the GPS L1 band is presented in this paper. The miniaturization and the right-hand circular polarization (RHCP) is obtained through a square-ring patch and truncated corners in which, respectively. The simulated results indicate that the impedance matching is excellent, the antenna has the characteristic of wide-angle circular polarization, and the radiation patterns of RHCP are excellent; In a wide elevation angle range, the RHCP level is at least 15dB higher than that of LHCP, which has a good suppression of interference from cross-polarization and multipath effect. Compared with the conventional GPS microstrip antenna with the same material, the characteristics of the proposed antenna are enhanced, the whole size is reduced, the area of the radiating patch is reduced by 26%, meanwhile,the cost of the antenna's substrate material is low, which satisfies with the requirements of GPS receiver for miniaturization, low-cost and compactness.

Keywords: GPS, Microstrip Antenna, Square-ring, RHCP, Miniaturization.

1 Introduction

The GPS applications have been melted into all aspects of national economic production and people's daily lives deeply, which has huge market potentialities and economic benefits. A GPS receiver is a terminal equipment which receives GPS signals transmitted from far away space, a GPS antenna is the entrance for radio waves to enter into a GPS receiver, so the performance of the GPS antenna has an important impact on the accuracy of navigation and positioning for a GPS receiver[1].

In order to make a GPS receiver work efficiently, a GPS antenna should supply some requirements[1][2]: ①Good radiation pattern in the upper half-plane, an ideal GPS receiver antenna should have a radiation pattern similar to a hemisphere; ② Right-hand circular polarization(RHCP): in order to eliminate Faraday rotation effect from ionosphere in the GPS signals, the GPS signals should be RHCP, meanwhile, GPS signals can shift into left-hand circular polarization(LHCP) when the GPS signal is reflected by ground or symmetrical objects, such as buildings, so RHCP can also have a good suppression of interference from cross-polarization and multipath effect;

D. Yang (Ed.): Informatics in Control, Automation and Robotics, Volume 2, LNEE 133, pp. 139–147.
springerlink.com

③In order to supply the requirements of developments for current portable electronic devices, a GPS antenna should be miniature and low-cost.

Microstrip antennas rank the first place in GPS receivers for several advantages, such as, its low-profile, small size, lightweight, easy integration, low cost and conformability to any structure. Currently, ceramic dielectric material with high permittivity is generalized used in order to achieve miniaturization, but high permittivity leads to very low gain and narrow impedance bandwidth, such as in the literatures[4][5],the circular polarization(CP) bandwidth is very narrow because of high permittivity, so it is difficult to achieve CP radiation and good impedance matching at the operating frequency, furthermore, GPS antennas are usually integrated with other equipments in practice, in all parameters of a microstrip antenna, impedance bandwidth is the most sensitive one to assembly errors, if impedance bandwidth is too narrow, the center operating frequency will shift because of assembly errors, which will make a GPS receiver unable to receive GPS signals, meanwhile, the price of substrate material with high permittivity is very high, which leads to the increase of cost for a GPS microstrip antenna. In the literature[6],the slots in the ground plane are used to achieve miniaturization and CP, but the backside levels become higher because of the slots in the ground plane. In the literature[7],it is the cross slots in the patch that are used to achieve miniaturization and CP, but the lengths of the two slots need only subtle differences in order to achieve CP, therefore, the requirements for fabrication tolerances are harsh. A GPS microstrip antenna which is used in the navigation ,communication and positioning fields is presented in this paper, the antenna is miniature, wider impedance bandwidth and low cost, and works at the GPS L1 frequency band(1.575G Hz).

2 Antenna Design

The geometry of the proposed antenna is plotted in Fig.1 (It will be referred as the design A below).

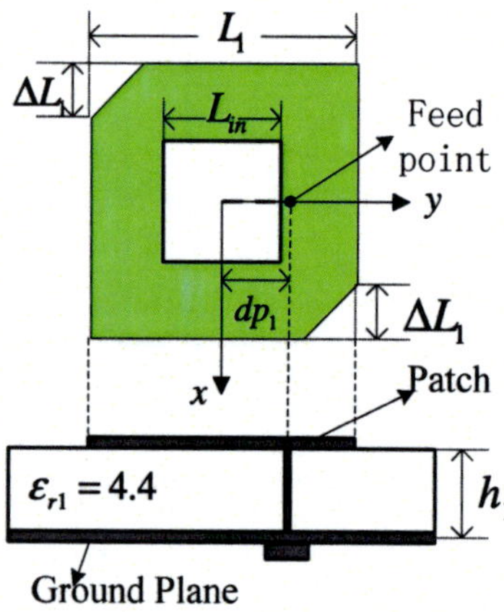

Fig. 1. Geometry of the proposed antenna (the design A)

In design A, a square-ring patch and two truncated diagonal corners are used to achieve miniaturization and RHCP operation, respectively; the proposed antenna is excited by a single feed, using coaxial cable. The square-ring microstrip antenna can be considered as a square microstrip antenna with a square slot. In Fig.1,the square-ring microstrip patch ,having an outer side length of L_1 and an inner side length of L_{in}, the slot in the patch makes the equivalent surface current path becomes longer, which not only lowers its resonance frequency, but also reduces its size. The parameter L_{in} which can control the resonance frequency, impedance performance, and the size of the patch is created, the new parameter L_{in} makes the design A more flexible, meanwhile, the antenna weight is also reduced.

In design A, the RHCP is achieved by a single feed and two truncated diagonal corners in the square-ring patch. A single feed microstrip antenna whose geometry is simple as well as whose cost is low can generate CP without any external phase shift and power divider network, which is suited for miniaturization. The side lengths of the two truncated diagonal corners which are also called near-degenerate mode splited unit are both ΔL_1, when the lengths of ΔL_1 as well as d_{p1} are chosen properly, two near-degenerate orthogonal modes of equal amplitudes and 90°phase difference for CP operation are generated. The dielectric material of the design A is FR4 whose price is low, which is helpful to reduce the cost of the design A .

The empirical equations of the Transmission Line Model for a microstrip antenna are shown as (1)~(4):

$$\varepsilon_e = \frac{\varepsilon_r + 1}{2} + \frac{\varepsilon_r - 1}{2}\left(1 + 12\frac{h}{W}\right)^{-\frac{1}{2}} \tag{1}$$

$$L = \frac{c}{2f_0}\frac{1}{\sqrt{\varepsilon_e}} - 2\Delta L \tag{2}$$

$$\Delta L = 0.412h\frac{(\varepsilon_e + 0.3)(W/h + 0.264)}{(\varepsilon_e - 0.258)(W/h + 0.8)} \tag{3}$$

$$W = \frac{c}{2f_0}\left(\frac{\varepsilon_r + 1}{2}\right)^{-\frac{1}{2}} \tag{4}$$

In the equations[8] above, c is the speed of light in free space, ε_r is the permittivity of the substrate material, $\varepsilon_r = \varepsilon_{r1} = 4.4$ in design A; ε_e is the equivalent permittivity; h is the thickness of the substrate, W and L is the width and length of the patch respectively; f_0 is the center frequency in free space, that is the GPS L1 frequency band here, so f_0=1.575G Hz.

3 Analysis and Simulation Results for Design A

In order to obtain the high performance of the proposed antenna, the electromagnetic modeling software, HFSS version 10.0, by Ansoft Corporation, was utilized to optimize all the parameters in Fig.1 repeatedly. The simulated return loss S11 of the proposed antenna with the variation of L_{in} is plotted in Fig.2, it can be seen that the resonant frequency decreases with increasing the inner side length L_{in} of the square-ring patch, this is because the equivalent surface current path becomes longer gradually when L_{in} increases gradually.

After optimization, the geometric dimensions are as follows: the ground plane is 60mm×60mm; the substrate is 60mm×60mm×3mm; the thickness h_1=3mm; the square-ring: the outer side length L_1=40mm, the inner side length L_{in}=13.6mm; the two truncated diagonal corners are both ΔL_1=5.4mm, and the feed point d_{p1}=8mm.

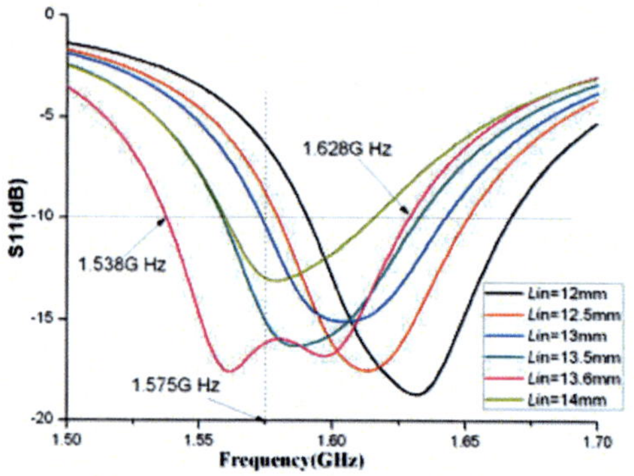

Fig. 2. Simulated results of S11 for the proposed antenna with the variation of L_{in}

The simulated results of the proposed antenna are as follows: the simulated return loss S11 are shown in Fig.3, the simulated return loss is –16.12dB at the GPS L1 frequency band (1.575G Hz),the impedance is well-matched, the -10dB bandwidth is approximately 90M Hz(1.538G Hz～1.628G Hz),which meets the requirements of the impedance bandwidth for a GPS receiver antenna(the general requirement is 2M Hz[9]), moreover, which can fully meet the requirements of the impedance bandwidth from assembly errors in practice. The axial ratio(AR) of RHCP in the broadside direction against frequency is plotted in Fig.4,the AR at the GPS L1 frequency band is 1.58dB, which meets the requirement of AR for a GPS receiver antenna(the general requirement is AR≤3dB at the center operating frequency),the CP bandwidth of 3dB axial ratio is 21M Hz (1.562G Hz～1.583G Hz); The airspace distribution characteristic of the axial ratio at the GPS L1 frequency band is plotted in Fig.5, it can been seen that the proposed antenna achieves CP radiation in both the upper half-plane and the lower half-plane, the GPS receiver can only receive satellite navigation and positioning signals in the upper half-plane($-90° \leq \theta \leq 90°$) in practice , the airspace coverage range determined from 3dB AR in the E-plane (φ=0°) is 142°(–74°～68°), which is 131°(–63°～68°) in the H-plane(φ=90°), so the proposed antenna has the characteristic of wide-angle right-hand circular polarization, which can receive GPS signals in a wide elevation angle range.

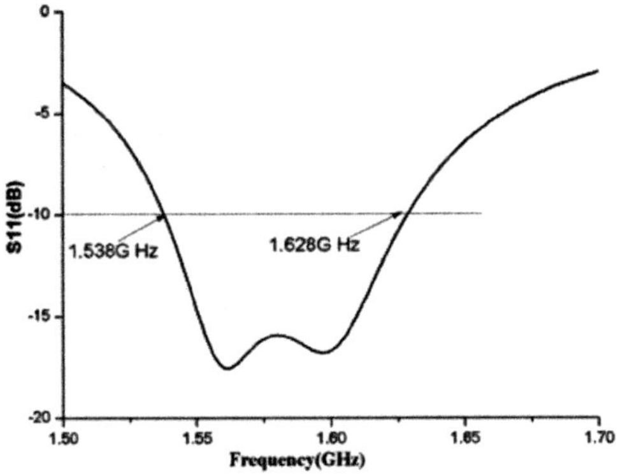

Fig. 3. Simulated results of S11 for the proposed antenna

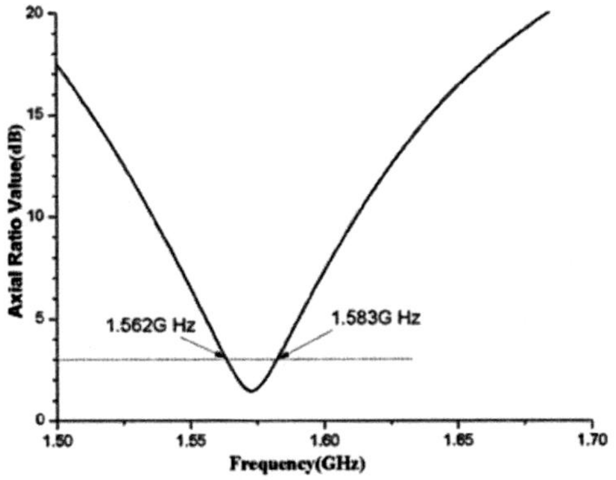

Fig. 4. Simulated results of AR for the proposed antenna

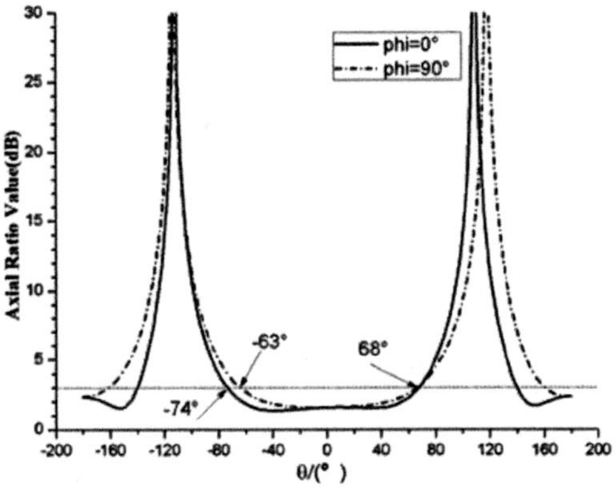

Fig. 5. The airspace distribution characteristics of the axial ratio in the working frequency of the proposed antenna

The radiation patterns of RHCP/LHCP in the E-plane and H-plane are plotted in Fig.6(a),(b), respectively, the gain of the proposed antenna is 2.6dB. The 3dB beamwidth of the E-plane is 101°(–49°～52°), that of the H-plane is 103°(–49°～54°), the RHCP characteristic is good within the 3dB beamwidth in both the E-plane and H-plane. In the E-plane, suppression of the LHCP gain by 21dB is achieved in the broadside direction, at least 15dB of gain suppression of the LHCP is achieved between the elevation angle range of $-74° \leq \theta \leq 67°$,the H-plane produces a similar result, so the proposed antenna has a good suppression of interference from cross-polarization and multipath effect(Greater than or equal to 15dB can be regarded as having a high suppression[10]). The simulated results indicate the performance of the antenna is good, which can be used in the GPS receiver.

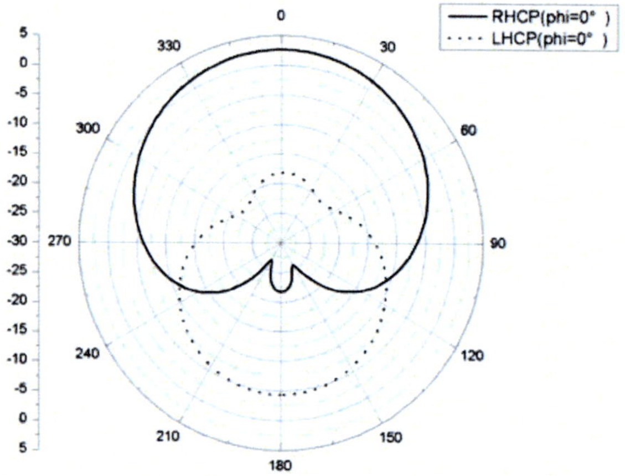

(a)

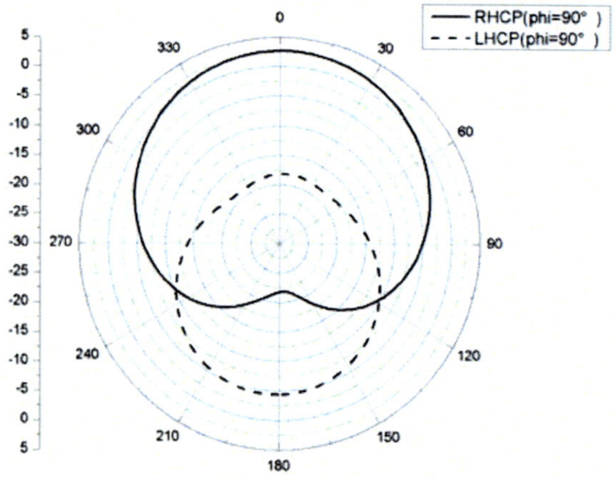

(b)

Fig. 6. Simulated radiation patterns for RHCP/ LHCP of the proposed antenna (a)E－plane/φ=0° (b)H－plane/φ=90°

As a comparison, a GPS microstrip antenna of conventional geometry is also modeled in HFSS, as shown in Fig.7,it will be referred as the design B below.

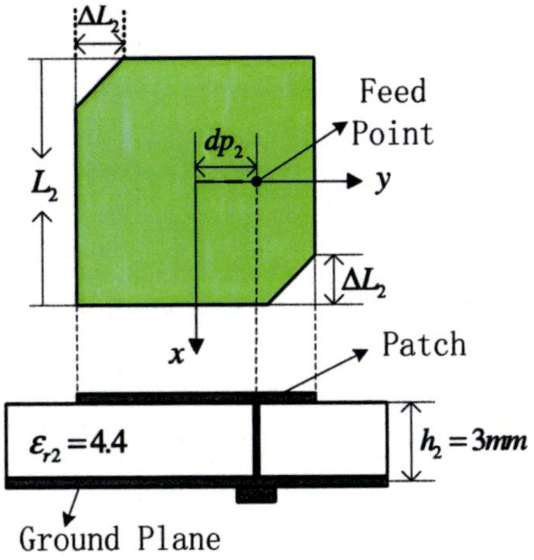

Fig. 7. Geometry of the conventional GPS antenna (the design B)

After optimization, the geometric dimensions of the design B are as follow: the ground plane is 64mmx64mm, the substrate is 64mmx64mmx3mm; the thickness and permittivity of the substrate in design B are both the same as those in design A, that is $h_1=h_2=3mm, \varepsilon_{r1}=\varepsilon_{r2}=4.4$; the side length of the patch in design B is $L_2=43.6mm$,the feed point is $d_{p2}=10mm$, the two truncated diagonal corners are both $\Delta L_2=5.9mm$.

The simulated results of both the design A and design B are set in table 1 as a comparison.

Table 1. The detailed comparison of simulated results for the design A and design B

Type	Impedance bandwidth (M Hz)	AR bandwidth (M Hz)	Gain (dB)	Beamwidth (°)
A	90	21	2.6	103
B	84	19	3.25	98

It can be seen from table1, the impedance bandwidth of the design A is 6M Hz more than that of the design B, this is due to the square-ring patch which equivalently leads capacitive effect into the surface of the patch, so part of the inductive effect from the feed probe is counteracted, therefore, the impedance bandwidth of the design A increases. The 3dB CP bandwidth of the design A is more than 2 M Hz than that of the design B; the gain of the design A is slightly 0.65dB less than that of the design B due to the decrease of the area of both the patch and the ground plane in design A. The 3dB beamwidth of the design A is 5°more than that of the design B, it can be seen that the overall performance of the design A is better than those of the design B except that the gain decreases slightly, furthermore, the side length of the ground plane in design A is 4mm less than that in design B, and the area of the patch in design A reduces by 26% than that in design B, which is helpful to be miniaturization and reduce the weight of the antenna.

4 Conclusions

The design of the proposed microstrip antenna (the design A) which works at the GPS L1 frequency band is completed with the help of electromagnetic simulation software HFSS in this paper. The miniaturization is obtained by the square-ring patch, the RHCP operation is obtained by two truncated diagonal corners which bring geometric perturbation in the square-ring patch as well as a single feed. The simulated results indicate the proposed antenna fully meets the requirements of a GPS antenna, furthermore, the proposed antenna has the characteristics of compactness, easily to be integrated, good electrical and mechanical properties, easily to be manufactured and so on. Compared with the GPS microstrip antenna of a conventional geometry (the design B), the characteristics of the proposed antenna are improved except that the gain decreases slightly, the whole size is decreased, the area of the patch is reduced by 26%.Meanwhile,the price of the substrate material in design A is low, which makes the cost of the proposed antenna low.

Hong-Gang Hao (male) was born in Yucheng, Shandong Province, in 1977. He is now a Associate Professor and Master Instructor. His research fields include optical inspection, electromagnetic theory and its applications. E-mail: haohg@cqupt.edu.cn

Hua-Xiao Lu (male) was born in Qingyang, Gansu Province ,in 1985. He is now a graduate student. His research fields include antenna & propagation and designs of microstrip antenna.for GPS. E-mail: princeandwealth@126.com

References

1. Liu, J.-Y.: The Principles and Methods for GPS satellite navigation, 2nd edn. Publishing House of Science, Beijing (2008) (in Chinese)
2. Gao, Y., Dong, S.-R., Wang, D.-M.: Development of Antennas for Global Position System. Wireless Commnication Technology (4), 34–39 (2008) (in Chinese)
3. Peng, X.-F., Zhong, S.-S., Xu, S.-Q., et al.: A Compact Dual-band GPS Micro strip antenna. Journal of Shanghai University (Natural Science) 11(1), 8–10 (2005) (in Chinese)
4. Chen, H.-M., Wang, Y.-K., Lin, Y.F., Pan, S.-C.: Microstrip-fed Circularly Polarized Square-Ring Patch Antenna for GPS Application. IEEE Trans. Antennas Propagation 36(4), 1264–1267 (2009)
5. Han, Y.-Y., Chen, Q., Ge, D.-Y., et al.: Development of minitype microstrip antenna for GPS application. Electronic Components Materials 27(9), 11–13 (2008) (in Chinese)
6. Zhao, H.-M., Niu, Y.-J., Liu, T.: A Novel Compact GPS Antenna. Telecommunication Engnineering 50(1), 107–109 (2010) (in Chinese)
7. Iwasaki, H.: A circuly polarized small-size microstrip antenna with a cross slot. IEEE Trans. Antenna Propragt. 44(10), 1399–1401 (1996)

8. Zhong, S.-S.: Microstrip Antenna Theory and Applications. Xidian University Press, Xi'an (1991) (in Chinese)
9. Geary, K., Schaffner, J.H., Hsu, H.-P., Song, H.J., Colburn, J.S., Yasan, E.: Single-Feed Dual-Band Stacked Patch Antenna for Orthogonal Circularly Polarized GPS and SDARS Applications. In: IEEE 68th Vehicular Technology Conference, VTC 2008-Fall, September 21-24, pp. 1–5 (2008)
10. Padros, N., Ortigosa, J.I., Baker, J., Iskander, M.F., Thornberg, B.: Comparative Study of High-Performance GPS Receiving Antenna Designs. IEEE Transactions on Antenna and Propagation 45(4), 698–706 (1997)

Camera Response Function Estimation and Application with a Single Image

Li Fu and Yue Qi

State Key Laboratory of Virtual Reality Technology and Systems,
Beihang University, Beijing 100191, China
{fuli,qy}@vrlab.buaa.edu.cn

Abstract. Camera response function (CRF) relates scene irradiance to image intensities. Estimation of CRF is a fundamental and necessary step in many computer vision applications such as the generation of high dynamic image, bidirectional reflectance distribution function (BRDF) estimation etc. In the paper, we compute CRF from a color image and edit images based on CRF to enhance contrast and change exposure of images. This method selects suitable edge windows with two uniform color regions, and is based on the empirical prior on camera response functions. The method uses only one color image rather than a set of images. The experiments show our estimation result is accurate.

Keywords: Camera response function, contrast, exposure.

1 Introduction

Most cameras are designed to have a non-linear mapping from the scene radiance to the image intensity. There might be several reasons, for example[1], the analog-digital converter might not work linearly, a gamma correction is applied, a correction is applied to make the image more visually appealing, and a white balance is applied to compensate for different light temperatures. The Camera Response Function (CRF) is used to describe the non-linear mapping.

Estimating the CRF is an essential work in computer graphics and computer vision. Many computer vision algorithms assume that image intensities are linearly related to scene irradiance. Some algorithms require precise measurements of scene irradiance to recover the scene properties, such as, construction of high dynamic range images [2, 3, 4], color constancy [5, 6], and recovery of BRDF from images [7]. The CRF is closely related to the image effects, for example, the contrast and exposure. The CRF can be edited to improve the image effects.

Our work is to estimate the camera response function from a single image captured from any camera under any light condition. Meanwhile, we improve the image effects based on the CRF. Most of previous methods require as input a set of registered images taken with varying camera exposures. Accurate calibration results can be obtained using these techniques; however, they require too limiting requirements.

D. Yang (Ed.): Informatics in Control, Automation and Robotics, Volume 2, LNEE 133, pp. 149–156.
springerlink.com

The remainder of the paper is organized as follows: in Sec.2, we review previous work on radiometric calibration. In Sec.3, we introduce our method to compute the camera response function. In Sec.4, we edit the image effects using CRF. Then we evaluate our method with experiments in Sec.5 and conclude with discussions about our algorithm and future work in Sec.6.

2 Related Work

Several different approaches for radiometric camera calibration are proposed before. Using the popular GretagMacbeth ColorChecker as the specific calibration object, CRF is estimated in [8]. Some methods use images acquired by a static camera with varying exposure times to compute the camera response function [9, 10].

In recent years, the methods of covering the camera response function using one image were proposed. These methods exploit image characteristics that reflect the non-linearity introduced by the camera response function. Lin et al. [19] perform the response function based on measured RGB distribution at color edges. Lin and Zhang [11] deal with a single grayscale image by using the histograms of edge regions. Matsushita and Lin [12] use the asymmetric profiles of noise distributions to compute the response function. Ng and Chang [13] use the geometry invariance theory to do this.

Several approaches estimate non-linear camera response functions from images acquired in uncontrolled conditions; further, the images are captured with different cameras. Diaz and Sturm [14] use an unstructured set of images to recover the camera's geometric calibration and a 3D scene model. With these inputs, the camera response function is estimated for each image. Luong et al. [15] also use a 3D model of the images scene, to compute the camera response function and estimate the illumination, and unlikely [14], they use a linear model for CRF's and require that the images be taken with the same camera under controlled conditions.

Different models for response function were proposed, such as gamma curve [16], polynomial [17], non-parametric [2], and PCA model [18].

3 Estimating the Camera Response Function with One Image

3.1 Background of CRF

The camera response function f relates the scene irradiance R to the image intensity I:

$$\mathrm{I} = f(R) \ . \tag{1}$$

In the paper, we deal with the image intensity rather than scene radiance, so we solve directly for the inverse response function $\mathrm{g} = \mathrm{f}^{-1}$. It is assumed reasonably that the response function f is invertible because the image intensity I increases monotonically with respect to scene radiance R.

3.2 Nonlinearity of Edge Colors

In a camera, CCD is the device that forms the image from scene radiance. Each array of CCD records a solid angle of the scene. Considering one channel, the irradiance Ir at x is:

$$\mathrm{Ir}(x) = \int_{\lambda} \int_{p\in S(x)} R(p,\lambda)\, q(\lambda) d_p d_\lambda \; . \tag{2}$$

Here, λ denotes the light wavelengths of the channel. The irradiance depends on the incoming scene radiances R and filter sensitivity q.

The image patches which contain two distinct uniform color regions are considered. The irradiance of the edge pixel is:

$$\begin{aligned}\mathrm{Ir}(x) &= \int_{\lambda} \int_{p\in S_1(x)} R_1(p,\lambda)\, q(\lambda) d_p d_\lambda + \int_{\lambda} \int_{p\in S_2(x)} R_2(p,\lambda)\, q(\lambda) d_p d_\lambda \\ &= a * \mathrm{Ir}_1(x) + (1-a) * \mathrm{Ir}_2(x) \; .\end{aligned} \tag{3}$$

where $a = \int_{p\in S_1(x)} d_p$.

It means the irradiance colors of the blended pixels lie on a line in RGB color space between the region irradiances Ir_1 and Ir_2 . Since f is nonlinear, the colors of edge pixels lie on a curve rather than a line. Inverse CRF g should transform non-linear edge color distributions to linear distributions. (Fig. 1).

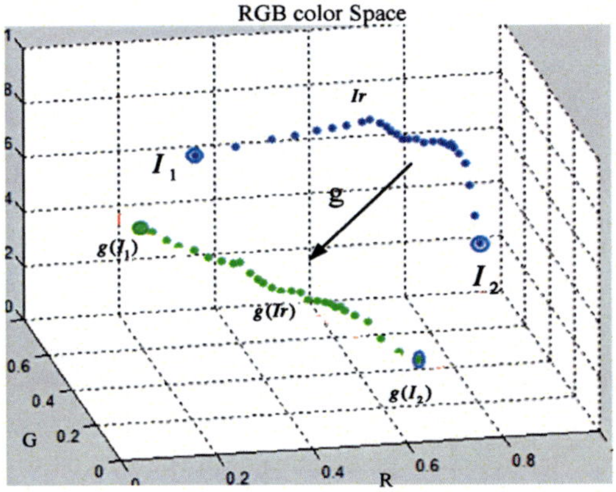

Fig. 1. Transformation of non-linear distributions of intensity to linear distributions of irradiance by the inverse response function g

3.3 Optimizing CRF Parameters

The prior $p(g)$ of inverse CRF is modeled from DoRF database [18], and can be formed as a Gaussian mixture model (K=5):

$$p(g) = \sum_{i=1}^{K} a_i N(g; u_i, \Sigma_i) \quad (4)$$

Grossberg and Nayar [18] proposed an Empirical Model of Response (EMoR), The model is a PCA based m^{th} order approximation:

$$g(I) = g_0(I) + \sum_i^m c_i h_i(I) \quad (5)$$

Where the g_0 is the mean function and the c_i 's are the coefficients for basis function h_i 's. With this prior information, we compute a Maximum a posteriori (MAP) estimation of CRF.

$$d = \frac{(2*|g(I_1)-g(I_2)|+|g(I_1)-g(I_r)|-|g(I_2)-g(I_r)|)}{g(I_1)-g(I_2)} \quad (6)$$

If $g(I_1)$, $g(I_2)$and $g(I_r)$ are on a line, d should be 1; if these three points form an equilateral triangle, d is 0.

$$D(g;\Omega) = \sum_{\Omega} d \quad (7)$$

We model the likelihood:

$$p(\Omega|g) = \frac{1}{Z}\exp(\lambda D(g;\Omega)) \quad (8)$$

Where Z is a constant, λ is empirically10^4.

$$g^* = \arg\max p(g|\Omega) = \arg\max p(\Omega|g)g(g) \quad (9)$$

The optimization is computed by the Levenberg-Marquardt method.

4 Editing Image Based on CRF

4.1 Enhance Contrast

Contrast can be defined as the brightness ratio of two areas of an image. The method for increasing the contrast is increasing the lighter pixels and decreasing the darker pixels. To increase the contrast, the camera response curves can be adjusted to the shape of ‘S’. To the contrary, for decreasing the contrast, the camera response curves can be adjusted to the shape of inverse ‘S’.

4.2 Adjust the Exposure

$$E = Rt \quad (10)$$

E is the exposure, t is time of exposure, and R is the image-plane irradiance.

Exposure is the product of irradiance and time of exposure. We can change the exposure by adjusting the shutter and aperture in cameras when capturing the images. The camera's non-linear response function will act on the exposure.

But with an image, instead of changing pixel values of the image directly, we firstly map the non-linear image intensity to linear irradiance, and then do multiplication and demultiplication on their radiance. Finally, we map the changed exposure to image intensity (Fig. 5.).

5 Experiment Results

We test the CRF estimation algorithm on images captured from two kinds of cameras, Nikon D100 and Samsung L830. For comparison, we recover the function using the method of Debevec and Malik[2], which use a set of images with varying exposure times.

We detect image patches containing edges between two uniform region colors (Fig. 2.). We use the Canny edge detection .For the uniform color regions, it is required that the color variance of each region is below a specified threshold, and the mean colors of two regions lie at least a specified distance from each others.

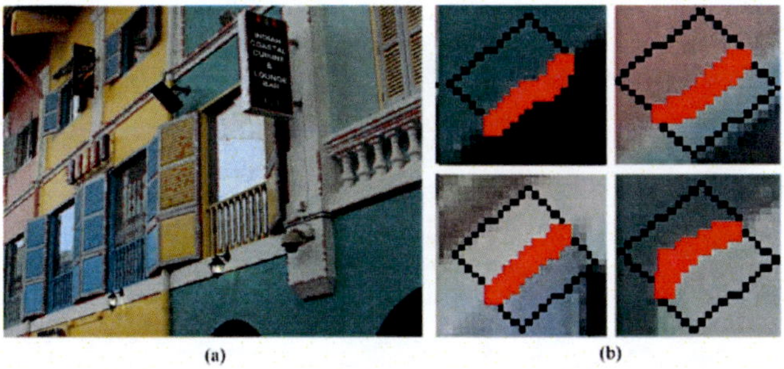

Fig. 2. Edge patches. (a) The whole image with edge patches. (b) Closeups of the edge patches. The thick lines mark the edge pixels

Table 1 shows the performance of the method in terms of RMS error. Fig.3 illustrates Inverse camera response functions for Nikon D100 and Samsung L830. Fig.4 presents enhancing the contrast of image using CRF. Fig.5 shows changing image exposure using CRF.

Table 1. RMSE of Inverse Response Function in RGB Channels

	RMSE(%) red	RMSE(%) green	RMSE(%) blue
Nikon D100	2.844	1.234	3.043
Samsung L830	1.513	3.269	3.923

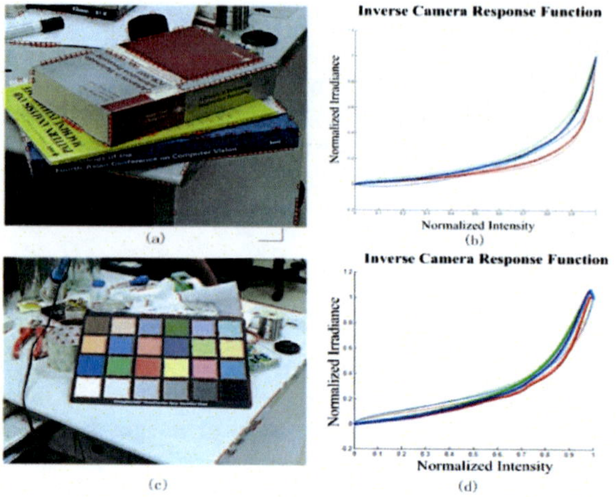

Fig. 3. Inverse camera response functions. (b, d) For Nikon D100 (a) and Samsung L830(c), the thin red, green and blue lines are results of Debevec & Malik's method [2], and the thick lines are the results of our method.

Fig. 4. Enhance the contrast of images with CRF. (a) The original image. (b) The CRF of the original image. (d) The new 'S' CRF. (c) New image with strong contrast, with the grass of background darker, and the flower hearts lighter.

Fig. 5. Change image exposure using CRF. (a) The original image. (b) The irradiance image mapped by inverse CRF. (c) The irradiance image which has higher brightness. (d) The high exposure image mapped by CRF.

6 Conclusion

In this paper, we use a single color image to recover camera response function. Comparing with some other methods, it does not needs a set of images and some limit in scene or illumination. We utilize the CRF estimated to edit image's contrast and exposure, improving the effect of the image.

Acknowledgments. The paper is supported by National Nature Science Foundation of China (No.61073078), and Beijing Municipal Natural Science Foundation (No.4102037). The authors wish to thank Shen Fangyang for discussion.

References

1. http://wiki.panotools.org/Camera_response_curve
2. Debevec, P., Malik, J.: Recovering high dynamic range radiance maps from photographs. In: Proc. SIGGRAPH 1997, pp. 369–378 (August 1997)
3. Mann, S., Picard, R.: Being 'undigital' with digital cameras: Extending dynamic range by combining differently exposed pictures. In: Proceedings of 46th Annual Conference on IS&T, pp. 422–428 (1995)
4. Mitsunaga, T., Nayar, S.K.: Radiometric Self Calibration. In: Proc. of CVPR, vol. 2, pp. 374–380 (June 1999)
5. Finlayson, G.D., Hordley, S.D., Hubel, P.M.: Color by Correlation: A Simple, Unifying Framework for Color Constancy. IEEE Trans. on PAMI 23(11), 1209–1221 (2001)
6. Land, E.H., McCann, J.J.: Lightness and Retinex Theory. JOSA 61(1), 1–11 (1971)
7. Debevec, P.E.: Rendering Synthetic Objects into Real Scenes. In: Proc. of ACM SIGGRAPH, pp. 189–198 (July 1998)

8. Chang, Y.-C., Reid, J.: RGB calibration for color image analysis in machine vision. IEEE Trans. on Image Processing, 1414–1422 (February 1996)
9. Mitsunaga, T., Nayar, S.K.: Radiometric self calibration. In: CVPR, vol. 1 (February 1999)
10. Kim, S.J., Frahm, J., Pollefeys, M.: Radiometric calibration with illumination change for outdoor scene analysis. In: CVPR (2008)
11. Lin, S., Zhang, L.: Determining the radiometric response function from a single grayscale image. In: CVPR (2005)
12. Matsushita, Y., Lin, S.: Radiometric calibration from noise distributions. In: CVPR (2007)
13. Ng, T.-T., Chang, S.-F., Tsui, M.-P.: Using geometry invariants for camera response function estimation. In: CVPR (2007)
14. Diaz, M., Sturm, P.: Radiometric Calibration using Photo Collections. In: ICCP (2011)
15. Luong, Q., Fua, P., Leclerc, Y.: The radiometry of multiple images. IEEE Trans. on PAMI, 19–33 (January 2002)
16. Mann, S., Picard, R.: On being 'undigital' with digital cameras: Extending dynamic range by combining differently exposed pictures. In: Proc. IS&T 46th Annual Conference, pp. 422–428 (March 1995)
17. Mitsunaga, T., Nayar, S.: Radiometric self-calibration. In: Proc. IEEE Conference on Computer Vision and Pattern Recognition, pp. 374–380 (July 1999)
18. Grossberg, M., Nayar, S.: Modeling the space of camera response functions. IEEE Transaction on Pattern Analysis and Machine Intelligence 26(10), 1272–1282 (2004)
19. Lin, S., Gu, J., Yamazaki, S., Shum, H.-Y.: Radiometric calibration from a single image. In: CVPR, pp. 2938–2945 (2004)

Automatic Extraction and Verification of State-Models for Web Applications

Andrey Zakonov and Anatoly Shalyto

State Univ. of Information Technologies, Mechanics & Optics, St. Petersburg, Russia
andrew.zakonov@gmail.com, shalyto@mail.ifmo.ru
http://www.ifmo.ru

Abstract. Complexity of Web applications and demand for their reliability have greatly increased over recent years as they have begun to be used in wide variety of areas, including control systems and enterprise applications. Due to short delivery times and changing requirements quality assurance of Web applications is usually an informal process. Meanwhile, formal methods have been proven to be reliable means for the specification, verification, and testing of systems. In this paper, we present an approach for automatic modelling of an existing web application using finite state machines. The paper describes a method to generate an application model by fully automatic dynamic analysis of any given existing web application combined with recorded user browsing sessions analysis. Method supports both applications with transitions between web pages and single-page applications with AJAX requests and dynamic DOM modifications. An algorithm is proposed that simplifies state model by merging similar states to achieve a human readable model even for complex real world web applications. The obtained model could be used to define formal requirements for the application, automatic model checking, documentation and test automation.

Keywords: Web Application, state model, model checking, formal requirements.

1 Introduction

Over the recent years web sites have transformed from a collection of static HTML pages to a complex interactive web applications, becoming more complicated and more popular, which is proved by examples like Facebook, GMail, etc. By a Web Application we understand a set of pages connected by hyperlinks. Each page could have a static content or could be a complex single page Ajax application, which Document Object Model (DOM) could be dynamically modified by application logic implemented in JavaScript.

While source code is the most accurate description of the behaviour of a web application, it is hardly suitable for high-level understanding of an application's intended behaviour. The goal the of presented research is to propose a method for automatic state model extraction of existing web applications, which would

D. Yang (Ed.): Informatics in Control, Automation and Robotics, Volume 2, LNEE 133, pp. 157–160.
springerlink.com

be suitable for writing formal specification requirements and their automatic verification using existing model checking tools.

First of all a method to discover web application states is developed. Automatic random-driven exhausting exploration of page states and page transitions is used together with the analysis of the collected user execution traces.

Secondly, an algorithm is proposed to discover similar states and to merge them. For complex applications considering each different DOM tree to be a different application state would lead to a model consisting of thousands of different states and transitions, which would make this model practically useless. Model simplification algorithm makes possible automatic generation of human readable models even for complex real world web applications.

2 Related Work

In [1] authors survey 24 different modelling methods used in web site verification and testing. Several techniques have been presented in the literature that propose automatic model extraction and verification of the existing web applications.

Page transitions are analysed in [3], but research is limited only to Web applications based on the Struts framework and JSP templates. Our approach support much wider range of web applications, due to support of Ajax Web applications that consist of a single page whose elements are updated in response to callbacks activated asynchronously by user or by a server message.

The work most similar to our approach is described in [4]. Paper proposes state-based testing approach for Ajax Web applications. State model is used to find semantically interacting events and generate tests. Approach is limited to single page applications with Ajax callbacks. Our approach handle all the possible state changes, which include page transitions and JavaScript DOM manipulation in event handlers triggered by user actions, as well as Ajax callbacks. Handling a Web application in whole makes it possible to apply our approach to real world application and to achieve more accurate model.

3 Automatic Model Extraction

Analysis of a given execution trace is done by employing the Selenium tool, which is able to replay all user actions. Before and after each action snapshot of the DOM state is recorded. Finally, execution trace is stored as a sequence of triads $< state1, action, state2 >$. Automatic analysis of a web application is done by the following method:

1. Page source code is analysed to get list of available actions *actionlist*, i.e. actions that trigger page transition or JavaScript code: *a*, *button*, *input*, elements with action handlers defined (*onlick*, *onmousedown*, jQuery handlers etc.).
2. Randomly select an action from *actionlist* and execute it. If action requires text input then value is generated randomly or selected from a supplied list (password/login/etc. values should be provided explicitly).

3. Triad $< state1, action, state2 >$ is added to the execution trace.
4. Go to step 1, unless last 20 iterations have discovered no new state (exploration has finished or looped).

To discover similar states, we introduce the following recursive definition of similarity: for DOM nodes A and B $similar(A, B) == True$ if and only if they have the same type, same set of attributes, same set of children, where each child of A is similar to the corresponding child of B.

We ignore text values of the elements, but compare only DOM structure, this means, that $< p >$ *text* $< /p >$ and $< p >$ *othertext* $< /p >$ nodes would be considered similar. Also, nodes of specific types like *link*, *script*, *meta*, etc. are filtered out of the DOM trees before comparison, as they do not directly affect the page state that the user could see.

An important feature of proposed algorithm is similar node "collapse" step. List of nodes $x_1, ..., x_n$ should be "collapsed" if $\forall i, j \in [1; n]$ $similar(x_i, x_j) ==$ $True$ && $x_i.parent == x_j.parent$. In this case list of nodes are replaced by one node, x_1. Due to "collapse" step pages like mail inbox or a task list, which often differ only by number of similar items, would become similar and the extracted model would make much more sense. "Collapse" algorithm steps:

1. Traverse the DOM tree, starting from its leafs.
2. For a given node fetch list of children nodes $list_c$.
3. Check all pairs $x_i, x_j \in list_c$ and if $similar(x_i, x_j) == True$ remove x_j node.

4 Model Checking and Testing

Extracted finite state model could be automatically converted into Promela format and served as an input to the Spin model checker. Properties to be verified could be expressed as Linear Temporal Logic (LTL) formulas. For example, navigational requirements could be conveniently formulated in LTL. Requirement "on all paths from page *Welcome* to page *Inbox*, page *Login* is present" would look like: [] (*Welcome* && ! *Inbox* → ((! *Inbox*) U ((*Login* && ! *Inbox*) || [] (! *Inbox*)))).

Testing can also be significantly automated by using finite state model. In [5] an approach for test generation from state model is introduced.

5 Case Study

A proof-of-concept tool was developed using Python 2.7 programming language, Selenium and Graphviz frameworks. Tool produces model as an XML file and a PNG image. XML description could be converted and used for model checking and test automation tools. PNG image contains a human readable representation of the model. Transitions contain description of the taken actions, like "click object *L*" or "type text *A* into field *B*". Object references are described using XPath language. Fig. 1 shows analysis of a simple execution trace of a Facebook

Fig. 1. Analysis of Facebook sending message execution trace

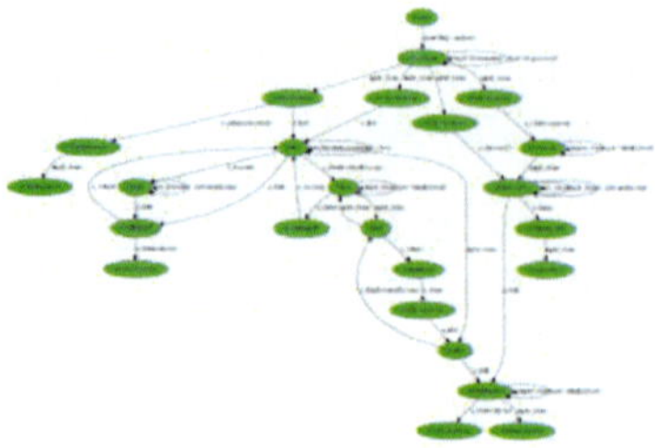

Fig. 2. Model automatically extracted for www.tadalist.com task list application

user. Fig. 2 presents model of a www.tadalist.com task list application, which was extracted completely automatically using exhaustive exploration technique.

Facebook send message model contained initially 9 different states and was automatically simplified to have only 5 states. Exhausting exploration of Tada list has discovered 54 different states in 6 minutes. After applying the simplifying algorithm a model containing only 25 different states were produced.

6 Conclusion

In this paper we presented an approach to extract human readable finite state models of the existing real world web applications. Generally model could not be 100% correct, as there would be no execution trace that explores each link, triggering each possible event handler on every page of the application. Nevertheless model that approximates application behaviour could be used for formal specification requirements and automated model checking and significantly improve software quality and defect detection rate.

References

1. Alalfi, M.H., Cordy, J.R., Dean, T.R.: Modelling methods for web application verification and testing: state of the art. Softw. Test., Verif. Reliab., 265–296 (2009)
2. Haydar, M.: Formal Framework for Automated Analysis and Verication of Web-Based Applications. In: ASE, pp. 410–413 (2004)
3. Kubo, A., Washizaki, H., Fukazawa, Y.: Automatic Extraction andVerication of Page Transitions in a Web Application. In: 14th Asia-Pacicfic Software Engineering Conference, APSEC 2007, pp. 350–357 (2007)
4. Marchetto, A., Tonella, P., Ricca, F.: State-Based Testing of Ajax Web Applications. In: ICST, pp. 121–130 (2008)
5. Zakonov, A., Stepanov, O., Shalyto, A.A.: GA-Based and Design by Contract Approach to Test Generation for EFSMs. In: IEEE EWDTS, pp. 152–155 (2010)

Recommender System to Analyse Students' Learning Productivity

Arturas Kaklauskas[1], Mark Seniut[1], Edmundas Kazimieras Zavadskas[1], Gintautas Dzemyda[3], Voitech Stankevic[2], Ceslovas Simkevicius[2], Sergejus Ivanikovas[3], Tomas Stankevic[2], Agne Matuliauskaite[1], and Lina Zemeckyte[1]

[1] Vilnius Gediminas Technical University Sauletekio Ave. 11, LT-10223 Vilnius, Lithuania
Arturas.Kaklauskas@vgtu.lt
[2] Semiconductor Physics Institute, Center for Physical Sciences and Technology, Vilnius University, Vilnius, Lithuania
[3] Vilnius University, Institute of Mathematics and Informatics, Vilnius University, Akademijos St. 4, Lt-08663 Vilnius, Lithuania

Abstract. A sufficient amount of studies worldwide prove an interrelation linking students' learning productivity, emotional and psychological state to physiological parameters. Emotional states and the interest in learning affect learning productivity, while physiological parameters demonstrate such changes. Since the research by the authors of the present article confirmed these interdependencies, a Recommender System to Analyse Student's Learning Productivity (Recommender System hereafter) has been developed. The Recommender System uses the Maslow's Hierarchy of Needs and the database of the best global practices to come up with recommendations for students how to improve their learning efficiency. Worldwide research includes various scientists who conducted in-depth studies on the different and very important areas in the field. We did not find any systems which would take physiological parameters of students, analyse their learning efficiency and, in turn, provide recommendations.

Keywords: Web-based recommender system, physiological techniques, emotional state, learning productivity, knowledge management, multiple criteria analysis methods, historical information, recommendations.

1 Recommender System

The Recommender System to Analyze a Student's Learning Productivity (i.e., Recommender System) consists of eight subsystems: e-Self-Assessment, Physiological computer mouse, Physiological Finger, Student's Physiological Database, Maslow's Pyramid Tables, Model-base Management and Model Base, Student's Interdependence between Physiological and Self-Assessment Parameters, Graphic Interface. The subsystems are briefly analyzed below.

Numerous studies have been conducted worldwide that analyze the reliability of self-assessment. A great many researchers reached reliable results proving that the

D. Yang (Ed.): Informatics in Control, Automation and Robotics, Volume 2, LNEE 133, pp. 161–164.
springerlink.com

reliability of self-assessment is sufficient. Due to the experiences of these authors, including the authors of this article, a decision was made to gain information about a student's learning productivity, interest in learning and emotional states among other matters by means of self-assessment. The authors herein based their developed Recommender System on the presumption that, by assigning a student self-assessment questions and then searching for the interdependency with that student's physiological parameters at that time, it is possible to determine this interlink rather accurately. Existing experiences and intuition give students a rather accurate indication about their personal learning productivity, interest in learning and emotional states under analysis. Before starting the work, a student must subjectively assess his/her own learning productivity, interest in learning and emotional states by using the *e-Self-Assessment questionnaire*.

Table 1. A main purpose of the Recommender System is to determine interdependency between the eleven pieces of data submitted on a student's own state of being and twelve of that student's physiological parameters at the time of self-assessment

Physiological parameters under measurement	Data on a student's state of being at the time of self-assessment: personal mood, productivity, interest in learning, stress, self-control, happiness, anger, fear, sadness, surprise, anxiety
Physiological computer mouse: temperature, humidity, skin conductance, touch intensity, heart rate **Physiological Finger:** humidity, electrogalvanic skin, conductance, skin temperature, heart rate **Blood pressure and heart rate data:** heart rate, systolic blood pressure, diastolic blood pressure	**Interdependency diagrams determined and outlined**

The student fills out the questionnaire and clicks on the button labeled "Start working". The *Physiological computer mouse* and the *Physiological Finger* begin selecting the physiological data on the student; these accumulate in the Student's Physiological Database and determine different interdependences (see Table 1).

This Physiological computer mouse is able to measure the temperature and humidity of a student's hand, skin conductance, touch intensity and heart rate. These physiological parameters give more information about the student and help to evaluate that student's learning productivity, interest in learning and emotional states. The schematic diagram of the sensors section of a Physiological computer mouse is presented in Fig. 1. The Physiological Finger collects data on the following parameters for correlations with a student's emotional state and work productivity: humidity (humidity sensor); electrogalvanic skin conductance (electrogalvanic skin response sensors); skin temperature (thermistor); heart rate (pulse sensor). The schematic diagram of the sensors section of the Physiological Finger is presented in Fig. 2. The signals from the skin temperature, resistance, humidity and heart rate sensors were read by the microcontroller PIC18F2458 with an internal 12-bit ADC. After reading these signals, the microcontroller calculates the values of skin temperature, humidity, resistance and pulse count and then sends the data to the computer via USB.

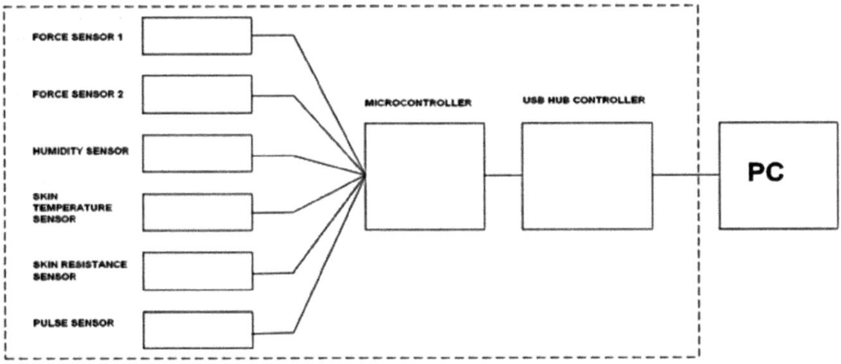

Fig. 1. Sensors section of the Physiological computer mouse

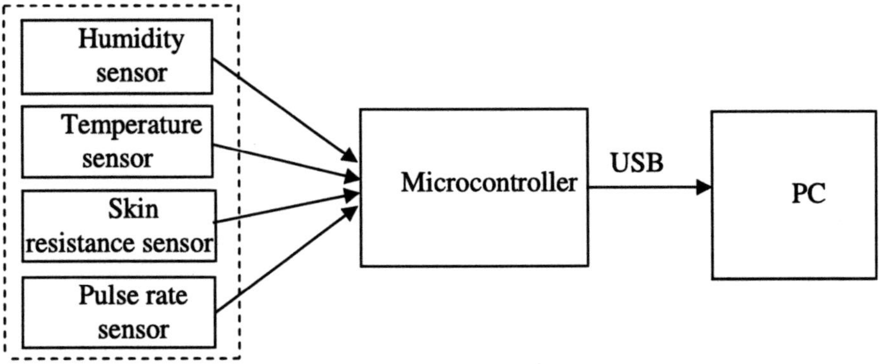

Fig. 2. Schematic diagram of the Physiological Finger

Once a student fills out the e-Self-Assessment questionnaire and clicks on the button labeled "Start working", the Physiological computer mouse and Physiological Finger begin to gather the physiological data of that student for storage in the Student's Physiological Database. Different physiological parameters have been tested for correlations with student's learning productivity, interest in learning and emotional states. According to the theory proposed by Maslow, there are five levels of needs, which determine human behavior: physiological, safety needs, social needs, the need for esteem and recognition and self-actualization (self-expression) need. The literature based on the Maslow Theory was used to compile the tables relative to Maslow's Pyramid. Maslow's Pyramid Tacit (abilities, skills, experience, ideals, emotions, values, learning culture, beliefs, images, perspectives) and Explicit (theoretical approach, problem solving, etc.) Knowledge Tables, based both on the developed Maslow's Knowledge Tables and on the models from the Model Base, provides recommendations on how to improve a student's learning productivity.

The *Model-base* consists of the following models: model to determine the correlation between the student's learning productivity, interest in learning, emotional states and the physiologicals parameters of the student's (Model for Dependence Analysis); model to automatically evaluate the student's learning productivity; model to determine initial criteria weights (data and student characteristics and recommendations) using expert evaluation methods; model to determine criteria weights (data from Maslow's Pyramid Tables and a student's physiological characteristics); model to design multi-variant recommendations; model to analyze multiple criteria and to prioritize recommendations; model to determine the utility degree of recommendations; model to deliver the recommendations. The methods developed by the authors herein provide the bases for the development of these models as follows [1]-[3]: a method for the complex determination of criteria weights; a multiple criteria method for a complex and proportional evaluation of alternatives; a method to define the utility degree and market value; a method of multi-variant design. The initial version of the Recommender System to Analyze a Student's Learning Productivity was developed and successful tested from 2006. Reliability of the System was proper.

2 Conclusions

The Recommender System developed by the authors determines a student's learning productivity based on the student's physiological parameters. It then generates thousands of alternative learning productivity recommendations based on the compiled Maslow's Pyramid Tables and selects out the most rational of these for the student's specific situation.

References

1. Kaklauskas, A., Zavadskas, E.K., Pruskus, V., Vlasenko, A., Seniut, M., Kaklauskas, G., Matuliauskaite, A., Gribniak, V.: Physiological and Intelligent Self-Assessment of Student Progress system. Computers & Education 55(2), 821–833 (2010)
2. Kaklauskas, A., Zavadskas, E.K., Pruskus, V., Vlasenko, A., Bartkiene, L., Paliskiene, R., Zemeckyte, L., Gerstein, V., Dzemyda, G., Tamulevicius, G.: Recommended Biometric Stress Management System. Expert Systems with Applications 38, 14011–14025 (2011)
3. Kaklauskas, A., Zavadskas, E.K., Seniut, M., Dzemyda, G., Seniut, M., Stankevic, V., Simkevicius, C., Stankevic, T., Paliskiene, R., Matuliauskaite, K.S., Bartkiene, L., Ivanikovas, S.A., Gribniak, V.: Web-based Biometric Computer Mouse Advisory System to Analyze a User's Emotions and Work Productivity. Engineering Applications of Artificial Intelligence 24, 928–945 (2011)

Intelligent Pupil Analysis of Student Progress System

Arturas Kaklauskas[1], Andrej Vlasenko[1], Vidas Raudonis[2], and Edmundas Kazimieras Zavadskas[1]

[1] Vilnius Gediminas Technical University, Sauletekio av. 11, Vilnius, LT-10223, Lithuania
[2] Kaunas University of Technology, K. Donelaicio St. 73, Kaunas, LT-44029, Lithuania

Abstract. Various methodologies are used in the effort to predict the knowledge of students. A sufficient amount of studies worldwide prove an interrelation between persons' cognitive load to pupil size. The Intelligent Pupil Analysis of Student Progress (IPASP) System for assessing student progress were developed on the basis of the aforementioned methodologies, the fifteen years of experience with distance learning and the seven years of experience with technology for pupil analysis by the authors of this article. The IPASP System developed by these authors is superior to the systems due to the use of pupil analysis and special algorithm which permits a more detailed analysis of the knowledge attained by a student.

Keywords: e-Learning, students, knowledge, pupil analysis, Intelligent system, e-Examination, Historical information.

1 Intelligent Pupil Analysis of Student Progress System

A sufficient amount of studies worldwide prove an interrelation linking persons' cognitive load to pupil size. This sensitivity of the pupil can give exhaustive data about the cognitive loads. Different research demonstrate that task-induced dilations can serve as reliable proxies for cognitive load and the sizes of blink pupil dilations reliably reflect diverse scale of difficulty within different activities, and have thus validated pupillary dilations. The original contribution of this article, compared to the research results published earlier, is as follows: the Intelligent Pupil Analysis of Student Progress (IPASP) System developed by the authors is superior to the traditional pupil analysis systems due to the use of pupil analysis and a special algorithm, which permits a more detailed analysis of the knowledge attained by a student. The IPASP System consists of the following components (see Fig. 1): Database Management Subsystem and Databases, Remote Eye Tracking Device, Model-base Management Subsystem and Model Bases, User Interface.

The stationary automatic eye tracking system is realized using (1) ordinary PTZ video camera which capture the eye image of the student (participant). The PTZ camera is placed on the table in front of the student at the distance of 1 meter. The system is used under the real exam conditions, so head movements of the human subjects were unrestricted. The camera is correctly positioned automatically using the servo motors of the PTZ camera in order to capture the eye image. The servo motors are controlled through RS232/RS485 converter.

D. Yang (Ed.): Informatics in Control, Automation and Robotics, Volume 2, LNEE 133, pp. 165–168.
springerlink.com

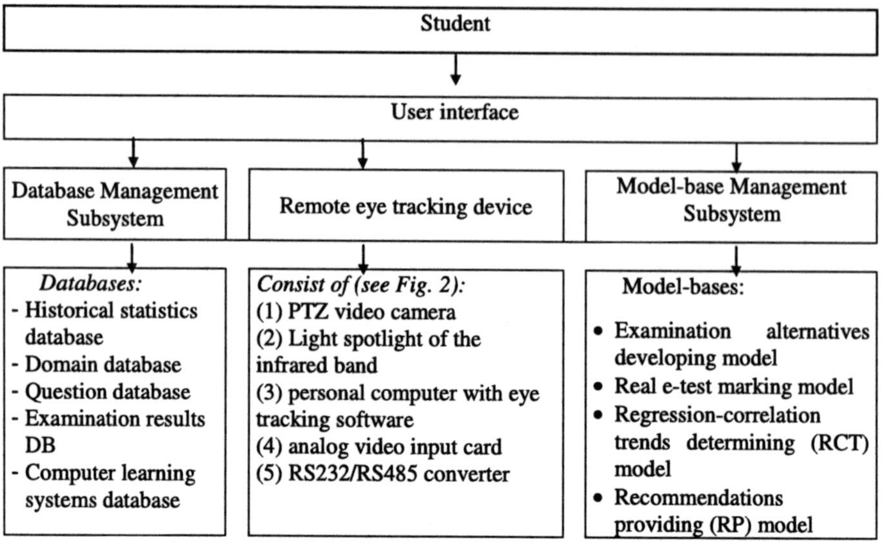

Fig. 1. Composite Parts of the Intelligent Pupil Analysis of Student Progress System

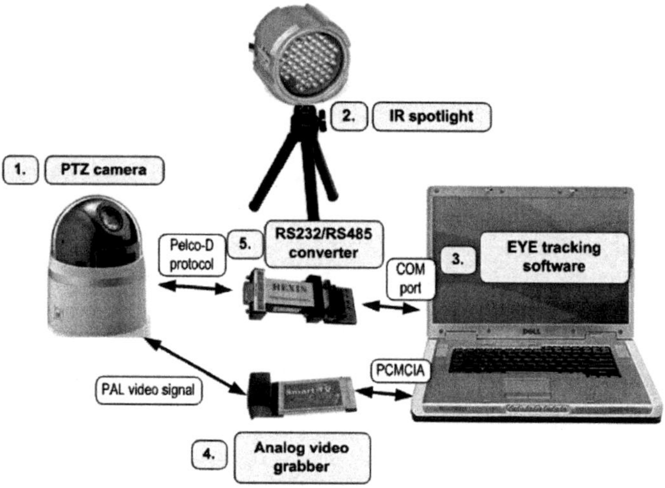

Fig. 2. Remote eye tracking device, were (1) PTZ video camera, (2) Light spotlight of the infrared band, (3) personal computer with eye tracking software, (4) analog video input card and (5) RS232/RS485 converter

The control commands are generated according Pelco-D protocol. One spotlight with diodes of the infrared band (2) is used in the system and it provide infrared brightening of the object (human face). PAL signal is transferred from the camera to the converter. Converted signal is directed to the computer PCMCIA input, and then processed by the special eye tracking software. The software ensures rapid and precise detection of the human head and eye pupil, its size, tracks pupil motion in image and captures its coordinates. The programme is design to follow subject's pupillary responses to the questions asked by the examinational system. Presented eye tracking system works on two different modes, i.e., face detection and eye pupil tracking modes. Our system uses PTZ camera with 10X optical zoom functionality. Autonomous system detects humans face in FOV. When student keep his head steady for certain time, the camera switches to ROI focusing mode in order to zoom upper left region of the human face (left eye). An active tracking mode consisting of a control loop with two interchangeable modules: camera control and tracking ones is then triggered to fallow the eye pupil by keeping it around the centre of an image.

During the exam, the changes of the eye pupil were recorded by a video camera. The taken photos were matched to the provided questions with reference to the time, and the average pupil diameter was calculated. Thus, a pupil diameter is determined, which corresponds to a particular question. According to them, it was possible to determine the student's reaction to each provided question. We may draw a conclusion that those students, whose pupil diameter underwent minor changes, were better prepared for the exam, questions did not seem so complicated to them and therefore their cognitive load is less. Figure 3 depicts the dependence of a pupil diameter on the complexity of questions.

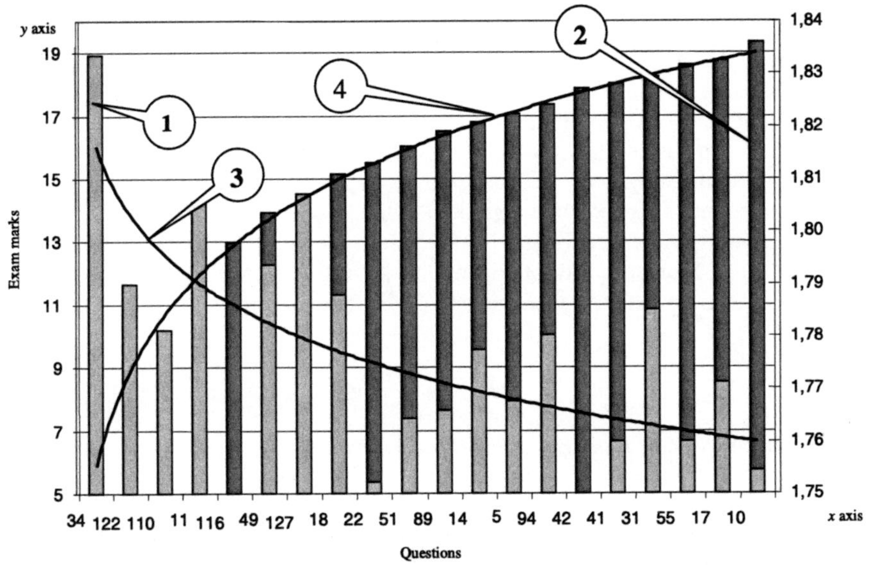

Fig. 3. Dependence of a pupil diameter on the complexity of questions: the x-axis indicates the numbers of the test questions, and the y-axis indicates the average of a pupil diameter (on the right) and the exam marks (on the left)

The x-axis indicates the ID numbers of the questions answered during the e-test and the y-axis indicates the average of a pupil diameter (on the right) and the exam marks (on the left). In case a student answers all the questions correctly, he/she may collect a maximum of 20 points. Figure 3 also shows two correlated curves obtained during the research that indicates direct dependence of a mark on the changes in a pupil diameter. For example, the bigger the change in a pupil diameter the worse the mark is, and vice versa.

Figure 3 show that the more complicated was the provided question the bigger the diameter of a pupil was. It widened most when answering the question No. 34. And actually, only 7 people out of 20 answered this question correctly. We may draw a conclusion that this question was most complicated. Moreover, a reciprocal process may be observed: the easier the question (more students answered it) the smaller the diameter of a pupil is. Fig. 3 obviously shows that the assumption was proved. The higher the pressure a person experiences (uncertain of his/her knowledge or sees the question for the first time) the wider the pupil of his/her eye becomes and its diameter increases. Currently student's knowledge can be evaluated automatically taking into consideration the obtained data and determined interdependences. For example, a lecturer gives several questions to a student about the level of preparation to the exam, and the pupil Recognition Analysis Subsystem, with regard to the changes in the average of the student's pupil, and taking into consideration the obtained data as well as determined interdependency.

2 Conclusions

All distance learning participants (students, professors, instructors, mentors, tutors, etc.) would like to know how well the students have assimilated the study materials being taught. A formal test or exam during the semester would cause needless stress for students. To resolve this problem, the authors of this article have developed an Intelligent Pupil Analysis of Student Progress (IPASP) System. In the future authors of the paper are intended to integrate IPASP System with Web-based Biometric Computer Mouse Advisory System [1], Recommended Biometric Stress Management System [2] and Biometric and Intelligent Self-Assessment of Student Progress system [3] which permits a more detailed analysis of the knowledge attained by a student.

References

1. Kaklauskas, A., Zavadskas, E.K., Seniut, M., Dzemyda, G., Stankevic, V., Simkevicius, C., Stankevic, T., Paliskiene, R., Matuliauskaite, A., Kildiene, S., Bartkiene, L., Ivanikovas, S., Gribniak, V.: Web-based Biometric Computer Mouse Advisory System to Analyze a User's Emotions and Work Productivity. Engineering Applications of Artificial Intelligence 24(6), 928–945 (2011)
2. Kaklauskas, A., Zavadskas, E.K., Pruskus, V., Vlasenko, A., Bartkiene, L., Paliskiene, R., Zemeckyte, L., Gerstein, V., Dzemyda, G., Tamulevicius, G.: Recommended Biometric Stress Management System. Expert Systems with Applications 38(11), 14011–14025 (2011)
3. Kaklauskas, A., Zavadskas, E.K., Pruskus, V., Vlasenko, A., Seniut, M., Kaklauskas, G., Matuliauskaite, A., Gribniak, V.: Biometric and Intelligent Self-Assessment of Student Progress system. Computers & Education 55(2), 821–833 (2010)

Research on "Multi-screen in One" Technology of Education Video Oriented U-learning*

Jun Xiao, Lamei Wang, and Xiaoxiao Zhu

Shanghai Distance Education Group, 10th Floor, Learning World, No.288 Guoshun Road, Yangpu District, Shanghai, China
xiaoj@shtvu.edu.cn, wanglamei@shtvu.edu.cn, zhuxx@shtvu.edu.cn

Abstract. On the research of the transcoding and the adaptive distribution and scheduling of education video which is compatible for various platforms, this paper proposes a technological solution of "multi-screen in one" technology oriented ubiquitous learning(u-learning) on the education video, which can realize the seamless joint of education video between different screens so as to ensure the continuation of the learning process, and achieve the intelligent recommendation of relevant video resource.

Keywords: Multi-screen in one, u-learning, continuous learning.

1 Introduction

In the 21 century, along with the rapid development of communication and network technology, the learning environment is changing dramatically and ubiquitous learning is also booming. In the flood of u-learning resources, education video resources account for a very large proportion. However, there are still some problems in the application of video resources. For example, there is insufficient support in the switch and adaptive distribution and scheduling of video resources between different devices, the search results can't meet the precise requirements of users and education video resources can only be used in separate devices and can't realize the seamless and continuous learning between different devices.

With the coming of the "Tri-network convergence, Multi-screen in one" era, multi-screen in one technology provides us a good solution to achieve continuous learning based on seamless migration between different devices. To solve the above problems, this paper presents a technological solution of "multi-screen in one" technology oriented u-learning on the education video, researches the transcoding and the adaptive distribution and scheduling of education video which is compatible for multi-platforms and, realizes the seamless joint of education video between different screens so as to ensure the continuous learning and, achieve the intelligent recommendation of relevant video resources. For this reason, this paper has an important theoretical and practical significance in the u-learning environment.

* The paper is supported by Innovation Program of Shanghai Municipal Education Commission.

D. Yang (Ed.): Informatics in Control, Automation and Robotics, Volume 2, LNEE 133, pp. 169–175.
springerlink.com

2 The Architecture of the "Multi-screen in One" Technology of Education Video

We have proposed the architecture of the "multi-screen in one" technology of education video oriented u-learning, which is shown in Fig. 1.

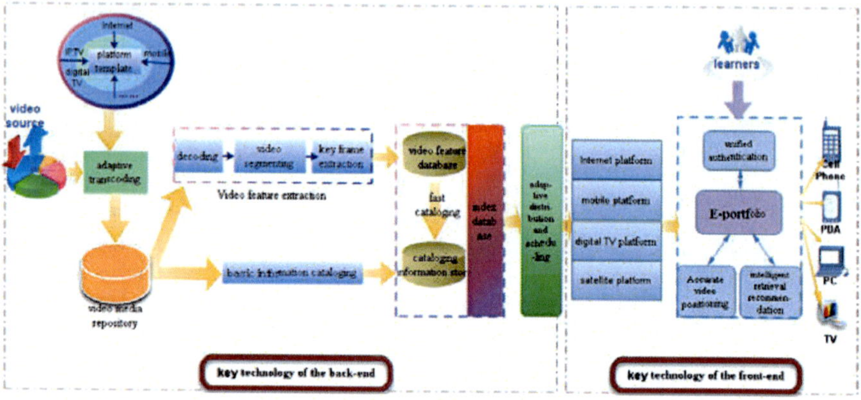

Fig. 1. The Architecture of "multi-screen in one" Technology of Education Video Oriented U-learning

From the above chart we can see that, first we establish a platform template based on a variety of platforms (including the Internet platform, the mobile platform, the digital television platform and others) at the back-end, and the template realizes the adaptive transcoding based different platforms and builds a video media resource database. Then we restructure these transcoded video resources by index, and in terms of the video features and non-video features, we build a video feature database and a video cataloging information database in order to achieve the intelligent cataloging of video resources and form the resource index database. Furthermore, the system distributes the transcoded learning resources through the adaptive scheduling list, and sends different video resources into relevant platforms based on the filter rules so that learners can acquire the appropriate video resources which are compatible with the network and the devices. And in the front-end, after the unified identity authentication based on the E-Portfolio, learners can continuously learn the various resources between different platforms.

In the learning process, the system will automatically track and accumulate learners' learning process (such as records of interrupted operations on learning resources),their behavior models and so on, and then save all the information into E-Portfolio databases of the remote service. The data can be exchanged between different platforms so as to realize the accurate positioning of the video resources. Considering the diverse search needs of learners, we allow the search model of key features, such as pictures and other multimedia, the system will transfer the features of the picture to the background video database, and through the association information the video database can return the results directly to the front end, and eventually to the learner.

3 Key Technology of the "Multi-screen in One" on Education Video

In order to achieve the "multi-screen in one" education video architecture, we need some key technological supports. The technological support can be divided into two areas: the back-end support and the front-end support.

3.1 Key Technology of the Back-End Support

The Adaptive Video Transcoding Technology. Video has unstructured data formats, 2making it incompatible between different networks, devices and data formats in the process of transmission and access. In response to this problem, through the adaptive video transcoding technology, the system provides the adaptive video format conversion based on different platforms (including the Internet platform, the mobile platform, the digital TV platforms, etc.).

Template Tlatform. Based on a variety of network platforms, we define the template platform, which specifies the video format that is compatible with corresponding platforms and has good scalability and reusability and can adapt to future network. According to the platform template, the system can convert the unstructured data formats, save them into the video media repository deposit and realize flexible transmission and access of video format.

Adaptive Transcoding. It realizes the concrete description of the video. The video files can be edited and synthesized, and the output can be a variety of audio and video formats, such as the Internet format, digital TV formats, mobile phone formats and so on.

The Intelligent Indexing of Video. Based on the intelligent indexing of video, we can realize such functions as video feature extraction and intelligent cataloging of the video resources, and finally establish the indexing center for all the video resources. The main features of a video can be devided mainly into the video information and non-video information (including audio information and other basic information.) For video information, you can attain the features by decoding, rendering,video segmenting, and key frame extraction, and then form a video features database. For non-video information, basic information can be used to form video cataloging information store. Based on the video feature database and the video cataloging information store we can realize a fast and intelligent cataloging of the video. By the micro-meta-data cataloging specifications we can achieve an accurate description of the video content, and ultimately form the index database.

Video Feature Extraction Technology. In content-based video retrieval, feature extraction occupies the most important position. It is the ability of describing the characteristics of the video that directly determines the retrieval performance as well as the computing performance. So the research of video feature extraction technology, the formation of the features database as well as the description of the visual characteristics(the color, texture, shape information),object information and sports features can provide basic information for the construction of video indexing and retrieval.

Construction Technology for Intelligent Video Indexing. The indexing structure is for video features, not for the video itself. When processing a search request, the system needs to calculate the similarity of the submitted content to the contents in the video database. So it needs to compare the submitted content and the content in the video database. Take picture indexing for an example.The picture is from the key frame of the video. when inserting a picture into the media database, you also insert the relevant characteristics into the indexing document; when deleting a picture, you also delete the relevant characteristics from the indexing document.

The Adaptive Distribution and Scheduling Technology of the Video. The networks and devices under the u-learning environment has characteristics of heterogeneity. The network includes the integration of a variety of network environments - digital television, Internet, mobile networks, satellite, etc. and it needs to address the discovery and adaptive issues of the equipment and network. Through the adaptive technology, learners can access the appropriate network resources. The adaptive distribution and scheduling technology has two functions: the generation of the adaptive scheduling task list and theadaptive distributed distribution.

The Generation of the Adaptive Scheduling Task List. The generation of the adaptive scheduling task list is based on two parts: the video list selected by the administrators from the most popular projects recently; the relevant video list generated by the terminal equipment (PC computer, mobile phones, TV, etc.) from the appointment or VOD video statistics by the learners.

First, the generation system of adaptive scheduling task lists receives the summary of the learning video appointment or demand information from various terminal platforms, and then it counts the hotspot and special cares of learners recently, and reports it to the backend database for records and management. The information receiving interface can use the Web-Service technology. Each application can access to Web Service by the network protocol or the prescribed standard data formats (Http, XML, Soap) and obtain the required results.

The Adaptive Distributed Distribution. Through the adaptive scheduling task list, the system can distribute the learning video, and through the distributed distribution technology, the system can manage the resources in a global way and dispatch the needed learning resources to terminals for learners. The scheduling process is transparent. Terminals store learners' E-portfolios and learning resources. When learners search the video resources, the system needs to determine the distribution objects.It conduct adaptive distribution from the normal storage area by the use of a set of filtering rules.These rules include: whether the video resources is usable in a certain mobile or desktop device, whether the video resource is beneficial to learners in a mobile or desktop computer situation. For example, the application of technical standards makes it unfeasible to download large files to a mobile device, some kind of video resources are only suitable for PC devices and unable to broadcast on the mobile equipment etc. Each video file corresponds to a metadata description file using XML technology. According to different templates, the system can transmit different data to the corresponding platform, thus this platform can release and retrieve the resources according to the metadata.The process is shown in Fig. 2.

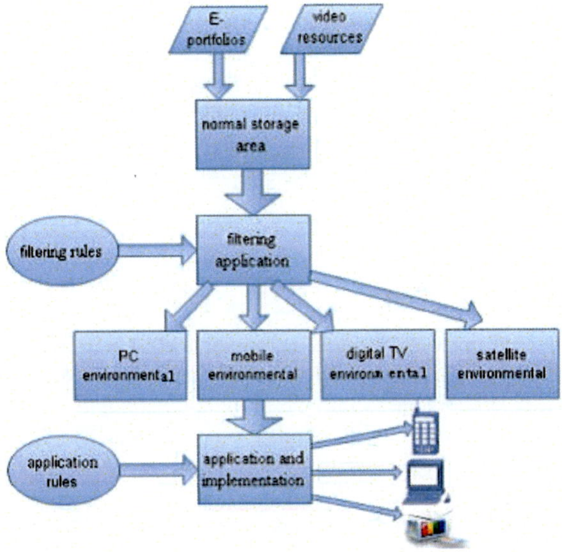

Fig. 2. The adaptive distribution and scheduling technology

3.2 Key Technology of Front-End Support

The Unified Authentication Technology for Learners. Unified authentication can solve such problems that the user need to memory multiple passwords to access multiple learning platforms or equipment. The system provides a unified identity authentication and comprehensive security service, in order to realize the unity of authentication on the Internet and the mobile platform. Through unified authentication, learners' learning process and records in different platforms can be called between other platforms.

Accurate Video Positioning Technologies. When learners are watching the video, they can add tags and collect it, as well as save the suspending information into the E-portfolio, and in another platform, they can quickly return to the needed video position and make accurate positioning. User authentication can also be inherited, which can help to realize the function of continuous learning based multi-screen.This requres that learning portfolios are able to record information in different screens, the records can work in different screens, and all screens have general interface standards, and thus can exchange data.

The Intelligent Retrieval Recommendation Technique. Facing huge amounts of video resources, in addition to realize full-text retrieval service, we must also focus on the intelligent video recommendation. For all the video, we can record the template by key figures in the picture or key characteristics. When importing pictures, the system can automatically contrast the templates, make records and identifications of those have the character's picture or key features, forming a classification. In a retrieval, after the key characteristics (can be key figures) images published online, the system will set the picture transfer to the background video database and the video center can directly return the results to the front end, and then recommend it to users.

And we also consider the following factors: those associated with the contents, the current equipment as well as learners' personal character (hobby information, learners' portfolios (such as learning process, access history)). Based on the above information, the system can intelligently recommend more individualized video resources.

4 Summary

In the U-learning environment, learning equipments change frequently with learners moving everywhere, so we need to find a solution for supporting the continuous learning effectively: learners' tasks may be suspended at anytime and anywhere, and then can be continued at any time in another place with another equipment, so that learners can't notice the noticeable differences despite changes of place, learning equipment and learning environment.

In a word, oriented u-learning "multi-screen in one" technology of education video proposed in the paper has made up for the research vacancy in the education field. The research of the "multi-screen in one technology" is not only to realize the share of the contents, but to achieve the seamless switching in the same learning resource on multi screens. Learners can access the needed learning resources using different devices at any time and place so as to ensure the continuation of the learning process.

References

1. Liu, X., Gong, Y.: Video Summarization and Retrieval Using Singular Value Decomposition. ACM Multimedia Systems 2, 157–168 (2003)
2. Li, B., Sezan, I.: Semantics ports video analysis: approaches and new applications. In: IEEE Proceedings of 2003 International Conference on Image Prodessing, pp. 17–20. IEEE Press, New York (2003)
3. Mezarls, V.: Real-time compressed-domain spatiotemporal segmentation and ontologies for video indexing and retrieval. In: IEEE Transactions on Circuits and Systems for Video Technology, pp. 359–362. IEEE Press, New York (2004)
4. Wengang, C., De, X.: Content-based video retrieval using audio and visual clues. In: IEEE Region 10 Conference on Computers, Communicationgs, Control and Power Engineering, pp. 586–589. IEEE Press, New York (2002)

5. Cheng, G., Yu, S., Yang, X.: Design and implementation of the "learning cell" running environment. Open Education Research 2, 27–36 (2009)
6. Wei, H.: Discussions on three screen syncretic technology. Journal of Information Communication 5, 51–53 (2009)
7. Hu, S., Li, J., Li, J.: Video retrieval based on the latent semantic analysis. Computer Engineering 13, 216–217 (2007)
8. Zhao, H.: Key technology analysis in the U-learning environment. China Educational Technology 7, 99–102 (2007)
9. Zhang, Z., Liu, H., et al.: Research on the video retrieval based on high-level semantics. Computer Engineering and Application 18, 168–170 (2007)

An Advanced Subspace Projection Method for the Weak Signal DOA Estimation in Presence of Coherent Strong Signals

Hou Shuai, Yang Jingshu, and Chen Gong

Laboratory of Information Countermeasure, Electronic Engineering Institute, Hefei, China
millstom@163.com

Abstract. For the problem of signal Direction of Arrival (DOA) estimation of the weak signals in the presence of coherent strong signals, a new method is proposed in this paper. First, a toeplitz de-correlation method using the maximum eigenvector is adopted to remove the coherence between strong and weak signal. Second, an advanced subspace projection method is to estimate the DOAs of weak signals. Compared with the existing algorithms such as JJM and RELAX, this method does not need the knowledge of the DOAs of strong signals and iterative steps. The performance of this method is demonstrated in the simulation results.

Keywords: DOA, weak signal, toeplitz matrix, subspace projection.

1 Introduction

Target orientation estimation which is called direction of arrival (DOA) estimation is a basic problem in array signal processing field. In many problems of practical interest, the receiving data of the array consists of weak and strong signals at the same time. The strong signals can either be strong jamming signal which is uncorrelated with the weak signal or be the powerful signal which is coherent with the weak signal.

If the strong signal is much more powerful such as more than 40dB than the weak signal, the spectrum peak of the weak signal is usually submerged by the peak of the strong signal when the spatial spectrum estimation methods such as MUSIC, CAPON etc. are adopted [2]. To solve this problem, many algorithms based on the idea of jam pre-processing have been proposed [3, 4]. However, these algorithms require the precise DOA knowledge of the strong signal which is hypothesized[3] or estimated beforehand[4]. So these algorithms have a relatively small application scale and large complexity.

Meanwhile, the performance of many classic spatial spectrum estimation methods such as eigen subspace algorithms represented by MUSIC algorithm will severely degrade or even be invalid if the receiving signals are coherent with each other. In order to solve this problem, many classic algorithms such as spatial smoothing [5], matrix decomposition [6] and vector singular value algorithm[7] have been proposed. However, these algorithms remove the signal coherence at the cost of losing the array aperture.

D. Yang (Ed.): Informatics in Control, Automation and Robotics, Volume 2, LNEE 133, pp. 177–183.
springerlink.com

This paper proposes an advanced subspace projection method to estimate the DOA of the weak signal in the presence of coherent strong signal. In this method, a toeplitz de-correlation method using the maximum eigenvector is adopted to remove the coherence between strong and weak signal. This method just needs the eigenvalue decomposition (EVD) twice and the spectrum peak searching only once. Meanwhile, it does not need the knowledge of the DOA of strong signal and which does not lose the array aperture during the signal de-correlation process.

This paper is structured as follows: in Section 2, the problem formulation is presented. In Section 3, the advanced MUSIC algorithm is proposed. Finally, in Section 4, the performance of the method is analyzed.

2 Problem Formulation

Consider N narrow-band far-field signals $\{s_n(t)\}_{n=1}^{N}$, arriving at an Uniform Linear Array(ULA) which consists of M omnidirectional sensors with the distance d between each other. After being corrupted by additive noise $n_m(t)$ which is uncorrelated with the signal source, the received data can be described by (1).

$$\mathbf{X}(t)=\mathbf{A}(\boldsymbol{\theta})\mathbf{S}(t)+\mathbf{N}(t) \tag{1}$$

Here, the received data of the array can be described as $\mathbf{X}(t)=[x_1(t),x_2(t),\cdots,x_M(t)]^T$, while $\mathbf{S}(t)=[s_1(t),s_2(t),\cdots,s_N(t)]^T$ and $\mathbf{N}(t)=[n_1(t),n_2(t),\cdots,n_M(t)]^T$.

In (1), $\mathbf{A}(\boldsymbol{\theta})$ is the manifold matrix, $\boldsymbol{\theta}$ is the DOA vector containing the directions of arrival of individual signals, i.e., the n-th component θ_n of $\boldsymbol{\theta}$ gives the DOA of the signal $s_n(t)$. The manifold matrix consists of the steering vectors $\{a(\theta_n)\}_{n=1}^{N}$:

$$\mathbf{A}=[a(\theta_1),a(\theta_2),\cdots,a(\theta_N)] \tag{2}$$

Where
$$a(\theta_n)=\left[1,\exp\left(-j\frac{2\pi d\sin\theta_n}{\lambda}\right),\cdots,\exp\left(-j\frac{2\pi(M-1)d\sin\theta_n}{\lambda}\right)\right]^T \tag{3}$$

According to (1), the covariance matrix of receiving data can be expressed as (4):

$$\mathbf{R}=E\{\mathbf{XX}^H\}=AE\{\mathbf{SS}^H\}\mathbf{A}^H+E\{\mathbf{NN}^H\}=\mathbf{AR_S A}^H+\mathbf{R_N} \tag{4}$$

$\mathbf{R_S}$, $\mathbf{R_N}$ in (4) are the covariance matrix of the signal and noise respectively. In the practical problems, $\mathbf{R}$ is usually represented by the sample covariance matrix $\hat{\mathbf{R}}$, namely:

$$\hat{\mathbf{R}}=\frac{1}{L}\sum_{i=1}^{L}\mathbf{XX}^H \tag{5}$$

L represents the snapshots number of the data. Assumes the arriving directions of strong signals are $\{\theta_n\}_{n=1}^{N_S}$ and weak signals are $\{\theta_n\}_{n=1}^{N_W}$, where N_S and N_W are the number of strong signals and weak signals respectively and $N_W + N_S = N < M$. The problem is to find $\{\theta_n\}_{n=1}^{N_W}$ from the receiving data $\mathbf{X}(t)$.

3 DOA Estimation Method

3.1 Toeplitz Decorrelation Method

When the signal sources are coherent with each other, all of the information in signals is contained in the maximum eigenvector of the covariance matrix of the receiving data. And the maximum eigenvector can be expressed as(6):

$$\mathbf{e} = \sum_{n=1}^{N} t_n a(h_n) = \mathbf{AT} \tag{6}$$

In (6), $\mathbf{T} = [t_1, t_2, \cdots, t_N]^T$ is linear combination factors, $\mathbf{e} = [e_1, e_2, \cdots, e_M]^T$. The toeplitz matrix $\mathbf{E}$ is defined by $\mathbf{e}$:

$$\mathbf{E} = \begin{bmatrix} e_1 & e_2^* & \cdots & e_m^* \\ e_2 & e_1 & \cdots & e_{M-1}^* \\ \vdots & \vdots & \cdots & \vdots \\ e_M & e_{M-1} & \cdots & e_1 \end{bmatrix} \tag{7}$$

According to the results in [9], the rank of the toeplitz matrix $\mathbf{E}$ is the number of signal sources, so using $\mathbf{E}$ to replace the covariance matrix $\mathbf{R}$ can get the correct signal subspace and noise subspace.

3.2 An Advanced Subspace Projection Method

The classic sort of subspace projection method such as MUSIC algorithm uses the orthogonality between the signal steering vectors and noise subspace[1]. The angles corresponding to the peaks of the spatial spectrum given by (8) are the signals' DOAs.

$$P(\theta) = \frac{1}{\mathbf{a}^H(\theta)\mathbf{U}_N\mathbf{U}_N^H\mathbf{a}(\theta)} \tag{8}$$

Where $\mathbf{U}_N = [e_{N+1}, \cdots, e_M]$ is the noise subspace whose columns are the $M-N$ noise eigenvectors obtained by computing the eigenvalue decomposition(EVD) of covariance matrix $\hat{\mathbf{R}}$ .As the steering vectors corresponding to the weak signals and the steering vectors corresponding to the strong signals are both orthogonal to the noise subspace, the spatial spectrum peaks of weak signals will be submerged by the peaks of strong ones if the classic MUSIC is used directly.

Assumes N_S strong signals and N_W weak signals which are coherent with each other are received by the array mentioned in *Section2*. First the coherence between the signals is eliminated by the method mentioned in Section3.2. Then computes the EVD of **E**. rearranges the eigenvalues of **E** from great to little and the eigenvectors correspond to the eigenvalues will construct the eigenspace **U** expressed by(9):

$$\begin{aligned}\mathbf{U} &= \left(u_1,\cdots,u_{N_S},u_{N_S+1},\cdots,u_{N_S+N_W},\cdots,u_{N_S+N_W+1},\cdots,u_M\right)\\ &= \left[\mathbf{U}_S,\mathbf{U}_W,\mathbf{U}_N\right]\end{aligned} \tag{9}$$

Where in (9), $\mathbf{U}_N$ is eigen subspace corresponding to the additive signals. The signal eigen subspace is divided into two parts, $\mathbf{U}_S$ is eigen subspace corresponding to the strong signals and $\mathbf{U}_W$ is the one corresponding to the weak signals and These three eigen subspace can be expressed by (10):

$$\begin{aligned}\mathbf{U}_S &= \left(u_1,\cdots,u_{N_S}\right)\\ \mathbf{U}_W &= \left(u_{N_S+N_W},\cdots,u_{N_S+N_W+1}\right)\\ \mathbf{U}_N &= \left(u_{N_S+N_W+1},\cdots,u_M\right)\end{aligned} \tag{10}$$

Constructs the projection to $\mathbf{U}_W$ for each steering vector $\mathbf{a}(\theta)$, the power of the projection can be expressed by (11):

$$P_J(\theta) = \mathbf{a}^H(\theta)\mathbf{U}_W\mathbf{U}_W^H\mathbf{a}(\theta) \tag{11}$$

Obviously, the power of the projection to $\mathbf{U}_W$ for the strong signals is much smaller than one for the weak signals and the works in[] have proved the result.

This method can be summarized in 4 steps:

Step 1. Estimates the covariance matrix of receiving data: $\hat{\mathbf{R}} = \frac{1}{L}\sum_{i=1}^{L}\mathbf{X}\mathbf{X}^H$

Step 2. Computes the EVD of $\hat{\mathbf{R}}$ and uses the maximum eigenvector of $\hat{\mathbf{R}}$ to construct a new covariance matrix **E** according to (7).

Step 3. Computes the EVD of **E**, and constructs the weak signal subspace $\mathbf{U}_W$ according to the eigenvectors of **E** and the number of weak signal N_W.

Step 4. Computes the power of the projection to $\mathbf{U}_W$: $P_J(\theta)$ and the angles corresponding to the peaks of $P_J(\theta)$ are the estimated DOAs of weak signals.

4 Simulation Results

Simulation 1:

Assumes the array consists of 8 sensors with the distance of half the signal wavelength. The number of the snapshots is 1024. Two signals which have the same carrier frequency $(f_c = 5MHz)$ and the same modulation (AM modulation with the modulation depth $M_b = 0.5$) are received by the array at the same time. The SNR of the strong signal and the weak signal are 50dB and 0dB respectively. Meanwhile, the direction of the strong signal and the weak signal are 30° and 50° respectively. First, the classic MUSIC algorithm is used to estimate the signal DOAs. The result in **Fig.1**(a) shows that the peak of the spectrum does not come out in the direction of weak signal. Then the proposed method is used to estimate DOAs, and the result in **Fig.1**(b) shows that the peak comes out in the direction of the weak signal.

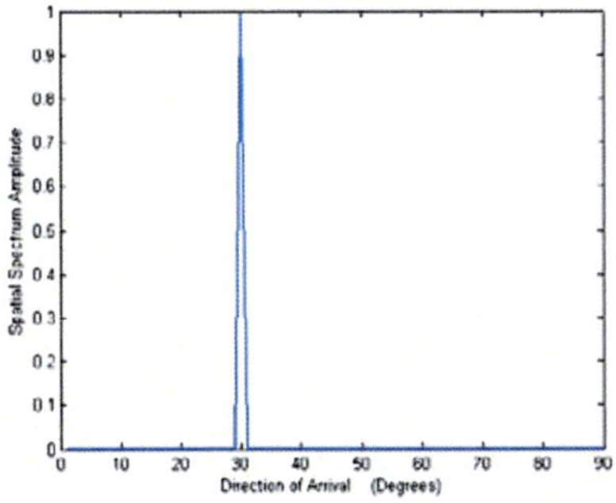

Fig. 1. (a) Weak signal DOA estimation by MUSIC algorithm

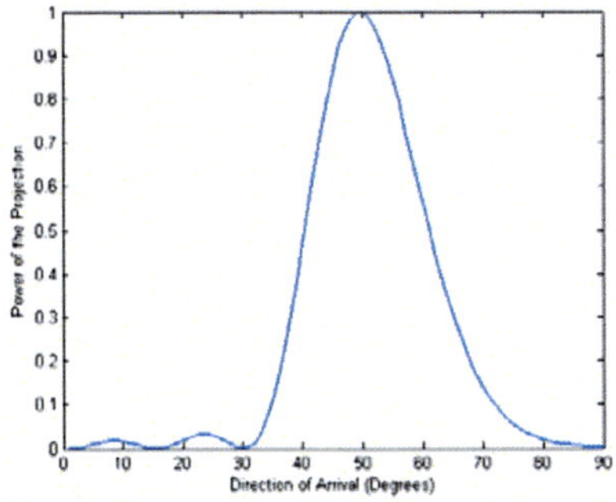

Fig. 1. (b) Weak signal DOA estimation by the proposed method

Simulation 2:

Here, we analyze the performance of the method using the root mean square error (RMSE) of weak signal DOA estimation. **Fig.2** shows the RMSE of weak signal DOA changed with the angle between strong and weak signal. In this figure, the SNR of weak signal is 0dB and the SNR of strong signal is 50dB.

Fig.3 shows the RMSE of weak signal DOA changed with the power ratio of strong to weak signal. In this figure, the angle between strong and weak signal is 20 degrees, the SNR of weak signal is 0dB while the SNR of strong signal changes from 0dB to 100dB.

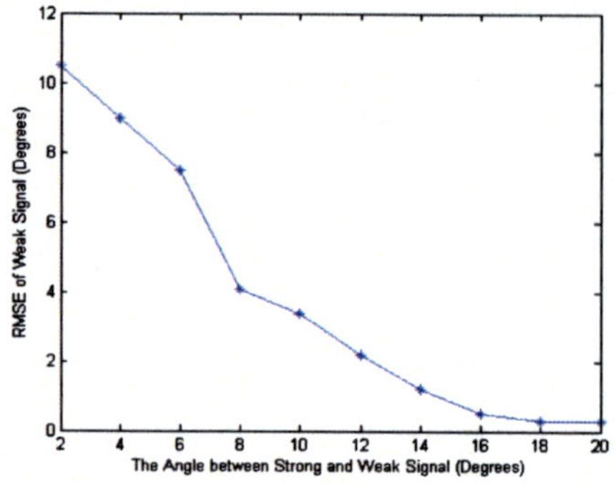

Fig. 2. RMSE of weak signal changed with the angle between strong and weak signal

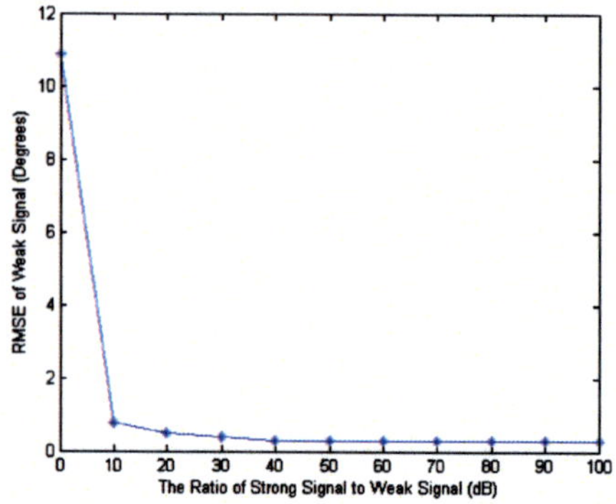

Fig. 3. RMSE of weak signal changed with the power ratio of strong to weak signal

5 Conclusions

An advanced subspace projection method has been proposed to estimate the DOA of weak signal in the presence of coherent strong signals. The simulation results have demonstrated the ability of this method to estimate the DOA of weak signal. The advantage of this method compared with existing algorithms is that it does not need the precise DOA knowledge of the strong signal. However, in this paper, we assume that the number of the strong and weak signals is known before the estimation. In future work, a more advanced method without the knowledge of the signal number will be researched.

References

1. Schimidt, R.O.: Multiple Emitter Location and Signal Parameter Estimation. IEEE Trans. on AP 34(3), 276–280 (1986)
2. Zhang, J., Liao, G.-S.: DOA Estimation Based on Extended Noise Subspace in The Presence of Strong Signals. Systems Engineering and Electronics 31(6), 1279–1283 (2009)
3. Chen, H., Su, H.-J.: A New Method to Estimate DOA in Presence of Strong Jamming /Signal Suppression. Acta Electronica Sinica 34(3), 530–534 (2006)
4. Wang, S., He, Z.-S.: A New Method to Estimate Weak Signal DOA in Presence of Wideband Strong Jamming Suppression. Modern Radar 28(9), 69–71 (2006)
5. Wang, B.-H., Wang, Y.-L., Chen, H.: Weighted Spatial Smoothing for direction-of-arrival estimation of coherent signals. In: Proceedings of IEEE Antennas and Propagation Society International Symposium (APSIS), San Antonio, Texas, USA, vol. (2), pp. 668–671. IEEE (2002)
6. Di, A.: Multiple Sources Location-a Matrix Decomposition Method. IEEE Trans ASSP 33(4), 1086–1091 (1985)
7. Gao, S.-W., Bao, Z.: Family of Data-based Matrix Decomposition for High-resolution Array Processing of Coherent Signals. Journal of China Institute of Communications 9(1), 4–13 (1988)
8. Xu, L., Zeng, C.: DOA and Source Number Estimation Method for Strong and Weak Signals Based on Eigen Beamforming. Journal of Electronics & Information Technology 33(2), 321–325 (2011)
9. Hu, X.-Q., Chen, H.: A Toeplitz Decorrelation Method Using the Maximum Eigenvector. Acta Electronica Sinica 36(9), 1710–1714 (2008)

Study of Node Localization Algorithm Based on Improved Particle Swarm Optimization and RSSI for WSNs

Shen Ming-yu, Lu Ya-jing, and Zhao Ming-shun*

School of Computer & Information Hefei University of Technology, Hefei 230000, China

Abstract. Based on RSSI ranging and trilateration positioning, an improved particle swarm optimization combined with RSSI self-correcting localization algorithm for Wireless Sensor Networks was proposed. Simulation results showed that the accuracy of this algorithm increased by almost 30% than trilateration, while improving the particle swarm optimization algorithm which may hardly converge to the optimal solution as falling into the local extreme area . This algorithm obtained better solution than node localization based on particle swarm optimization for WSNs, and then enhance the stability and convergence of the algorithm.

Keywords: Wireless Sensor Network, RSSI ranging, Particle Swarm Optimization, localization accuracy.

1 Introduction

With the rapid development of microelectronic, sensor, and wireless communication technology, a new type of network technology is developed——Wireless Sensor Network[1,2], which is based on node localization and has the functions of data collection, processing, communication and management. Node localization is the key technology in wireless sensor networks. It can not only monitor the location, but keep watching on moving and tracking of the targets, assist network management, and predict the target trajectory. Based on the need of measuring the distance, the node localization algorithm in WSNs could be divided into two categories: range-free and range-based localization [3]. To some extent, the range-free location simplifies the steps of localization, but generally has low accuracy unless there is higher density of node deployment. So it is not practical.

Distance ranging methods of Range-based localization include Time of Arrival (TOA), Time Difference of Arrival (TDOA), Angle of Arrival (AOA) and Received Signal Strength Indicator (RSSI), etc. Compared with the previous locations, RSSI measurements can get the distance though the wireless channel model without additional hardware. It acheives easily and has low cost relatively and does not require time synchronization strictly. Based on the RSSI measurements, this dissertation introduces the algorithm, which is combined with RSSI self-correction mechanism and an improved particle swarm optimization, to optimize the nodes localization for WSNs.

---

* Hefei University of Technology Ph.D. Special Fund (GDBJ2009-005).

D. Yang (Ed.): Informatics in Control, Automation and Robotics, Volume 2, LNEE 133, pp. 185–193.
springerlink.com

2 Related Work

1) RSSI Measurement

RSSI indicates the electromagnetic energy in current medium, the value is decreasing with the increasing distance. Because of reduced signal strength in the communication process, the distance between transmitter and receiver can be estimated according to the signal strength of the receiver. Most of the RSSI measurements use a signal propagation model that maps RSSI values to distance estimates. The most popularly used signal propagation model is the log-normal shadowing model [4, 5]:

$$RSSI(d) = P_T - PL(d_0) - 10n\log(\frac{d}{d_0}) + X_\sigma \tag{1}$$

Here, P_T is the transmit power, $PL(d_0)$ is the path loss for a reference distance d_0, n is the path loss exponent and environment dependant. X_σ is a Gaussian random variable with zero mean and σ 2variance that models the random variation of the RSSI value.

2) Trilateration Measurement

Trilateration measurement is a common method of plane localization.The principle is as follows: the unknown node can be determined if the distance between the unknown node and three anchor nodes is known. For example, the coordinate values of the three anchor nodes A_1, A_2, A_3 are (x_1, y_1), (x_2, y_2), (x_3, y_3), the distances to the node X are $\lambda_1, \lambda_2, \lambda_3$, so the coordinate value of X that can be computed form formula (2) is the formula (3):

$$\begin{cases} (x-x_1)^2 + (y-y_1)^2 = \lambda_1^2 \\ (x-x_2)^2 + (y-y_2)^2 = \lambda_2^2 \\ (x-x_3)^2 + (y-y_3)^2 = \lambda_3^2 \end{cases} \tag{2}$$

$$\begin{bmatrix} x \\ y \end{bmatrix} = \begin{bmatrix} 2(x1-x3)2(y1-y3) \\ 2(x2-x3)2(y2-y3) \end{bmatrix}^{-1} \begin{bmatrix} x_1^2 - x_3^2 + y_1^2 - y_3^2 - \lambda_1^2 + \lambda_3^2 \\ x_2^2 - x_3^2 + y_2^2 - y_3^2 - \lambda_2^2 + \lambda_3^2 \end{bmatrix} \tag{3}$$

3) RSSI Localization Algorithm Based on Self-correction

In the process of RSSI-based algorithm for trilateral localization, the selection of anchor nodes has different effects on node localization to some extent. This paper selects anchor nodes based on RSSI, and adopts self-correcting mechanism of anchor nodes.The process is as follows:

STEP 1: Start the localization mechanism, the anchor nodes radio data packet including their own ID and coordinate information, then the unknown nodes within the communication area will receive this packet. And firstly, compared with the existed ID numbers, if the ID is new, the unknown nodes calculate the RSSI and save it. If not, the new RSSI of anchor node compares with the existed RSSI, if the former is bigger than the latter, takes the place of the existed one, otherwise drops it.

STEP 2: The unknown node orders the received RSSI of the anchor nodes in descending order, takes the former three to participate in trilateral localization, and obtains the coordinate value of the unknown node, (x, y).

STEP 3: The anchor nodes $A_1, A_2, A_3 ...... A_n$ are positioned as the unknown nodes. For instance, on the basis of the above algorithm, the coordinate value of anchor node $A_i (i = 1, 2, 3, ..., n)$ is $(\hat{x_i}, \hat{y_i})$. As the actual coordinate is already known, so the location error of each node can be calculated, and then the average error is obtained:

$$\begin{cases} \overline{ERROR_x} = \frac{1}{n}\sum_{i=1}^{n}(x_i - \hat{x_i}) \\ \overline{ERROR_y} = \frac{1}{n}\sum_{i=1}^{n}(y_i - \hat{y_i}) \end{cases} \tag{4}$$

Combined with the coordinate value of the unknown nodes form formula (3) and (4), the final coordinate value is:

$$\begin{cases} \hat{x} = x + \overline{ERROR_x} \\ \hat{y} = y + \overline{ERROR_y} \end{cases} \tag{5}$$

To some extent, the self-correcting mechanism improves the localization accuracy, but it still has error which is defined as (6):

$$Error_i(\hat{x}, \hat{y}) = \sqrt{(x_i - \hat{x})^2 + (y_i - \hat{y})^2} - d_i \tag{6}$$

Here d_i is the measured distance between the node i and the unknown node, (x_i, y_i) and $(\hat{x}, \hat{y})$ are the actual and test coordinate value of the unknown node. To improve the localization accuracy, the localization problem turns into the optimization problem.

4) Particle Swarm Optimization Algorithm

Particle Swarm Optimization (PSO) [6] is a swarm bionic optimization algorithm, which models the behavior of flocks of birds and fish. Its process does not depend upon the quality of the objective function, and then converge to the most optimal solution in a larger probability. So it is commonly used to solve the optimization problems.

The basic PSO algorithm can be described as follows:

Each individual called "Particle" in the group represents a feasible solution of the problem. In D-dimensional searching space, the coordinate value of the t generation of the particle i is $X_i^t = (x_{i1}^t, x_{i2}^t, \cdots, x_{iD}^t)$ $i = 1, 2, \cdots, N$ and the flight speed is $V_i^t = (v_{i1}^t, v_{i2}^t, \cdots, v_{iD}^t)$. The objective function value corresponding to the particles' location is regarded as the fitness value which is used to evaluate the merits of particle. In the process of iteration, each particle dynamically tracks the two extreme values to update its position and velocity. One extreme value called individual extreme value is the current best optimal solution of the particle itself, the location is $P_i^t = (p_{i1}^t, p_{i2}^t, \cdots, p_{iD}^t)$, the other is the best solution in the particle swarm, called global extreme value: $P_g^t=(p_{g1}^t, p_{g2}^t, \cdots, p_{gD}^t)$. The particles update their own locations and velocities in the light of the following formulas.

$$V_i^{t+1}=V_i^t+c_1r_1(P_i^t-X_i^t)+c_2r_2(P_g^t-X_i^t) \tag{7}$$

$$X_i^{t+1}=X_i^t+V_i^{t+1} \tag{8}$$

Where t is the number of iterations, c_1 and c_2 are known as the acceleration constants, usually $c_1=c_2$=2; r_1 and r_2 are random numbers uniformly distributed in [0, 1]. In the right of formula (7), V_i^t is the previous velocity of the particle, $c_1r_1(P_i^t-X_i^t)$ and $c_2r_2(P_g^t-X_i^t)$ indicate the particle's updating. The algorithm runs after initializing a swarm of particles randomly, and the particles update themselves through formula (7) and (8) in each iteration until it meets the condition of stopping iteration [7-8].

However, the above algorithm has shortcomings of slow convergence and poor accuracy. In order to improve it, Shi and Eberhart[9] have introduced the concept of inertia weight, and proposed a strategy of linearly decreasing weight (LDW), the updating formula (9) is:

$$V_i^{t+1}=wV_i^t+c_1r_1(P_i^t-X_i^t)+c_2r_2(P_g^t-X_i^t) \tag{9}$$

$$w=w_{\max}-\frac{(w_{\max}-w_{\min})\times iter}{iter_{\max}} \tag{10}$$

Here, w is the inertia weight that controls the exploration of local and global search space, $w_{\max}$ is the maximum inertia weight, $w_{\min}$ is the smallest inertial weight, $iter$ is the current iteration number, $iter_{\max}$ is the maximum iteration number, the inertia weight w will linearly decrease with the increasing iteration. Large number of experiments show that the algorithm, which is called Linearly Decreasing Weight Particle Swarm Optimization [10, 11], can obtain best test results when w is decreasing from 0.9 to 0.4. When algorithm evolves into the latter process, the particle is easy to fall into the local minimum area, and it is difficult to converge to the better solutions. To optimize this problem, an improved particle swarm optimization is proposed in this dissertation.

5) Improved Particle Swarm Optimization

Improved Particle Swarm Optimization (IPSO) is mainly to initialize particles which fell into the local minimum area by combining with their own information to prevent particles falling into the local minimum area. The algorithm can improve the degree of convergence and stability on the basis of obtaining optimal solution.The main process is as follows:

STEP 1: The fitness values of particles in the t generation of the swarm are $f_1^t, f_2^t, \cdots f_N^t$, the average fitness value of current generation is as follows:

$$f_{avg}^t = \sum_{i=1}^{N} f_i^t / N \tag{11}$$

The absolute deviation value between particle i and f_{avg}^t is as follows:

$$\delta_i^t = \left| f_i^t - f_{avg}^t \right| \tag{12}$$

The test value is as follows:

$$\lambda_i^t = f_{avg}^t / N \tag{13}$$

Where N is the number of the particles.

STEP 2: If λ_i^t is bigger than δ_i^t, the particle falls into the local minimum area. To maintain the stability of the algorithm, the particle which fell into local minimum will be processed.The operation is as follows:

$$X_i^t = \frac{\delta_i^t}{f_i}(Xup - Xdown) + Xdown \times (-\delta_i^t) \tag{14}$$

Here, Xup and $Xdown$ are the upper and lower boundary of the solution in the searching area. The algorithm uses step2 to determine whether the particle fell into the local minimum area, and then re-initializes the particle. Thus, the problem of local extreme area is solved and the algorithm can converge to better solution.

3 Localization Algorithm Based on IPSO for Wireless Sensor Networks

1) Computation of Fitness Value

PSO algorithm determines the individual and global extreme according to the fitness value of the particle, the smaller fitness value represents the higher fitness value. The following formula is used to calculate the fitness value:

$$fit_i = d_i Error_i(\hat{x}, \hat{y})^2 \tag{15}$$

Here d_i is the distance between the unknown node and the i anchor node, to some extent, d_i is inversely proportional to the received RSSI. The algorithm adopts d_i to supplement the minimized errorinsufficiently, and to calculate the fitness value by using the formula (15).

2) Algorithm Process

STEP 1: Generate a random population of N particles, a random initial set of velocities, P_g^0 and P_i^0.

STEP 2: For each particle of the population, compute the fitness value fit_i from (15).

STEP 3: Repeat step 2 for all the N particles in the population.

STEP 4: Compare each particle's fitness value with its P_i^0. If it is better, then its position and value are updated. The best fitness value among P_i^0 is stored as the P_g^0.

STEP 5: Compute the weight, velocity and position of the particle form (10), (9),(8). If the current location of the particle exceeds the upper bound, the location is the new upper bound; On the contrary, it is the lower bound.

STEP 6: Compute the f_{avg}^t, δ_i^t, λ_i^t of the particle according to (11), (12), (13), and determine whether the particle falls into the local minimum area. If it is, use formula (14) to initialize the particle randomly, and proceed to step 5; if not, proceed to step 7.

STEP7: Substitute the particle's coordinate (x_i, y_i) that correspond to the current P_g^0 into formula $\left|\sqrt{(x-x_i)^2+(y-y_i)^2}-d_i\right|$, if $\left|\sqrt{(x-x_i)^2+(y-y_i)^2}-d_i\right|<\mu$ or $iter > iter_{max}$, proceed to step 8, or $iter$ is plus one, and then repeat step 2.

STEP 8: The result P_g^0 gives the desired and estimated location of the deployed sensor.

4 Simulation and Analysis

1) Designing Simulation

To test and verify the feasibility and effectiveness of the algorithm, the simulation uses the Matlab combined with C++ to evaluate and compare the performance of our new scheme to other localization schemes including trilateration and PSO-based localization. Simulation runs are conducted mainly in a 100m×100m wireless sensor environment, including 50 unknown nodes and different number of anchors that depends on different condition. Table 1 lists typical values of simulation parameters. The evaluation criteria of the algorithm is $AVE=\frac{1}{N}\sum_{i=1}^{N}\sqrt{(x_i-\hat{x}_i)^2+(y_i-\hat{y}_i)^2}$

Table 1. Typical value of simulation parameters

Communication radius	$10m$
Attenuation index n	3.4
Standard deviation σ	4
Reference distance d_0	1m
Inertia weight w	$w_{max}=0.9$
	$w_{min}=0.4$
Maximum iteration	$iter_{max}=100$
The particle scale N	$N=50$
Maximum updating velocity	$V_{max}=kX_{max}(0.1<k<1)$

Here, $(\hat{x_i}, \hat{y_i})$ is the test location of unknown node, (x_i, y_i) is the actual position, N is the number of the unknown nodes.

To verify the searching performance of IPSO algorithm, four test functions are chosen to test PSO and IPSO algorithm. The functions are as follows:

Table 2. Test functions

Functions	Boundary
$f_1(x)=\sum_{i=1}^{n} x_i^2$	[-100,100]
$f_2(x)=\sum_{i=1}^{n}\left[100(x_{i+1}-x_i^2)^2+(x_i-1)^2\right]$	[-100,100]
$f_3(x)=\sum_{i=1}^{n}\left[x_i^2-10\cos(2\pi x_i)+10\right]$	[-5.12,5.12]
$f_4(x)=(x_2-\frac{5.1}{4\pi}x_1^2+\frac{5}{\pi}x_1-6)^2+10(1-\frac{1}{8\pi})\cos x_1+10$	$x_1\in[-5,10]$ $x_2\in[0,15]$

2) The Simulation Result

Because the distance measured from RSSI is different in vary condition, and could get different ranging error. So one condition of the simulation is based on different ranging error, the experiment compares localization based on IPSO to trilateration measurement and PSO. The result is as fig.1, the average location errors of the three localizations are increasing with ranging error becoming bigger in general. When the ranging error is 50%, trilateration localization produces average location errors top to 10m, the PSOs' is almost 8m, but the average errors of IPSO localization is lower than 6m. And the average location error of IPSO algorithm is increasing slower with the increasing of ranging error than the former two. Our scheme actually generates the smallest location errors by avoiding the sensors to fall into the local minimum area.

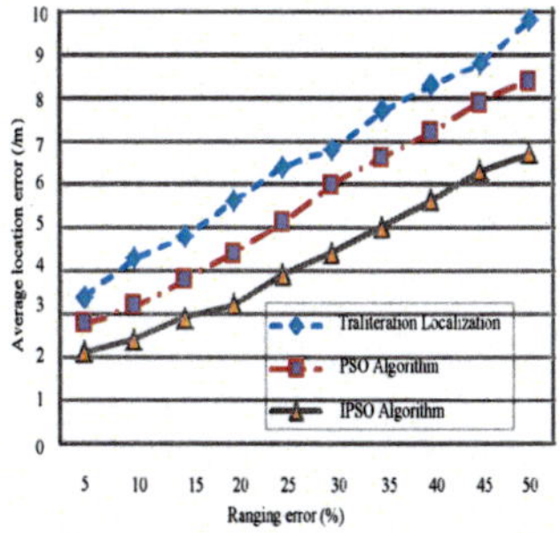

Fig. 1. Localization errors vs. different ranging errors

Fig.2 shows number of anchor nodes vs. location errors, the location accuracy is proportional to the number of anchor nodes, but the number is up to certain degree and the algorithm become relative stable. PSO algorithm has significant improvement than the trilateration localization, but IPSO has smaller average location error on the basis of PSO algorithm in the same condition, optimizes the accuracy of nodes location.

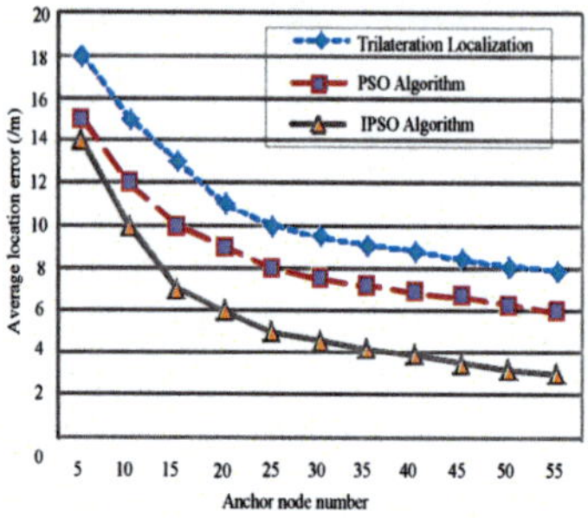

Fig. 2. Number of anchor nodes vs. location errors

Through the test result of the functions as showed in table 3, the average optimal value of IPSO algorithm is improved almost 40% than PSO, and as the fig.3 displaying, the PSO algorithm iterates fifty times to get the optimal solution, but our algorithm just needs thirty times. Because IPSO algorithm not only improves the global searching capability, but optimizes the problem that the particle could not achieve the optimal solution easily.

Table 3. Test results of the functions

Functions	Average optimal value	
	PSO algorithm	IPSO algorithm
f_1	1.3972	0.9345
f_2	0.9217	0.4259
f_3	3.5283	2.0847
f_4	3.7149	2.2635

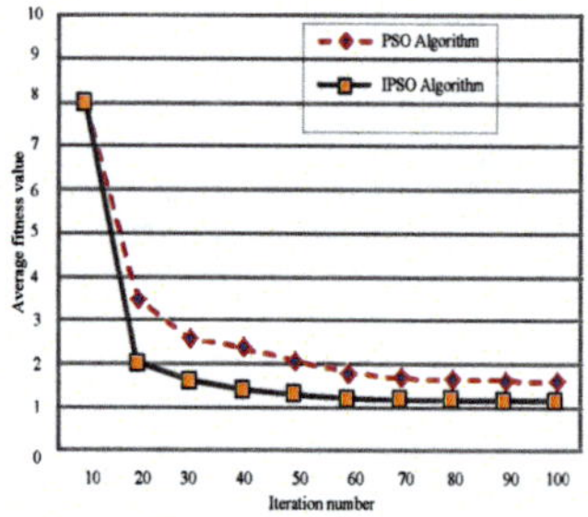

Fig. 3. Iterations number vs. average fitness value

5 Conclusion

This dissertation presents node localization algorithm based on improved particle swarm optimization algorithm for Wireless Sensor Networks. In this algorithm, the RSSI self-correcting mechanism is introduced to improve the accuracy of ranging, and the IPSO algorithm is proposed to optimize the problem of local extreme value. Our localization scheme could obtain better and stable overall performance. But this is just the theoretical study for particle swarm optimization applied to nodes localization for WSNs, the next step is to prove the algorithm and to study how to save the energy.

References

1. Pottie, G.J.: Wireless Sensor Networks. In: IEEE Information Theory Workshop, pp. 139–140 (1998)
2. Culler, D.: Wireless Sensor Networks-the ITRevolution., `http://www.cs.berkeley.edu/culer/talks/KES2004.ppt`
3. Li, C., Ju, H., Yong, M.: Study on the proceeding of Wireless Sensor Networks. Journal of Computer Research and Development 42(1), 163–174 (2005)
4. Liu, C., Wu, K., He, T.: Sensor Localization with Ring Overlapping Based on comparison of Received Signal Strength Indicator. In: 2004 IEEE International Conference on Mobile AD-hoc and Sensor System (2004)
5. Fang, Z., Zhao, Z., Geng, D., Xuan, Y., Du, L., Cui, X.: RSSI Variability Characterization and Calibration Method in Wireless Sensor Network. In: 2010 IEEE International Conference and Automation (2010)
6. Kulkarni, R.V., Venayagamoorthy, G.K.: Particle Swarm Optimization in Wireless Sensor Networks: A Brief Survey. IEEE Trans. On System, Man, and Cybernetics-Part C: Applications and Reviews (2010)
7. Yang, W., Li, Q.: A Brief of Particle Swarm Optimization. Engineering Sciences 6(5), 87–94 (2004)
8. Dortgo, M., Maniezzo, V.: Ant System: Optimization by A Colony Cooperating Agents. IEEE Trans. on System, Man, and Cybernetics 26(1), 29–41 (2006)
9. Kennedy, J., Eberhart, R.C.: Particle swarm optimization. In: Proceedings of IEEE International Conference on Neural Networks
10. Shi, Y., Eberhart, R.C.: A Modified Patricle Swarm Optimizer. In: Proceeding of the IEEE Congress on Evolutionary Computation, pp. 69–73 (1998)
11. Jiao, B., Lian, Z.G., Gu, X.S.: A dynamic inertia weight particle swarm optimization algorithm. Chaos, Solitions and Fractals 37, 698–705 (2008)

Research on a Memory-Weighted Learning Model for Fictitious Play

Chun-yan Yu, Xiao-sheng He, Yu-qi Lin, and Feng-qin Chen

College of Mathematic and Computer Science, Fuzhou University,
Fuzhou 350108, Fujian, China

Abstract. Each adaptive agent always has bounded rationality and incomplete knowledge in complex adaptive system. And, agents with self-learning ability can modify their behaviors constantly to evolve adaptively according to the feedback of interaction. To address this issue, a memory model was introduced into fictitious play, then a memory-weighted learning model was proposed used for directing the strategy selection of adaptive agents, and finally the learning convergence in different types of games was observed based on swarm platform. The experience results show that although the limits of forgetting and local interaction exist, the agents could adjust strategy according to the information observed continuously, and lead to the realization of adaptive learning.

Keywords: Fictitious Play, Memory, Complex Adaptive System, Adaptive Agent.

1 Introduction

A Complex Adaptive System (CAS) is constituted by numerous adaptive agents which interact and affect each other [1]. In CAS, each agent can't sufficiently aware of its surroundings, but can only acquire part of the information around it, thus has strong locality. And when an agent interacts with the surroundings or other agents, it will continuously adjust its behavior according to the scattered observed information, with strong adaptability and autonomy.

Fictitious play and its variant are widely used faith learning models. In fictitious play, each participant is firmly convinced that its opponents' strategies obey a constant and unknown distribution, then games with other players around according to this selection strategy [2]. Similar to the basic ideas of CAS, each participant of fictitious play continuously adjust its faith according to the observation of opponents' behavior. Meanwhile, participant's current strategy will influence opponents' current state and faith, then influence opponents' strategy selection of next stage. This kind of interaction that influences and restricts each other, forms a kind of feedback, then plentiful operation of this local feedback will emerge from the bottom up to a global dynamic, and consequently implement participators' adaptive learning of the model.

Virtual action process has been demonstrated to achieve Nash equilibrium quickly, and it is often used to explain how participants realize cognitive evolution based on

D. Yang (Ed.): Informatics in Control, Automation and Robotics, Volume 2, LNEE 133, pp. 195–202.
springerlink.com

the actual behavior. Traditional virtual action process implicitly supposes that the participants think that the environment is constant, and asks that participants have complete memory at the same time. However, the participants' memories often concentrate on just a few of fragmentary and incomplete information. And compared with the earlier memory, participants often give higher weight to the latest observed information [3]. According to the characteristics of the CAS and the limits above, this paper introduces memory model into traditional virtual action process, and proposes a kind of virtual action memory-weighted learning model, then conducts a simulation research on the participant's behavior under the swarm platform, and combining the characteristics of the adaptive agents observes the influence of incomplete information on behavior decision-making from the memory model perspective.

2 Fictitious Play

During fictitious play process, participants always remember relative frequency of each strategy that the opponent adopted. At each stage, participants compute the expected payoff of all strategies taken by themselves according to the opponents' strategy distribution, and choose a strategy with the highest expected payoff [2-4].

Traditional fictitious play gives same weight to all the information observed by the participants, while ignores the fact that the information observed too early may become "obsolete" for participants, so that may be ignored totally[5]. An improvement is that assuming participants to discount the importance of the past information, that is, compared with the prior information, to give recent observations a higher weight, thus creating a mixed form called as "Weighted fictitious play" [3]. A typical case is the three-parameter model proposed by Cheung and Friedman [5].

Another shortage of traditional fictitious play process is that it demands the participants to track the entire history when selecting the strategy every time [3-5], it is unrealistic in many models. A variant of this is a random sampling model, that is, the participants independently and randomly sample from the information observed in the past, and choose the optimum reaction according to the selected samples. According to whether the participants have sampled from the whole previous information or not, it can form two models: partial sampling model with unbounded memory and partial sampling model with limited memory [7]. Kaniovski and Young use stochastic approximation theory to analyze and study the above two models separately, and verify their convergence [6, 8].

However, the weighted fictitious play still requires the participants owning all the information and being fully rational. Although the partial sampling model with unbounded memory is added with part of sampling disturbance, the participants still retain all historical information that may be sampled at any time. The partial sampling model with bounded memory just keeps the recent historical information simply, while ignoring the characteristics of the phenomenon of memory and oblivion. The process of fictitious play is the process that the participants accumulate the experience, while the process of the experience accumulation always based on memory [9]. In order to better simulate the behavior evolution of the participants in the complex adaptive system, we have the necessary to consider the memory's impact on learning according to the characteristics of adaptive agents. Withal, this paper introduces memory model and local limits into the fictitious play process and observes the learning situation of the participants based on the swarm simulation platform.

3 Behavior Memory

Memory, which comes from cognitive psychology, is considered to be inseparable with learning: Learning is a memory-based experience and rules-accumulation process, and it performs on the changes of the agents' behavior [9]; Memory's retaining state directly affects the participant's behavior-selection and learning situation. Concerning the memory-retaining characteristics, the Forgotten Curve concluded by Ebbinghaus is the most famous research in the history of experimental psychology [10]. The curve in figure 1 shows that forgetting speeds in the beginning, then gradually changes slowly.

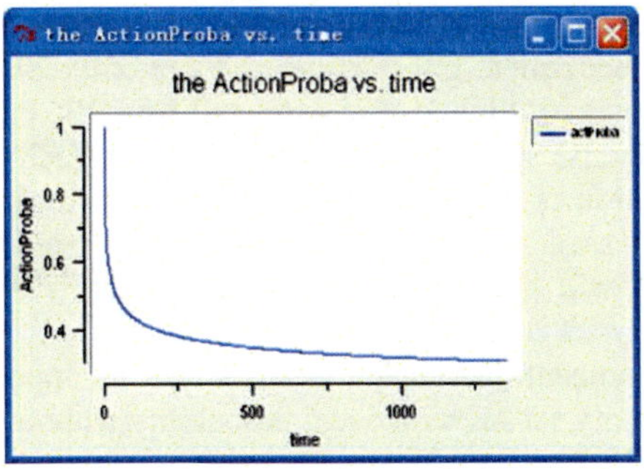

Fig. 1. Ebbinghaus forgotten curve, shows relationship between memory stock and time

The curve can be represented by the formula (1) approximately, where, b means a stock, t means a time interval, k and c are constants, whose approximate estimated values in psychology are k = 1.84, c = 1.25.

$$b = \frac{k}{k + (\log t)^c}, t \in N_+ \quad (1)$$

From the formula (1), we can get the retentive probability of some information with the τ time interval.

$$p = \begin{cases} 1 & if\ \tau = 1 \\ \dfrac{k + (\log(\tau - 1))^c}{k + (\log \tau)^c} & if\ \tau \neq 1\ and\ \tau \in N_+ \end{cases} \quad (2)$$

From the formula (1) and (2), it can be found that, as time goes on, the memory concentrating information will continuously grow.

In addition, we can consider the influence of other phenomena in cognitive psychology upon the behavior memory, such as stimulating strength, memories, and recognition's influence on memory retaining, and the participant's "depressed" phenomenon in recalling, and so on [11]. For simplicity's sake, this paper only considers the influence of the forgetting phenomenon upon the participants' behavior selection: each player has an unbounded memory, but the event of every stage may be forgotten as time goes by, and the rate of forgetting is shown as formula (2). According to this characteristic, this paper puts forward a kind of Memory-Weighted Learning Model (MWLM), and observes the participants' learning convergence when there exist forgetting phenomenon.

4 MWLM Model

To simulate the agents' learning in complex flexibility system, we introduce the feature of adaptive agents in the process of fictitious play. Different from other fictitious play, the key feature of every participant in MWLM model is that: each participant only has incomplete information, and can only conduct an expectation of other agents' strategies with limited data extracted from the memory, and then make restricted rational response upon it.In the MWLM model, each participant will encode the observed information of various stages as memory, and store in own memory set M. However, these messages are not stored permanently, from the beginning of every moment, participants will get into oblivion, and the oblivion probability of every element in the memory set is shown as formula (2).

Suppose that at t moment, participant i games with its opponents. First, drawing k samples from its memory set Mi with equal probability without replacement to form a sample set C, $C \subseteq M$. If M is insufficient for sampling, that is, when K is greater than the size of the memory set, the sampling set will be equal to the memory set. Participants according to the extracted sample compute weighted function k_t^i of opponents' strategy space S_i^{-i} : $S_i^{-i} \to R_+$. Taking into consideration that the records of different time have different importance upon current strategy, we assign a weight for each record according to the record's time interval. Defining recording weight of the τ time interval record to $\phi^{\tau-1}$, where $0 \le \phi \le 1$, the calculation of the weight $k_t^i(s^{-i})$ of the opponents' strategy s^{-i} is shown as formula (3):

$$k_t^i(s^{-i}) = \sum_{c \in C} I(c.s^{-i}, s^{-i})\phi^{c.\tau-1} \quad (3)$$

Where $I(c.s^{-i}, s^{-i}) = \begin{cases} 1 & if\ c.s^{-i} = s^{-i} \\ 0 & others \end{cases}$.

We can predict the opponents' distribution strategy γ_t^i at the time t using k_t^i . At the time t, participant i deems the probability of the strategy s^{-i} adopted by its opponent as follows:

$$\gamma_t^i(s^{-i}) = \begin{cases} \dfrac{k_t^i(s^{-i})}{\sum_{\tilde{s}^{-i}\in S_i^{-i}} k_t^i(\tilde{s}^{-i})} & if\ s^{-i}\in S_i^{-i} \\ 0 \quad if\ s^{-i}\notin S_i^{-i} \end{cases} \tag{4}$$

According to the information of the calculated strategy distribution γ_t^i and the payment matrix, we can calculate that the participant i takes different strategy s^i at the time t as follows:

$$\bar{u}_t^i(s^i) = \sum_{s^{-i}\in S_i^{-i}} \gamma_t^i(s^{-i}) u_t^i(s^i, s^{-i}) \tag{5}$$

Participant checks all optional strategies and estimates the expecting payment of every strategy. For the strategy $s^i \in S^i$, if $\bar{u}_t^i(s^i)$ is higher than the expecting payment ρ_t^i of the participant i at the time t, then s^i will be added into the candidate strategy set D_t^i. If $D_t^i \neq \varnothing$, participant i will select the strategy randomly from it. The probability of the strategy s^i be selected is as follows:

$$\lambda_t^i(s^i) = \begin{cases} \dfrac{\bar{u}_t^i(s^i)}{\sum_{\tilde{s}^i\in D_t^i} \bar{u}_t^i(\tilde{s}^i)} & if\ s^i\in D_t^i \\ 0 \quad if\ s^i\notin D_t^i \end{cases} \tag{6}$$

If $D_t^i = \varnothing$, that is when expecting payments of all the strategies of the participant i at the t time are lower than the expected payoff, participant i carries the optimal reaction, choosing the strategy with the highest expecting payment.

Participant i chooses its own strategy according to the rules above to interact with its opponent. After the game, participant i observes opponent's strategy at that time and its own payoff u_t^i, and store them into the memory. Finally, dynamically updating the expecting payment of the participant i as follows:

$$\rho_{t+1}^i = \begin{cases} (1-w_+^i)\rho_t^i + w_+^i u_t^i & if\ u_t^i \geq \rho_t^i \\ (1-w_-^i)\rho_t^i + w_-^i u_t^i & if\ u_t^i < \rho_t^i \end{cases} \tag{7}$$

Where w_+^i (w_-^i) separately represents the effecting weight of positive (negative) payment discrepancy on the expecting payment of participant i. The greater w_+^i (w_-^i) is, it means, the participant i reacts more sensitive to the expecting payment when it obtains positive (negative) payment discrepancy. In addition, when $w_+^i > w_-^i$, it means, participant i is an optimistic one ; when $w_+^i < w_-^i$, participant i is a pessimistic one.

5 Game Experiment Model

A game model constitutes of a large number of participants as well as their surroundings, and each participant exists as others surroundings. Each participant will be considered as an adaptive agent, thus Game model can be represented as follows: Mod=<Env, A>, here, Env is external surroundings and A is agents' set.

The external environment of a model is the external environment where all the principals exist. In our daily life, interaction of the participants will be often affected by geographical, cultural, professional, and many other factors. A grid of $n \times n$ can be used to simulate spatial environment of the real world. We set mountains, rivers and other obstacles or roads to restrict movement and perspective of the agents, influence the agents' interactive topology [12], and thus influence the matching selection of the agents' game. For simplicity's sake, this paper supposes every agent has a limited horizon, and can only interact with other agents in its horizon.

For each agent $a \in A$, it can be expressed by an eight-tuples: a=<Pos, S, u, M, φ, w+, w-, vis>. Where Pos represents the current location of a, and S represents the limited strategy space of a; u: $S^a \times S^{-a} \rightarrow R+\eta^a$ is the payoff function of a, where η^a is an absolute continuous random vector about Lebesgue Measure on the finite interval, where different agents have different random payment disturbance[13]; M represents the memory set of participant a; and φ is discount factor of a about historical information, the greater φ is, it means the weight given to the historical information by a is greater; w+ and w- represent the expecting payoff's updating weight of the participant a, the greater w+ and w- are, the expecting payoff of a is closer to the payment gotten at the latest moment; vis represents visual field of the participant a, which can only game with other agents among the field.

For the test of MWLM model, this paper adopts different payment matrixes under various different conditions, as shown in table 1(the numbers in brackets represents the equilibrium of this type of the game), then observes the influence of different information conditions upon the participant's learning behavior.

Table 1. Various game types of payoff matrix.

Game types	payoff matrix	
Prisoners'	0.7, 0.7	0, 1
game(0)	1, 0	0.2, 0.2
Hawk-Dove	0, 0	1, 0.4
game 1	0.4, 1	0.8, 0.8
(0.333)		
Hawk-Dove	0, 0	1, 0.2
game 2	0.2, 1	0.5, 0.5
(0.714)		
Stag hunt	1, 1	0, 0.3
(1)	0.3, 0	0.3, 0.3
Stag hunt 2	1, 1	0, 0.5
(1、0或0.5)	0.5, 0	0.5, 0.5

The experiment model is a grid of 50 * 50, where randomly placed 50 participants. Each player randomly moves, and in its horizon randomly selects another participant for matching. Every turn, each participant matches only once at the most. If finds its opponent to game with, then use MWLM model to choose its own strategy. When interaction ends, each player observes opponent's strategy at this moment, its own strategy and obtained pay, then deposits it into the memory set. Figure 2 represents various game types of average marginal experience distribution in MWLM model.

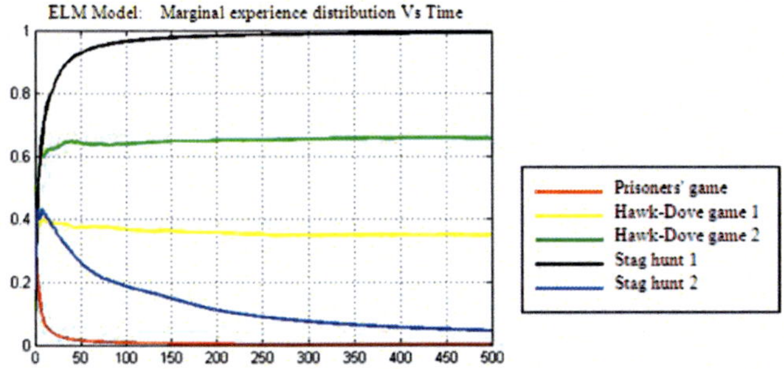

Fig. 2. Marginal experience distribution of various games in MWLM model

In prisoner game, due to the relative low expecting payoff of participants and the non-optimal reactions in early time, a certain probabilistic cooperation is needed. As participants' strategies of cooperation can affect their opponents' beliefs, then the opponents have the possibility of cooperation in the rest rounds. But along with the elevation of the expecting payoff and the refinement of the participants' information, the probability of the participants' cooperation falls rapidly and finally approaches to the equilibrium point of (betray, betray). In eagle-pigeon game, the distribution of participants' strategies doesn't rigidly converge to the Nash equilibrium. When facing the distribution of the opponents' strategies with mixed strategies equilibrium, the participants will select two strategies with equal probability; while the equal probability selection will also make a deviation of the mixed strategies equilibrium. In the first situation of the deer-hunting game, the participant's strategy distribution quickly converges to the Nash equilibrium. However in the second case, the participant's strategy distribution converges to the (rabbit rabbit) combination without any exception. We can think that when the expecting payment of the participant is less than 0.5, it has certain probability to choose the rabbit, while choosing rabbit will reduce the expecting payment of the opponent using deer hunting strategy, making the opponents have greater probability to select hunting rabbits in the next stage, and therefore cause the whole group converge to the combination of (rabbit rabbit).

From the above experimental results, we can see that because the participant has less available information, every memory will have important influence upon the strategies selection of the participant. But along with the increasing of the memory set, participant has more and more complete information; the influence of the noise in the MWLM model on the participant's strategies selection gradually disappears, and still can quickly converge to equilibrium.

6 Conclusion

This paper, from fictitious action process of game learning model, considered the participants' learning situation when oblivion phenomenon exists, meanwhile added partial limits into fictitious play process based on the adaptive agents' characteristics, and proposed a fictitious play weighted-memory learning model. Based on the Swarm platform observed that through a lot of individual participants' adaptive activities, it emerged the whole system's regularity. The simulation results showed that when the participants has less available information in the earlier stage, the oblivion phenomenon and the local limits will cause an obvious influence on participants' behavior choice; But as the memory set got complete, the effect of noise gradually disappeared, and gradually approached to an equilibrium point. The introduction of the memory model and the adaptive agents' characteristics can be better to simulate the participants' learning behavior, and have stronger realistic significance.

Acknowledgments. This work is sponsored by the National Natural Science Foundation of China under Grant No.60805042, the Natural Science Foundation of Fujian Province under Grant No.2010J01329, the Industry-University-Research Major Project of Fujian Province under Grant No.2010H6012 and the Program for New Century Talents of Fujian Province under Grant No.XSJRC2007-04.

References

1. Holland, J.H., Zhou, X.-M., Han, H.: Hidden Order: Adaptive make complexity, pp. 5–10. Shanghai Scientific and Technological Educational Publishing House, Shanghai (2000)
2. Fudenberg, D., Levine, D.K.: Learning in Games. European Economic Review, 631–639 (1998)
3. Moller, C.K., He, J.-T.: Behavior game: An experimental research on strategy interaction, pp. 267–315. China Renmin University Press, Beijing (2006)
4. Drew, F., Levin, D.K., Xiao, Z.-Y., Hou, C.-Q.: Game learning theory, pp. 33–59. China Renmin University Press, Beijing (2004)
5. Cheung, Y.W., Friedman, D.: Individual Learning in Normal Form Games: Some Laboratory Results. Games and Economic Behavior, 46–76 (1997)
6. Kaniovski, Y.M., Young, H.P.: Learning Dynamics in Games with Stochastic Perturbations. Games and Economic Behavior, 330–363 (1995)
7. Young, H.P., Wang, Y.: Personal strategy and social structure: the theory of the system evolution, pp. 106–114. Shanghai People's Publishing House, Shanghai (2004)
8. Young, H.P.: The Evolution of Conventions. Econometrica, 57–83 (1993)
9. Wilhite, S.C., Payne, D.E.: Learning and Memory: The Basis of Behavior. Allyn and Bacon, 218–226 (1992)
10. Ebbinghaus, H.: Memory: A Contribution to Experimental Psychology, pp. 77–79. Teachers College, Columbia University, New York (1913)
11. Kuhn, Zhen, G.: An introduction of psychology: the understanding road of thought and behavior, pp. 301–381. China Light Industry Press, Beijing (2004)
12. Deng, H.-Z., Tan, Y.-J.: The simulation analysis of multi-person mixed game. Journal of Management Sciences in China, 77–82 (2002)
13. Fudenberg, D., Kreps, D.M.: Learning Mixed Equilibria. Games and Economic Behavior, 320–367 (1993)

Improved S Surface Controller and Semi-physical Simulation for AUV

Chong Lv and Xinhua Zhang

Beijing Institute of Automatic Control Equipment, 100074 Beijing, China
lvchong6@yahoo.com.cn

Abstract. S surface controller is proved to be effective in underwater vehicle motion control. But it is difficult to adjust parameters. How to choose optimum control parameters for special underwater vehicle is a important problem for control effect. In order to solve this problem, a modified Particle Swarm Optimization (MPSO) algorithm is proposed to optimize control parameters of S surface controller. It realized rapid convergence of S surface controller parameters and avoided loitering near local peak value. This paper also designed a semi-physical simulation system for the control system of a certain AUV. The total structure and operational principle of this semi-physical simulation system were explained in detail. The 6-DOF mathematic model of AUV moving in underwater space was constructed. Results of simulation reflect the motion of the AUV realistically and can provide important reference for the real underwater experiments.

Keywords: AUV, improved S surface control, modified Particle Swarm Optimization, control system, semi-physical simulation.

1 Introduction

Autonomous Underwater Vehicle (AUV) has been greatly used in civil field, military field and scientific research [1]. In the complicated ocean environment, nonlinear, high coupling of 6-DOFs' motion make AUV's dynamic model very complex, especially for the AUV with control surface (rudder, wing, etc). So design the controller for AUV must considered the nonlinear problem, the controller is independent of model. Dr. Liu Xuemin presented S surface control[2] which combines fuzzy control and PID control. It predigests the controller's design and guarantees the control effect. But the control parameters completely depend manual adjustment; it made the work cockamamie and can not guarantee the good robust. So the Modified Particle Swarm Optimization (MPSO) algorithm is proposed to optimize control parameters of S surface controller. And the semi-physical simulation system for the control system of a certain AUV is designed to verify the controller.

D. Yang (Ed.): Informatics in Control, Automation and Robotics, Volume 2, LNEE 133, pp. 203–210.
springerlink.com

2 S Surface Control

Form the rules of general fuzzy controller we can see the change of the control output is disciplinary. Values in the leading diagonal can be connected a polygonal line. And we can use a smooth curve to fit this polygonal line. Actually, this curve can be seen as innumerability polygonal lines whose lengths tend to zero. Usually we adopted the form of sides loose and center dense, this form accords with the change of Sigmoid function. Accordingly Sigmoid function embodied fuzzy concept in a certain extent. Equally the surface of polygonal line of the whole fuzzy control rules can be replaced by the smooth curve surface.

Generally Sigmoid function is:

$$y = 2.0/\left(1.0 + e^{-kx}\right) - 1.0 \tag{1}$$

So the Sigmoid curve surface function is:

$$z = 2.0/\left(1.0 + e^{(-k_1 x - k_2 y)}\right) - 1.0 \tag{2}$$

The designed S surface controller model is:

$$u = 2.0/\left(1.0 + e^{(-k_1 e - k_2 \dot{e})}\right) - 1.0 \tag{3}$$

Where e and $\dot{e}$ are the control inputs, and they represent the normalized error and error change rate. u is the normalized force of moment in each degree of freedom. k_1 and k_2 are control parameters of error and error change rate, and we only need adjust k_1 and k_2 to control the AUV. Due to the ocean current and other unexpected disturbances, we need to eliminate the steady-state errors and reduce the oscillation by shifting the sigmoid surface. We adopt the model shown as follows:

$$u = 2.0/\left(1.0 + e^{(-k_1 e - k_2 \dot{e})}\right) - 1.0 + \Delta u \tag{4}$$

Where Δu is the disturbance compensation obtained by self adapting. The step of self adapting is following:

① Judge the kinematic velocity of AUV whether less than a set threshold. If true go to②, if false go to③;
② Set a array to store the deviant of this degree of freedom and put the set counter add one. If counter reach a set value then go to④;
③ Put the value of array forward one decimal location and detract counter one, go to ①;
④ Get the weighted average of this array and use it to calculate the side-play amount of control output, adjust the controller's out put by self in order to eliminate fastness deviation. Meanwhile put the counter and array return to zero, go to next circle.

According above stated, a simple and applied controller for AUV is designed; it can be satisfying the work request in the complex ocean circumstance. But the control parameters of S surface controller completely depend manual adjustment, so we hope get the control parameters by self adapting using MPSO.

3 MPSO Optimized Parameters of Control

3.1 Introduction of MPSO Algorithm

PSO is an evolution calculation technology developed by Kennedy et al. in 1995, stemming from the simplification and imitation of a social model. Compared with the traditional optimization method, PSO demands nothing about the shape of the optimal object function, and it is even able to describe the problem hardly expressed by the traditional method as the object function, so that it is of wider use. Its validity in dealing with certain kind of complex optimization problem had been proved by practices.

Formula(4) can get a good control effect by adjusting k_1 and k_2. However with the uncertain factors increasing, adjusting quantization factor and scale factor only by manual experience is harder. It is hard to get optimal parameters. Here modified PSO algorithm is presented to determine k_1 and k_2.

MPSO updates velocity and position of particles by these two equations:

$$\begin{aligned} v_{id} &= wv_{id} + c_1 r_1 (p_{id} - x_{id}) + \\ & c_2 r_2 (p_{gd} - x_{id}) \end{aligned} \tag{5}$$

$$x_{id} = x_{id} + \eta v_{id} \tag{6}$$

Where v_{id} denotes the d-th dimension velocity of the i-th particle which is champed by $V_{\max}$ in case that the particle might leave the search space; v_{id} represents the current d-th dimension position of the i-th particle, c_1 and c_2 are acceleration coefficients; r_1 and r_2 are values drawn form the uniform random distribution on the interval [0, 1]; P_{id} is the d-th dimension component of the particle's personal best position, while P_{gd} is the d-th dimension component of the swarm's best position; w is inertia weight, and η is the restraint factor.

In formula(5, 6), the parameters to be determined are inertia weight w, acceleration constants c_1, c_2, constraint factor η and maximum velocity of particle $V_{\max}$. In formula(5), as the effect of w and $V_{\max}$ is maintenance balance of global and local searching ability, when $V_{\max}$ increased we can minish w in order to reach the searching balance. Therefor fix the range of $V_{\max}$, use nonlinearly decreasing weight strategies[3] and a opening upward parabola to adjust w.

$$w=\frac{(w_{start}-w_{end})t^2}{t_{\max}^{\ 2}}+\frac{2t(w_{end}-w_{start})}{t_{\max}}+w_{start} \tag{7}$$

At the beginning convergence rate of PSO is very fast, so MPSO introduced this strategy in order to make algorithm into local searching quickly by accelerating the decrease speed of w.

Generally give $c_1 = c_2 = 0.4$, but the motion state of the particles is uncertain during iterative process. So we need to adjust η according to environmental change of particle swarm. MPSO introduces immune algorithm[4] into PSO to adjust η dynamically:

$$D_i = \sum_d \sqrt{\sum_j (x_{id} - x_{jd})^2} \tag{8}$$

Formula(26) means the distance between particle i and other particles, define the concentration of particle i in swarm:

$$A_i = 1/(1+D_i) \tag{9}$$

So

$$\eta = A_i\sqrt{1-F_i} \tag{10}$$

Where Fi is fitness of particle i.

In formula(10) the immune principle shows that if concentration of certain particle is larger, affinity is smaller and deviation the optimal solution. So constraint factor η should be larger and vice versa.

3.2 Optimize k_1 and k_2 Using MPSO

The number of control parameters is 2. These parameters are position vector of particles in the swarm in each dimension. The number of dimension of search space is 2. Search for the optimal position of the swarm means search for the optimal parameters of the controller. According to experience, define wstart=0.9, wend=0.4, $V_{\max} = 10$, tmax is maximal simulation time, t is current simulation time. The size of swarm is m =80.

The objective function of particle is used to evaluate the capability of controller, then obtain the particle's fitness, it is the evidence of adjusting controller's parameters. $J = \sum_{i=1}^{k} T^2 |e_i|$ is used to evaluate the capability of controller where T denotes the computer sampling step size (T =0.5s), e_i represents the error at i-th time, and k denotes simulation times. The fitness of particle is

$$F = 1/(\alpha + J) \tag{11}$$

Where α is a small constant and take α =0.1 in practice. J is the smaller, the capability of controller is the better.

The following is flowchart of MPSO as in Figure 1.

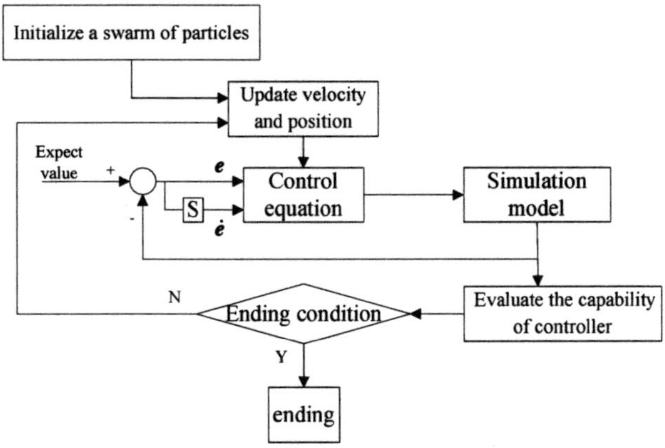

Fig. 1. Flowchart of MPSO

Step1: Initialize a swarm of particles (swarm scale m =80) including stochastic position and velocity.

Step2: Evaluate every particle's fitness in terms of formula (11)

Step3: In the case of each particle, compare the current fitness with the fitness corresponding to the best position P_i; if the current one is better, update the index number of P_i

Step4: In the case of each particle, compare the current fitness with the fitness corresponding to the global best position P_g; if the current one is better, update the index number of P_g

Step5: According to formula (5,6,7,10), update the particles' velocities and positions.

Step6: If the ending condition (reaching a set maximal generation Gmax=600) is not satisfied, then return to Step 2.

The simulation time step is set as 0.5s. There are ten groups input data for learning, each group data is used for each particle of each generation. The simulation time has 400 steps (this time can ensure the system reaches the steady-state). Putting the deviation of every learning into the objective function we can get object function values of particles and then we can get the fitness of every particle by formula(11). When the training is over, we get ten optimal positions. We put these into a new generation particle swarm and search optimization again. The result of this searching is used as controller's parameters.

4 Semi-physical Simulation for AUV

4.1 The Structure of Semi-physical Simulation System

The research object was a certain AUV with rudder and wing operation. The motion control system is based on the use VxWorks real-time operating system. This control system includes AMD Geode LX800 CPU kernel module, multi-board embedded control system base on PC104 buss , power module, impulse board module and etc. The structure of system is Figure. 2.

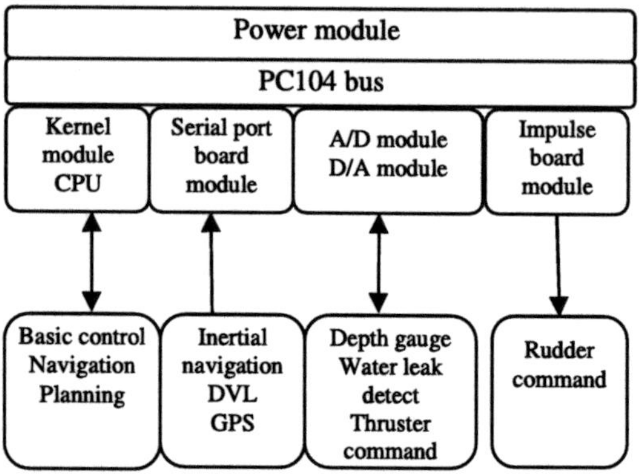

Fig. 2. Structure of AUV motion control system

In the semi-physical simulation, the gesture sensors of AUV (inertial navigation, optical fibre gyro and etc) need rotating table to simulate. But this way cost too much and was difficult to actualize. Therefore data interface PC is used to transform the outputs of the relative motion sensors and the outputs of the controller, through which the semi-physical simulation is constructed.

4.1.1 The Hardware of Simulation System

The hardware of simulation system includes motion simulation solver, superjacent bed supervisor, data interface message processor, view simulation processor, basic control, planning and navigation processor.

Motion simulation solver adopts PC which uses Window XP operation system. Based on VC++ plant the motion information of AUV is obtained by solving the mathematic model of AUV using Rubge-Kutta method. View simulation processor adopts OCTANE active station, uses MultiGen Creator and Vega build virtual-reality-system of AUV movement. Data interface message processor adopts X86 industrial control computer which has eight channels PCI serial port, A/D, D/A transfer card. Basic control, planning and navigation processor adopts SCM/LX-3072 PC104 CPU module which integrated PS/2 keyboard, PS/2 mouse, CRT, IDE, USB 2.0, 10/100BaseT Ethernet and some other interfaces. The operation system is real-time embedded operation system VxWorks.

The hardware structure of semi-physical simulation system was Figure 3.

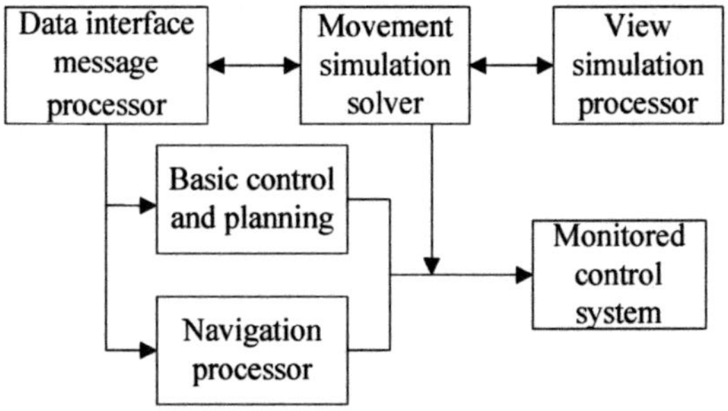

Fig. 3. Connection of hardware

4.1.2 Data Transmission of Simulation System

Through motion simulation solver we get current kinematic velocity, gesture, depth, latitude and longitude and some other information of AUV. Base on TCP/IP protocol the information was transferred to data interface message processor and view simulation processor by network in real-time. The data were translated into some certain formats of sensors by data interface message processor and imported to motion control processor through RS232 serial port. The data of sensors mainly include velocity afforded by Doppler velocimeter, gesture angle and accelerated velocity afforded by inertial navigation, depth afforded by depthometer and so on.

The output commands of control system include rudder command, thruster voltage command. These commands were transferred to motion simulation solver through data interface processor to simulate AUV in real-time. The simulation gave motion information, basic control, planning, navigation and transfers these to the monitor. The monitored control system was designed by VC++ plant and communicated with other systems by network.

4.2 Result of Simulation

Base on the semi-physical simulation system make motion simulation, prove the motion control system, reliability of hardware and software before launch, and analyze capability of control system[5].

Simulation gave the heading control curve and depth control curve when AUV's cruise velocity was 1m/s. the result was showed in view simulation and the simulation result of AUV during the long distance cruise was also given.

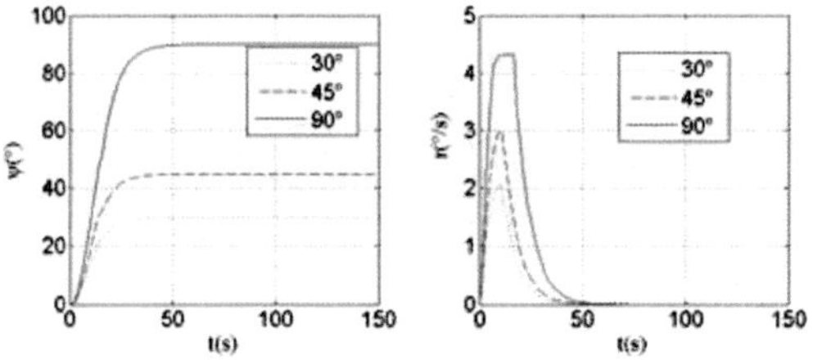

Fig. 4. Curves of heading angle and velocity of heading angle in direction finding

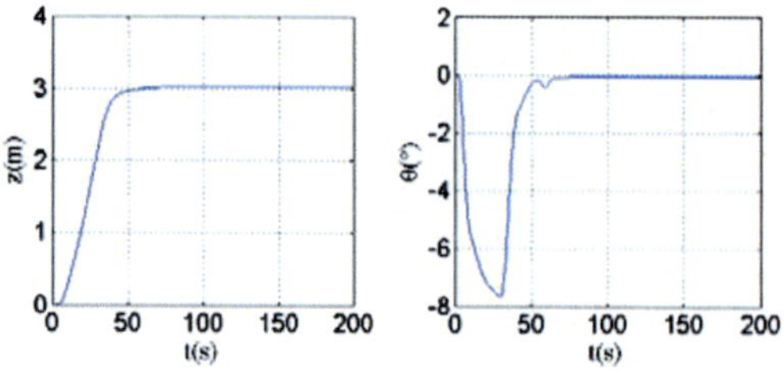

Fig. 5. Curves of depth and trim angle in depth finding

From Figure 4 we can see AUV can rapidly turn into stabilization and has no overshoot basically. It presented the good capability of circumgyration in horizontal plane. The change of heading angle was bigger, the velocity of heading angle was bigger. Figure 5 showed the change of AUV depth was consecutive and smooth; the steady-state error was less than 1%. The angle of trimming rise to 8° firstly, along with the decrescence of depth error, θ decreased too and finally stabilized about zero.

5 Conclusion

MPSO algorithm was adopted to determine the control parameters of S surface. Parts of hardware of control system were put into simulation loop; through data interface processor simulate some motion sensors. Create semi-physical simulation flat of AUV control system to prove hardware system, software structure and data interface. The result showed the controller had a good control effect. Proceeding semi-physical simulation of AUV control system is a reliable evaluation method before the actual trial.

References

1. Xu, Y.-R., Pang, Y.-J., Gan, Y., et al.: AUV- state-of-the-art and prospect. CAAI Transactions on Intelligent Systems 1(1), 9–16 (2006)
2. Liu, X., Xu, Y.: S control of automatic underwater vehicle. Ocean Engineering 19(3), 81–84 (2001)
3. Chen, G.-M., Jia, J.-Y., Han, Q.: Study on the strategy of decreasing inertia weight in particle swarm optimization algorithm. Journal of Xi'an Jiaotong University 40(1), 53–56 (2006)
4. Luo, F., He, M.-Y.: Multilayer feed-forward neural networks based on immune genetic algorithm. Computer Applications 25(7), 1661–1662 (2005)
5. Meng, T., Yu, F., Jin, Z., Han, K.: Hardware-in the-Loop Simulation Method for Pico-satellite Attitude Determination and Control System. Journal of Astronautics 28(5), 1156–1160 (2007)

Optimal Battery Energy Storage System Charge Scheduling for Peak Shaving Application Considering Battery Lifetime*

Xuzhu Dong[1], Guannan Bao[2], Zhigang Lu[1], Zhichang Yuan[2], and Chao Lu[2]

[1] Electric Power Research Institute of China Southern Power Grid, Guangzhou, 510080
[2] Department of Electrical Engineering , Tsinghua University, Beijing, 100084

Abstract. Peak shaving is one of the most important applications of battery energy storage system. In order to prolong battery life or to study the relationship between the battery lifetimes, the charge-discharge cycles and the depth of discharge, constraints concerning charge-discharge cycles and depth of discharge should be added to the optimization model. Therefore, algorithms based on continuous models fail to solve this kind of problem. This paper presents a dynamic programming algorithm containing multi-modules of states. The remaining capacity is discretized and modularized. Tests are carried out using a set of predicted load data. The results show the effectiveness of the method.

Keywords: Battery energy storage system, peak shaving, dynamic programming.

1 Introduction

Battery energy storage system (BESS) can be used to shave the peak by discharging during peak periods and charging during valley periods. For grid owners, the benefits of peak shaving include increasing the utilization factor of facilities, putting off the upgrade of facilities, and saving renewal expenses. For customers, peak shaving brings economic benefits because of the power rates difference between peak periods and valley periods. Many large scale BESS are in operation in the world. In China, a MW level battery energy storage system demonstration project is carried out by China Southern Power Grid. The method presented in this article is used in the demonstration project.

According to the predicted daily load data, the day-ahead optimization block of peak shaving gives the optimal 24-hour charge strategy of BESS before a new day. The real-time control block of peak shaving gives the charge instructions to the power conversion system (PCS) according to the charge strategy and the real-time data such as real-time load and the state-of-charge (SOC) of batteries. The relationship between day-ahead optimization block and real-time control block of peak shaving is shown in figure 1.

---

* This work was supported by National Natural Science Foundation of China (51037002) and the Major State Basic Research Development Program of China (2012CB215206).

D. Yang (Ed.): Informatics in Control, Automation and Robotics, Volume 2, LNEE 133, pp. 211–218.
springerlink.com

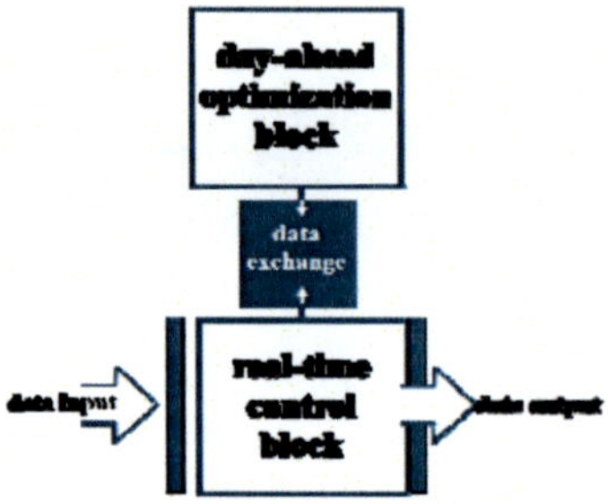

Fig. 1. The relationship between day-ahead optimization block and real-time control block of peak shaving

Preceding paper [1] solves the day-ahead optimization problem with sequential quadratic programming method in order to achieve the maximum profits of the BESS; [2]-[8] solve the optimization problem with intelligent algorithm including evolutionary programming, particle swarm optimization approach and simulated annealing technique; [9] and [10] present dynamic programming (DP) to solve the day-ahead optimization. One advantage of DP is that the nonlinear constraints such as the internal loss constraints are easy to handle. Another advantage is that DP doesn't require continuous functions and it's easy to implement using computer program.

In this article, considering the effects of charge-discharge cycles and depth of discharge (DOD) on battery lifetime, we put the cycle constraint and DOD constraints into the optimization model and solve the problem with DP containing multi-modules of states. The following Section introduces the day-ahead model for peak shaving; Section 3 describes DP containing multi-stage states; finally, calculation results are presented in Section 4.

2 Day-Ahead Optimization Model

2.1 Assumptions

(1) The battery power can change instantaneously, i.e. ramp rate constraints are negligible.
(2) The battery internal losses are negligible.

2.2 Notations

To describe the mathematical model, the following notation is first introduced:

N: number of load data in one day;
Δt: time interval of load data;
D(i), i=1,2,…,N: the ith predicted load data;
b(i), i=1,2,…,N: the ith output power of BESS, positive for charge and negative for discharge;
s(i), i=0,1,2,…,N: the ith remaining capacity;
S: the total battery capacity;

S_{low}, S_{high}: the lower limit and upper limit of remaining capacity;
$S_{initial}$, S_{final}: the initial value and final value of remaining capacity;
P_{max}: the maximum power of charge and discharge;
k: limit number of charge-discharge cycles;
$DOD(j), j = 1,2,\ldots,k$: limit value of DOD of the jth discharge.

2.3 Objective Function

In mathematics, variance reflects the degree of deviation from the random variables to their mean value. The objective function is the variance of the load data to reduce the volatility of the load curve.

$$\min\ f(b) = \frac{1}{N}\sum_{i=1}^{N}\left(\big(D(i) + b(i)\big) - \frac{1}{N}\sum_{i=1}^{N}\big(D(i) + b(i)\big)\right)^2 \tag{1}$$

2.4 Constraints

(1) Capacity constraints

$$S_{low} \le s(i) \le S_{high}, \qquad i = 0,1,2,\ldots,N \tag{2}$$

where

$$s(0) = S_{initial} \tag{3}$$

$$s(i) = s(i-1) + b(i) \times \Delta t, \qquad i = 1,2,\ldots N \tag{4}$$

$$s(N) = S_{final} \tag{5}$$

Because the initial value $S_{initial}$ and the final value S_{final} are determinate, $\frac{1}{N}\sum_{i=1}^{N}\big(D(i) + b(i)\big)$ is constant, the objective function (1) can be simplified to:

$$\min\ f(b) = \sum_{i=1}^{N}\big(D(i) + b(i)\big)^2 \tag{6}$$

The above objective function is separable which meets the requirement of DP method.

(2) Power constraints

$$-P_{max} \le b(i) \le P_{max}, \qquad i = 1,2,\ldots,N \tag{7}$$

Battery lifetime is closely related to charge-discharge cycles and DOD. In order to prolong battery life or to study the relationship between the battery lifetimes, the charge-discharge cycles and the DOD, constraints concerning charge-discharge cycles and DOD should be taken into account.

(3) Charge-discharge cycles constraint

There are three different battery states: charge, discharge and idle. Idle can be seen as charge or discharge with zero power. The process 'charge-idle-charge' is regard as a charge stage and 'discharge-idle-discharge' is regard as a discharge stage. Idle in process 'charge-idle-discharge' is regard as charge with zero power while idle in 'discharge-idle-charge' is regard as charge with zero power. Therefore, one day is divided into several stages in which charge stage and discharge stage appear alternately. The charge-discharge cycle constraint is

$$\text{charge-discharge cycles=k} \quad \text{or} \quad \text{charge-discharge cycles} \leq \text{k} \tag{8}$$

(4) DOD constraints

$$s(i) \geq S * DOD(j), j = 1,2,\ldots,k, i \in j^{th} \text{ discharge stage} \tag{9}$$

Constraints (3) and (4) are not continuous expressions, so algorithms based on continuous models fail to solve this kind of problem. This paper presents a DP algorithm containing multi-modules of states which can solve the model successfully.

3 DP Algorithm Containing Multi-modules of States

The theorem of optimality is the basis of DP: 'an optimal policy must contain only optimal sub-policies' [11]. One day is divided into N intervals. Each time interval is Δt. There are M states in every interval. We choose remaining capacity as state, discretizing it with increment Δs. The battery initial state is in interval number zero and the final state is in interval number N. The optimal policy is the optimal path from initial state to final state. In an interval, M states are divided into $2 * k$ modules. Assume that charge module comes first. The states and possible paths are shown in figure 2.

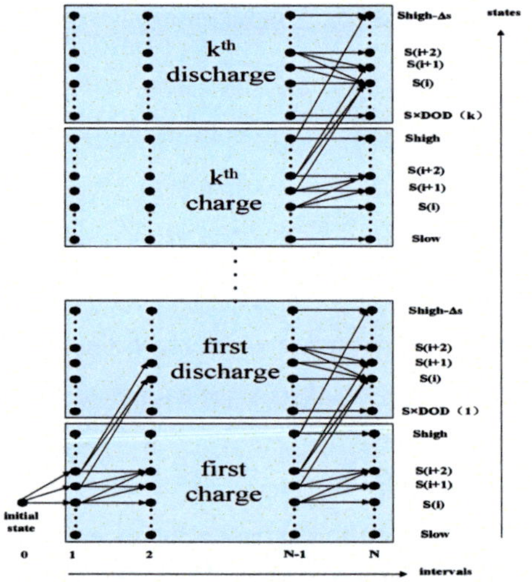

Fig. 2. DP algorithm containing multi-modules of states

In charge modules, possible paths include battery idle paths, battery charge paths and battery discharge paths to the discharge module above it; In discharge modules, possible paths include battery idle paths, battery discharge paths and battery charge paths to the charge module above it. In both charge modules and discharge modules, all the states meet the requirements of S_{low} and S_{high}. In the jth discharge modules, states also meet the requirement of DOD(j). In figure 2, an arrow represents a possible step. The difference of the remaining capacity connected by an arrow is less than $P_{max} * \Delta t$. Go forward along the path indicated by arrows, the four constraints——capacity constraints, power constraints, charge-discharge constraint and DOD constraints are guaranteed to be satisfied.

There are many possible decisions starting from one state or pointing to one state. The starting points of arrows pointing to state m of interval n make up a set $E(n, m)$. The charge power from state r of interval n-1 to state m of interval n is P_{rm}, and the single-step decision-making indicator is $v_n(r, m) = (D(n) + P_{rm})^2$. The optimal index function from initial point to state m of interval n is $f(n, m) = \min\{r \in E(n, m) | f(n-1, r) + v_n(r, m)\}$.

An optimal solution set with different final value S_{final} can be solved simultaneously in just one call of DP. If there is no final value constraint $s(N) = S_{final}$ in the model, we can select the solution with the minimal index function from the optimal solution set.

4 Numerical Simulation

We use the predicted load data of Biling substation in Shenzhen. Δt=5min and N=288. The total battery capacity is 20MW·h. Maximal charge and discharge power is 5MW. The sum of squares of predicted load data $\sum_{i=1}^{N} D(i)^2$ is 218096. Initial value $S_{initial}$, final value S_{final} and DOD limitations $DOD(j), j = 1,2, \ldots, k$ are zeros. Set the capacity increment Δs from 1MW× Δt gradually reduced to 0.1MW× Δt, and k=1. The optimization results are shown in table1. It is clear that with the decrease of capacity increment Δs, the objective function decreases while the computing time increases. The smaller the capacity increment Δs we set, the more accurate the results we get, but the greater the amount of calculation in DP. The optimal time meets the practical needs.

Table 1. Comparison of optimal results with different capacity increments

Capacity increment	Objective function	Computing time
1MW	213778	93ms
0.5MW	213751	343ms
0.2MW	213740.078125	1872ms
0.1MW	213739.4375	7325ms

Let k increases from 1 to 4 when capacity increment is $1MW \times \Delta t$. The optimization results are shown in table 2.

Table 2. Comparison of optimal results with different cycle limits

Cycle limits k	Objective function	Computing time	figure
1	213778	93ms	figure 3
2	213698	202ms	figure 4
3	213676	320ms	figure 5
4	213676	430ms	figure 5

The more charge-discharge cycles, the longer time it takes to compute. The load curves with different charge-discharge cycles are shown from figure3 to figure5. The dashed curve in figure above is predicted load curve. The solid curve in figure above is the load curve after peak shaving. The curve in figure below is the battery power curve.

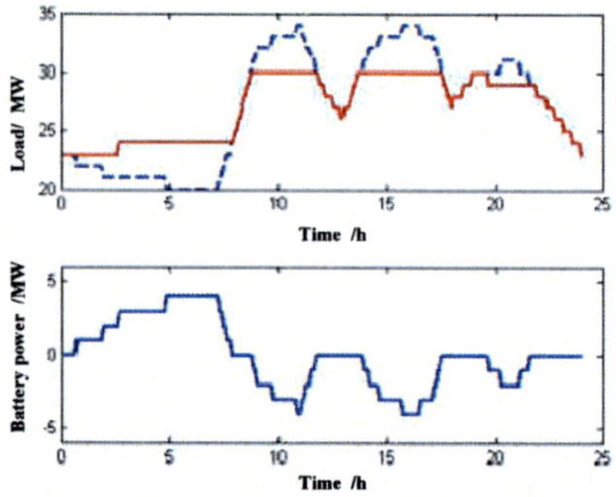

Fig. 3. Load curves and battery power curve when k=1

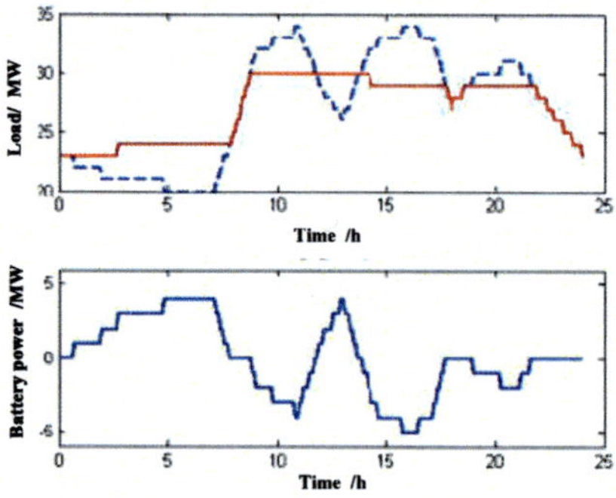

Fig. 4. Load curves and battery power curve when k=2

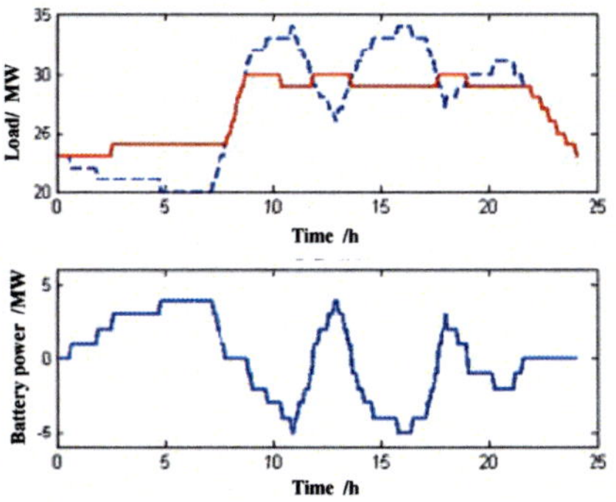

Fig. 5. Load curves and battery power curve when k=3 and k=4

The optimal solution when k=3 is the same as the optimal solution when k=4, indicating that the optimal charge-discharge cycles is 3. It can be understood that the charge and discharge power is zero in the fourth cycle when we set k=4. Although the optimal solutions are the same, the computing time when k=4 is longer than the computing time when k=3 because there are more states. The objective functions when k=1 and k=2 are larger than the objective function when k=3 or k=4, because there are charge-discharge cycles constraints. It is the price to pay to extend battery lifetime.

5 Conclusion

The DP containing multi-modules of states we presented here can not only solve the day-ahead optimization problem effectively which shaves the load peak, but also satisfy the charge-discharge cycles constraint and DOD constrains which prolong battery life. The computing time meets the practical needs. The smaller the capacity increment Δs we set, the more accurate the results we get, but the greater the amount of calculation in DP. The larger the charge-discharge cycles k we set, the longer computing time it takes to optimize the problem. If the optimal charge-discharge cycles are less than the k we set, it can be understood that the charge and discharge power is zero in several cycles. The objective function may be larger when the charge-discharge cycles constraint added in, which is the price to pay to extend battery lifetime.

References

1. Weihao, H., Zhe, C., Birgitte, B.-J.: Optimal Operation Strategy of Battery Energy Storage System to Real-Time Electricity Price in Denmark. In: IEEE Power and Energy Society General Meeting, pp. 1–7 (2010)
2. Gaul, J., Handschin, E., Hoffmann, W., Lehmkoster, C.: Establishing a Rule Base for A Hybrid ES/XPS Approach to Load Management. IEEE Transactions on Power Systems 13(1), 86–92 (1998)

3. MacGill, I.F., Kaye, R.J.: Decentralised Coordination of Power System Operation Using Dual Evolutionary Programming. IEEE Transactions on Power Systems 14(1), 112–119 (1999)
4. Cau, T.D.H., Kaye, R.J.: Multiple Distributed Energy Storage Scheduling Using Constructive Evolutionary Programming. In: Proc. IEEE Power Eng. Soc. Int. Conf. Power Ind. Comput. Appl., pp. 402–407 (2001)
5. Cau, T.D.H., Kaye, R.J.: Evolutionary Optimization Method for Multistorage Hydrothermal Scheduling. IEE Proc. Gener. Transm. Distrib. 149(2), 152–156 (2002)
6. Chacra, F.A., Bastard, P., Fleury, G., Clavreul, R.: Impact of Energy Storage Costs on Economical Performance in A Distribution Substation. IEEE Trans. Power Syst. 20(2), 684–691 (2005)
7. Lee, T.-Y.: Operating Schedule of Battery Energy Storage System in a Time-of-Use Rate Industrial User with Wind Turbine Generators: A Multipass Iteration Particle Swarm Optimization Approach. IEEE Trans. Energy Conv. 22(3), 774–782 (2007)
8. Fung, C.C., Ho, S.C.Y., Nayar, C.V.: Optimisation of A Hybrid Energy System Using Simulated Annealing Technique. In: Proc. IEEE Int. Conf. Comput. Commun. Control Power Eng., vol. 5(0), pp. 235–238 (1993)
9. Maly, D.K., Kwan, K.S.: Optimal Battery Energy Storage System (BESS) Charge Scheduling with Dynamic Programming. In: IEE Proc.-Sci. Meas. Technol. vol. 142(6) (1995)
10. Oudalov, A., Cherkaoui, R., Beguin, A.: Sizing and Optimal Operation of Battery Energy Storage System for Peak Shaving Application. In: IEEE Proceedings of Power Tech 2007, Lausanne, Switzerland (2007)
11. Bellman, R.E., Dreyfus, S.E.: Applied Dynamic Programming. Princeton University Press, Princeton (1962)

Applications of Data Sharing Approaches in P2P Environments

Mutaz Beraka, Sofien Gannouni, and Hassan Mathkour

Computer Science Department, College of Computer and Information Sciences,
King Saud University, Riyadh, Kingdom of Saudi Arabia
mutaz999@hotmail.com,
{gnnosf,mathkour}@ksu.edu.sa

Abstract. During the last decade, many research and development efforts have been deployed with the aim of proposing new approaches for accessing remote, heterogeneous and autonomous data sources to share them across the network. These approaches vary in terms of concepts and standards used in the implementation of the approach. From our review of the literature, we identified the following four approaches: Transaction Processing Monitors (TPM), Tuplespace, Resource Description Framework (RDF), and Data Service Approach (DSA). We also provide an overview of the P2P systems that have been developed based on each of these approaches.

Keywords: Data sharing, TPM, Tuplespace, RDF, DSA, P2P systems.

1 Introduction

Peer-to-Peer (P2P) architecture presents advantages over centralized approaches [1]. It avoids both computational performance and information update bottlenecks, and provides robustness, reliability, and scalability [1]. In P2P environments, sharing data sources in a distributed manner has huge value for businesses and users.

Transaction Processing Monitors (TPM) [2] [3], Tuplespace [4], Resource Description Framework (RDF) [5] [6], and Data Service Approach (DSA) [7] are approaches that allow users to share data in P2P environments. These approaches, proposed during different periods of time, were influenced by the views of data and data sharing at that time. Many decentralized distributed systems (P2P systems) were developed to implement these approaches in the respective periods. Nowadays, the demand for sharing data stored in distributed data sources has gained further importance and attracted the attention of many researchers.

This paper describes the TPM, the Tuplespace, the RDF, and the DSA, approaches. It then describes their applications in various P2P systems that have been developed for each of these approaches during the last decade.

D. Yang (Ed.): Informatics in Control, Automation and Robotics, Volume 2, LNEE 133, pp. 219–224.
springerlink.com

2 Transaction Processing Monitors

TPM facilitates developing and administering complicated transaction processing systems with a large number of users and multiple servers. It is considered as a standard interface that allows users to submit their queries and then encapsulates them as transactions to process and execute them across various DBMS servers. Each transaction must satisfy ACID properties to be completed successfully as a whole unit; it cannot remain in an intermediate state [2]. These properties prevent any transaction from being in a partially completed state, an unknown state, or in an inconsistent state. In addition, they guard against hardware and software errors [3]. Systems that implement the TPM approach in P2P environments include: the Peer Agent System (PAS) and the P2P database network (P2PDBN).

The Peer Agent System was proposed by Penserini et al. [8]. It's a new form of distributed and cooperative information system based on the P2P architecture. This system provides the ability to exchange data and solves the problems associated with heterogeneous data management. A PAS consists of a Multi Agent System (MAS) with four core agents for each peer: a wrapper, mediator, facilitator, and planner [8]. These agents allow a peer to play one or more roles within the system.

The P2P database network (P2PDBN) [9] is an implementation of the TPM approach in a P2P environment. The P2PDBN is a network of peers, where each peer has its own database and participates in this network to exchange and share data with other peers [9]. In the P2PDBN, peer databases are created autonomously because each peer has its own database schemas and maintains data independently, although these data may be semantically related to each other. A model of the peer database network is shown in Figure 1 [9].

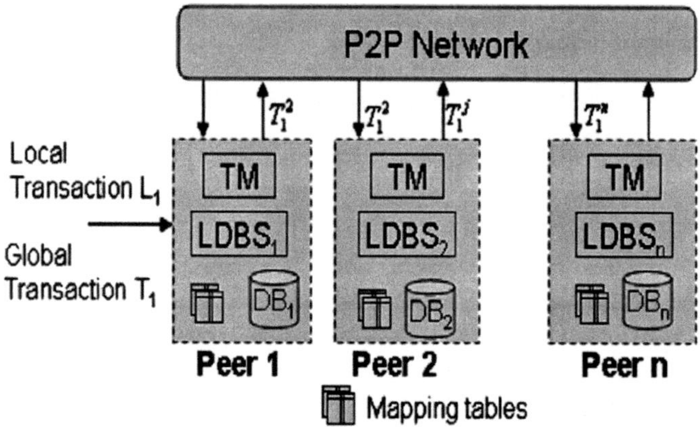

Fig. 1. Peer Database Network Model

3 Tuplespace

The Tuplespace was proposed by Gelernter and Carriero [4] as a coordination model that provides the ability to build and develop centralized and decentralized applications. It is based on the concept of a virtual shared memory [4]. It allows data providers to insert and publish their data as tuples in a shared space, and data consumer to fetch and retrieve data from the shared space that matches certain pattern. In Tuplespace, users interact only by exchanging tuples, which are ordered collections of data items. A tuple is either simple or complex [10]. There are many systems that implement the Tuplespace approach in P2P environments. These include TIBCO Active space and Grinda.

TIBCO Active space is a distributed P2P in-memory data grid using a virtual shared space for sharing, exchanging, and processing data in real time [11]. It presents a unified interface that provides messaging and database functionalities. In addition, it provides transactional support and concurrent access control [11]. ActiveSpaces is a platform-independent system that can be used as middleware to exchange data between heterogeneous platforms [11]. Its architecture is shown in Figure 2 [11].

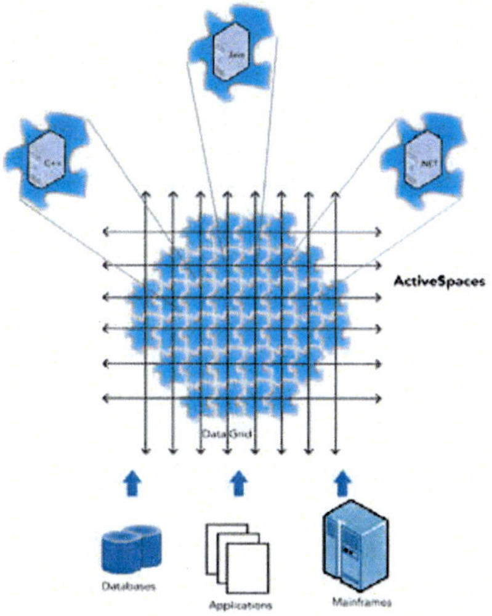

Fig. 2. Architecture of TIBCO ActiveSpaces

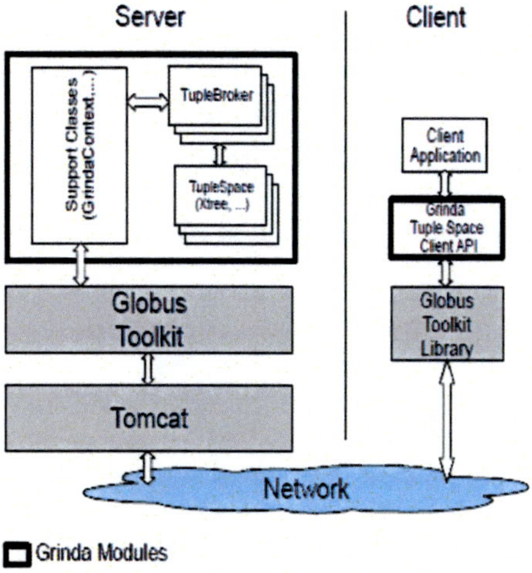

Fig. 3. General Architecture of Grinda Modules

Grinda uses the concept of a tuple space to implement a Grid service (Grid + Linda). Grinda is a tuple space system designed to be deployed in the Globus Toolkit and to serve as the coordination framework for applications deployed on the generic tuple space infrastructure [4]. In other words, Grinda is an implementation of distributed tuple spaces for grid-link environments in structured P2P networks [4]. The general architecture of Grinda modules is shown in Figure 3 [4].

4 Resource Description Framework

RDF is a W3C specification. It was originally designed as a metadata data model to represent and store any specific data as Web resources over the Internet [5]. Being a semantic web technology, RDF supports the exchange associated with data as well as knowledge on the web. RDF is an application of XML that imposes important structural constraints to present unambiguous techniques of expressing semantics [12]. Resources of RDF are recognized by Uniform Resource Identifiers (URIs) on the web [6]. There are many systems that implement the Tuplespace approach in P2P environments. These include Edutella, Piazza and GridVine.

The Edutella is based on W3C standards for RDF and RDF schema, and built on JXTA technology to provide an infrastructure for RDF-based metadata for P2P applications [13]. Edutella provides metadata and semantic services for Edutella peers, with these services built on JXTA technology [14]. Examples of these services are: querying, replication, mapping, mediation, and so on [14].

Piazza is a peer data management system that allows users to share heterogeneous data in a P2P environment [14] [15]. In the Piazza system, each peer contributes schema, mappings, data and/or computation. Capabilities of query answering are provided over a distributed collection of local schemes with pair-wise schema mappings between them [14]. It supports data integration over a P2P system by using the mechanism of decentralized schema mediation [14] [15].

The GridVine system uses a P2P overlay network (P-Grid) to realize semantic overlays [14] [16]. This realization of GridVine is supporting by separating the logical layer from the physical layer, applying the well-known database principle of data independence [14]. The operations on logical layer support semantic interoperability, while operations on the physical layer exploit the structured overlay network [14].

5 Data Service Approach

DSA embodies the concepts found in SOA to supply an abstract layer named data service layer (DSL). DSL is responsible for accessing structured, semi-structured and unstructured data stored in heterogeneous data sources, and exposing them as data services [7]. These services are loose-coupled and reusable facilitating thereby the development of a new application that uses such services to integrate data from multiple data sources [7] [17]. Implementations of the DSA approach in a P2P environment are rare and almost nonexistent. However, we give an overview of the only system (WSO2) that implements the DSA approach and approximates a P2P system.

WSO2 is software built on top of WSO2 Carbon, a lightweight high-performance platform for taking data stored in data sources and making them available as Web services. It uses Axis2 as the underlying SOAP processing engine [18]. The architecture of WSO2 is shown in Figure 4 [18]. WSO2 supports relational databases, CSV files, and Microsoft Excel files [18].

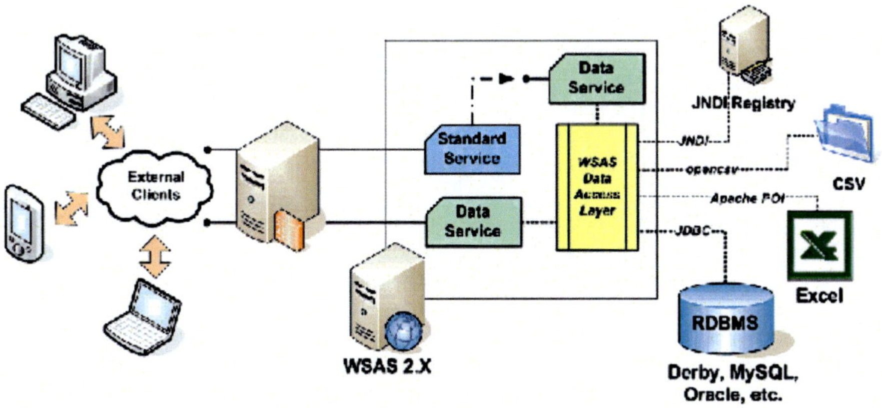

Fig. 4. Architecture of WS02

6 Conclusion

P2P systems provide greater benefits for implementation of data sharing approaches, including greater stability, reliability, and so on. In this paper, we outlined the four proposed approaches supporting data sharing. Moreover, we overviewed the implementations of these approaches in P2P architectures (P2P systems). These systems allow peers to exchange and share distributed data across the network. In addition, these systems must be accurate because they are used many times throughout the day, and must mask the heterogeneity between peer platforms and data sources.

Acknowledgment. This work was supported by the Research Center and College of Computer and Information Sciences, King Saud University, Riyadh, Saudi Arabia.

References

1. Haase, P., Schnizler, B., Broekstra, J., Ehrig, M., Harmelen, F., Menken, M., Mika, P., Plechawski, M., Pyszlak, P., Siebes, R., Staab, S., Tempich, C.: Bibster – A Semantics-Based Bibliographic Peer-to-Peer System. In: McIlraith, S.A., Plexousakis, D., van Harmelen, F. (eds.) ISWC 2004. LNCS, vol. 3298, pp. 122–136. Springer, Heidelberg (2004)
2. Alonso, G.: Transaction Processing Monitors (TP-monitors). Computer Science Department, Swiss Federal Institute of Technology (ETHZ), `http://masteritgov.dia.uniroma3.it/didattica/MW/Middleware-04-TP-Monitors.ppt`
3. Browne's, C.: Transaction Processing Monitors, `http://linuxfinances.info/info/tpmonitor.html`
4. Capizzi, S.: A Tuple Space Implementation for Large-Scale Infrastructures. PhD thesis, Department of Computer Science, University of Bologna, Italy (2008)
5. Alexander, N., Ravada, S.: RDF Object Type and Reification in the Database (2004)

6. Zhou, S.: Exposing Relational Database as RDF. In: 2nd International Conference on Industrial and Information Systems (2010)
7. Bloomberg, J., Goodson, J.: Best Practices for SOA: Building a Data Services Layer. SOA World Magazine 8(5) (2008)
8. Penserini, L., Panti, M., Spalazzi, L.: Agent-Based Transactions into Decentralised P2P. Preliminary Report, Computer Science Institute, University of Ancona. ACM Publisher, Italy (2002)
9. Masud, M., Kiringa, I.: Transaction processing in a peer to peer database network. Data & Knowledge Engineering 70(4), 307–334 (2010)
10. Omicini, A., Denti, E.: From tuple spaces to tuple centres. Science of Computer Programming 41(3), 277–294 (2001)
11. TIBCO ActiveSpaces, `http://www.tibco.com/products/soa/in-memory-computing/activespaces-enterprise-edition/default.jsp`
12. Miller, E.: An Introduction to the Resource Description Framework. Bulletin of the American Society for Information Science and Technology, Dlib Magazine 25(1), 15–19 (1998)
13. Nejdl, W., Wolf, B., Qu, C., Decker, S., Sintek, M., Naeve, A., Nilsson, M., Palmèr, M., Risch, T.: EDUTELLA: A P2P Networking Infrastructure Based on RDF. In: 11th International World Wide Web Conference, Honolulu, Hawaii, USA, pp. 7–11 (2002)
14. Agneeswaran, V.S.: A Survey of Semantic Based Peer-to-Peer Systems. Technical report, LSIR, communicated to International Journal of Computer Science and Software Technology (2007)
15. Tatarinov, I., Ives, Z., Madhavan, J., Halevy, A., Suciu, D., Dalvi, N., Dong, X., Diyska, Y., Miklau, G., Mork, P.: The Piazza Peer Data Management Project. ACM SIGMOD Record 32(3), 47–52 (2003)
16. Aberer, K., Cudré-Mauroux, P., Hauswirth, M., Van Pelt, T.: GridVine: Building Internet-Scale Semantic Overlay Networks. In: McIlraith, S.A., Plexousakis, D., van Harmelen, F. (eds.) ISWC 2004. LNCS, vol. 3298, pp. 107–121. Springer, Heidelberg (2004)
17. Goundar, K., Singh, S., Ye, X.: An Investigation into Concurrency Control Mechanisms in Data Service Layers. In: 14th Asia-Pacific Software Engineering Conference, pp. 318–325. IEEE (2007)
18. Rubasinghe, S., Anandagoda, A.: WSO2 Data Services An Executive Overview. White paper (2008)

Rolling Process Automation for Instant Ramen Noodles

Byung Kook Yoo[1] and Dong Sang Yoo[2]

[1] Engineering Team, Nongshim Co. Ltd. 153 Sinsohyeon-Dong, Anseong, 456-782 Korea
y2335@hanmail.net
[2] Dept of EE, Hankyong National University, 327 Chungangro, Anseong, Gyeonggido, 456-749 Korea
dsyoo@hknu.ac.kr

Abstract. A noodle making process is composed of mixing, rolling, boiling, frying, cooling, and packing processes. For uniform products, the roller gap in the rolling process needs to keep the uniform thickness of dough sheets in spite of different kinds of raw materials. In this paper, we propose a rolling process automation using a PLC and an AC servomechanism to contribute to process standardization by transferring from tacit knowledge of skilled workers to explicit knowledge. The proposed system organizes to adjust the roller gap and visualizes gap values in a touch panel.

Keywords: Automation, Rolling Process, Roller Gap Adjustment, Instant Ramen Noodle.

1 Introduction

In 1958, the first instant ramen in the world went on sale in Japan. This product consisted of noodles that had been seasoned, fried, and then dehydrated. Momofuku Ando, the founder of food manufacturer Nissin, began developing instant ramen in the early years following World War II.

Instant ramen noodles are made with wheat flour, water, salt, and alkaline water that adds elasticity to the noodles. First, the ingredients are kneaded together to make a dough. Next, this dough is rolled out and cut into thin noodles. The noodles are then steamed and are finally packaged after dehydration [1][2].

In the rolling process, when the dough is rolled out, depending on some characteristics of wheat flour such as moisture and protein content, harvest season and region, etc., the thickness of dough sheets varies at the same roller gap and workers frequently should adjust the roller gap to maintain the uniform thickness. But, the roller gap adjustment is a very difficult job for the workers because a worm gear which adjusts the roller gap has a backlash not to be controlled and there is no display device to estimate the roller gap.

In this paper, using an AC servomechanism and a PLC with a display device, we aim to improve a roller gap adjustment system in the rolling process which can contribute to ease of operation and process standardization.

D. Yang (Ed.): Informatics in Control, Automation and Robotics, Volume 2, LNEE 133, pp. 225–229.
springerlink.com

2 The Proposed System for the Rolling Process

As shown in figure 1, the rolling process is composed of complex rollers and continuous rollers.

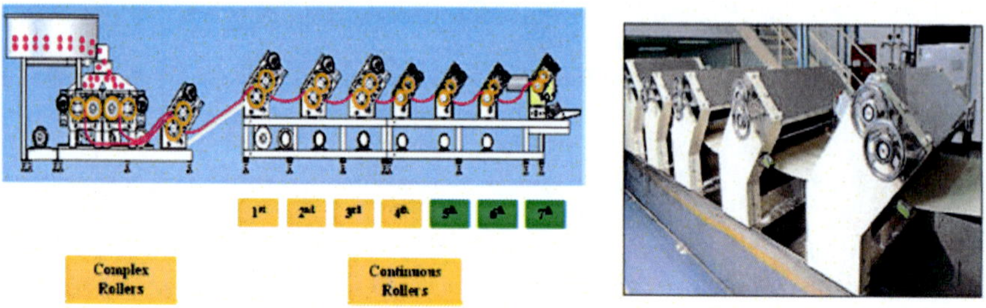

Fig. 1. Rolling process composed of complex rollers and continuous rollers

In the rolling process, even though the roller gap is set to the same value, depending on some characteristics of wheat flour such as moisture and protein content, harvest season and region, etc., the viscosity and stiffness of the dough vary and the fluctuation of the roller gap is necessarily occurred when the dough is rolled out. Therefore, workers should frequently adjust the roller gap to maintain the uniform thickness. But, the roller gap adjustment is a very difficult job for the workers because a worm gearbox in the conventional system which adjusts the roller gap has a backlash not to be controlled and there is no display device to estimate the roller gap. Because of these reasons, whenever product items in the same production lines are changed, the roller gap adjustment of each product depends on the experience of the worker.

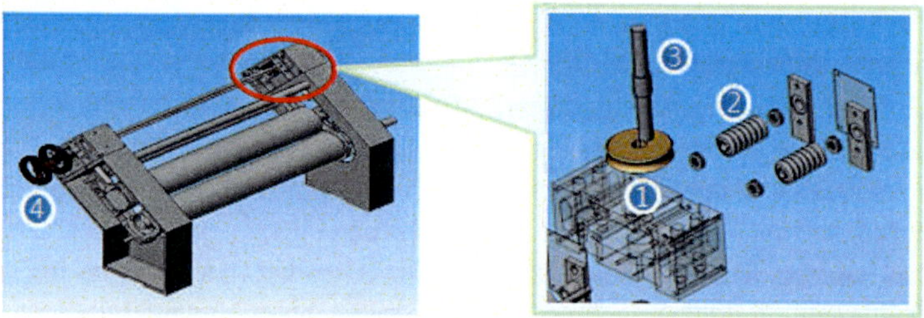

Fig. 2. A worm gearbox in the conventional system. ① wheel ② worm ③ shaft ④ handle

Figure 2 shows a worm gearbox for the roller gap adjustment being used in the conventional production line.

In general, since the 5th through 7th stages in the continuous rollers have more influence on the thickness quality of the dough sheets than the 1st through the 4th stages, we propose the roller gap adjustment system for the 5th through 7th stages using an AC servomechanism and a PLC as shown in figure 3.

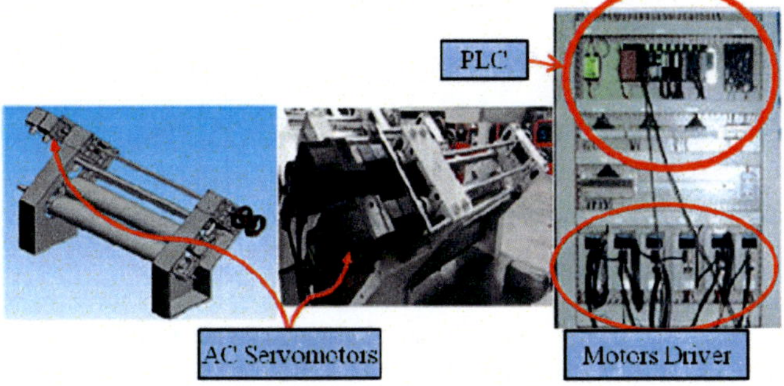

Fig. 3. The Proposed Adjustment System

The proposed system has the following specification:

a. Worm shaft stroke per one revolution of worm wheel: 6mm
b. Worm gear ratio: 1/30
c. Gear reduction ratio of AC servomotor (500W): 1/40
d. Worm shaft pitch per one revolution of motor: $6/(30\times40)=0.005$mm

In addition, a touch panel with PLC interface is used as an input-and-display device for the roller gap adjustment and whenever workers click once on the touch panel, the roller gap is controlled in units of 0.01mm as well as the current gap is displayed on the touch panel screen. Consequently, the proposed system may improve the workability and correspond to the change of product items by recalling the data from DB.

3 Application Results

During the normal production for a specific instant ramen "O", we compared the roller gap displacements in the 6th stage of the continuous rollers between the conventional system and the proposed system in figure 4.

As shown in figure 4, the fluctuation of the roller gap of the proposed system using the AC servomechanism was less than the conventional system.

The measurement results according to the changes of the product items in the mixed noodle production lines for "O" and "N" noodles were depicted in figure 5. As shown in figure 5, when the roller gap was automatically adjusted by the AC servomechanism, the measurement results of each stage were almost closed to the reference values and the standard deviations were greatly reduced over manual adjustments. This means that the proposed system can be improve the thickness quality of dough sheets in the rolling process.

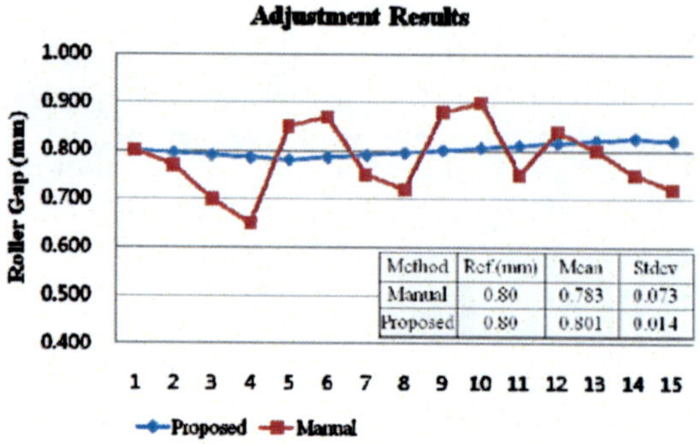

Fig. 4. Application Results in the 6th stage of the continuous rollers for an Instant Ramen "O"

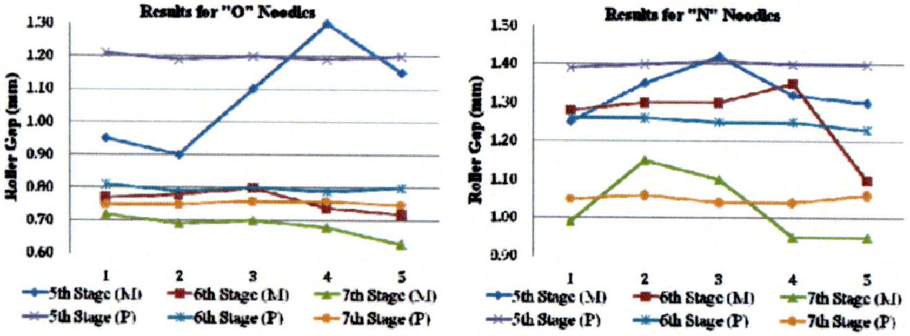

stage	Ref. (mm)	Manual (mm)		Proposed (mm)	
		Mean	Stdev	Mean	Stdev
5	1.20	1.080	0.160	1.198	0.008
6	0.80	0.762	0.032	0.798	0.008
7	0.76	0.684	0.034	0.754	0.005

stage	Ref. (mm)	Manual (mm)		Proposed (mm)	
		Mean	Stdev	Mean	Stdev
5	1.40	1.328	0.063	1.400	0.007
6	1.25	1.266	0.096	1.250	0.012
7	1.05	1.028	0.092	1.050	0.010

Fig. 5. Application Results for Mixed Noodle Production Lines

4 Conclusion

The roller gap adjustment in the conventional rolling process relied on the knowhow of a skilled worker because a worm gearbox by which they adjusts the roller gap has a backlash not to be controlled and there is no display device to estimate the roller gap.

In this paper, we implemented the automated roller gap adjustment system using an AC servomechanism and a PLC with a touch panel which can display the roller gap. The proposed system can adjust the roller gap in units of 0.01mm and accommodate various product items which have different thickness specification by recalling the presetting values of the desired thickness from database.

References

1. Hou, G.G.: Asian Noodles: Science, Technology, and Processing. Wiley (2010)
2. Web Japan, http://web-japan.org/kidsweb/hitech/ramen/

The Pressure Detection and Control for Gas Tightness Test of Metal Capillary Tube

Yang Yu, Min Zhao, and Yongheng Yang

College of Information and Technology, DongHua University, Songjiang District, China, 201620

Abstract. The paper introduced the high pressure detection and control for gas tightness test of metal capillary tubes used in aircraft and airspace fields. The pressure sensor was connected in the input-output high- pressure pipeline system, and the circulating pipeline system was controlled by PID. Thereby, the stability and precision for detection of high pressure output system could be ensured, and thus the control system property would be approved.

Keywords: Single Chip Computer(SCC), PID control, A/D conversion, gas tightness test for High pressure, metal capillary tube.

1 Introduction

The metal capillary tubes (namely the tube with thinner diameter (D≤3mm) and thickness (T ≤0.5mm) have been widely used in circulating pipeline systems of heat exchangers etc. in aircraft and airspace fields. The gas tightness test for high pressure (≤20MPa) have to be performed strictly before they are used in pipeline systems by welding or connecting methods. The precise detection of high-pressure gas tightness of metal capillary tubes is a key factor to find and overcome the inside-defects in order to avoid the larger disaster.

The PID Control is one of practical method in industry control fields with simple, stable and reliability and stronger robustness etc. According to PID Control theory, the pressure detection and control system by means of Single Chip Computer (SCC) is designed in order to collect the signal data from pressure sensor and promptly control or adjust the gas pressure output. In this paper, the hardware of SCC (such as AT89C51), conventional pressure sensor (such as MPX2100) and A / D converter are chosen to connect in close-loop system of high-pressure gas tightness test of metal capillary tubes to perform the collecting and converting for pressure message. The pressure data collected was controlled and revised by incremental PID algorithm to improve the stability and precision for detection.

D. Yang (Ed.): Informatics in Control, Automation and Robotics, Volume 2, LNEE 133, pp. 231–236.
springerlink.com

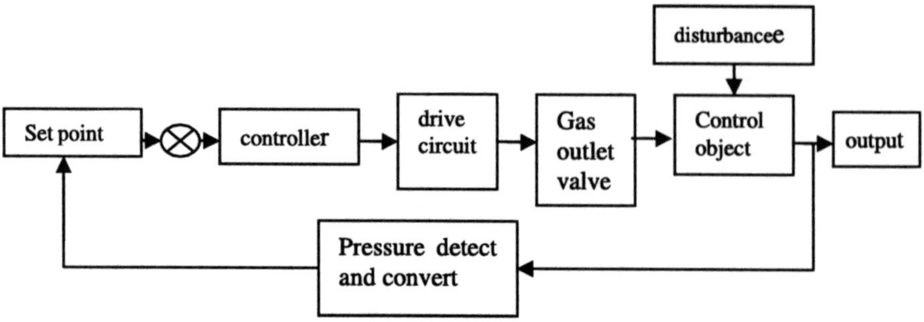

Fig. 1. Structure diagram of the system

2 System Hardware Design

2.1 Theory of the System Structure

The system hardware frame structure is shown in Fig.2. The work process of the system is followed as: The information of the gas pressure is converted by the pressure sensor to be the voltage signal which is linear shifted with the variation of the gas pressure. The range of original signal of the voltage is quite micro compared with that of the A/D input voltage, so the signal must be amplified by the signal processing circuit and then converted by the A/D converter to be the digital signal that will be send into the SCC AT89C51 and compare with the set point (the target value of the pressure, for example 20MPa), then the deviation will be calculated by the PID program to output the controlled values so that the power of the circuit output can be adjusted and controlled effectively.

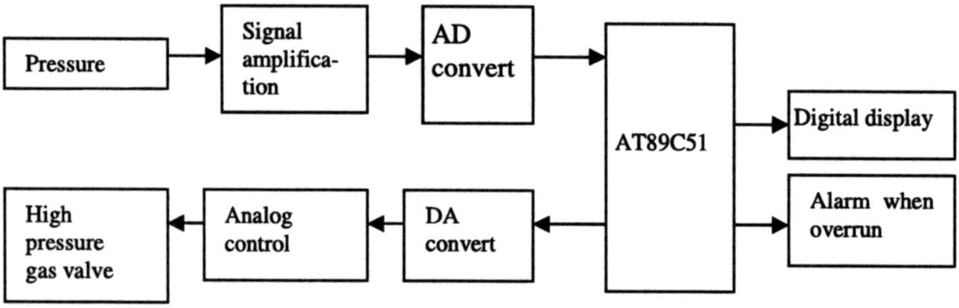

Fig. 2. Structure diagram of the system hardware

2.2 Hardware of System

In this design, data acquisition and processing is that the current pressure values from the sensor are converted to the corresponding voltage values so that the SCC can compare these values with the target points, and then get the deviation signal for the control calculating. Taking all kinds of factors into account, our design expands a data acquisition system with the AT89C51, some I/O connectors and TLC 2543 serial A/D converter.

AT89C51 is an 8-bit SCC with lower power consumption and higher performance. The device was made by the ATMEL which is of high density and nonvolatile, and compatible with standard MCS-51 instruction system and 8051 pin structure. AT89C51 which integrate the general 8-bit CPU and ISP Flash memory cell was widely used in many kinds of embedded systems. AT89C51 combine FLASH, RAM, serial I/O connectors, so it could make up the intelligent pressure control system with a small quantity of peripheral circuit. This terse structure is helpful to reduce the power dissipation and enhance the reliability of the system.

TLC2543 is the 12-bit serial pattern A/D converter of the TI Company, which has wider application in instrument and meter with higher resolution ration and moderate price. Its AD converting process is finished by one by one approaching of switch electric capacity which the serial structure could retrench the I/O resource of the I/O of SCC.

We choose the pressure sensor of the system from the Motorola Company MPX2100 which has the high accuracy, linearity and flexibility. The analog pressure value from the appliance was sent to the AT89C51 after amplifying to convert into the digital signal by A/D converter with stronger anti-interference ability. The converted digit amount was then calculated by the SCC and compensation software etc. The result also could be shown and uploaded through the serial connectors for the human-computer exchange. The composition principle diagram of pressure sensor MPX2100 is shown in Fig.3.

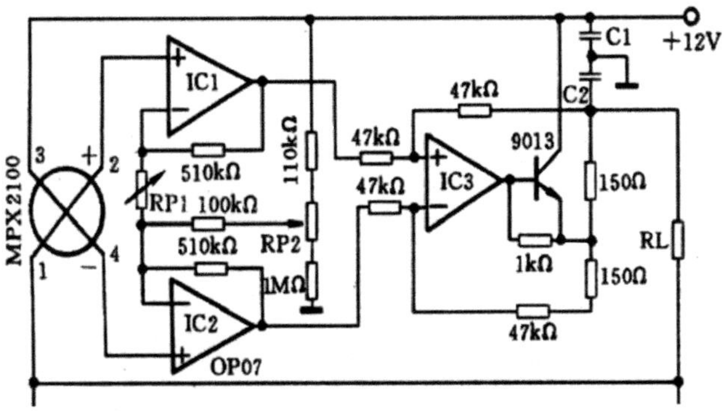

Fig. 3. The composition principle diagram of pressure sensor MPX2100

2.3 PID Algorithm

As a conventional system control method, The PID algorithm was widely used in the process control with its micro calculated amount, better real-time character and easier achievement. We choose the incremental PID control to make sure that the output of the gas pressure stays at the target value quickly and accurately. Because of its increment of controlled quantity for PID, the influence of the accidents will be slight if the computers have breakdowns. The actuator has the memory function and could stay at the normal position, thereby it corresponds with the industry process because the range of AD converter is always smaller and not easy to reach increment saturation.

According to the incremental theory, we can conclude as followed:

$$u(k)=Kp\left[e(k)+\frac{T}{Ti}\sum_{i=0}^{k}e(i)+\frac{Td}{T}\big(e(k)-e(k-1)\big)\right] \quad (1)$$

$$\Delta u(k)=u(k)-u(k-1)=q_0 e(k)+q_1 e(k-1)+q_2 e(k-2) \quad (2)$$

In formula $$q_o = K_p(1+\frac{T}{T_i}+\frac{T_d}{T}) \quad (3)$$

$$q_1 = -K_p(1+2\frac{T_d}{T}) \quad (4)$$

$$q_2 = K_p\frac{T_d}{T} \quad \text{T is cycle period} \quad (5)$$

Thus, if the PID control parameters K_p, T_i, T_d are determined, q_0,q_1,q_2 can be calculated.

3 Design of the System Software

3.1 Main Program

The main program is the system monitor routine programmed by C language in which the initialization designing etc. of SCC will be accomplished. The automatic control flow chart is shown in the figure 4.

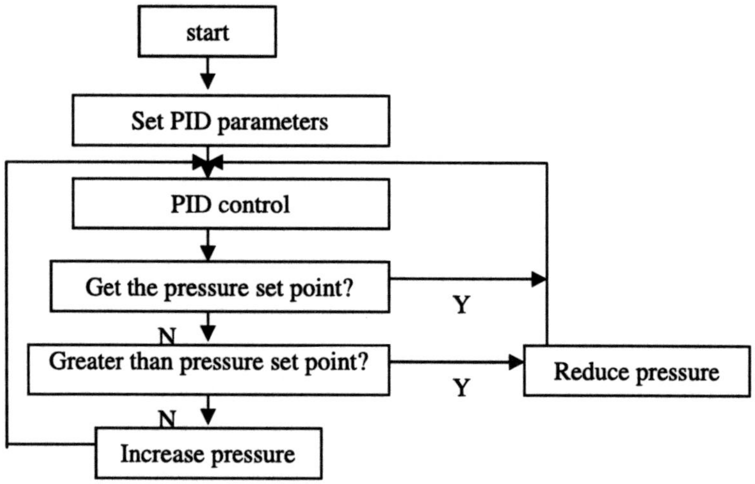

Fig. 4. The flow chart of automatic control

3.2 A/D Conversion Program

The main function of the A/D conversion program is to send the sampled signal to TLC2543 for the A/D conversion when the interruption appear, and send the results to AT89C51.For the TLC2543, the first data should be abandoned for the internal adjustment. The program flow chart is shown in the figure 5 as followed.

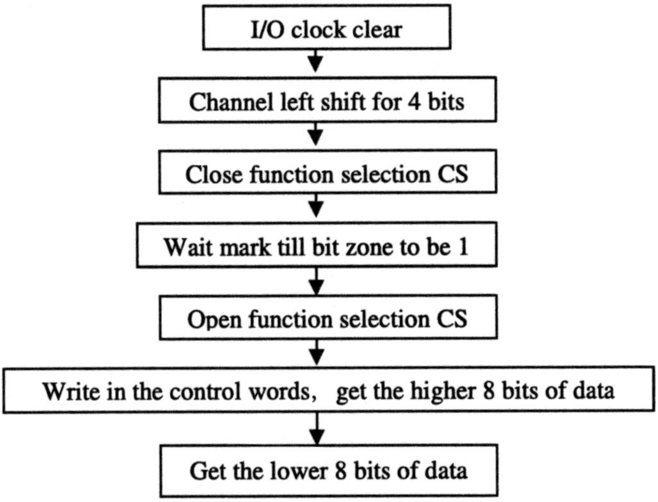

Fig. 5. The program flow chart

3.3 PID Control Subroutine

This PID control algorithm is designed to enhance the accuracy. And the signal is latched in the incremental PID algorithm. The incremental quantities are related to the last three collected samples. Thus it could reduce the influence of the disturbance.

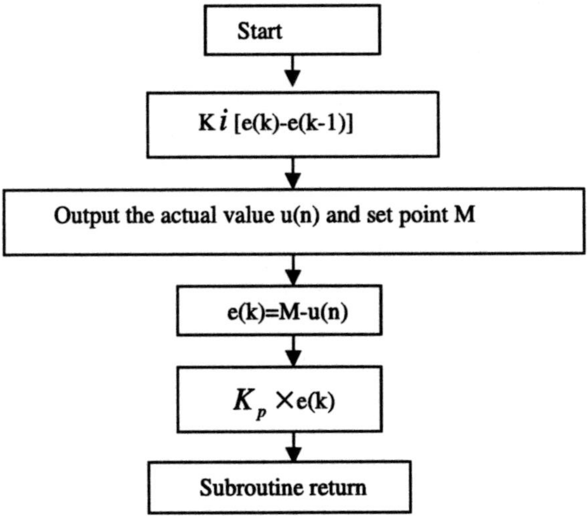

4 Conclusions

The gas pressure control system for gas tightness test of metal capillary tubes, which is higher reliability and simpler structure, was designed with Single Chip Computer as the main controller accompanying with other fittings such as pressure sensor etc. The PID algorithm and digit control, which is advantageous in control system of digit circuit, was successfully applied in above-mentioned pressure control system with quicker system responding rate velocity, smaller overshoot and better stability. The novel gas pressure control system would be helpful in practical application of high pressure detection and control for gas tightness test of metal capillary tubes used in aircraft and airspace fields.

References

1. Chen, X., Zhang, L.: Application of the digital PID algorithm in the automatic test system Foreign electric measurement technique (04) (2006)
2. Fu, X., Jiang, B.: Design of the digital PID controller for DSP. Modern Electronic Technique (2007)
3. Tan, H., Chen, K., Tu, T., Xiao, R.: Design of the data acquisition system in AT89C51 Mechatronics (06) (2005)
4. Li, F.: Research on digital PID control algorithm. Journal of Liaoning University 32(4) (2005)

Combined Vision and Frontier-Based Exploration Strategies for Semantic Mapping

Islem Jebari, Stéphane Bazeille, and David Filliat

Electronics & Computer Science Laboratory, ENSTA ParisTech, Paris, France

Abstract. We present an approach to multi-objective exploration whose goal is to autonomously explore an unknown indoor environment. Our objective is to build a semantic map containing high-level information, namely rooms and the objects laid in these rooms. This approach was developed for the Panoramic and Active Camera for Object Mapping (PACOM)[1] project in order to participate in a French exploration and mapping contest called CAROTTE[2]. To achieve efficient exploration, we combine two classical approaches: frontier-based exploration for 2D laser metric mapping and next-best view computation for visual object search. Based on a stochastic sampling strategy, this approach looks for a position that maximizes a multi-objective cost function. We show the advantage of using this combined approach compared to each particular approach in isolation. Additionally, we show how an uncertainty reduction strategy makes it possible to reduce object localization error after exploration.

Keywords: SLAM, multi-objective exploration, semantic mapping.

1 Introduction

The Panoramic and Active Camera for Object Mapping (PACOM)[1] project aims at developing a robotic system dedicated to indoor autonomous exploration and mapping. In particular, we seek to extract high-level semantic information that is easy to understand and important to the robot users such as surrounding objects and the environment structure.

The semantic map we build is based on a 2D occupancy grid to which the location of objects detected using vision is added. The exploration strategy should therefore both ensure that the coverage of the occupancy grid is complete and that all the objects of the environment have been perceived. We initially considered a purely vision-based exploration strategy that gave overall correct results and covered almost all the area using vision in most cases but had disadvantages such as incoherent and unnecessary long trajectory of the robot. In order to solve these problems, we chose to adopt a multi-objective exploration strategy in which we fuse a frontier-based exploration criteria dedicated to occupancy grid construction, a vision-based criteria

---

[1] The PACOM project is supported by DGA in the frame of the "CAROTTE" competition and funded by ANR under the subvention 2009 CORD 102. <http://cogrob.ensta.fr/pacom/>

[2] CAROTTE is organized by the French research funding agency (ANR) and the French armament procurement agency (DGA). <http://www.defi-carotte.fr>

D. Yang (Ed.): Informatics in Control, Automation and Robotics, Volume 2, LNEE 133, pp. 237–244.
springerlink.com

dedicated to object search and a distance criteria in order to shorten exploration trajectories.

This paper is organized as follow: next section overviews the related work. Section 3 describes the semantic map construction process. The multi-objective exploration is described in Section 4 and Section 5 gives experimental results.

2 Related Work

Autonomous exploration techniques can be classified according to two distinct approaches. In the first one, exploration strategies make use of predefined trajectories [1] which are static and defined off-line, exploiting some a priori information about the environment. In the second approach, exploration is viewed as a sequence of steps, each one composed of a movement towards a location and of an observation with which the robot acquires data about the environment. The exploration of an unknown environment using a 2D scanning laser sensor is often performed using the frontier-based exploration algorithm [2] or an exploration strategy choosing the next best position for the robot given the utility of this position for the mapping problem [3]. In the context of our project, the exploration should take into account the need to search for objects. This general problem of optimally covering the whole environment with several viewpoints is referred to as the "art gallery problem" [4] and is known to be NP-hard. However, several practical solutions exist, for example for optimizing the probability of finding the target given a limited number of actions [5].

3 Semantic Mapping

The semantic map build by our robot integrates the 2D metric map of the environment, the location of objects detected using vision and the segmentation of the free space into distinct rooms. The object recognition algorithm we developed is able to detect initially known objects in the robot's visual environment and to estimate their location in the camera reference frame (see [6] for details). Object detection is performed using a two-step method: first, a fast extraction of the salient regions [7] is performed in order to segment multiple possible objects inside an image. Second, a local features extraction [8] and a bag of visual words algorithm [9] is used to recognize the segmented objects. Once an object is detected, its distance from the camera is estimated using the scale difference between current and reference image, using SIFT keypoint matching between the two images.

The position, name and images of the detected objects along with the segmented rooms are added to the 2D occupancy grid to create the semantic map. Rooms are detected in the occupancy grid through an algorithm that detects doors and analyses the resulting connected components of open space, recognizing as rooms the areas with a sufficient size. The positions in the absolute reference frame of the detected objects are estimated using a Kalman filter for each object.

4 Multi-objective Exploration for Semantic Mapping

The main goal of our project is to build such a semantic map to describe as much of the environment as possible, as quickly as possible. Exploration is therefore required to completely map the environment using the laser scanner and to detect by image processing using a pan-tilt camera the objects laid on the floor. We took advantage of the fact that the two sensors have similar field of view (a semi-circle in front of the robot) to integrate these two objectives into a single function. Additionally, we used a map improvement strategy in order to improve the precision of object localization after exploration is completed.

Computing an optimal set of positions to cover the environment is a NP-hard problem. We therefore use a stochastic sampling strategy inspired by the Randomized Art-Gallery Algorithm [10] to search for the next robot position that discovers the more unmapped area. For each sampled position, a score is computed integrating several criteria:

$$S = \lambda_{obj} S_{obj} + \lambda_{map} S_{map} + \lambda_{dist} S_{dist} \quad (1)$$

where S_{obj} is a score giving the area potentially discovered by vision, S_{map} is a score related to unknown environment discovery using laser and S_{dist} is a score giving the cost of going to this position from the current position. Each score is normalized between 0 and 1 in order to remove the influence of their different orders of magnitude. The sampled position with the highest score is taken as the next position the robot should reach using the Navigation subsystem while performing SLAM (Simultaneous Localization and Mapping) and object detection. When the position is reached, the camera is swept from left to right to perform object detection and a new exploration position is computed. The exploration is performed until a given percentage (95% in this paper) of the free area has been observed by the camera.

For object search, the score S_{obj} is computed as the number of cells of the open area visible through the camera that has not been observed yet. The visible area takes the pan-tilt capability of the camera into account to compute the left and right limit of the field of view. The minimal resolution needed for object recognition and the minimal object size are taken into account to compute the minimum and maximum perception distance. The unobserved area is computed using a 2D view-map that records the area previously perceived by the robot camera. The view-map is updated during robot movements using the camera angle of view, the camera being oriented in front of the robot. When the robot reaches an exploration goal, the area perceived by moving the camera left to right is recorded. Once the view map is updated, the connected components of the unobserved open areas are computed (see the colored areas in the right part of Fig. 1 for example).

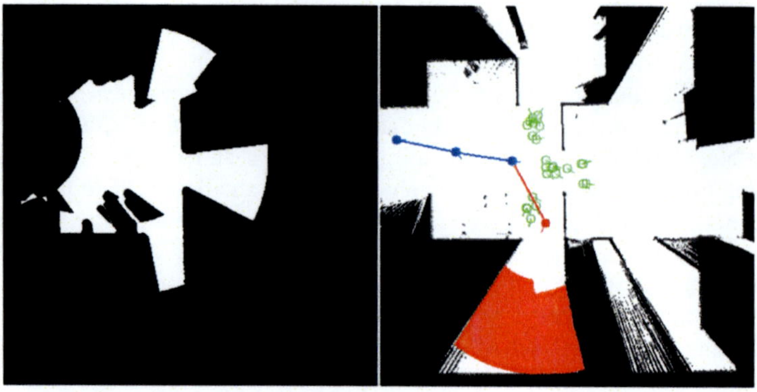

Fig. 1. Left: View-map recording already seen areas. Right: Exploration trajectory with the sampled positions *(in green)*, the selected next position and the next observed area *(in red)*.

The score S_{map} is linked to frontiers discovery. The idea behind frontier-based exploration [11] is to gain new information about the world by moving to the boundary between open space and unexplored space. Then, the robot can see into unexplored space and add the new information to its map. Any open cell adjacent to an unknown cell is labeled as frontier cell. Connected frontier cells are grouped into frontier regions that will be used in the sampling strategy. The score S_{map} is computed as the number of frontier cells between free and unknown area that are visible through the laser sensor from the sampled position.

The score S_{dist} is computed as the opposite of the travel distance from the current robot position. The distance of all grid cells to the current robot position is pre-computed before sampling the positions using the Dijkstra algorithm in order to rapidly evaluate this criterion for any sampled position.

As our approach is a greedy method that selects the next best position, but does not take the whole trajectory into account, the sampling strategy used to select the positions to be evaluated (50 in this paper) may have an influence on the resulting exploration trajectory. In order to favor more continuous trajectories, the positions are first sampled inside the free space in a semi-circular area in front of the robot, with random orientations. If no reachable position can be found in this area or if all the positions have an exploration score below a threshold, positions are sampled from the whole map in order to ensure global covering. These positions are sampled from the connected components of unseen areas and frontier cells by computing robot positions inside the open space that will observe these areas.

A complementary strategy has been implemented to reduce the error position of the detected objects at the end of the multi-objective exploration. We select in our map the objects whose uncertainty is above a threshold and we sample uncertainty reduction positions along the small axis of the object uncertainty ellipses. We then select the position that is the closest to the current robot position. When the position is reached, the Kalman filters are updated to reduce this error if the object is correctly detected. If the object is not detected, it is removed from the semantic map.

5 Experimental Results

Experiments have been performed in a 63 m^2 indoor environment composed of 2 rooms and 2 corridors. We use a Pioneer 3 dx mobile base, with a SICK Laser Range Finder and a Pan-Tilt-Zoom camera. We performed several experiments using vision criteria, frontier criteria and both in order to demonstrate the interest of our approach.

5.1 Vision-Based Exploration

The vision-based exploration strategy using parameters $(\lambda_{obj}, \lambda_{map}, \lambda_{dist}) = (1, 0, 0.6)$ is the first approach we explored because it should be sufficient for correct mapping. Indeed, the laser range is larger than the camera range. So, by trying to cover visually the environment to detect objects we should also discover the whole map using laser.

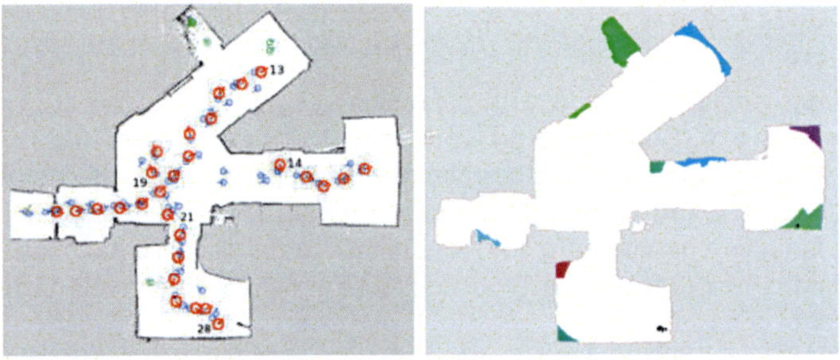

Fig. 2. Left: Vision-based exploration map: red circles correspond to exploration points with the robot orientation. Right: view-map showing areas perceived by the camera *(white areas)*.

Moreover, using this strategy, images are taken mostly when the robot is stopped, resulting in good image quality. However, as seen on the left of Fig. 2, the trajectory generated is disordered and not optimal. Indeed, as we are looking for the larger unseen area, the robot travels more than it needs. As a result, we obtain a complicated trajectory and a slow exploration. A second limitation is that the map doesn't completely cover the area to be explored (the room on the top left wasn't explored) because the robot did not get close enough to discover the room using the laser.

5.2 Frontier-Based Exploration

The frontier-based exploration strategy using parameters $(\lambda_{obj}, \lambda_{map}, \lambda_{dist}) = (0, 1, 0.6)$ is the most suitable to give a complete laser map. It is the fastest way to discover the environment and the resulting trajectory is consistent because exploration points are limited. As an example, since we are looking for edges in the laser map it is not possible to miss the top left room (see Fig. 3). However, this method presents some limitations. Using only this strategy many parts are not explored visually and as images are all recorded during motion between exploration points, the object detection is sometimes missed or less accurate because of blurry images.

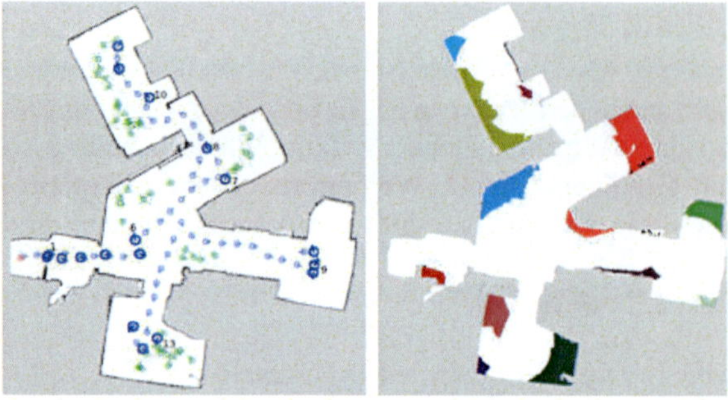

Fig. 3. Left: Frontier-based exploration map: dark blue circles correspond to exploration points with the robot orientation. Right: view-map showing areas perceived by the camera *(white areas)*.

5.3 Multi-objective Exploration Strategy

Fig. 4 shows a map generated with parameters (λ_{obj}, λ_{map}, λ_{dist}) = (1, 1, 0.6). This multi-objective exploration strategy combines the advantages of the two previous methods. As they are complementary the resulting semantic map is complete from a laser point of view, the vision covers nearly the entire laser map and the exploration trajectory is rather consistent. In addition, thanks to the distance parameter, the robot avoids coming back to a previously visited area (a room is completely visited the first time the robot comes inside).

As it is shown on Fig. 4, the multi-objective strategy can switch automatically between visual exploration and laser exploration. During the exit of the small starting room, the score of the first exploration points are mostly due to the frontier criteria because frontier edges are very large. After that, the laser exploration score gets smaller and vision is privileged in order to complete the map. When every near area is seen, frontier exploration becomes stronger again and lead to discover a new room.

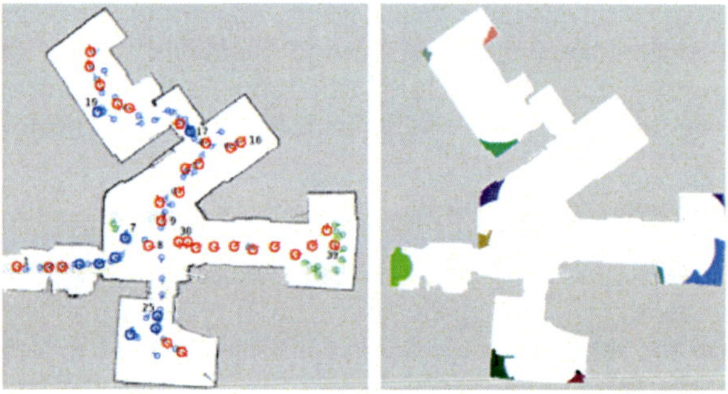

Fig. 4. Left: Frontier & Vision-based exploration map: dark blue circles correspond to exploration points where frontier score is higher than vision score; red points correspond to positions where vision score is higher. Right: view-map showing areas perceived by the camera *(white areas)*.

5.4 Semantic Map Improvement

The final result corresponding to the semantic map obtained at the end of the exploration run is shown on Fig. 5 (left). Green color corresponds to a correct detection with a correct label. Red color refers to a false detection and orange color refers to missed objects. Some images of the detected objects are shown with frames using the same code color as the ellipse of uncertainty. Room detection algorithm successfully detected the rooms and the main corridor. The map improvement module was developed to verify the object detection. That is to say, to delete false detection and to reduce the ellipses of uncertainty. Purple points in Fig. 5 (left) show the positions the robot reached in order to reduce the ellipse of uncertainty, along with some images taken by the robot from these positions. Fig. 5 (right) shows the semantic map after the map improvement module. The ellipses of uncertainty are smaller and one false detection is deleted. Room detection algorithm successfully detects the rooms and the main corridor.

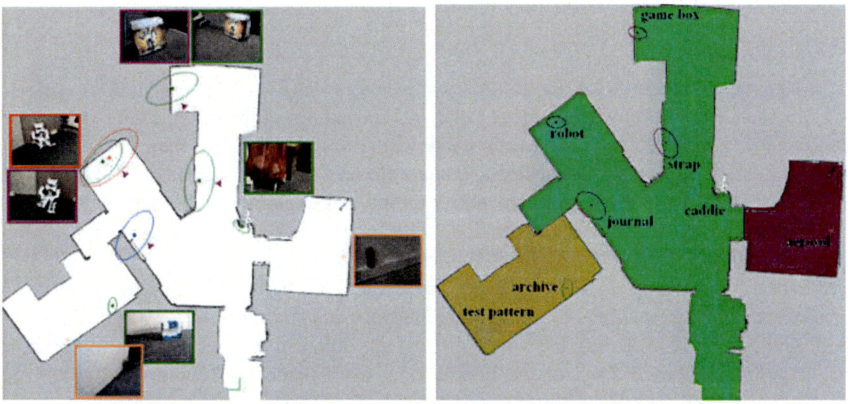

Fig. 5. Left: Semantic map showing the environment structure and the objects position with their ellipses of uncertainty. Right: Semantic map after the map improvement module.

6 Conclusion and Perspectives

In this paper, we present our novel multi-objective autonomous exploration strategy in order to build a complete semantic map of an unknown indoor environment. Our multi-objective strategy provides better performances than single-objective explorations since it successfully integrates both frontier-based and vision-based strategies to benefit from their advantages. Our solution leads to complete laser-based exploration, complete vision-based area coverage, a consistent and nearly optimal trajectory and a better quality of detection as we added a map improvement module. It has to be noticed that our strategy is based on random sampling and may produce results that are intuitively non-optimal, depending on the choice of the multi-objective function parameters. Experimental tuning of the parameters was done for this paper to make it possible to reach good behaviors, but we are developing quantitative experiments using a simulator to have a better estimate of the pros and cons of the proposed approach. However, a difficulty is to find objective performance metrics for the semantic mapping problem.

Currently, our final map improvement strategy is a separate behavior executed after exploration is complete. In future work, we plan to integrate this objective directly in the multi-objective function in order to give a more accurate position of the object after exploration. This should make the robot to move and to take more images of an object after the first detection and should avoid coming back later.

References

1. Bourque, E., Dudek, G.: Viewpoint selection-an autonomous robotic system for virtual environment creation. In: IROS, vol. 1, pp. 526–532 (1998)
2. Yamauchi, B.: Frontier-based exploration using multiple robots. In: Proceedings of the Second International Conference on Autonomous Agents, Minneapolis, Minnesota, United States (1998)
3. Tovar, B., Muñoz-Gomez, L., Murrieta-Cid, R., Alencastre-Miranda, M., Monroy, R., Hutchinson, S.: Planning Exploration Strategies for Simultaneous Localization and Mapping. Journal on Robotics and Autonomous Systems 54(4), 314–331 (2006)
4. Shermer, T.C.: Recent results in art galleries. Proceedings of the IEEE, 1384–1399 (1992)
5. Tsotsos, J.K., Shubina, K.: Attention and Visual Search: Active Robotic Vision Systems that Search. In: The 5th International Conference on Computer Vision Systems (2007)
6. Jebari, I., et al.: Multi-sensor semantic mapping and exploration of indoor environments. In: Proceedings of the IEEE International Conference on Technologies for Practical Robot Applications (TePRA 2011), Massachusetts, USA (2011)
7. Butko, N.J., Zhang, L., Cottrel, G.W., Movellan, J.R.: Visual Saliency Model for Robot Cameras. In: International Conference on Robotics and Automation (2008)
8. Filliat, D.: A visual bag of words method for interactive qualitative localization and mapping. In: Proceedings of the International Conference on Robotics and Automation (ICRA) (2007)
9. Sivic, J., Zisserman, A.: Video Google: A Text Retrieval Approach to Object Matching in Videos. In: Proceedings of the International Conference on Computer Vision (2003)
10. Gonzalez-Banos, L.J.: A randomized Art-Gallery Algorithm for Sensor Placement. In: Annual Symposium on Computational Geometry (2001)
11. Yamauchi, B.: A frontier-based approach for autonomous exploration. In: Pproceedings of the 1997 IEEE International Symposium on Computational Intelligence in Robotics and Automation (1997)

Research and Application of Optimization Extreme of Nonlinear Function Based on Particle Swarm Optimization Algorithm

Qi Han* and Li Zhang*

School of Electronics and Information,
Xi'an Polytechnic University, China, 710048
hanqilin1116@163.com

Abstract. A hybrid Particle Swarm Optimization algorithm was improved the performance of the basic PSO that is easily sank into local optimum and has a slower convergence rate based on the idea of variation of Genetic Algorithm. The hybrid PSO would be used for the extreme optimization of nonlinear function and simulation in MATLAB. By the way of changing the value of inertia weight, we have a conclusion that the algorithm can have a faster convergence velocity and better accuracy .Comparing the simulation result to the basic PSO, algorithm in this paper has a faster convergence velocity avoid sinking into local best.

Keywords: Particle Swarm Optimization (PSO), global optimization, adaptive inertia weight.

1 Introduction

Most optimization problem is multi-objective optimization. Such as the multi-targeted of elevator group control system. The method of evaluate function is current mainstream approach. The multiple-target optimization is translated into the single-target problem. Most function involved in the optimization of the nonlinear function, a key problem is the optimization extreme of nonlinear function .In the field of intelligent control , a problem often encountered in fitting the experimental data. How to solve the problem of fitting data depends on the difficulty of the spatial dimension and nonlinear characteristics of the model .Generally speaking, the more parameters are, the more strong nonlinear and the more time of fitting data is. The accuracy of fitting data is lower. Therefore, we are not certain whether the training process tend to global optima. Particle swarm optimization algorithm is possessed of parallel processing and good robustness. It is used to solve the optimization problems of nonlinear, non-Differential and Many-Valued function. Particle Swarm Optimization makes the test program more succinct, but largely improved the design efficiency.

---

* Qi Han, Postgraduate; Li Zhang, PhD.
Foundation item: This paper is supported by special item fund of education department of Shanxi provincial government(2010JK570).

D. Yang (Ed.): Informatics in Control, Automation and Robotics, Volume 2, LNEE 133, pp. 245–252.
springerlink.com

A new proposed evolutionary computation called Particle swarm Optimization is applied to obtain parameters of the extreme optimization in nonlinear function. In the paper, a hybrid PSO algorithm with mutation is used and this revised version is proved to be more probable to find a global optimal solution and to achieve faster convergence. Simulated result using MATLAB and the experimental result is also on the paper.

2 The Basic PSO Algorithm and Evolution

2.1 The Basic PSO Algorithm

The presented PSO algorithm is an evolutionary computation technique developed by Dr. Eberhart and Dr. Kennedy in 1995, which uses direct encoding method. PSO algorithm was an optimized algorithm based on swarm intelligence. It roots in research on colony movement of bird and fish. It is a kind of evolutive calculational methods based on group brainpower. This algorithm can be described as follows: In the D-dimensional space, we choose any n particles which are regarded as a new swarm X （X=X1,X2,...,Xn）, let i_th particle be a vector in the D-dimensional space $Xi=(X_{i1,} X_{i2}...X_{iD})^T$,it represents the position of i_th particle in the D-dimensional space. According to the Objective function, adaptive value of i_th position was calculated. The speed of i_th particle can be represents as follows: $Vi=（V_{i1,} V_{i2}, ..., V_{iD}）^T$; individual extreme: $Pi=（P_{i1,} P_{i2}...P_{iD}）^T$; global extreme of the swarm: $Pg=（P_{g1,} P_{g2}...P_{gD}）^T$.

The formula for updating speed and position of the particle can be described as follows:

$$V_{id}^{k+1}=w\times V^k{}_{id}+c_1\times r_1\times(P^k{}_{id} - X^k{}_{id})+c_2\times r_2\times(P^k{}_{gd} - X^k{}_{id}) \quad (1)$$

$$X^{k+1}{}_{id}=X^k{}_{id}+V_{id}{}^{k+1} \quad (2)$$

In the formula, w represents inertia weight; d=1,2,...,D, i=1,2,...,n; k represents iteration times at present;c1and c2 represents non-negative constant, they are called acceleration factor;r1 and r2 are random numbers between 0 and 1; Pi(Pi=Pi1, Pi2,..., PiD)represents i_th particle best position by searching whole space so far; Pg (Pg =Pg1, Pg2,...,PgD) is swarm best position by searching whole.

2.2 Selection of the Inertia Weight

Selection of the inertia weight is about the extent to which the speed of current particle inherits previously. Shi. Y introduced the inertia weight to the PSO algorithm. They also analyzed that a large inertia weight is help for the global search and a small inertia weight is help for the local search. Now, we improve the algorithm, which could strengthen the local search, and also considered the global search simultaneously. In the PSO algorithm, the inertia weight is embedded into the adaptive optimization algorithm to improve its search performance.

The expression is of the form:

$$w(k) = w_{max} - (w_{max} - w_{end})(T_{max} - k) / T_{max} \tag{3}$$

Wmax is the largest inertia coefficient which we can select;
Wmin is the smallest inertia coefficient which we can select;
k is the current iteration times;
Tmax is the largest iteration times.

Generally speaking, algorithm can have better performance when Wmax is 0.9 and Wmin is 0.4.The hybrid algorithm takes full advantage of the largest inertia coefficient to improve the ability of global search of PSO algorithm. With iteration times increases, the smallest inertia coefficient improved the ability of local search. Non-linear inertia coefficient was used to avoid the algorithm into a local optimum.

There are several commonly used expressions:

$$w(k) = w_{min}(w_{max} / w_{min})^{1/(1+ck/Tmax)} \tag{4}$$

2.3 Tracking of Sensitive Particle

PSO algorithm is widely used to solve the static optimization in many fields. However, the effects of the environment in practical application, the problems are more complex and time-varying. In other words, problems are dynamic. Based on the best fitness of the particles, the ranges of both search space and velocity of the particles are contracted dynamically with the evolution of particle swarm. To track dynamic extremes, we can carry out optimization with sensitive particle PSO algorithm in dynamic environment. We should calculate the fitness value of sensitive particles in each of iteration. When the fitness values changes, we can judge that the environment has changed.

Improved formula for the position:

$$x(t+1) = x(t) + r \times v(t+1) \tag{5}$$

r is random numbers between 0 and 1;

2.4 Mutation Operation

The speed of PSO algorithm was reduced by the use of mutation. A eugenic evolution strategy was proposed to improve the efficiency of the basic particle swarm optimization algorithm searching. Variation operation was embedded in the evolution process to speed up the global searching. This strategy expanded the shrink search space with the increasing of the iteration times. Particle can jump out the optimal position by means of searching previously. Corresponding MATLAB code is as follows:

```
If   rand•0.9
K=ceil (2*rand);
Pop (j,k)=rand;
End
```

3 Application of the Extreme Optimization of Non-linear Function

3.1 Figure of Algorithm

The figure of the extreme optimization of nonlinear function base on the PSO algorithm is as follows:

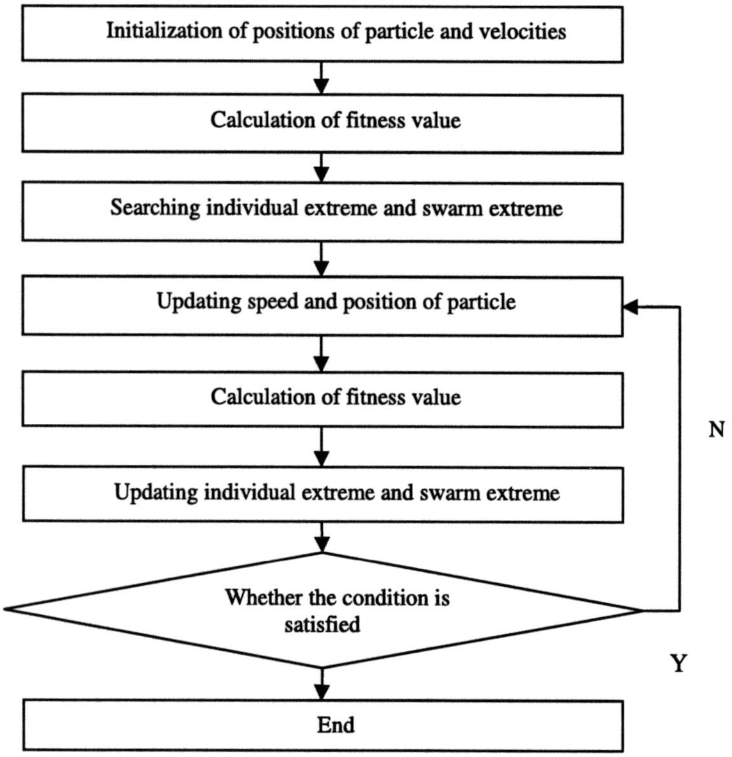

Fig. 1. Flow chart of algorithm

Ackley function was chosen as the expression of the fitness function.
The expression is of the form:

$$y = -c_1 \exp\left[-0.2\sqrt{1/n\sum_{j=1}^{n} x_j^2}\right] - \exp\left(1/n\sum_{j=1}^{n}\cos(2\pi x_j)\right) + c_1 + e \quad (6)$$

The particle number of the swarm is 20; the dimension of each particle is 2; 200 was chosen as the largest number of iteration. First, the initialization process is completed. The effect of the initial information is assign random numbers to the position and velocity of the particle. Then, we should calculate the fitness-value of the particle. Individual extreme and swarm extreme can be determined according to the fitness-value. To update the velocity and position of the particle according to the (1) (2)steps. Finally, new individual extreme and swarm extreme should be determined according to the fitness-value of particle of the new swarm.

3.2 Part Code of Algorithm

It is important to achieve this process with the code in MATLAB.
Part of the MATLAB code is as follows:

```
Initialization of positions of particle
For i=1; sizepop
%random swarm
Pop (i, :) =5*rands(1,2);
V (I, :) =rands(1,2);
%calculation of the fitness-value of the particle
Fitness (i)=fun(pop(i,:));
End
The code of the initial extreme values
[bestfitness bestindex]=min(fitness);
Zbest=pop (bestindex,:);
Gbest=pop;
Fitnessgbest=fitness;
Fitnesszbest=bestindex;
```

The basic PSO algorithm and improved PSO algorithm have been simulated respectively in MATLAB, and the theory of improved PSO algorithm has been studied. The results for fitting of nonlinear function are analyzed and compared.

4 Simulation and Analysis of Data

In the last passage we had discussed the process of two algorithms. Best individual fitness of every generation was plotted in MATLAB. The figures are as follows:

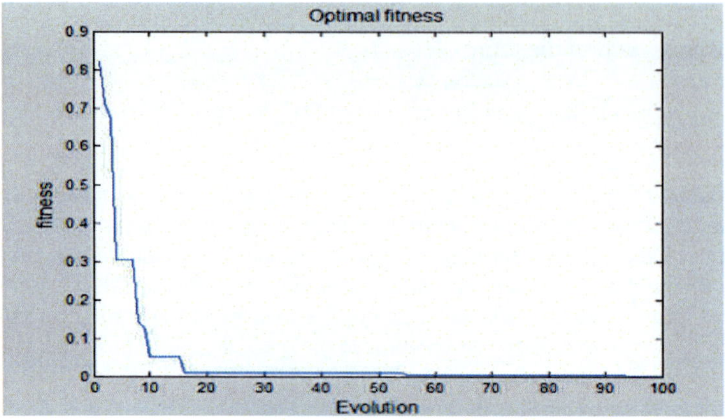

Fig. 2. Optimization fitness value (1)

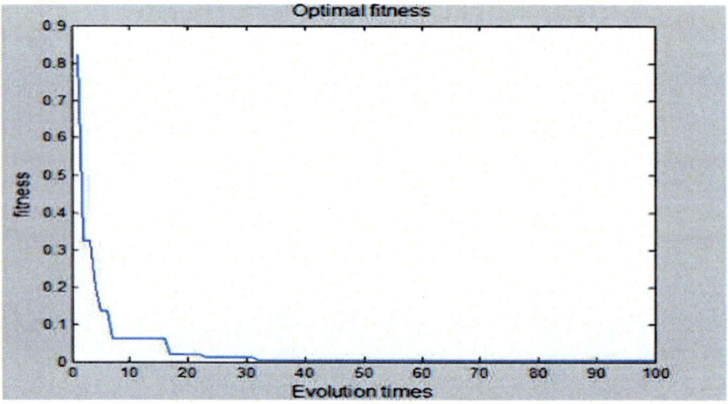

Fig. 3. Optimization fitness value (2)

From the end result, simulation shows that the algorithm has weak point. The basic algorithm is an effective and high veracity method, but it is easy to get into the local best. In other words, algorithm has a fast convergence but not get the global optimal solution in a theory. The global extreme (gbest) by searching swarm is a local optimal solution. Algorithm appeared a phenomenon that is known as precocity. Variation operation was introduced to the algorithm for searching global extreme. This operation changed the direction of the particle and make the particle has an opportunity to searching other areas .New individual and global extreme could be found by this method. Then the algorithm can continue searching whole space. According to **Fig. 1.** We can get the best individual fitness-value (0.0329), corresponding to the position (0.0083,-0.0068) of particle. The actual optimal value can get close to the optimal value by the experiment. According to **Fig. 2.** We can get the best individual fitness-value (0.0063), corresponding to the position (0.0020, 0.0008).Compared to the basic PSO algorithm, improved algorithm by introducing variation operation can get out of the local minimum extreme points for better solution.By comparing and analyzing, we can embed variation operation in the basic PSO algorithm avoid sinking into local optima. This method is simple and fast in speed of operation. Improved PSO algorithm can get better solution by appling to fitting of nonlinear function.The effect of changing inertia weight is obvious. Let do an experiment to confirm the idea.We chose an ordinary function, the expression is as follows:

$$F(x) = 100 * (x_i - x_j^2)^2 + (x_j - 1)^2 \tag{7}$$

i and j represent integral;
the figures are as follows:

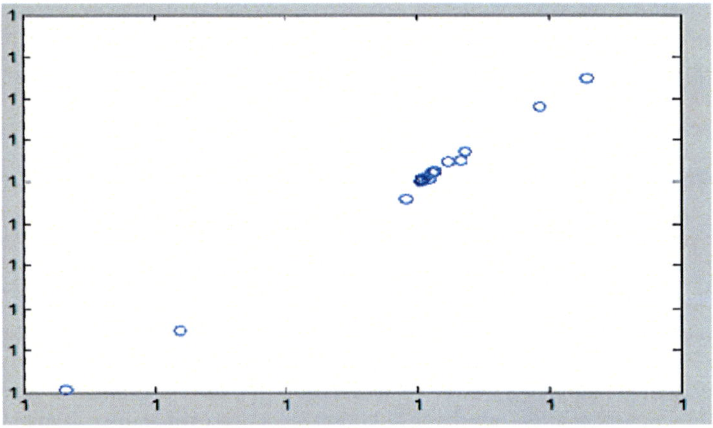

Fig. 4. The curve of extreme optimization (1)

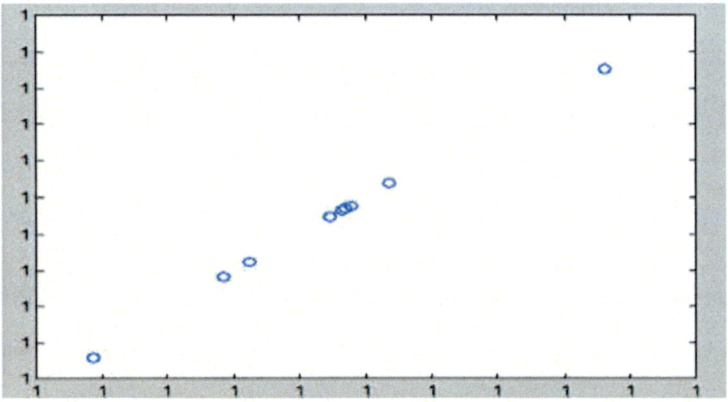

Fig. 5. The curve of extreme optimization (2)

Using a PSO algorithm coded by MATLAB, the fitting of nonlinear function problem was calculated. Assuming the maximum number of iterations is 200.We set different inertia weight value for algorithm .Two algorithms have been simulated respectively. A kind of inertia weight value is conventional, another kind is nonlinear degressive. By comparing and analyzing to the two charts, right result was certificated. According to **Fig. 3.**, there are many local optimization values. We can find that a lot of points around the global optimum. Comparing to **Fig. 6.**, there are many local optimum but not many points around the global optimum. From the chart4-3, we can judge that the algorithm of nonlinear inertia weight value has a better ability of searching whole space. Other selection is more easily lead the algorithm into the local best. In this procedure, the expression of the inertia weight is of the form:

$$W(k)=Wmax-(Wmax-Wmin)*(k/Tmax)^2 \tag{8}$$

Based on the above content, It has a better ability of searching by introducing the variation operation. The MATLAB simulation shows the proposed method is valid. To change the value of inertia weight has improved the ability of searching for global extreme avoid sinking into local optima.

5 Conclusion

In this paper, the basic particle swarm optimization algorithm was described. Secondly, by making use of the variation operation, the whole space search is carried out in the whole solution space to avoid earliness and convergence. Application of the extreme optimization for nonlinear function has proved that the improved algorithm is feasible. Then, we set the different value of the inertia weight to discuss the effects of the algorithm. Finally, the result of simulation shows the improved algorithm is practicable and its optimal results take the precedence over the traditional methods. There is multi-objective particle swarm optimization algorithm based on fuzzy control. The development of hybrid PSO algorithm is the direction of the future.

References

1. Ded, K., Pratap, A., Agarwal, S., et al.: A fast and elitist multi-objective genetic algorithm: NSGA2. IEEE Transactionon Evolutionary Computation 6(2), 149–172 (2002)
2. Kennedy, J., Eberhart, R.C.: Particle Swarm Optimization. In: Proceedings of IEEE International Conference on Neural Networks, pp. 1942–1948
3. Shi, Y., Eberhart, R.C.: Empirical Study Optimization. In: The Proceedings of the IEEE Congress on Evolutionary Computation, pp. 1945–1950 (1999)
4. Shi, Y., Eberhart, R.C.: Comparing inertia weights and constriction factors in particle swarm optimization. In: The Proceedings of the IEEE Congress on Evolutionary Computation, pp.84–88 (2000)
5. Hua-Liang, W., Billings, S.A., Yifan, Z., Lingzhong, G.: Lattice Dynamical Waveler Neural Networks Implemented Using Particle Swarm Optimization for Spatio-Temporal System Identification. IEEE Trasactions on Neural Networks 20(1), 181–185 (2009)
6. Li, X.-D.: A non-dominated sorting particle swarm optimizer for multi-objective optimization. In: Proc of International Conference on Genetic and Evolutionary Computation, pp. 37–48 (2003)

Design of Temperature Measuring and Controlling System Based on STM32

Xiangtong Kong, Chunping Wang, and Shuying Sun

Department of Optical and Electronic Engineering,
Ordnance Engineering College, Shijiazhuang, 050003, China
kongxiangtong28@163.com

Abstract. In the DIP devices' quick release system's design, the temperature adjustment of the solder is needed to protect the circuit. In this paper, a intellective temperature control and measure system is designed, which is based on the STM32 series micro-controllers. The structure of the system's hardware is introduced detailed. For the Fuzzy-PID temperature control strategy, the system can adjust the temperature of the solder accurately.

Keywords: DIP, temperature control, STM32, Fuzzy-PID.

1 Introduction

For most of the DIP package devices' disassembly tools, misconduct will result in a tilt of the circuit board copper foil or the damage to the peripheral devices, even cause the circuit board scrap accident[1]. For the above defects, we designed a tin bath for the disassembly technique based on local heating. As the disassembly device needs accurate temperature of the solder, while the traditional analog temperature control method's accuracy is not higher enough to meet the requirements, therefore, to design a convenient and accurate temperature measuring and controlling system is very important.

The design of the temperature control system based on MPU can significantly improve the efficiency and accuracy of temperature control. The microcontrollers of STM32 family is produced by STMicroelectronics which is a type of 32-bit, sustain real-time simulation and tracking, based on the ARMv7 architecture. It also has the advantages of high performance, low cost and low power consumption [2].

This paper describes a design of temperature measurement and control system based on the microprocessor STM32F103R8. By using a simple touch-screen-based interactive systems, we can facilitate the achievement of real-time display and precise control the temperature.

2 The Structure of the DIP Rapid Removal Tools

The traditional commonly used removal tools, such as the hot air gun, tin stove ,have the problems of the heating area is too large ,So it's easily to damage the surrounding devices. In order to solve these problems, The designed heated Tin bath tooling is shown in Figure 1.

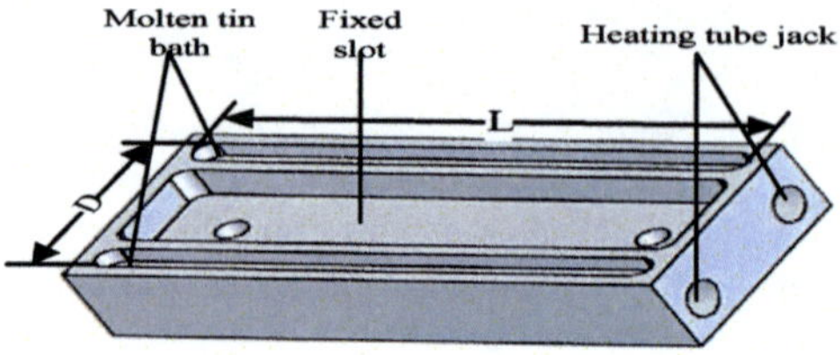

Fig. 1. The Schematic diagram of heated tin bath tooling

The jack of the Heating tube is located in the below of the molten tin bath, by inserting the Heating tube to heat up the solder. The model of the Heating tube is BaKon1323. The structure is shown in Figure 2.

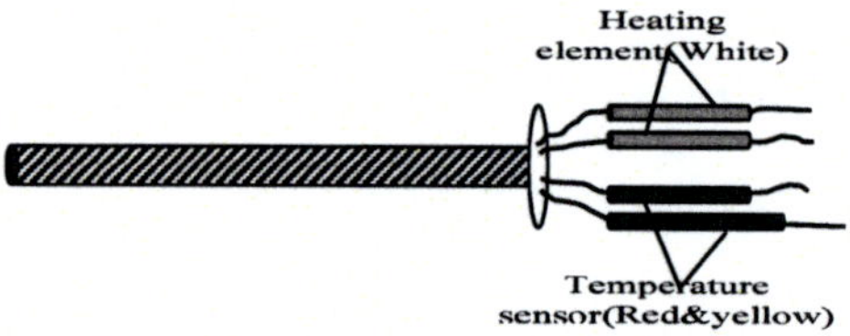

Fig. 2. The Schematic diagram of the BaKon1323

3 Temperature Control System Hardware Design

The system mainly consists of the micro-controller module, the data acquisition module, the temperature control module, and interactive module and other components. The overall structure of the system is shown in Figure 3.

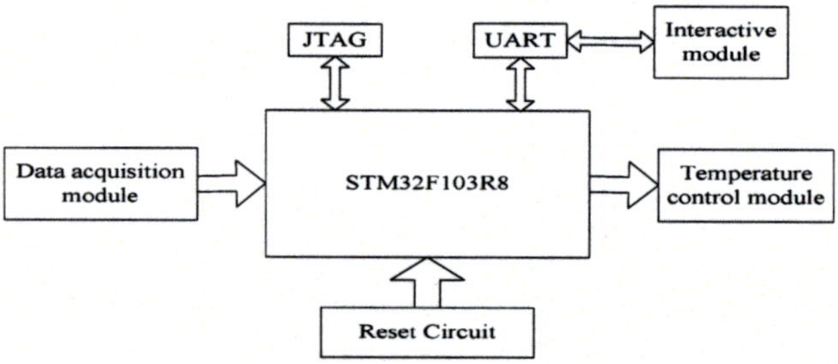

Fig. 3. System block diagram

3.1 The Micro-controller Module

The Micro-controller module is made up of the STM32F103R8 and its peripheral circuit, mainly used for the data transmission and processing, which is the core of the system. The operating frequency of this type microprocessor is 72MHz, with 128K bytes of Flash memory and 20K bytes of SRAM and a wealth of GPIO and flexible 7-channel general purpose DMA, and a rich communication interface (USTAR,I2C,SPI,CAN,USB) in it[3]. The two 12-bit ADC which rates can up to 1MHz, which is an important factor for the system chose this device as the core controller. The STM32F103R8 microprocessor is chosen the LQFP-64 package.

3.2 The Data Acquisition Module

The Data acquisition module is mainly made up of The BaKon1323 Replacing the Heating Element and The Dual operational amplifier LM358. When operating, the voltage signal that is generated by the temperature sensor can reflect the temperature of the solder. After signal conditioning (mainly filtering) and amplification, the signal can be processed by the Microcontroller. The schematic of the Data acquisition module circuit is shown in Figure 4.

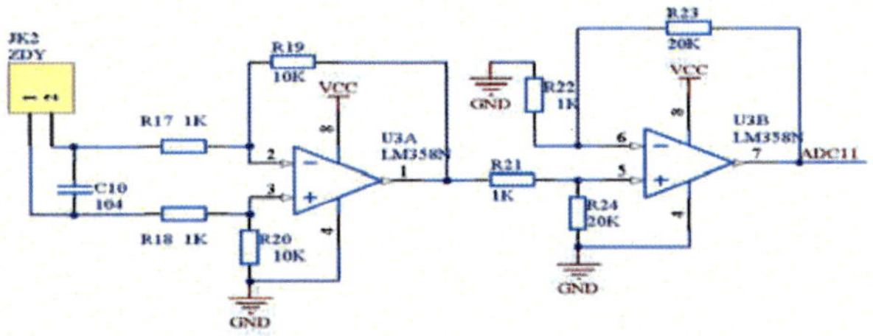

Fig. 4. The schematic of the Data acquisition module circuit

According to the experiment, the relationship of the BaKon1323's voltage signal and the temperature is shown in Figure 5, to meet the linear relationship.

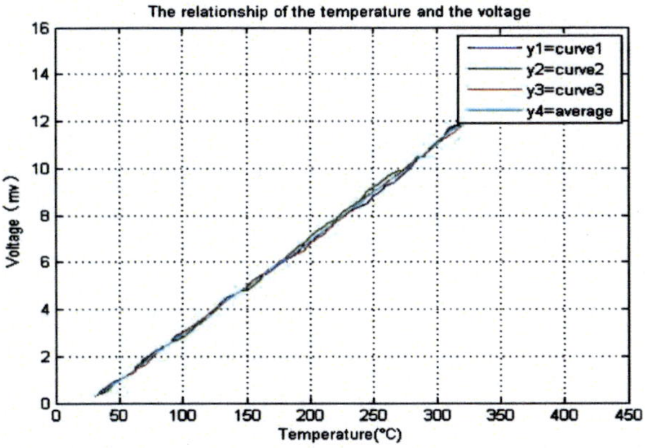

Fig. 5. The diagram of the relationship of the voltage and the temperature about the sensor

The amplifier selected LM358, which consists of two independent, high gain, internally frequency compensated dual operational amplifiers[4]. LM358 is +5 V single power supply mode.

3.3 Temperature Control Module

The Temperature control module is mainly made up of the MOSFET IRF3205 and NPN transistor. By controlling the relevant signals to control FET's breaking up to realize the purpose of flexible control of the heating power. The Heating module as in Figure 6. Cooling time can be used in the natural way of the air.

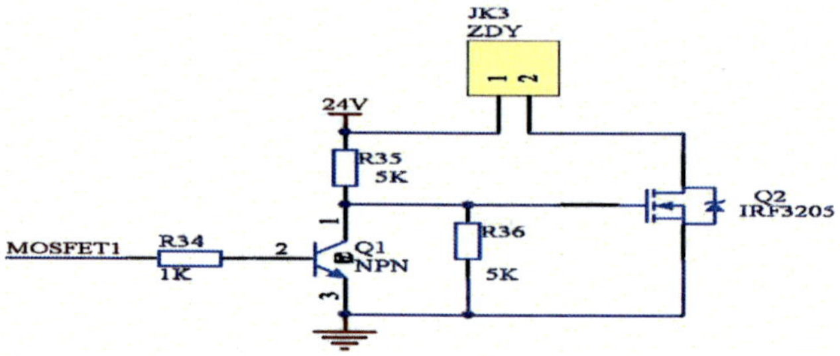

Fig. 6. The schematic of the Heating module circuit

3.4 Human-Computer Interaction Module

The Human-Computer Interaction module mainly consists of the DWIN LCD touch screen and serial communication modules. As DWIN LCD touch screen uses the RS-232 communication standard, the system designed two RS-232 serial communication interfaces. The communication port is shown in figure7.

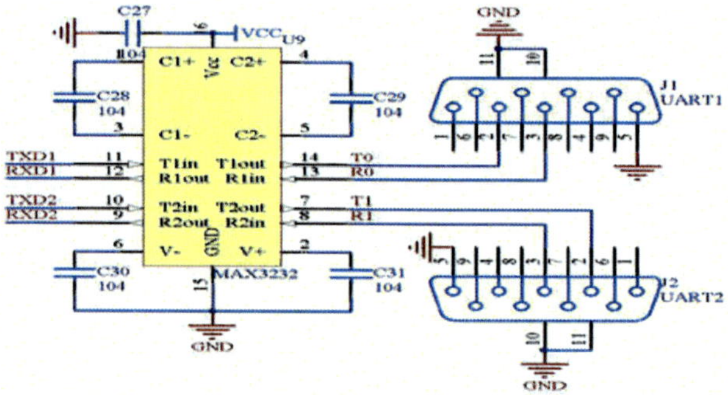

Fig. 7. The schematic of the serial port circuit

4 Software Design

4.1 Software Design of the Temperature Control System

After the software initialization, the processor firstly processes the data transmitted by the temperature sensor, the current temperature will display on the LCD touch screen, the operator touch the corresponding button on the touch screen to input the set temperature. Then, the microcontroller will adjust the system's temperature. In the adjustment process, the microcontroller will control the heating power according to the relationship of the temperature sensor's information and real-time temperature and the selected temperature control strategy. Eventually, the closed-loop control system will reach the quickly and accurately change about the temperature. The total software flow chart of the system is shown in Figure 8.

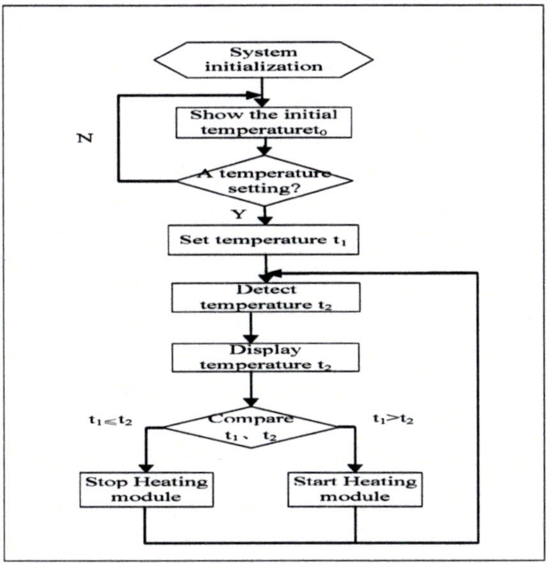

Fig. 8. The total software flow chart of the system

4.2 Fuzzy-PID Control

Fuzzy control theory provides a feedback control method which treats the computer-constitute fuzzy controller as the core and the simulation of the human brain way of thinking as the basic starting point[5]. The simulation curves of the Fuzzy PID and the general PID are shown in Figure9. The curve 1 is an ordinary PID control performance curve, 2 is the effect of the Fuzzy-PID control. From the curves, we know that the Fuzzy-PID temperature control strategy has the characteristics of the smaller overshoot, shorter response time and better robustness.

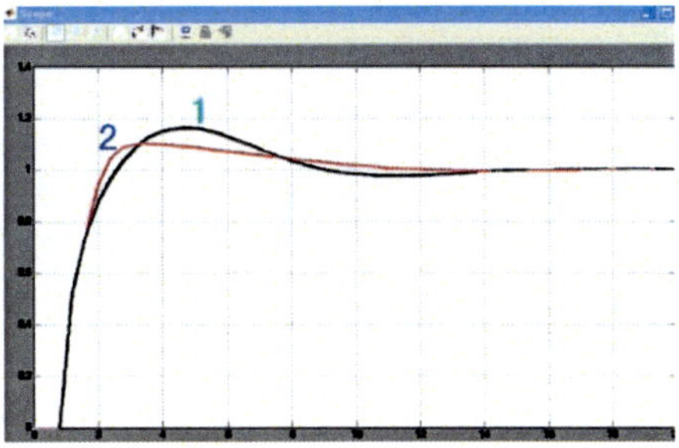

Fig. 9. Simulation curves of the Fuzzy-PID and PID control strategy

The temperature control strategy using the Fuzzy-PID closed-loop control strategy. To set the difference of the set temperature and real temperature as the Fuzzy-controller input, the rate of the temperature change as the input rate of the Fuzzy-PID controller[6].

5 Conclusion

Because of the needs of the DIP devices' Quick release system design, this paper designed a temperature measurement and control system based on the ARMv7 architecture stm32 processor, used the Fuzzy-PID control strategy to achieve the accurate temperature control. The experiment shows that the heating system is rapid heating and low temperature overshoot, at the same time, the system can keep the temperature constant, which can meet the design requirements. When using in the DIP package device quick disassembly system, the heating effect is obvious. The system can also be extended to other applications that require precise temperature control.

References

1. Lou, L.: Research on digital PCB and IC replacing maintenance technology based on FPGA/CPLD, pp. 59–60. Ordance Engineering College, Shijiazhuang (2010)
2. Wang, Y., Xu, W., Hao, L.: ARM Cortex-M3 Microcontrollers Principles and Practice of STM32 Series, pp. 22–26. Beihang University Press, Beijing (2008)
3. Sun, S., Chen, Z., Kou, C.: The Developing and Application of the New Generation of Embedded MPU-STM32F103. Micro Computer Applications 12, 59–63 (2010)
4. ST Microelectronics. Rev 8, http://www.ti.com
5. Zhu, J.: Fuzzy control theory and system principle, pp. 20–25. China Machine Press, Beijing (2005)
6. Ishibuchi, H., Yamamoto, T.: Fuzzy rule selection by multi-objective genetic local search algorithms and rule evaluation measures in data mining. Fuzzy Sets and Systems 141(1), 23–37 (2004)

Expert-System-Based Design of Large Networks of Deployable Mechanisms

Bingyin Ma[1], Zongquan Deng[2], Hailin Huang[1], Bing Li[1], and Rongqang Liu[2]

[1] Shenzhen Graduate School, Harbin Institute of Technology,
518055 Shenzhen, China
[2] School of Mechatronics Engineering, Harbin Institute of Technology,
150001 Harbin, China
bingyin_ma@hotmail.com, denzq@hit.edu.cn, huanghl@hitsz.edu.cn,
libing.sgs@hit.edu.cn, liurq@hit.edu.cn

Abstract. Design of large networks of deployable mechanisms often involves simultaneously multiple considerations. This paper presents an expert system, the reasoning approach of which combines rule-based reasoning (RBR) and fuzzy comprehensive evaluation method, for generating all the available deployable candidates that satisfying the requirements specified a prior by the designers. The expert system can also automatically evaluate the performances of the generated mechanisms, which is the basis for the selection between all available candidates.

Keywords: Rule-based reasoning, Fuzzy Comprehensive Evaluation, Large networks of deployable mechanisms (LNDM).

1 Introduction

Deployable mechanisms, which are widely used in aerospace industries and general civil engineering, are the special type of engineering mechanisms whose geometry can be changed from small volume into big volume to meet practical requirements [1], [2]. However, the design of large networks of deployable mechanisms involves simultaneously multiple considerations so that it is difficult.

Expert systems have been regarded as a powerful tool for creating intuitive human computer interfaces [3]. It is also a convenient tool for disseminating rare or costly expertise available, such as design and evaluation of LNDMs. However, the design of LNDMs often involves multiple considerations, many of which may be contradict with each other. Hence, the reasoning with IF-THEN only is not suitable for this type of problems.

In order to acquire and evaluate the potential schemes, this paper proposed a new reasoning approach which combines RBR and fuzzy comprehensive evaluation [4]. This approach can automatically select all potential schemes and locate the optimal case. By integrating this approach, our expert system can provide an effective computerized environment for users to design and evaluate LNDMs.

D. Yang (Ed.): Informatics in Control, Automation and Robotics, Volume 2, LNEE 133, pp. 259–262.
springerlink.com

2 Design of Large Networks of Deployable Mechanisms

Designing LNDMs is a very complex process. The fully deployed configuration of LNDMs may range from planar configuration to spatial configuration. However, this paper only focuses on planar configuration. Specifically, if the fully deployed configurations of LNDMs are connected to the different planar deployed profiles, such as triangle, hexagon, etc, then the fully deployed configuration of LNDM can be regarded as a plane which consists of those deployed profiles. Fig. 1 shows an example of LNDM; the fully deployed configuration can be considered to be a plane which constitutes of triangle profiles.

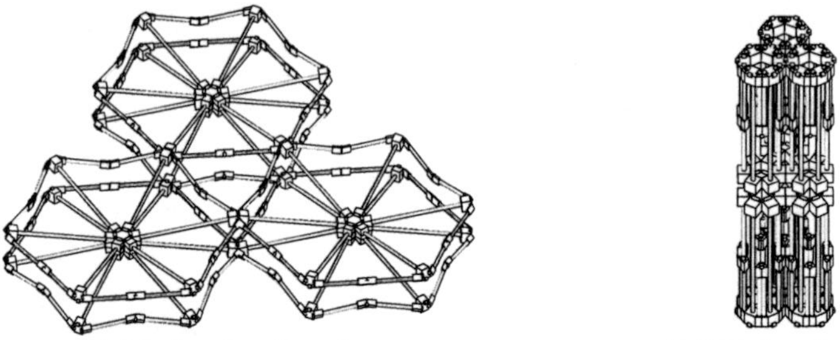

(a) Deployed configuration (b) Folded configuration

Fig. 1. Example of large networks of deployable mechanisms

In the following sections, we will give a detailed introduction on the hybrid reasoning process of RBR and fuzzy comprehensive evaluation.

3 Reasoning Process

Fig. 2 shows the overview of the reasoning process. Firstly, the system loads the information that the designers have been selected, then RBR reasons all possible deployable schemes, which constitute the available candidate scheme set. If this set is null or has only one solution, then continue another reasoning process or obtain the only solution. Otherwise, by using fuzzy comprehensive evaluation method, the optimal solution could be obtained according to the references the designer has defined.

3.1 RBR

In this expert system, we first collect all feasible LNDMs that have been reported in the Ref. [2], and then classify them based on the deployed profile of the single loop deployable units. To sum up, there are three categories of deployed profiles in the system, i.e. triangle, quadrilateral, hexagon. Other relevant criteria, such as degree of freedom (DOF), type of the joint, also have been considered.

Example rule is as follow:

```
IF (DOF=1) & (type_joint=Revolute) & (profile=triangle)
Then LNDM='Bricard' or 'Myard'
```

'Bricard' and 'Myard' in last rule stand for the set of deployable Bricard mechanisms and deployable Myard mechanisms, i.e., all possible triangle deployed profile of LNDMs are based on Bricard and Myard mechanism.

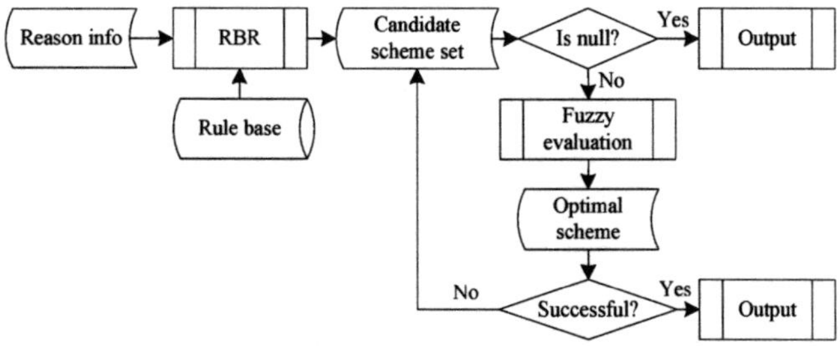

Fig. 2. Overview of the reasoning process

3.2 Fuzzy Comprehensive Evaluation

To reduce redundant computational efforts, the minimal assembly unit is defined as a unit when: (a) having the same degrees of freedom with the LNDM; (b) connected by the minimal number of deployable units; (c) using the same connection method, the unit can be extended to the whole LNDM. All criteria can be tested based on the minimal assembly unit. To implement fuzzy comprehensive evaluation, the following steps have to be taken.

(1) Get candidate scheme set $U = \{u_1, u_2, \ldots, u_m\}$.

By using RBR, the candidate deployable scheme set has been located.

(2) Define the influencing criterion set $V = \{v_1, v_2, v_3, v_4\}$.

The evaluation is influenced by many criteria, four criteria for LNDM are proposed in the system: the deployable ratio v_1, the numbers of links per unit area v_2, the numbers of joints per unit area v_3 and the degree of freedom v_4.

Deployable ratio. The deployable ratio is defined as the ratio of the minimum enveloping area of LNDM divided by the maximum deployed area of LNDM.

Numbers of links and joints per unit area. The numbers of links and joints per unit area are defined as ratios of the numbers of links and joints divided by the area of the fully deployed minimum assembly unit.

(3) Compute the influencing criterion of each candidate schemes, and then by comparison with each value under the same criterion to decide the performance norms (r_{ij}) of LNDM. To deployable ratio and degree of freedom, the more numbers a candidate has, the more likely to be taken an optimal case. On the other hand, the less numbers the other two references have, the more likely to be taken an optimal case. Lastly, form the evaluation matrix R.

$$R=\begin{bmatrix} r_{11} & r_{12} & ,\cdots, & r_{1m} \\ r_{21} & r_{22} & ,\cdots, & r_{2m} \\ r_{31} & r_{32} & ,\cdots, & r_{3m} \\ r_{41} & r_{42} & ,\cdots, & r_{4m}, \end{bmatrix}=\left(r_{ij}\right)_{4\times m}.$$

(4) Determine weight set for each influence factor.

According to analytic hierarchy process (AHP) method [4], the weight set can be obtained by following steps:

Firstly, construct the judgment matrix A according to AHP judgment table.

$$A=\begin{bmatrix} 1 & 7 & 7 & 9 \\ 1/7 & 1 & 1 & 5 \\ 1/7 & 1 & 1 & 5 \\ 9 & 1/5 & 1/5 & 1 \end{bmatrix}.$$

Secondly, compute and construct the weight set W.

$W = (w_1, w_2, w_3, w_4) = (0.672, 0.142, 0.142, 0.043)$.

Thirdly, determine consistency for the judgment matrix.

Since $CR=0.091<0.1$, the judgment matrix satisfied consistency.

(5) Synthesize the above procedures to find the optimal solution.

On the basis of the weight set W and the evaluation matrix R, the sorted scheme set can be obtained by the fuzzy operation: $B = W \bullet R = \left(b_1, b_2, \cdots, b_m\right)$. Therefore the optimal scheme is obtained based on the sequenced results. $b_k = \max_{j=i}^{m}\left\{b_j\right\}$.

4 Conclusion

This paper has presented an expert system with hybrid reasoning approach of RBR and fuzzy comprehensive evaluation that can assist the design of LNDMs.

According to the criteria that we have defined, the presented approach provides a novel method to evaluate deployable mechanisms and select the optimal solution. With the input of requirements information, our expert system could automatically locate the optimal LNDMs and evaluate its performance.

References

1. Chen, Y.: Design of Structural mechanism. University of Oxford (2003)
2. Deng, Z.Q., Huang, H.L., Li, B., Liu, R.Q.: Synthesis of Deployable/Foldable Single Loop Mechanisms with Revolute Joints. ASME Journal of Mechanisms and Robotics (2011)
3. Peng, G.L., Chen, G.F., Wu, C., Xin, H., Jiang, Y.: Applying RBR and CBR to develop a VR based integrated system for machining fixture design. Expert Systems with Applications 38, 26–38 (2011)
4. Peng, Z.Z., Sun, W.Y.: Fuzzy Mathematics and its Applications, pp. 78–114. Wu Han University Press, Wu Han (2007)

A Heuristic Algorithm for the Initial Codebook Design in Vector Quantization

Xu Yatao*, Lin Changqing, Yuan Qin, and Zheng Deshun

Academic Research Department, Logistics Academy, Beijing, 100858, China
Logistics Command Department, Logistics Academy, Beijing, 100858, China
Academic Research Department, Logistics Academy, Beijing, 100858, China
Technology Research Department, Piaoxun Network Limited Company,
Nanjing, 210093, China
yataoxu@gmail.com

Abstract. In this article we propose a heuristic scheme to design VQ initial codebook. Following this scheme we design a concrete algorithm. Our main idea is to construct a weight function as the heuristic function, according to which we divide the training vectors set into many objective subsets dynamically, and finally we generate codes in each objective subset at random. Training this initial codebook by GLA algorithm, we achieve an approximate global optimal one.

Keywords: initial codebook design, Vector Quantization, maximum descent, split algorithm, global optimal.

1 Introduction

GLA algorithm[1] is a milestone of Vector quantization (VQ)[2] research. Its performance relies on the initial codebook design closely. There are several initial codebook generation methods such as random coding algorithm[3], pair wise nearest neighbor (PNN) algorithm[4], splitting algorithm[1] and maximum descent (MD) method[5]. The latter two are widely used in picture compression and speech recognition, but they usually trap into local optima. In this article we propose a heuristic scheme for initial codebook design, called division and distribution (DD) scheme, to achieve an approximate global optimal codebook. Here is our main idea and motivation on DD scheme: (1) Identify the training vectors set as the current set. (2) Divide the current set into several subsets. (3) Calculate the required code number for each subset according to the weight function. (4) For each subset, test whether the required code number is lower then a given threshold. If so, design the initial codes of this subset by Monte Carlo method. Otherwise, identify this subset as current set and go to Stage 2.

Why we divide the training vectors set? It has been proved that the work of searching for the optimal codebook is NP-hard[6], even for the dimension as low as two. If we divide the training vectors into several small sets, and search for the codes

* Author sponsored by NSFC11101212.

D. Yang (Ed.): Informatics in Control, Automation and Robotics, Volume 2, LNEE 133, pp. 263–271.
springerlink.com

in every set separately, the work becomes much easier. In the ultimate situation, there is only one code in a set, and the centroid is the best code. But on the other hand, division plane will break the optimal cluster.

How to construct weight function? Weight function plays the most important role in DD scheme. Our construction is based on the concept of distortion, which is defined in section 2. How to decide recursion stop condition? Since too much division will break the optimal cluster seriously. We should set a stop condition. After several iterations, the training set is divided into many small subsets. When the training vectors in a small subset obey uniform distribution, we can use random coding method to generate codes, which has a better performance for the uniform signal source. According to the experimental results of random coding algorithm, we usually stop the recursion for one subset when its required code number w nears 5% of the given number. How does it look like after division? Regard the square in figure 1 as a projection of training set in a plane, and the gray stands for the density of training vectors. Every grid stands for a subset. Figure 1 shows the division effect.

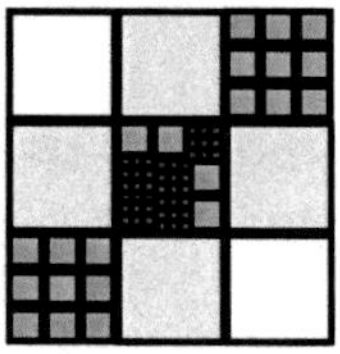

Fig. 1. Divide a training set into several subsets

This article is organized as follows: In the next section, we introduce some notations and mathematical tools to be used in the latter. Section 3 shows the whole construction and complexity of our algorithm. The conclusion is drawn in the last section.

2 Notations and Mathematical Tools

2.1 Notations

In the following of this article we denote the training vectors set by D=(D1,…,Dj)T, where $D_i (1 \le i \le j)$ is a k-dimension vector.

Definition 1. Let D be the training vectors set, A= (a1,…,ak) a training vector and $\hat{A} = (\hat{a}_1, ..., \hat{a}_k)$ a reconstructed vector of A. Now we give two standards, squared error and mean squared error, to measure the distortion [2]:

$$\text{Squared error (SE): } d(A, \hat{A}) = \sum_{i=1}^{k} (a_i - b_i)^2$$

$$\text{Mean squared error (MSE): } E[\frac{1}{k} d(A, \hat{A})] = \frac{1}{jk} \sum_{i=1}^{j} d(D_i, \hat{D}_i)$$

Definition 2. Let row vectors A= (a1,...ak), B=(b1,...,bk), D=(D1,...Dj)T. Define covariance as follows[7]:

Covariance between A and B:

$$\sigma_{AB}^2 = \frac{\sum_{i=1}^{k}(a_i - \bar{a})(b_i - \bar{b})}{n-1} = \frac{1}{n-1} AB^T$$

Covariance matrix of D:

$$C = \frac{1}{n-1} DD^T = \begin{pmatrix} \sigma_{D_1D_1}^2 & \sigma_{D_1D_2}^2 & \cdots & \sigma_{D_1D_j}^2 \\ \sigma_{D_2D_1}^2 & \sigma_{D_2D_2}^2 & \cdots & \sigma_{D_2D_j}^2 \\ \vdots & \vdots & \ddots & \vdots \\ \sigma_{D_jD_1}^2 & \sigma_{D_jD_2}^2 & \cdots & \sigma_{D_jD_j}^2 \end{pmatrix}_{j \times j}$$

In section 3, we divide the current set into n subsets, denoted by S1, ..., Sn. Each subset Si corresponds to a k-dimension hypercube Hi and a 2-dimension block Bi. With i as the index of i-th subset, we use lij to denote the length of Hi in j-th dimension, gi a kind of volume function of Hi, ni the number of training vectors in Si, wi weight function of Hi, xi, mi respectively the theoretical and practical required code number of Hi, si the overall SE distortion for Si, tip the p-th training vector in Si, and we use tipj to denote the j-th dimension of tip.

Furthermore, by si(1), si(xi) we denote the overall SE distortion for Si with 1 and xi codes, respectively. For the case of 1 code, set the code at the center of the hypercube Hi, i.e. $(\frac{l_{i1}}{2},...,\frac{l_{ik}}{2})$ .For the case of xi codes, according to the distribution characteristic of signals, we set the codes at $(\frac{l_{i1}}{2x_i},...,\frac{l_{ik}}{2x_i})$, $(\frac{3l_{i1}}{2x_i},...,\frac{3l_{ik}}{2x_i})$,..., $(\frac{(2x_i-1)l_{i1}}{2x_i},...,\frac{(2x_i-1)l_{ik}}{2x_i})$. Each code of them corresponds to a small k-dimension hypercube $H_{ij}(1 \le j \le x_i)$.

2.2 Principal Components Analysis

Principal Components Analysis (PCA)[8] is a linearly dimension-reducing technique in Data Analysis. It is used to reduce the dimensionality of the specific data set and identify new meaningful underlying variables for the set. Concretely speaking, PCA transforms a number of correlated variables into much less uncorrelated variables called principal components. The first principal component accounts for as much of the variability in the data as possible, and each succeeding component accounts for as much of the remaining variability as possible.

Lemma 1. If C is a real symmetric matrix of order n, then there exists a orthogonal matrix P such that $P^{-1}CP = diag(\lambda_1,...,\lambda_n)$, where $\lambda_1,...,\ \lambda_n$ are eigenvalues of C.

Lemma 1 Proof: see [9].

According to Lemma 1, we calculate the eigenvalues of the covariance matrix C and choose the two greatest eigenvalues of them, say λ1, λ2. Then we calculate their eigenvectors, v1, v2, respectively. Finally, we orthonormalize these two eigenvectors to be e1, e2. Now (e1, e2) become a 2-dimension normalized orthogonal base, whose elements represent the first and second most significant principal components of the current set.

2.3 Monte Carlo Method

We generate random codebook by Monte Carlo method, details see [11].

3 DD Algorithm

3.1 Dividing the Current Set

Every k-dimension training vector in the current set has a 2-dimension projection with respect to (e1, e2). According to the projection, we divide the current set into n subsets such that all the subsets occupy congruent (e1, e2)-blocks:

Definition 3. Let $D_{proj} = D(e_1,e_2)$ =(D1,D2,…,Dj)T(e1,e2)= $\begin{pmatrix} D_1e_1 & D_1e_2 \\ D_2e_1 & D_2e_2 \\ \vdots & \vdots \\ D_je_1 & D_je_2 \end{pmatrix}_{j\times 2}$ Define the bound of the hypercube of D as follows:

1. $Floor_1 = \min_{1\le i\le j}\{D_ie_1\}$
2. $Floor_2 = \min_{1\le i\le j}\{D_ie_2\}$
3. $Upper_1 = \max_{1\le i\le j}\{D_ie_1\}$
4. $Upper_2 = \max_{1\le i\le j}\{D_ie_2\}$

Let n = r ∗ c. For any training vector A =(a1, …, ak), we calculate which block it belongs to like this:

$$\begin{cases} i = \begin{cases} \left\lceil r * \dfrac{Ae_1 - Floor_1}{Upper_1 - Floor_1} \right\rceil & Ae_1 > Floor_1 \\ 1 & Ae_1 = Floor \end{cases} \\ j = \begin{cases} \left\lceil c * \dfrac{Ae_2 - Floor_2}{Upper_2 - Floor_2} \right\rceil & Ae_2 > Floor_2 \\ 1 & Ae_2 = Floor_2 \end{cases} \end{cases} \tag{1}$$

Where $\lceil . \rceil$ is the upper truncated operation. Thus, the training vector A belongs to (e1, e2)-block Bij.

3.2 Weight Function

Definition 4. Let si(xi) be the SE distortion of Hi with xi codes, sij(1) the SE distortion of Hij with 1 code, see section 2. Define their relation as follows:

$$s_i(x_i) = \sum_{j=1}^{x_i} s_{ij}(1)$$

Lemma 3. Let gi=g(li1,…, lik)= $\sum_{j=1}^{k} l_{ij}^2$, then

$$s_i(x_i) = \frac{n_i g_i}{12 x_i^2}$$

Proof 1

$$s_i(1) = \int_0^{l_{i1}} \cdots \int_0^{l_{ik}} \frac{n_i}{l_{i1} \cdots l_{ik}} [(s_1 - \frac{l_{i1}}{2})^2 + \cdots + (s_k - \frac{l_{ik}}{2})^2]ds_1 \cdots ds_k = \frac{n_i}{12}(l_{i1}^2 + \cdots l_{ik}^2) = \frac{n_i g_i}{12}$$

Similarly

$$s_{ij} = \frac{n_{ij} g_{ij}}{12}$$

By Definition 4 we have

$$s_i(x_i) = \sum_{j=1}^{x_i} s_{ij}(1) = \frac{x_i}{12} \frac{n_i}{x_i} \frac{g_i}{x_i^2} = \frac{n_i g_i}{12 x_i^2}$$

Theorem 1. For the conditional extremum problem

$$\begin{cases} \min \sum_{i=1}^{n} s_i(x_i) \\ \sum_{i=1}^{n} x_i = m \\ 0 < x_i < m \end{cases} \tag{2}$$

When

$$x_i = \frac{m \sqrt[3]{n_i g_i}}{\sum_{j=1}^{n} \sqrt[3]{n_i g_i}}$$

$\sum_{i=1}^{n} s_i(x_i)$ has the least value. Then the weight function is

$$w_i = \sqrt[3]{n_i g_i}$$

Proof 2. Step 1: Calculate the stationary points.

We define a Lagrange function[10] as follows,

$$L(x_1,\cdots,x_n) = \sum_{i=1}^{n} s_i(x_i) + \lambda(\sum_{i=1}^{n} x_i - m) = \sum_{i=1}^{n} \frac{n_i g_i}{12x_i^2} + \lambda \sum_{i=1}^{n} x_i - \lambda m$$

Let

$$L_{x_1} = -\frac{n_i g_i}{6x_i^3} + \lambda = 0 \qquad (1)$$

$$\vdots \qquad\qquad \vdots$$

$$= L_{x_n} = -\frac{nigi}{6x_i^3} + \lambda = 0 \qquad (n)$$

$$\sum_{i=1}^{n} x_i = m \qquad (n+1)$$

Transform expressions (i) and (j) as follows:

$$\frac{n_i g_i}{6x_i^3} = \lambda \qquad (i')$$

$$\frac{n_j g_j}{6x_j^3} = \lambda \qquad (j')$$

$(i')/(j')$, we have

$$\frac{x_i}{x_j} = \frac{\sqrt[3]{n_i g_i}}{\sqrt[3]{n_j g_j}}$$

Then the weight function

$$w_i = \sqrt[3]{n_i g_i}$$

Combining with (n+1), we conclude

$$x_i = \frac{m\sqrt[3]{n_i g_i}}{\sum_{j=1}^{n} \sqrt[3]{n_j g_j}}$$

Now P=(x1, …, xn) is a stationary point of the Lagrange function L(x1, …, xn).

Step 2: Check if P is the minimal point.

For the Lagrange function $L(x_1,\cdots,x_n) = \sum_{i=1}^{n} \frac{n_i g_i}{12x_i^2} + \lambda \sum_{i=1}^{n} x_i - \lambda m$, we calculate its Jacobi Matrix

$$J = \left(\frac{\partial^2 L}{\partial x_i x_j}(x_1,\cdots,x_n,\lambda)\right)_{n\times n} = \begin{pmatrix} Lx_1x_1 & Lx_1x_2 & \cdots & Lx_1x_n \\ Lx_2x_1 & Lx_2x_2 & \cdots & Lx_2x_n \\ \vdots & \vdots & \ddots & \vdots \\ Lx_nx_1 & Lx_nx_2 & \cdots & Lx_nx_n \end{pmatrix}_{n\times n}$$

$$= \begin{pmatrix} \frac{n_1 g_1}{2x_1^4} & 0 & \cdots & 0 \\ 0 & \frac{n_2 g_2}{2x_2^4} & \cdots & 0 \\ \vdots & \vdots & \ddots & \vdots \\ 0 & 0 & \cdots & \frac{n_n g_n}{2x_n^4} \end{pmatrix}_{n\times n}$$

We see that J is a positive definite matrix, hence P is the least point, and the minimal point since it is the unique stationary point. Next we continue to deal with the decimal point:

Rearrange $\{x_i - \lfloor x_i \rfloor\}$ to be a decreasing sequence $\{x_{k_i} - \lfloor x_{k_i} \rfloor\}$, such that

$$x_{k_i} - \lfloor x_{k_i} \rfloor \geq x_{k_{i+1}} - \lfloor x_{k_{i+1}} \rfloor, i \in \{1,\cdots,n\}.$$

Let

$$c = m - \sum_{i=1}^{n} \lfloor x_i \rfloor$$

Then

$$x_{k_i} - \lfloor x_{k_i} \rfloor \geq \cdots \geq x_{k_c} - \lfloor x_{k_c} \rfloor \geq \cdots \geq x_{k_n} - \lfloor x_{k_n} \rfloor.$$

Here $\lfloor . \rfloor$ is the lower truncated function.

Definition 5. Define the practical required code number of subset Si as follows:

$$m_i = \begin{cases} \lfloor x_i \rfloor + 1 & x_i - \lfloor x_i \rfloor \geq x_{k_c} - \lfloor x_{k_c} \rfloor \\ \lfloor x_i \rfloor & otherwise \end{cases} \quad (3)$$

3.3 Algorithm

Let k= 16, n = r*c=3*3, w=16, that is, suppose the current set is divided into 9 subsets and recursion stops for some subset when it requires less than 16 codes.

Algorithm:

Step 1: Identify the training vectors set D as the current set.

Step 2: Calculate the corresponding normalized orthogonal base (e1, e2) of the current set through PCA method.

Step 3: Divide the current set into 3*3 subsets, such that all the subsets occupy congruent (e1 , e2)-blocks. Each subset corresponds to not only an (e1, e2)-block, but also a 16-dimension hypercube.

Step 4: Identify the position of every training vector in the current set. For any training vector A, it belongs (e1, e2)-block Bij, where i, j are expressed in (1).

Step 5: Calculate the code number each subset deserves. For subset Si($1 \le i \le 9$), by weight function $w_i = \sqrt[3]{n_i g_i}$, where gi= $\sum_{j=1}^{k} l_{ij}^2$. Here, we use tolerance formula to evaluate lij, $l_{ij} = 6 * \sqrt{\frac{\sum_{p=1}^{n_i}(t_{ipj} - \bar{t}_{ij})}{n_i - 1}}$, where $\bar{t}_{ij} = \frac{\sum_{p=1}^{n_i} t_{ipj}}{n_i}$. We have xi, from which mi is calculated, see (3).

Step 6: for(i = 1; i ≤ 9; i + +){
if (mi≥ 16)
{identify subset Si as the current set, and go to Step 2;}
else
{identify subset Si as an objective set, and by Monte Carlo method, choose mi training vectors at random from Si, as the initial codes of this hypercube.}}

Step 7: Train the initial codebook by GLA algorithm, and finally get an approximate global optimal codebook.

3.4 Complexity

Suppose the initial training set contains N training vectors, and the dimension of the training vectors is k, the size of codebook is M, every current training set is divided to n subsets in every recursion. The main steps of DD algorithm can be concluded as: The Division of Set and The Distribution of Codes. And the two steps are repeated by subset until the stop condition is met. The major time consuming step of Set's Division is Principle Component Analysis, which needs to solve the covariance of training vectors, the time complexity of which is O($C_k^2 N$). The major time consuming step of the distribution of codes is to solve the Weight function. To evaluate the length in every dimension of every subset, it need a time complexity of O(kN). So one recursion of the Division and Distribution process for N training vectors of k dimension is corresponding to the time complexity of O($C_k^2 N$). The exact time complexity is related with recursion depth, which is related with data distribution. For signal source of uniform distribution, codes of current set are distributed into n subsets equally. If the stop condition is no more than one code in each subset, the recursion depth is $\sqrt[n]{M}$, and the time complexity is O($\sqrt[n]{M}\ C_k^2 N$).

4 Conclusions

In this article we propose a heuristic scheme, DD scheme, to design initial codebook. Using mathematical tools we construct a heuristic weight function, by which we make a dynamic division for the training vectors set.

References

1. Linde, Y., Buzo, A., Gray, R.M.: An algorithm for vector quantizer design. IEEE Transactions on Communications COM-28, 84–95 (1980)
2. Gersho, A., Gray, R.M.: Vector Quantization and Signal Compression. Kluwer, Boston (1992)
3. Gray, R.M.: Source Coding Theory. Kluwer, Boston (1990)
4. Equitz, W.H.: A new vector quantization clustering algortihm. IEEE Transactions on ASSP 37(10), 1568–1575 (1989)
5. Chan, C.K., Ma, L.K.: A fast method of designing better codebooks for image vector quantization. IEEE Trans. Communications COM-42, 237–242 (1994)
6. Drineas, P., Frieze, A., Kannan, R., Vempala, S., Vinay, V.: Clustering in large graphs and matrices. In: Proc. 10th Annual ACM-SIAM Symposium on Discrete Algorithms, SODA (1999)
7. Wonnacott, T.H., Wonnacott, R.J.: Introductory Statistics. Wiley, New York (1990)
8. Dunteman, G.H.: Principal Components Analysis. Sage, London (1989)
9. Zhou, B.X.: Introduction to Advanced Algebra. Higher Education (1989)
10. Brannan, D.A.: A First Course in Mathematical Analysis. Cambridge University Press, Cambridge (2006)
11. Frenkel, D.: Introduction to Monte Carlo methods. NIC Series 23, 29–60 (2004)

Efficient SAT-Based Verification of Asynchronous System

Xu Yatao*, Lin Changqing, Yuan Qin, and Zhou Conghua

Academic Research Department, Logistics Academy, Beijing, 100858, China
Logistics Command Department, Logistics Academy, Beijing, 100858, China
Academic Research Department, Logistics Academy, Beijing, 100858, China
Computer Science Department, Jiangsu University, Zhenjiang, 212013, China.
yataoxu@gmail.com

Abstract. Bounded model checking has been recently introduced as an efficient verification method for reactive systems. This technique reduces model checking of linear temporal logic to propositional satisfiability. However this method does not work well for asynchronous systems. In this paper we propose a new SAT-based verification method for asynchronous systems. Our method can reduce verification time by representing the behavior by very succinct formulas.

Keywords: Bounded Model Checking, Symbolic Model Checking, SAT.

1 Introduction

Model Checking [1] is a powerful technique for verifying systems and detecting errors at early stages of the design process, which is obtaining wide acceptance in industrial setting. In model checking, the specification is expressed in temporal logic-either Computation Tree Logic(CTL)[2] or Linear Temporal Logic(LTL)[3]-and the system is modeled as a finite state machine(FSM). A traversal algorithm verifies exhaustively whether the FSM satisfies the property or not. For realistic designs, the number of states of the system can be very large and the explicit traversal of the state space becomes infeasible. Symbolic model checking [4, 5], with boolean encoding of the finite state machine, can handle more than 10^{20} states. BDDs[6], a canonical form for boolean expressions, have traditionally been used as the underlying representation for symbolic model checker[5]. Model checkers based on BDDs are usually able to handle systems with hundreds of state variables. However, for larger systems the BDDs generated during model checking become too large for currently available computers. In addition, selecting the right ordering of BDD variables is very important. The generation of a variable ordering that results in small BDDs is often time consuming or needs manual intervention. For many examples no space efficient variable ordering exists.

* Author sponsored by NSFC11101212.

D. Yang (Ed.): Informatics in Control, Automation and Robotics, Volume 2, LNEE 133, pp. 273–281.
springerlink.com

Recently a new approach for symbolic model checking has been proposed called Bounded Model Checking(BMC)[7], which is based on SAT techniques. Given a FSM M and an LTL specification f, the idea is to look for counterexamples of maximum length k, and to generate a boolean formula which is satisfiable if and only if such counterexample exists. The boolean formula then is given as input to a SAT solver. If the formula is satisfiable, the satisfying assignment returned is converted into a counterexample execution path. SAT procedures do not suffer from the potential space explosion of BDDs and can handle propositional satisfiability with thousands of variables.

However, bounded model checking does not work well for asynchronous systems like Petri nets, because the encoding scheme into propositional formulas is not suited for such systems; it would require a large formula to represent the transition relation, thus resulting in large execution time and low scalability. In this paper we propose a method to address this issue. Our basic idea is that for each reachable state in the asynchronous system we impose a transition: each reachable state is reachable from itself in one step. Our method uses a much succinct formula compared the existing method. This shortens the execution time of a SAT procedure. Therefore Our method enhances the effectiveness of SAT-based verification.

The rest of the paper is organized as follows: in the next Section we simply introduce asynchronous system, linear time temporal logic LTL , bounded model checking. In Section 3, we show our new encoding method. Section 4 are some related works. In Section 5, we give some conclusions and directions for future research.

2 Formal Preliminaries

2.1 Asynchronous System

We use a type of state transition graph called a Kripke structure[1] to describe the behavior of asynchronous systems. Let AP be a set of atomic propositions. A Kripke structure M over AP is a four tuple $M = (S, S_0, R, L)$ where S is a finite set of states, $S_0 \subseteq S$ are the initial states, $R \subseteq S \times S$ is a transition relation that must be total, that is, for every state $s \in S$ there is state $s' \in S$ such that $R(s, s')$, $L : S \rightarrow 2^{AP}$ is a function that labels each state with the set of atomic propositions true in that state. In a Kripke structure M, a path π is an infinite sequence of states, $\pi = s_0 s_1 \cdots$ such that $(s_i, s_{i+1}) \in R$ for every $i \geq 0$. π_i denotes the suffix of π starting at s_i, and $\pi(i)$ denotes the state s_i. Let $V = \{v_1, \ldots, v_m\}$ be the set of system variables and the variables in V range over a finite set D. A *valuation* for V is a function that associates a value in D with each variable v in V. A state of a concurrent system is just a valuation $s : V \rightarrow D$ for the set of variables in V. We think of the variables in V as present state variables and the variables in V' as next state variables. The transition relation for asynchronous system is most naturally expressed as the disjunction of the transition relations, that is, $R(V, V') \equiv R_1(V, V') \vee \cdots \vee R_n(V, V')$.

2.2 Bounded Model Checking

Linear time temporal logic LTL formulas are defined recursively: if p is an atomic proposition then $p, \neg p$ are in LTL; if f, g $\in$LTL then so are $Xf, Ff, Gf, fUg, fRg, f \wedge g, f \vee g$.

Definition 2.1. *(Semantics of LTL) Let M be a Kripke structure , π be a path in M, and f, g be formulas of LTL. $M, \pi \models f$ denotes that f is true in the path π in M. The relation $\models$ is defined inductively as follows:*
$M, \pi \models p$ *iff* $p \in L(\pi(0))$, $M, \pi \models \neg p$ *iff* $p \notin L(\pi(0))$
$M, \pi \models f \wedge g$ *iff* $M, \pi \models f$ *and* $M, \pi \models g$. $M, \pi \models f \vee g$ *iff* $M, \pi \models f$ *or* $M, \pi \models g$
$M, \pi \models Xf$ *iff* $M, \pi_1 \models f$. $M, \pi \models Gf$ *iff* $\forall i \geq 0, M, \pi_i \models f$.
$M, \pi \models Ff$ *iff* $\exists i \geq 0, M, \pi_i \models f$
$M, \pi \models fUg$ *iff* $\exists i \geq 0 (M, \pi_i \models g$ *and* $\forall 0 \leq j < i, M, \pi_j \models f)$
$M, \pi \models fRg$ *iff* $(\forall i \geq 0, M, \pi_i \models g$ *or* $\exists j \geq 0, (M, \pi_j \models f \wedge \forall 0 \leq l \leq j, M, \pi_l \models g))$

An LTL formula f is universally valid in a Kripke sturcture M, denoted as $M \models Af$, iff for any path π with $\pi(0) \in S_0$ in M, $M, \pi \models f$. An LTL formula f is existentially valid in a Kripke sturcture M, denoted as $M \models Ef$, iff there exists a path π such that $\pi(0) \in S_0$ and $M, \pi \models f$. Logic LTL$_{-X}$ is a subset of LTL. The only diffence between LTL$_{-X}$ and LTL is that there is no temporal operator X in LTL$_{-X}$.

We briefly recall some basic notions of the bounded model checking as proposed in [7]. For $l \leq k$ we call a path π a $(k,l)-$*loop* if $R(\pi(k), \pi(l))$ and $\pi = u \cdot v^{\omega}$ with $u = (\pi(0), \ldots, \pi(l-1))$ and $v = (\pi(l), \ldots, \pi(k))$. We call π a $k-$ *loop* if there exists an integer l with $k \geq l \geq 0$ such that π is a $(k,l)-$ *loop*.

Definition 2.2. *(Bounded Semantics for a Loop) Let $k \geq 0$ and π be a $k-$loop. Then an LTL formula f is valid along the path π with bound k (in symbols $\pi \models_k f$) iff $\pi \models f$.*

Definition 2.3. *(Bounded Semantics without a Loop)Let $k \geq 0$ and let π be a path that is not a $k-$loop. Then an LTL formula f is valid along π with bound k(in symbols $\pi \models_k f$) iff $\pi \models_k^0 f$ where*
$\pi \models_k^i p$ *iff* $p \in L(\pi(i))$ $\pi \models_k^i \neg p$ *iff* $p \notin L(\pi(i))$
$\pi \models_k^i f \wedge g$ *iff* $\pi \models_k^i f$ *and* $\pi \models_k^i g$. $\pi \models_k^i f \vee g$ *iff* $\pi \models_k^i f$ *or* $\pi \models_k^i g$.
$\pi \models_k^i Gf$ *is always false.* $\pi \models_k^i Ff$ *iff* $\exists j, i \leq j \leq k, \pi \models_k^j f$.
$\pi \models_k^i Xf$ *iff* $i \leq k$ *and* $\pi \models_k^{i+1} f$.
$\pi \models_k^i fUg$ *iff* $\exists j, i \leq j \leq k[\pi \models_k^j g$ *and* $\forall n, i \leq n < j . \pi \models_k^n f]$.
$\pi \models_k^i fRg$ *iff* $\exists j, i \leq j \leq k[\pi \models_k^j f$ *and* $\forall n, i \leq n \leq j . \pi \models_k^n g]$.

Lemma 2.4. *Let k be an LTL formula and π a path, then $\pi \models_k f \Rightarrow \pi \models f$.*

An LTL formula f is bounded existentially valid with respect to the bound k in a Kripke structure M, denoted as $M \models_k Ef$, iff there exists a path π such that $\pi(0) \in S_0$ and $M, \pi \models_k f$.

Theorem 2.5. *Let f be an LTL formula and M be a Kripke structure. Then $M \models Ef$ iff there exists $k \geq 0$ such that $M \models_k Ef$.*

Given an LTL property ϕ, a Kripke structure M and a bound k, bounded model checking is performed by generating and solving a propositional formula $[[M]]_k \wedge [[\phi]]_k$ where a path of length k is encoded by $[[M]]_k$ and the LTL specification is encoded by $[[\phi]]_k$.

Definition 2.6. *(Unfolding of the Transition Relation). For a Kripke structure M, $k \geq 0$,*

$$[[M]]_k := I(s_0) \wedge \bigwedge_{i=0}^{k-1} R(s_i, s_{i+1}).$$

There are several known methods for generating $[[\phi]]_k$. Since in this paper we focus on the transition relation, the definition of $[[\phi]]_k$ wiil not be introduced here.

3 Efficient Bounded Model Checking for Asynchronous System

In this section we first recall the stuttering equivalent and the bounded stuttering equivalent. Two infinite paths π, π' are stuttering equivalent[1], denoted $\pi \sim_{st} \pi'$, if there are two infinite sequences of integers $0 = i_0 < i_1 < i_2 < \cdots$ and $0 = j_0 < j_1 < j_2 < \cdots$ such that for every $k \geq 0$, $L(\pi(i_k)) = L(\pi(i_k + 1)) = \cdots = L(\pi(i_{k+1} - 1)) = L(\pi'(j_k)) = L(\pi'(j_k + 1)) = \cdots = L(\pi'(j_{k+1} - 1))$. In [1] they have proved that any LTL_{-X} property is invariant under stuttering. We call a finite sequence of identically labelled states a *block* and use the notation B_j^π to represent the jth block: $\pi(i_j), \pi(i_j + 1), \ldots, \pi(i_{j+1} - 1)$ in the path π. And call the set $\{B_0^\pi, B_1^\pi, \ldots\}$ is a partition of π. For each *block* B_i^π we define $Size(B_i^\pi)$ is the number of states in B_i^π. And use the notation $B_i^\pi(j)$ to represent the jth state in B_i^π.

Definition 3.1. *(Bounded Stuttering Equivalent[11]) For two integers $m \geq 0, n \geq 0$, two bounded paths π_m, π'_n are bounded stuttering equivalent, denoted $\pi_m \sim_{bst} \pi'_n$, if and only if one of the following two conditions holds.*

(1) For each $0 \leq l \leq m$, if $(\pi(m), \pi(l)) \in R$ then there exists an integer h with $0 \leq h \leq n$ such that $(\pi'(n), \pi'(h)) \in R$ and the paths $\pi(0), \ldots, \pi(l-1)(\pi(l), \ldots, \pi(m))^\omega$, $\pi'(0), \ldots, \pi'(h-1)(\pi'(h), \ldots, \pi'(n))^\omega$ are stuttering equivalent. For each $0 \leq h \leq n$, if $(\pi'(n), \pi'(h)) \in R$ then there exists an integer l with $0 \leq l \leq m$ such that $(\pi(m), \pi(l)) \in R$ and the paths $\pi(0), \ldots, \pi(l-1)(\pi(l), \ldots, \pi(m))^\omega$, $\pi'(0), \ldots, \pi'(h-1)(\pi'(h), \ldots, \pi'(n))^\omega$ are stuttering equivalent.
(2) π_m is not a $m-$loop path, π'_n is not a $n-$loop path and there are two finite sequences of integers $0 = i_0 < i_1 < i_2 < \cdots < i_l$ and $0 = j_0 < j_1 < j_2 < \cdots < j_l$ such that for every $0 \leq x < l$, $L(\pi(i_x)) = L(\pi(i_{x+1})) = \cdots = L(\pi(i_{x+1} - 1)) = L(\pi'(j_x)) = L(\pi'(j_{x+1})) = \cdots = L(\pi'(j_{x+1} - 1))$ and $L(\pi(i_l)) = L(\pi(i_l + 1)) = \cdots = L(\pi(m)) = L(\pi'(j_l)) = L(\pi'(j_l + 1)) = \cdots = L(\pi'(n))$.

Theorem 3.2. *Let π, π' be two paths and two bounded paths π_m, π'_n are their prefixes. If π_m, π'_n are bounded stuttering equivalent then for any LTL_{-X} formula ϕ, $\pi \models_m \phi$ iff $\pi' \models_n \phi$.*

In this section for the convenience of discuss we assume that the transition relation R is the disjunction of n transition relations, that is $R = R_1 \vee \ldots \vee R_n$. And the state variable V is m tuple boolean vector, that is $V = \{v_1, \ldots, v_m\}$. Let $I = \{1, \ldots, n\}$.

Definition 3.3. *For each transition relation R_i we define a new transition relation $R'_i = R_i \vee (\bigwedge_{i=1}^{m} v_i \leftrightarrow v'_i)$.*

Intuitively the difference of R_i and R'_i is that in R'_i for each reachable state we impose a transition to itself. Since the transition relation of the system is changed some properties holding in the new system may not hold in the original system. In the following we will show that the properties expressed with LTL_{-X} are invariant.

Definition 3.4. *(New Unfolding of the Transition Relation) For a Kripke structure M, $k \geq 0$, we define $[[M]]'_k := I(s_0) \wedge \bigwedge_{i=0}^{k-1} R'(s_i, s_{i+1})$, where $R'(s_i, s_{i+1}) = R'_1(s_i, s_{i+1}) \wedge \ldots \wedge R'_n(s_i, s_{i+1})$.*

Theorem 3.5. *Given a Kripke structure M, an integer $k \geq 0$, an LTL_{-X} formula ϕ, if $[[M]]'_k \wedge [[\phi]]_k$ is satisfiable then $M \models E\phi$.*

In order to present the proof of the above theorem we introduce the following definition.

Definition 3.6. *If a sequence of states $s_0 \ldots s_j$ satisfying that for each $0 \leq i < j$, $R'(s_i, s_{i+1})$ holds, then we define a sequence of states $comp(s_0 \ldots s_j) := s'_0 \ldots s'_l$ inductively :*

- $s'_0 = s_0$
- *If $s'_x = s_y$ and $R(s_y, s_{y+1})$ does not hold, then let $y = y + 1$.*
- *If $s'_x = s_y$ and $R(s_y, s_{y+1})$ holds, then let $s'_{x+1} = s_{y+1}$ and $x = x + 1, y = y + 1$.*

Lemma 3.7. *If a sequence of states $s_0 \ldots s_j$ satisfying for each $0 \leq i < j$, $R'(s_i, s_{i+1})$ holds, then $s_0 \ldots s_j$ and $comp(s_0 \ldots s_j)$ are bounded stuttering equivalent.*

The proof of the theorem 3.5 is directly from theorem 3.2 and lemma 3.7. Now we compare our encoding method with the general encoding method. For comparison's fairness we takes standard clause form translation. And for simplicity we assume for each transition $R_i (1 \leq i \leq n)$ and any two states s, s', the size of the conjunctive normal form(CNF) of $R_i(s, s')$ is h. Then the size of CNF of $R(s, s')$ is h^n. Therefore the size of CNF of $R(s_0, s_1) \wedge \ldots \wedge R(s_{k-1}, s_k)$ is kh^n. That is it's size is exponential in the size of $R_i(s, s')$. It is clear the size of CNF of $R'_i(s_j, s_{j+1})$ is $2mh$. Then the size of CNF of R' is $2knmh$ and is linear in the size of $R_i(s, s')$. It is clear our method uses more succinct formula.

4 Optimization

Our encoding framework assumes that in each step all transition relations can be enable. However, in fact in each step there are only few transition relations to be enable. For example, Figure 1 is a simple Petri net PN_1, Figure 2 is its reachability. In PN_1 there are 7 transitions, however from PN_1's reachability graph we can see that in each step there are most four transitions to be enable. In the following we propose an algorithm to find all enable transition relations for each step. We first introduce some terms in our algorithm.

- M_0 is the initial state of an asynchronous system, $Tran = \{R_1, \ldots, R_n\}$ is the set of its transition relations.
- $v_1, v_2, \ldots$ are k-tuple boolean vectors. T is a set of k-tuple boolean vectors.
- $v|i = (v(1), \ldots, v(i))$.
- $M \xrightarrow{R_i} M'$ represents that in the state M transition R_i is enable and if fired leads to the state M'.

Algorithm 4.1. *ComputeEnalbeTransitionrRelations*$(M_0, Tran)$ {
$v_1 = (0, \ldots, 0), T = \{v_1\}, i = 1, j = 1$
while$(i \leq k)$
For the state M_{i-1}, *look for a transition* $R_i \in Tran$ *such that* R_i *satisfying the following condition:*
R_i *is enable in* M_{i-1} *and* $\forall v \in T, (v_j|_{i-1}, i) \neq v|_i$.
Then there exists three cases to be considered:
Case 1. If there is no transition enable in M_{i-1}, *then return and report there is a deadlock.*
Case 2. If M_{i-1} *is not a deadlock state and no transition satisfies the condition (1), then we consider two cases:*
(1) if $i = 1$ *then return.*
(2) if $i > 1$ *then compute the state* M_{i-2} *such that* $M_{i-2} \xrightarrow{R_{v_j(i-1)}} M_{i-1}$. *Let* $i = i - 1$.
Case 3. If find a transition R_m *satisfying the condition (1), then we consider two cases:*
(1) if $i = k-1$, *then let* $v_{j+1}|_i = v_j|_i, v_{j+1}(i+1) = m, j = j+1, T = T \cup \{v_{j+1}\}$.
(2) if $i < k - 1$, *then let* $v_j(i + 1) = m, i = i + 1$. }

Given a system, assume by the algorithm 5.1 we have compute the set T, then we define $m \in T_i$ iff there exists a vector $v \in T$ such that $v(i) = m$. That is, T_i is all the transitions which are enable in i step. For example, in PN_1, $T_1 = \{1, 2\}, T_2 = \{3, 4, 5, 6\}, T_3 = \{3, 4, 5, 6\}, T_4 = \{7\}$. Assume we have computed the sets $T_1, \ldots, T_k$,then we redefine transition unfolding.

Definition 4.2. *For a Kripke structure* M, $k \geq 0$, *we define* $[[M]]''_k := I(s_0) \wedge \bigwedge_{i=0}^{k-1} R'(s_i, s_{i+1})$, *where* $R'(s_i, s_{i+1}) = \bigwedge_{j \in T_i} R'_j(s_i, s_{i+1})$).

It is clear that $[[M]]''_k$ is more succinct and the satisfiability of $[[M]]'_k \wedge [[\phi]]_k$ implies that $E\phi$ is satisfied by M.

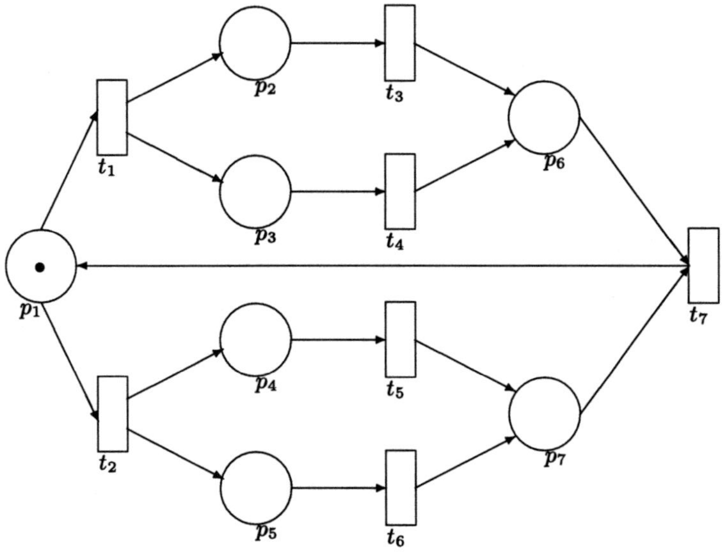

Fig. 1. A Petri net PN_1

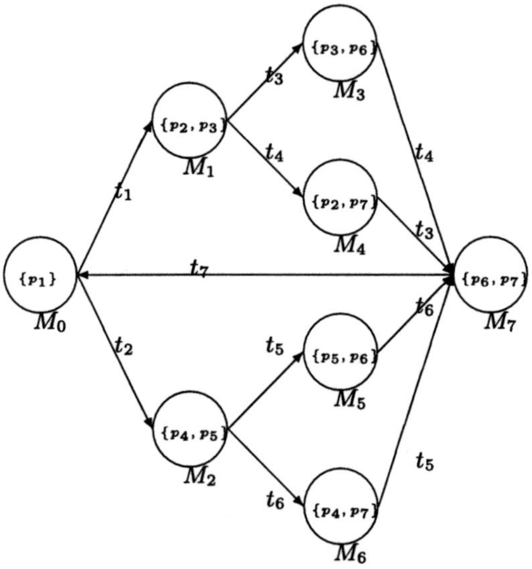

Fig. 2. Reachability Graph

Theorem 4.3. *Given a Kripke structure M, an integer $k \geq 0$, an LTL_{-X} formula ϕ, if $[[M]]''_k \wedge [[\phi]]_k$ is satisfiable then $M \models E\phi$.*

5 Related Work

As we know, bounded model does not work well for asynchronous systems, because the encoding scheme into propositional formulas is not suited for such systems; it would require a large formula to represent the transition relation, thus resulting in large execution time and low scalability. To address this issue, in [12] they propose a new SAT-based verification method for safe Petri nets which is very similar with ours. Their main idea is that for safe petri nets they impose a transition relation such that each reachable state is reachable from itself in one step. Their method has two drawbacks: (1) The state space that can be explored by their method depends on the order of the transitions; (2) For any LTL properties checking, their method is not suit because then the encoding of LTL specification is very large. In [13, 14] approaches that use other techniques than SAT decision procedures have been proposed. These approaches allow bounded model checking of Petri nets by using answer set programming and by using Boolean circuit satisfiability checking.

6 Conclusions

In this paper we proposed a new method for bounded model checking. By exploiting the interleaving nature of asynchronous systems, our method generates much more succinct formulas than ordinary bounded model checking, thus resulting in high efficiency. There are several directions in which further work is needed. First although our technique in theory is feasible, we also we need to show that our theory works in practice. Second a more comprehensive comparison is needed with existing verification methods. Especially, comparison with bounded reachability checking proposed by Shough ogata[12].

References

[1] Clarke, E.M., Grumberg, O., Peled, D.: Model checking. MIT Press (2000)
[2] Ben-Ari, M., Manna, Z., Pnueli, A.: The temporal logic of branching time. Acta Information 20, 207–226 (1983)
[3] Pnueli, A.: A temporal logic of concurrent programs. Theoretical Computer Science, 13, 45–60
[4] McMillan, K.L.: Symbolic model checking. Kluwer Academic Publishers, Dordrecht (1993)
[5] Burch, J.R., Clarke, E.M., McMillan, K.L.: Symbolic model checking: 10^{20} states and beyond. Information and Computation 98, 142–170 (1992)
[6] Bryant, R.E.: Graph-based algorithms for boolean function manipulation. IEEE Transactions on Computers 35, 677–691 (1986)
[7] Biere, A., Cimatti, A., Clarke, E.M., Zhu, Y.: Symbolic model checking without BDDs. In: Cleaveland, W.R. (ed.) TACAS 1999. LNCS, vol. 1579, pp. 193–207. Springer, Heidelberg (1999)
[8] Godefroid, P., Wolper, P.: A partial approach to model checking. Information and Computation 110(2), 305–326

[9] Murata, T.: Petri Nets: Properties, analysis and application. Proc. of the IEEE 77(4), 541–574 (1989)
[10] van der Aalst, W.M.P.: Verification of Workflow Nets. In P. Azema and G. Balbo, editors, Application and Theory of Petri Nets 1997. In: Azéma, P., Balbo, G. (eds.) ICATPN 1997. LNCS, vol. 1248, pp. 407–426. Springer, Heidelberg (1997)
[11] Zhou, C.: SAT-based Compositional Verification Strategy for Concurrent Software with States, Events. Journal of Software 20(6), 1414–1424 (2009)
[12] Ogata, S., Tsuchiya, T., Kikuno, T.: SAT-Based Verification of Safe Petri Nets. In: Wang, F. (ed.) ATVA 2004. LNCS, vol. 3299, pp. 79–92. Springer, Heidelberg (2004)
[13] Heljanko, K.: Bounded reachability checking with process semantics. In: Larsen, K.G., Nielsen, M. (eds.) CONCUR 2001. LNCS, vol. 2154, pp. 218–232. Springer, Heidelberg (2001)
[14] Heljanko, K., Niemelã, I.: Bounded LTL model checking with stable models. Theory and Practice of Logic Programming 3(4&5), 519–550 (2003)

Three Dimensional Optimal Guidance Law for Micro Missiles to Attack Maneuvering Targets

Zhigang Wang and Wei Li

College of Astronautics, Northwestern Polytechnical University, Xi'an 710072, China
viperyuqi@163.com

Abstract. In order to adapt to the developmental trend of micro missiles, three-dimensional optimal guidance law design for air-launched micro missiles to attack ground armed vehicles. Relative motion equations of micro missile and target are established. The performance index function is designed to reach undershoot quantity least and overload did not exceed a certain value. The optimal guidance equations are detailed derived in the three-dimensional space. We did the simulation which optimal and proportional guidance law are adopted for micro missiles to attack ground maneuvering armed vehicles. The simulation results illustrate that the optimal guidance law compared to the proportional guidance law has following advantages: high precision-guided and low overload characteristics.

Keywords: Micro Missiles, Optimal Guidance Law, Principle of the Minimum.

1 Introduction

In recent years, micro missiles become a new direction of precision-guided weapons because of its advantages: low price, light weight, powerful, high precision characteristics. As a man portable weapon, weaponry or tactical UAVs, micro missiles will be used for critical situations, in order to destroy enemy vehicles and key personnel of the armed light aircraft.

In order to reduce collateral damage when UAV attack ground armed vehicles, micro missiles are required a very small miss distance, or even directly impact the target, which bring a new legal challenge for guidance design. In traditional aircraft guidance, proportional navigation and its extended form has been widely studied and applied[1]. This guidance system can meet the requirements when targets in the case of uniform motion, but in attacking maneuvering targets, its low accuracy [2]. At present use of modern control theory methods to derive the optimal guidance law study received widespread attention, the optimal guidance law to improve the accuracy of the missile's guidance has significant effect[3,4] When it attack high-speed maneuvering targets.

D. Yang (Ed.): Informatics in Control, Automation and Robotics, Volume 2, LNEE 133, pp. 283–286.
springerlink.com

2 Micro-missiles and Targets Motion Model

Select the ground coordinate system $Axyz$ define is in ref[5]. Micro missiles and targets relative motion relationships shown in ref[5].

Micro missile and target relative motion equation is:

$$\dot{r}_x = V_{rx} \quad \dot{r}_y = V_{ry} \quad \dot{r}_z = V_{rz} \quad \dot{V}_{rx} = a_{Mx} - a_{Tx} \quad \dot{V}_{ry} = a_{My} - a_{Ty} \quad \dot{V}_{rz} = a_{Mz} - a_{Tz} \tag{1}$$

Eq.(1), r_x、r_y、r_z are the projection for the relative distance $\boldsymbol{r}$ in the ground coordinate system; V_{rx}、V_{ry}、V_{rz} are the projection for the relative velocity in the ground coordinate system; a_{Mx}、a_{My}、a_{Mz} are the projection for the acceleration $\boldsymbol{a}_M$ in the ground coordinate system; a_{Tx}、a_{Ty}、a_{Tz} are the projection for the acceleration $\boldsymbol{a}_T$ in the ground coordinate system.

Eq.(1)describes the system is a linear time-invariant second-order system, the system is written in the form of a matrix:

$$\dot{\boldsymbol{x}} = \boldsymbol{A}\boldsymbol{x} + \boldsymbol{B}\boldsymbol{u} \tag{2}$$

3 Optimal Guidance Law Derivation

3.1 The Establishment of Performance Index Function

The performance index function is designed to reach undershoot quantity least and overload did not exceed a certain value. The system designed to facilitate the state variable feedback control, using the following formula for the performance index [6]:

$$J = \frac{1}{2}\boldsymbol{x}^T(t_f)\boldsymbol{S}\boldsymbol{x}(t_f) + \frac{1}{2}\int_{t_0}^{t_f} \boldsymbol{R}\boldsymbol{u}^2 dt \tag{3}$$

3.2 Guidance Equation Derivation

From Eq.(2) and Eq.(3) can write the Hamiltonian as:

$$H = \frac{1}{2}\boldsymbol{R}\boldsymbol{u}^2 + \boldsymbol{\lambda}^T(\boldsymbol{A}\boldsymbol{x} + \boldsymbol{B}\boldsymbol{u}) \tag{4}$$

Canonical equation is:

$$\dot{\boldsymbol{\lambda}} = -\frac{\partial H}{\partial \boldsymbol{x}} = -\boldsymbol{A}^T\boldsymbol{\lambda} \tag{5}$$

The minimum conditions can be obtained from optimal control theory: $\boldsymbol{R}\boldsymbol{u} + \boldsymbol{B}^T\boldsymbol{\lambda} = 0$. The optimal control variables as:

$$\boldsymbol{u}^*(t) = -\boldsymbol{R}^{-1}(t)\boldsymbol{B}^T(t)\boldsymbol{\lambda}(t) \tag{7}$$

Consistent cut-off conditions:

$$\boldsymbol{\lambda}(t_f) = \frac{\partial[\frac{1}{2}\boldsymbol{x}^T(t_f)\boldsymbol{S}\boldsymbol{x}(t_f)]}{\partial \boldsymbol{x}(t_f)} = \boldsymbol{S}\boldsymbol{x}(t_f) \tag{8}$$

As the optimal control Eq.(7) linearly dependent $\lambda(t)$, $\lambda(t)$ and terminal state vector have a linear relationship Eq.(8). For optimal control of linear dependent on the state vector, we can assume that $\lambda(t)$ with the linear relationship between the state vector $\boldsymbol{x}(t)$:

$$\lambda(t) = \boldsymbol{P}(t)\boldsymbol{x}(t), \quad t \in [t_0, t_f] \tag{9}$$

Substituting Eq.(9) into Eq.(7) can be obtained

$$\boldsymbol{u}^*(t) = -\boldsymbol{R}^{-1}(t)\boldsymbol{B}^{\mathrm{T}}(t)\boldsymbol{P}(t)\boldsymbol{x}(t) \tag{10}$$

$\boldsymbol{P}(t)$ can be obtained by solving the Riccati matrix differential equation, so we can get the optimal control variables as:

$$\boldsymbol{u}^* = \begin{bmatrix} -\dfrac{k_1(t_f - t)r_x}{k_2 r_1 + (t_f - t)^{27}} & -\dfrac{k_1(t_f - t)r_y}{k_2 r_2 + (t_f - t)^{27}} & -\dfrac{k_1(t_f - t)r_z}{k_2 r_3 + (t_f - t)^{27}} \end{bmatrix}^{\mathrm{T}} \tag{11}$$

To estimate the remaining flight time, you need to know relative speed $\dot{r}$ and relative distance r. Remaining flight time is shown as follows: $\tau = t_f - t = r / |\dot{r}|$.

4 Simulation

Air-launched Micro missiles to attack ground maneuvering armed vehicles. Armed vehicles acceleration $a_T = 25\sin(0.5t)\mathrm{m/s^2}$, velocity $V_T = 11\mathrm{m/s}$, armed vehicles center of mass of the starting position coordinates $x_T = 3500\mathrm{m}$, $y_T = 1.5\mathrm{m}$, $z_T = 0\mathrm{m}$. The initial simulation parameters of missile are: $V_0 = 20\mathrm{m/s}$、$\theta_0 = -5°$、$\psi_{V0} = 0°$、$x_0 = 0\mathrm{m}$、$y_0 = 500\mathrm{m}$、$z_0 = 0\mathrm{m}$, using designed optimal guidance law and the proportional guidance law for simulation, proportional navigation coefficient 3. Simulation curve as follow:

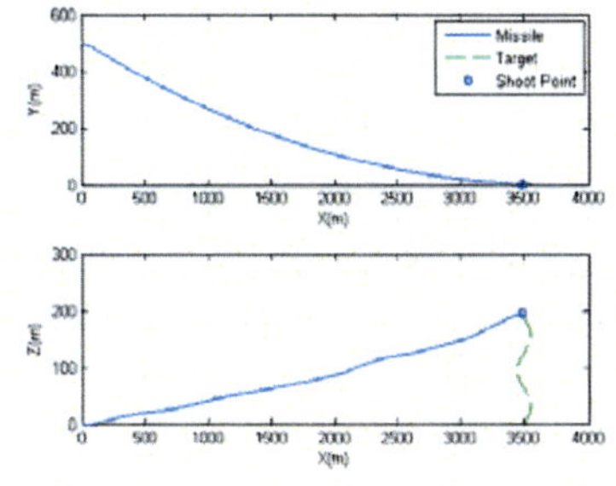

Fig. 1. ON Trajectory Curve

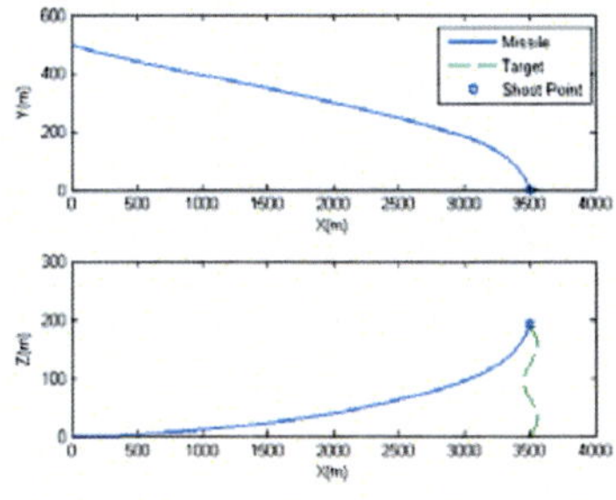

Fig. 2. PN Trajectory Curve

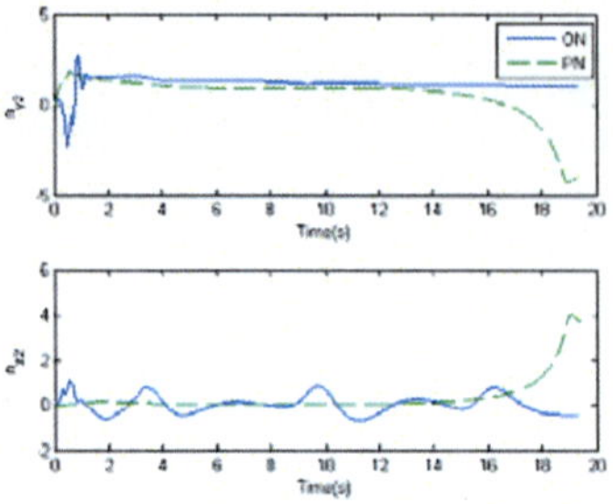

Fig. 3. Overload Curve

Curve ON denote the simulation results of the optimal guidance law, PN denote the simulation results of the proportional guidance law. The miss distance simulation result is $\delta R_{ON} = 0.8\text{m}$ 、 $\delta R_{PN} = 3.8\text{m}$.

Figure 1 and Figure 2 shows that the designed optimal guidance law trajectory curve more flat than the proportional guidance law. The optimal guidance law's miss distance is smaller than proportional guidance law, the miss distance of the optimal guidance law at less than 1m. From Figure 3, in the end portion of the optimal guidance law trajectory curve, the overload is smaller than the proportional navigation law.

5 Conclusion

From the three dimensional relative motion model of micro missile and target, using optimal control theory and flight mechanics principle, an optimal guidance law is designed. Simulation results show that The trajectory curve of the optimal guidance law compared to the proportional guidance law has higher precision and lower overload characteristics in the end portion of trajectory.

References

1. Poh, E.K., Li, D., Chai, J.S.: Near Optimal Midcourse Guidance Law for Flight Vehicle. In: AIAA 98-0583 (1998)
2. Bhattacharyya, A., Bhattacharjee, R.N.: Seeker Based Optimal Guidance Law,a Few Issues. In: AIAA 2003-5794 (2003)
3. Pint, L.: Nonlinear Optimal Guidance. In: AIAA 2006-6079 (2006)
4. Shaferman, V., Shima, T.: Linear Quadratic Guidance Laws for Imposing a Terminal Intercept Angle. Journal of Guidance, Control, and Dynamics 31(5), 1400–1412 (2008)
5. Li, X.-G., Fang, Q.: Winged missile flight dynamics, pp. 28–36. Northwestern Polytechnical University Press, Xi'an (2004)
6. Chen, S.-L.: Modern Aircraft Flight Mechanics, pp. 166–188. Northwestern Polytechnical University Press, Xi'an (1987)

Optimal Maneuver Technology for Satellite Formation Flying

Zhigang Wang and Xiang Chen

College of Astronautics , Northwestern Polytechnical University, 710072, Xi'an, China
sbeotal@nwpu.edu.cn

Abstract. This article produces a method to settle the optimal maneuver problem for satellite formation flying. Based on Hill's equations and Pontryagin's maximum principle, the optimal maneuver problem can be described as a state transition problem. Then, the optimal solution can be found by solving the state transition equations. The simulations show that this method is effective for optimal maneuver technology.

Keywords: Formation Flying, Optimal Maneuver, Minimum-time Maneuver.

1 Introduction

Formation flying is of many advantages compared with single satellite[1]. Usually, orbit maneuver is needed for formation redressal and formation keeping during the flying in space. Therefore, a method for settling the optimal maneuver is produced. This method changes the optimization problem into solving nonlinear equations[2] to avoid the disadvantages of the traditional method. The simulation results show that the method is effective for optimal maneuver technology. Besides, this method can also produce a time reference for minimum-fuel maneuver.

2 Mathematics Model

2.1 Dynamics Model

Relative satellite dynamics[3], as defined with a circular reference orbit, are shown as Eq.(1). They are convenient to work with because the cross-track motion Z decouples from the in-plane motion X and Y. In the development, only thrust in the Y and Z directions is considered because they are the most common[4].

$$\begin{cases} \ddot{x} = 3n^2 x + 2n\dot{y} + \Delta a_x \\ \ddot{y} = -2n\dot{x} + \Delta a_y \\ \ddot{z} = -n^2 z + \Delta a_z \end{cases} \tag{1}$$

D. Yang (Ed.): Informatics in Control, Automation and Robotics, Volume 2, LNEE 133, pp. 287–290.
springerlink.com

2.2 Minimum-Time Maneuver

Hill's equations depict a linear system. It is well known that the minimum-time solution is bang-bang for the system[5]. It follows that the form of the minimum-time control is shown in Fig. 1.

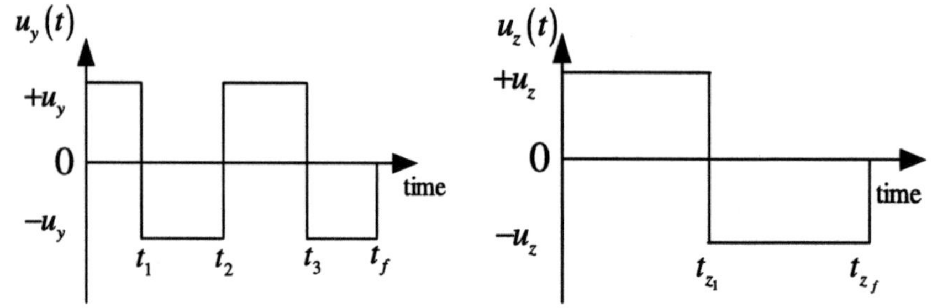

Fig. 1. Assumed switching functions for minimum-time maneuver

Minimum time maneuvers are characterized by Pontryagin's maximum principle with a cost defined as

$$J_{time} = \int_{t_0}^{t_f} dt \tag{2}$$

Using this cost function, the Hamiltonian can be defined as

$$H_{time}(t) = 1 + \boldsymbol{\lambda}^T(t)\left[\mathbf{AX}(t) + \mathbf{Bu}(t)\right] \tag{3}$$

where $\boldsymbol{\lambda}(t)$ is the costate and is defined as

$$\dot{\boldsymbol{\lambda}}(t) = -\mathbf{A}^T\boldsymbol{\lambda}(t) \tag{4}$$

The extremum condition is shown as Eq. (5). The t_i is defined as switching time.

$$\boldsymbol{\lambda}^T(t_i)\mathbf{B} = 0, \forall t_i \tag{5}$$

There is also an additional constraint equation on the Hamiltonian for minimum-time commands, which can be shown as $H_{time}(t) = 0$.

With the form of the minimum-time control and the dynamics functions, solving the optimal problem can be changed into solving nonlinear equations which are shown as Eq. (6) and Eq. (7) for the linear system described as Hill's equations.

$$\begin{bmatrix} 4-3\cos(nt_f) & 0 & -\sin(nt_f)/n & 2(1-\cos(nt_f))/n \\ 6(nt_f-\sin(nt_f)) & 1 & 2(\cos(nt_f)-1)/n & (3nt_f-4\sin(nt_f))/n \\ -3n\sin(nt_f) & 0 & \cos(nt_f) & -2\sin(nt_f) \\ 6n(\cos(nt_f)-1) & 0 & 2\sin(nt_f) & 4\cos(nt_f)-3 \end{bmatrix} \begin{bmatrix} x_f \\ y_f \\ \dot{x}_f \\ \dot{y}_f \end{bmatrix} = \begin{bmatrix} x_0 \\ y_0 \\ \dot{x}_0 \\ \dot{y}_0 \end{bmatrix} + \frac{u_{y_0}}{n^2} \begin{bmatrix} 2c_1-2c_2 \\ 4c_3+1.5c_4 \\ 2nc_3 \\ -3nc_1+4nc_2 \end{bmatrix} \tag{6}$$

$$\begin{cases} z_f\cos(nt_f) - \dot{z}_f\sin(nt_f)/n - z_0 - \left[2\cos(nt_{z_1}) - (1+\cos(nt_f))\right]u_{z_0}/n^2 = 0 \\ z_f\sin(nt_f) + \dot{z}_f\cos(nt_f) - \dot{z}_0 - (2\sin(nt_{z_1}) - \sin(nt_f))u_{z_0}/n = 0 \end{cases} \tag{7}$$

where $c_1 = n\left(2t_1 - 2t_2 + 2t_3 - t_f\right)$, $c_2 = 2\sin(nt_1) - 2\sin(nt_2) + 2\sin(nt_3) - \sin(nt_f)$, $c_3 = 2\cos(nt_1) - 1 - 2\cos(nt_2) + 2\cos(nt_3) - \cos(nt_f)$, $c_4 = n^2\left(2t_1^2 - 2t_2^2 + 2t_3^2 - t_f^2\right)$.

Then the optimal solutions of the minimum time maneuver can be found by solving the nonlinear equations.

3 Simulations

As an example, we suppose the reference orbit height is 1000km. The state transition course is from initial state($r_0=[180,-350,100]^T$m, $v_0=[-0.15,-0.40,-0.60]^T$ms^{-1}) to final state($r_f=[0,200,0]^T$m, $v_0=[0,0,0]^T$ms^{-1}). The initial control accelerations in the Y and Z directions are both 6×10^{-5}ms^{-2}. The simulation ignores the variety of mass.

3.1 Simulation Results

In-plane minimum-time constraints are solved in Eq. (6). The three switching times and final time are $[t_1, t_2, t_3, t_f]$ = [1531, 3777, 5537, 5891]s. Finally, the position states are $[0.06,200.02]^T$m, and the velocity states are $[-7.84\times10^{-5}, 9.35\times10^{-5}]^T$ms^{-1}.

Out-plane minimum-time constraints are solved in Eq. (7), the switching time and final time are $[t_{z1}, t_{zf}]$ = [2545, 3009]s. Finally, the position is 0m, and the velocity is 0ms^{-1}.

The time history of position states and velocity states are shown as Fig. 2 and Fig. 3. Figure 4 show the scaled input, the Hamiltonian, and the switching function $\boldsymbol{\lambda}^T\mathbf{B}$.

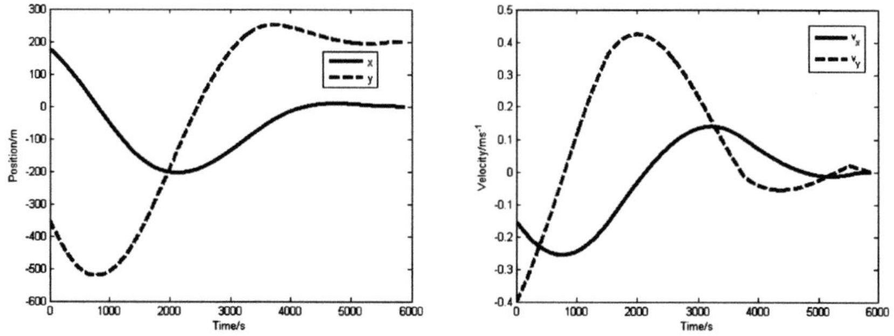

Fig. 2. Time history of in-plane position states and velocity states

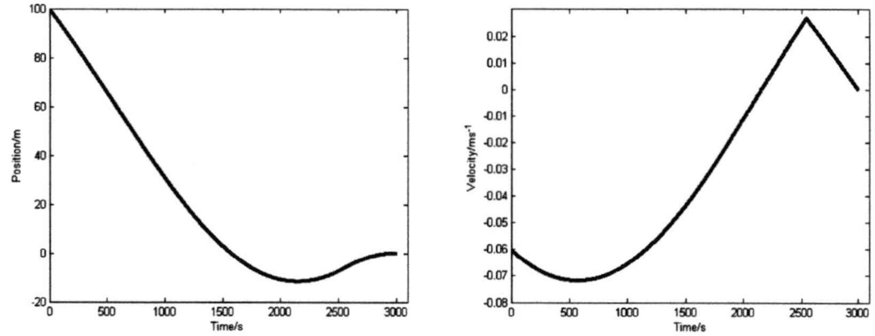

Fig. 3. Time history of out-plane position state and velocity state

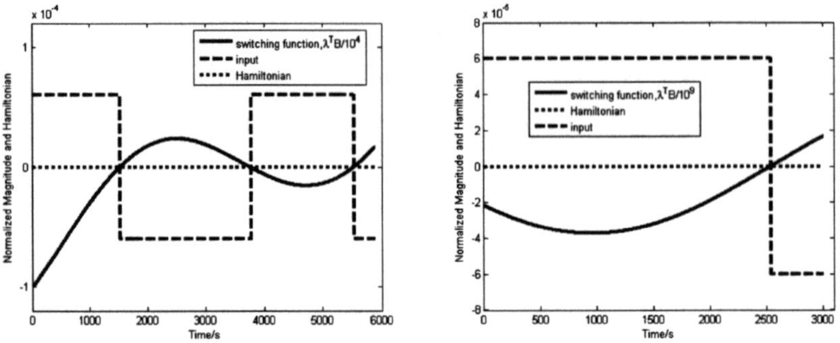

Fig. 4. Time history of input, the Hamiltonian, and the switching function

3.2 Analysis

As shown by simulation results, the maneuver starts and ends at the correct conditions. The fuel cost is 0.3535ms^{-1} for in-plane maneuver. Besides, the fuel cost is 0.1527ms^{-1} for out-plane maneuver. Figure 4 shows each of the switching constraints, given as $\boldsymbol{\lambda}^T\mathbf{B}=0$, are met for all three switching times. The in-plane and out-plane Hamiltonian are met for the desired value. The Hamiltonian and the switching constraints prove that we have found the optimal solutions for minimum-time maneuver.

4 Conclusion

This article produces a method to settle the optimal maneuver problem for satellite formation flying. The optimal solutions can be found through solving the state transition equations. Therefore we can avoid the disadvantages of the tradition optimization method. This method can produce optimal solutions which can be used for reference for practice. It can also produce a time reference for minimum-fuel maneuver.

References

1. Vadali, S. R., Vaddi, S. S., Naik, K. Control of Satellite Formations. AIAA Paper 2001–4028 (August 2001)
2. Campbell, M.E.: Planning Algorithm for Multiple Satellite Clusters. Journal of Guidance, Control, and Dynamics 26(5) (September-October 2003)
3. Wiesel, W.E. (ed.): Spaceflight Dynamics, 2nd edn., pp. 80–85. McGraw-Hill, New York (1989)
4. Carpenter, R.: Decentralized Control of Satellite Clusters. International Journal on Nonlinear and Robust Control 12(2-3), 141–161 (2002)
5. Kirk, D.E.: Optimal Control Theory: An Introduction, pp. 240–258. Prentice-Hall, Upper Saddle River (1970)

Comparisons between Two Kinds of Star Sensor Measurement Models by Attitude Estimation Accuracy

Zhigang Wang and Yifan Deng

College of Astronautics, Northwestern Polytechnical University, Xi'an 710072, China
ivan.dunne@163.com

Abstract. Compare two kinds of star sensor measurement models: vector observation model and quaternion observation model, which used in the attitude estimate algorithm based on star sensor and gyro by accuracy. Built the attitude kinematical model, sensors measurement model, and extended Kalman filtering (EKF) model. The simulation shows that the vector observation model is more accurate.

Keywords: Attitude Estimation, Star Sensor, Measurement Model, Extended Kalman Filter(EKF).

1 Introduction

Star sensor with gyro is a widely used scheme for high accuracy attitude determination, with extended Kalman filter(EKF)[1,2,3]. There are two kinds of measurement models of star sensor for filter, one of them use vector information as measurement, the other use attitude quaternion obtained by star sensor. This paper compared those two kind of models through computer simulation.

2 Coordinate Systems and Attitude Quaternion Kinematical Equation

Define the earth center inertial coordinate system (ECI) and spacecraft body-fixed system as usual. Assume the rotation order from ECI system to body system is 3-2-1, corresponding attitude quaternion is $q = [q_0 \quad q_1 \quad q_2 \quad q_3]^T$. The spacecraft's angular velocity under body system is $\boldsymbol{\omega} = [\omega_x \quad \omega_y \quad \omega_z]^T$. We have the quaternion kinematical equation

$$\dot{q} = \frac{1}{2} q \otimes \omega \tag{1}$$

D. Yang (Ed.): Informatics in Control, Automation and Robotics, Volume 2, LNEE 133, pp. 291–294.
springerlink.com

3 Sensors' Models

3.1 Gyro Measurement Model

Assume the gyro is fixed with body system. The gyro measurement model can be written as

$$\boldsymbol{\omega}_g = \boldsymbol{\omega} + \mathbf{b} + \mathbf{d} + \mathbf{n}_g \tag{2}$$

Where $\boldsymbol{\omega}_g$ is the gyro output, $\boldsymbol{\omega}$ is the real angular velocity, $\mathbf{b}$ is the gyro's constant bias, the gyro's time variation drift $\mathbf{d}$ is modeled as first order marcov process, gyro measurement noise $\mathbf{n}_g$ is white Gauss noise.

3.2 Star Sensor's Vector Observation Model

Assume the place matrix of the star sensor is unit matrix, then the direction of star light vector in the body system $\mathbf{r}_b$ is related to the direction in the ECI system $\mathbf{r}_0$ according to

$$\mathbf{r}_b = \mathbf{A}(q)\mathbf{r}_0 + \Delta\mathbf{r}_s \tag{3}$$

Where $\mathbf{A}(q)$ is the attitude matrix $\Delta\mathbf{r}_s$ is the white Gauss noise of the star sensor.

3.3 Star Sensor's Quaternion Observation Model

The star sensor's quaternion observation model can be described as

$$q_s = q \otimes n_q \tag{4}$$

Where q_s is the quaternion obtained by star sensor, q is the real attitude and n_q is the measurement error.

4 EKF Model

4.1 State Equation

Linearize the kinemics' equation (1), describe the attitude quaternion error as a rotation quaternion whose scalar part is quite close to 1 and all attitude information is contained in the 3 dimensional vector part[4].

We define the 9 dimensional error state vector $\Delta\mathbf{X} = \begin{bmatrix} \delta\bar{\mathbf{q}}^T & \Delta\mathbf{b}^T & \Delta\mathbf{d}^T \end{bmatrix}^T$, then we have the filtering state equation as follows.

$$\Delta\dot{\mathbf{X}}(t) = \mathbf{F}(t)\Delta\mathbf{X}(t) + \mathbf{G}(t)\mathbf{w}(t) \tag{5}$$

Where $\hat{\boldsymbol{\omega}}\times$ is the cross multiplication matrix.

Then we can obtain the dispersed state equation.

4.2 Measurement Equation

Assume that the two vectors in the vector observation model obtained from the star sensors are along the Y and Z axis of the body system, and the quaternion in the quaternion observation model is calculate from those two vectors by deterministic algorithm.

In the vector observation model, choose the two observed vectors as the measurement state $\mathbf{Z}=\begin{bmatrix}\mathbf{r}_{b1}^T & \mathbf{r}_{b2}^T\end{bmatrix}^T$, projecting the measurement equation to the plane which is perpendicular to the vectors in order to avoid the singular problem[5]. The linearized measurement equation of the vector observation model can be written as follows.

$$\mathbf{P}\Delta\mathbf{Z}(t)=\mathbf{P}\mathbf{H}(t)\Delta\mathbf{X}(t)+\mathbf{V}(t) \tag{6}$$

Where $\mathbf{P}$ is the projection matrix.

Then we can obtain the dispersed measurement equation.

In the quaternion observation model, choose the vector part of the quaternion as the measurement state $\mathbf{Z}=[\mathbf{q}_s]$, we have

$$\delta\mathbf{Z}=\delta\mathbf{q}+\mathbf{n}_q \tag{7}$$

Then we can obtain the dispersed measurement equation.

According to the state equation and the measurement equation, the filtering can be performed.

5 Simulation Results

Set the simulation parameters as follows: star sensor noise $\sigma_s=1''$, gyro constant bias $\mathbf{b}=5^\circ/h$, gyro measurement noise $\sigma_g=0.1^\circ/s$. State error covariance matrix original value $\mathbf{P}_0=diag\begin{bmatrix}1 & 1 & 1 & 1 & 1 & 1 & 1 & 1 & 1\end{bmatrix}$.

Set the filtering step as 1sec and the simulation time as 1day. The simulation results can be shown in fig. 1.

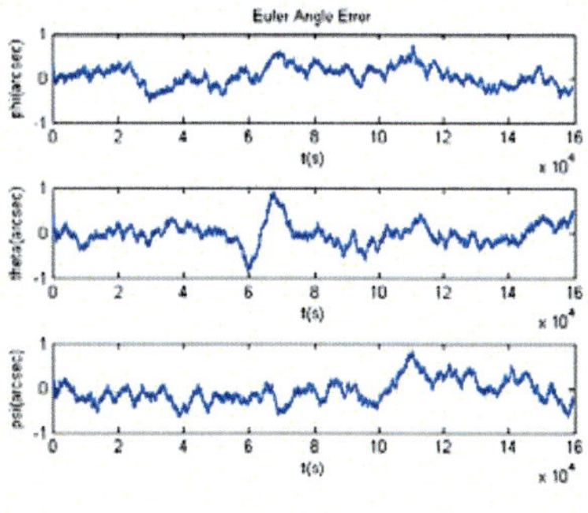

(a) vector observation model

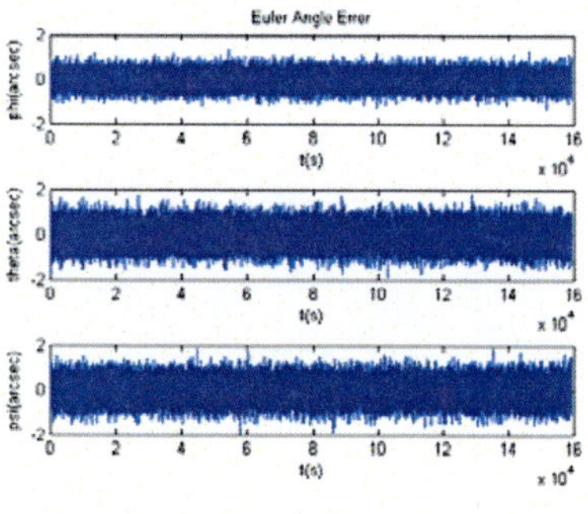

(b) quaternion observation model

Fig. 1. Attitude estimation error

From the simulation results we can see clearly that the attitude estimation through EKF from vector observation model is more accurate than that from quaternion observation model.

6 Conclusion

This paper discussed two kinds of measurement models of star sensor for filter: the vector observation model and the quaternion observation model, in the attitude estimation algorithm, and compared the attitude estimation accuracy through simulation.

Through the computer simulation, we get the conclusion that the attitude estimation through EKF from vector observation model is more accurate than that from quaternion observation model.

References

1. Shuster, M.D., Oh, S.D.: Three-axis attitude determination from vector observations. AIAA 81-4003
2. Markley, F.L.: Attitude determination and parameter estimation using vector observation: theory. Journal of Astronautical Sciences 37(1) (1989)
3. Markley, F.L.: Attitude determination and parameter estimation using vector observation: application. Journal of Astronautical Sciences 39(3) (1991)
4. Lefferts, E.J., Markley, F.L., Shuster, M.D.: Kalman filtering for spacecraft attitude estimation. AIAA 82.0070R
5. Shuster, M.D.: Kalman filtering of spacecraft attitude and the QUEST model. The Journal of Astronautical Sciences 38(3), 377–393 (1990)

Hypersonic Vehicle Attitude Single-Channel Controller Design Based on Spatially Optimal Rotation Vector

Zhigang Wang and Yu Ning

School of Astronautics, Northwestern Polytechnical University, Xi'an 710072, China
sbeotal@nwpu.edu.cn

Abstract. A novel single-channel control method for hypersonic vehicle is developed based on the idea of rigid body spatially optimal rotation vector. A double-loop, single-channel attitude controller, which solved the seriously coupling problem among channels existing in the traditional three-channel attitude control system, is designed under such method. Outer loop tracks attitude angles by motivating vehicle to rotate in terms of spatially optimal vector, while inner loop tracks angular velocity by producing a statement moment vector. Appropriate control system parameters are chosen through 6-DOF simulation. The simulation results show that this single channel attitude controller provides robust tracking of the desired mission angle profiles.

Keywords: hypersonic vehicle, attitude control, single-channel, spatially optimal rotation vector.

1 Introduction

Hypersonic vehicles are much more sensitive to aerodynamics than normal aircrafts. The traditional three-channel controller is no longer satisfying the requirements of such situation. Heller and Sachs designed longitudinal robust controller [1]. Georgie and Valasek investigated re-entry attitude control problem of X-38 [2]. Johnson and Calise used neural networks for adaptive guidance and control [3]. Richter and et al. utilized variable structure systems theory for the vehicles [4]. In this paper, a novel control method for hypersonic vehicle is developed. The idea of using rigid body spatially optimal rotation vector replaces the three-channel uncoupled control method.

2 Single-Channel Attitude Controller Design Idea

A single-channel attitude controller for hypersonic vehicle is designed as shown in Fig. 1. This is an inner-outer double-loop controller design. Outer loop tracks attitude angles by motivating vehicle to rotate in terms of spatially optimal vector, while inner loop tracks angular velocity by producing a statement moment vector. Such design solves the seriously coupling problem among channels existing in the traditional three-channel attitude control system. Definitions of coordinate systems and symbols are same as in Ref. [5].

D. Yang (Ed.): Informatics in Control, Automation and Robotics, Volume 2, LNEE 133, pp. 295–298.
springerlink.com

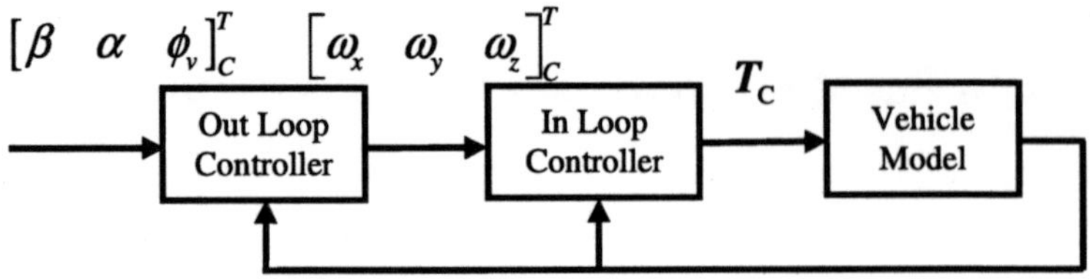

Fig. 1. Double-loop, single-channel attitude control system flow chart

3 Vehicle Motion Equations

The motion of vehicle is described by centroid movement and attitude rotation. The dynamic equations for centroid movement with respect to earth center is

$$\frac{d\boldsymbol{V}_k}{dt} = -\frac{\mu \boldsymbol{r}}{r^3} + \frac{\boldsymbol{P}+\boldsymbol{A}}{m} - 2\boldsymbol{\omega}_E \times \boldsymbol{V}_k - \boldsymbol{\omega}_E \times (\boldsymbol{\omega}_E \times \boldsymbol{r}) \tag{1}$$

Utilizing spherical coordinates r, λ, ϕ_c to express location of vehicle, we obtain

$$\dot{r} = -V_{\mathrm{k,uz}}, \ \dot{\lambda} = V_{\mathrm{k,uy}} / (r\cos\phi_c), \ \dot{\phi}_c = V_{\mathrm{k,ux}} / r \tag{2}$$

The dynamic equations for attitude rotation is

$$(\mathbf{I})_b \left[d(\boldsymbol{\omega})_b / dt \right] + (\boldsymbol{\omega})_b^{\times} (\mathbf{I})_b (\boldsymbol{\omega})_b = (\mathbf{M})_b \tag{3}$$

The kinematic equations is shown by quaternion.

4 Attitude Controller Design

Here the spatially optimal rotation vector for the vehicle is defined. Assuming two reference frames will be identical after a rotation along certain axis by angle σ, and the angles between the axis and the frame are $\beta_1, \beta_2, \beta_3$, the relative directions between can be settled by

$$\mathbf{Q} = q_0 + q_1\mathbf{i} + q_2\mathbf{j} + q_3\mathbf{k} \tag{4}$$

So the rotation from body system S_b to expected frame $S_{b'}$ is

$$\mathbf{Q}_{b'b} = \mathbf{Q}_{b'u} \circ \mathbf{Q}_{ub} = \mathbf{Q}_{b'u} \circ \mathbf{Q}_{bu}^* \tag{5}$$

where $\circ$ represents quaternion multiplication, and $\mathbf{Q}_{ub} = \mathbf{Q}_{bu}^*$. Then the vectorial part of $\mathbf{Q}_{b'b}$ is the spatially optimal vector in system S_b, and the minimum rotation angle is

$$\sigma = 2\arccos\left(q_{b'b0}\right) \tag{6}$$

where $q_{b'b0}$ is the scalar part of $\mathbf{Q}_{b'b}$.

4.1 Outer Loop Controller Design

While the vehicle flying, there will be an expected attitude at each moment. The function of an attitude control system is to adapt the real vehicle attitude to the expected one by control moments. Considering the rotation quaternion $\mathbf{Q}_{b'b}$ as attitude error, the expected angular velocity vector $\boldsymbol{\omega}_c$ provides gradual tracking of variety of expected target angle $\boldsymbol{\gamma}_{\mathrm{d}} = \left[\beta_{\mathrm{d}} \quad \alpha_{\mathrm{d}} \quad \phi_{\mathrm{vd}}\right]^{\mathrm{T}}$.

In system S_b, there is

$$\boldsymbol{\omega}_{\mathrm{d}} = \mathbf{Q}_{b'b}^{*} \circ \boldsymbol{\omega}_{\mathrm{d}}' \circ \mathbf{Q}_{b'b} \tag{7}$$

The control law is

$$\boldsymbol{\omega}_{\mathrm{c}} = k_1 \sigma \mathbf{q}_{b'b} + \boldsymbol{\omega}_{\mathrm{d}} \tag{8}$$

4.2 Inner Loop Controller Design

In the body system S_b, the angular momentum is

$$\mathbf{H}_{\omega_c} = \mathbf{I}\boldsymbol{\omega}_{\mathrm{c}} \tag{9}$$

Assume that $\mathbf{u}_{\mathrm{E}}$ is velocity vector of the end point of angular momentum, and $\boldsymbol{M}$ is the main moment of outer forces, then $\mathbf{u}_{\mathrm{E}} = \boldsymbol{M}$. Assume that

$$\mathbf{u}_{\mathrm{E}} = k_2\left(\mathbf{H}_{\omega_c} - \mathbf{H}_{\omega}\right) \tag{10}$$

Then the control moment will be

$$\mathbf{T}_{\mathrm{c}} = k_2\left(\mathbf{H}_{\omega_c} - \mathbf{H}_{\omega}\right) \tag{11}$$

5 Numerical Tests

In the computer simulation, the standard earth assumption and model is adopted. The parameters of vehicle are identical to space shuttle. The aerodynamic coefficients are obtained by linear fitting. The results are shown in Fig. 2. The controller rapidly converged after the disturbances and the attack angle well followed the expected one.

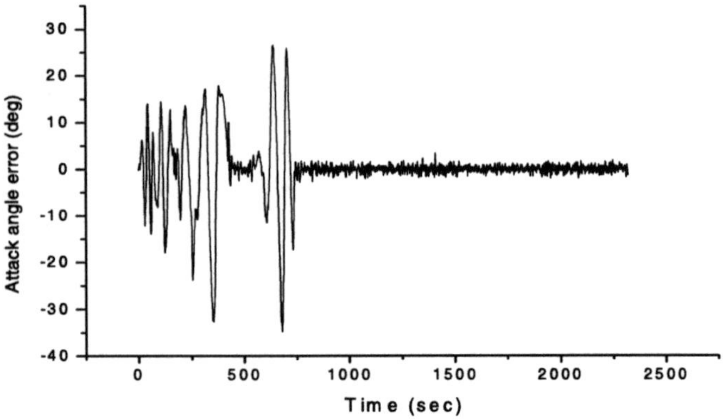

Fig. 2. Attack angle error

6 Conclusion

The idea of single-channel control method for hypersonic vehicle is explored. The method is based on the idea of rigid body spatially optimal rotation vector. A double-loop, single-channel attitude controller is designed. The outer loop tracks attitude angles by motivating vehicle to rotate in terms of spatially optimal vector, and the inner loop tracks angular velocity by producing a statement moment vector. The method solves the seriously coupling problem among channels existing in the traditional three-channel attitude control system. The simulation results show that this single channel attitude controller provides robust tracking of the desired mission angle profiles.

References

1. Heller, M., Sachs, G.: Flight dynamics and robust control of a hypersonic test vehicle with ramjet propulsion. In: AIAA International Space Planes and Hypersonic Systems and Technologies Conference, Norfolk, US, April 27-30 (1998)
2. Georgie, J., Valasek, J.: Selection of longitudinal desired dynamics for dynamic inversion controlled re-entry vehicles. In: AIAA Guidance, Navigation, and Control Conference and Exhibit, Montreal, Canada, August 6-9 (2001)
3. Johnson, E.N., Calise, A.J.: Reusable launch vehicle adaptive guidance and control using neural networks. In: AIAA Guidance, Navigation, and Control Conference and Exhibit, Canada, August 6-9 (2001)
4. Brinker, J.S., Wise, K.A.: Stability and flying qualities robustness of a dynamic inversion aircraft control law. Journal of Guidance, Control and Dynamics 19(6), 1270–1277 (1996)
5. Keshmiri, S., Colgren, R.: Six DOF nonlinear equations of motion for a generic hypersonic vehicle. In: AIAA Atmospheric Flight Mechanics Conference and Exhibit, South Carolina, US, August 20-23 (2007)

Shield Attitude Rectification Decision Function Based on Support Vector Data Description*

Guo Zheng-gang**, Wang Feng-tao, and Sun Wei

School of Mechanical Engineering, Dalian University of Technology,Dalian 116023, China
gzg1700w@126.com

Abstract. A shield attitude rectification decision function is proposed to reduce snake track phenomenon and rectification failure problem during shield attitude rectification. A method of shield attitude rectification working condition evaluation based on support vector data description(SVDD) is introduced. Shield attitude mechanics model containing priori knowledge assists to feature selection. SVDD handles the one class classification problem. Decision function for attitude rectification is proposed based on working condition evaluation. The result of engineering data processing indicates that the evaluation method is stable and practical for shield attitude rectification decision.

Keywords: shield, attitude rectification, SVDD, rectification decision.

1 Introduction

It is important to keep the shield actual axis on the designed axis during shield excavation. But there will always be deviation between actual and expected attitude during shield excavation. So the attitude rectification should be carried out when the deviation is over certain threshold. The shield driver observes show value of actual attitude, and then operates the jack to provide reverse moment for attitude rectification. But the way of attitude rectification droved only by attitude deviation always results in snake track phenomenon and rectification failure problem[1].

So a valid shield attitude rectification decision function should be established to stop attitude rectification before working condition being deteriorated, considering shield attitude is affected by relevant working condition.

Shield attitude rectification working condition must be evaluated accurately before establishing rectification decision function. Building a shield attitude mechanics model is a helpful way for working condition evaluation. When the deviation between actual and modeled attitude is over certain threshold, working condition deterioration is deemed to take place. But quite a few parameters of the model are difficult to acquire, and the selection of the ground constitutive model is not confirmed. But the

* Received date: 2011 -
Supported by Chinese National Basic Research Program (973 Program) (No.2007CB714006).
** Corresponding author.

D. Yang (Ed.): Informatics in Control, Automation and Robotics, Volume 2, LNEE 133, pp. 299–307.
springerlink.com

mechanics model can retrogress to a reference of working condition selection to solve the problem of working condition selection.

The other way is to analyses working condition historical sample, and then carry through the evaluation . The evaluation problem is classification problem, essentially. There are many effective arithmetic for dealing with classification problem, including cost sensitive support vector machines[2] [3], cost sensitive probabilistic neural network[4],condensed nearest neighbor data domain description[5], etc. Cost sensitive support vector machines and cost sensitive probabilistic neural network need two class samples at least, but there are always one class normal sample could be acquired during attitude rectification. Condensed nearest neighbor data domain description can use one class sample for classification, but over fitting can not be avoided. Tax proposed support vector data description(SVDD) based on support vector machine to solve one class classification[6] [7]. This method go well using small sample, and have powerful generalization ability. So SVDD is introduced to establish shield attitude working condition evaluation mechanism, and an improved way is introduced for unseen sample classification.

Method of shield attitude rectification working condition evaluation is proposed based on support vector data description. Shield attitude mechanics model containing priori knowledge assists in feature selection. SVDD handles the one class classification problem. A decision function for attitude rectification is proposed. Shield attitude is composed of yawing, pitching and rotation attitude. Only the yawing attitude is discussed and expressed as shield attitude in the article.

2 Attitude Rectification Working Condition Selection

Attitude rectification working condition selection means select several working conditions which have more influence on attitude rectification among numerous working conditions during shield excavation. The selected working conditions are used as sample feature to build sample vector. Descending dimension method (e.g. PCA) and experience method can handle feature selection respectively. Experience method based on shield attitude mechanics model is adopted for feature selection. Feature selection has dual significance. On the one hand, priori knowledge is introduced to reveal key factors of class difference of shield attitude rectification working condition. On the other hand, feature selection descends input space dimension, downscales the problem, and reduces the computational complexity. Effectiveness and availability should be considered on feature selection.

Mitsutaka Sugimoto has developed a model of shield mechanics [8] [9]. The model points out that the shield attitude is affected by the following five loads: force at face, jack force, force acting on shield periphery, forces on shield tail and self weight of shield. In depth analysis of Mitsutaka Sugimoto's model, the working conditions of shield attitude are selected, as illustrated in Table 1.

Table 1. Selection of shield attitude working condition

Load	Load decompose	Availa bility	Select or not	Reasons	The value involved in computing
Force at face	Earth pressure	easy	yes	Related to shield attitude closely. Represented by the readout of earth pressure sensors on the chamber board.	Readout of earth pressure sensors
	Frictional force	hard	no	Unavailable。 Concern with earth pressure at face.	-
Jack force	-	easy	yes	Related to shield attitude closely.	Jack oil pressure
Force acting on shield peripher y	Earth pressure	hard	no	Respond to other selected working conditions.	-
	Frictional force	hard	no	Unavailable。 Concern with earth pressure on shield periphery.	-
Forces on shield tail	-	hard	no	Ignorable when the shield tail is not contacts the segment ring. When the shield tail is contacts the segment ring, the force is not ignorable. But in this situation, the existing monitor system will alarm as abnormity.	-
Self weight of shield	-	easy	no	Since the sample comes from historical data of target shield, this value is a constant.	-

According to Table 1, attitude angle θ, readout of earth pressure sensors and jack oil pressure are selected as shield attitude working conditions.

The following will make some instruction and transformation of the selected working conditions for building sample vector.

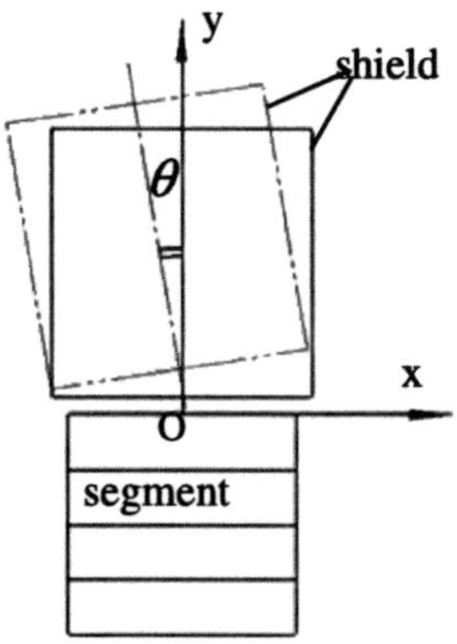

Fig. 1. Shield attitude angle

Shield attitude angle θ is the angle between shield axis and y axis as illustrated in Fig.1. Coordinate system xoy fixed on the segment. The origin point o is been set on center of the segment end plane which shield jack props on. The plane xoy is parallel to horizontal plane. The coordinate system xoy moves to new place when a new segment is set.

Difference between right and left side of earth pressure sensors readout, expressed as δ, conducts as a dimension of sample vector.

As to jack oil pressure, should be transformed to moment τ.

$$\tau = \sum_{i=1}^{N} y_i p_i l_i, \quad y_i \in \{-1,1\} \tag{1}$$

where N is the quantity of jack; $y_i = 1$,when ith jack is located in x axis negative direction, otherwise positive direction; p_i is the jack oil pressure of ith jack; l_i is the arm of ith jack.

So the sample vector x can be expressed as $x = [\theta, \delta, \tau]^T$.

Many other researchers also made great efforts for shield attitude model. Loganathan proposed a tunnel ground displacement model based on gap parameter, and verified the model with engineering data [10]. Wei Gang carried on an improvement to the Loganathan's model in that proposed a model including all kinds of solid [11] [12]. But there are large deviations between obtained results and the actual results, because of multiple uncertain parameters involved, so these models is not suitable for working condition selection.

3 Support Vector Data Description

Support vector data description defines a closed boundary around the normal sample: a hypersphere. The sphere is characterized by center a and radius R. We minimize the volume of the sphere by minimizing R^2, and demand that the sphere contains all training objects x_i. To allow the possibility of outliers in the training set, the distance from x_i to the center a should not be strictly smaller than R, but larger distances should be penalized. Therefore we introduce slack variables $\xi_i \geq 0$, $i = 1,2,\ldots,l$ and the penalty variable C. The parameter C controls the trade-off between the volume and the errors. Analogous to the Support Vector Classifier, a kernel function $K(x_i \cdot x_j) = \langle \varphi(x_i), \varphi(x_j) \rangle$ can be introduced to the algorithm. An ideal kernel function would map the target data onto a bounded, spherically shaped area in the feature space and outlier objects outside this area. According to the literature[5], a Gaussian kernel is selected as $K(x, y) = \exp(-\|x-y\|^2 / 2\sigma^2)$. As to this kernel, $K(x_i, x_i) = 1$. The original problem can be described as:

$$\begin{aligned} \min \quad & R^2 + C\sum_{i=1}^{l}\xi_i \\ s.t. \quad & \|\varphi(x_i) - a\| \le R^2 + \xi_i \\ & \xi_i \ge 0, \forall i = 1,\cdots,l \end{aligned} \quad . \tag{2}$$

where l is the quantity of the samples.

The Lagrangian is:

$$\begin{aligned} L(R,a,\alpha_i,\beta_i,\xi_i) = & R^2 + C\sum_{i=1}^{l}\xi_i \\ & -\sum_{i=1}^{l}\alpha_i\left\{R^2 + \xi_i - \left[\|\varphi(x_i)\|^2 - 2a\cdot\varphi(x_i) + \|a\|^2\right]\right\} - \sum_{i=1}^{l}\beta_i\xi_i \end{aligned} \quad . \tag{3}$$

where α_i and β_i are Lagrange multipliers.

Setting $\partial L/\partial R = 0$, $\partial L/\partial a = 0$ and $\partial L/\partial \xi_i = 0$,and then substituting the results back into $L(\cdot)$, and solve the corresponding QP problem.

The SVDD sphere center a is expressed as:

$$a = \sum_{i=1}^{l}\alpha_i\varphi(x_i) \ . \tag{4}$$

According to the Kuhn-Tucker conditions, the target training data can be classified into three categories. The samples with $C > \alpha_i > 0$ are called unbounded support vectors(UBSVs). Radius R can be determined by taking any $x_k \in UBSVs$ and calculating the distance from its image to the center a:

$$R^2 = 1 - 2\sum_i \alpha_i K(x_i \cdot x_k) + 2\sum_{i,j}\alpha_i\alpha_j K(x_i \cdot x_j), \forall x_k \in SV \ . \tag{5}$$

For an unseen sample z, its distance to the center a is:

$$\|\varphi(z) - a\|^2 = K(z\cdot z) - 2\sum_{i=1}^{l}\alpha_i K(z\cdot x_i) + \sum_{i=1}^{l}\sum_{j=1}^{l}\alpha_i\alpha_j K(x_i \cdot x_j) \ . \tag{6}$$

The difference of Eq.(6) of Eq.(5) yields the decision function:

$$D(z) = 1 - R^2 + \sum_{i=1}^{l}\sum_{j=1}^{l}\alpha_i\alpha_j K(x_i \cdot x_j) - 2\sum_{i=1}^{l}\alpha_i K(z\cdot x_i) \ . \tag{7}$$

If $D(z)\le 0$, z is accepted as a target sample; otherwise it is rejected as an outlier. Thus the problem of sample processing and evaluation criterion has been solved.

Take notice of the first three items on the right side of equal sign of Eq.(7) is constant(does not contain z), the amount of calculation of unseen sample classification rest on the fourth item. If the amount of calculation could be reduced, the calculation of unseen sample classification will be speed up.

4 Improved Way for Unseen Sample Classification

When the Gaussian kernel is used, all $\varphi(x)$ must be on the unit ball centered at the origin O_F in feature space, because of $K(x,x)=1$, illustrates as Fig.2.

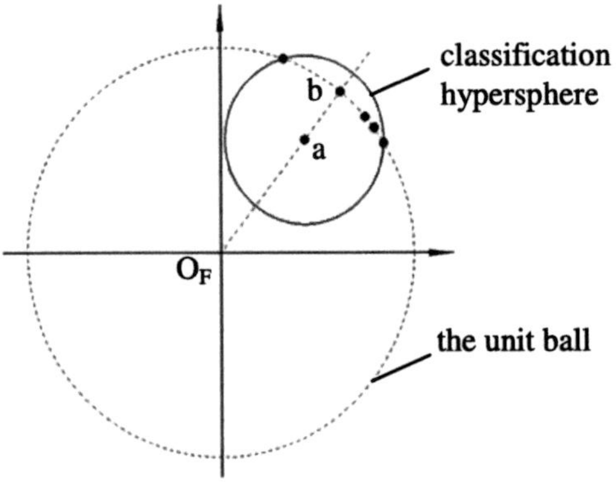

Fig. 2. Classification hypersphere in feature space

Extension line of $O_F a$ and the unit ball intersect at point b . b has the same direction with a , so

$$b = \gamma a = \gamma \sum_{i=1}^{l} \alpha_i \varphi(x_i) \ . \tag{8}$$

For a fixed σ andC, γ is a constant.

$$\gamma = \frac{1}{\|a\|} \ . \tag{9}$$

Suppose that the preimage of b is $\hat{x}$, hence

$$a = \frac{b}{\gamma} = \frac{\varphi(\hat{x})}{\gamma} \ . \tag{10}$$

Substituting Eq.(10)back into Eq.(6), obtain the new decision function:

$$\begin{aligned} D(z) &= \|\varphi(z)-a\|^2 - R^2 \\ &= 1 - R^2 + 1/\gamma^2 - \frac{2}{\gamma} K(z,\hat{x}) \end{aligned} . \tag{11}$$

Notice that the fourth item on the right side of equal sign of Eq.(11) just needs to evaluate the kernel function one time, so the calculation of unseen sample classification will be speed up comparing Eq.(7).

5 Decision Function for Attitude Rectification

One sample set is composed of the data of one segment excavation, and the classification hypersphere is calculated. The amount of sample in the sample set is N_s. The real time samples are treated as unseen sample for classification during excavation. The amount of abnormal sample is N_a. Then the attitude rectification decision function is:

$$\frac{N_s}{N_a} \times \frac{\theta}{\theta_e} \leq \eta \quad ,\eta \in [0,1] . \tag{12}$$

Where θ is the attitude angle; θ_e is the expected angle for this segment. η is set artificially. The lower η is been set, the less tolerance to working condition deteriorating will have.

If the Eq. (12)is satisfied, the attitude keep going, otherwise should be stopped.

The steps of the shield attitude rectification decision are:

Step 1: initialize C and σ.
Step 2:solve the QP problem of Eq.(3).
Step 3:compute the sphere radius R.
Step 4:compute the scaling factor γ.
Step 5:find the preimage $\hat{x}$ of b.
Step 6:classification decision.
Step 7:attitude rectification decision.

6 Application

Tunnel excavation data from Guangzhou is introduced to verify algorithm. The data is extracted from tunnel horizontal curve alignment with a radius of 500*m*. A sample set is composed of 100 samples from one segment excavation for the sphere radius computation. Three continuous sample sets sphere radius are computed for stability verification. If the stability is good, the algorithm in this article can run as the evaluation criterion well.

Verification concentrates on preprocessing of engineering data. The original engineering data illustrated as Fig.3.

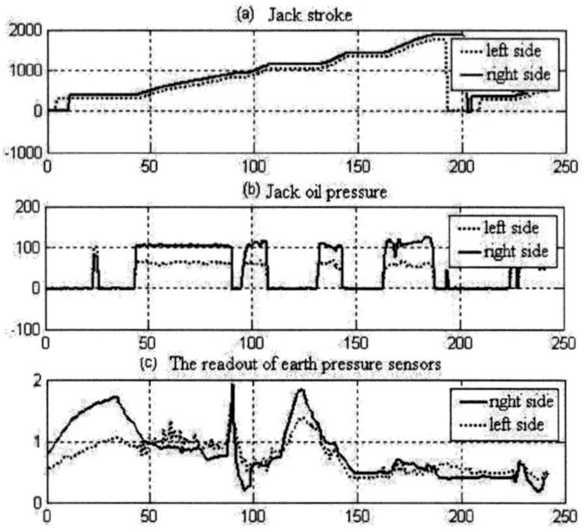

Fig. 3. Original engineering data of one segment

The original engineering data includes the readout of earth pressure sensors, jack oil pressure and jack stroke.

The steps of engineering data preprocessing are:

Step 1: delete the data which the jack oil pressure approaching to zero.
Step 2: remove the idle stroke
Step 3: subtract the readout of left earth pressure sensors from right.
Step 4: processing jack oil pressure data according to Eq.(1).
Step 5:attitude angle θ computing. Attitude angle θ is corresponding rigidly to the difference between right side jack stroke and left side jack stroke, in addition the value of a small attitude angle is equal to the sine of θ approximately, so the attitude angle θ can be computed by ratio of jack stroke difference between right and left to shield diameter.
Step 6: extract 100 rows uniformly-spaced data and normalize the data to build up a sample set.

After preprocess and process the data of three segment excavation, the hypersphere radius are 0.48,0.52and 0.49 separately. The stability of R is good, so the algorithm in this article can run as the evaluation criterion well.

7 Conclusions

Evaluation of shield attitude rectification working conditions is helpful for rectification decision, and could be described as a one class classification problem in most cases. SVDD method is introduced to handles the one class classification problem. Decision function for attitude rectification is proposed. The result of engineering data processing indicates that the evaluation method is stable and practical for shield attitude rectification working condition evaluation and rectification decision. Some working conditions not be selected, which also are closely related to shield attitude rectification, for hard to be acquired. If they are be collected in the future, they could be added to the algorithm easily to improve the decision function.

References

1. Chen, K., Hong, K.: Shield construction technique. China Communications Press (2009) (in Chinese)
2. Zhen, E.H., Li, P., Song, Z.H.: Cost sensitive support vector machines. J. Control and Decision 21, 473–476 (2006)
3. Muntean, M., Ileana, I., Rotar, C.: Improving classification with cost-sensitive approach and support vector machine. In: 9th RoEduNet IEEE International Conference, vol. 21, pp. 180–185 (2010)
4. Tang, M.Z., Yang, C.H., Gui, W.H.: Cost-sensitive probabilistic neural network and its application in the fault diagnosis. J. Control and Decision 25, 1074–1078 (2010)
5. Angiulli, F.: Condensed nearest neighbor data domain description. J. Pattern Analysis and Machine Intelligence 29, 1746–1758 (2007)
6. Tax, David, M.J., Duin, R.P.: Support vector data description. J. Machine Learning 54, 45–66 (2004)
7. Ypma, A., Tax, David, M.J., Duin: Robust machine fault detection with independent component analysis and support vector data description. In: Neural Networks for Signal Processing-Proceedings of the IEEE Workshop, pp. 67–76 (1999)
8. Sugimoto, M., Sramoon, A.: Theoretical Model of Shield Behavior During Excavation I:Theory. Journal of Geotechnical and Geoenvironmental Engineering 128, 138–155 (2002)
9. Sugimoto, M., Sramoon, A.: Theoretical Model of Shield Behavior during Excavation II: Application. Journal of Geotechnical and Geoenvironmental Engineering 128, 156–165 (2002)
10. Loganathan, N.: Analytical Prediction for Tunneling Induced Ground Movement in Clays. Journal of Geotechnical and Geoenvironmental Engineering 124, 846–856 (1998)
11. Wei, G.: Establishment of Uniform Ground Movement Model for Shield Tunnels. Journal of Geotechnical Engineering 29, 554–559 (2007)
12. Wei, G.: Prediction of ground deformation induced by shield tunneling construction. Journal of Rock Mechanics and Engineering 28, 418–424 (2009)

Embedded Microscope Vision Based Mechanical Platform for LED Wafer Automatic Inspection

Xinyan Gao[1,*], Ning Zhou[2], Dakui Li[1], and Yuan Yue[3]

[1] School of Software, Dalian University of Technology, Dalian, China
Tyler.gxy@hotmail.com, ldk@dl.cn

[2] School of Computer and Information Technology, Beijing Jiaotong University, Beijing, China
Tomzhoun@gmail.com

[3] School of Mathematics and Computer Science Institute, Northwest University for Nationalities, Lanzhou, China
helloyy2@gmail.com

Abstract. In this paper, we propose a novel technique solution towards LED wafer defects automatic full inspection using neural network chip array to assure defect-free outgoing dies. Our research intends to develop an automatic inspection system for defect pattern recognition in order to substitute human visual judgement. This solution mainly includes a three degree-of-freedom precise mechanical positioning stage and an automatic robot arm working with an embedded microscope vision system. A built-in parallel neural network chip array acts as the recognition engine instead of traditional software approach. Meanwhile, the mechanical motion control is also based on neural network method. This solution will benefit greatly from hardware engine acceleration as for performance improvement.

Keywords: LED, Automatic Inspection, Neural Network, Microscope Vision, Mechanical Platform.

1 Introduction

With the rapid development of LED industry, the yield ratio of LED wafer manufacturing has drawn more and more attention. In particular, in most cases, wafer defect inspection is still conducted manually, so that the full inspection process can incur significant personnel costs and lower yield ratio due to human fatigue.

Consequently, an automatic inspection process is required to maintain a high yield and consistent inspection results in LED defect inspection or wafer manufacturing which benefits from low-cost and acceptable reliability.

* The project is supported by "the Fundamental Research Funds for the Central Universities" (1600-852014) and partly supported by the National Natural Science Foundation of China under Grant No. 60873118 and 60973147 and partly supported by Grants (HCIC201102) of Guangxi Key Laboratory of Hybrid Computational and IC Design Analysis Open Fund.

D. Yang (Ed.): Informatics in Control, Automation and Robotics, Volume 2, LNEE 133, pp. 309–316.
springerlink.com

At present, on the one hand, several neural network based approaches for inspecting LED defect have been proposed [2,3,7,1]. In [5], an automatic solution using neural networks and statistical approaches for defect inspection of LEDs was studied. All these proposed methods are either based on learning vector quantization, or using statistical approach, or utilizing geometric and textural features, or by efficient knowledge acquisition. However, essentially, they are software based methods for neural networks implementation and suffer from lower efficiency and speed.

On the other hand, the neural network based motion control system has recently been widely investigated [8,9]. The ability to learn and high performance make neural network based method become an attractive subject. We aim to develop a control method for auto-focusing and motion error compensation by using neural network approach.

Therefore, in this paper, we attempt to investigate a novel visual inspection solution for LED wafer images based neural network hardware chip array and develop an automatic mechanical and electronic integrated platform based on this idea.

2 Platform Architecture

In this section, we will firstly give an overview of the inspection platform.

As shown in Fig.1, the embedded computer system is composed of a SOC (System-On-Chip) processor based on ARM11 core with one 32KB data cache and one 32KB instruction cache acting as the core processor to support Linux system. Microscope vision system with an integrated advanced camera system can be used to control external auto focus and perform image processing. Mechanical stage is used to hold wafer for scanning and positioning. The neural network chip array works as the pattern recognition engine under the control of software.

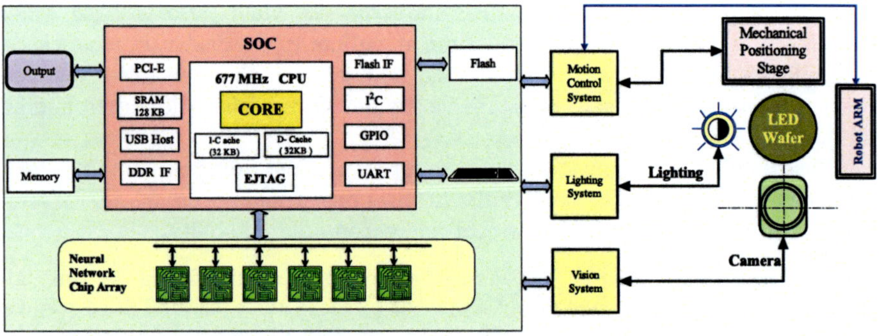

Fig. 1. Embedded Vision Inspection Platform

The platform mainly include 4 types of sub-system and the core components can be listed as the following:

(1) ARM11 based Embedded System
- one USB 2.0 host controller interface connected with USB camera
- one flash memory interface connected with NOR flash chip(32MB)
- one I^2C bus interface connected to neural network chip
- one DDR2 interface connected with 256MB memory module
- neural network chip array (up to 16 chips)

(2) Mechanical Positioning Stage and Robot Arm
- X-Y positioning stage
- rotary positioning stage
- driver and motion controller
- 3 degree-of-freedom (X-Y & Z) robot arm

(3) Microscope Vision Sampling System
- Z linear positioning stage
- microscope for machine vision [4] system
- 1.3 Megapixel CCD camera with USB 2.0 interface

(4) Lighting System
- LED lighting equipment
- illumination adjustment module

When inspecting, the target wafer is held in a special holder and is moved by a commercial X-Y precision stage driven by two stepper motors. The planar range of motion of this stage is 80×80 mm^2. The speed of movement is maximum to about 5mm per second. In this project, a neural network chip array module (CM1K[6], from "Cognimem" company) is integrated and used as the recognizing engine. Our project implementation is done on an embedded Linux distribution running a 2.6.29 kernel.

3 Preprocessing and Experiments

3.1 Auto-Focusing

For vision system, the magnification is greater than 100 times and focus position can heavily impact the quality of image acquired. Thus, an auto-focusing system should be developed for this platform.

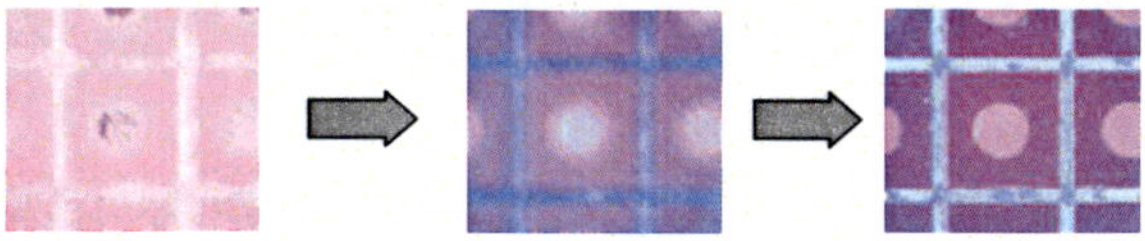

Fig. 2. Focus Position Finding

Fig.2 shows the focus position locating process, the auto-focusing operates by calculating the standard deviation of all grey levels of each image as it steps through the focal plane. By the "local maximum" algorithm, the focus position is found by camera movement with Z-axis linear stage under the control of software. The focus position is then used for capturing images with the CCD camera.

3.2 Lighting

Lighting is an important factor in viewing objects through a microscope. For microscope, good lighting is the cornerstone of professional imaging. Thus, a specified lighting system is used to adjust light on a microscope to obtain maximum results. In our platform, the microscope works with external illumination light sources of LED ring equipment.

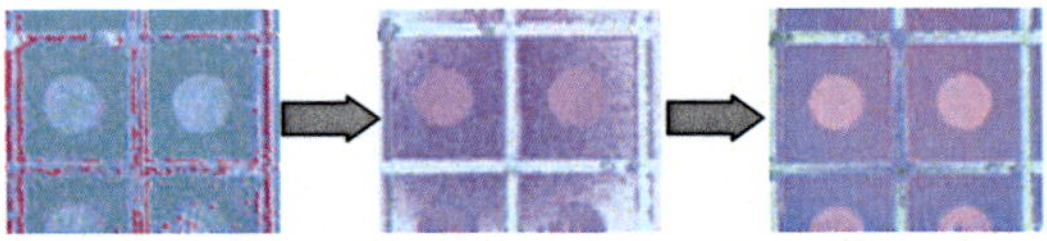

Fig. 3. Lighting Adjustment

As show in Fig.3, the left picture shows the object cannot be seen clearly when the light is too dark while the right picture shows a clear image can be yielded if the light is properly adjusted.

4 Inspection

4.1 Inspection Flow

As shown in Fig.4-(1), each LED wafer contains thousands of dies while each LED die consists of 2 key components: light-emitted area and P-electrode. The right part Fig.4-(2) is captured from microscope vision system by zooming in 100 times.

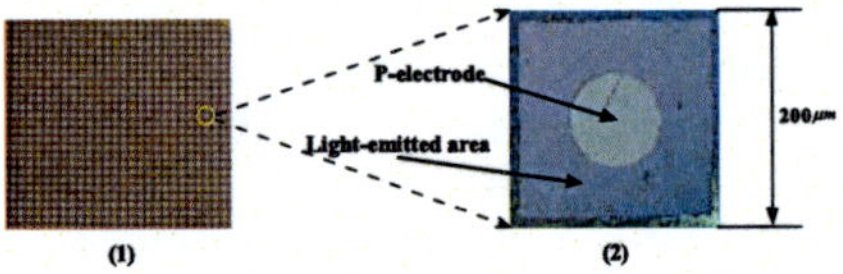

Fig. 4. LED Wafer and Die

Thus, an automatic approach which can locate, inspect, and mark individual dies is proposed to inspect the post-sawing LEDs. The basic inspection process consists of 3 steps: geometry inspection, light-emitted area inspection and

P-electrode inspection. Thus, we have the following elementary checking flow ***FLOW.1*** based on hardware neural network algorithm.

```
FLOW.1: "Inspection Flow "
BEGIN

  /* Step 1: Preprocessing */
    Powered On;
    System Inintialization(CPU, Cache, DDR, GPIO, ...);
    Move to original position;
    Adjust Light;
    Move Z − axis for finding focus position;
  /* Step 2:  Inspecting the whole wafer */
  /* Step 2.1: Initializing Neural Network chip and reset the category */
    Init_NN_Array( );
  /* Step 2.2: Learning vectors */
    Learn_Vectors( );
  /* Step 2.3: Scanning each block of the whole wafer and inspecting */
    while(!Scan_End){
        Taken snapshot and get sub − image; /* Image acquirement */
        Geometry_Measure(width, height); /* appearance inspection */
        / * Block Area Inspection * /;
        Feature_Extracting(Light_emitted_Area);
        Nearual_Network_ChipArray(vector);
        pass = Defect_inspection( ); /* call "Recognizing( )" */
        if(!pass);{
            Make a mark at current position;}
        / * P − electrode Inspection * /;
        Feature_Extracting(P − electrode);
        Nearual_Network_ChipArray(vector);
        pass = Defect_inspection( ); /* call "Recognizing( )" */
        if(!pass);{
            Make a mark at current position;}
        Move_to_Next( ); /* Motion control and move to next block */
     } /* End of while */

END;
```

4.2 Algorithm Framework

In this section, we will discuss the defect inspection algorithm framework based on *Neural Network Chip* architecture.

As shown in Fig.5-(2), a software hierarchy including core algorithms is depicted. The low-level routines for accessing neurons have been implemented and wrapped in kernel driver module while high-level algorithm functions are organized as a user layer software package.

Next, three core procedures: feature extraction, learning a vector, and recognizing a vector will be discussed respectively.

1) Feature Extraction

Generally, CM1K is used as a high-performance pattern recognition engine featuring a network of 1024 neurons expandable through its parallel architecture. The chip can classify a digital signal received directly from a sensor. The feature extraction embedded in the CogniMem chip is called sub-sampling and used 6 parameters:

- NWidth and NHeight (or the ROI (Region of Interest) nominal size)
- BWidth and BHeight (or the size of the primitive blocks inside the ROI)
- LEFT and TOP

According to specifications[6], the ROI nominal size must fit a number of primitive blocks less than or equal to 256. The relationship between the parameters is:

$$NWidth = n{\times}BWidth,\quad NHeight = m{\times}BHeight \quad (n{\times}m \leq 256) \qquad (1)$$

As shown in Fig.5-(1), the pixels of each block are used to produce each component of the signature vector.

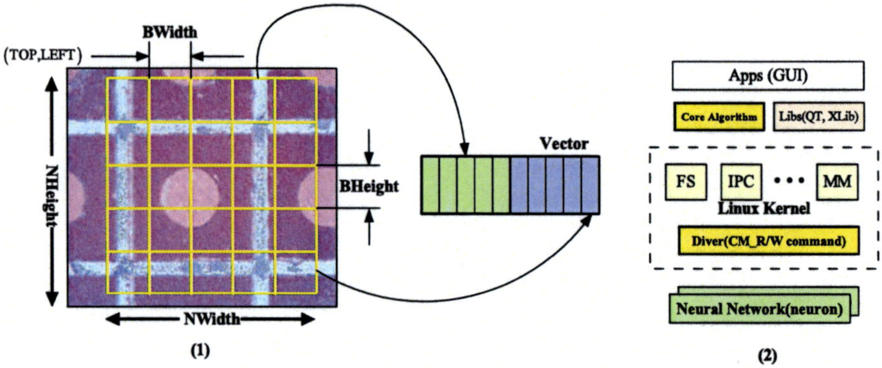

Fig. 5. Image Sub-Sampling

When needed, the following basic functions can be used to access the registers of CogniMem chip.

- ▷ $CM\_Write(byte\ Reg,\ word\ data);$
- ▷ $CM\_Read(byte\ Reg,\ word\ *data);$
- ▷ $Camera\_Write(0x01,\ byte\ Reg, word\ data);$
- ▷ $Camera\_Read(0x01,\ byte\ Reg, word\ *data);$
- ▷ $CM\_ReadProto(word\ *Vector,\ word\ VecLength);$
- ▷ $CM\_WriteProto(word\ *Vector,\ word\ VecLength);$

From CogniMem technology reference guide [6], a group of registers, such as "CM_NCR", "CM_COMP", "CM_LCOMP", "CM_DIST", "CM_CAT", and etc, are provided for operating neural network for learning and recognizing.

2) Learning a Vector

Learning a vector is the first step for recognizing, which consists of broadcasting a vector to the neurons and assigning it a category value. The following pseudo-codes can describe this function.

```
void Learning(Vector[ ], int CatVal){
    For(i = 0; i < N − 1; i + +){
        CM_Write(CM_COMP, Vector[i]); // Broadcast the N-1 components
    }// End of For
        CM_Write(CM_LCOMP, Vector[N]);
        /* Broadcast the last component(which updates the ID, UNC flags) */
        /* Broadcast the category of the vector */
        CM_Write(CM_CAT, CatVal);
}// End of Learning
```

3) Recognizing a Vector

Recognizing a vector is our final target which consists of broadcasting a vector to the neurons and reading the distance and category registers successively. Then a category value will be return in the end. The function can be described by the following pseudo-codes.

```
int Recognizing(Vector[ ]){
    For(i = 0; i < N − 1; i + +){
        CM_Write(CM_COMP, Vector[i]); // Broadcast the N-1 components
    }// End of For
        CM_Write(CM_LCOMP, Vector[N]);
        /* Broadcast the last component(which updates the ID, UNC flags) */
        /* Broadcast the category of the vector */
    do{
        CM_Read(CM_DIST, Out_dis);
        CM_Read(CM_CAT, Out_cat);
    }while(Out_dis! = 0xFFFF)
    return Out_cat;
}// End of Recognizing
```

5 Conclusion

In this paper, we presented a solution for the automated post-sawing inspection of LED defects. This solution involves the applications of computer vision technique and image processing algorithms development based on neural network chip to identify and classify LED defects. The performance of this solution will be about 100 times faster than a computer running at 2.4 Ghz.

We are now developing the K-Nearest Neighbor (KNN) classifier inspection programm based on our current solution. Higher level motion control system software based on neural network for interfacing and positioning will be developed in the near future.

The advantage of our solution is that the performance in terms of accuracy and speed can be improved greatly due to neural network chip array. The current preprocessing experiments show the proposed solution achieves pretty satisfied results and it is a promising method should be particularly explored in the future.

Therefore, future work also involves continuing this research and further develop a specified algorithm based on the built-in RBF(Radial Basis Function) neural network of the chip. We will then consider using simple but powerful features input vector to optimize this solution.

Acknowledgments. The project is supported by "the Fundamental Research Funds for the Central Universities" (1600-852014) and partly supported by the National Natural Science Foundation of China under Grant No. 60873118 and 60973147 and partly supported by Grants (HCIC201102) of Guangxi Key Laboratory of Hybrid Computational and IC Design Analysis Open Fund. The authors would like to thank their colleagues for participating in the research. They also appreciate the anonymous reviewers for their helpful comments.

References

1. Lin, H.D.: Automated defect inspection of light-emitting diode chips using neural network and statistical approaches. Expert Syst. Appl. 36(1), 219–226 (2009)
2. Su, C.T., Yang, T., Ke, C.M.: A neural-network approach for semiconductor wafer post-sawing inspection. IEEE Trans. Semi. Man. 15(2), 260–266 (2002)
3. Li, C., Chang, C., Jeng, M.: Applying regional level-set formulation to postsawing four-element led wafer inspection. IEEE Transactions on Systems, Man, and Cybernetics, Part C: Applications and Reviews (99), 1–12 (2010)
4. Zhang, J.M., Lin, R.M., Wang, M.J.: The development of an automatic post-sawing inspection system using computer vision techniques. Comput. Ind. 40(1), 51–60 (1999)
5. Lin, H.D.: Automated defect inspection of light-emitting diode chips using neural network and statistical approaches. Expert Syst. Appl. 36(1), 219–226 (2009)
6. http://www.cognimem.com
7. Chang, C.Y., Li, C.H., Chang, C.H., Jeng, M.: Learning vector quantization neural networks for LED wafer defect inspection. Int. J. Innov. Comp. Inf. Control 4(10), 2565–2579 (2008)
8. Corradini, M., Ippoliti, G., Longhi, S.: Neural Networks Based Control of Mobile Robots: Development and Experimental Validation. Journal of Robotic Systems 20(10), 587–600 (2003)
9. Hendzel, Z.: Adaptive Critic Neural Networks for Motion Control of Wheeled Mobile Robot. Nonlinear Dynamics 50(4), 849–855 (2007)

A Selective Encryption Scheme for H.264/AVC Video Coding

Wei Huang[1], Wenqing Fan[1], and Tingting Zhang[2]

[1] School of Computer Science, Communication University of China, 100024, Beijing, China
[2] Information Security Center, Beijing University of Posts and Telecommunications, 100876 Beijing, China
ztt861215@hotmail.com

Abstract. H.264/AVC is the most advanced standard for video compression while the confidentiality of the content encoded by H.264/AVC is not well implemented in the coding standard. In this paper, we proposed a selective encryption scheme which makes effective use of the data structure of NAL in H.264/AVC. We used AES encryption algorithm to encrypt important I frames data, thus it can keep the advantage of selective encryption algorithm while provide strong enough confidentiality. The experiments demonstrated that our scheme can meet the requirements of real-time processing and effective confidentiality.

Keywords: H.264/AVC, video encryption, selective encryption.

1 Introduction

H.264/AVC is the latest video coding standard, which is currently one of the most commonly used formats for the recording and compression. H.264/AVC can support a wide range of applications, such as digital TV broadcasting, real time video communication, media streaming and so forth. With the development of H.264/AVC, the design and implementation of an efficient encryption platform based on H.264/AVC becomes an urgent issue recently.

Many researches have been conducted on video data encryption, but most of them focused on the old standards, such as MPEG-2, MPEG-4. Although these algorithms have a high confidentiality level, they can't meet the real-time processing requirement due to the large computational overhead.

To solve the above issues, we propose a selective encryption scheme based on the data structure of NAL in H.264/AVC. In this scheme, only part of data, such as I frames, sequence parameter set and picture parameter set, are chosen to be encrypted. This scheme can substantially reduces the computational overhead and improves the efficiency of the scheme.

The rest of this paper is organized as follows. The selective encryption algorithms based on H.264/AVC are described in Section 2. Then the design of the new selective encryption scheme is given in Section 3. The performance of the selective encryption

D. Yang (Ed.): Informatics in Control, Automation and Robotics, Volume 2, LNEE 133, pp. 317–323.
springerlink.com

algorithm and experimental results are given in detail in Section 4. Finally, conclusion and future work are presented in Section 5.

2 Related Work

Depending on the different encryption algorithms and compression processes, video encryption algorithms have been classified into three categories [1]: complete encryption algorithm, selective encryption algorithm and encryption algorithm with compression function. Complete encryption algorithm has the highest confidentiality level, but encrypting all of the video data brings high computational overhead. Encryption algorithm with compression function combines encryption process into compression coding, this algorithm solves of the problem of computational overhead, though it is less confidentiality than the other two algorithms and has effect on the compression ratio. Selective encryption algorithm utilizes the characteristics of video data, and encrypts only the important data. It becomes the major research field in video content confidentiality based on H.264/AVC. In the past few years, several different selective encryption algorithms were proposed [2]-[7] which can be divided into partial encryption algorithm and DCT coefficients encryption algorithm.

MPE algorithm [3] is the most commonly used in partial encryption algorithm. In view of inter picture prediction technology, MPE algorithm encrypts the data which is important for image reconstruction and I frames are generally chosen. The three major picture types used in the H.264/AVC videos are I frames, P frames and B frames. I frames are the least compressible, P frames use data from previous frames (I frames) to decompress, B frames use both previous and forward frames (I frames and P frames) to get the highest amount of data compression. Only I frames don't require other frames to decode. If I frames are encrypted, the attackers can't get right I frames even they get P frames and B frames, and they can't use motion compensation to predict the image. MPE algorithm usually uses the traditional block cipher (DES, AES and IDEA) or stream cipher (RSA) to encrypt, these cryptographic algorithms have been proven that they can resist all kinds of known attack, therefore MPE is considered as a algorithm with high confidentiality level.

H.264/AVC standard uses Discrete Cosine Transform (DCT) and the results are quantized and entropy coded to achieve higher compression ratios. DCW algorithm [5] uses a random permutation list to replace the Zig-Zag order to map the DCT coefficients. The algorithm is easy to operate, but it makes a great effect on compression ratio. Encryption DCT coefficients has destroyed the statistical characteristic which influences the entropy encoding and so the compression ratio is changed. At the same, DCW algorithm confidentiality is not satisfactory. VEA algorithm [6] uses a secret key randomly changing the sign bits of DCT coefficients of I frames. MVEA algorithm use a key randomly changing the sign bits of all DCT coefficients and the sign bits of motion vectors. These two algorithms encrypt only a small part of data, have little effect to real-time property and compression ratio, but they don't have strong resistance for the known plaintext attack. In order to improve the security, the third algorithm called RVEA [6] is proposed. RVEA algorithm use cryptography algorithm to encrypt selected sign bits of the DCT coefficients and motion vectors. RVEA algorithm has excellent confidentiality feature as the traditional cryptography. On the other hand, selection data and encrypting operation increase delay-time.

In conclusion, the advantage of selective algorithm is reducing the computational overhead and meeting the real time requirement, but it affects the confidentiality and compression ratio. So, confidentiality and compression ratio are the key problem which urgently waits to be solved in selective encryption algorithm.

3 The Proposed Selective Video Encryption Scheme

The key point of selective encryption algorithm is to select the proper data to encrypt. According to the layered structure of H.264/AVC, the proposed selective encryption algorithm encrypts some important NAL units thus achieves higher efficiency and lower cost.

In the H.264/AVC standard, the codec specification itself distinguishes conceptually between a video coding layer (VCL) and a network abstraction layer (NAL). The VCL uses inter picture prediction with motion compensation, transformation, quantization and entropy coding to compress and encode video data. The NAL encoder encapsulates the slice output of the VCL encoder into Network Abstraction Layer Units (NALU) [8], which are suitable for transmission over packet network. A NALU consists of a one byte header and the payload byte string. The NALU header has the following format:

F: forbidden_bit(1 bit), the value indicates whether the NAL unit type octet and payload contain bit errors or other syntax violations. The value of 0 means no error in payload.

NRI: nal_ref_idc(2 bits), the component signals the relative importance. The value of 11 indicates that the content of the NAL unit is the most important to reconstruct reference pictures for inter picture prediction.

TYPE: nal_unit_type(5 bits), the component specifies the NAL unit payload type.

For NAL units having TYPE to 5(indicating an IDR picture), 7(indicating a sequence parameter set) or 8(indicating a picture parameter set), an H.264 encoder should set the values of NRI to 11 in binary format. IDR picture is actually an I frame which only contains I slice or SI slice. This also shows that I frames, SPS and PPS play a major role in reconstructing pictures at the decoder.

NAL units are encapsulated in RTP (Real-Time Transport Protocol) packages to transmission on the network. The RTP .payload format defines three different payload structures:

Single NAL Unit Packet: Contains only a signal NAL unit in the payload.

Aggregation Packets: Aggregates multiple NAL units into a signal RTP payload.

Fragmentation Units: Fragments a single NAL unit over multiple RTP packets.

No matter use which kinds of structure, the first byte of RTP payload header serves as a NAL unit. If the RTP payload is IDR picture, the first byte of RTP payload equals to 01100101 (F = 0, NRI = 11, TYPE= 00101) in binary that converts to a decimal number is 101.SPS and PPS are expressed as 103 and 104 in decimal.

Sometimes P frames and B frames are intra prediction coded, in this case encrypting only I frames does not provide a satisfactory confidentiality level. To achieve the higher level of video encryption, the proposed scheme uses AES algorithm in OFB mode to encrypt I frames, SPS and PPS By the determining the first byte of RTP payload equals to 101,103 or 104, the proposed scheme can select the proper RTP payload to encrypt. The overview of the proposed method is shown in figure 1.

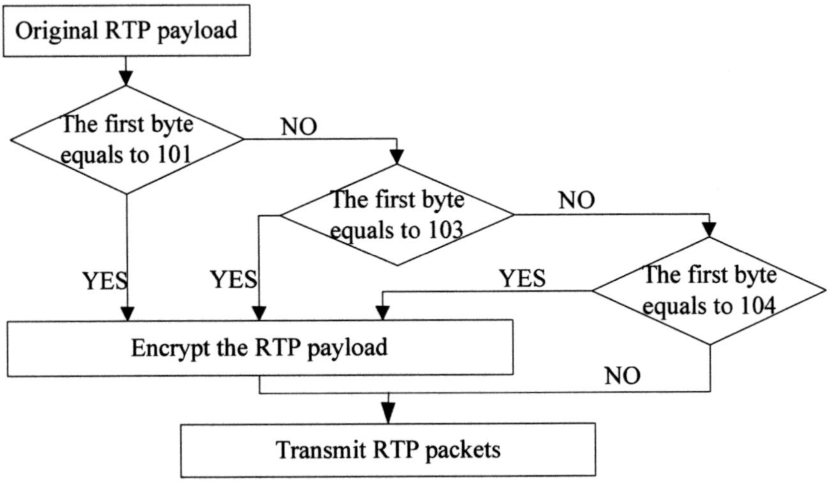

Fig. 1. The overview of the proposed method

4 Experimental Results

In order to evaluate the performance of the proposed scheme, we use several typical videos for testing. Each of them represents different motion complexity, luminance and chrominance. The comparisons of encryption effect are shown in Figure 2. In figures, the left is original images, the middle is only I frames encrypted and the right is I frames, SPS, PPS encrypted by our scheme. We can see from Figure 2 that using our scheme video images are encrypted better.

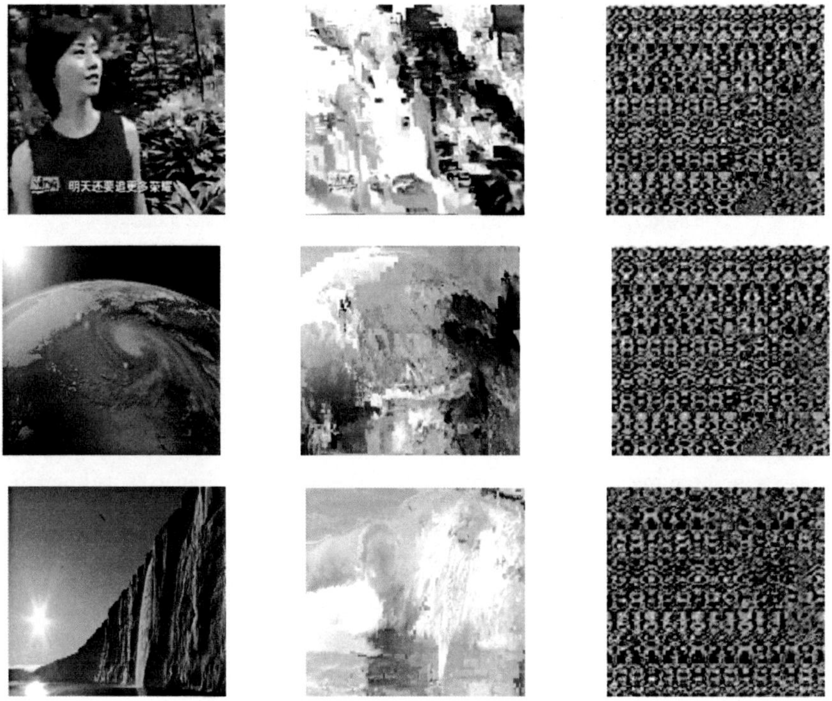

Fig. 2. The original and encrypted video images

4.1 Security Analysis

AES is a very powerful standard cipher and no effectively attack has been found on AES. The research [9] suggests that the design and strength of all key lengths of the AES algorithm are sufficient to protect classified information. Meanwhile, the proposed scheme encrypts parameter sequence, attackers can't get the parameters of the decoder. I frames are completely encrypted that damage the synchronous video sequence and lead to subsequent data frames can't be decoded. So the proposed scheme in this paper is generally considered to be security enough and experiments show the proposed scheme gets obviously degraded images and good encryption effect. Moreover, the scheme has resistance for the exhaustion attack, since it is impossible to completely get plaintext which needs about 2^{128} attacks.

4.2 Computational Complexity

The proposed scheme uses AES to encrypt, the data can't be encrypted until get a 128-bit block, and that would increase encryption and decryption delay. Considered the speed of the scheme, we adopt OFB mode which has the feature of stream cipher and encrypts less than 128-bit block in time. Meanwhile, AES algorithm has three different key sizes of 128, 192, or 256 bits and has 10 rounds for 128 bits, 12 rounds for 192 bits, and 14 rounds for 256 bits. The proposed scheme chooses the 128-bits key size to reduce the rounds of implementation. Moreover, we improve AES algorithm, in every round the operations are only shifting cyclically, table look-up and bitwise XOR. As a result, the speed of the proposed scheme will decline obviously.

To demonstrate the efficiency of our proposed scheme, we have compressed some H.264/AVC videos with different definition of 480P, 720P and 1080P. The change of code length by the selective scheme and encryption proportion are given in Table1.

Table 1. The encryption proportion of three H.264/AVC videos

Test video	Video length (byte)	Encryption length(byte)	Encryption proportion
480-H.264	126,750,540	3,850,240	3.04%
720-H.264	466,184,540	27,973,241	6.00%
1080-H.264	3,596,753,772	343,849,660	9.56%

One can note that with definition increases, the quality of the encrypted video increases, and the maximum of encrypted data is 9.56%, less than 10%. Although the length of an I frame is longer than a P frame or B frame, the amount of I frame is the least, so only a very small number of data to be encrypted. Meanwhile, the scheme encrypts the sensitive video data is fast enough, that means it can meet the requirement of real time encoder and decoder work for H.264/AVC videos.

4.3 Compression Ratio

In this selective video encryption scheme, we use AES algorithm to encrypt, the length of bit streams obtained is not affect. So there is no change between the compression ratio of original video and encrypted video. That means the proposed selective encryption scheme doesn't effect on the compression ratio at all.

5 Conclusions and Future Work

In this paper, the selective encryption algorithm has been proposed base on the H.264/AVC video coding standard. Experimental results have shown that the proposed algorithm can achieve good security, high efficiency with low computational complexity and not impact on the video length. So it is suitable for real-time secure video transmission based on H.264/AVC.

In this paper, we don't consider key management which is one of the kernel problems in video encryption technology. At the same time, the using of chaotic stream ciphers in video encryption is becoming more and more because of its high security and efficiency. Future research will focus on the combination of chaotic stream ciphers with current encryption algorithm and adding key management to insure the security of key storing, backuping and restoring.

References

1. Liu, S.L., Sun, J., Wang, Z.: A Fast Encryption Scheme Based-on Chaos. In: The 8th International Conference on Control, Automation, Robotics and Vision (ICARCV), pp. 126–131 (2004)
2. Jiang, J., Liu, Y., Su, Z.: An improved selective encryption for H. 264 video based on intra prediction mode scrambling. Journal of Multimedia. 5, 464–472 (2010)
3. Agi, I., Gong, L.: An empirical study of MPEG video transmissions. In: Proceedings of the Internet Society Symposium on Network and Distributed System Security, pp. 137–144 (1996)
4. Varalakshmi, L.M.: H.264 based selective video encryption for mobile applications. International Journal of Computer Applications 17, 13–20 (2011)
5. Tan, L.: Method for encrypting and decrypting MPEG video data efficiently. In: Proceedings of the Fourth ACM International Multimedia Conference, pp. 219–230 (1996)
6. Shi, C., Bharat, B.: A fast MPEG video encryption algorithm. In: Proceedings of the Sixth ACM International Multimedia Conference, pp. 81–88 (1998)
7. Wang, Y., Cai, M., Tang, F.: Design of a new selective video encryption scheme based on H.264. In: 2007 International Conference on Computational Intelligence and Security (2007)
8. RFC 6184 - RTP Payload Format for H.264 Video, `http://www.packetizer.com/rfc/rfc6184`
9. Hathaway, L.: National Policy on the Use of the Advanced Encryption Standard (AES) to Protect National Security Systems and National Security information, `http://csrc.nist.gov/groups/ST/toolkit/documents/aes/CNSS15FS.pdf`
10. Schneier, B., Kelsey, J., Whiting, D.: The Twofish Team's Final Comments on AES Selection, `http://www.schneier.com/paper-twofish-final.pdf`

Independent Component Analysis Based on Smooth Discrete Wavelet Coefficients

JingHui Wang and YuanChao Zhao

Tianjin Key Laboratory of Intelligence Computing and Novel Software Technology
Tianjin University of Technology
TianJin, China 300191
csr_dsp@sina.com

Abstract. In this paper, a novel Independent Component Analysis(ICA) using Smooth Discrete Wavelet Coefficients is presented. Discrete Wavelet transform is an important time-frequency analysis tool and it particularly suitable for nonlinear signal. Independent Component Analysis using Smooth Discrete Wavelet Coefficients is suitable for dealing with non-stationary signal. And simulation results have shown the method is feasible.

Keywords: Blind Separation Signal, Discrete wavelet.

1 Introduction

Blind source separation (BSS) is an approach that recovering a set of unknown mutually independent source signals from their observed mixtures without knowing the mixing matrix[1].In recent years, the works of BSS based on Independent Component Analysis(ICA). ICA plays an important role in a variety of signal processing applications like blind source separation (BSS). ICA searches for a linear or nonlinear transformation which minimizes the statistical dependence between components [2].

ICA is an optimization problem. Researchers focused on blind signal separation, blind deconvolution and blind equalization. Those methods involving statistical signal processing and adaptive signal processing. Now, the ICA models and its algorithm are closely related to a specific application. When ICA algorithm migrated from one background to other applications, it's a difficult work.

In order to improve migration and adaptive, researchers started from time domain and time domain to the time-frequency domain analysis. Time-frequency analysis became an important tool, because most practical situation, the input signals are time-varying signal. ICA based on time-frequency approaches is well-suited to non-stationary signals[3].

Now, there are three classes time-frequency methods emerge. (1)Those methods related to classical BSS approach, such as joint diagonalization of a combined set of spatial time-frequency distributions (TFD). (2) Those method based on the ratios of

D. Yang (Ed.): Informatics in Control, Automation and Robotics, Volume 2, LNEE 133, pp. 325–331.
springerlink.com

time-frequency transformations of observe signals. (3) Those methods based on the independent signals have independent time-frequency transformation or time frequency distribution.

From time domain and frequency domain, wavelet transform is an important tool. The method reflects the signal feature in time domain and scale domain. The scale domain and frequency are closely related. Wavelets are functions that cut up data into different frequency components, and each component with a resolution matched to its scale. The wavelets have advantages over traditional Fourier methods in analyzing physical situations, especially where the signal contains discontinuities and local features.

In this paper, we provided an independent component analysis approach based on smooth discrete wavelet transform. This approach exploits the difference wavelet coefficients between difference groups. By smoothing the wavelet coefficients and blind source separation criterion, we get mixed matrix. Compared directly separation signals in time domain, the method is suitable for blind separation of non-stationary signals.

2 Blind Sources Separation Model

Blind sources separation (BSS) is m sensors receiving an instantaneous linear mixture of signals emitted from n sources. The m vector $x(t)$ denotes the output of the sensors at time instant t which may be corrupted by an additive noise $n(t)$. The generative BSS model can be presented as follows[4,5]:

$$x(t) = As(t) + n(t) \tag{1}$$

The $m \times n$ matrix A is called the 'mixing matrix'. The n source signals are collected in a $n \times 1$ vector. $s(t)$ is the source signal vector. The sources are assumed to have different structures in the time frequency domain. The mixing matrix A is full column rank and it is unknown.

The ICA unmixes matrix W is show as follows

$$S = WX \tag{2}$$

where $W = [w_1, w_2, ..., w_m]^T$,and $w_i = [x_{i1}, x_{i2}, ..., x_{in}]$, i =1, ...,m.

The main work of ICA is to recover source signal S from observation X by estimating the weight matrix W.

In the real situation, the mixing model is more complex, such as instantaneous mixing, anechoic mixing and echoic mixing. Sometimes, it's a difficult work to distinguish blind sources models.

3 ICA Using Smooth Wavelet Coefficients

3.1 Wavelet Transform and Smooth Wavelet Coefficients

Wavelets provide a description data that localizes information at a range of scales and positions[6,7]. The discrete wavelet transform decomposes a function as a sum of basis functions called wavelets. It known as the scaling function ϕ and the wavelet function ψ.

An orthonormal wavelet ψ is defined as a Hilbert basis, that is a complete orthonormal system, for the Hilbert space $L^2(R)$. The Hilbert basis is constructed as the family of functions $\{\psi_{jk} : j, k \in Z\}$.

These translates and dilations are defined as follows:

$$\begin{aligned} \phi_{j,k}(x) &= 2^{j/2}\phi(2^j x - k) \\ \psi_{j,k}(x) &= 2^{j/2}\psi(2^j x - k) \end{aligned} \tag{3}$$

And $j, k \in Z$. The family is an orthonormal system if it is orthonormal under the inner product,

$$< \psi_{jk}, \psi_{lm} >= \delta_{jl}\delta_{km} \tag{4}$$

where δ_{jl} is the Kronecker delta.

The index j defines the dilation or level while the index k defines the translation. Generally, sums of the $\phi_{i,k}(x)$ capture low frequencies and sums of the $\psi_{j,k}(x)$ represent high frequencies in the data. More precisely, for any suitable function $f(x)$.

$$f(x) = \sum_k c_k^{j_0}\phi_{j_0,k}(x) + \sum_{j \geq j_o}\sum_k d_k^j \psi_{i,k}(x) \tag{5}$$

where the c_k^j and d_k^j are known as the scaling coefficients and the detail coefficients respectively.

When we get the wavelet coefficients, can be smoothed in various ways, such as the group average, sliding smoothly, minimum variance, least-squares method. Also, we can remove the coefficient below the threshold, it's means pre-filter.

3.2 Blind Wavelet Coefficients Separation Criteria

If the original signal is independent, the wavelet coefficients reflect independence of different scales and translation wavelets. There are many separation criteria of independent components. Some of them come from entropy[8]. It defined as:

$$H(p_x) = -\int p_x(u)\log p_x(u)du \tag{6}$$

Because the source signals contain the least Gaussian components, a measure of nongaussianity is the key for estimating the weight matrix, and correspondingly, the independent components. The classical measure of nongaussianity is kurtosis, which is the fourth order statistics and has zero value for Gaussian distribution[9]. However, kurtosis is sensitive to outliers. The negentropy is then used as a measure of nongaussianity since Gaussian variable has the largest entropy among all random variables of equal variance[10],

$$J(Y) = H(Y_{gauss}) - H(Y) \tag{7}$$

where Y is a random variable, $H(Y_{gauss})$ is the entropy of a Gaussian random variable with the same covariance matrix as Y. Amd the negentropy is difficult to compute, an approximation is given as follows

$$J(Y) \approx \{E[G(Y)] - E[G(Y_{gauss})]\}^2 \tag{8}$$

where $G(Y)$ is a non-quadratic function. In the ICA algorithm, $Y = w^T X$. Obviously, by choosing wisely, we obtain better approximation of negentropy. The following choice of G has been proved useful.

$$G(u) = \frac{1}{a}\log\cosh(au) \tag{9}$$

where $1 \le a \le 2$, and often $a = 1$.

In order to maximize the objective function (7), we can apply FastICA algorithm. Specific process is as follows

$$\begin{aligned} w_i^+ &= E\{Xg(w_i^T X)\} - E\{g'(w_i^T X)\}w_i \\ w_i &= \frac{w_i^+}{\left\|w_i^+\right\|} \end{aligned} \tag{10}$$

where g denotes the derivative of the non-quadratic function G in Eq.8, and $g(u) = \tanh(au)$. The decorrelation process keeps different weight vectors from converging to the same maximal. It decorrelates the $(p+1)th$ weight vector from the preceding p weight vectors.

$$\begin{aligned} w_{p+1}^+ &= w_{p+1} - \sum_{i=1}^{p} w_{P+1}^T w_i w_i \\ w_{p+1} &= \frac{w_{p+1}^+}{\left\|w_{p+1}^+\right\|} \end{aligned} \tag{11}$$

3.3 Summary Algorithm

The separation algorithm based on smoothing discrete wavelet transform is summarized as follows:

1. Grouping the input data X . An input data is X mixed signal group, group for a power of 2.
2. Calculate the wavelet coefficients of different packet data.
3. Smoothing those wavelet packet's coefficients.
4. Calculate the mixing matrix based on the independent criterion of wavelet packet coefficients.
5. Solving the source signal.
6. Calculate the wavelet coefficients of the next group.
7. Update the smooth wavelet coefficients.
8. If it is not the last set of data, turning to (4).

4 Simulation and Synthetic Signals

The following experiment was divided into two parts, first part is the speech-music mixed and its independent component analysis, the second part is a mixture of random signals. In the simulation, the proposed method is compared with the Jade and FRFT method.

The data for reference only, with a minimum variance smooth wavelet coefficient, the different results come from different simulation data.

Speech - Music Separation

The speech-music mixing data come form[11]:

http://cnl.salk.edu/~tewon/Blind/blind_audio.html.

A speaker has been recorded with two distance talking microphones (sampling rate 16kHz) in a normal office room with loud music in the background. The distance between the speaker, cassette player and the microphones is about 60cm in a square ordering. (All files are in 16kHz wav-format).

From the hearing point of view, we heard two different voices clearly, that mean other sound dampening effect is also very good.

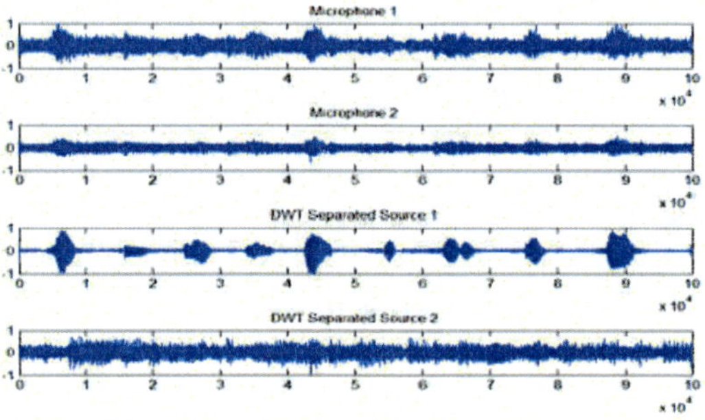

Fig. 1. Seperated Signal Speech-Music Separation

Random Signal Separation

The following is a Gaussian random signal generation and its mixing signals mixed by random matrix, using a smooth discrete wavelet transform independent component analysis can be reconstructed signal.

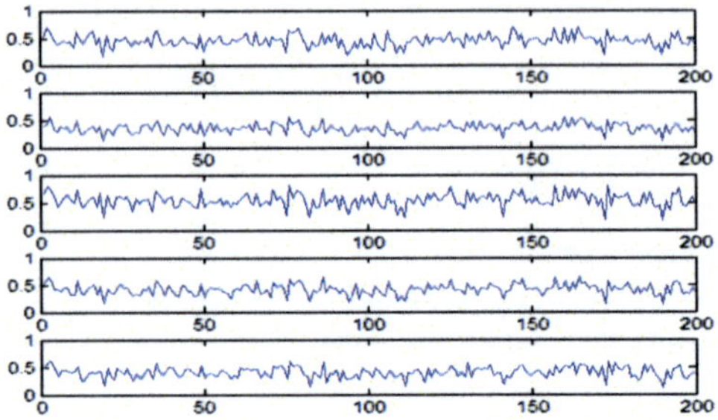

Fig. 2. Mixed Gaussian Random Signal

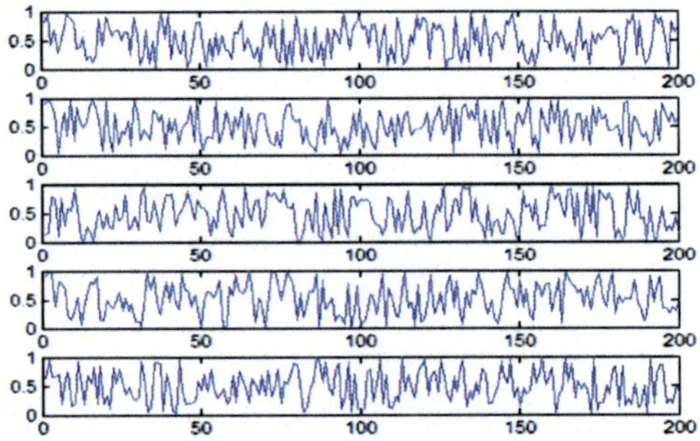

Fig. 3. Seperated Signal

Table 1. JADE \FRFT ICA and SDWT(Random Signal)

Method	SDR	SIR	SNR
JADE	5.96	7.43	18.32
FRFT(256)	3.22	4.63	12.98
SDWT	4.12	3.30	10.20

5 Conclusion

In this paper, we have presented blind sources separation algorithms based on discrete wavelet coefficients. The method comes from wavelet coefficients independence. With the input signal change, the coefficients independence adaptive. The method suitable for nonlinear signal and time-varying signal.

Simulation results have shown that the method is feasible. The method extends the relationship between time and frequency to the scale and frequency. When dealing with signals buried in heavy noise, the method can be removed those coefficients which less than the threshold. This efficiency is very high. The future work is to improve the decomposition efficiency.

Acknowledgment. This work is supported by National Natural Science Foundation of China (No.61001174).

References

1. Hyvärinen, A., Oja, E.: A Fast Fixed-Point Algo-rithm for Independent Component Analysis. Neural Computation 9, 1483–1492 (1997)
2. Belouchrani, A., Amin, M.G.: Blind source separation based on time-frequency signal representations. IEEE Trans. Spch. Aud. Proc. 46, 2888–2897 (1998)
3. Grinsted, Moore, J.C., Jevrejeva, S.: Application of the cross wavelet transform and wavelet coherence to geophysical time series. In: Nonlinear Processes in Geophysics, pp. 561–566. European Geosciences Union (2004)
4. Wang, J.-H., Kong, G.-Q., Liu, C.-H.: A parallel independent component implement based on learning updating with forms of matrix transformations. In: Huang, D.-S., Heutte, L., Loog, M. (eds.) ICIC 2007. LNCS (LNAI), vol. 4682, pp. 202–211. Springer, Heidelberg (2007)
5. Donoho, D.L., Johnstone, I.M.: Ideal spatial adaptation by wavelet shrinkage. Biometrika 81, 425–455 (1994)
6. Lindsay, R.W., Percival, D.B., Rothrock, D.A.: The discrete wavelet transform and the scale analysis of the surface properties of sea ice. IEEE Trans. Geosci. Remote Sens. 34, 771–787 (1996)
7. Weng, H., Lau, K.-M.: Wavelets, period doubling, and time-frequency localization with application to organization of convection over the tropical western. Pacific. J. Atmos. Sci. 51, 2523–2541 (1994)
8. Shan, Z.: Aviyente Source Separation in the Time-Frequency Domain by Maximizing an Information-Theoretic Criterion. In: 2006 IEEE International Conference on Electro/information Technology, May 7-10, pp. 43–48 (2006)
9. Reju, V.G., Koh, S.N., Soon, I.Y.: Underdetermined Convolutive Blind Source Separation via TimeCFrequency Masking. IEEE Transactions on Audio, Speech, and Language Processing, 101–116 (2010)
10. Shan, Z., Swary, J., Aviyente, S.: Underdetermined Source Separation in the Time-Frequency Domain. In: International Conference on Acoustics, Speech and Signal Processing, pp. 945–948 (2007)
11. `http://en.wikipedia.org/wiki/Wavelet_transform`

Nonlinear Control Design of a Hypersonic Aircraft Using Sum-of-Squares Method*

Zhiling Yang and Hongfei Sun

Department of Automation, Xiamen University, Fujian Province, P.R. China
Yangzl0613@hotmail.com

Abstract. The main focus of this paper is on developing a state feedback nonlinear controller for a hypersonic aircraft using the Sum-Of-Squares (SOS) approach. The SOS technique allows the joint search of a Lyapunov function and a nonlinear controller using semidefinite programs. Different from the aproximate linearization and feedback linearization methords, the approach proposed in this paper cause less conservatism. Simulation results are presented to evaluate the stability and performance of the controller.

Keywords: hypersonic aircraft, Sum-Of-Squares (SOS), Lyapunov function, semidefinite programs.

1 Introduction

In view of the particular characteristics of multi-input multi-output, strong coupling, and high nonlinearity, analysis and contro l of hypersonic vehicle systems are of great challenges [1]. In recent years, a variety of linear or nonlinear control approaches have been used to deal with nonlinear systems [2, 3]. Approximate linearization [9, 11] is a widely used method for analysis and synthesis of nonlinear systems, but this will lose much nonlinear characteristics thus definitely cause low accuracy. Nonliner control seems to be more suitable for hypersonic vehicles because of the wide range of the speed and altitude which gives rise to large variations of the flight parameters . Feedback linearization (FBL), in its various forms, is perhaps the most commonly employed nonlinear control method in flight control [4, 5, 6]. When applied to flight control problems, the underlying method is often referred to as "dynamic inversion" (DI) [7, 10]. As is well known, the prospective benefits of DI lie in the straightforward manner in which a controller structure is obtained, the possibility to effectively account for system nonlinearities and the inherent gain scheduling. However, FBI or DI method can not guarantee satisfactory robustness because it relies heavily upon the nonsingularity of a diffeomorphism.

To overcome deficiencies above, this paper introduces the recently developed sum-of-squares (SOS) techniques into the study of hypersonic vehicles. To the authors' knowledge, the combination of SOS and hypersonic vehicle control has not been extensively addressed in the literature. Compared with the other control techniques, the

* This work is supported by the research of Spatial Intelligence Research Key Laboratory and Xiamen University.

D. Yang (Ed.): Informatics in Control, Automation and Robotics, Volume 2, LNEE 133, pp. 333–342.
springerlink.com

main advantages of SOS techniques for nonlinear systems analysis and design are that 1) the use of SOS techniques can convert many non-convex nonlinear programming problems into convex programming problems; 2) the use of SOS techniques does not require much simplification of the nonlinear model which means the model error is small; 3) the designed controller is the polynomial function of the state variables which would facilitate engineering implementation; 4) the use of SOS techniques does not require the diffeomorphism which would cause better robustness. In the lately years, SOS techniques has been drawn more and more attention, for more imformation, the readers can refer to [13, 14]. This paper attempts to apply the SOS design strategy of reference [13] to the hypersonic vehicle proposed in [12]. From the simulation results, each state can converge to the trim condition within a short time, and control inputs are within the admissible ranges.

The remainder of the paper is organized as follows: in Section 2 the hypersonic aircraft model is introduced and the control odjective is stated. Section 3 presents some preliminary results concerning the sum of squares decomposition and its application to solve state dependent LMIs, while in Section 4 simulation results are given. Finally, conclusions are drawn in Section 5.

2 Hypersonic Aircraft Model

The rigid-body vehicle dynamics considered in this study is given as follows [12]

$$\begin{aligned} \dot{V} &= \frac{T\cos\alpha - D}{m} - g\sin\gamma \\ \dot{\gamma} &= \frac{T\sin\alpha + L}{mV} - \frac{g}{V}\cos\gamma \\ \dot{\alpha} &= Q - \dot{\gamma} \\ \dot{Q} &= \frac{M}{I_{yy}} \end{aligned} \tag{1}$$

This model comprises four state variables $x = [V, \gamma, \alpha, Q]^T$ and two control inputs $u = [\Phi, \delta_e]^T$ which affect (1) through the thrust T, the pitching moment about the body y-axis M, lift L, and drag D. The meaning of the state variables and the input vector is given in Table 1. Following [8], approximations of the forces and moments to be employed for control design and stability analysis have been derived as follows

$$\begin{aligned} & L \approx \bar{q}SC_L(\alpha, \delta_e), \qquad D \approx \bar{q}SC_D(\alpha, \delta_e) \\ & T \approx C_T^{\alpha^3}\alpha^3 + C_T^{\alpha^2}\alpha^2 + C_T^{\alpha}\alpha + C_T^0, \\ & M \approx Z_T T + \bar{q}S\bar{c}\left[C_{M,\alpha}(\alpha) + C_{M,\delta_e}(\delta_e)\right] \end{aligned} \tag{2}$$

where $\bar{q} = 1/2\rho V^2$ denotes dynamic pressure, and Z_T is a known quantity.

The forms for the functions appearing in Eqs.(2) and the forces are selected as

$$C_L = C_L^{\alpha}\alpha + C_L^{\delta_e}\delta_e + C_L^0,$$

$$C_D = C_D^{\alpha^2}\alpha^2 + C_D^{\alpha}\alpha + C_D^{\delta_e^2}\delta_e^{\,2} + C_D^{\delta_e}\delta_e + C_D^0$$

$$C_{M,\alpha} = C_{M,\alpha}^{\alpha^2}\alpha^2 + C_{M,\alpha}^{\alpha}\alpha + C_{M,\alpha}^0, \qquad C_{M,\delta_e} = C_e\delta_e$$

$$C_T^{\alpha^3} = \beta_1\Phi + \beta_2, \qquad C_T^{\alpha^2} = \beta_3\Phi + \beta_4$$

$$C_T^{\alpha} = \beta_5\Phi + \beta_6, \qquad C_T^0 = \beta_7\Phi + \beta_8$$

Table 1. Admissible Ranges For States, Input, and Variables of Interest

Var		Min Value	Max Value
V	Vehicle Velocity	7000 ft/s	11000 ft/s
γ	Flight-Path Angle (FPA)	-5 deg	5 deg
α	Angle-of-Attack	-10 deg	10 deg
Q	Pitch Rate	-15 deg/s	15 deg/s
Φ	Fuel-to-air Ratio	0.2	1.2
δ_e	Elevator Deflection	-20 deg	20 deg
$\bar{q}$	Dynamic Pressure	500 psf	2000 psf

Because the ranges of α, γ are very small, we can let $\sin\alpha \approx \alpha$, $\cos\alpha \approx 1$, $\sin\gamma \approx \gamma$, $\cos\gamma \approx 1$, then system (1) can be transformed into the following form

$$\dot{V} = f_{11}(\alpha,\gamma) + f_{12}(\alpha)\Phi + f_{13}\delta_e + f_{14}\delta_e^2,$$

$$V\dot{\gamma} = f_{21}(\alpha) + f_{22}(\alpha)\Phi + f_{23}\delta_e,$$

$$V\dot{\alpha} = f_{31}(\alpha,V,Q) + f_{32}(\alpha)\Phi + f_{33}\delta_e,$$

$$I_{yy}\dot{Q} = f_{41}(\alpha) + f_{42}(\alpha)\Phi + f_{43}\delta_e, \tag{3}$$

where

$$f_{11}(\alpha,\gamma) = \frac{1}{m}\left[\left(\beta_2\alpha^3 + \beta_4\alpha^2 + \beta_6\alpha + \beta_8\right) - \bar{q}S\left(C_D^{\alpha^2}\alpha^2 + C_D^{\alpha}\alpha + C_D^0\right)\right] - g\gamma,$$

$$f_{12}(\alpha) = \frac{1}{m}\left(\beta_1\alpha^3 + \beta_3\alpha^2 + \beta_5\alpha + \beta_7\right), f_{13} = -\frac{1}{m}\bar{q}SC_D^{\delta_e}, f_{14} = -\frac{1}{m}\bar{q}SC_D^{\delta_e^2},$$

$$f_{21}(\alpha)=\frac{1}{m}\left[\left(\beta_2\alpha^4+\beta_4\alpha^3+\beta_6\alpha^2+\beta_8\alpha\right)+\bar{q}S\left(C_L^{\alpha}\alpha+C_L^0\right)\right]-g,$$

$$f_{22}(\alpha)=\frac{1}{m}\left(\beta_1\alpha^4+\beta_3\alpha^3+\beta_5\alpha^2+\beta_7\alpha\right), f_{23}=\frac{1}{m}\bar{q}SC_L^{\delta_e},$$

$$f_{31}(\alpha,V,Q)=VQ-f_{21}(\alpha), f_{32}(\alpha)=-f_{22}(\alpha), f_{33}=-f_{23}$$

$$f_{41}(\alpha)=Z_T\left(\beta_2\alpha^3+\beta_4\alpha^2+\beta_6\alpha+\beta_8\right)+\bar{q}S\bar{c}\left(C_{M,\alpha}^{\alpha^2}\alpha^2+C_{M,\alpha}^{\alpha}\alpha+C_{M,\alpha}^{0}\right),$$

$$f_{42}(\alpha)=Z_T\left(\beta_1\alpha^3+\beta_3\alpha^2+\beta_5\alpha+\beta_7\right), f_{43}=\bar{q}S\bar{c}C_e$$

3 Controller Design Using SOS Method

Definition 1. [13] A multivariate polynomial $f(x)$ (where $x\in R^n$) is a sum of squares if there exist polynomials $f_1(x),\cdots,f_m(x)$ such that $f(x)=\sum_{i=1}^{m} f_i^2(x)$. This can be shown equivalent to the existence of a special quadratic Form stated in the following proposition.

Proposition 2. [13] Let $f(x)$ be a polynomial in $x\in R^n$ of degree $2d$. In addition, let $Z(x)$ be a column vector whose entries are all monomials in x with degree no greater than d. Then $f(x)$ is a sum of squares iff there exists a positive semidefinite matrix Q such that

$$f(x)=Z^T(x)QZ(x) \tag{4}$$

Now let's consider the system $\dot{x}=f(x)+g(x)u$, which written in the following state dependent linear-like representation:

$$\dot{x}=A(x)Z(x)+B(x)u \tag{5}$$

where $A(x)$ and $B(x)$ are polynomial matrices in x, and $Z(x)$ is and $N\times 1$ vector of monomia- ls in x satisfying the following assumption.

Assumption 3. [13] $Z(x)=0$ iff $x=0$.

In addition, define $M(x)$ to be a $N\times n$ polynomia1 matrix whose (i,j)-th entry is given by $M_{ij}(x)=\frac{\partial Z_i}{\partial x_j}(x)$

for $i=1,\cdots,N, j=1,\cdots,n$. Finally, let $A_j(x)$ denotes the j-th row of $A(x)$, $J=\{j_1,j_2,\cdots,j_m\}$ denote the row indices of $B(x)$ whose corresponding row is equal to zero, and define $\bar{x}=(x_{j_1},x_{j_2},\cdots,x_{j_m})$.

In view of our objective is to find a state feedback controller $u=k(x)=F(x)Z(x)$ which renders the equilibrium $x=0$ stable, here we have not included any performance objective in the synthesis. The following lemma is useful in the proof of the main result for the solution to the synthesis problem.

Lemma 4. [13] For a symmetric polynomial matrix $P(x)$ which is nonsingular for all x, we have

$$\frac{\partial P}{\partial x_i}(x)=-P(x)\frac{\partial P^{-1}}{\partial x_i}(x)P(x) \tag{6}$$

Proof: Since $P(x)$ is non-singular, we have $P(x)P^{-1}(x)=I$. Differentiating both sides with res- pect to x_i yields

$$\frac{\partial P}{\partial x_i}(x)P^{-1}(x)+P(x)\frac{\partial P^{-1}}{\partial x_i}(x)=0$$

This immediately gives (6).

Theorem 5. [13] For the system (5), suppose there exist an $N\times N$ symmetric polynomial matrix $P(\bar{x})$, an $n\times N$ polynomial matrix $K(x)$, a constant $\varepsilon_1>0$, and a sum of squares $\varepsilon_2(x)$, such that the following two expressions

$$v^T\left(P(\bar{x})-\varepsilon_1 I\right)v, \tag{7}$$

$$-v^T\Big(P(\bar{x})A^T(x)M^T(x)+M(x)A(x)P(\bar{x})+K^T(x)B^T(x)M^T(x)\cdots$$
$$+M(x)B(x)K(x)-\sum_{j\in J}\frac{\partial P}{\partial x_i}(\bar{x})\left(A_j(x)Z(x)\right)+\varepsilon_2(x)I\Big)v \tag{8}$$

are sums of squares, where $v\in\mathbb{R}^N$. Then the state feedback stabilization problem is solvable, and a controller that stabilizes the system is given by:

$$u(x)=K(x)P^{-1}(x)Z(x)$$

Furthermore, if (8) holds with $\varepsilon_2(x)>0$ for $x\neq 0$, then the zero equilibrium is asymptotically stable, and $P(\bar{x})$ is a constant matrix, then the stability holds globally.

Proof: See [14].

Remark 6. [13] If all rows of $B(x)$ are not equal to zeros, then $P(\bar{x})$ can be a symmetric positive definite polynomial constant matrix.

4 Simulation Results

Suppose the trim condition is $(x^*, u^*) = (V^*, \gamma^*, \alpha^*, Q^*, \Phi^*, \delta_e^*)$, and let the states error $e = x - x^* := (e_V, e_\gamma, e_\alpha, e_Q)$, inputs error $e_\Phi = \Phi - \Phi^*$ and $e_{\delta_e} = \delta_e - \delta_e^*$. In view of the range of $1/V$ is very small ($1.4286\times10^{-4} \sim 9.0909\times10^{-5}$), we can let $1/V^*$ instead of $1/V$, then system (3) can be converted into the following

$$\dot{e_V} = F_{11}(e_\alpha, e_\gamma) + F_{12}(e_\alpha)e_\Phi + F_{13}e_{\delta_e} + F_{14}e_{\delta_e}^2,$$

$$V^*\dot{e_\gamma} = F_{21}(e_\alpha) + F_{22}(e_\alpha)e_\Phi + F_{23}e_{\delta_e},$$

$$V^*\dot{e_\alpha} = F_{31}(e_\alpha, e_Q) + F_{32}(e_\alpha)e_\Phi + F_{33}e_{\delta_e},$$

$$I_{yy}\dot{e_Q} = F_{41}(e_\alpha) + F_{42}(e_\alpha)e_\Phi + F_{43}e_{\delta_e}, \tag{9}$$

The coefficients $F_{ij}(i = 1,\cdots,4, j = 1,\cdots,4)$ are related to $f_{ij}(i = 1,\cdots,4, j = 1,\cdots,4)$ above. System (9) can be written in the following state dependent linear-like representation for the trim condition by Table 2

Table 2. Trim condition for the truth model

State	Value	State	Value	Input	Value
V	7702.0808 ft/s	α	1.5127 deg	Φ	0.2514
γ	0 deg	Q	0 deg/s	δ_e	11.4657 deg

$$\dot{e} = A(e)Z(e) + B(e)u \tag{10}$$

where $Z(e) = \begin{bmatrix} e_V & e_\gamma & e_\alpha & e_Q \end{bmatrix}^T, u = \begin{bmatrix} e_\Phi & e_{\delta_e} \end{bmatrix}^T$

$$A(e) = \begin{bmatrix} 0 & -32.1522 & -10.7676 - 729.8360e_\alpha - 439.9507e_\alpha^2 & 0 \\ 0 & 0 & 0.0687 - 0.0021e_\alpha - 0.0571e_\alpha^2 - 0.0202e_\alpha^3 & 0 \\ 0 & 0 & -0.0687 + 0.0021e_\alpha + 0.0571e_\alpha^2 + 0.0202e_\alpha^3 & 1 \\ 0 & 0 & 2.9454 + 6.9233e_\alpha - 2.2068e_\alpha^2 & 0 \end{bmatrix},$$

$$B(e)=\begin{bmatrix} 24.4285+120.5655e_\alpha\text{-}10.1295e_\alpha^2\text{-}1.2564\text{e+}003e_\alpha^3 & -18.6258 \\ \text{-}2.0287\text{e-}006\text{-}1.1582\text{e-}004e_\alpha\text{-}0.0017e_\alpha^2\text{-}0.0092e_\alpha^3\text{-}0.0161e_\alpha^4 & 0.0112 \\ 2.0287\text{e-}006+1.1582\text{e-}004e_\alpha+0.0017e_\alpha^2+0.0092e_\alpha^3+0.0161e_\alpha^4 & -0.0112 \\ 0.1225+0.6048e_\alpha\text{-}0.0508e_\alpha^2\text{-}6.3023e_\alpha^3 & 0.1225 \end{bmatrix},$$

according to Assumption 3, $M=\begin{bmatrix} 1 & 0 & 0 & 0 \\ 0 & 1 & 0 & 0 \\ 0 & 0 & 1 & 0 \\ 0 & 0 & 0 & 1 \end{bmatrix}$.

All rows of $B(e)$ are not equal to zeros, then $P(e)$ can be a symmetric positive definite polynomial constant matrix. By Theorem 5, (7), (8), the state feedback stabilization problem is solvable, and a controller that stabilizes the system is given by:

$$u(e)=K(e)P^{-1}(e)Z(e)$$

To test the performance of controller, simulations have been performed on the nonlinear vehicle model described in system (10), which is a nominal error system. The trim condition refers to Table 2, and the initial state $e(0)=(0.01ft/s, 0.05\deg, 0.06\deg, 0.05\deg/s)^T$. Application SOSTOOLS packa- ge to solve the problem in Definition 1, the resulting controller is as follows

$$P=\begin{bmatrix} 9.0917 & 0.0110 & 0.0015 & -0.0082 \\ 0.0110 & 0.35\times10^{-4} & 0.41\times10^{-6} & 0.46\times10^{-5} \\ 0.0015 & 0.41\times10^{-6} & 0.28\times10^{-5} & -0.77\times10^{-5} \\ -0.0082 & 0.46\times10^{-5} & -0.77\times10^{-5} & 0.52\times10^{-4} \end{bmatrix}$$

$$K=\begin{bmatrix} -0.34\times10^{-10}+0.96\times10^{-9}e_\alpha & 0.10\times10^{-6}-0.54\times10^{-10}e_\alpha & 0.43\times10^{-7}-0.52\times10^{-10}e_\alpha & \cdots \\ 0.31-0.056e_\alpha & 0.32\times10^{-3}-0.28\times10^{-4}e_\alpha & -0.18\times10^{-3}-0.75\times10^{-4}e_\alpha & \end{bmatrix}$$

$$\begin{matrix} 0.62\times10^{-7}+0.40\times10^{-10}e_\alpha \\ -0.24\times10^{-2}+0.48\times10^{-3}e_\alpha \end{matrix}\Bigg]$$

$$u=\begin{bmatrix} \left(-0.52\times10^{-5}+0.62\times10^{-10}e_\alpha\right)e_V+\left(0.37\times10^{-2}-0.30\times10^{-5}e_\alpha\right)e_\gamma+\cdots \\ \left(-0.0177+0.0117e_\alpha\right)e_V+\left(12.0127-6.0345e_\alpha\right)e_\gamma+\cdots \end{bmatrix}$$

$$\begin{matrix} \left(0.0293-0.26\times10^{-4}e_\alpha\right)e_\alpha+\left(0.44\times10^{-2}-0.19\times10^{-5}e_\alpha\right)e_Q \\ \left(-286.8341+0.2044e_\alpha\right)e_\alpha+\left(-92.638811.6340e_\alpha\right)e_Q \end{matrix}\Bigg]$$

The simulation results by using Simulink toolbox in MATLAB is shown below

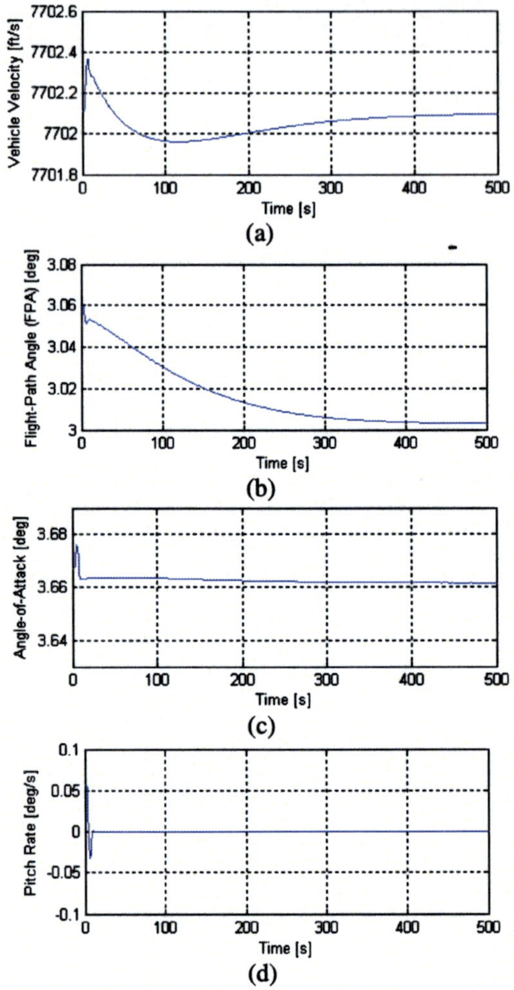

Fig. 1. States response of the SOS Controllers

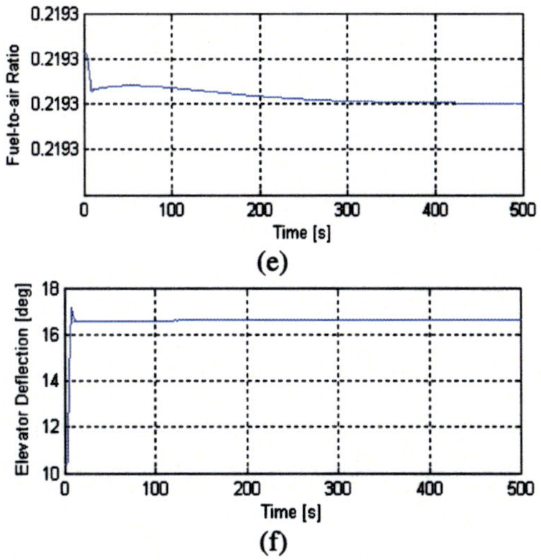

Fig. 2. Control Inputs response of the SOS Controllers

From Figure 1 and Figure 2 we can see that the velocity, flight path angle, angle of attack , pitch rate, fuel-to air ratio, and elevator deflection are asymptotically stable at the trim condition.

5 Conclusion and Future Work

In this paper, we have designed a nonlinear controller using the sum-of-squares techniques combined with the Lyapunov stability theorem for the longitudinal dynamics of a hypersonic aircraft model, the effectiveness and feasibility of the controller design by using SOS method is shown. In the future, we intend to consider some performance objective in the synthesis and incorporate robustness systematically in the controller design.

References

1. Fidan, B., Mirmirani, M., Ioannous, P.A.: Flight Dynamics and Control of Air-Breathing Hypersonic Vehicles: Review and New Direction. In: AIAA International Space Planes and Hypersonic Systems and Technologies, Nofolk, Virginia, pp. 107–118 (2003)
2. Khalil, H.K.: Nonlinear Systems, 2nd edn. Prentice-Hall, Inc., Upper Saddle River (1996)
3. Krstic, M., Kanellakopoulos, I., Kokotovic, P.: Nonlinear and Adaptive Control Design. John Wiley & Sons, New York (1995)
4. Bugajski, D.J., Enns, D.F., Elgersma, M.R.: A Dynamic Inversion Based Control Law with Application to High Angle of Attack Research Vehicle. AIAA-90-3407-CP, pp. 826–839 (1990)
5. Lane, S.H., Stengel, R.F.: Flight Control Design Using Nonlinear Inverse Dynamics. Automatica 31(4), 781–806 (1988)

6. Menon, P.K.A., Badget, M.E., Walker, R.A., Duke, E.L.: Nonlinear Flight Test Trajectory Controllers for Aircraft. Journal of Guidance, Control and Dynamics 10(1), 67–72 (1987)
7. Richard, J.A., Siva, S.B.: Robust Flight Control Design Using Dynamic Inversion and Structured Singular Value Synthesis. IEEE Transactions on Control Systems Technology 1(2), 80–92 (1993)
8. Parker, J.T., Serrani, A., Yurkovich, S., Bolender, M.A., Doman, D.B.: Control-oriented modeling of an air-breathing hypersonic vehicle. Journal of Guidance, Control and Dynamics 30(3), 856–869 (2007)
9. Nam, K., Lee, S., Won, S.C.: A Local Stabilizing Control Scheme Using an Approximate Feedback Linearization. IEEE Transactions on Automatic Control 39(11), 2311–2314 (1994)
10. Hovakimyan, N., Lavretsky, E., Cao, C.Y.: Adaptive Dynamic Inversion via Time-Scale Separation. In: IEEE Conference on Decision & Control, pp. 1075–1080 (2006)
11. Kanellakopoulos, I., Kokotovic, P., Morse, A.S.: Systematic Design of Adaptive Controllers for Feedback Linearizable Systems. IEEE Transactions on Automatic Control 36(11), 1241–1253 (1991)
12. Bolender, M.A., Doman, D.B.: A nonlinear longitudinal dynamical model of an air-breathing hypersonic vehicle. Journal of Spacecraft and Rockets 44(2), 374–387 (2007)
13. Prajna, S., Papachristodoulou, A.: Wu. F.: Nonlinear Control Synthesis by Sum of Squares Optimization: A Lyapunov-based Approach. In: Asian Control Conference, pp. 157–165 (2004)
14. Parrilo, P.A.: Structured Semidefinite Pmgrams and Semialgebraic Geomety Methods in Robustness and Optimization. PhD thesis, California Institute of Technology, Pasadena, CA (2000)

3D Flight Simulation Based on Virtools

Ying Xiao[1], Wei Mei[1], and Xiaowei Zhang[2]

[1] Department of Optics and Electronic Engineering, Ordnance Engineering College, Shijiazhuang 050003
[2] PLA Representation Office in Xi'an
xiaobengying@163.com

Abstract. There are lots of flight scene simulation ways about air targets, and common way is simulation to kinematics and dynamics of targets, which is of high precision, but enormous computation, complicated modeling. so this simulation system mainly adopts segment outputting and random combination method to create many complicated flight track, finally builds a flight control system based on Virtools, in which we can control 3D aircraft model to finish all sorts of flight actions by mouse and keyboard.

Keywords: Segment outputting, Random combination, Virtools.

1 Mathematic Modeling of Flight Segments

Airway is decomposed into four segments: level flight、 level acceleration or deceleration maneuver、 dive or climb maneuver and level turn maneuver.

1.1 Level Flight

Level flight means an aircraft is making horizontal and straight flight. What we are going to study in this section is steady level flight, that is, an aircraft makes uniform rectilinear flight without any bank or sideslip in a horizontal plane.

Track equations are as follow:

$$\begin{bmatrix} \dot{X}_h \\ \dot{Y}_h \\ \dot{Z}_h \end{bmatrix} = \begin{bmatrix} V_0 \\ 0 \\ 0) \end{bmatrix} \quad \begin{bmatrix} X_h \\ Y_h \\ Z_h \end{bmatrix} = \begin{bmatrix} V_0 * t \\ 0 \\ 0 \end{bmatrix} \quad \begin{bmatrix} \gamma \\ \theta \\ \psi \end{bmatrix} = \begin{bmatrix} 0 \\ 0 \\ \psi_0 \end{bmatrix} \tag{1}$$

1.2 Level Acceleration or Deceleration Maneuver [1]

In each sampling interval Δt ,target velocity V_t can be expressed as follow.

$$V_t = V_{t-1} + a_t \cdot \Delta t \tag{2}$$

Thereinto, a_t is target acceleration. We analyse and find that target acceleration a_t is not a sudden change, but a gradual increase course in the course of accelerated motion;

D. Yang (Ed.): Informatics in Control, Automation and Robotics, Volume 2, LNEE 133, pp. 343–349.
springerlink.com

With the increase of velocity, air resistance is also increasing, acceleration will gradually reduce to zero, and in the new force balance, target will fly at a even speed again. According to this, the connection of target acceleration a_t and time t can be approximatively expressed by parabola.

$$a_t = a_{\max} - a^* \cdot (t - t_a)^2 \tag{3}$$

Thereinto, $0-2t_a$ is accelerated time, $a_{\max}$ is, a^* is acceleration change rate. For expediently create random airway, we can give target accelerated time $2t_a$ and maximum acceleration $a_{\max}$ in the simulation, so the unknown variable a^* can be solved using acceleration $a_t = 0$ when accelerated beginning time $(t=0)$ and ending time $(t=2t_a)$.

$$a^* = a_{\max} / t_a^2 \tag{4}$$

So $a_t = a_{\max} - a_{\max} \cdot (t - t_a)^2 / t_a^2$ (5)

As a result of velocity change

$$\Delta V = \int_0^t a_t dt, (0 \le t \le 2t_a) \tag{6}$$

Integral and get

$$\Delta V = \frac{a_{\max}}{t_a^2} \cdot (t_a \cdot t^2 - \frac{t^3}{3}) \tag{7}$$

The final track equations are as follow:

$$\begin{bmatrix} \ddot{X}_h \\ \ddot{Y}_h \\ \ddot{Z}_h \end{bmatrix} = \begin{bmatrix} \frac{a_{\max}}{t_a^2} * (2 * t_a * t - t^2) \\ 0 \\ 0 \end{bmatrix} \begin{bmatrix} \dot{X}_h \\ \dot{Y}_h \\ \dot{Z}_h \end{bmatrix} = \begin{bmatrix} V_0 + \frac{a_{\max} * t^2}{t_a^2} (t_a - \frac{t}{3}) \\ 0 \\ 0 \end{bmatrix} \begin{bmatrix} X_h \\ Y_h \\ Z_h \end{bmatrix} = \begin{bmatrix} V_0 * t + \frac{a_{\max} * t^3}{t_a^2} * (\frac{t_a}{3} - \frac{t}{12}) \\ 0 \\ 0 \end{bmatrix} \tag{8}$$

1.3 Dive or Climb Maneuver

Dive is the reverse course of climb, their mathematic modeling method is the same, so we only describe track equation of climb. Climb is decomposed into three stages: entry、line and recovery stages, T is the total climb or dive time.

1.3.1 Entry Stage

During climb, the pliot should maintain the AOA (hold the stick) and only work on the throttle. If he advances the throttle, the excess thrust will increase, and so the climb rate; if he retards the throttle, the excess thrust will decrease, and so the climb rate.

Track equation is as follow (we take 30 degree's climb angle as example here) :

$$\begin{bmatrix} \ddot{X}_h \\ \ddot{Y}_h \\ \ddot{Z}_h \end{bmatrix} = \begin{bmatrix} 0 \\ 0 \\ \frac{V_0 * \pi}{2\sqrt{3} * T} * \sin(\frac{\pi * t}{T}) \end{bmatrix} \begin{bmatrix} \dot{X}_h \\ \dot{Y}_h \\ \dot{Z}_h \end{bmatrix} = \begin{bmatrix} V_0 \\ 0 \\ \frac{V_0}{2\sqrt{3}} * (1 - \cos(\frac{\pi * t}{T})) \end{bmatrix} \begin{bmatrix} X_h \\ Y_h \\ Z_h \end{bmatrix} = \begin{bmatrix} V_0 * t \\ 0 \\ \frac{V_0}{2\sqrt{3}} * (t - \frac{T}{\pi} \sin(\frac{\pi * t}{T})) \end{bmatrix} \tag{9}$$

Climb angle $\theta = \arctan(\frac{\dot{Z}_h}{\dot{Y}_h}) = \arctan(\frac{1-\cos(\frac{\pi * t}{T})}{2\sqrt{3}})$ (10)

Angular velocity $\omega_y = \dot{\theta} * \frac{\pi * \cos(\pi * t / T)}{(1+k^2)} \quad k = \frac{1-\cos(\frac{\pi * t}{T})}{2\sqrt{3}}$ (11)

1.3.2 Line Stage

After the aircraft is stabilized, the lift coefficient will not change because the AOA did ont chage;and the lift required will be almost the same because the climb angle was almost the same. Therefore, we can think the climb speed will be basically the same.

In this stage, aircraft climbs to needed height at an even velocity, and $\omega_y = 0$.

So we can get the equation.

$$\begin{bmatrix} \ddot{X}_h \\ \ddot{Y}_h \\ \ddot{Z}_h \end{bmatrix} = \begin{bmatrix} 0 \\ 0 \\ 0 \end{bmatrix} \quad \begin{bmatrix} \dot{X}_h \\ \dot{Y}_h \\ \dot{Z}_h \end{bmatrix} = \begin{bmatrix} V_0 \\ 0 \\ \frac{V_0}{\sqrt{3}} \end{bmatrix} \quad \begin{bmatrix} X_h \\ Y_h \\ Z_h \end{bmatrix} = \begin{bmatrix} V_0 * t \\ 0 \\ \frac{V_0}{\sqrt{3}} * t \end{bmatrix} \tag{12}$$

1.3.3 Recovery Stage

Recovery stage is the reverse of entry stage.

$$\ddot{Z}_h = -\frac{V_0 * \pi}{2\sqrt{3} * T} * \sin(\frac{\pi * t}{T}) \tag{13}$$

$$\dot{Z}_h = \frac{V_0}{2\sqrt{3}} * (1 + \cos(\frac{\pi * t}{T})) \tag{14}$$

$$Z_h = Z_0 + \frac{V_0}{2\sqrt{3}} * (t + \frac{T}{\pi} \sin(\frac{\pi * t}{T})) \tag{15}$$

$$\theta = \arctan(\frac{\dot{Z}_h}{\dot{Y}_h}) = \arctan(\frac{1+\cos(\frac{\pi * t}{T})}{2\sqrt{3}}) \tag{16}$$

1.4 Level Turn Maneuver

Aircraft turn is supposed to be coordinated and no sideslip, and the track is in the same level. We take right turn for an example to analyse turn radius and angular velocity in the course of turning.

Assuming turn velocity is $\dot{Y}_h$, turn radius is R, angular velocity is $\omega_x = \dot{Y}_h / R$, turn centripetal force is offered by declining component of gravity, acceleration is $\ddot{X}_h$.

So

$$\ddot{X}_h = -R * (\omega_x)^2 = -(\dot{Y}_h)^2 / R = -g * \tan\gamma \tag{17}$$

$$R = (\dot{Y}_h)^2 / g * \tan\gamma \tag{18}$$

$$\omega_x = \dot{Y}_h / R = g * \tan\gamma / \dot{Y}_h \tag{19}$$

Turn course is also decomposed into three sections: aircraft change roll angleand enter turn section from level flight, keep roll angle and turn at constant angular velocity, and recovery section after turning. Assuming that bank angle γ and angular velocity ω_x are both linearly increase at the beginning of turning, and they are both linearly decrease at the end of turning.

1.4.1 Entry Stage

In this section, aircraft is flying at constant angular velocity and roll angle is adjusted to needed value. Assuming transition time is 2s, roll angle is $\pi/6$, total turn time is Ts, the formulas are as follow:

$$\dot{\gamma}=\pi/12 \quad \gamma=\pi*t/12 \tag{20}$$

$$\omega_x=0.5*\omega*t \tag{21}$$

$$\psi=0.25*\omega*t^2 \tag{22}$$

1.4.2 Hold-on Stage

$$\dot{\gamma}=0 \quad \gamma=\pi/6 \tag{23}$$

$$\omega_x=-g*\tan(\pi/6)/V_x \tag{24}$$

$$\psi=\omega*(t-1) \tag{25}$$

1.4.3 Recovery Stage

$$\dot{\gamma}=-\pi/12 \quad \gamma=\pi*(T-t)/12 \tag{26}$$

$$\omega_x=\omega-0.5*\omega*(t-T+2) \tag{27}$$

$$\psi=\omega*(T-3)+\omega*(t-T+2)-0.25*\omega*(t-T+2)^2 \tag{28}$$

In a word, the final track equation is as follow:

$$\begin{bmatrix} X_h \\ Y_h \\ Z_h \end{bmatrix}=\begin{bmatrix} X_z+R(\cos\psi_t-\cos\psi_z) \\ Y_z+R(\sin\psi_0-\sin\psi_z) \\ Z_z \end{bmatrix} \tag{29}$$

X_z, Y_z, Z_z, ψ_z are initial position and drift angle of turning.

Because aircraft unconsciously maneuver、air turbulence and other random factors can all influence flight track, we can use random numbers $v_x \sim N(0,\sigma_x^2)$、$v_y \sim N(0,\sigma_y^2)$、$v_z \sim N(0,\sigma_z^2)$ to simulate process yawp on the three coordinate axes, and superimpose them into flight track, then we can get a more vivid track.

2 System Design and Implementation

2.1 Virtools Introduction

Virtools is an extensive collection of technologies for 3D visualization. Virtools is an authoring application that allows you to quickly and easily create compositions full of rich, interactive, 3D content. Industry standard media such as models, animations, images and sounds are brought to life by Virtools' behavior technologies.

Virtools provides an extensive collection of reusable behaviors in the form of Behavior Building Blocks (BBs) that allow you to create almost any type of content through a simple, graphical interface – without writing a single line of code.

Virtools is a Software Development Kit (SDK) that provides access to the Behavioral Engine and the Render Engine. With the SDK, you can create new behaviors or modify existing ones, create new parameter types, create media plugins to read any media of your choice, replace the Virtools render engine with a render engine of your choice, create a custom executable file (.exe), modify and extend the Virtools render engine – full source code to the rendering engines is provided.

2.2 3D Model

3D aircraft model is produced by 3DS MAX software. Size scale is 1:1,namely, each geometry unit represent one meter, airframe principal axis is parallel to X axis, airhead faces to forward direction of X axis, vertical symmetry is parallel to Y axis, confirming Z axis according to left-hand helix rule, and this is Virtools tolerant 3D coordinate direction, as figure 1 shows.

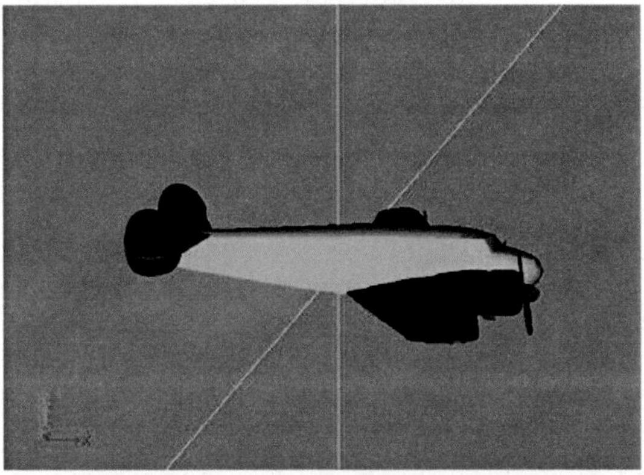

Fig. 1. 3D aircraft model based on 3DS MAX

After finishing 3D aircraft model, it is converted to .nmo file and imported Virtools platform, then we can combine building blocks according to our need.

2.3 System Module Design and Implementation

After comprehensive analysis to system function, we put system into the follow modules:

1. Initial module: finish initial configuration at the beginning of system running.
2. Aircraft position、 attitude computation module: according to initial information、 maneuvering mode and mathematic model of flight, working out real-time aircraft position and attitude information.

3. Real-time display module: using Text Display module which is Virtools' own module to real-time display aircraft position and attitude information on the screen.
4. Control module: implementing maneuvering actions choosing and running、 pause and restoration fuctions in the course of flight.

Flow chart of flight simulation system is shown in Figure 2. System running interface is shown in Figure 3. the top half of Figure 3 is scene area,the bottom half is building blocks (flight control、 computation and maneuver switch) adding area.

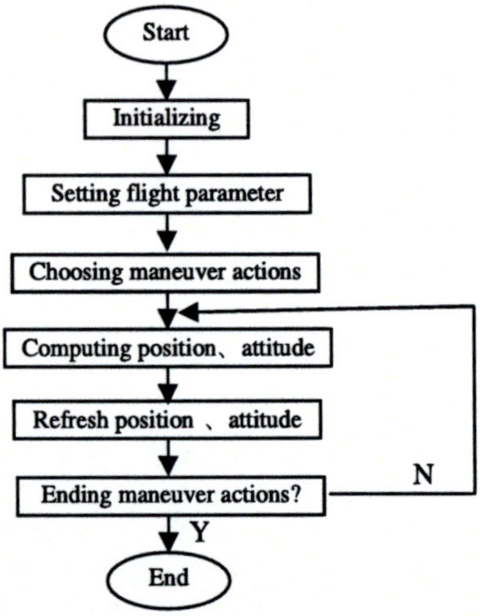

Fig. 2. System running flow chart

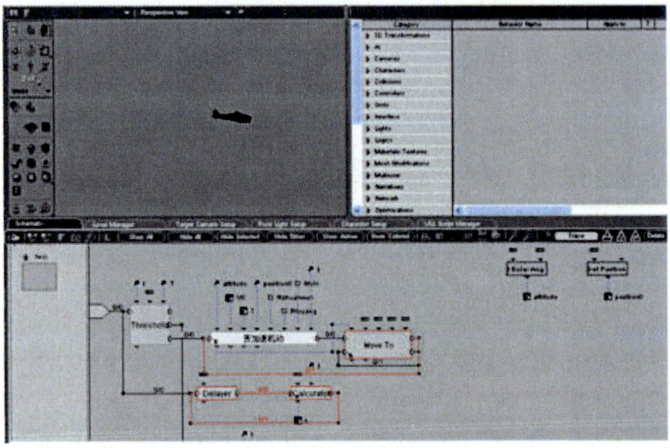

Fig. 3. System running interface

3 Conclusion

Flight simulation system development is finished based on Virtools platform, which decomposes the full flight track into four sections, and arbitrarily combines them by man, then creates quite vivid flight track. At the same time, it can directly show basic flight actions and real time output aircraft position and attitude information. Further more, it can also provide validation technique for antiaircraft gun air defense effectiveness evaluation of simulation games.

References

1. Li, Z., Li, Z.: The Automatic Creating Technique of a Flight Track. Fire Control & Command Control (2002)
2. Teng, Y., Zheng, J., Gao, Z.: Design and implementation of Interactive 3D Scenes Based on Virtools. International Forum on Computer Science-Technology and Applications (2009)
3. Virtools Dev User Guide (2001)
4. Yang, H., Guo, A., Dong, W., Yang, Q.: Design and Realization of Flight Simulation Based on 3DS. Computer Simulation (2004)
5. Eichhenberger, I.A.: Your Pilot's License (2003)

Design of Large-Scale PV Power Station Supervisory Control and Data Acquisition System Based on Fieldbus and Industrial Ethernet

Yao Hongchun[1] and Xu Ming[2]

[1] State Grid Electric Power Research Institute
No. 8 Nanrui Road, Nanjing, China, 210003
[2] Zhangjiakou Wind and Solar Power Energy Demonstration Station CO. LTD.
No. 15 Sifangtaigou Road, Zhangjiakou, China, 075000

Abstract. According to the development status of large-scale PV power station, this paper expatiates the requirement of supervisory control and data acquisition (SCADA) system of large-scale PV power station, and presents the design method and application technique of the system based on fieldbus and industrial ethernet. The system is flexible and reliable, and will promote the efficiency of operation and maintenance of large-scale PV power station.

Keywords: Fieldbus, Industrial Ethernet, PV Power Station, SCADA.

1 Technical Requirement of the SCADA System

With the support of policies of "Solar Roof Plan" and "Golden Sun Works", the quantity and scale of the finishing, under construction and planning solar energy PV power stations are rising sharply each year. According to the differences of grid voltage level and installed capacity, PV power stations can be divided into small-scale, medium-scale and large-scale; according to different forms of construction, PV power stations can be divided into off-grid and grid-connected. At present, large-scale grid-connected PV power stations account for a significant proportion of the finishing or under construction PV power stations in China.

Large-scale grid-connected PV power station must be equipped with automatic operation, fully functional SCADA system. This SCADA system is different from conventional power plant control system or substation automation system, relatively speaking, the types of equipments are less than the conventional power station, and the control process is not very complicated within the large-scale grid-connected PV power station. But the power station is typically characterized by large capacity (10MW and above), wide area (150 acres or more), geographic isolation, and the maintenance staff is relatively small, so requires the concentration of operation, equipments monitoring, environmental monitoring and security, and is able to adapt to the scattered locations and flexible configuration features. The SCADA system based on fieldbus and industrial ethernet meets these requirements.

D. Yang (Ed.): Informatics in Control, Automation and Robotics, Volume 2, LNEE 133, pp. 351–357.
springerlink.com

2 LonWorks Technology Features

LonTalk protocol used in LonWorks technology follows the ISO/OSI reference model, and is defined to achieve the OSI 7-layer protocol. LonWorks technology is the only 7-layer protocol fieldbus.

The Neuron Chip is the core technology of LonWorks bus, and LonTalk protocol is completely embedded in the chip. Thus, as long as the neuron chip is embedded in the device, the device can be installed in LonWorks network. Neuron Chip includes three 8-bit CPU working in parallel: the first CPU working as media access controller deals with the first and second layer of LonTalk protocol; the second CPU working as the network processor deals with the third to sixth layer of LonTalk protocol; the third CPU working as the application processor deals with user code and operating system service.

When applied to large-scale grid-connected PV power station SCADA system, LonWorks bus technology has significant advantages, mainly reflected in:

(1) LonWorks technology can choose twisted-pair, power lines, fiber optic and other physical media, and have a smart transceiver chip which combines neuron chip and physical layer. Smart Transceiver includes two types of physical layer, twisted-pair and power line. The electrical performance of Twisted-pair can achieve 2km/78kbps; power line can be coupled to power supply line, and the sensitivity can achieve 80dB, and can use two bands respectively with the communication rate of 3.6kbps and 5.4kbps.

(2) LonWorks based devices can use distributed peer communication to construction the distributed control system. Each network device can receive, send and process the network information independently without other devices, so each LonWorks device can execute autonomous decision-making and information processing, rather than relying on the main CPU. Any node failure will not cause the system to a standstill.

(3) Flexible network topology. The bus-shaped, star, ring and other network topology are all applicable. This characteristic greatly reduces the workload of the construction and maintenance.

(4) LonWorks network using logical connections between nodes. It is easy to add and remove nodes, and easy to adjust and expand the system.

(5) LonTalk protocol can be configured according to different application environments, and is completely open.

3 Industrial Ethernet Technology Features

Industrial ethernet is technically compatible with commercial ethernet (IEEE802.3 standard), and is widely used in industrial control systems. However, in product design, industrial ethernet should be able to meet the needs of the industrial field in material selection, product strength, real-time, reliability and environmental applicability. Its main technical features are as follows:

(1) Wide range of applications. Industrial ethernet has many characteristics of low price, various transmission media options, high speed, easy-to-network, and has been applied in the resource management layer and manufacturing execution layer of integrated automation systems in industrial enterprises. Recently, the development trend of industrial Ethernet in field control is presenting.

(2) High speed of communication. Industrial ethernet communication speed range is from 10/100Mbps to today's 10Gbps, of which 10/100Mbps has been widely used, and 1000Mbps and 10Gbps Ethernet technology is gradually mature.

(3) Security of control network. Industrial ethernet uses anti-virus measures, intrusion detection, network isolation, message encryption and other measures, has proposed a series of control network security models.

(4) Supports a variety of physical media and topology. Industrial ethernet supports a variety of transmission media, supports bus and star topologies, and multiple redundant connections can be used to improve network performance.

4 System Framework

Large-scale PV power station SCADA system is made up of monitoring center, fiber optic ring network and control subnet. The control subnet includes routing unit, control unit, fieldbus network and security system components. The overall system topology is shown in Figure 1.

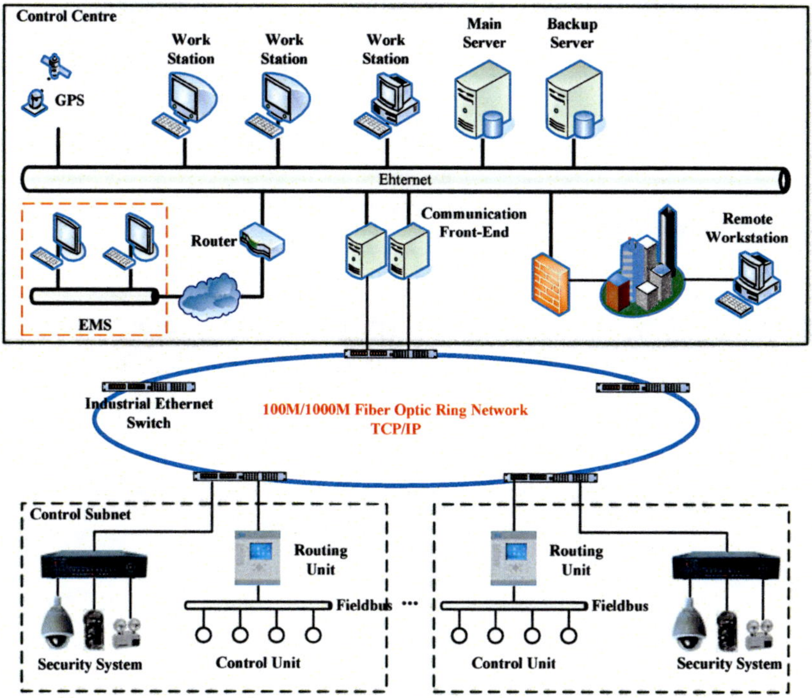

Fig. 1. Overall system topology

The control center is the core of the SCADA system. It exchanges information with on-site systems through the communication network, completes the operation monitoring, order dispatching, data analysis, status display, statistical analysis and other functions. It receives up-level dispatching instructions and controls active/reactive power of PV power station.

Fiber optic ring network is the backbone communications network of the SCADA system, which utilizes single mode optical fiber as physical media and uses TCP/IP protocol. The network with high speed, real-time, high reliability can be configured as a single ring or loop form. The control center and the on-site system access the network through industrial ethernet switch.

The control subnet includes routing unit, control unit, fieldbus and security module, and gets access to fiber optic ring network through industrial ethernet switch. Large-scale PV power stations generally set 500kW or 1MW capacity as a PV power cell, the range of functions of the control subnet is defined by the PV power cell. The control subnet topology is shown in Figure 2. The control unit of PV convergence box utilizes DC power line as carrier communication cable, and the control unit in DC power distribution cabinet receives the data, and sends it via twisted pair to the fieldbus.

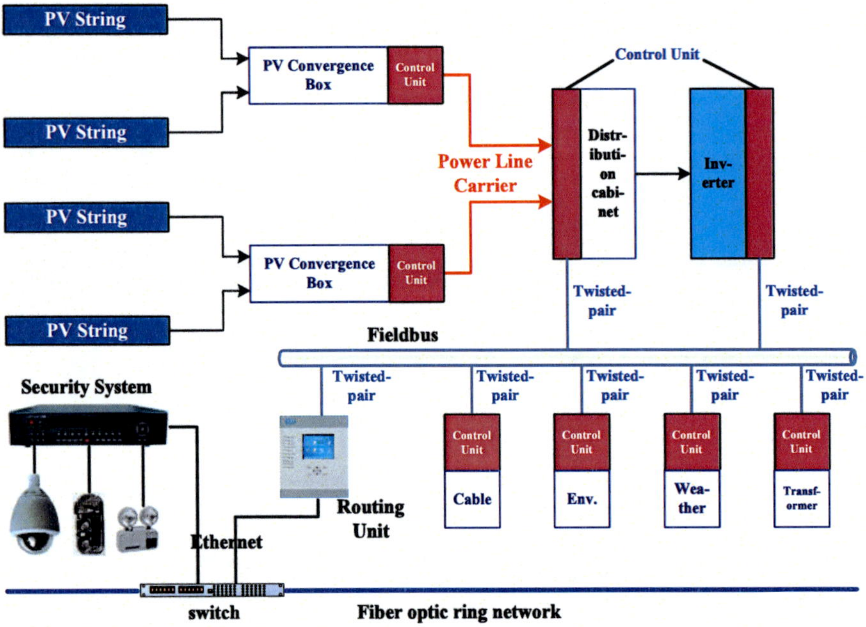

Fig. 2. The control subnet topology

The routing unit which is responsible for data pre-processing and protocol converter is the key equipment for transmitting the information. The fieldbus protocol and other communication protocol of the control subnet will be transferred in the TCP/IP network by the technique of protocol converting and tunneling.

The control unit is the basic equipment of the SCADA system. The SCADA system is responsible for the data acquisition and command distribution, which interacts with the routing unit. According to monitoring and control objects, processing ability and different communication media, the control unit is also divided into different types, which can achieve flexible deployment and lower costs.

5 Main Functions of the System

5.1 SCADA

The SCADA functions of large-scale PV power station can be divided into operational class and management class. Data collection, status monitoring, device control, event logging are belong to operational class; user management, security management, system setup, time synchronization are belong to management class. The more important features include:

(1) Data acquisition and processing. The system receives real-time data of the routing unit through communication frond-end server, and does pre-processing according to the different types of analog, digital and electrical measurement.

(2) Condition monitoring and evaluation. The control center monitors the main operating parameters and operating status of PV power station. And all details of the station through the system configuration diagram, electrical diagrams, process graphs and statistical reports are presented. Based on abundant information, it will evaluate the status of system.

(3) Equipment operation and control. The control center can control the main equipments of PV power station, and the control commands are issued by the routing unit.

(4) System alarms and diagnostics. Alarm systems can be divided into two types, one is accident alarm, and the other is early warning alarm. Generally, the former includes the inverter fault information and unplanned breaker tripping, etc; the latter includes general equipment deformation, abnormal data collection and trend alarms.

(5) Record and recall events. When an accident of PV power station or power grid happened, the SCADA system automatically extracts relevant information before and after the specified period of time for analysis. Recalling of the accident can be automatically triggered by the critical fault conditions, or can be started by a user-specified way.

5.2 Active/Reactive Power Automatic Control

The active power automatic control dispatches the active power output of the whole system and each inverter according to the grid's dispatching commands and system's operational status, to meet the requirements of grid scheduling, and optimize system efficiency.

The reactive power automatic control dispatches the reactive power output of SVC/SVG and each inverter which can be used in 0.95 (leading) to 0.95 (lagging) power factor of the system according to the grid dispatching commands and grid status. At this rate, the power requirements of the voltage of common coupled point will be fulfilled.

5.3 Environmental Monitoring

Large-scale PV power station covers large area, deploys less staff, and is maintained difficultly. Especially when the surface of PV cells was polluted, power generation is greatly influenced. Environmental monitoring functions within the power station can monitor weather status and air quality continuously, and will effectively improve system power efficiency and reduce maintenance workload.

5.4 Security Features

Security features include video surveillance, infrared detection, alarm and emergency lighting, and can complete blind spot coverage of the station. The staff can improve the security level and reduce the intensity of work with this function. This function also provides the technical means for the environment monitoring and equipment surveillance.

6 The Basic System Components

6.1 Control Unit

The control unit completes data acquisition and control commands assigned tasks for monitoring and controlling objects directly. The core of the control unit is microprocessor and smart transceivers. According to the application environment, microprocessor should use ARM9. Specifically, microprocessor is charged with data acquisition, preprocessing, control command recognition and other functions. And smart transceiver is responsible for sending and receiving data. The basic structure of the control unit is shown in Figure 3.

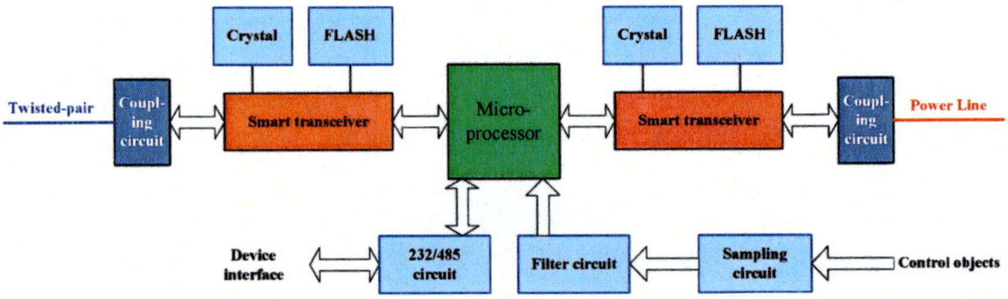

Fig. 3. The basic structure of the control unit

6.2 Routing Unit

The routing unit is constructed by PowerPC embedded platform and smart transceiver. The unit supports LonTalk protocol, IEC60870, CDT and other communication protocol, and has multiple communication interfaces of LonWorks twisted-pair, ethernet, RS232 and RS485. So it can communicate with multiple control unit through different media for information exchange, and can correctly receive, identify, process, execute remote command sent by the control center. The basic structure of the routing unit is shown in Figure 4.

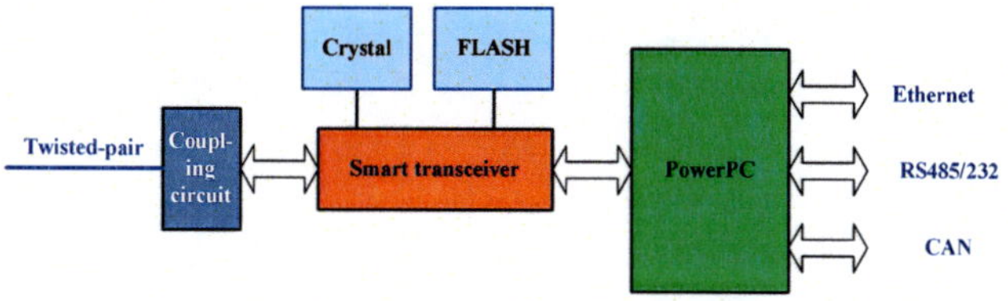

Fig. 4. The basic structure of the routing unit

6.3 Control Centre

The control center software adopts the cross-platform and distributing system architecture. The support platform system consists of the message/service soft-bus, system management service, security service, real-time data service, history data service, business operations service and public service. The public service includes data service, model service, file service, picture service, alarm service, access service, reporting service and logging service. All SCADA and other applications run on the support platform.

7 Conclusion

LonWorks and industrial ethernet technology achieves significant benefits in real-time, reliability, maintenance, and other aspects. The large-scale PV power station SCADA system based on this technology can effectively improve the efficiency of operation and maintenance of PV power station, because of so many advantages of full-featured, network flexibility, easy installation, stability, reliability and quick reaction.

References

1. Liu, W., Bu, H., Qing, T., Liu, X.: The Application of Lonworks Fieldbus Technology in Environment Inspect System. Logistics Sci-Tech. 4, 114–116 (2007)
2. Hu, G., Lu, L., Huang, L.: Development of Substation Automation Based on the LonWorks control System Technology. Proceeding of the CSU-EPSA 17(1), 83–87 (2005)
3. Jiang, L., Xu, A., Song, Y., Wang, J.: Research and Realization of High Availability Industrial Ethernet Technology. Computer Engineering 35(11), 260–262 (2009)
4. Wang, Q.: Design and application of fiber industrial ethernet switch based on IEC61850. Power System Protection and Control 38(13), 113–115 (2010)

Design and Implementation of a SOLR Plug-in for Chinese-English Cross-Language Query Expansion Based on SKOS Thesauri

Wei Sun[1], Fabrizio Celli[2], Ahsan Morshed[2], Yves Jaques[2], and Johannes Keizer[2]

[1] Institute of Agriculture Information, Chinese Academy of Agricultural Sciences, Beijing 100081, P.R. China
sunwei@caas.net.cn
[2] FAO of the United Nations, Rome, Italy
Fabrizio.Celli@fao.org,
Ahsan.Morshed@fao.org,
Yves.Jaques@fao.org,
Johannes.Keizer@fao.org

Abstract. Given that existing studies for query expansion techniques for Chinese-English are relatively few and their level of standardization low, in order to improve efficiency of Chinese-English cross-language retrieval, this paper discusses the design and implementation of a SOLR plug-in for Chinese-English cross-language query expansion based on SKOS thesauri and used within the AGRIS agricultural bibliographic system. The paper also elaborates the key techniques involved in the plug-in. Finally, taking the AGRIS data resources as an example, the paper shows application examples for segmentation of mixed Chinese and English, user query parsing and AGRIS retrieval system etc., techniques that have improved the Chinese-English cross-language retrieval efficiency to a certain extent, and laid a technical foundation for research about knowledge retrieval and discovery in related fields.

Keywords: SOLR, SKOS, Linked open data, Index, Query Expansion.

1 Introduction

With the rapid expansion of networked information and the continuous enrichment of multilingual information, cross-language retrieval has become a key factor in global knowledge sharing while user demand for retrieval efficiency are also escalating. Query expansion techniques can improve recall and precision rates for information retrieval, and hold out the hope of solving not only the problems of language extension but semantic extension as well. Although methods of semantic concept expansion can make up for the inherent limitations of mechanical string expansion that focuses on query words, most of the existing related techniques cover query expansion among western languages [1-3], while few techniques cover those between Chinese and English, particularly based on standard resource models such as

D. Yang (Ed.): Informatics in Control, Automation and Robotics, Volume 2, LNEE 133, pp. 359–367.
springerlink.com

SKOS[4-6]. In addition, semantic and cross-language query expansion will increase the indexing workload, and the indexing quality is an important indicator to measure retrieval efficiency. Therefore, how to build a unified standard for semantic expansion technologies and how to index efficiently and qualitatively for meeting users' retrieval demands within a wider scope while improving search efficiency are still urgent problems in the field of information retrieval.

AGRIS (International Information System for the Agricultural Sciences and Technology) is a public domain Database with nearly 3 million structured bibliographical records on agricultural science and technology [7] developed by FAO (Food and Agriculture Organization of the United Nations) since 1974 to make agricultural information world-wide available. Its open database covers many aspects of agriculture, including forestry, animal husbandry, aquatic sciences and fisheries, and human nutrition, and its content is provided by more than 150 participating institutions from 100 countries. AGRIS' focus is on solving problems concerning global agricultural science and technology information resources sharing [7]. It is thus clearly useful for such a system to solve cross-language semantic retrieval problems based on a unified standard and to improve its efficiency.

Recently emerging "Linked Data" technologies can use a lightweight, scalable and extensible dynamic mechanism to achieve knowledge organization of semantic associations among dynamic, heterogeneous, and distributed data[8]. As a technical standard for resource description proposed by W3C, RDF is the basic language for Linked data, while SKOS is a common data model for sharing and linking knowledge organization systems via the Semantic Web [9]; Apache Solr is an open source search server written in Java and based on the Apache Lucene search library: it has REST-like HTTP/XML and JSON APIs that make it easy to use from any programming language [10]. Indexes created by Solr are fully compatible with the Lucene search engine library, but with more powerful full-text search capabilities and high scalability.

Based on Linked data, the paper describes the design and implementation of the SOLR plug-in for Chinese-English cross-language query expansion or Solr Query expander (SQE) - which is independent of retrieval systems - and also reveals the extended results by taking the AGRIS system as an example.

2 Retrieval System Architecture Based on SQE

To realize the SQE plug-in retrieval system based on SOLR technology, we need to begin by clarifying the data flow between the SOLR server and the retrieval system, and the role of SQE during the process of indexing and retrieval. The paper elaborates this by using the AGRIS system as an example. As shown in **Fig. 1**, the complete retrieval in AGRIS based on SOLR technology is based on three processes: initialization, indexing, and searching.

2.1 The Initialization of SOLR

When Tomcat starts up, it initiates by reading some related configuration documents including solr.xml, solrconfig.xml, schema.xml and other related xml configuration files, and will then finish the loading of the related SOLR handler, and the configuration for the searching and indexing processes.

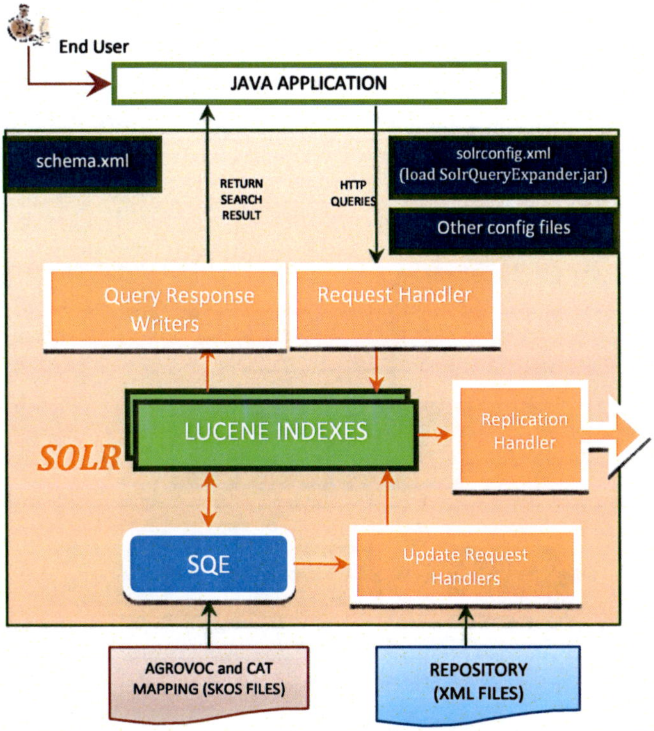

Fig. 1. AGRIS architecture based on SQE

During the process of indexing configuration, the SOLR Query Expander will index the "AGROVOC-CAT mapping (SKOS file)" which begins with a Jena analysis of the SKOS file prior to Lucene indexing. The index result is then updated by the SOLR Update Request Handler.

2.2 Indexing

During the process, first an XML document needs to be parsed and then imported into SOLR by a SOLR server update. Then, during the indexing process of the "REPOSITORY (XML FILES)" by Lucene, the SOLR Query Expander searches the indexed SKOS file and adds the expanded terms to the index, thus producing a new index via the Update Request Handler.

2.3 Searching

This process doesn't need SQE. After SOLR receives a request, it analyzes the user query with a sequence of analyzers, and then looks for the result in the constructed index to finish the search. The search result will be returned with the query response writers.

3 Logical Structure of SQE

The SQE logical structure (**Fig. 2**) contains two processes, "SKOS load and Index" and "Muti-term Expansion and Index Updating", which are marked by "①" and "②" separately.

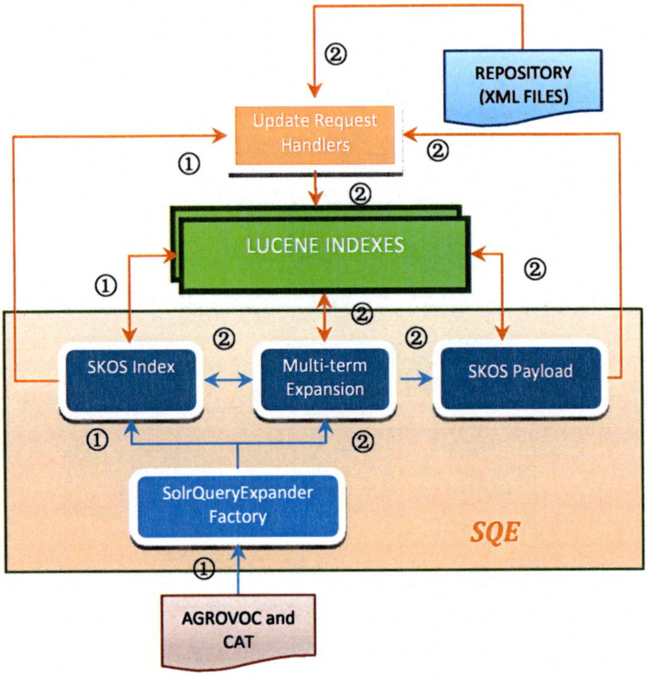

Fig. 2. Logical structure of SQE

3.1 The Logical Process of Loading and Indexing SKOS

In this process, "AGROVOC-CAT mapping (SKOS file)" is first loaded by an operation of the SOLR Query Expander Factory. This SKOS file is then indexed with Lucene by producing a SKOS engine class. The indexed result will be stored in the specific directory and updated by the Update Request Handler.

3.2 The Logical Process of Multi-term Expansion and Index Updating

In the second process, "THE REPOSITORY (XML FILES)" can be loaded, overriding the SOLR method in the SOLR Query Expander Factory. During the process of analysis of XML files, multi-term expansion takes each analyzed term as input and searches the constructed SKOS index to match concepts. If a match is found, it adds the concept's label to the constructed index. Then, it implements the index updating via operation of the SKOS payload and the Update Request Handler.

4 Key Techniques of SQE

4.1 Techniques for Producing SKOS

SQE is a plug-in based on a SKOS thesaurus document. The document is generated from the linked data version of the agricultural Chinese and English thesauri CAT and AGROVOC. Once the thesauri have been published, their related resources can be searched and browsed, and associations among them can be created with namespaces. The SKOS document contains associations between phrases in AGROVOC and CAT, which are generated by searching the SKOS with a Jena query. A fragment of the SKOS document is shown in **Fig. 3**.

```
<skos:Concept rdf:about="http://aims.fao.org/aos/agrovoc/c_5410">
        <skos:prefLabel>ornamental birds</skos:prefLabel>
        <skos:altLabel>观赏禽</skos:altLabel>
        <skos:narrower rdf:resource="http://aims.fao.org/aos/agrovoc/c_14425"/>
        <skos:broader rdf:resource="http://aims.fao.org/aos/agrovoc/c_935"/>
</skos:Concept>
<skos:Concept rdf:about="http://aims.fao.org/aos/agrovoc/c_5411">
        <skos:prefLabel>Ornamental bulbs</skos:prefLabel>
        <skos:altLabel>Flowering bulbs</skos:altLabel>
        <skos:altLabel>球根花卉</skos:altLabel>
        <skos:broader rdf:resource="http://aims.fao.org/aos/agrovoc/c_5417"/>
        <skos:related rdf:resource="http://aims.fao.org/aos/agrovoc/c_1144"/>
</skos:Concept>
```

Fig. 3. A fragment of SKOS

4.2 Techniques for Updating and Controlling Indexes

In order to meet the needs of system administrators and make cross-language retrieval, indexing, and updating of semantic expansion more flexible, the control on SQE using a SOLR configuration document can make SQE loading more flexible, so that the goals of controlling query expansion are achieved.

Fig. 4 is a part of the configuration for indexing and retrieving SOLR configuration documents. SQE plug-in controls SKOS loads with an expand variable. When the expand feature is on, SOLR will load the SKOS document during the indexing process and use SKOS-based concept expansion to update the index. On the contrary, when the expand feature is off, the SKOS file will not be loaded, and concept expansion won't be performed during indexing. Thus, in the case the SKOS document has not been changed, it can be indexed once, and not re-indexed until it has been changed. Thus the efficiency of indexing is improved to some extent.

```
<fieldType name="html_st_text" class="solr.TextField" positionIncrementGap="100">
    <analyzer type="index">
        <tokenizer class="org.wltea.analyzer.solr.IKTokenizerFactory" isMaxWordLength="false"/>
        <filter class="solr.StopFilterFactory" ignoreCase="true" words="stopwords.txt"
enablePositionIncrements="true" />
        <filter class="fao.agris.luceneSKOS.solr.SKOSFilterFactory" skosFile="ChEnmapping.rdf"
expand="on"/>
        <filter class="solr.LowerCaseFilterFactory"/>
    </analyzer>
    <analyzer type="query">
        <tokenizer class="org.wltea.analyzer.solr.IKTokenizerFactory" isMaxWordLength="true"/>
        <filter class="solr.StopFilterFactory" ignoreCase="true" words="stopwords.txt"
enablePositionIncrements="true" />
        <filter class="solr.LowerCaseFilterFactory"/>
    </analyzer>
</fieldType>
```

Fig. 4. Screenshots of SOLR configuration document

4.3 Analysis Techniques of Mixed Chinese-English

As can be seen from **Fig. 4** (a fragment of the SOLR configuration document), an analyzer is used during both indexing and retrieval. IK analyzer is a lightweight Chinese analyzer kit based on Java [11] [12]. IK analyzer 3.2.0 - or later versions - provides expansion for Solr 1.4 at API level. The new version of IK Analyzer uses a unique algorithm, "the most fine-grained segmentation algorithm of forward iteration". The algorithm supports two kinds of segmentation mode: the maximum word length and the fine-grained. This algorithm has a high processing speed, and also supports segmentation of letters, numbers and Chinese phrases. But it is only a Chinese analyzer kit, and it can only segment English by whitespace: so it doesn't support mixed Chinese and English segmentation and English phrase segmentation. So, a technical issue of the analyzer is to solve mixed Chinese and English segmentation and to improve segmentation efficiency.

As the mapping database based on AGROVOC and CAT contains a large number of Chinese and English phrases, SQE plug-in development required the usage of techniques of mixed Chinese and English segmentation. Analysis of mixed Chinese and English in SQE is transformed by the IK Analyzer package, which takes the phrases in SKOS as a user dictionary, uses the "Most fine-grained segmentation method" and adds analysis of mixed Chinese and English on the basis of the IK Analyzer.

Firstly, English letters are included in pre-segmentation texts, and segmented using the most fine-grained technique by taking a letter as a unit. Then, the analysis results are filtered with a letter as a unit (**Fig. 5**). Thus, the segmentation problem of mixed Chinese and English is solved.

```
//Outputing unknown phrases by a word
  for(int i = uBegin ; i <= uEnd ; i++){
newLexeme = new Lexeme(context.getBuffOffset() , i , 1   ,
Lexeme.TYPE_CJK_UNKNOWN);
     if(!(segmentBuff[i] >= 'a' && segmentBuff[i] <= 'z')&&!(segmentBuff[i] >= 'A'
&& segmentBuff[i]<='Z') && segmentBuff[i]!=' ')
context.addLexeme(newLexeme);
 }
```

Fig. 5. Code fragment 2 of mixed Chinese and English segmentation

5 Examples Demonstrating Key Techniques of SQE

5.1 Analysis Examples of Mixed Chinese and English

SQE plug-in is based on an analysis of mixed Chinese and English. In the AGRIS system, we use "most fine-grained segmentation" to analyze mixed Chinese and English texts, segmenting English not only by referring to white space but also based on SKOS labels. The examples of mixed Chinese and English analysis are as follows:

- Pre-analyzed text: Welcome to the Institute for Development Strategy.生产费用可以按照与产品的关系分类。
- Text analyzed: Welcome | institute | development strategy | development | strategy | 生产费用 | 生产费 | 生产 | 费用|可以|按照|产品|关系|分类

5.2 Demonstration of Query Expansion Results

When we search "rice" in the SOLR analysis page, we will get 8 expanded terms, basmati rice, broken rice, 谷物, small grain cereals (grain), cereals, paddy, 稻米, 大米 with SQE.

5.3 Example of User Query Parsing

For different query expansion content, parsing methods are also different. Parsing techniques for user queries in AGRIS system add new and corresponding parsing content based on SOLR techniques[13]. Given differences between Chinese and English parsing techniques, parsing techniques in AGRIS system based on SQE also add new content.

In the SKOS, Chinese "生产费用" is an exactMatch term for the English "Operating costs". Here we take the query expression related to "生产费用" and "Operating costs" as an example to illustrate the decomposed query expression result.

Except for Solr query syntax, the search result should be in accordance with the decomposed query expression in **Table 1** below.

Table 1. Parsing result of user query

No.	Query Expression	Decomposed Query Expression
1	生产费用	"生产费用" or "Operating costs"
2	Operating costs	"Operating" or "costs" or "生产费用"
3	"生产费用"	"生产费用" or "Operating costs"
4	"Operating costs"	"Operating costs" or "生产费用"

In order to improve the retrieval precision, Chinese phrases are segmented using the maximum matching method, while English words and phrases are all segmented into a user query according to SOLR parsing rules.

5.4 Example of the AGRIS System

The paper has made search testes by taking Chinese and English data from 2008 to 2010 as data sources in AGRIS. Specific operations are to first search three pairs of Chinese and English corresponding phrases, and then classify statistically the search result. Initial test results are shown in **Table 2**.

Table 2. Retrieval result before and after query expansion

query	Before expansion	After expansion
Rice	5 classes, 447 results	5 classes, 464 results
大米	2 classes, 9 results	5 classes, 461 results
生产费用	0 classes, 0 results	0 classes, 0 results
Operating costs	13 results	3 classes, 54 results
生产成本	11 results	1 classes, 45 results
Production costs	5 classes, 355 results	5 classes, 943 results

As can be seen from **Table 2**, recall rate of most of the search result has been improved significantly not only from the field angle but also from the number of records retrieved. In addition, although the initial indexing time at server startup before and after expansion is 54 and 345 seconds respectively, the difference in actual search retrieval time before and after expansion is measured in milliseconds.

6 Conclusion

In summary, based on an analysis of mixed Chinese and English, the SQE plug-in designs and implementation that this paper discusses has realized Chinese-English cross-language expansion retrieval. Analysis also shows that the recall rate of retrieval is improved to a certain extent according to the search test. As SQE is based on SKOS thesauri and independent of retrieval systems, it can provide technical support for knowledge organization and discovery research in a wide variety of other situations.

The improvement of retrieval recall rate is limited by the thesauri. Although indexing is only performed at startup and the affect on retrieval efficiency is not apparent, data management efficiency can be furthered by improving indexing efficiency. Therefore, the next step is to work on index updating efficiency by improving the underlying SKOS thesauri. A second improvement could be to configure the system such that two indexes are created at startup, one expanded and one not, allowing queries to be dynamically routed to either one of the two indexes as query expansion can in some cases create large, unwanted result sets.

References

1. Vossen, P.: EuroWordNet a Multilingual Database with Lexical Semantic Networks. Computational Linguistics 25(4), 628–630 (1999)
2. CINDOR, `http://www.cindorsearch.com/`
3. Lexical Database Homepage About WordNet, `http://wordnet.princeton.edu/`
4. Wu, D., Wang, H.: The Mechanism of Ontology Applied to Cross Language Information Retrieval. Library and Information Service 9, 10–13 (2006)
5. Wu, D.: Ontology Driven Cross Language Information Retrieval. New Technology of Library and Information Service 5, 22–26 (2006)
6. Chen, S.J., Zhang, Y.: English-Chinese Cross-language Information Retrieval Using Lucene System. Computer Engineering 13, 62–64 (2005)
7. AGRIS System, `http://agris.fao.org/`
8. Linked Data - Design Issues, `http://www.w3.org/DesignIssues/LinkedData.html`
9. SKOS Simple Knowledge Organization System RDF Schema, `http://www.w3.org/TR/2008/WD-skos-reference-20080829/skos.html`
10. Lucene Apache Solr., `http://lucene.apache.org/solr/`
11. ik-analyzer, `http://code.google.com/p/ik-analyzer/`
12. IKAnalyzer Chinese Analyzer V3.1.6 user handbook, `http://wenku.baidu.com/view/b4598a5abe23482fb4da4c84.html`
13. Apache Lucene - Query Parser Syntax, `http://lucene.apache.org/java/2_4_0/queryparsersyntax.html`

A Study on the Implementation of Data Storage and Data Curve in Monitoring and Control System Software

Weiqiang Zhao[1], Yongxian Liu[2], Mowu Lu[1], and Qingjun Guo[1]

[1] Engineering Training Center, Shenyang Aerospace University, Shenyang Liaoning, 110136, China
[2] School of Mechanical Engineering & Automation, Northeastern University, Shenyang Liaoning, 110004, China
zwq7618@yahoo.com.cn, yxliu@mail.neu.edu.cn, lumowu@163.com, gqj81@tom.com

Abstract. This paper introduces a new method for the implementation of data storage and data display in monitoring and control software, provides an effective real-time data storage method using dynamic link to actualize real-time and historical data curve dynamic display based on Windows API. As an real example, a real monitoring and control system software developed by the author using this method is discussed in detail. And also, the key source code is given.

Keywords: Monitoring and control system, Data curve, Dynamic link.

1 Introduction

The implementation of data storage and data curve is very important for industrial monitoring and control software development. The design of data storage has significant influence on the program universality and efficiency.

Data curves include real-time data curve and historical data curve. Real-time data curve reflects current value and the variation trend of the monitoring parameters. Real-time data curve is an intuitive description of the current working condition.

Historical data curve displays historical data in a graphic interface and reflects the variation trend of the monitoring parameters in a past period of time. Historical data curve is a effective tool for working condition analysis and fault diagnosis of the system.

The working condition monitoring and fault diagnosis system for aircraft ground hydraulic oil pump vehicle developed by the author has functions including data acquisition, data storage and data curve dynamic display to actualize real-time monitoring for oil pump vehicle working condition. In what follows, the implementation method of the data storage and data curve in this system is introduced.

D. Yang (Ed.): Informatics in Control, Automation and Robotics, Volume 2, LNEE 133, pp. 369–375.
springerlink.com

2 Data Storage Method

Data storage function is necessary for monitoring and control system. The stored data can be used to establish historical record of working condition for the system being monitored. The historical record is useful for achieving accurate result for system condition evaluation, fault diagnosis or remainder life time prediction. Data storage module of monitoring and control system stores the acquired data on storage media in a certain pattern. There are two questions in this process, one question is how data be expressed and managed in memory, another question is how data be managed on storage media.

2.1 Data Expression and Management in Memory

Traditional data expression and management method uses variables to store data. This method is involved with massive disk operation and is ineffective. In this paper, dynamic link structure is used to store data in memory. According to this method, a node class module is constructed to store a data node (including 18 data) from one acquisition. For each acquisition, the Add method of CNodes is used to allocate memory locations. By detecting whether the Count property of cNodes reaches the preset value, system determines whether need transfer data to storage media or not. This dynamic link structure reduces memory requirements and disk operation. Structure of CNodes is shown in Fig.1.

Data1	...	Data18	Pointer

Fig. 1. Node structure

2.2 Data Expression and Management in Storage Media

Data stored on storage media should be well indexed so that data can be easily retrieved. In this paper MS Access2000 database management system is used. It is fully compatible with VB6.0. Access database can be operated easily by the controls in VB6.0. This system stores data in Access database and operates database by using ADO(ActiveX Data Object). Data acquired during an acquisition period is stored into a new data table in database as a single acquisition. The name of the new data table is the current date and time. Meanwhile, information of this acquisition is added into a preset acquisition record data table in the database. This preset table stores information of each acquisition including data table name, time of acquisition start and stop. Therefore it is useful for information retrieval. The detailed method and source code are as follows:

Add the following code to the Click event of the Start acquire command button:

```
StartTime=Time                              'record acquisition start time
FormName=Format(Date, "YY/M/D") & StartTime      'form name
Set CMD=New ADODB.Command                          'set new data table
```

```
CMD.CommandText="create table " & FormName & "(Field Name,
Field Type, ......)"
Set CMD.ActiveConnection=DB
CMD.Execute
Set CMD=Nothing
Set RS=New ADODB.Recordset      'open record set to store data
RS.CursorType=adOpenKeyset
RS.LockType=adLockOptimistic
RS.Open FormName, DB, , adCmdTable
```

The code above establishes new data form for the current acquisition. Add the following code to the Click event of stop acquire command button:

```
RS.close:Set RS=Nnothing
Set RS=New ADODB.Recordset      'open record set for new data
RS.CursorType=adOpenKeyset
RS.LockType=adLockOptimistic
RS.Open "Acqirement record", DB, , adCmdTable
RS.AddNew
RS("Start time")=StartTime 'record acquisition information
RS("Form name")=FormName
RS("Finish time")=Format(Time, "HH:MM:SS")
RS.Update
RS.Close
Set RS=Nothing
```

The code above is used to establishes acquisition information record table. DB in the above code is the database link (Private DB as adodb.connection). Add the following code to the Form_Load event to establish database link:

```
Set DB=New ADODB.Connection                'set database link
DB.CursorLocation=adUseClient
DB.Open "Provider=microsoft.jet.oledb.3.51;" & "data
source=Database name"
```

Add the following code in the acquisition timer for storage of each acquisition data node:

```
RS.AddNew
RS("Field name")=field value               'record information
...
RS.Update
```

3 Implementation of Data Curve

VB6.0 offers controls and methods for graph and image processing. The PictureBox control is used to display data curves in this paper. There are 18 parameters to be monitored in this system and among which, there are 12 parameters need to be displayed in curves. So, a control array including 13 PictureBoxes, pic(0) to pic(12), is added on the control panel Form. Some properties of these PectureBoxes are set as follows: BackColor is set to vbBlack, AutoRedraw is set to True and ScaleMode is set to 0-User. The Visible property of pic(0) is set to True. It is used as main screen to display background grids and draw curves. The Visible properties of pic(1) to pic(12) are set to False. These PictureBoxes are used to draw corresponding parameter curves. The control array of checkboxes on the Form are used to control which parameter curves should be displayed on the main screen pic(0). Explanations for the module level variables used in display process are as follows:

nW: nW=pic(1).ScaleWidth, nH: nH=pic(1).ScaleHeight, pNumber is the number of points to displayed on one screen, xSpace is X-coordinate space between two adjacent points, xSpace=nW/pNum, G(1 to 12) and B(1 to 12) are arrays used for curves scaling, CurveSelector(1 to 12) is the array used for curves display selection information.

The following code actualizes curve display:

```
Private Sub ShowWave(ByVal data As cNode)        ' current point
  Static N As Integer                          'current point number
  Static lx(1 to 12) As Integer                    'last point array
  Static ly(1 to 12) As Single
  Static llx(1 to 12) As Single
  Static lly(1 to 12) As Single
  Dim cy(1 to 12) As Single                    'current point array
  Dim cx(1 to 12) As Single
  Dim i As Integer
  For i=1 to 12                  'read current point and transform
   cy(i)=B(i) - G(i) * data.Item(i)
  Next i
  If N < pNumber Then              'curve does not reach right
boundary, do not need curve translation
   N=N + 1
   For i=1 to 12                'set x-coordinate of current point
    cx(i)=N * xSpace
```

```
    Next i
    For i=1 to 12        'draw line connecting last point and
  current point using different colors
     pic(i).Line (lx(i),ly(i))-(cx(i),cy(i)),QBColor(i + 9)
     If N < pNumber - 1 Then
       pic(i).Circle (cx(i), cy(i)), 2, QBColor(i + 9)
         'draw a circle flag on point
     End If
    Next i
    If N=pNumber - 1 Then
     For i=1 To 4
      llx(i)=cx(i): lly(i)=cy(i)
     Next i
    End If
Else 'curve reaches right boundary, and makes left translation
 For i=1 to 12
   BitBlt pic(i).hdc, 0, 0, pic(1).ScaleWidth - xSpace, nH,
  pic(i).hdc, xSpace, 0, vbSrcCopy
   pic(i).Line (nW, ly(i))-(llx(i), lly(i)),
  pic(i).BackColor
     'erase the last section of the curve using backcolor
  pic(i).PSet (llx(i), lly(i)), pic(i).BackColor
     'erase the crossing point
  pic(i).Line (nW - xSpace,ly(i))-(nW,cy(i)),QBColor(i + 9)
     'draw a new line
  pic(i).Circle (nW-2 * xPace, lly(i)),2,QBColor(i + 9)
     'draw a circle flag on point
 Next i
 For i=1 to 12
  llx(i)=nW - xSpace
  lly(i)=ly(i)
 Next i
End If
   For i=1 to 12                'set coordinate of the last point
     lx(i)=cx(i): ly(i)=cy(i)
```

```
  Next
  pic(0).Cls
    'the main picture box cls and ready for new pictures
  For i=1 to 12
   If Curveselector(i)=1 Then    'display the selected curves
    BitBlt pic(0).hdc,0,0,nW,nH,pic(i).hdc,0,0, vbSrcPaint
   End If
  Next i
End Sub
```

Fig.2 shows the interface of the monitoring and control system.

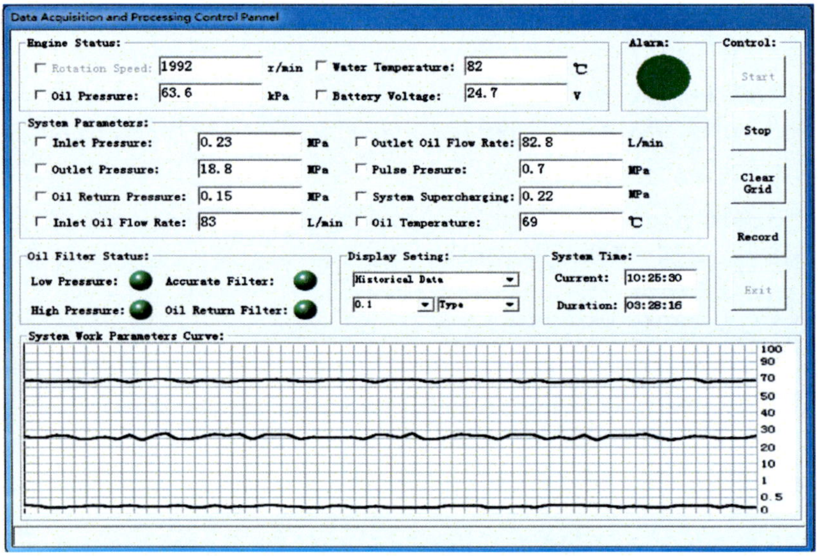

Fig. 2. Data acquisition and data curve display interface

Summary

According to the dynamic link data storage and data curve implementation method introduced in this paper, multi-curves dynamic display and zoom are actualized. This method has successfully applied to the working condition monitoring and fault diagnosis system of aircraft ground oil pump vehicle and proved effective. This method has broad applicability and fits for data storage and data curve dynamic display in various monitoring and control software.

References

1. Su, B.H.: Remote monitoring and control system based on socket technology used in coal mine. Industry and Mine Automation, China (2003)
2. Guo, J.H.: The realization of real-time communication between PC and PLC based on VB6.0. Mine Machinery, p. 56 (March 2006)
3. Gong, P.Z.: A simple tutorial of visual basic programming. High Education Press, China (2003)
4. Yuan, H.Y.: A new real-time ECG detection algorithm. Journal of Electronic Measurement and Instrument 23(9), 79–83 (2009)

Vibration Analysis of a Certain Type of Aero-engine Turbine Blade Based on UG

Weiqiang Zhao[1], Yongxian Liu[2], Mowu Lu[1], and Qingjun Guo[1]

[1] Engineering Training Center, Shenyang Aerospace University, Shenyang Liaoning, 110136, China
[2] School of Mechanical Engineering & Automation, Northeastern University, Shenyang Liaoning, 110004, China
zwq7618@yahoo.com.cn, yxliu@mail.neu.edu.cn, lumowu@163.com, gqj81@tom.com

Abstract. This paper introduces the FEA method for a certain type of aero-engine turbine blade and makes a vibration characteristics analysis to this aero-engine turbine blade based on this method. The vibration characteristic of this aero-engine turbine blade is studied and the natural modal of the turbine blade is calculated based on UG software. The first six natural frequencies and mode shapes are given. According to the analysis results the dynamic characteristics of the blade are discussed. The analysis method and results in this paper can be used for further study on optimal design and vibration safety verification for the blade.

Keywords: Turbine blade, Finite element method, Modal analysis.

1 Introduction

Blades are important components in aero-engine, and they are easily damaged because they work under a condition of high centrifugal loads, aerodynamic loads and alternate loads caused by vibration when working. According to statistics, 60 percent aero-engine failures are vibration failure and 70 percent vibration failures are blades vibration failure[1-3].

Vibration fatigue damage of blade is an seriously problem for aero-engine. Thus, in order to avoid vibration failure, the vibration natural characteristic analysis including natural frequencies analysis, vibration modes analysis and vibration stress distribution using finite element method (FEM) is necessary for blade design and developing process.

In this paper, the 3-demention model of a certain type of aero-engine turbine blade is established using UGNX/CAD module, and the vibration characteristics of this blade is analyzed using UGNX/CAE module. The natural frequencies and corresponding vibration modes of this blade are obtained and also the vibration characteristics are analyzed. The analysis results are helpful to make further study on optimal design and vibration safety verification.

D. Yang (Ed.): Informatics in Control, Automation and Robotics, Volume 2, LNEE 133, pp. 377–382.
springerlink.com

2 FEA Model for Turbine Blade

2.1 FEA Equations of Modal Analysis for Turbine Blade

The purpose of modal analysis is to calculate natural frequencies and vibration modes, and this problem comes down to eigen value and eigen vector problem. The FEM basic equation of dynamics problem for non damping free vibration is:

$$M\ddot{\delta}(t)+C\dot{\delta}(t)+K\delta(t)=P_f(t)\cdot \tag{1}$$

Let the damping force and the excitation force in eq.1 be zero:

$$M\ddot{\delta}(t)+K\delta(t)=0 \; . \tag{2}$$

Free vibration of any elastic body can be resolved into a superposition of a set of harmonic vibrations. Let the harmonic vibrations solution of eq.2 be:

$$\delta(t) = \delta_0 \sin \varpi t \; . \tag{3}$$

Substituting eq.3 into the basic equation of free vibration, we can get eq.4:

$$(K-\varpi^2 M)\delta_0 = 0 \; . \tag{4}$$

Since amplitudes of nodes in structure δ_0 are not all zero, value of coefficients determinant must be zero:

$$|(K-\varpi^2 M)| = 0 \; . \tag{5}$$

Since the stiffness matrix K mass matrix M are n order matrix, eq.5 is a n order equation of ϖ^2. The natural frequencies are generalized eigenvalue and can be solved by Rayleigh quotient method:

$$\varpi_i = \frac{\delta_{0i}^T K \delta_{0i}}{\delta_{0i}^T M \delta_{0i}} \; . \tag{6}$$

Natural frequencies:

$$f_i = \frac{\varpi_i}{2\pi} \; . \tag{7}$$

2.2 Solid Model of Turbine Blade

This blade is the first level turbine blade of a certain type of aero-engine. The blade body is torsional curved surface and the blade root is firry tenon. The solid model is set up using UG. UG software has effective solid modeling function and the UG/CAE module can use the solid model for FEA analysis without any transformation.

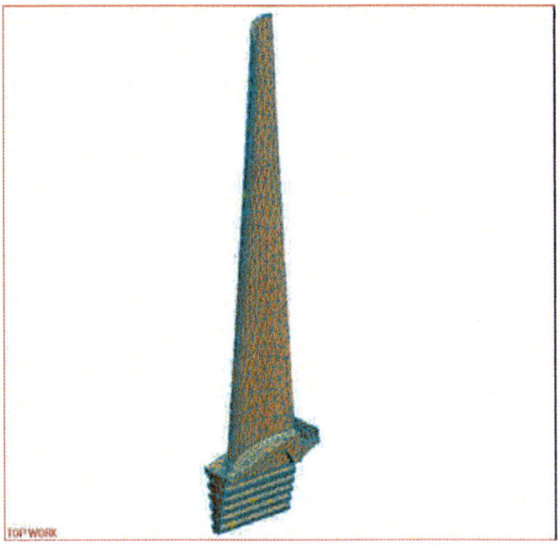

Fig. 1. FEA mesh of turbine blade

2.3 Material and Physical Characteristic Parameters of Turbine Blade

The blade material is GH4049, elastic modules is E=224GPa, poisson ratio is 0.295 and density is 8.44g/cm^3.

2.4 FEA Element and Meshing

The UG/CAE module supports 4-nodes and 10-nodes tetrahedron elements. In this paper, 10-nodes tetrahedron element is used and the global size is set to 5 according to the recommended value of the automatic element sizing function. This global size is a parameter which controls the edge length of tetrahedron element in the whole solid.

UG/CAE module supports three solid meshing algorithms, Facet, Parametric and Sequence. Facet algorithm is a flexible meshing algorithm and generates high quality grid for almost all geometry. Parametric algorithm will be used when Facet algorithm fails. While Sequence algorithm is a synthesis of the two algorithms above. This algorithm uses Facet algorithm fist and will start up Parametric algorithm automatically if the Facet algorithm fails. Sequence algorithm is default option.

In UG/CAE module, there are three methods controlling the mapping of nodes in 10-nodes element to geometry, they are Mixed method, Curved method and linear method. Mixed method maps element middle nodes whose Jacobi value less than critical value to geometry. Curved method maps all middle nodes to geometry without regard to the influence to element mass. While Linear method maps all middle nodes to a line connecting the two points.

In this paper, Mixed method is used to control element quality and solid element is used for the whole model to improve analysis accuracy. Meanwhile, swept mesh with different density is used for the blade body and the blade root respectively to improve analysis efficiency.

There are 8893 nodes and 15926 elements in the whole blade. The finished FEA meshing result is shown in Fig.1.

2.5 Boundary Conditions and Solving Method

The blade root is a firry tenon. The clamped plane usually sets on the root section on the first tongue of the firry tenon. In this paper, the boundary condition is that the blade is cantilevered clamped on the pitch line of the first tongue. The solving method uses Structures P.E solver to make modal analysis based on Lanczos eigenvalue extraction algorithm.

3 Calculation Results and Vibration Characteristics Analysis

3.1 Modal Analysis Results

UG/CAE module has effective post process function and outputs analysis summary and nodes displacements, stress and strain diagrams. Table.1 and Fig.2 to Fig.4 show the first six order natural frequencies and corresponding mode shapes.

3.2 Vibration Characteristics Analysis of the Bade

As shown in Fig.2 to Fig.4, the first order mode shape is first order circumferential bending vibration, the second order mode shape is first order axial bending vibration, the third to sixth order mode shape are first order torsional vibration, second order axial bending vibration, second order torsional vibration and second order axial bending vibration.

Table 1. Natural modal of turbine blade

Order	Frequency[Hz]	1st circumferential bending
1	4.557e+002	1st axial bending
2	1.359e+003	1st torsional
3	1.968e+003	2nd axial bending
4	2.032e+003	2nd torsional
5	3.500e+003	2nd circumferential bending
6	4.379e+003	1st circumferential bending

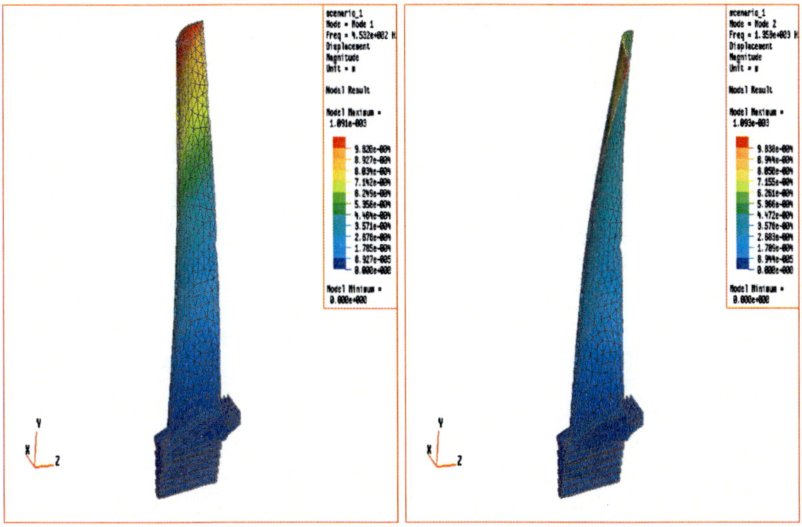

Fig. 2. The 1st and 2nd mode shape

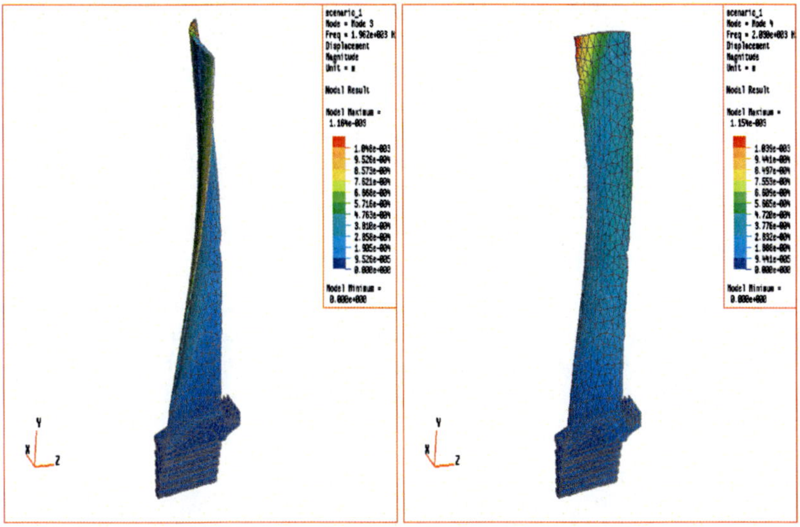

Fig. 3. The 3rd and 4th mode shape

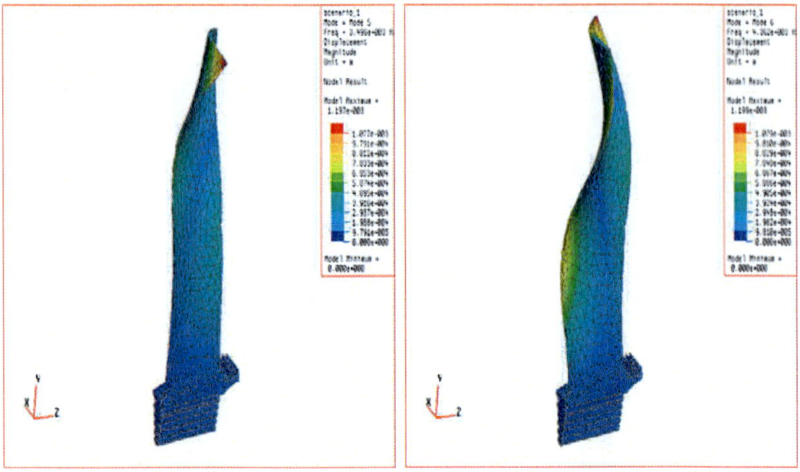

Fig. 4. The 5th and 6th mode shape

4 Summary

The UG software has effective solid modeling and FEA functions. Thus, it is very suit for solid modeling and FEA analysis for aero-engine turbine blade which has very complex geometry shape. In this paper, the FEA model of a certain type of aero-engine turbine blade is built and calculated based UG software. The first six natural frequencies and mode shapes are given. The analysis results can be used as reference for optimization design and vibration safeness verification for this turbine blade.

References

1. Zhang, J., Liu, X.P.: Principle and Numerical Methods of Modal Analysis to Turbo-machines, pp. 333–369. National Defence Industry Press, China (2001)
2. Tao, C.H., Zhong, P.D., Li, R.Z.: Failure Analysis and Prevention for Rotor in Aero-engine, pp. 102–163. National Defence Industry Press, China (2000)
3. Shlyannikov, V.N., Iltchenko, B.V., Stepanov, N.V.: Engineer Failure Analysis 8(5), 461–475 (2001)
4. Thomas, D.L.: Int. J. Num. Meth. Engineering 14(8), 561–564 (1979)
5. Shen, D.K.: Strength Calculation of Aero-engine, pp. 1–30. National Defence Industry Press, China (1990)
6. Zhang, J., Wang, W.L., Chen, X.J.: Acta Mechanica Solida Sinica (English Edition) 1(1), 61–70 (1988)
7. Zhou, C.Y.: Journal of Harbin Institute of Technology 12(10), 24–28 (2000)

Research and Implement of Distributed Network Motion Control System

ChongQuan Zhong[1,2], HaiBo Jin[1], YeChen Han[1], and ZhanMeng Wang[2]

[1] School of Electronics and Information Engineering, Dalian Univ. of Technology, LiaoNing Dalian 116024, China
[2] DUT Computer Control Engineering CO, LTD, LiaoNing Dalian 116045, China
zhongcq@dlut.edu.cn

Abstract. In order to control the motion of multi-axis motor in robotic arm, high precision clock synchronization technology based on PTP of IEEE1588 and NURBS interpolation algorithm are realized in distributed network motion control system we designed which is composed with industrial computer, embedded controller, ac servo driver and motor, industrial Ethernet and robotic arm. Clock synchronization technology whose principle is to exchange messages consecutively to determine the offset between controllers is used for controlling synchronization between multi-axis motors. NURBS interpolation algorithm is used for controlling motion track of each single motor. Finally, experiment on precision of clock synchronization and interpolation algorithm is conducted, experiment result shows that the error of clock synchronization is no more than 1us and the error of interpolation is no more than 3um.

Keywords: high precision clock synchronization, NURBS interpolation algorithm, distributed network motion control systems.

1 Introduction

Motion control system, which is essentially electrical system achieving the preconceived track by controlling motor's torque, speed and angle[1], has extensive application in traditional industry such as CNC machining, robot and high speed printing. In motion control system, there are three basic types of motion controllers, namely computer standard bus-based motion controller, soft open motion controller and embedded motion controller[2]. Computer standard bus-based motion controller has open architecture and is independent of computer. This kind of controller using DSP or microprocessor chip as CPU can achieve motion planning, high-speed and real-time interpolation, servo driving and can also be used as standard common interface between I/O device. Soft open motion controller, as another type controller, provides maximum flexibility to user, because software of controller which is installed in computer entirely is independent of hardware only as standard interface between computer and servo driver. User can develop required control function using open motion control kernel built in operating system. Embedded motion controller can run independently because its control instructions are stored in controller itself through computer's downloading and it can connect to superior computer or a control

D. Yang (Ed.): Informatics in Control, Automation and Robotics, Volume 2, LNEE 133, pp. 383–391.
springerlink.com

panel through communication interface such as Industrial Ethernet, RS485 and Profibus. In addition, embedded motion controller can be provided with hard disk drivers, even get remote diagnosis through the Internet. So the embedded motion controller is more suitable to multi-axis motion controlling. With the mature of network technology, the network motion control system comprised with embedded controllers has become a trend. Industrial Ethernet has been applied to motion control system because of its low cost, high transmission speed, strong compatibility, which accelerates the development of distributed network motion control system. However, the Ethernet CSMA/CD mechanism which leads to synchronous precision cannot meet to requirement of motion control system. So how to improve synchronous precision become important and urgent.

In this paper, distributed network motion control system is designed and in this system two key technologies which are clock synchronization for multi-axis motor and interpolation algorithm for each single-axis motor are discussed deeply. The rest of this paper is organized as fallows. Section 2 describes the network motion control system for multi-axes motion controlling our designed. Section 3 discusses the high precision clock synchronization for synchronizing multi-axis motors. Section 4 discusses the NURBS interpolation algorithm for controlling each single motor's track. Section 5 is experiment and result analyzing. Section 6 is conclusion.

2 Network Control System

Our designed control system including industrial control computer, embedded motion controllers (a kind of PLC), servo drivers is essentially distributed network integration control system as shown in Fig. 1.

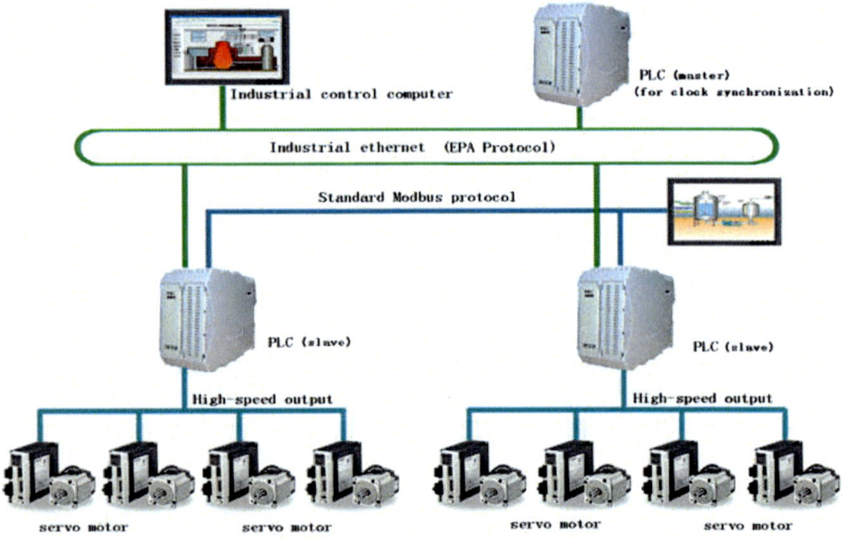

Fig. 1. Structure of distributed network control system

In this system, the industrial computer's main work is (1) monitoring running state of the whole system. (2) Generating motion track according to graphic in AutoCAD and communicating with PLC including download, upload and diagnosis. Embedded motion controllers we used are programmable logic controllers (PLC) which has four channels high-speed output ports. Master PLC is responsible for synchronizing time for each PLC clock and slave PLC is responsible for computing single-axis motor's track according to NURBS interpolation algorithm and output pulse to ac servo motor. Ac servo motor are used to converting pulse to actual displacement of the robotic arm. The connection between industrial control computer, master PLC and slave PLC is EPA protocol fieldbus which is stable industrial Ethernet.

3 Clock Synchronizing

There are two kinds of clock synchronization protocol frequently-used, which are network time protocol (NTP) and precision time protocol (PTP)[3] defined by IEEE 1588. Compared with NTP, PTP provides a standard method to synchronize devices on a network with sub microsecond precision and ensures that events and timestamps in all devices use the same time base[4]. PTP is optimized for user-administered, distributed systems, minimal use of network bandwidth and low processing overhead. So in this paper, the clock synchronization between master and slave realized in this system is based on PTP whose principle is to exchange messages consecutively to determine the offset between master and slave but the message transit delay exists through the network (see figure 2).

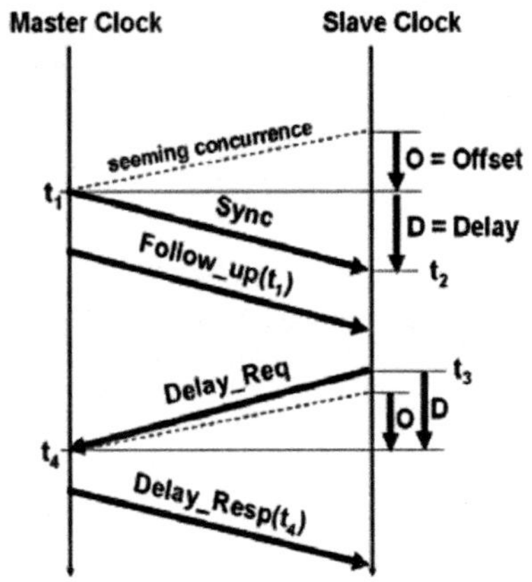

Fig. 2. Message exchange between master and slave

In order to correct offset, the master cyclically transmits Sync messages to the slave clock at defined intervals. A time stamping mechanism determines the exact transmission time t_1 as precise as possible and sends it down to the slave on behalf of a second message called the Follow_up message. The slave clock measures the exact reception time t_2 of the Sync message. On reception of the Sync and the corresponding Follow_up message, the slave clock calculates the offset correction in relation to the master clock. The deviation between master and slave is the transit delay the Sync message has experienced when traveling through the network. Furthermore, consecutive Sync measurements allow the slave's frequency drift to be compensated. Another message exchange is required to measure the delay between slave and master. For this purpose, the slave clock sends a so-called Delay_Req packet to the master. The exact transmission and reception time t_3 and t_4 of this message are measured. The Delay_Resp message carries the measured value t4 to the slave, where now delay time and offset time can be calculated through equations.

$$\begin{cases} T_{delay} + T_{offset} = t_2 - t_1 \\ T_{delay} - T_{offset} = t_4 - t_3 \end{cases} \tag{1}$$

Calculating equations (1), we can get:

$$\begin{cases} T_{delay} = \dfrac{(t_2 - t_1) + (t_4 - t_3)}{2} \\ T_{offset} = \dfrac{(t_2 - t_1) - (t_4 - t_3)}{2} \end{cases} \tag{2}$$

Clock offset comprises static offset which caused by different starting time and dynamic offset which caused by crystal oscillation. Static offset is invariant, but it is note that the dynamic offset will be accumulated linearly with the passage of time. So the offset exhibits linear variation with the passage of time approximately. Therefore, we establish linear module to clock offset:

$$T_{offset} = kt + \mathrm{b} \tag{3}$$

Where, k is coefficient of crystal oscillation offset, b is invariant offset.

In order to calculate k, we need to measure clock offset twice.

$$\begin{cases} T_{offset1} = kt_1 + \mathrm{b} \\ T_{offset2} = kt_2 + \mathrm{b} \end{cases} \tag{4}$$

Calculating equations (4), we can get:

$$k = \frac{T_{offset2} - T_{offset1}}{t_2 - t_1} \tag{5}$$

Synchronizing period is 1s, namely $t_2 - t_1 = 1$, so equation (5) is simplified as:

$$k = T_{offset2} - T_{offset1} \tag{6}$$

The unit of system timer clock is 1us, so crystal oscillation offset k is equal to difference of master and slave's counting in 1s. The clock synchronizing offset of PTP adopted in this system is no more than 5 us.

4 NURBS Interpolation Algorithm

In each single-axis motor's track controlling, track precision is determined by interpolation algorithm directly. In our system, NURBS interpolation algorithm[5] is adopted, because NURBS has various useful properties such as smoothness and the possibility of local modifications, which facilitate the representation of general freeform curve.

A NURBS curve is defined by:

$$P(u) = \frac{\sum_{i=0}^{n} w_i d_i N_{i,k}(u)}{\sum_{i=0}^{n} w_i N_{i,k}(u)} \tag{7}$$

Where, $\{w_i\}$ are the weights , the $\{d_i\}$ are the control points, and $\{N_{i,p}(u)\}$ are the normalized B-spline basis functions of p degree. The weight w_i determines the influence of the ith control vector d_i on the curve. The ith basis function $N_{i,p}(u)$ defined on a knot vector $U = \{u_0, u_1 \cdots u_{n+k+1}\}$ is recursively defined by

$$N_{i,k}(u) = \frac{u - u_i}{u_{i+p} - u_i} N_{i,k-1}(u) + \frac{u_{i+p+1} - u}{u_{i+p+1} - u_{i+1}} N_{i+1,p-1}(u) \tag{8}$$

Where , $N_{i,0}(u) = \begin{cases} 1 & \text{if } u_i \le u \le u_{i+1} \\ 0 & \text{otherwise} \end{cases}$.

For simplify equation (7), all weights are set to 1. Equation (7) is simplified:

$$P(u) = \sum_{i=0}^{n} d_i N_{i,k}(u) \tag{9}$$

To determine a NURBS curve interpolation (m+1) data points $\{Q_k\}$, it suffices to have (m+1) independent conditions. In this paper, we assign a parameter value $\bar{u}_k$ to each data point, and select a knot vector U and a weight vector W, then we set up the curve interpolation problem[6][7] as

$$Q_k = C(\bar{u}_k) = \sum_{i=0}^{m} P_i R_{i,p}(\bar{u}_k) \qquad for\, k = 0,\cdots,m \tag{10}$$

Where the rational functions $\{R_{i,p}(u)\}$ are defined by

$$R_{i,p}(u) = \frac{w_i N_{i,k}(u)}{\sum_{i=0}^{m} w_i N_{i,k}(u)}$$

Now, the core problem is how to determine parameter value $\bar{u}_k$. In this paper, we adopt Centripetal method:

$$\begin{cases} \bar{u}_0 = 0, \quad \bar{u}_n = 1 \\ \bar{u}_i = \bar{u}_{i-1} + \dfrac{|Q_i - Q_{i-1}|^{\frac{1}{2}}}{\sum_{j=0}^{n} |Q_j - Q_{j-1}|^{\frac{1}{2}}} \end{cases} \tag{11}$$

For the Centripetal method, the knot vector U is defined by

$$U = \{0,0,\cdots 0, u_{k+1}, \cdots u_n, 1,1,\cdots 1\} \tag{12}$$

Where $u_{j+k} = \frac{1}{k}\sum_{i=j}^{j+p-1} \bar{u}_i \quad for\; j = 1,\cdots n-k$.

Using Eqs. (7)-(11), the set of track points $\{P_i\}$ can easily be calculated from the data points $\{Q_k\}$.

5 Experiment on Synchronization and Interpolation Precision

In experiment, the industrial control computer and PLC in our system are programmable automation controller (PAC) and DEM3000 respectively. The control object is robotic arm which has 4-Degree of freedom (DOF). All of above are shown in Fig.3.

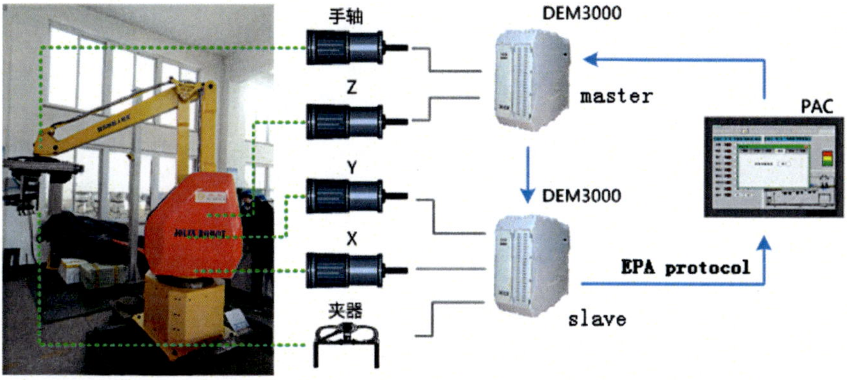

Fig. 3. Realistic controller and robotic arm adopted in experiment

In clock synchronization experiment, synchronization time is triggered by timer in each DEM3000. Timer was set to be triggered at different 10 times and triggered time was recorded in PAC as shown in Table 1. The first column data is regular time and second is triggered time, the third is triggered error.

Table 1. Triggered time for synchronization

Regular time	Triggered time	Trigger error
500000	500000.046	0.046
1500000	1500000.077	0.077
3000000	2999999.986	-0.014
5000000	5000000.046	0.046
7500000	7500000.001	0.001
10500000	10500000.020	0.020
14000000	13999999.970	-0.030
18000000	18000000.060	0.060
22000000	22000000.000	0.000
27500000	27500000.020	0.020

The precision of triggered time is no more than 0.1us in view of the third column data in Tab.1.

Next, we experiment on synchronized time between master and slave controller based on trigger mechanism by two-channel DS5062 oscillograph with 60M bandwidth in each channel. Experiment result is shown as Fig. 4.

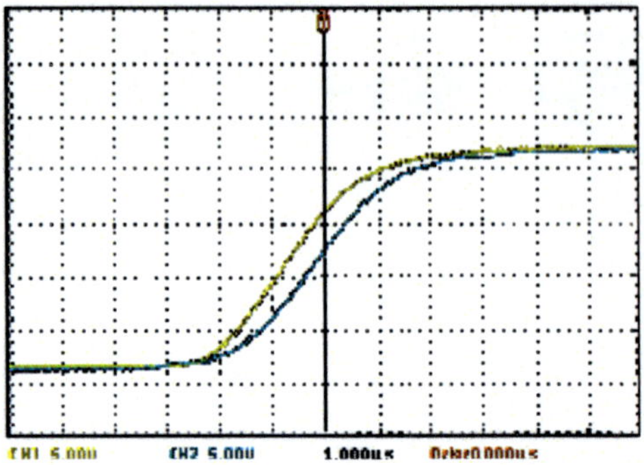

Fig. 4. Synchronized time between master and slave controller

The precision of synchronization is no more than 1us in view of Fig.4.

In interpolation experiment, we construct a free curve with 300 points which are motion track of robotic arm and then fit it into B-spine curve. Stepping parameters Δu are set to 0.001, pulse equivalence is set to 0.001 and motion speed is 1000 mm/sec. error limited is 0.005mm. Motion track includes line, arch, circle, ellipse, spline, solid, trace and polyline. Experiment result about error of interpolation algorithm is shown in table 2.

Table 2. The error of interpolation algorithm

GE type	Error (pulse number)	GE type	Error (pulse number)
LINE	± 1	SPLINE	± 1
ARC/CIRCLE	± 1	SOLID/TRACE	± 3
ELLIPSE	± 1	POLYLINE	± 3

Experiment result shows that interpolation error, on simple graphic such as line, arc and ellipse is no more than 1 um, on complex graphic such as solid, trace and polyline is no more than 3um, as view from these errors.

6 Conclusions

Distributed network motion control system was designed using industrial computer, motion controller, ac servos motors. In this system, each controller's clock was synchronized precisely by high-precision clock synchronization based on PTP of IEEE 1588 so that all motors were able to run simultaneously. Therefore, the synchronization control of multi-motors was solved and the automation, efficiency and robustness in this system were also enhanced. The track of robotic arm was generated according to given graphic by NURBS interpolation algorithm. Finally experiments on precision of clock synchronization and interpolation algorithm were conducted. Experiment result show that the error of clock synchronization is no more than 1us, the error of interpolation algorithm is no more than 3 um, which both can meet the requirement of industry robot.

References

1. Xi, Z., Zhou, H.: Development and status of motion controller. Electrical Drive Automation 27(3), 5–10 (2005)
2. Xi, C., Yanbo, C., Cheng, K.W.E.: Motion Controller and the Application of PMAC in AC Servo CNC System. In: International Conference on Power Electronics System and Application (2009)
3. IEEE Std. 1588-2002, IEEE Standard for a Precision Clock Synchronization Protocol for Networked Measurement and Control Systems
4. Neagoe, T., Cristea, V., Banica, L.: NTP versus PTP in Computer Networks Clock Synchronization. Industrial Electronics 01, 317–362 (2006)
5. Jung, H.B., Kim, K.: A New Parameterisation Method for NURBS Surface Interpolation. Int. J. Adv. Manuf. Technol. 16, 784–790 (2000)
6. Liu, B.: Degree reduction of NURBS curves based on genetic algorithm. Computer Engineering 34(14), 194–196 (2008)
7. Dong, Y.: Research on approaching and NC machining technology based on NURBS surface. North China Electric Power University, 20–29 (2007)

On the Optimization of Dijkstra's Algorithm

Seifedine Kadry, Ayman Abdallah, and Chibli Joumaa

American University of the Middle East, Egaila, Kuwait
{Seifedine.kadry,Ayman.Abdallah}@aum.edu.kw,
Chibli.joumaa@acm.edu.kw

Abstract. In this paper, we propose some amendment on Dijkstra's algorithm in order to optimize it by reducing the number of iterations. The main idea is to solve the problem where more than one node satisfies the condition of the second step in the traditional Dijkstra's algorithm. After application of the proposed modifications, the maximum number of iterations of Dijkstra's algorithm is less than the number of the graph's nodes.

Keywords: Dijkstra's algorithm, directed graph, shortest path.

1 Introduction

The shortest-route problem determines a route of minimum weight connecting two specified vertices, source and destination, in a weight graph (digraph) in a transportation network. Other situations can be represented by the same model like VLSI design, equipment replacement and others. There are different types of shortest path algorithm [5] to find the shortest path of any graph. Most frequently encountered are the following:

- Shortest path between two specified vertices
- Shortest path between all pairs of vertices
- Shortest path from a specified vertex to all others

The most efficient algorithm used to find the shortest path between two known vertices is Dijkstra's algorithm [1]. Some improvements on Dijkstra algorithm are done in terms of efficient implementation [3] and cost matrix [4]. In this paper, we propose some improvement in order to reduce the number of iterations and to find easily and quickly the shortest path.

2 Dijkstra's Algorithm

The problem of finding the shortest path from a specified vertex s to mother t can be stated as follows:

A simple weighted digraph G of n vertices is described by a n by n matrix $D=[d_{ij}]$, where, d_{ij} = length (or distance or weight) of the directed edge from vertex i to vertex j:

D. Yang (Ed.): Informatics in Control, Automation and Robotics, Volume 2, LNEE 133, pp. 393–397.
springerlink.com

$$d_{ij} = \begin{cases} > 0, \ if\ i \neq j \\ = 0, if\ i = j \\ = \infty\ (large\ number), if\ there\ is\ no\ edge\ from\ i\ to\ j \end{cases}$$

Dijkstra's algorithm labels the vertices *of* the given digraph, At each stage in the algorithm some vertices have permanent labels and others temporary labels. The algorithm begins by assigning a permanent label 0 to the starting vertex *s,* and temporary label infinity to the remaining n-1 vertices. From then on in each iteration, another vertex sets a permanent label, according to the following rules:

a. Every vertex j that is not yet permanently labeled gets a new temporary label whose value is given by min[old label of j, (old label of i + d_{ij})], where i is the latest vertex permanently labeled, in the previous iteration, and d_{ij} is the direct distance between vertices i and j. If i and j are not joined by an edge, then d_{ij}=infinity.

b. The smallest value among all the temporary labels is found, and this becomes the permanent label of the corresponding vertex. In case of more than one shortest path, select any one of the candidates for permanent labeling. Steps a and b are repeated alternately until the destination vertex t gets a permanent label. The first vertex to be permanently labeled is at a distance of zero from s. The second vertex to get a permanent label (out of the remaining n-1 vertices) is the vertex closest to s From the remaining n-2 vertices, the next one to be permanently labeled is the second closest vertex to s. And so on. The permanent label of each vertex is the shortest distance of that vertex from s. This statement *can* be proved by induction.

3 Modified Dijkstra's Algorithm

In this section, we will present a modified version of Dijkstra's Algorithm in order to reduce the total number of iteration by optimized the situation of many shortest paths:

a. Every vertex j that is not yet permanently labeled gets a new temporary label whose value is given by min[old label of j, (old label of i + d_{ij})], where i is the latest vertex permanently labeled, in the previous iteration, and d_{ij} is the direct distance between vertices i and j. If i and j are not joined by an edge, then d_{ij}=infinity [6].
b. The smallest value among all the temporary labels is found, and this becomes the permanent label of the corresponding vertex. In case of more than one shortest path, select all of them for permanent labeling.
c. Steps (a) and (b) are repeated alternately maximum n-1 times until the destination vertex t gets a permanent label.
d. Le T=[t_{ij}] is the shortest path matrix that we can build it form D, as follows

$$t_{ij} = \begin{cases} 0, if\ d_{ji} = 0\ or\ d_{ji} = \infty \\ d_{ji}, otherwise \end{cases}$$

4 Illustrative Example

The network in the following figure (figure 1) gives the distances in miles between pairs of cities 1,2,... and 8. For efficiency purpose, we will find the shortest route between cities 1 and 8 using the traditional and modified version of Dijkstra algorithm:

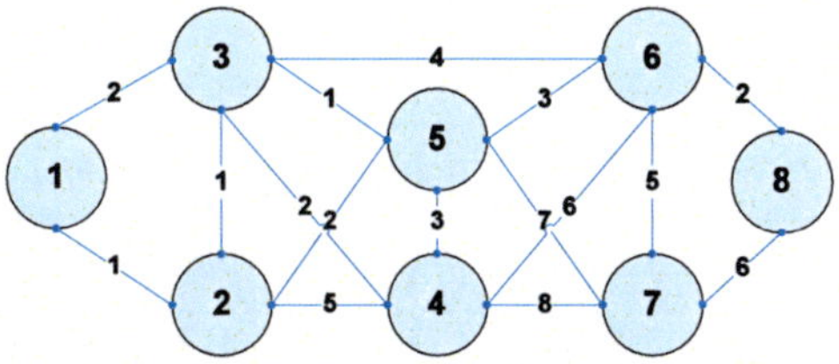

Fig. 1. Network of 8 nodes

Dijkstra algorithm

The number of iterations to find the shortest route between cities 1 and 8 by using Dijkstra algorithm is 8 (figure 2), this is the result of TORA [1] software:

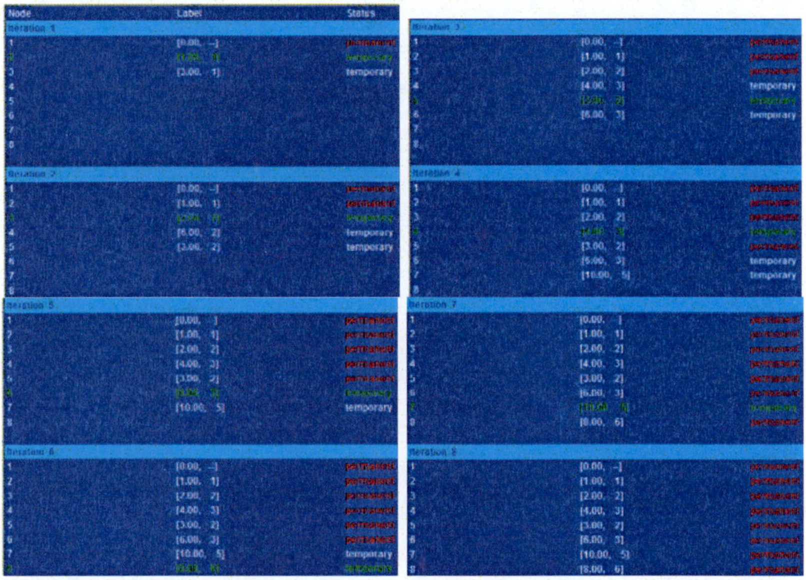

Fig. 2. Iterations 1 to 8 of Dijkstra algorithm

Modified Dijkstra algorithm

By using the modified version of Dijkstra algorithm, 5 iterations only are required to get the shortest route between cities 1 and 8:

Initially

$$D=\begin{bmatrix} 0 & 1 & 2 & \infty & \infty & \infty & \infty & \infty \\ 1 & 0 & 1 & 5 & 2 & \infty & \infty & \infty \\ 2 & 1 & 0 & 2 & 1 & 4 & \infty & \infty \\ \infty & 5 & 2 & 0 & 3 & 6 & 8 & \infty \\ \infty & 2 & 1 & 3 & 0 & 3 & 7 & \infty \\ \infty & \infty & 4 & 6 & 3 & 0 & 5 & 2 \\ \infty & \infty & \infty & 8 & 7 & 5 & 0 & 6 \\ \infty & \infty & \infty & \infty & \infty & 2 & 6 & 0 \end{bmatrix}$$

and the set of permanent vertices is P={1}.

Iteration 1:	$P=\{1,2\}, d_{1j}=\{0,1,2,4,\infty,\infty,\infty,\infty\}$
Iteration 2:	$P=\{1,2,3\}, d_{1j}=\{0,1,2,4,3,6,\infty,\infty\}$
Iteration 3:	$P=\{1,2,3,5\}, d_{1j}=\{0,1,2,4,3,6,10,\infty\}$
Iteration 4:	$P=\{1,2,3,4,5,6\}, d_{1j}=\{0,1,2,4,3,6,10,8\}$
Iteration 5:	$P=\{1,2,3,4,5,6,7,8\}, d_{1j}=\{0,1,2,4,3,6,10,8\}$

$$T=\begin{bmatrix} 0 & 1 & 0 & 0 & 0 & 0 & 0 & 0 \\ 0 & 0 & 1 & 0 & 2 & 0 & 0 & 0 \\ 0 & 0 & 0 & 2 & 0 & 4 & 0 & 0 \\ 0 & 0 & 0 & 0 & 0 & 0 & 0 & 0 \\ 0 & 0 & 0 & 0 & 0 & 0 & 7 & 0 \\ 0 & 0 & 0 & 0 & 0 & 0 & 0 & 2 \\ 0 & 0 & 0 & 0 & 0 & 0 & 0 & 0 \\ 0 & 0 & 0 & 0 & 0 & 0 & 0 & 0 \end{bmatrix}$$

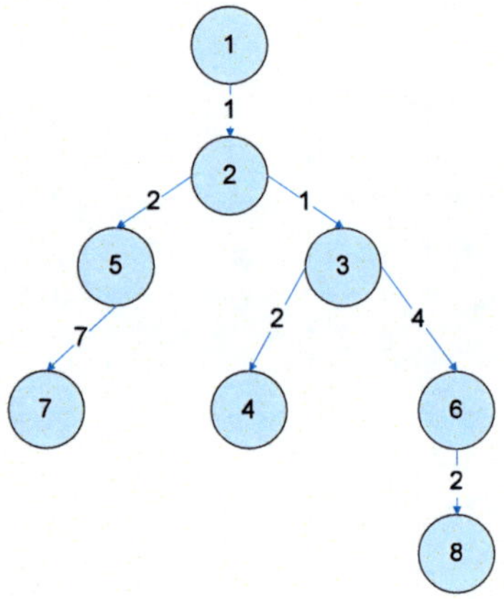

Fig. 3. Tree of shortest route

Based on the previous tree (figure 3), the shortest route and distance between cities 1 and 8 are: 1-2-3-6-8 and the distance is 8 miles.

5 Conclusion

In this paper an optimization of Dijkstra algorithm is presented. The main idea is to deal with the second stage of the traditional version of this algorithm where the shortest routes from a node to its successors are many and to reduce the number of iterations. The main feature of the proposed amendment is the remarkable reduction in the number of iterations and the ease in finding the shortest route from any city to any city. In the future work, we will implement the proposed modifications then to compare it with the existing one in terms of complexity.

References

1. Taha, H.: Operations research an introduction, 9th edn. Pearson publisher (2011)
2. Nar, D.: Graph theory with applications to engineering and computer science. Prentice Hall (1997)
3. Sneyers, J., Schrijvers, T., Demoen, B.: Dijkstra's Algorithm with Fibonacci Heaps: An Executable Description in CHR. In: 20th Workshop on Logic Programming (WLP 2006), Vienna, Austria (February 2006)
4. Brendan, H., Gordon, J.: Generalizing Dijkstra's Algorithm and Gaussian Elimination for Solving MDPs. In: Conference: International Conference on Automated Planning and Scheduling/Artificial Intelligence Planning Systems - ICAPS (AIPS), pp. 151–160 (2005)
5. Ravi, N., Sireesha, V.: Using Modified Dijkstra's Algorithm for Critical Path Method in a Project Network. International Journal of Computational and Applied Mathematics 5(2), 217–225 (2010)
6. Sudhakar, T.D., Vadivoo, N.S., Slochanal, S.M.R., Ravichandran, S.: Supply restoration in distribution networks using Dijkstra's algorithm. In: International Conference on Power System Technology (PowerCon 2004), Signapure (2004)

Design and Performance Study of Smart Antenna Systems for WIMAX Applications

Ayman Abdallah, Seifedine Kadry, and Chibli Joumaa

American University of the Middle East, Egaila, Kuwait
{Ayman.abdallah,Seifedine.kadry}@aum.edu.kw,
Chibli.joumaa@acm.edu.kw

Abstract. In this paper we propose an approach that uses homodyne receivers to design smart antenna systems. The receivers functions are to detect angles of arrivals of seven incoming RF signals using MUSIC or ESPRIT algorithms. The characteristics of each algorithm are critical for the system's precision as well as receivers types. Results are deduced from the simulation of each system, using the Advanced Design System (ADS) and MATLAB. These are compared to results deduced from real systems in the WIMAX (3.5GHz) domains.

Keywords: Smart Antennas, DOA, Spatial Smoothing, ESPRIT, MUSIC, Adaptive Antennas.

1 Introduction

The problem of localization of sources radiating energy by observing their signal received at spatially separated sensors is of considerable importance, occurring in many fields, including radar, sonar, mobile communications, radio astronomy, and seismology.[1]

Thus a variety of methods, using heterodyne or homodyne techniques, for the Detection Of Arrivals (DOA) estimation are used including spectra estimation, minimum-variance distortion less response estimator, linear prediction, maximum entropy, and maximum likelihood [2]. In addition to previous methods the most famous methods used in DOA are Eigen structure methods, including many versions of MUSIC algorithms [3] [9], minimum norm method [4], ESPRIT method [5] [10], and the weighted subspace fitting method.

The main goal of this paper is to provide a step by step methodology in simulation and in conception of smart antenna systems that use 6-port or 5-port reflectometers [6] [7] [8]. The purpose of the simulation is to compare the quality of the detected angles of both circuits, especially when the number of the incoming RF signals increases till seven. A comparison is also made to a smart antenna system that uses either ESPRIT or MUSIC algorithms. This paper consists of seven main sections.

Section 2 presents a general introduction of smart antennas. Section 3 explains the theory of the MUSIC [9] and ESPRIT [10] algorithms. Whereas section 4 describes

D. Yang (Ed.): Informatics in Control, Automation and Robotics, Volume 2, LNEE 133, pp. 399–406.
springerlink.com

the experimental results of the implementation of a smart antenna system using 5-ports reflectometer .Finally the conclusions are summarized in section 5.

2 Smart Antennas Systems

Smart antennas are essential parts of the new generation of mobile systems. They can be used to exploit spatial and spectral characteristics of the incoming signals to provide highly accurate location information [1]. Before implementing such systems, a simulation phase must be present in order to optimize its efficiency. A smart antennas system is made of two main parts which are the hardware part and signal processing part. The hardware part is responsible of capturing RF signals and converting them to low frequency (LF) signals. The signal processing part, section three, detects the complex envelopes of the LF signal and provides an estimation of the arrival angles. A physical smart antenna is composed of a set of receiving antenna injects the RF signals into Low Noise Amplifiers (LNA). The issued signals are down-converted to LF signals [6] as shown in Figure 1.

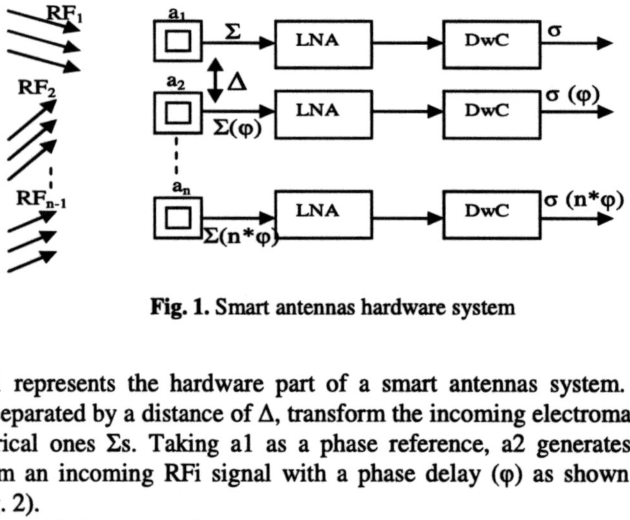

Fig. 1. Smart antennas hardware system

Figure1 represents the hardware part of a smart antennas system. "n" coplanar antennas separated by a distance of Δ, transform the incoming electromagnetic signals into electrical ones Σs. Taking a1 as a phase reference, a2 generates an electrical signal from an incoming RFi signal with a phase delay (φ) as shown in the figure below (fig. 2).

The electrical phase delay between two consecutive antennas φio may be computed as $\varphi_i = \frac{2\pi\Delta\sin(\theta_i)}{\lambda}$. θ_i Is the RF shift angle from the normal of the antennas, λ is the wave length of the central frequency RF.

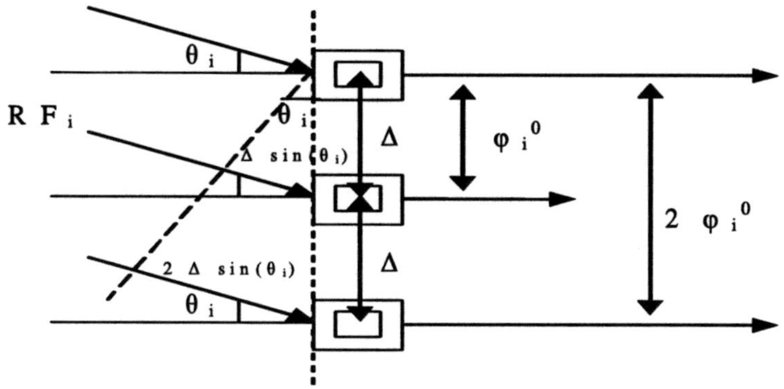

Fig. 2. Geometry structures in coplanar antennas

LNAs amplify RF signals and create an additional phase delay. In this paper, we will consider that the software phase delay calculations overcome the later problem by calibrating the down-converters.

Down converters are responsible of transforming RF signals into LF which are easy to manipulate by the software phase. Two major types of down-converter are used in smart antennas, Microwave mixers and Homodyne receivers. In this paper, we will use homodyne receivers and especially six-port reflectometers.

The signal processing phase is dedicated for the Detection of Arrivals DOA and Signal to Noise Ratios (SNR) determination. The used software programs can be applied in both of real and simulated systems. The smart antenna circuits, including antennas, low noise amplifiers LNA and homodyne receiver junctions are calibrated to provide the required data for the MUSIC and ESPRIT algorithms.

3 Smart Antenna in Simulation

In order to stimulate a smart antenna system we utilized the advanced design system (ADS) as a circuit simulation and Matlab tool as signal processor [11]. The characteristics of the smart antenna are derived from the following simulation.

3.1 Performance Comparison of MUSIC and ESPRIT

Music & Esprit utilize the $\frac{a_2}{a_1}$ ratio of each reflectometer (5or 6-port) in order to estimate the angle of arrivals. We simulated the smart antenna to detect the following theoretical AOA: 51.10, 19.50, 00, -19.50,-51.10. The result of the simulation is shown in the Table 1.

Table 1. DOA with MUSIC and ESPRIT

ESPRIT		MUSIC	
DOA	S dB	DOA	S dB
51.0	53.1	51.1	21.5
19.4	33.4	19.5	24.95
-0.0	32	0	25.9
-19.5	31.7	- 19.5	23.34
-51.0	30.4	-51.7	23.2

The above table demonstrates that we can use either MUSIC or ESPRIT algorithms to get a precise result. Note that the simulation was done in a noiseless environment.

3.2 Performance Comparison of Five and Six Port Reflectometers

In order to compare their performances in DOA, we simulated seven signals to be detected by a smart antenna using six port reflectometers and another one using five ports reflectometers. Both systems perform well as shown in the following figure (fig. 3).

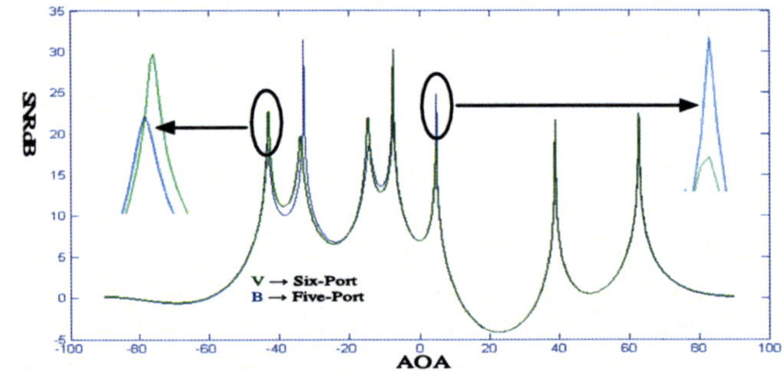

Fig. 3. DOA of seven RF signals

In the case of instability of the LO power of the five port reflectometer, the results become inaccurate, as shown in Figure 4.

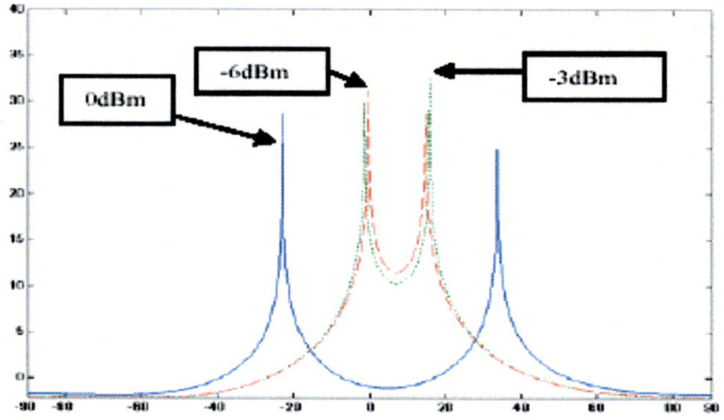

Fig. 4. DOA of 2 signals (-230 and 340) in instable five ports environments

4 Experimental Results

A feasibility study has been made by the conception of a smart antenna demonstrator as shown in figure 5 [12].

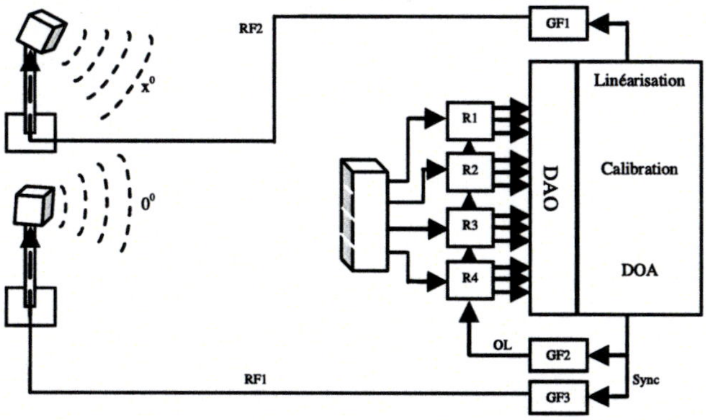

Fig. 5. Smart antenna system demonstrator

The smart antenna demonstrator uses 3 RF generators synchronized together representing the LO source and the RF sources conducted to the antenna emitters. The system is calibrated using Neveux's method.

4.1 Experimental Comparison of MUSIC and ESPRIT

In order to demonstrate the good performance of the DOA algorithms, MUSIC&ESPRIT, we utilized the complex envelopes provided by the demonstrator. We generated 2 RF signals of AOA (0° and 34°) uncorrelated in frequency, we deduced the good performance of our algorithm as shown in figure 6.

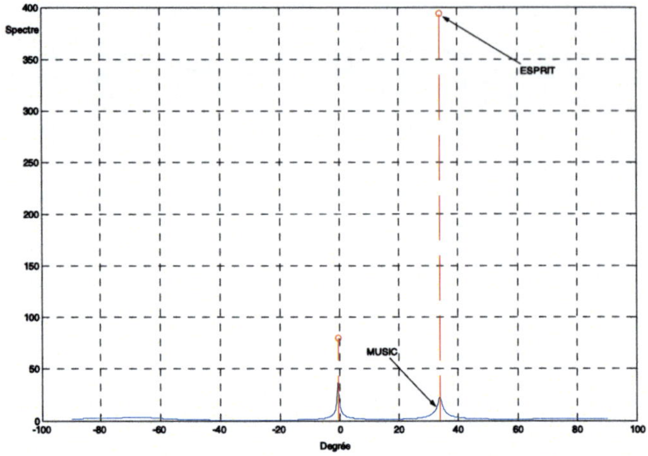

Fig. 6. DOA of 0° and 34° with MUSIC and ESPRIT

4.2 Noisy Environment

The second measurement is made to compare the efficiency of both algorithms in a noisy environment. As shown earlier, both algorithms have a good performance in a noiseless environment. Whereas, in a noisy environment (SNR decreased by 20dB from the initial experiment), we observed a degradation in the performance of the MUSIC algorithm, while the ESPRIT algorithm maintain an acceptable one, as shown in figure 7.

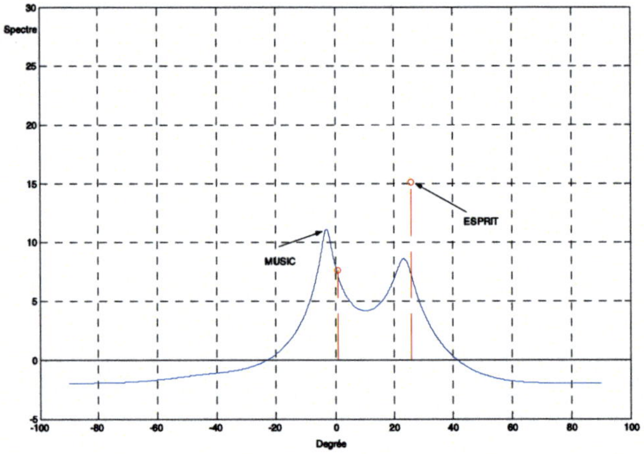

Fig. 7. Decrease of SNR by 20dB

4.3 Resolution

The resolution is calculated by a progressive reduction of the angular difference of the two AOA. In simulation both MUSIC and ESPRIT provide a 10 resolution. The experimental measurements showed that when we had angular separations that are below 50 both algorithm fail, as shown in figure 8.

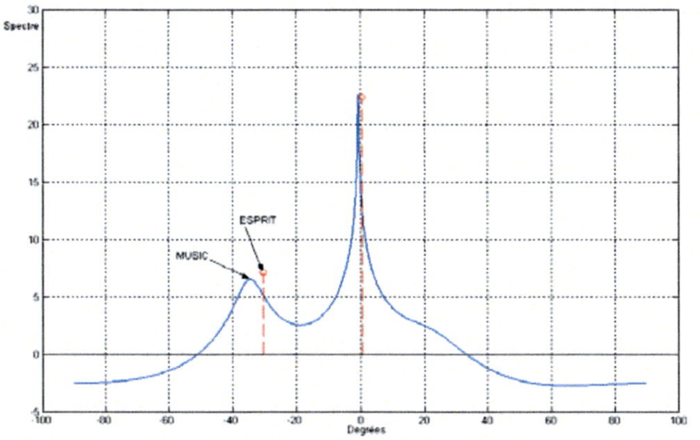

Fig. 8. DOA of 2 RF separated by less than 50

4.4 Multipath

To overcome the problem of multipath that is present in an environment with high level of signal correlations, we had to pre-treat the signals with a smoothing algorithm. In order to demonstrate the efficiency of this algorithm, we used 2 RF signals with 00 and 270 AOA respectively, having the major power in the 270 signal. The results provided by MUSIC & ESPRIT without preprocessing, demonstrates the failure of the system. The result provided by the system when applying the preprocessor algorithm is shown in the following figure (fig. 9).

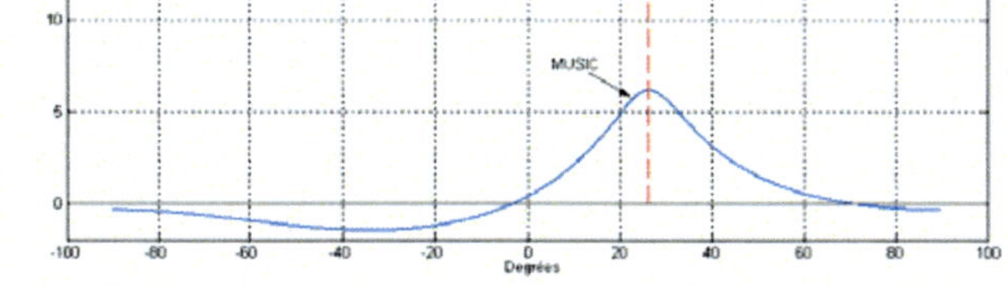

Fig. 9. Spatial smoothing for multipath rejection

The above figure shows that the spatial smoothing algorithm has eliminated the interferer RF signal and detected the desired signal.

5 Conclusion

This paper gives a clarified idea about simulation and design of smart antenna systems. A performance comparison of two receivers six port and five ports reflectometers was presented. A performance comparison for two DOA algorithms which are ESPRIT and MUSIC was also presented. The homogeneity of the results in simulations and in experiments was demonstrated in terms of accuracy, noise resolution and multipath.

References

1. Godara, L.C.: Application to antenna arrays to mobile communications. Part II: Beamforming and direction of arrival considerations. IEEE Proc. 85, 1195–1247 (1997)
2. Godora, L.C.: Smart Antennas. CRC Press LLC (2004)
3. Schmidt, R.O.: Multiple emitter location and signal parameter estimation. IEEE Trans. Antenna and Propag. 34, 276–280 (1986)
4. Buckley, K.M., Xu, X.L.: Spatial spectrum estimation in a location sector, IEEE Trans. Acoust. Speech Signal Process. 38, 1842–1852 (1990)
5. Yuen, N., Friedlander, B.: Asymptotic performance analysis of ESPRIT, higher order ESPRIT, and virtual ESPRIT algorithms. IEEE Trans. Signal Process. 44, 2537–2550 (1996)
6. Garcia, A.: Conception d'un radar d'aide à la conduite automobile utilisant un système discriminateur de fréquence type six-port, Thèse ENST (2003)
7. Weidmann, F.: Développements pour des applications grands publiques du réflectometre six-portes: algorithme de calibrage robuste, réflectometre à très large bande et réflectometre intégré MMIC. Thesis ENST Paris (1997)
8. Neveux, G.: Demodulateur directe de signaux RF multi-mode et multi-bande utilisant la technique "cinq-port", these ENST Paris (2003)
9. Urazghildiiev, I.: MUSIC with Outlier Rejection: A New Tool for Improving the Estimation Performance at the Threshold Region of the SNR. In: European Radar Conference, Amsterdam (2004)
10. Roy, R., Kailath, T.: ESPRIT: estimation of signal parameters via rotational invariance techniques. IEEE Trans. Acoust. Speech Signal Process. 37, 984–995 (1989)
11. Abdallah, A.: Stability study of smart antenna systems using homodyne receivers for DOA estimation, Wseas (2006)
12. Abdallah, A.: Contribution à l'étude d'une antenne adaptative et à la conception d'un démonstrateur, thèse XLIM Limoges (2007)

Research on the Evaluation of Feature Selection Based on SVM

Yongjun Ma* and Linqiang Zhan**

College of Computer Science and Information Engineering, Tianjin University of Science & Technology, Tianjin, China, 30022
yjma@tust.edu.cn, zhanlinqiang@yahoo.com.cn

Abstract. Feature selection is an important step of pattern recognition. An effective algorithm of feature selection can improve the rate and efficiency of recognition. There are many standards of evaluation of feature selection, but most of them depend on algorithm and effect of the classification. Therefore, most of standards have limitations. The paper applies SVM method to evaluate feature selection. The main influence factors of effect of feature selection in classification are analyzed in the paper as well. The paper proposed the evaluation criterions of feature selection. The new criterions can be applied to feature selection. And then we give an example to show the Feasibility of new criterions.

Keywords: SVM algorithm, the measure of the correlation of feature, margin distance between clusters.

1 Introduction

The major method of feature selection is to utilize certain mathematical tools to decrease pattern dimension in order to seek for the most effective feature-composition pattern vector with low dimension, which is applied to pattern recognition [1].

The pattern recognition is rather crucial in the selection of feature. To some extent, the effect of feature selection refers to the performance and design style of grader.

Feature selection needs to solve two problems. One is to select proper feature selection algorithm. The other is to find out the standard of assessment as the standard to measure that whether the feature combination is optimum.

---

* Yongjun Ma, vice president of College of Computer Science and Information Engineering at Tianjin University of Science & Technology, tutor for graduates; engineering Doctor of University of Science & Technology of China.

** Linqiang Zhan, Master graduate student of Computer Science and Information Engineering at Tianjin University of Science & Technology;research direction is Image processing and pattern recognition.

D. Yang (Ed.): Informatics in Control, Automation and Robotics, Volume 2, LNEE 133, pp. 407–414.
springerlink.com

2 Introductions to Statistical Learning Theory and SVM

2.1 Empirical Risk Minimization and Structural Risk Minimization Theory

Now, the recognition of two patterns will be considered. Given decision function set is $\{f_\lambda(x):\lambda\in\Lambda\}, f_\lambda(x):R^N\rightarrow\{-1,1\}$ .Where Λ is parameter set. Suppose a group of sample submitting to an unknown distribution P(x,y). Our purpose is to find out the function $f_\lambda(x)$ among the given decision function set, which makes the expected risk minimum, in other words, $R(\lambda)=\int|f_\Lambda(x)-y|P(x,y)dxdy$ is least.

Generally, we name the function $f_\lambda:R^N\rightarrow\{-1,1\}$ making the expected risk minimum that we are seeking for as the supposing function, and the aggregate $H=\{f_\lambda(x):\lambda\in\Lambda\}$ is named as supposing space. $R_{emp}(\lambda)=\frac{1}{l}\sum_{i=1}^{l}|f_\lambda(x_i)-y_i|$ is Empirical risk. Because the distribution P(x,y) is unknown, $R(\lambda)$ cannot be calculated. Therefore, the expected risk mentioned above cannot be minimized. Some sample points of P(x,y) are given. When the amount 1 of the sample point tends to infinite, the empirical risk tends to expected risk. The empirical risk minimization principle means to estimate the minimum value of the expected risk through the minimum value of the empirical risk.

In 1971, Vapnik put forward that the minimum value of empirical risk is not necessarily a converge to the minimum value of expected risk [2]. Besides, the VC dimension of the supposing space $\{f_\lambda(x):\lambda\in\Lambda\}$ is limited [3]. Through the research, Vapnik and Chervonenkis found that the empirical risk $R_{emp}(\lambda)$ and real risk $R(\lambda)$ have the following relation, namely their probability of $1-\eta$ satisfies [4]:

$$R(\lambda)\leq R_{emp}(\lambda)+\sqrt{\frac{h(\ln\frac{2l}{h}+1)-\ln\frac{\eta}{4}}{l}}\ . \tag{1}$$

Where, h is the VC dimension of the supposing space. It is found in (1) that the minimum value of empirical risk $R_{emp}(\lambda)$ can not ensure the minimum value of $R(\lambda)$. In order to solve this problem, Vapnik put forward structural risk minimization theory, namely to try to make the two items at the right side of the less than or equal sign minimum at the same time. The empirical risk $R_{emp}(\lambda)$ and VC dimension of supposing space will be minimum at the same time, through which, ensure the minimum value of expected risk $R(\lambda)$.

2.2 Brief Introduction on Support Vector Machine (SVM) Algorithm

Suppose two data collections H_1 and H_2. Suppose they are linear-separable. In d dimensional space, the general pattern of linear discriminant function is $g(x)=\vec{w}\cdot\vec{x}+b$. The sorting-surface equation is $\vec{w}\cdot\vec{x}+b=0$.

Considering the two sample sets, $\vec{w}\cdot\vec{x}_i+b\geq 1, \forall x_i\in H_1$ and $\vec{w}\cdot\vec{x}_i+b\leq 1, \forall x_i\in H2$.

Our purpose is to seek for the optimum sorting function, which can make the optimum separation of H_1 and H_2. Suppose that the space is made up of indicator function set $sign(\vec{w}\cdot\vec{x}_i+b)$ and fulfills $\min_{i=1,2,\ldots,l}\left|\vec{w}\cdot\vec{x}_i+b\right|=1$. The hyperplanes that meet the condition is named as typical hyperplane. Vapnik proves that the VC dimension of typical hyperplane is $N+1$ [5].

Statistical learning theory[5] points out that in N dimensional space, suppose that all the samples are in the sphere with the radius of R. The VC dimension of indicator function set $f(\vec{x},\vec{w},b)=\mathrm{sgn}\{\vec{w}\cdot\vec{x}_i+b\}$ (where $\vec{w}$ fulfills $\|\vec{w}\|\leq A$) fits the following inequality[6].

$$h\leq\min([R^2A^2],N)+1\ . \tag{2}$$

The distance from x, a point in the space to the hyperplane confirmed by $(\vec{w},b)$ is

$$d(x,w,b)=\frac{\left|\vec{w}\cdot\vec{x}_i+b\right|}{\|\vec{w}\|}\ . \tag{3}$$

Therefore, getting the optimum $(\vec{w},b)$ can be summarized as a problem of quadratic programming (linear-separable, namely the empirical risk is zero)

$$\min_{\vec{w},b}\frac{1}{2}\|\vec{w}\|^2\ . \tag{4}$$

$$s.t.\ y_i(\vec{w}\cdot\vec{x}_i+b)-1\geq 0, i=1,2,\cdots l\ . \tag{5}$$

Fulfillment of the constraint conditions of the quadratic programming problem mentioned above means linear-separable, namely the empirical risk is zero.

Supposing the space fulfills $\{f_{w,b}=sign(\vec{w}\cdot\vec{x}_i+b)\big|\|\vec{w}\|\leq A\}$, it is found in (2) and (5) that when the empirical risk is zero, if the minimum value of $\|\vec{w}\|^2$ is obtained, it can ensure that the upper bound of the VC dimension is the minimum. Thereby, the VC dimension can be minimized. However, in the point set (x_i,y_i), the points that fulfill the constraint condition of the planning problem are nearest to the hyperplane. They determine the distance between two data convex hulls. Taking $\min_{i=1,2,\dots,l} |\vec{w}\cdot\vec{x}_i + b| = 1$ into (3), get $d(x,w,b) \le \frac{1}{\|\vec{w}\|} = d_1 = d_2$.

Therefore, the assorting hyperplane that makes the $\|\vec{w}\|^2$ minimum and the largest between class distances achieves the largest distance among each assorting convex hull that is perpendicular to its own direction. Besides, the distance from the two assorting consecutive points is equal to the distance to the assorting hyperplane. Thus, SVM algorithm is also named as maximum edge algorithm.

3 Analysis of the Major Assessment Factors of the Feature Selection

3.1 Correlation among the Features

The larger the correlation among the features is, the more redundant the information is, and in the pattern recognition, the more complex the hardware achievement and the algorithm of the grade are. Therefore, feature selection algorithm should ensure the minimum correlation among the features. Based on the feature, select the algorithm, such as principal component analysis, etc. It starts with the decreasing the correlation of the feature. Generally speaking, there's no quantitative standard for the measurement of the correlation among the features. However, we can apply canonical correlation analysis and other methods to make investment on the correlation among the features of the feature vectors obtained after feature selection. The definition of the correlation of two features include micro-correlation (related coefficient 0-±0.25), relative correlation (related coefficient ±0.25-±0.4), solid correlation (related coefficient ±0.4-±0.6), significant correlation (related coefficient ±0.6-±0.85), and highly correlation (related coefficient ±0.85-±1).

3.2 Dimensionality

From the formula (2), it is found that the upper bound of h of VC dimension replies on R^2A^2 and N, while R^2A^2 depends on the maximum value of $\|w\|^2$. N stands for the dimensionality of the feature vector. Thus, the larger N is, the larger the upper bound of h is.

3.3 Error Sorting Ratio

Error sorting ratio is an effective method to measure the sorting algorithm. After completing the feature selection, corresponding grader is needed to make the sorting learning. In order to obtain a uniform assessment standard, we adopt SVM algorithm to sort the samples and calculate the error sorting ratio of the sorting. To some extent, the indicator can reflect the effect of feature selection algorithm.

3.4 Class-Among Margin Distance

3.4.1 Positive Class-Among Margin Distance

Definition: Here we define the positive class-among margin distance between two sorting convex hull is $2 \cdot \min(d1, d2)$. In other words, it is two times of the smaller distance from the sorting hyperplane to the two sorting margins. If the sorting hyperplane is obtained based on the SVM algorithm, the positive class-among margin distance defined by us is two times of the maximum distance from sorting hyperplane to the convex hull margin, namely $2 \cdot \max(d(x, w, b))$. It is also the practical class-among margin distance.

3.4.2 Negative Class-Among Margin Distance

What we discussed above is based on the supposing of linear-separable condition. In the following, we will discuss the linear-inseparable condition.

Consider in (5), based on the linear-inseparable condition, a slack variable σ ($\sigma \geq 0$) is introduced. (4) can be expressed as

$$y_i(w \cdot x_i + b) - 1 + \sigma \geq 0, i = 1,2,\cdots,l \ . \tag{6}$$

The quadratic programming problem mentioned above can be expressed as $\min\limits_{\vec{\omega},\vec{b}} \frac{1}{2}\|\vec{\omega}\|^2 + C\left|\sum \sigma_i\right|$ and $s.t. \ y_i(w \cdot x_i + b) - 1 + \sigma \geq 0, i = 1,2,\cdots,l$.

Therefore, we should consider the maximum class-among sorting distance and the minimization of the error sorting samples at the same time. Besides the positive class-among margin distance mentioned above, we also need to define a volume to measure the error sorting of the samples. In accordance to the positive class-among margin distance, we define the negative class-among margin distance.

Definition: Suppose that the maximum distance of two error sorting samples with the sorting hyperplane is $\Delta 1$ and $\Delta 2$ respectively. The negative class-among margin distance is defined as the sum of the two distances, namely $\Delta 1 + \Delta 2$.

4 Evaluation of Feature Selection

Through these indicators, we can summarize the assessment method of feature selection.

1. Minimum dimension assessment principle: based on the condition with same other indicators, the smaller the dimension is, the better the effect of feature selection algorithm are.
2. Maximum positive class-among margin distance assessment principle: based on the condition with same other indicators, the larger the positive class-among margin distance are, the better the effect of feature selection algorithm is.

3. Minimum negative class-among margin distance assessment principle: based on the condition with same other indicators, the less the negative class-among margin distance is, the better the effect of feature selection algorithm is.
4. Minimum feature correlation assessment principle: based on the condition with same other indicators, the less the feature correlation is, the better the effect of feature selection algorithm is.
5. Minimum error sorting ratio assessment principle: based on the condition with same other indicators, the less the error sorting ratio of the SVM algorithm is, the better the effect of feature selection algorithm is.

5 Simple Application Example of Feature Selection Evaluation

Indicate the application of feature selection interpretational criteria through recognition of handwritten numeral 6 and 8, while 84×84 dot matrix handwritten numeral is existing. Adopt algorithm of singular value decomposition(SVD) [7] to decompose image matrix A into weighted sum of 84×84 order matrix with rank 1.

$$U^{T}AV = \Lambda^{\frac{1}{2}} . \tag{7}$$

where U and V are 84×84 order unitary matrix, so

$$\Lambda^{\frac{1}{2}} = \begin{bmatrix} \lambda^{\frac{1}{2}}(1) & & & \vdots & \\ & \ddots & & \vdots & 0 \\ & & \lambda^{\frac{1}{2}}(R) & \vdots & \\ \cdots & \cdots & \cdots & \cdots & \cdots \\ & 0 & & \vdots & 0 \end{bmatrix} . \tag{8}$$

where $\lambda^{\frac{1}{2}}(R)$ is singular value, so we can set up feature vector $H_i = [\lambda^{1/2}(1), \lambda^{1/2}(2), \cdots, \lambda^{1/2}(R)]$. Make feature extraction for a handwriting sample "6" in this way, then get the feature vector [40.2991 19.6839 13.3665 10.7331 … 0.0017 0.0015], which is 74×1 vector. If take this feature to make pattern recognition, calculated amount is largish. According to related discussion of "curse of dimensionality", it will take a lot of resource and time cost. Therefore it needs the feature extraction.

Table 1. Handwritten numeral recognition feature extraction

U	X1	X2	X3	X4	X5	X6	X7	X8	...	Y
"6"Sample1	38.90	19.64	10.85	8.68	7.43	3.43	2.19	1.24	...	A
"6"Sample2	40.30	19.68	13.37	10.73	7.91	3.83	1.93	1.46	...	A
"6"Sample3	39.73	19.79	11.78	9.45	7.67	3.64	2.03	1.29	...	A
"6"Sample4	39.13	19.97	12.87	10.07	7.52	3.33	1.89	1.39	...	A
"8"Sample1	46.63	17.32	13.62	8.88	5.01	2.42	1.30	1.00	...	B
"8"Sample2	47.55	19.44	15.51	10.44	5.55	2.89	1.92	1.40	...	B
"8"Sample3	47.68	18.75	14.72	9.48	5.62	2.62	1.66	1.29	...	B
"8"Sample4	48.23	19.27	16.17	10.13	5.19	2.78	1.34	1.47	...	B

Assuming by the two algorithm feature selection, we get feature vectors (X1 ,X2) and (X1 ,X5), and here we evaluate and compare the two selection methods.

SVM-based maximum positive margin distance of feature vector (X1,X2) is 3.1685, while SVM-based maximum positive margin distance of feature vector (X1,X5) is 3.0984. Dimensionalities of these two feature vectors is equal and they a re both linearly separable, so SVM-based classified misclassification rate and negative margin distance are zero. Through correlation analysis, both are micro- related. According to maximum positive margin distance criterion we defined before, the selection result of feature vector (X1,X2) is better.

And then we take other 30groups of sample to test, recognition accuracy based on feature (X1,X2) is 86.67%, while recognition accuracy based on feature (X1,X5) is 73.33%, which is confirm to the result above.

6 Summary

SVM algorithm is applied to feature selection evaluation in this article, and related definition as well as interpretational criteria that affect feature selection are provided. In addition interpretational criteria are further applied to process of pattern recognition feature selection, so as to achieve purpose of further screening for feature combination. In this article we only give the comparison between single evaluation elements, while the comparison and measurement between different evaluation elements need further study.

Acknowledgement. This research is sponsored by a grant of Tianjin Science & Technology Development Foundation of High School under contract 20090805.

References

1. Wen, X., Hu, Q.: Patter recognition and status monitoring. National University of Defense Technology Press, China (1997)
2. Mertens, S., Engel, A.: Vapnik-Chervonenkis Dimension of Neural Net works With Binary Weights. Physical Review E 55(4), 4478–4488 (1997)

3. Vapnik, V., Chervoknenkis, A.Y.: On The Uniform Convergence of Relative Frequencies of Events to Their Probabilities. The Prob. and Its Applications 17(2), 264–280 (1971)
4. Osuna, E.E., Freund, R., Girosi, F.: Support Vector Machines: Training and Applications. A.I.Memo 1602, Mit Artificial Intelligence Laboratory (1997)
5. Vapnik, V.: The nature of statistical learning theory. Springer, Heidelberg (1995)
6. Zhang, X.: On statistical learning theory and support vector machine. Automatic Chemistry Newspaper (1), 32–42 (2000)
7. Jing, R., Ye, X.: Computer image processing, pp. 128–130. Zhejiang University Press, China (1990)

A Study on the Impact of Photovoltaic Power Network-Connected on Voltage and Power Loss of Distribution Network

Yi Zhao, Li Liu, and Gang Wang

Shenyang Institute of Engineering, Shenyang 110136, Liaoning Province, China

Abstract. The quantity and direction of active and reactive power in transmission of distribution circuit have changed after introducing 10kV low-voltage distribution grid into photovoltaic power supply. This paper researches the changes of voltage amplitude and total power loss in each load node after the radial distribution grid in different locations incorporated into photovoltaic power supply of different capacities. The author conducted an example simulation on 33 node distribution grid, and then conducted the theoretical analyses on characteristics of voltage distribution, interconnected location of PV power and the biggest interconnected level limit.

Keywords: distribution network, flow calculation, voltage distribution, power loss.

1 Introduction

With more and more distributed photovoltaic power accessing to distribution network, the analysis methods of traditional distribution network are receiving new challenges.

At present most of the power distribution systems in urban and rural areas in China still give priority to radial structure. With the introduction of the distributed generation technology, the distribution network reliability and economical efficiency will be improved. But the power loss and voltage distribution have great influence, for example, the trend and distribution of flow and power loss and node voltage will be changed by the insert position and capacity of photovoltaic power.

This paper, which based on the constant power load model, studies on radial distribution network. With the 33 nodes standard distribution network test system, changing the position and capacity of distributed photovoltaic power, using the forward and backward substitution method .Through quantitative analysis, we get datas from power losses and node voltage of distribution system. Then we compare change rate of node voltage with previous rate and simulate all sorts of conditions. At last , discuss the influence law on power losses and node voltage of distribution system when photovoltaic power accessing.

D. Yang (Ed.): Informatics in Control, Automation and Robotics, Volume 2, LNEE 133, pp. 415–422.
springerlink.com

2 Influence on Node Voltage

As shown in figure 1, distribution line has been sectioned in n part, every impedance is $R_i + jX_i$, it`s number is 1, 2...m, ...n along the line. Where, node 1 is source bus and its power is $P_2 + jQ_2$, $P_3 + jQ_3$, ... $P_m + jQ_m$, ... $P_n + jQ_n$.

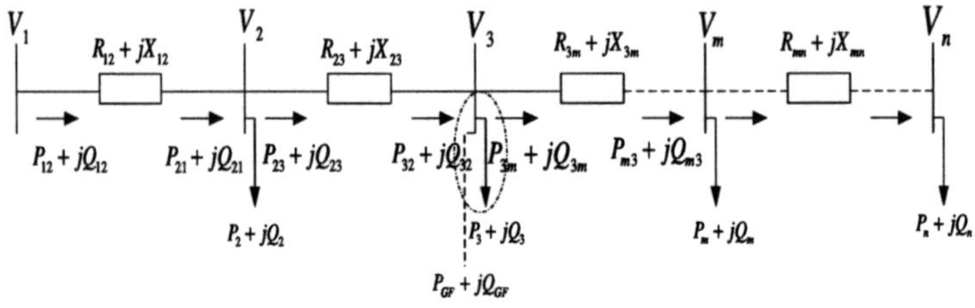

Fig. 1. Radial distribution grid model

Insert photovoltaic power in node 3 which capacity is $P_{GF} + jQ_{GF}$ and consider it as a negative load. Every node before photovoltaic power accessed is calculated as:

$$V_2 = V_1 - \Delta V_{12} = V_1 - \frac{P_{12}R_{12} + Q_{12}X_{12}}{V_1}$$

$$V_3 = V_2 - \Delta V_{23} = V_2 - \frac{P_{23}R_{23} + Q_{23}X_{23}}{V_2}$$

$$V_n = V_{n-1} - \Delta V_{n-1,n} = V_{n-1} - \frac{P_{n-1,n}R_{n-1,n} + Q_{n-1,n}X_{n-1,n}}{V_{n-1,n}}$$

As discussed above, node voltage is related to transmission power on lines, while transmission power is related to load power, so when photovoltaic power accessing, transmission power on lines and node voltage will be changed.

Centralized power supply of distribution networks is character with radial structure. Voltage is gradually reducing along the feeder while power system is running in steady-state operation. After photovoltaic power accessed, load node voltage will be raised because of the decrease of transmission power on feeder. It may lead to load node voltage excursion exceed standard. How much of the voltage rise is related to the insert position and capacity of photovoltaic power.

Setting reasonable operation mode of photovoltaic power for the distribution network voltage adjustment is very important.

When it is sunny at noon, photovoltaic power supply's output is bigger. If the line is light load, photovoltaic power will raise the access node voltage obviously. If the access node is at the end of feeder, access node voltage may cross the upper limit, at that time must set reasonable operation mode of photovoltaic power. Such as: photovoltaic power must participate in adjusting voltage in order to absorb the excess reactive power on line.

When it is heavy load at night, photovoltaic power always work without the output, but it can also supply reactive power in order to improve Voltage quality.

3 Influence on Power Loss

The power loss mainly depends on the system flow the photovoltaic power has influence on both the flow distribution of the system and the power loss of distribution network. The photovoltaic power access may increase or decrease the power loss, this mainly depends on the interconnected location of the photovoltaic power, the capacity , the relative size of the payload , the network topology and other factors.

After accessing photovoltaic power into distribution network, the traditional condition of transmitted power single phase flow may has a change while the flow calculation is the main tool for quantitative analysis. As shown in figure 1, the radial distribution network without the photovoltaic power, the power of node I is:

$$P(i)=\begin{cases}\sum_{j=i}^{n}P_j+\sum_{j=i+1}^{n}\Delta P_j(1\le i<n)\\ P_n(i=n)\end{cases}$$

$$Q(i)=\begin{cases}\sum_{j=i}^{n}Q_j+\sum_{j=i+1}^{n}Q_j(1\le i<n)\\ Q_n(i=n)\end{cases}$$

Power loss of the branch line i is:

$$\Delta P(i)=\frac{(P(i)^2+Q(i)^2)R_i}{U_i^2}$$

$$\Delta Q(i)=\frac{(P(i)^2+Q(i)^2)X_i}{U_i^2}$$

Node K access photovoltaic power, node i power is:

$$P(i)=\begin{cases}\sum_{j=i}^{n}P_j+\sum_{j=i+1}^{n}\Delta P_j-P_{GF}(1\le i\le m)\\ \sum_{j=i}^{n}P_j+\sum_{j=i+1}^{n}\Delta P_j(k<i<n)\\ P_n(i=n)\end{cases}$$

$$Q(i)=\begin{cases}\sum_{j=i}^{n}Q_j+\sum_{j=i+1}^{n}\Delta Q_j-Q_{GF}(1\le i\le m)\\ \sum_{j=i}^{n}Q_j+\sum_{j=i+1}^{n}\Delta Q_j(k<i<n)\\ Q_n(i=n)\end{cases}$$

Considering the load node as PQ node and the photovoltaic power as a negative load, we use the flow of distribution network algorithm, named the back/forward sweep method. The back/forward sweep method is suitable for the distribution network with a single power, just on the branch line via the KVL and KCL theory to get the result, which has the huge superiority about the easy programming realizing, efficient calculation and high speed, while the negative point is its inconvenient for the network with PV node. If the network with distribution power, it needs to be revise. At the same time by the calculation result, we can analyse the original network impact after introducing distribution grid into photovoltaic power, it means that we can through the transformer capacity , Line load rate and other factors to make sure of a node which can introduce the capacity of photovoltaic power.

4 Example Analysis

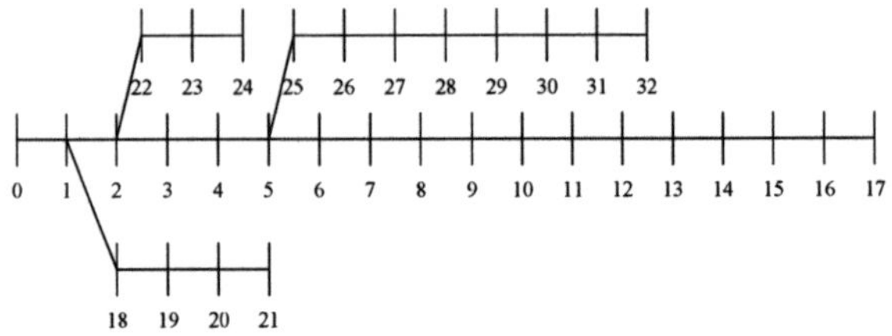

Fig. 2. 33 Bus testing system

This test system using 33 bus testing system as shown in figure 2 , load a constant power model, total power load of the system is 3715.0kW Total reactive termination is 2300.0kvar.Three-phase power benchmark values $S_B = 10000kVA$, Line voltage reference value $U_B = 12.66kV$, Convergent precision is $1.0 \times e^{-5}$. Branch parameter and bus-load of data in table 1 below.

Table 1. Branch parameters and Bus load data

Branch number	First node	End node	R (Ω)	X (Ω)	Terminal active (kW)	Terminal reactive (kvar)
1	0	1	0.0922	0.0470	100.0	60.0
2	1	2	0.4930	0.2511	90.0	40.0
3	2	3	0.3660	0.1864	120.0	80.0
4	3	4	0.3811	0.1941	60.0	30.0
5	4	5	0.8190	0.7070	60.0	20.0
6	5	6	0.1872	0.6188	200.0	100.0
7	6	7	0.7114	0.2351	200.0	100.0
8	7	8	1.0300	0.740	60.0	20.0
9	8	9	1.0440	0.740	60.0	20.0
10	9	10	0.1966	0.065	45.0	30.0
11	10	11	0.3744	0.1238	60.0	35.0
12	11	12	1.4680	1.1550	60.0	35.0
13	12	13	0.5416	0.7129	120.0	80.0
14	13	14	0.5910	0.5260	60.0	10.0
15	14	15	0.7463	0.5450	60.0	20.0
16	15	16	1.2890	1.7210	60.0	20.0
17	16	17	0.7320	0.5740	90.0	40.0
18	1	18	0.1640	0.1565	90.0	40.0
19	18	19	1.5042	1.3554	90.0	40.0

Table 1. *(continued)*

20	19	20	0.4095	0.4784	90.0	40.0
21	20	21	0.7089	0.9373	90.0	40.0
22	2	22	0.4512	0.3083	90.0	50.0
23	22	23	0.8980	0.7091	420.0	200.0
24	23	24	0.8960	0.7011	420.0	200.0
25	5	25	0.2030	0.1034	60.0	25.0
26	25	26	0.2842	0.1447	60.0	25.0
27	26	27	1.059	0.9337	60.0	25.0
28	27	28	0.8042	0.7006	120.0	70.0
29	28	29	0.5075	0.2585	200.0	600.0
30	29	30	0.9744	0.9630	150.0	70.0
31	30	31	0.3105	0.3619	210.0	100.0
32	31	32	0.3410	0.5302	60.0	40.0

4.1 Voltage Distribution

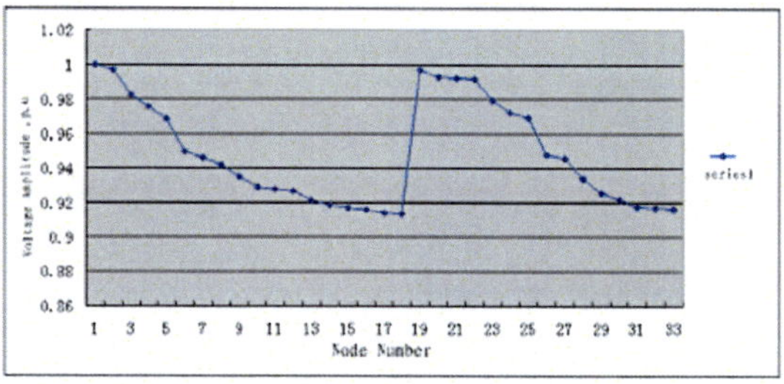

Fig. 3. Each node voltage distribution curve of 33 node test system without photovoltaic power connection

1) Access location and voltage distribution

Node1, Node8, Node17 access photo voltaic power of which capacity is $1000+j500kVA$, Considering the node type as PQ node carry through flow calculation, the results as shown in Figure 4. It is easy to find out from the diagram that photovoltaic power access can improve the entire voltage level of the system. The access position closely associated with the node voltage amplitude. Photovoltaic power the more closer to the power supply the access node voltage impacts smaller, the more closer to the end of the line voltage, the access node voltage impacts greater.

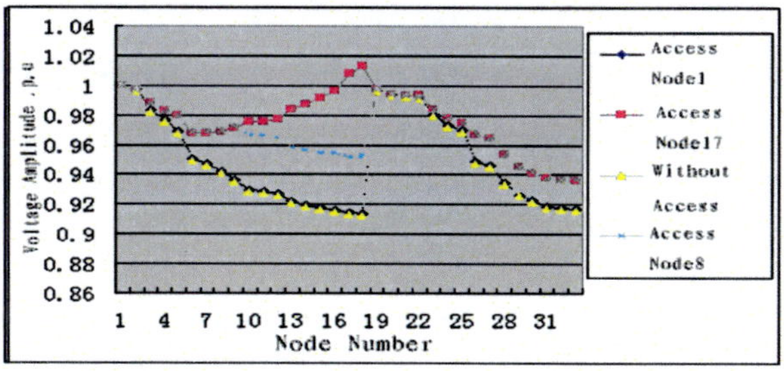

Fig. 4. Photovoltaic power supply access position and voltage amplitude change contrast

2) Access capacity and voltage distribution

In the node access photovoltaic power and change its capacity, specific parameters shown in a table 2 , the flow calculation results entry table 2 And figure 5.

Table 2. The data of photovoltaic power supply capacity and node voltage

NUM	Access capacity (kVA)	Node voltage distribution (.p.u)				
		node1	node8	node17	node24	node32
1	0	0.997	0.935	0.913	0.969	0.916
2	45+j20	0.997	0.937	0.918	0.97	0.918
3	90+ j 40	0.997	0.938	0.923	0.97	0.919
4	150+ j 60	0.997	0.941	0.929	0.97	0.919
5	210+ j 80	0.997	0.943	0.934	0.971	0.921
6	390+ j 170	0.997	0.95	0.953	0.972	0.925
7	510+ j 230	0.998	0.954	0.965	0.972	0.927
8	670+ j 300	0.998	0.96	0.98	0.973	0.93
9	1190+ j 550	0.998	0.977	1.028	0.976	0.939
10	1400+ j 670	0.998	0.984	1.047	0.976	0.943
11	1500+ j 730	0.998	0.987	1.056	0.976	0.945

Analysis shows that in the constant position the access node capacity greater the voltage higher.

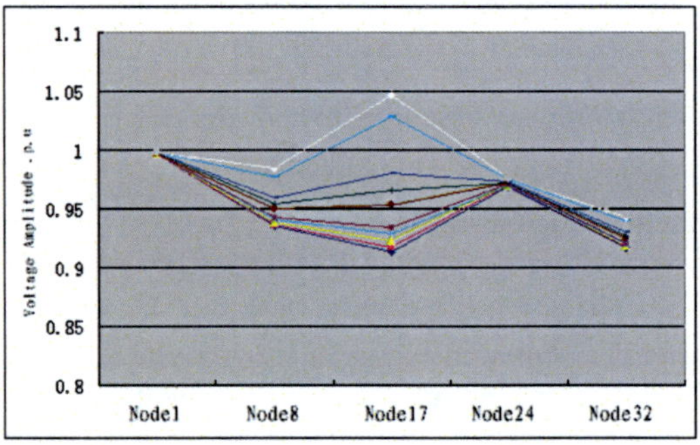

Fig. 5. Photovoltaic power supply capacity and voltage amplitude access change contrast

4.2 System Network Loss

When no photo voltaic power access, system total loss is 205kW, loss rate is 5.23%

1) Access location and network loss

Node 1-32 access capacity is $1000+j500kVA$ Photovoltaic power respectively, considering its node type is PQ node, carry through flow calculation, the results as shown in Figure 6.

Visible system network losses and access capacity of Photovoltaic power are relevant, on the feeder 1-17 the system's loss goes down firstly, then up its minimum value occurs in the Central feeder.

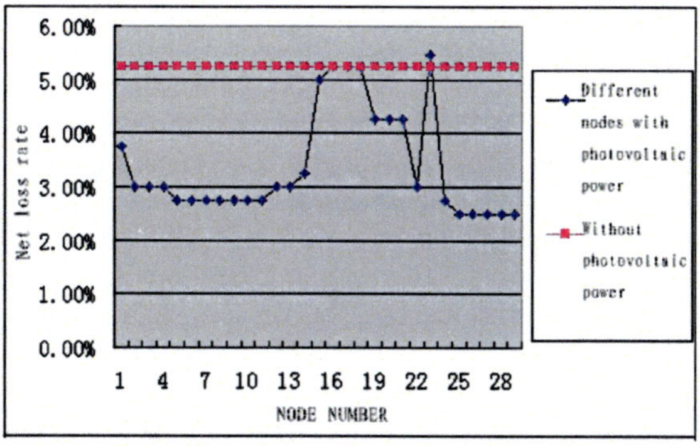

Fig. 6. Photovoltaic power supply position and system access grid loss rate contrast

2) Access capacity and network loss

Node 17 access photovoltaic power, its node type is PQ node, the capacity shown as table 3 below.

Table 3. The data of photovoltaic power supply capacity and the system total access network loss

NUM	Access capacity (kVA)	power losses (kW)	NUM	Access capacity (kVA)	power losses (kW)
1	0	205	6	390+ j 170	145
2	45+j20	190	7	510+ j 230	135
3	90+ j 40	185	8	670+ j 300	125
4	150+ j 60	175	9	1400+ j 670	135
5	210+ j 80	165	10	1500+ j 730	145

Flow calculation the results as shown in figure 7. In the access location unchanged, capacity increases, variation of the total system loss is reduced firstly then increased. The maximum access capacity should not be greater than the load capacity on the system.

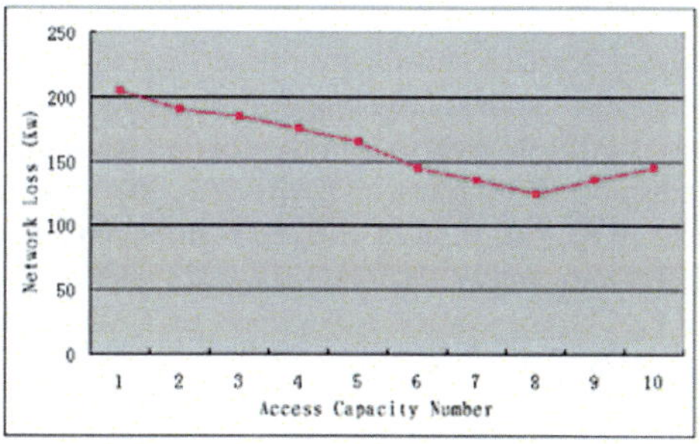

Fig. 7. Photovoltaic power supply capacity and the system total access grid loss contrast

5 Conclusions

As traditional energy sources increasingly exhausting, increasingly serious environmental pollution appears. Solar, wind and other renewable energy development and utilization for the sustainable development is of great significance. Photovoltaic power grid-connected to a certain extent reduces the problem of energy crisis. In order to better utilize photovoltaic power, we must analyze the influence of grid-connected on the existing distribution network. This article based on 33 node test system, by photovoltaic power access locations and the capacity to change, analyzes voltage distribution and total loss of system. Come to the following conclusion:

a. As photovoltaic power capacity constantly , each accessing a different node, Access location near feeder end, Each node voltage value rises more; access location near the feeder first end, Node voltage's change is smaller.

b. When the photovoltaic power position is constantly, accessing different capacity. When the larger capacity, contribute more, equivalent to a smaller load, each node voltage rises more.

c. When photovoltaic power capacity is large enough, flow-backward, Local access point voltage formation system Voltage Max.

d. When photovoltaic power accesses, it can effectively reduce system losses. Reduction of loss associated with photovoltaic power accesses location and capacity.

This constant power load model and load node and photovoltaic power grid-connected nodes are considered as PQ node, thus it has limitations on the application, but the resulting general laws have the certain reference value.

Acknowledgments. The work is supported by the Shenyang Science and Technology Plan Projects of China (Grant No. F10-205-1-80).

References

1. 2010-2015 China solar photovoltaic power industry depth assessment and investment prospects forecast
2. Wang, S.-X., Wang, C.-S.: Modern distribution system analysis. Higher education press, Beijing (2007)
3. Liu, L., Jian, G., Peng, J.-C.: Distributed power generation to distribution grid loss and the influence of the voltage distribution. The Computer Simulation 27(4), 279–283 (2010)
4. Goswami, S.K., Basu, S.K.: A New Algorithm for the Reconfiguration of Distribution Feeders for Loss Reduction. IEEE Trans. on PWRS 7(3), 1484–1491 (1992)
5. Civanlar, S., Grainger, J.J., Yin, H., Lee, S.S.H.: Distribution Feeder Reconfiguration for Loss Reduction. IEEE Trans. on PWRD 3(3), 1217–1223 (1988)
6. Baran, M.E., Wu, F.F.: Optimal Capacitor Placement on Radial Distribution System. IEEE Trans. on PWRS 4(1), 725–734 (1989)

Research of Double-Bar Parallel COSA Finger for Humanoid Robot Hands

Han Liu and Wenzeng Zhang

Dept. of Mechanical Engineering, Tsinghua University, Beijing 100084, China;
Key Lab for Advanced Materials Processing Technology, Ministry of Education, China
liuhan09@mails.tsinghua.edu.cn, wenzeng@tsinghua.edu.cn

Abstract. A novel Coupled and Self-adaptive (COSA) finger with double-bar parallel mechanism is proposed. As a further attempt to achieve the efficient combination of coupled and self-adaptive motion, the finger conducts humanoid grasp with self-adaptation. Concise in structure, compact in conformation, low in manufacturing cost, and smooth in transmission, the finger is competent for the effect of multiple DOF and high adaptability of humanoid robot hands. This paper is composed of the introduction and comparison with existing technology, the principle of the parallel double-bar COSA finger, a feasible specific design based on the concept, and finally the analysis of the feasibility.

Keywords: humanoid robot hand, coupling, self-adaption, under-actuation, double-bar, parallel.

1 Introduction and the Principle of the COSA Finger

An under-actuated hand is a design of which the actuators are less than the DOF[1]. Its grasping contains two patterns : the coupled grasping and the self-adaptive grasping. The coupled grasping robot hands, such as TBM Hand[2], MANUS-Hand[3], SDM Hand[4] and iCub Hand[5], operate in a humanoid pattern. When grasping objects, as the joints are coupled in a fixed way, they bend proportionally. Insufficiency is that, objects would be nipped only with distal phalanx. The self-adaptive grasping robot hands, such as Laval Hand[6], LARM Hand[7], TH-3R Hand[8,9] and GCUA Hand[10], operate in a more adaptive pattern. The mechanism realizes the special effect of double-bar parallel under-actuated grasping while deficiency is that the finger keeps straight before touching objects. Hence, the new concept, COSA (coupling and self-adaption)--combination, instead of simply addition, of the two patterns--sets a milestone in the development of robot hands design. This paper presents a design of a two-linkage mechanism with double bars.

The coupling pattern and the self-adaption pattern have respective deficiencies, while they found the base for the COSA finger, which can achieve the humanoid grasping as well as adaption to the figure of objects. As an embodiment of the COSA finger, the double-bar parallel COSA finger achieves the effect of a coupled rotation integrated with a self-adaptive rotation. The structure is compact, transmission smooth, manufacturing and maintenance costs low, the appearance similar to human fingers: it can be used in the robot hand as the fingers or as a part of fingers.

D. Yang (Ed.): Informatics in Control, Automation and Robotics, Volume 2, LNEE 133, pp. 423–430.
springerlink.com

2 Feasible Specific Design Based on the Concept

2.1 Components

As a embodiment of the COSA finger, the device is composed of a base, a proximal shaft, a proximal phalanx, a distal shaft, a distal phalanx, a motor, a retarder, springs and two parallel transmission mechanisms: a crank-driving transmission mechanism and a bar-slider mechanism.

2.2 Relative Setting Positions

Structure of the double-bar parallel COSA finger device is shown in Fig. 1. It includes a base, a proximal shaft, a proximal phalanx, a proximal pulley, a distal shaft, a distal phalanx, a distal pulley, a driving plate, a connecting bar, a crank, the first spring, the second spring, a slider, the first bevel gear, the second bevel gear, a motor, a retarder.

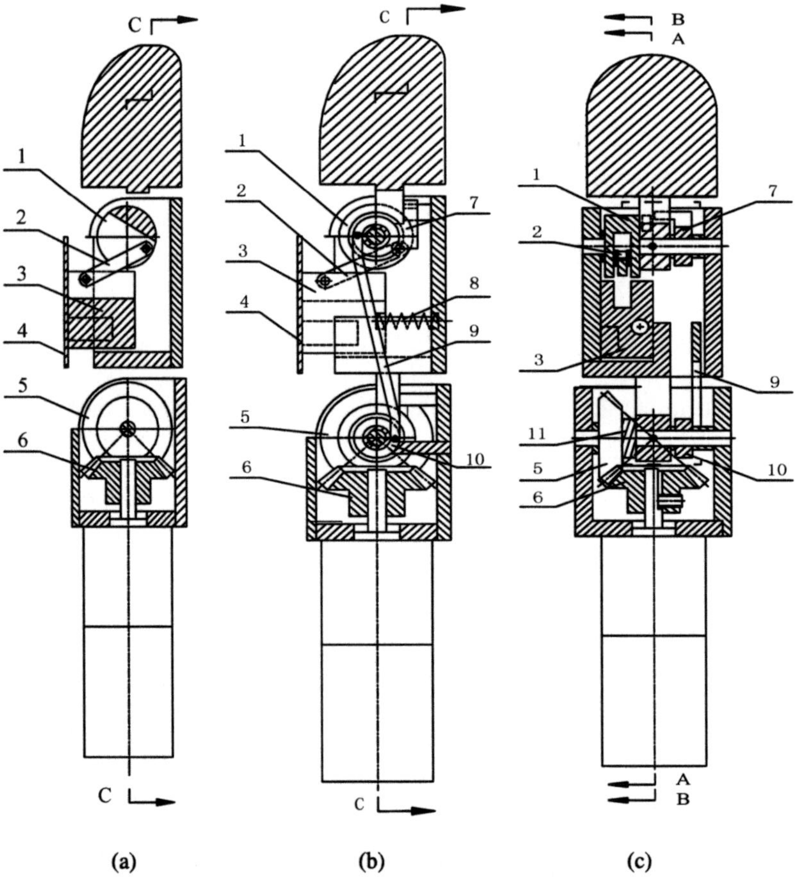

Fig. 1. Views. (a) The crank-driving transmission mechanism; (b) The bar-slider mechanism; (c) Cutaway view from the front; (d) Front view: (e) Right view; (f-g) Stereo view. 1-the driving plate, 2-the connecting bar, 3-the slider, 4-the slider panel, 5-the second bevel gear, 6-the first bevel gear, 7-the distal pulley, 8-the first spring, 9-the crank, 10-the proximal pulley, 11-the second spring.

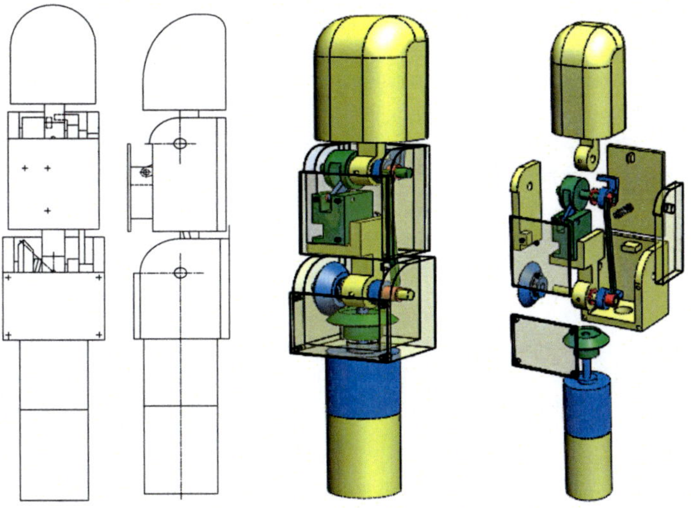

(d) (e) (f) (g)

Fig. 1. (*continued*)

The proximal phalanx is fixed to the proximal shaft; the proximal pulley is fixed to the proximal shaft; the distal pulley is fixed to the base; the slider is inlayed in the sliding chute of the proximal pulley; the distal pulley protrudes a driving bar which touches the distal phalanx; the driving plate is fixed to the distal shaft; ends of the connecting bar are hinged respectively to the slider and the driving plate; the first bevel gear is fixed to the output shaft of the retarder; the second bevel gear meshes the first bevel gear, and is fixed to the proximal shaft. The slider panel is covered with a special material, which appears tough along the axial direction and smooth perpendicular to the axial direction.

2.3 Grasping Process

The initial state of the robot finger is as shown in Fig. 3a. When the finger does not touch the object, it keeps straight. The bump in the base withstands the proximal phalanx for fear of reverse bending of the finger. The spring on the distal shaft, exactly a compression spring, forces the distal phalanx straight. The bump in the proximal phalanx withholds the distal phalanx for fear that the finger would rotate reversely.

The two grasping patterns of the device are as follows:

a. The coupled grasping:

When the robot finger is to grasp the object: the motor rotates, driving the first bevel gear via the retarder, and then the second bevel gear, which brings the proximal shaft to rotate forward, driving the proximal phalanx forward around the axis of the proximal shaft. The proximal pulley encases the proximal shaft and is fixed to the base; thus, rotation of the proximal phalanx would make the crank hinged with the proximal pulley and the distal pulley rotates a certain angle around the proximal pulley. The distal pulley would be pulled by the crank to rotate forward. The driving bar of the distal pulley would push the distal phalanx, so that the distal phalanx rotates forward relatively to the proximal phalanx, bending to the object. The process is like this, until the distal phalanx comes into contact with objects, as is shown in Fig. 2a-2b, with the characteristic that $\alpha_1 \approx \beta_1$.

Releasing process is similar to the grasping process: the motor reversal would drive the proximal phalanx and the distal phalanx simultaneously, releasing the object to achieve the final straight as is back to the initial state of the finger.

b. The under-actuated grasping:

There are two forms of under-actuated grasping:

1. The first under-actuated grasping process: as other fingers or an external force pushes the object in the direction of the arrow, the object extrudes the slider, triggering the under-actuated grasping, finally the final distal phalanx fastening the object. Specifically, when the slider of the proximal phalanx contacts the object, the distal phalanx doesn't touch the object; as the object pushes the slider under other fingers or an external force, the slider slips internally, driving the connecting bar so as to drive the driving plate forward around the axis of the distal shaft, thereby enabling the distal phalanx forward until contact the object to achieve grasping. Specific movement is shown in Fig. 2c-2d, with features that $\alpha_2 > \beta_2$, $\alpha_3 > \beta_3$.

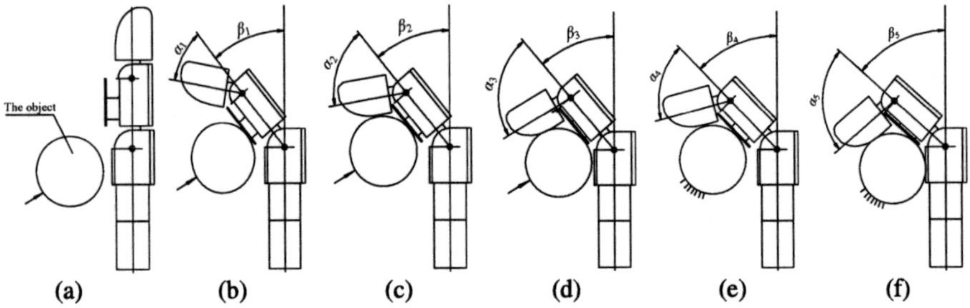

Fig. 2. Under-actuated grasping process. (a) The initial state; (b) The coupled grasping; (c) The under-actuated grasping of the first pattern; (d) The final state of the first pattern; (e) The under-actuated grasping of the second pattern; (f) The final state of the second pattern.

2. The second under-actuated grasping process: the object is stationary or fixed with other fingers or external constraints, then the finger continues to rotate, making the slider pushed into the proximal phalanx, resulting in the under-actuated grasping, and finally, the distal phalanx fasten the object. Specifically, when the slider on the proximal phalanx does, while the distal phalanx doesn't touch the object, the slider is blocked by the object because that the object is constrained immobile. The second process is shown in Fig. 2e-2f, with features that $\alpha_4 > \beta_4$, $\alpha_5 > \beta_5$.

Summing up a. and b. processes, this embodiment achieves a particular under-actuated grasping with the characteristic of coupling followed by self-adaption, with the decoupling in a natural way, consuming little power.

3 Analysis of the Progress Concerned Feasibility

Specific analysis would be found here, including geometric relations and force equations, setting that: l_1=50mm, r_{11}=r_{12}=5mm; r_2=7.5mm, l_4=22.5mm, α_{01}=α_{02}=30°; h_1=35mm, h_2=20mm; k=0.5N/mm; T_m=5000N·m.

3.1 Geometric Relations

In the coupled grasping process, when it comes to the relation between the rotating angle of the distal shaft caused by coupling and the rotating angle of the proximal shaft, geometric relation is show in Fig.3a. Here follows the mathematical analysis.

According to the Pythagorean theorem and the cosine law, one can get,

$$l_2^2 - l_1^2 = (r_{11} + r_{12})^2, \tag{1}$$

$$l_3^2 = r_{11}^2 + l_1^2 + 2r_{11}l_1 \sin\theta_1, \tag{2}$$

$$\cos\alpha_1 = \frac{l_1^2 + l_3^2 - r_{11}^2}{2l_1 l_3}, \tag{3}$$

$$\cos\alpha_2 = \frac{r_{12}^2 + l_3^2 - l_2^2}{2r_{12}l_3}. \tag{4}$$

Then,

$$\theta_{20} = \alpha_2 - \alpha_1 = \arccos\frac{l_1 r_{11}\sin\theta_1 - r_{11}r_{12}}{r_{12}l_3} - \arccos\frac{l_1 + r_{11}\sin\theta_1}{l_3} \tag{5}$$

The relation is shown in Fig. 4a.

In the self-adaptive grasping process, the distance that the slider moves along the direction of the proximal phalanx, results in the rotating angle of the distal shaft caused by self-adaption. Geometric relation is shown in Fig. 3b. Here comes the mathematical analysis.

According to position relation, one can get:

$$\alpha_0 = \theta_{21} - \alpha_{01},\ l_5 = r_2\sin\alpha_{01} + l_4\sin\alpha_{02}, \tag{6}$$

$$h_0 = r_2\sin\alpha_0, \tag{7}$$

$$s_0 = r_2(\cos\alpha_0 - \cos\alpha_{01}), \tag{8}$$

$$s_1 = l_4\cos\alpha_{02}, \tag{9}$$

$$s_2 = \sqrt{l_4^2 - (l_5 + h_0)^2}. \tag{10}$$

Then,

$$\begin{aligned}\Delta s = {} & l_4\cos\alpha_{02} + r_2[\cos(\theta_{21} - \alpha_{01}) - \cos\alpha_{01}] \\ & -\sqrt{l_4^2 - [(r_2\sin\alpha_{01} + l_4\sin\alpha_{02}) + r_2\cos(\theta_{21} - \alpha_{01})]^2}\end{aligned} \tag{11}$$

The relation is shown in Fig. 4b.

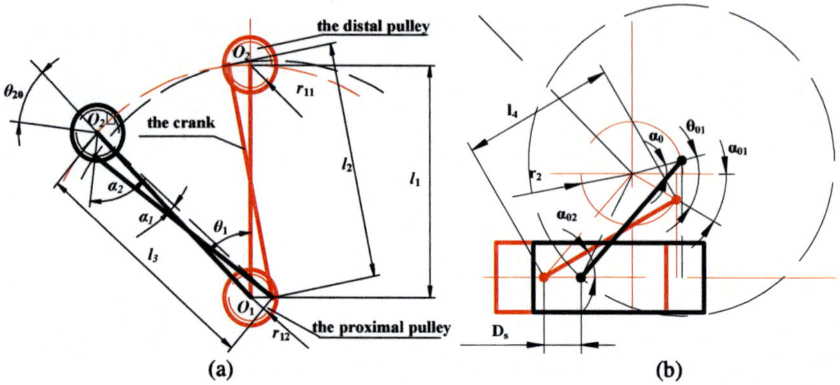

Fig. 3. Geometric relations. (a)The coupled grasping; (b) The self-adaptive grasping.

The f_1, f_2 are the grasping forces on the proximal phalanx and the distal phalanx, respectively; θ_1 is the rotating angle of the proximal shaft; θ_2 is the rotating angle of the distal shaft; $\theta_2 = \theta_{20} + \theta_{21}$;$\theta_{20}$ is the rotating angle of the distal shaft caused by coupling; θ_{21} is the rotating angle of the distal shaft caused by self-adaption; Δs is the distance the slider moves along the direction of the proximal phalanx; T_m is the torque of the motor. Others are distances and angles used in calculating, which are all marked in Fig. 4.

3.2 Force Equations

Force balance and moment balance act on the following mathematical analysis, when it comes to the relation among the grasping forces and the rotating angle of the proximal shaft and the distal shaft.

As for the whole, according to equilibrium and the Hooke's law, one can get,

$$T_m = f_1 h_1 + f_2 (h_2 + l_1 \cos\theta_2), \tag{12}$$

$$f_2 h_2 = f_t l_5 + f_t \frac{l_5 + h_0}{s_2}(s_2 - s_0), \tag{13}$$

$$f_1 - f_t - f_s = 0,\ f_s = k\Delta s. \tag{14}$$

Then

$$f_1 = \frac{T_m + k\Delta s[2l_5 + h_0 - \frac{s_0}{s_2}(l_5 + h_0)]\frac{l_1\cos\theta_2 + h_2}{h_2}}{h_1 + [2l_5 + h_0 - \frac{s_0}{s_2}(l_5 + h_0)]\frac{l_1\cos\theta_2 + h_2}{h_2}} \tag{15}$$

$$f_2 = \frac{T_m - f_1 h_1}{l_1\cos\theta_2 + h_2} \tag{16}$$

Finally, one can get relations between f_1, f_2 and θ_1,θ_2, as shown in Fig. 5a-5b.

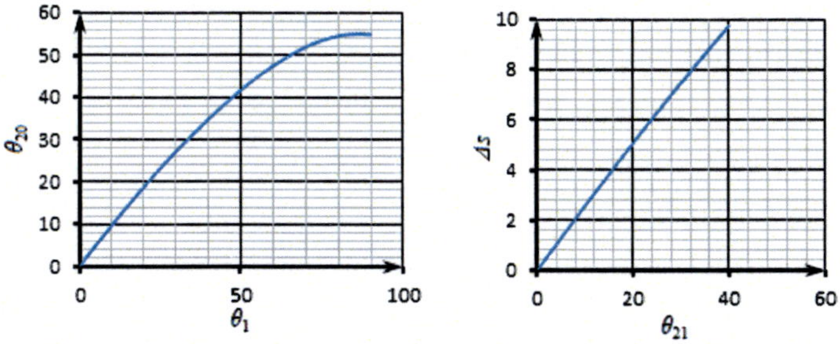

(a) (b)

Fig. 4. Relations among angles and displacement. (a) Relation between θ_{20} and θ_1; (b) Relation between Δs and θ_{21}.

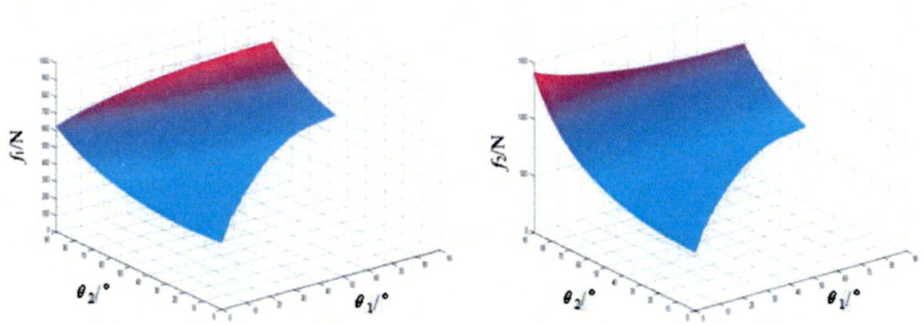

(a) (b)

Fig. 5. Force analysis of the grasping process. (a) f_1 as a function of $\theta_1$and θ_2; (b) f_2 as a function of θ_1 and θ_2.

Allowing for all above, in coupled grasping process, when the proximal shaft rotates an angle θ_1, the distal shaft would rotate an angle θ_{20}, which is not equal to θ_1; in self-adaptive grasping process, when the slider moves a distance Δs , the distal shaft would rotate another angle θ_{21}, and θ_{21} is approximately of direct proportion to Δs. In the whole process, the distal shaft rotates θ_2=θ_{20}+θ_{21} totally, as least θ_{20}, which is determined by the proximal shaft. Forces f_1 and f_2 are corresponding to angles θ_1, θ_2, thus θ_1, θ_2 are determined by the shape of the object as well as the distance between the object and fingers.

As shown in Fig. 5, if θ_1 increases, then the f_1 and f_2 will change slightly; if θ_2 increases, then the f_1 and f_2 will increase distinctly. The f_1 and f_2 will always keep positive and large enough to keep the finger being effective to grasp objects stably.

4 Conclusion

A novel design of double-bar parallel COSA was introduced in this paper. Compared with traditional coupled or self-adaptive gasping patterns, the design showed its ascendancy in many aspects: concise in structure, low in manufacturing cost, compact in conformation, and smooth in transmission. Nevertheless, the design still has its deficiency: as the distal phalanx is driven indirectly, the grasping performed through the distal phalanx appears less powerful, which needs further improvement.

Acknowledgement. This paper was supported by the NSF of China (No.50905093).

References

1. Laliberte, T., Gosselin, C.: Simulation and Design of Under-actuated Mechanical Hands. Mech. Mach. Theory 33(1/2), 39–57 (1998)
2. Dechev, N., Cleghorn, W., Naumann, S.: Multiple finger, passive adaptive grasp prosthetic hand. Mechanism and Machine Theory 36, 1157–1173 (2001)
3. Dubey, V.N.R., Crowder, M.: Grasping and control issues in adaptive end effectors. In: Proc. of ASME on Design Engineering Technical Conf. and Computers and Information in Engineering Conf., pp. 1–9 (2004)
4. Dollar, A.M., Howe, R.D.: The SDM Hand as a Prosthetic Terminal Device: A Feasibility Study. In: IEEE 10th Int. Conf. on Rehabilitation Robotics, Noordwijk, Netherlands, pp. 978–983 (2007)
5. Davis, S., Tsagarakis, N.G., Caldwell, D.G.: The Initial Design and Manufacturing Process of a Low Cost Hand for the Robot iCub. In: IEEE Humanoids, 40–45 (2008)
6. Birglen, L., Gosselin, C.M.: Kinetostatic Analysis of Underactuated Fingers. IEEE Trans. on Robotics and Aut. 20(2), 211–221 (2004)
7. Wu, L., Ceccarelli, M.: A Numerical Simulation for Design and Operation of an Under-actuated Finger Mechanism for LARM Hand. Mechanics Based Design of Structure and Machines 37(1), 86–112 (2009)
8. Zhang, W., Che, D., Liu, H., et al.: Super Under-actuated Multi-fingered Mechanical Hand with Modular Self-adaptive Gearrack Mechanism. Industrial Robot: An Int. J. 36(3), 255–262 (2009)
9. Zhang, W., Che, D., Chen, Q., et al.: A dexterous and self-adaptive humanoid robot hand: gesture-changeable under-actuated hand. In: Proc. 2nd Int. Conf. on Intelligent Robotics and Applications, Singapore, pp. 515–525 (December 2009)
10. Zhang, W., Che, D., Chen, Q., et al.: Study on Gesture-Changeable Under-actuated Humanoid Robotic Finger. Chinese J. of Mechanical Engineering 23(2), 142–148 (2010)

Research of Passive Grasp Multi-fingered Robot Hand

Jiangxia Shi[1,2] and Wenzeng Zhang[1]

[1] Dept. of Mechanical Engineering, Tsinghua University, Beijing 100084, China
[2] Dept. of Automobile Engineering, Beihang University, Beijing 100191, China
shijiangxia@ae.buaa.edu.cn, wenzeng@tsinghua.edu.cn

Abstract. This paper presents a robot hand, PG Hand, which is able to perform many operations with passive grasp. The PG Hand is designed by imitating the human hand, and can either be used as one part of robots or the handicapped. All fingers in the hand are similar in their grasping principle. Each finger has two or three joints with a tendon-spring mechanism. All joints in the hand are actuated by only one motor embedded into the palm. There is an incomplete gear in the middle of the transmission path. There are three stages: fingers are bent initially due to springs; fingers are opened by the motor before grasping; fingers grasp quickly once none tooth part in the incomplete gear is actuated into transmission path. Bearing the characteristics of under-actuation and self-adaption, the device can open and close fingers efficiently and robustly with tendon-spring mechanisms by only one motor.

Keywords: Humanoid robot hand, Under-actuation, Self-adaption, Passive grasp, Tendon-spring mechanism.

1 Introduction

Humanoid robot hands are important sensing and acting part of robots. They can serve as object graspers, the medium between the robot body and the environment, or tactile sensors of robots. According to the number of joints, the field of humanoid robot hands has developed into two major directions: dexterous hands and under-actuated hands.

Research on humanoid robot hands was first focused on the area of dexterous hands in the 1970s, and the Okada Hand [1] from Japan was a masterpiece. To meet the need for the robot to interact with the environment like human beings, dexterous hands were designed to accomplish complicated manipulations via multiple active joints. As industry developed, its demand for robot hands grew rapidly in the 1980s and since then, research on dexterous hands has thrived for decades, with such extraordinary works as the Utah/MIT Hand [2], the Robonaut Hand [3] and the DLR/HIT Hand II [4]. However, aiming at imitating human hands and their functions, dexterous hands have to bear too many DOFs with complicated drive and control systems, making them into hands with huge volumes.

Later, another kind of hands—under-actuated hands appeared on stage and became the concentration of research of robot hands. With less actuators—the number of which is smaller than that of DOFs—and less weight and volume, under-actuated

D. Yang (Ed.): Informatics in Control, Automation and Robotics, Volume 2, LNEE 133, pp. 431–437.
springerlink.com

hands are able to accomplish certain manipulations that can satisfied the demand of robots enough. W. Zhang et. al. in Tsinghua University have worked on under-actuated hands for years and contributed a lot to their development[5-6]. The robot hand in this paper, which consists of multiple tendons and springs but only one motor, is designed to develop the advantages of under-actuated hands.

2 The Humanoid Under-Actuated PG Hand

2.1 Design of the Typical Finger

One typical finger of the hand is showed in Fig. 1a-1c, containing three phalanxes and three joints—each has a joint shaft. There is a spring mounted on each joint shaft, connecting the lower phalanx and the higher phalanx. When the springs are loose, the finger is in the closed position, as showed in Fig. 2a-2d. There is a pulley mounted on each joint shaft, and the lowest one is connected with the lowest joint shaft together. A tendon, whose one end is connected with the lowest pulley together and the other end with the highest phalanx together, goes through the lowest pulley and phalanx, the middle pulley and phalanx, and finally the highest pulley and phalanx all the way.

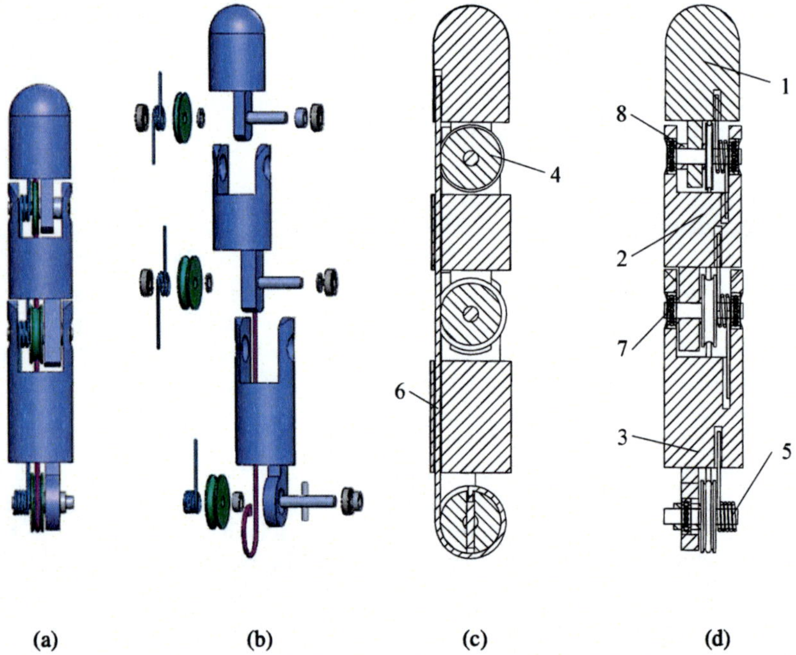

Fig. 1. The finger with three joints. 1-highest finger; 2-middle finger; 3-lowest finger; 4-pulley; 5-spring; 6-tendon; 7-bearing; 8-spacer.

When the tendon is pulled downwards, it moves across the pulleys and pulls the whole finger to an open position, making the springs into tense state as well. The tendon, then, if released, would allow the finger close rapidly under the drive of the springs, thus the finger could finish a typical grasping process.

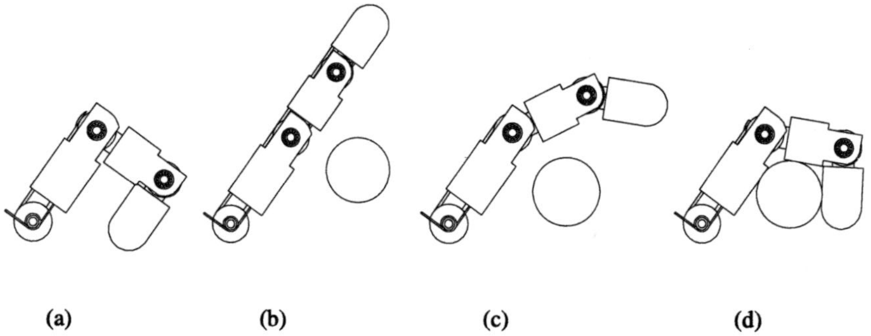

Fig. 2. A typical finger's grasping process

2.2 Formulation of the Whole Hand

The whole hand is designed to imitate the human hand thus containing five fingers named the thumb, first finger, middle finger, third finger and little finger. Besides the thumb that has two phalanxes, the five fingers are of the same structure and principle as illustrated above and are mounted onto the palm by sleeving the lowest joint shaft into the palm. Specifically, the lowest shafts of the last four fingers are made into one shaft called the center shaft in this example.

A center pulley is mounted on the center shaft and is connected with one end of the tendon of the thumb. The tendon then passes two small pulleys—located in the palm and not showed in Fig. 3—and straight to the lowest pulley of the thumb.

As showed in Fig. 3, there is a motor located in the palm, which is connected with a reducer. An incomplete gear is mounted on the output shaft of the reducer, and meshes with a complete gear that is connected with the center shaft together. There is a notch inside the incomplete gear, as illustrated in Fig. 4.

A proximity sensor is located in the palm, perpendicular to the end of the gear, and it can sense whether the notch is passing through its direction. The angle of the notch is intentionally designed according to the angle of the teeth. To be specific, the notch starts earlier than the teeth and ends also earlier than the teeth so that when the hand reaches the open position, the sensor senses the end of the notch and the left one or two teeth help keep the position still. Then if the sensor senses the notch beginning again, the control system will stop the motor—to end up the loop.

All the pulleys of the last four fingers have the same diameter, so that when each phalanxes move from the state where all are perpendicular to each other to the state of open position, it will cost the complete gear about 270 degrees to rotate. As to the thumb, adjustment of the diameter of the center pulley would achieve the same result.

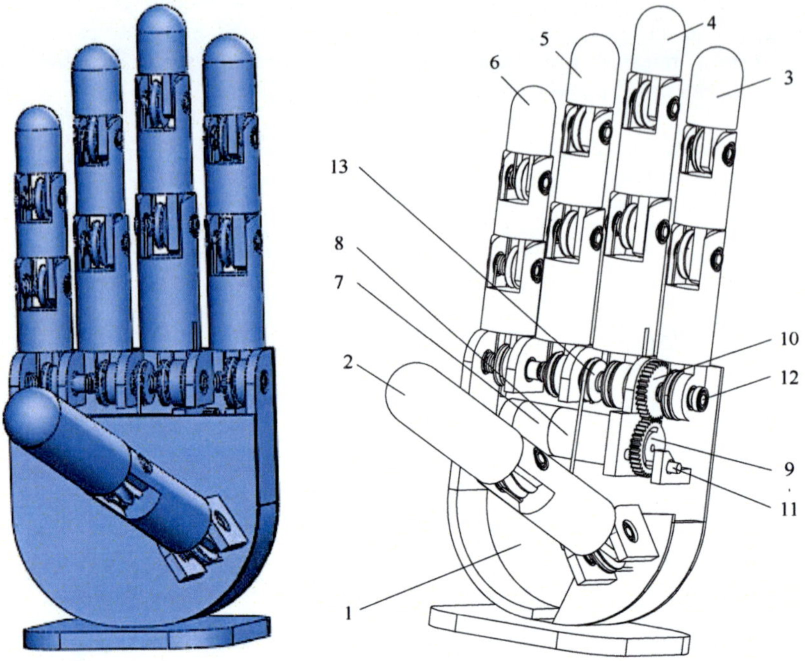

(a) (b)

Fig. 3. The formulation of PG Hand. 1-palm; 2-thumb; 3-first finger; 4-middle finger; 5-third finger; 6-little finger; 7-motor; 8-reducer; 9-incomplete gear; 10-complete gear; 11-proximity sensor; 12-center shaft; 13-center pulley.

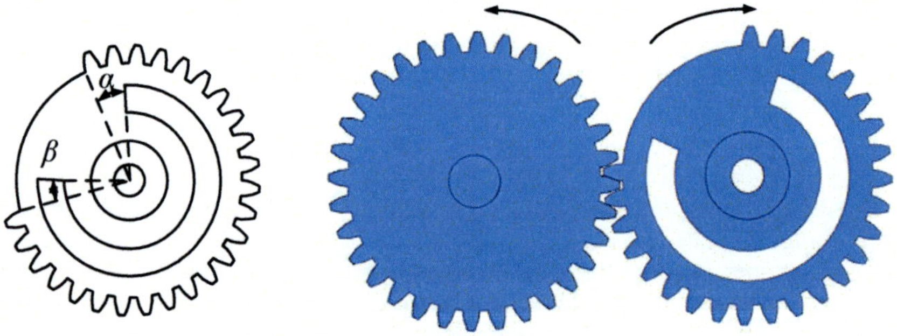

(a) (b)

Fig. 4. The incomplete gear mechanism

2.3 The Grasping Progress and Its Control System

A brief flow chart of the control system is given in Fig. 5, showing how a typical grasping process goes. There is a MCU taking charge of the whole process, and there are two buttons on it. Button 1 is for starting the motor, and Button 2 is for starting the grasp movement. As soon as we press Button 1, the motor is turned on and it makes the reducer function. Then the incomplete gear starts to drive the complete one to roll, thus driving the center shaft to rotate. With the aid of pulleys connected to the center shaft and tendons that are fastened to the pulleys, the whole hand can be driven to an open position.

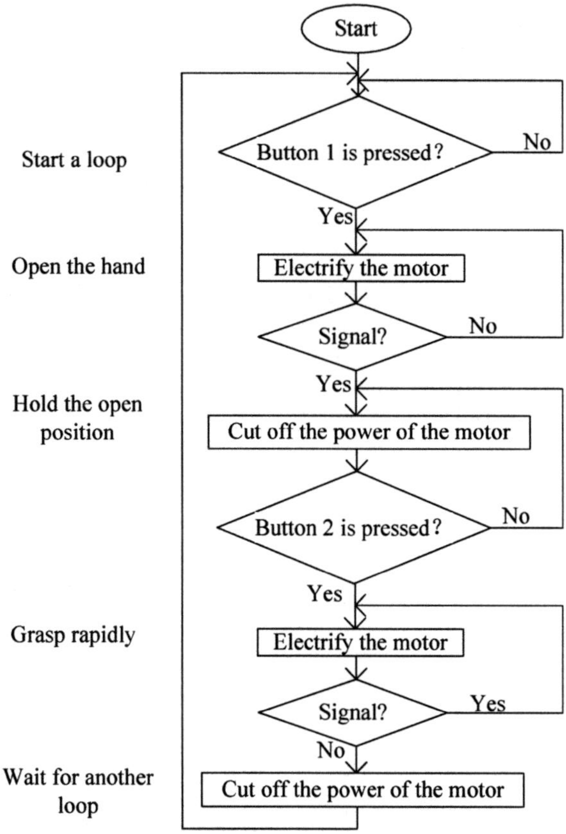

Fig. 5. The flow chart of control system

Then the proximity sensor senses the notch ending and will send a signal to the MCU. When the MCU receives the signal, it cuts off the electricity of the motor. As mentioned above, there is still one or two teeth on the incomplete gear meshing with the other gear, thus holding the fingers at the open position still. After moving the hand near the object, press Button 2. The motor is electrified again, while there is still a signal sent by the proximity sensor, which hasn't sense the notch coming yet. The gears will continue to rotate and the complete gear will be released due to the absence of the teeth of the incomplete gear, so will the tendons. As soon as the notch comes, the signal disappears and the motor is power off. At this time, the teeth of complete gear just haven't run into teeth of the incomplete gear, so this grasp ends with the whole hand closed by the drive of springs. With such mechanism of tendons and springs, the five fingers can wrap around objects of various shapes with self-adaption and passive behavior. To put down the object, we just need to fulfill another loop.

3 Extensions

This paper gives an example of how this kind of humanoid robot hand works. However, such principles can lead to endless applications. Each finger could be of four or even more joints, according to the size and shape of the object. Besides, we can choose less fingers—for example, three fingers are possible if they are enough to grasp small and focused object—or more, like six or more fingers, if needed.

4 Conclusion

The PG Multi-fingered Robot Hand showed in this paper can accomplish various manipulations with 14 joints driven by only one motor, thus decreasing the volume and weight of the hand significantly. Meanwhile, the tendon-spring combination endows the hand with passive behavior and self-adaption, since it can adapt to different shapes easily and flexibly. The tense springs can also make the grasping movement rapidly and efficiently. Its structure is simple and useful, but the PG Hand is able to finish movements like grasping, lifting and carrying objects reliably, and it makes the research of under-actuated robot hands a step further.

Acknowledgment. This research was supported by National Natural Science Foundation of China (No.50905093).

References

1. Okada, T.: Computer Control of Multijointed Finger System for Precise Object-Handing. IEEE Transactions on Systems, Man and Cybernetics, 289–299 (1982)
2. Jacobsen, S.C., Iversen, E.K., Knutti, D.F., Johnson, R.T., Biggers, K.B.: Design of the Utah/MIT Dexterous Hand. In: 2001 IEEE Inter. Conf. on Robotics and Automation, San Francisco, USA, pp. 1520–1532 (1986)
3. Lovchik, C., Aldridge, H., Diftler, M.: Design of the NASA Robonaut Hand. J. ASME Dynamics and Control Division. DSC 67, 813–830 (1999)
4. Liu, H., Wu, K., Meusel, P.: Multisensory Five-finger Dexterous Hand: The DLR/HIT Hand II. In: 2008 Inter. Conf. on Intelligent Robots and Systems, Nice, France, pp. 3692–3697 (2008)

5. Zhang, W., Che, D., Liu, H., et al.: Super Under-actuated Multi-fingered Mechanical Hand with Modular Self-adaptive Gearrack Mechanism. Industrial Robot: An Int. J. 36(3), 255–262 (2009)
6. Zhang, W., Che, D., Chen, Q., et al.: Study on Gesture-Changeable Under-actuated Humanoid Robotic Finger. Chinese Journal of Mechanical Engineering 23(2), 142–148 (2010)
7. Bicchi, A., Kumar, V.: Robotic Grasping and Contact: A Review. In: 2000 IEEE International Conference on Robotics and Automation, San Francisco, CA, USA (2000)
8. Li, Z., Hsu, P., Sastry, S.: Grasping and Coordinated Manipulation by a Multifingered Robot Hand. The International Journal of Robotics Research 8(4), 33–50 (1989)
9. Hirose, S., Umetani, Y.: The Development of Soft Gripper for the Versatile Robot Hand. Mechanism and Machine Theory 13(3), 351–359 (1978)
10. Kobayashi, H.: Control and Geometrical Considerations for an Articulated Robot Hand. The International Journal of Robotics Research 4(1), 3–12 (1985)

Analysis of Coupled-Active Fingers with Gear Mechanism

Mengyang Liang[1] and Wenzeng Zhang[2]

[1] Dept. of Automotive Engineering, Tsinghua University, Beijing 100084, China
[2] Dept. of Mechanical Engineering, Tsinghua University, Beijing 100084, China
liangmengyang@hotmail.com, wenzeng@tsinghua.edu.cn

Abstract. This paper proposes a novel structural design of the coupled-active (CA) finger. The CA finger combines coupled under-actuated finger with active finger, adopting a new control strategy to let these two kinds of grasping pattern working effectively in different grasping steps. Therefore, this new design contains the advantages of both easy control and self-adaption. To further analyze the CA finger mechanism, this paper considers some other parameters, including phalanx forces, the speed ratio and constraining conditions. With physical and mathematical models, we have come to theoretical results of this novel humanoid finger.

Keywords: Robotic hand, coupled finger, active finger, self-adaptive grasp, under-actuation.

1 Introduction

In the field of humanoid robot, there have been a great number of effective designs, including MANUS-Hand[1], TBM Hand[2] and Southampton Hand[3], belonging to the coupled under-actuated (UA) hand, and Utah/MIT Hand[4], Robonaut Hand[5] and DLR-II Hand[6], belonging to active hand.

As a traditional UA hand style, the coupled UA hand has a settled proportion between the rotational angle of the first joint and that of the 2nd joint. However, the two joints can only move in a fixed rotational volocity without any shape self-adaption. This character can be illustrated in Figure 1 where $\theta_1 : \theta_2 = const$. In Fig. 1c, an evidence is also provided that coupled UA hand sometimes cannot successfully grasp objects.

On the other hand, though active hand has a relatively large working space within which it could also accomplish dexterous movements, it brings, at the same time, a complex control system and large mass caused by the motors controlling every joint (or every degree of freedom).

Therefore, in this paper we are going to discuss a combination of active hand and coupled UA hand, in order to simplify the algorithm and transmission chain, achieve shape self-adaption and accomplish required agility. Synchronously, we displace the motors to the base so that we could minimize the mass in the moving elements.

D. Yang (Ed.): Informatics in Control, Automation and Robotics, Volume 2, LNEE 133, pp. 439–446.
springerlink.com

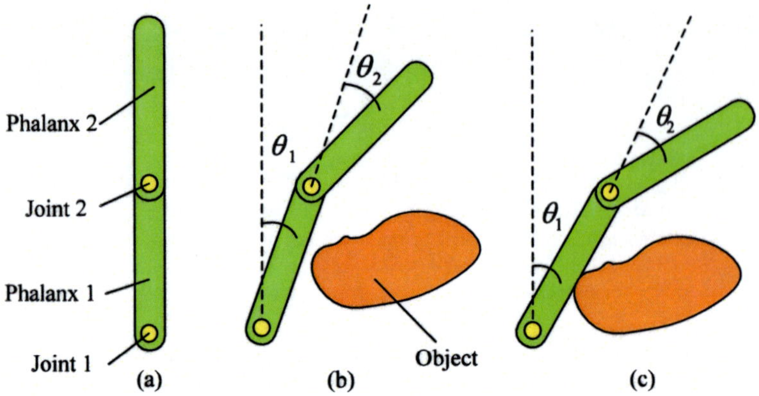

Fig. 1. Grasping process of 2-joint coupled finger

2 Mechanical Design

The novel Coupled-active (CA) finger structure is designed beased on coupled UA hand, adding active module with pulley-belt transmission to ensure the disposal of motors on the base. Fig. 2 provides a structural configuration of this CA finger.

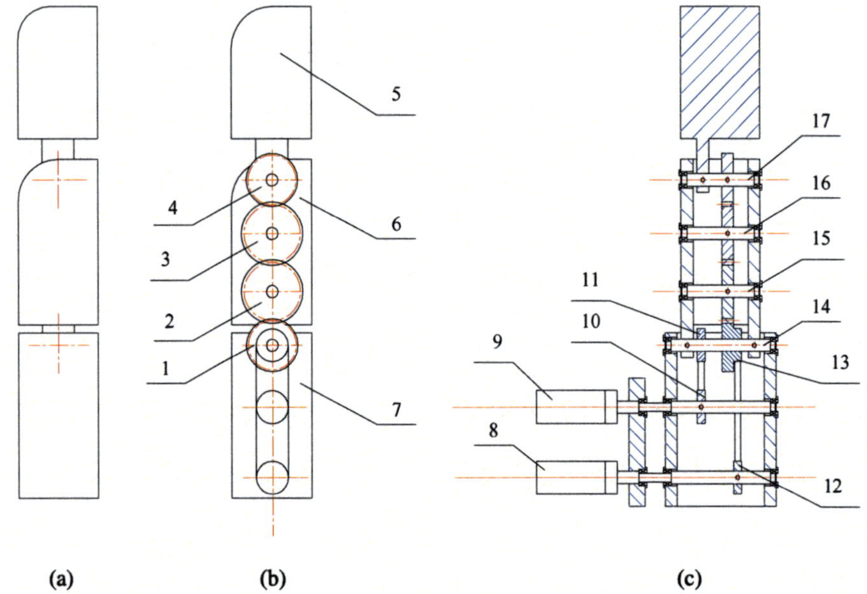

Fig. 2. Structure design of CA hand. 1-first gear; 2-second gear; 3-third gear; 4-forth gear; 5-first phalanx; 6-second phalanx; 7-base; 8-second motor; 9-first motor; 10-first pulley; 11-second pulley; 12-third pulley; 13-forth pulley; 14-first shaft; 15-second shaft; 16-thrid shaft; 17-forth shaft.

The coupled motion is accomplished by holding the 2nd motor still and starting the 1st motor. As shown in Fig. 3(a), the 1st gear is fixed still with the 2nd motor. Then, when the 2nd gear rotates a certain angle caused by the rotation of the 1st phalanx around the 1st joint (θ_1 in Fig. 3), the gear mechanism will convey this rotational motion to the 4th gear, resulting in a rotational angle (θ_2 in Fig. 3) of the 2nd phalanx. With reference point A and B, we can clearly see the rotation of the 2nd and 4th gear.

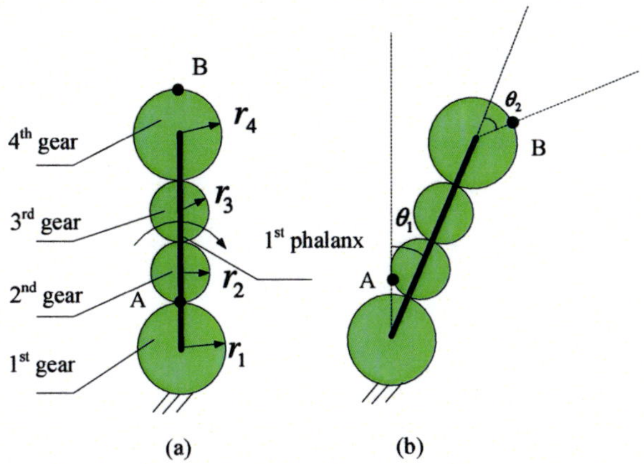

Fig. 3. Coupled movement of the CA finger

Besides the coupled motion, this novel design can also perform active motion, which is achieved by holding the 1st motor still and activating the 2nd motor to rotate the 1st gear. Then the 4th gear will rotate around the 2nd joint independently.

3 Grasping Process of 2-Joint CA Finger

As mentioned and discussed above, the novel CA finger is derived from coupled UA finger and active finger to eliminate their shortcomings. So the working process of CA finger can be divided into 2 stages: the coupled step and the active step.

The grasping process starts with the coupled step, which brings the two phalanges move in proportional rotational velocity. In ordinary cases, the 1st phalanx is to touch the object before the 2nd phalanx. Then the coupled step is ended and the active step is started, in which the 2nd motor drives the 2nd phalanx to move. The whole process is illustrated in Fig. 4. In other cases, however, the 2nd phalanx touches the object first in the coupled step. This is called the pinch mode, in which the grasping process is terminated without actuating active step.

This combination of the two moving patterns brings another concern regarding sensing system. We will accomplish this function by installing moment sensor in the joints. When the 1st phalanx touches the objects, the moment in the 1st joint is going to rise severely; also, the moment in the 2nd joint is going to increase when the 2nd phalanx touches the object. Therefore, by estimating the signal of these sensors, we could judge the condition of the phalanges.

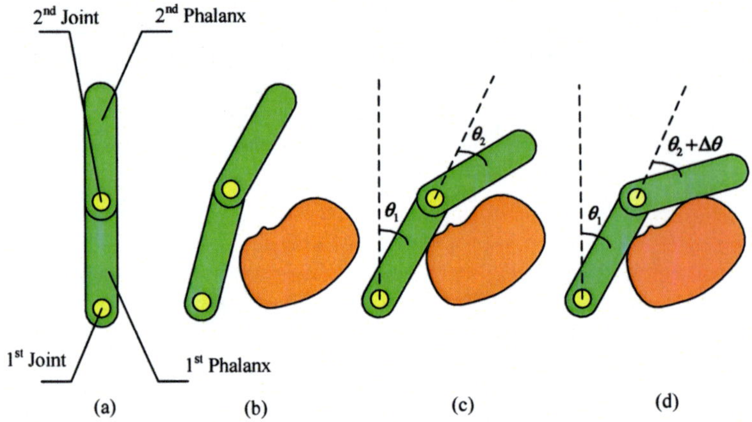

Fig. 4. Grasping Process of the CA finger. (a)-(c), coupled step; (d), active step.

4 Kinematic Analysis

Let the teeth numbers of the 4 gears be n_1, n_2, n_3 and n_4. Then, since all the gears share the same modulus, we could safely come to the equation:

$$r_i = {mn_i}/{2} \tag{1}$$

which could be transformed into:

$$r_1 : r_2 : r_3 : r_4 = n_1 : n_2 : n_3 : n_4 \tag{2}$$

Then as illustrated in Fig. 3, the relation between the rotational angels is

$$\theta_2 = {r_1\theta_1}/{r_4} \tag{3}$$

The equation above points out an essential kinematic relation in coupled step that the rotational speed ratio between the 1st and the 2nd phalanx is utterly determined by the radius of the 1st and 4th gear, which is to say that we don't have to consider the radius of other gears when designing the speed ratio.

4 Mechanical Analysis

Since the dexterous hand with CA fingers mainly works in conditions where it is applied in holding object steadily, this paper particularly focuses on the statics analysis of CA finger design, including functional forces, moments and stability of grasping objects. Although, in the progress of further structure design and optimization, dynamics analysis is fatally critical, this further estimation and consideration will not be included in this paper.

4.1 Analysis of a General Process

A general process is defined here as the situation in which both of the phalanges touch the object, as shown in Fig. 5. Note that this structure has 2 degrees of freedom (DOF). Therefore, one can assume 2 generalized displacement variables as $\delta\theta_1$ and $\delta\theta_2$. Utilizing these two parameters we could have the sum of the virtual work:

$$\sum \delta A = \left(F_1 h_1 + F_2\left(h_2 + l_1 \cos\theta_2\right) - M_1\right)\delta\theta_1 + \left(F_2 h_2 - M_2 \frac{r_4}{r_1}\right)\delta\theta_2 \tag{4}$$

where two generalized forces are:

$$Q_1 = F_1 h_1 + F_2\left(h_2 + l_1 \cos\theta_2\right) - M_1 \tag{5}$$

$$Q_2 = F_2 h_2 - M_2 \frac{r_4}{r_1} \tag{6}$$

The term in the principle of virtual work can be expressed as

$$Q_1 = Q_2 = 0,$$

then we can solve eq.(4) and eq.(5), resulting:

$$F_2 = \frac{M_2 r_4}{h_2 r_1} \tag{7}$$

$$F_1 = \frac{M_1 - \frac{M_2 r_4}{h_2 r_1}\left(h_2 + l_1 \cos\theta_2\right)}{h_1} \tag{8}$$

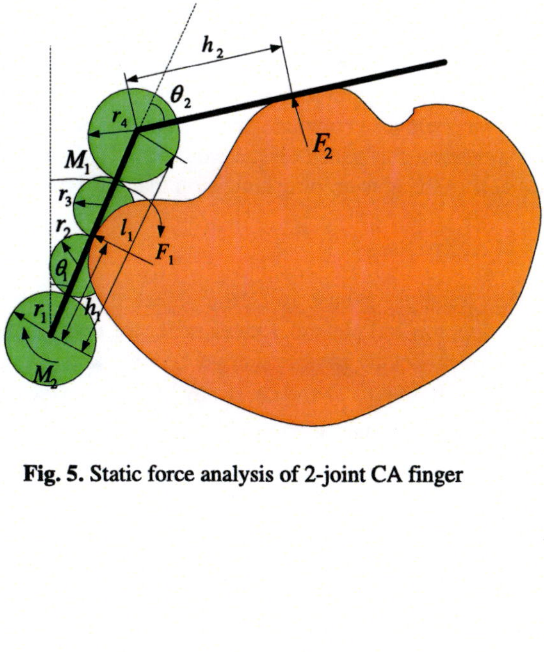

Fig. 5. Static force analysis of 2-joint CA finger

4.2 Analysis of Coupled Step

By a further study of the general model above, adding extra constrains and parameters, we could come to unique equations for coupled step. Assume a special condition in which the two phalanges touch the object simultaneously at the end of coupled step. Then, the relation between the two generalized displacement variables is similar to eq. (3) that can be expressed as:

$$\delta\theta_2 = {r_1\delta\theta_1}/{r_4} \tag{9}$$

Then apply the function above with eq. (4), one could get the one-DOF equation:

$$\sum \delta A = (F_1h_1 + F_2(h_2 + l_1\cos\theta_2) - M_1)\delta\theta_1 + \left(F_2h_2 - M_2\frac{r_4}{r_1}\right)\frac{r_1\delta\theta_1}{r_4} \tag{10}$$

Also using the principle of virtual work, similarly we could arrive at:

$$M_2 = F_1h_1 + F_2(h_2 + l_1\cos\theta_2) - M_1 + F_2h_2\frac{r_1}{r_4} \tag{11}$$

with which we could transform eq. (7) and eq. (8) into functions utterly based on the variable of M_1:

$$F_2 = \frac{\left(F_1h_1 + F_2(h_2 + l_1\cos\theta_2) - M_1 + F_2h_2\frac{r_1}{r_4}\right)r_4}{h_2r_1} \tag{12}$$

$$F_1 = \frac{M_1h_2r_1 - \left(F_1h_1 + F_2(h_2 + l_1\cos\theta_2) - M_1 + F_2h_2\frac{r_1}{r_4}\right)(h_2 + l_1\cos\theta_2)}{h_1h_2r_1} \tag{13}$$

In this way, we have finally come to the function of forces between phalanges and the object, only regarding to the output torque of the 1st motor. Meanwhile, we should note that from eq. (11) we could also conclude that though the 2nd motor is not actuated, the minimum of its quiescent resistance torque should be larger than that in eq. (11), by which we could ensure the constraining condition that the 1st gear is hold still.

4.3 Analysis of Active Step

In most cases, the phalanges cannot touch the object at the same time in the coupled step. So in active step, the 2nd phalanx rotates under the control of the 2nd motor after the coupled step is over. Let the rotational angel of the 2nd phalanx in this step be $\Delta\theta$.

By replacing all the θ_2 in eq. (7) and eq. (8) by $\theta_2 + \Delta\theta$, we could have the functions of forces applied on the phalanges in the active step:

$$F_2 = \frac{M_2r_4}{h_2r_1} \tag{14}$$

$$F_1 = \frac{M_1 - \frac{M_2 r_4}{h_2 r_1}(h_2 + l_1 \cos(\theta_2 + \Delta\theta))}{h_1} \tag{15}$$

where F_2 can be explained as the active force between the 2nd phalanx and the object caused by the rotational movement; meanwhile, F_1 is caused by the object on the 1st phalanx because the 1st phalanx is hold still by the 1st motor. Therefore, such F_1 is a constraining force that cannot be solved with the previously assumed parameters in this paper; and the force on the phalanx in the active step cannot be utterly expressed in terms of M_2.

Also, to ensure the 1st phalanx to be still, we should note that the minimum of the quiescent resistance torque of the 1st motor should be larger than

$$M_1 = F_1 h_1 + F_2(h_2 + l_1 \cos\theta_2) \text{ ,} \tag{16}$$

which is a transformation of eq. (8).

4.4 Analysis of Pinch Mode

The pinch mode refers to the situation that the 2nd phalanx touches the object before the 1st one, as shown in Fig. 6. However unique, this is also a special case of the general model. The fundamental difference is that the F_1 in eq. (7) and eq. (8) equals to zero in this case. Moreover, in this mode, M_2 is a constraining torque. Then we would expect to express F_2 utterly based on M_1, by which we could control the F_2 by modulating M_1.

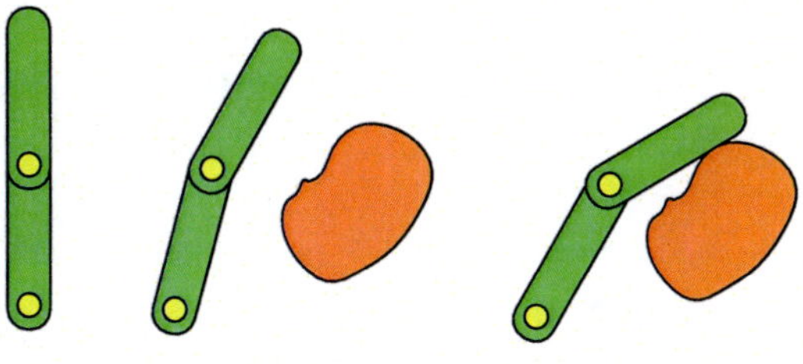

(a) (b) (c)

Fig. 6. Pinch mode of the finger

It is clear that this mode is within the coupled step, which means that eq. (11) and eq. (12) can also be applied in the pinch mode. Setting

$$F_1 = 0 \text{ ,}$$

we could transform eq. (12) into

$$F_2 = \frac{\left(F_2(h_2 + l_1 \cos\theta_2) - M_1 + F_2 h_2 \frac{r_1}{r_4}\right) r_4}{h_2 r_1} \tag{17}$$

which is the force on the 2nd phalanx in pinch mode. Furthermore, after setting

$$F_1 = 0 \quad ,$$

eq. (11) is reshaped into

$$M_2 = F_2(h_2 + l_1 \cos\theta_2) - M_1 + F_2 h_2 \frac{r_1}{r_4}$$

showing the minimum required quiescent resistance torque of M_2 to ensure the 1st gear being still.

5 Conclusion and Future Work

Though the novel design of CA finger has many advantages as shown above, there are also weaknesses that we need to improve in the future. For example, the gear structure may bring extra mass to the phalanx, which may increase the system inertia, and thus, impede the agility of the humanoid hand. Moreover, this paper only concerns about the phalanx force without discussing the stability of grasping. In fact, besides the magnitude of the force, the vector directions, sensors and torque distribution between the two motors could all affect whether the humanoid hand could hold objects successfully.

Above all, this paper proposes a design of novel CA finger with gear mechanism. Focusing on this particular case, its dynamic patterns and control modes to accomplish missions of dexterous movements are estimated. The relation between the phalanx force and motor torque is discussed, so that the phalanx forces from motors can be controlled well.

Acknowledgement. This paper was supported by the NSF of China (No.50905093).

References

1. Pons, J.L., Rocon, E., Ceres, R., et al.: The MANUS-HAND dexterous robotics upper limb prothesis: mechanical and manipulation aspects. Autonomous Robots 16, 143–163 (2004)
2. Dechev, N., Cleghorn, W., Naumann, S.: Multiple finger, passive adaptive grasp prosthetic hand. Mechanism and Machine Theory 36, 1157–1173 (2001)
3. Dubey, V.N.R., Crowder, M.: Grasping and control issues in adaptive end effectors. In: Proc. of ASME on Design Engineering Technical Conf. and Computers and Information in Engineering Conf., pp. 1–9 (2004)
4. Jacobsen, S., Iversen, E., Knutti, D., et al.: Design of the Utah/MIT Dexterous Hand. In: Proc. IEEE Inter. Conf. on Robotics and Automation, pp. 1520–1532. IEEE Computer Society Press, San Francisco (1986)
5. Lovchik, C., Aldridge, H., Diftler, M.: Design of the NASA Robonaut Hand. In: Proc. ASME Dynamics and Control Division, DSC, vol. 67, pp. 813–830. American Society of Mechanical Engineers, New York (1999)
6. Butterfass, J., Grebenstein, M., Liu, H., et al.: DLR-Hand II: Next generation of a dexterous robot hand. In: Proc. of IEEE Inter. Conf. on Robotics and Automation, pp. 109–104 (2001)

Hybrid Under-Actuated Robotic Finger with Triple Pulley-Belt Mechanism

Yifan Liu[1] and Wenzeng Zhang[2]

[1] Dept. of Industrial Engineering, Tsinghua University, Beijing 100084, China
[2] Dept. of Mechanical Engineering, Tsinghua University, Beijing 100084, China
liuyifan.thu@gmail.com, wenzeng@tsinghua.edu.cn

Abstract. This paper presents a hybrid under-actuated grasping mode, which includes two stages: grasping with coupled mode first and then grasping with self-adaptive mode. A novel double-joint robotic finger mechanism unit with triple pulley-belt mechanism, TPB finger, is designed based on the hybrid under-actuated grasping mode. The TPB finger has many advantages: better humanoid grasping, more dexterous actions, more stable in grasping, self-adaptive grasping to objects with different sizes and shapes, lower requirements for control, easier to use, more compact, more efficient, cheaper.

Keywords: Multi-fingered robotic hand, Under-actuated mechanism, Coupled grasp, Self-adaptive grasp, Hybrid grasping mode.

1 Introduction

Hand is vital to human beings, thus developing some robotic hands to help peoplein various fields arouses great interests. Strengthening the functions of robotic hands has a significant impact on the usage ofhumanoid robots.In the recent three decades, a plenty of arts in dexterous hands, such as Utah/MIT Hand[1], DLR II Hand[2], Robonaut Hand[3], have developed. On the one hand, increasing the number of joints or actuators enables the robotic hands more precise and dexterous; on the other hand, the more complexityin the structure, the more negative effect on the sensing and control. Meanwhile, the robotic hands need to be small enough with sufficient grasping force.

Under-actuated mechanism is the one number of actuators is less than DOFs. Generally speaking, there're two kinds of under-actuated finger hands: the self-adaptive one, for example,SARAH Hand[4], TH-3R Hand[5], and the coupled one, like Southampton Hand, MANUS-Hand, etc.

2 The Hybrid Grasping Mode

The coupled mechanism is the one that different joints rotate simultaneously and there's a certain mathematical relationship between the rotating angles. The predictability of the fixed grasping gesture of the finger makes the control of the finger clearer. What's more, the coupled grasping finger can use the fingertip to grasp small objects.However,disadvantages also exist. The initial gesture of the finger is fixed

D. Yang (Ed.): Informatics in Control, Automation and Robotics, Volume 2, LNEE 133, pp. 447–452.
springerlink.com

before contacting, which differs from the real human fingers. What's more, the coupled grasping mode would result in the escape of objects.

The self-adaptive grasping mode is the one that the gesture would wrap the object so that the grasping is more stable. Its advantage is that it introducesself-adaptive grasp, which greatly lowers the requirements for the sensing and real-time control, simplifying the mechanism, thus reducing its cost. Due to the merits, in the recent ten years the under-actuated hands have been prevailing in robotic hands.

The hybridunder-actuated grasping mode is the combination of the coupled mode and the self-adaptive mode. It overcomes the demerits of the two independent grasping modes while gaining their merits together in the same finger.The whole grasping process consists of two stages; the first stage is coupling, following by the second stage of self-adaptive grasping motion.At the first stage, the finger performs coupled grasping mode, during which different parts of the phalanges of the finger rotates at certain angles, respectively. At the second stage, the finger conducts the self-adaptive mode, which enables the finger behaves like a real human finger, whose gesture resembles the shape of the object, so that the grasping would be more stable.

3 The TPB Finger with 2 Joints

Based on the principle of coupled and self-adaptive grasping mode, a novel double-joint robotic finger with triple pulley-belts is designed, TPB finger for short.The TPB Finger consists of the following parts, shown in Fig. 1a-c.

At the 1st stage, before the finger contacts with the objects to be grasped, the motor torque is transmitted to the gear transmission mechanism through the conical gear mechanism, then to the proximal Joint Shaft. Because the right proximal pulley is fixed with the middle Joint Shaft with its radius the same as the right middle proximal pulley, the middle phalanx would rotates the same degree as the proximal Joint Shaft under the action of the spring. Since the left proximal belt is fixed with the proximal Joint Shaft, the left proximal pulley rotates with the proximal Joint Shaft at the same rate. According to the ratio between the left proximal pulley and the middle pulley (here in this finger, the ratio is 2), the middle Joint Shaft would rotates a certain degree, two times of the degree of the proximal Joint Shaft. So the torque makes the middle phalanx and the distal phalanx bent toward the object with the same rotating angles until the middle phalanx touches the object. Then it stops its movements. The overall process is the coupling grasping motion.

At the 2nd stage, although the middle phalanx stops, the left proximal Belt still rotates while the middle Joint Shaft doesn't rotate because of the spring, which also gives rise to the grasping force of the middle Joint Shaft, so the distal phalanx would go on rotating until it also contacts with the object. The torque transmitted to the distal phalanx exerts some proper grasping force on the objects. The above process is the self-adaptive grasping motion.

So under the two stages, the object is grasped by the finger firmly and stably. Fig. 1e shows the workspace of the TPB finger, which also indicates the grasping process of the TPB finger. The middle and the distal phalanges bend towards the object till the position of ABC. Then the middle phalanx stops while the distal phalanx continues bending toward the object until it also contacts the object. The TPB finger perfectly wraps the object. The forces exerted on the TPB finger when grasping an object is shown in Fig. 1d.

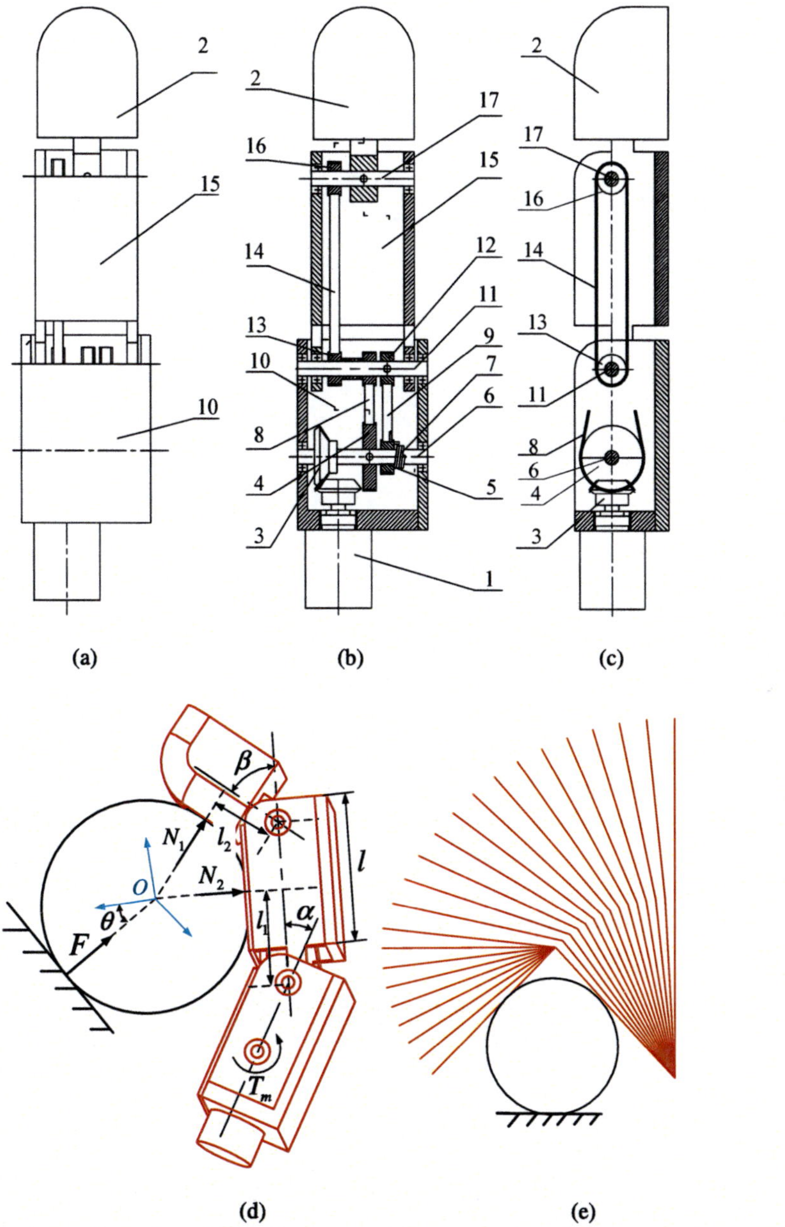

Fig. 1. Mechanical design of the TPB finger.(a)Front view; (b)Frontcutaway view; (c)Side cutaway view; (d)Force analysis; (e)Workspace. 1-Motor; 2-Distal Phalanx; 3-Conical Gear; 4-Left Proximal Pulley; 5-Right Proximal Pulley; 6-Proximal Joint Shaft; 7-Spring; 8-Left Proximal Belt; 9-Right Proximal Belt; 10-Proximal Phalanx; 11-Middle Joint Shaft; 12-Right Middle Pulley; 13-Left Middle Pulley; 14-Middle Belt; 15-Middle Phalanx; 16-Distal Pulley; 17-Distal Joint Shaft.

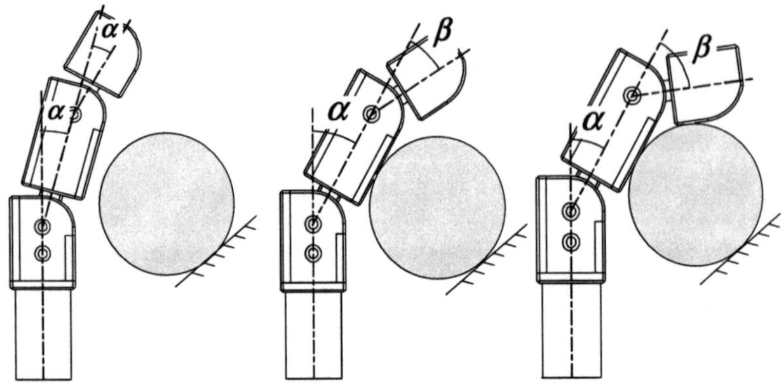

Fig. 2. The Grasping process of the TPB finger

The analysis is based on the assumption that there's no friction between the object and the finger. Where,

N_1: the force exerted on the middle phalanx, N;
N_2: the force exerted on the distal phalanx, N;
F: the force exerted on the object to be grasped, N;

θ: the angle between F and the horizontal line, °;
α: the rotation angle of the proximal phalanx relative to the base, °;
β: the rotation angle of the distal phalanx relative to the proximal phalanx, °;

T_M: the torque of the motor, Nm;
l: the length of the middle phalanx, mm;
l_1: distance between the middle joint shaft and N_1action spot, mm;
l_2: distance between the distal joint shaft and N_2action spot, mm.

$$\begin{cases} F\cos\theta = N_1\cos\alpha + N_2\sin\left(\frac{\pi}{2} - \alpha - \beta\right) \\ F\sin\theta = N_1\sin\alpha + N_2\cos\left(\frac{\pi}{2} - \alpha - \beta\right) \end{cases} \tag{1}$$

What's more, using the balancing condition for the torque of the finger, there is:

$$T_M = N_1 l_1 + N_2(l_2 + l\cos\beta) \tag{2}$$

From the above equations, one get the following expressions for $N_1$and N_2:

$$N_1 = \frac{T_M}{l_1 + (l_2 + l\cos\beta)\frac{(\tan\theta\cos\alpha - \sin\alpha)}{\sin(\alpha+\beta) - \tan\theta\cos(\alpha+\beta)}} \tag{3}$$

$$N_2 = \frac{T_M}{l_1\frac{(\sin(\alpha+\beta) - \tan\theta\cos(\alpha+\beta))}{\tan\theta\cos\alpha - \sin\alpha} + (l_2 + l\cos\beta)} \tag{4}$$

When $T_M = 5000\text{mNm}, l = 4.15\text{mm}, l_1 = 3.25\text{mm}, l_2 = 1.00\text{mm}, \theta = 60°$, the relationship between N_1, N_2 and θ,β respectively, are shown in Fig. 3.Because of the mechanism of the TPB finger, in the graph, only the $\alpha \le \beta$ part is shown.

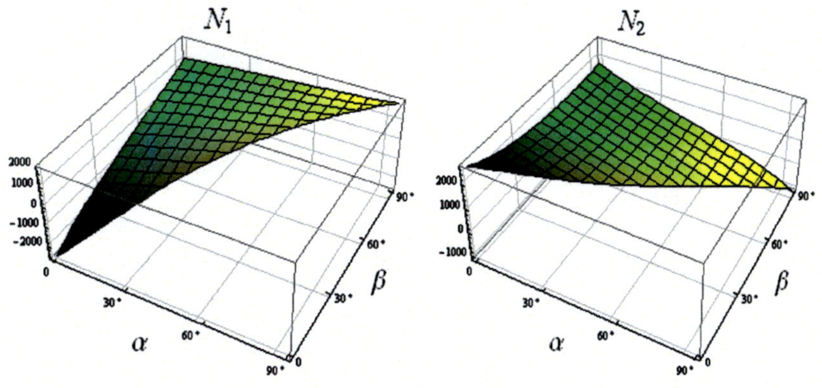

(a) (b)

Fig. 3. The relationship between forces and angles of the TPB finger

From the graphs, conclusions are drawn as below:

When the rotating angel of the middle phalanx α is relatively small, indicating that the finger is grasping a large object or the object is really close to the finger, the force exerted on the middle phalanxN_1, would increasedrastically with the increasing of the rotating angel of the distal phalanx β, while the force exerted on the distal phalanx N_2, would be almost the same; Whenα is big, indicating that the finger is grasping a small object or the object is a little bit far from the finger, N_1 would increase slightly with the increase of the rotating angel of the distal phalanx β, while N_2 would still remain quite small.

From the above analysis, the advantage of TPB finger is obvious:

- When the finger grasps a small object and the object is not far away from the finger, which means that α and β are both very big and there's a difference between the two angels, N_1 is very large while $N_2$is almost nullified, indicating that the grasping force is mainly exerted by the middle phalanx and the distal phalanx only functions as supporting the object to make the grasping more stable.
- When the object is small and is far away from the finger (for example, the object lies between the two phalanges), the finger would wrap the object, indicating that α and β are both big and almost the same, N_1 and N_2 are both of intermediate level of their effective grasping force range, and the magnitudes of the two forces are the same. Under this situation, the coupled grasping motion takes the main role.

- When the finger grasps a large object and the object is not far away from the finger, which means that α and β are both very small, N_1 is very small while N_2 is very large, indicating that the distal phalanx takes the main role of imposing the grasping force onto the object while the middle phalanx is almost "useless". Under this circumstance, the self-adaptive motion plays the main role.

These simulation results demonstrate that the TPB finger can perform stably and effectively in grasping objects.

4 Conclusion

This paper presents a hybrid under-actuated grasping mode, which include two stages: grasp with coupled mode first and then grasp with self-adaptive mode. A novel double-joint robotic finger with triple pulley-beltmechanism, TPB finger, is designed based on the hybrid under-actuated grasping mode. The TPB finger has many advantages: better humanoid grasping, more dexterous actions, more stable in grasping, self-adaptive grasping to objects with different sizes and shapes, lower requirement for control, easier to use, more compact, more efficient, cheaper.

Acknowledgements. This research was funded by National Natural Science Foundation of China (No. 50905093).

References

1. Jacobsen, S.C., Wood, J.E., Knutti, D.F., et al.: The Utah-MIT dextrous hand: work in progress. Int. J. of Robotics Research 3(4), 21–50 (1984)
2. Lovchik, C., Aldridge, H., Diftler, M.: Design of the NASA Robonaut Hand. In: Proc. of ASME Dynamics and Control Division, DSC, vol. 67, pp. 813–830. American Society of Mechanical Engineers, New York (1999)
3. Butterfass, J., Grebenstein, M., Liu, H., et al.: DLR-Hand II: Next generation of a dexterous robot hand. In: Proc. IEEE Int. Conf. on Rob. and Auto., pp. 109–104 (2001)
4. Laliberte, T., Gosselin, C.: Simulation and Design of Under-actuated Mechanical Hands. Mechanism and Machine Theory 33(1/2), 39–57 (1998)
5. Zhang, W., Che, D., Liu, H., et al.: Super Under-actuated Multi-fingered Mechanical Hand with Modular Self-adaptive Gear-rack Mechanism. Industrial Robot: An Int. J. 36(3), 255–262 (2009)

Robot Remote Control Using Bluetooth and a Smartphone Augmented System

Gaowei Chen, Scott A. King, and Michael Scherger

School of Engineering and Computing Sciences, Texas A&M University –Corpus Christi, Corpus Christi, TX, 78412, USA
gchen@islander.tamucc.edu, scott.king@tamucc.edu, michael.scherger@tamucc.edu

Abstract. In this paper, we present an innovative robotic system with augmented smartphones. We propose a smartphone accelerometer controlled differential steering algorithm and implement it on a two-motor driven robotic vehicle with real-time videorendering. In the system design, we create two self-validating protocolsthat build upon our Bluetooth communication network. The overall performance is evaluated by two projects, MOCOVE and RIZZY. The proposed system can be used in robotic remote control, autonomy, robotic localization and human-robot interaction.

Keywords: mobile robot, smartphone, remote control, human-robot interaction.

1 Introduction

Research on robotics has gone on for several decades. Despitedissimilar appearances, robots all share a common mechanical and moveable structure under three distinct phases of control: perception, processing, and action. A robot uses sensors to perceive the environment. Consumer robots tend to have controllers with limited computational power and small amounts of memory. Sensors for consumer robots are alsoquite costly. For instance, a HiTechnic NXT Acceleration/Tilt sensor is sold for $55 at its official website, and a Lego®Mindstorms™ sound sensor sells for $30. In addition, it takes microcontroller resources to manage these mounted sensors. With the advancement of contemporary smartphonetechnology, many common sensors are already embedded into a smartphone. Given the limited processing and high cost of individually packaged sensors, we chose to augment a consumer robot with a smartphone gaining the advantage of more computational power combined with numerous sensors.

The greatest advantage of using a mobile phone to remotely control a robot is location independence[12].There are a number of existing articlesstudying the development of communication models between a cellular phone and a robot. Smartphonestend to bepowerful platforms for roboticautonomyand remote control. A contemporary smartphone possesses many auxiliary features (e.g. digital compass, accelerometer, audio, microphone and camera) and many convenient communication interfaces (e.g. Wi-Fi, Bluetooth, etc.).There is no doubt that smartphone processorshave higher computational power and more memory than common

D. Yang (Ed.): Informatics in Control, Automation and Robotics, Volume 2, LNEE 133, pp. 453–460.
springerlink.com

microcontrollers. Thismotivatesusto integrate a smartphone into robotic systems in order to tackle complex tasks and improve the overall performance.In this research, we implemented a differential steering system with smartphone motion control. In addition, we develop two self-validatingprotocols for Bluetooth communication. The overall performance of the proposed system is evaluated by two projects we designed, MOCOVE (Motion Control Vehicle) and RIZZY. Our system can be used in remote control, robotic autonomy, robotic localization and human-robot interaction.

Section 2 discusses the related research. Section 3 explicitly depicts our system design and implementation. Section 4 evaluates our work and Section 5 concludes.

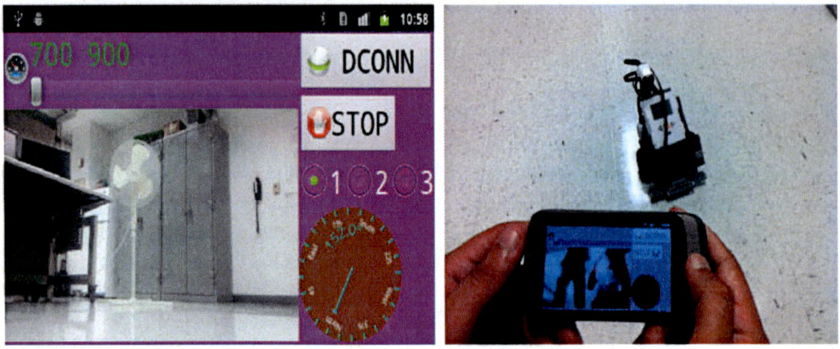

MOCOVE Project screenshots

2 Related Research

Previous articles focus on remotely controlling a robot using cellular phones. Some communication models were proposed, especially through wireless connection. There are mainly two service platforms utilizedfor building the wireless communication between a cellular phone and a robot: to use it as a mobile service platform[6] [7][8][9] [11][12] or to communicate through the networks[3] [4] [10].

One way to achieve remote control is through telephony service. Dual tone multi frequency signaling (DTMF) is used for telecommunication and telephone signaling over the line in the voice-frequency band between phones and the switching center. [12] DTMF assigns each key on the dial pad a specific dual tones frequency. Based on DTMF, amicrocontroller can easily identify the input key (i.e. pressing '5' results in sending a tone made by adding 1336 Hz and 770 Hz) [9].Although it becomes relatively simple for signaling, it would be difficult to develop a complicated protocol using DTMF.

A mobile phone is not only limited to voice calls. For instance, Simple Message Service (SMS) can be utilized for building a communication network. SMS is a mobile service transmitting a short message to a service-activated mobile phone. One communication model illustrates the process of how a SMS message is used to remotely control a robot[12]. The message is first sent and analyzed on an external server, which connects to a local IP network. The content of the SMS message contains a special protocol. After validation and verification of the message by the server, the exact control command extracted from the protocol will be sent to the robot over the IP network. Using mobile services, low cost, large service location coverage and wide support by most service providers are the prominent advantages. However, because of the nature of the discrete SMS messages sending process, it is impossible to develop a continuous control flow on the robot.

Wireless application protocol (WAP) is an open and international standard commonly used for mobile phones. WAP mobile phones can be used for a location independent remote robot control system. However, WAP is not really desirable due to its comparatively high cost, low data rate, high latency and uncomfortable data input[3].

Additionally, in order to develop a voice-controlled robot, the function of speech recognition (and speech understanding) is necessary[8]. As well as voice control, a robot can detect and even recognize human faces with facial detection functionality.

3 System Design and Implementation

3.1 Challenges and Problems

On one hand, smartphone hardware and its operating system exert computational influence on problem solving. On the other hand, the communication latency between the smartphone and the microcontroller sets up a performance gap. Real-time remote control and synchronized data surveillance is a problem we encountered in this research.In addition, our proposal of having the smartphone augmented robot is relatively new. Until now, there are few mature solutions for the integration of a robot and smartphone. One solution is Cellbots [2], an open source project launched by Google Cloud Robotics, where robotic hobbyists and researchers are promoting a similar concept.

3.2 System Design and Implementation

This innovative integration builds upon contemporary smartphone technology. To conquer the challenges, our methodology is to split problems into individual pieces, and solve them separately. This section describes the design and implementation of our system.

The microcontroller is programmed with triggering logic under a thread-parallel structure. We implemented a motion control differential steering system based on the smartphone's accelerometer and achieved real-time remote control. Additionally, we adopt Bluetooth technology in our system design and implementation. Rather than SMS, DTMF, WAP, Bluetooth is more competent and fairly compatible. There are some existing projects using wired connection through USB interface. However, this structure requires assembling an extra break-out board. In contrast, Bluetooth does not require any tangible modification of hardware. In addition, the advantages such as robust connection, fast and continuous data flow propagation, low power consumption, and cheap cost are highly appreciated. Moreover, self-validating protocols are designed for building a reliable communication with as little overhead as possible. Reliability implies a system with a robust continuous control flow. Security is one concern. By nature, Bluetooth is broadcasting. If a malicious user hacks in the system the robot can be kidnapped. Security is beyond the scope of this project, but we have active research projects in our department for securing Bluetooth communication that could be utilized if security was an issue. We built two projects to evaluate the proposed system. MOCOVE is a two-motor-driven robot vehicle that is remote controlled by a smartphone. MOCOVE has real-time video rendering on the controlling device. The motion control functionality is implemented so as to be similar to steering a car wheel. It uses an accelerometer sensor on a smartphone and is controlled by a human operator. Another smartphone mounted on MOCOVE transfers back real-time video from its camera. RIZZY is a university mascot robot that interacts with humans. The "brained" humanoid robot detects a human face and is able to make a simple scripted conversation with the person.

3.3 Object Oriented Concepts in Robotics

The microcontroller is designed in a thread-parallel structure where multiple threads are working concurrently. Each thread is defined to conquer one task. This structure allows threads to be added for any incremented tasks. The main program is running in a loop that is ended either when the user presses the exit button or exits from the smartphone interface. Threads will start when the program is instructed to run. Separate threads for Bluetooth communication, locomotion control, and sensor reading are used in MOCOVE. BluetoothThread listens for incoming connections and continuously receives data from the established socket. LocomotionThread implements differential steeringalgorithm to perform movement operations. SensorReadingThread monitors sensors that are used in the project. All common data is exchanged through an object named CommData among threads. This is implemented on a customized firmware, leJOS [6].The following pseudocode describes this structure.

```
  BluetoothThread.start();//start a thread
  BluetoothThread.setDataEntity(CommData);//DATA exchange
  LocomotionThread.start();
  SensorReadingThread.start();
…//add threads here that performs other tasks
  While(CommData.getExit() or Button.pressExit())
  Thread.yield();
```

3.4 Smartphone Accelerometer Controlled Differential Steering Algorithm

Our robot MOCOVE has a simple and reliable two-wheel-driven propulsion system. It is atypical example of a differentially steered drive system. In Lucas's paper[5], a kinematics approach is used. The mathematical equations from his paper are especially useful for predicting two-wheel-robot's trajectory.

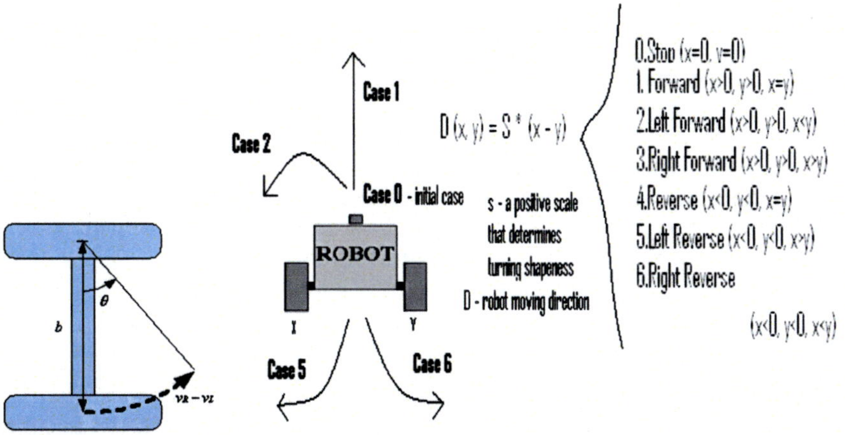

Fig. 1. On left, a simplified two-wheel-robot with different velocities, figure retrieved from [7].On right, a differential steering system with the input motor speeds from smartphone accelerometer is presented.

Where,

θ	is the change in orientation
V_R	is theright wheel speed
V_L	is theleft wheel speed

Differential steering algorithm is intuitive. If $V_R$is equal to V_L, the robot moves in a straight line. The Robot moves along a curved path when V_R and V_L are different. Occasionally, the robot pivots if $V_R$is equal to the opposite of V_L.

According to differential steering algorithm, the control of a robot is achieved byvarying the rotationalspeed of the two driving motors. We propose an accelerometer controlled differential steering system, where the wheels' speeds areobtained and scaledfrom the smartphone's accelerometer data. To simplify possible scenarios, we developedthe equation below for path explanation

$$D\,(x, y) = S * (x - y) \tag{1}$$

Where,

D	is the robot's actual moving direction
S	is a positive constant number used to scale turning speeds
x	is the integer that represents left motor speed
y	isthe integer that represents right motor speed

As displayed in the right hand of figure 1, robot locomotion is conducted by motor speeds x and y. Sevenscenarios could happen after directly changing the motor speeds. The initial scenario is when both x and y are 0 (case 0). The constant S scales the turning sharpness and partially determines the final moving speed. When x and y are the same positives, it is reflected as moving forward in differential drive algorithm (case 1). When xis less than y and both are positives, it's making a left forward turn (case 2). If both x and y are negative, and x is greater than y, the robot would be reversing towards the left (case 5).We use Sto scale eachmotor'sspeed to be within the power range. In the actual implementation, the maximum power for a motor speed isup to 900 (8 volts).

Each Android device has a coordinate systemdefined relative to the device screen [1] and accelerometer data is returned in this 3D space. By specifying a delivery rate to the sensor manager, we can acquire accelerometer data and filterimproper dataout from sharp flipping. In MOCOVE, only y and z are used and scaled properly to be the left and right motor speed. By sending speed integers continuously from smartphone to the microcontroller, a real-time remote control is accomplished.

3.5 Bluetooth Communication and Self-validating Protocols

The Bluetooth protocol is based on a master-slave structure where data is transferred as packets. Atlow-level transport protocol, RFCOMM (Radio Frequency Communication)creates a virtual serial stream that allows binary data to be transported. Each service is recognized with a unique UUID (Universally Unique Identifier) in a form of 16 bits. Thus, each socket connection isguaranteedunique.

Mindstorms™robotic kits include a powerful but unreliable Bluetooth protocol, which is explained in its BDK (Bluetooth Development Kit). Three channels can be used, whereas at one time a robot can only talk to one Bluetooth device. Nevertheless, we use leJOS to build a customized Bluetooth Client-Server structure. Processing large packets by the microcontroller is not recommended. The protocol should have small overhead but be able to validate itself. Therefore, we assume no packet is larger than 256 bits, which is the length of the buffer used to receive data from the socket.

Table 1. Protocol A used between microcontroller and smartphone

Name	Packet length	Task ID	Actual data
Data type	unsigned int8	unsigned int8	
Data length	8 bits	8 bits	0 ~240 bits
Value range	0 ~ 255	0 ~ 255	

Table 2. Protocol B used between smartphones

Name	Packet length	Fragment sequence	Fragment total	Task ID	Actual data
Data type	unsigned int8	unsigned int8	unsigned int8	unsigned int8	
Data length	2 * 8 bits	8 bits	8 bits	8 bits	0 ~472 bits
Value range	0 ~ 511	0 ~ 255	0 ~ 255	0 ~ 255	

We propose the followingprotocol, Protocol Abetween the smartphone and the robot, and it is illustrated in table 1.One byte is sacrificed for packet validation. When either the microcontroller or the smartphone receives this protocol, it calculates the whole packet length and converts an unsigned integer value from the head 1 byte. If both values are the same, the received packet is assumed correct. Otherwise, it indicates the packet is not legal in our network and the packet will be ignored. In a common case, data is 32 bits that represent left and right motor speeds. Therefore, a common packet is 48 bits.

In table 2, we introduce the protocol B used between smartphones. In this case, larger data packets need to be transmitted. And large datasetsare split into fragments at the sender side. All fragments are put into a queue and recovered at the receiver side. For instance, images taken by MOCOVE's mounted smartphone are sent to the user end smartphone.

4 Evaluation of Proposed Work

4.1 MOCOVE

This section discusses the MOCOVEproject, a two-wheel-driven robot controlled by HTC 2.1controllerthrough Bluetooth.A HiTechnic compass sensor and a Kyocera Echo brainare assembled on the robot for real-time geographical positioning and surveillance vision acquisition. Control and sensor data between smartphone and robot is transmitted using the protocol A. Visual data and voice commands are sent via the protocol B.Video vision from Kyocera is rendered asynchronously on the controller. MOCOVEalso sends back its GPS location to the controller, which drawsa graph ofits path trace. The smartphone accelerometer eventsare delivered to the sensor listener in nearly real-time, depending on the delivery rate.

4.2 RIZZY

RIZZY is a project that demonstrates the proposed system in human-robot interaction.The "talking" and walkinghumanoid robot is assembled a Motorola™ XOOM head that hears voice commands from users. The robot will respond to the human talker when hearing special keywords. The first circumstance is the robot detects a human face. When realizing the talker is a human, it reacts. For example, when it hears"go ahead", itmakes the corresponding action. This feature is accomplished using Google™ Face Detection, by processing a face image. In addition, the feature of text to speech is also accomplished. RIZZY talks under certain circumstances. For instance, when it hears a command it cannot understand, it will say "Sorry, I do not understand." The proposed system and multithreading structure enable robot to perform biped walking, data receiving, face detection, speech recognition and speech synchronization.

5 Conclusions and Future Work

In this paper, we propose an innovative robotic system with augmented smartphones. In future work, we consider increasinga robot's intelligence by using a smartphone processing heavy image-or algorithms-based problems such as visual gesture control and robotic localization. We foresee the trend that smartphones are growing to be a popular platform for the IT industry. Developing a visual programming language for smartphone is also promising for this project. Instead of having protocols, robot program codes generated by the smartphone using a visual language would give users a complete flexibility to do everything. In future, this could be used to program a home service robot to do a customized routine task such as cleaning a kitchen.

Acknowledgements. This work was partially funded by NSF Grant #CNS-0708573.

References

1. Android SDK, `http://developer.android.com`
2. Cellbots: Using Cellphones as Robotic Control Platforms, `http://www.cellbots.com`
3. d'Angelo, P., Corke, P.: Using a WAP phone as robot interface. In: IEEE International Conference on Robotics and Automation, Proceedings ICRA 2002, vol. 2, pp. 1173–1178 (2002)
4. Hebah, N., Amjad, A.: Smartphone control robots through Bluetooth. IJRRAS 4(4) (September 2010)
5. Lucas, G.W.: A tutorial and elementary trajectory model for the differential steering system of robot wheel actuators. The Rossum Project (2001), `http://rossum.sourceforge.net/papers/DiffSteer/DiffSteer.html`
6. Ji-Dong, Y., Sungkuk, C., Keechul, J., Shaw, C.D.: Development of Communication Model for Social Robots Based on Mobile Service. In: 2010 IEEE Second International Conference on Social Computing (SocialCom), August 20-22, pp. 57–64 (2010)
7. Kubik, T., Sugisaka, M.: Use of a cellular phone in mobile robot voice control. In: Proceedings of the 40th SICE Annual Conference on International Session Papers SICE 2001, pp. 106–111 (2001)
8. Manikandan, D., Pareek, P., Ramesh, P.: Cell phone operated robot. In: 2010 International Conference on Emerging Trends in Robotics and Communication Technologies (INTERACT), December 3-5, pp. 183–184 (2010)
9. Miyazaki, F., Matsubayashi, S., Yoshimi, T., Arimoto, S.: A new control methodology toward advanced teleoperation of master-slave robot systems. In: Proceedings of 1986 IEEE International Conference on Robotics and Automation, vol. 3, pp. 997–1002 (April 1986)
10. Rui, Z., Dongying, J.: Based on Mobile Device Multi-Biped Robot Monitor and Coordinated Control. In: Second International Conference on Innovative Computing, Information and Control ICICIC 2007, September 5-7, p. 25 (2007)
11. Sekmen, A., Koku, A.B., Zein-Sabatto, S.: Human robot interaction via cellular phones. In: IEEE International Conference on Systems, Man and Cybernetics 2003, October 5-8, vol. 4, pp. 3937–3942 (2003)
12. Yun Chan, C., Jae Wook, J.: Remote robot control system based on DTMF of mobile phone. In: 6th IEEE International Conference on Industrial Informatics, INDIN 2008, July 13-16, pp. 1441–1446 (2008)

The Proportion Integral Control of Torque with AC Servo System

Xiaoyuan Wang, Pingxin Wang, Peng Gao, and Zhe Yang

School of Electrical Engineering &Automation, Tianjin University Tianjin 300072 China
xywang62@tju.edu.cn

Abstract. The defects of the servo unit are occurred with the increasing application of servo motor and control requirements of the specific application system. Combining with the characteristics of servo system, the practical application of servo unit under loading cyclical output is studied for the demand in this paper. PI control module is designed and completed on the base of principle of proportional-integral control. The control system is completed combining with the basic control characteristics of servo unit. The inadequate of control is made up for the servo unit by PI module. The need of practical applications is finally met. The system is confirmed with good tendency, static property and good control performance through test results.

Keywords: Proportion integral controller, Cycle loading control of the pulling force, Servo motor, Servo unit.

1 Introduction

Since the 1980s, the technology of servo drive has a rapid development with the development of the integrated circuits, power electronics and AC variable speed drive technology. A AC digital control system is used as the servo system [1]. Now, the traditional servo unit can not meet the needs of practical applications because there are some special requirements in application systems [2-3]. In order to simulate the force wave of the output in system, the PI control module is designed in this paper. The results show that the designed PI control module can make the servo system operate smoothly and the output tension of cyclic loading is achieved.

2 System Components

The system is required to output the specified tension or pressure among a small circle according to the given signal in practical applications. The PI strategy is used to control the torque in order to meet the simulation needs.

2.1 Principles of Systems

The schematic diagram of the system is shown in Figure 1.

D. Yang (Ed.): Informatics in Control, Automation and Robotics, Volume 2, LNEE 133, pp. 461–467.
springerlink.com

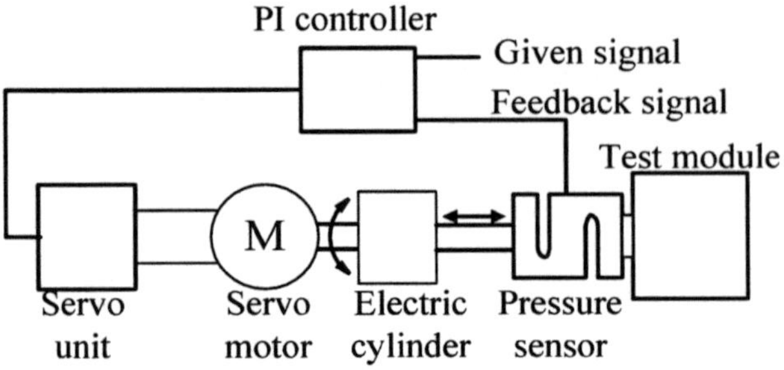

Fig. 1. Basic Block Diagram

The system consists of the control part, the movement execution part and the test part. The control part includes the PI controller and the servo unit. The movement execution part includes the servo motor and electric cylinder. The test part is used to test modules.

The system reference signal is the sinusoidal voltage signal as shown in Figure 1. The frequency of the system is 0.1Hz to 1Hz. The reference signal and the feedback signal returned by the pressure sensor are calculated in the PI module. The limit speed and the protection measures must be set in the servo unit to prevent the rapid movement. The cyclic loading is achieved from servo motor outputs. The motor torque is converted into the line pressure or tension signal with the electric cylinder. The feedback signal is generated from the sensor connected to the drive shaft. The system is designed so that the specified cyclic loading tension in the smaller displacement is get.

2.2 The Test Platform

The system includes Yaskawa servo motor and servo unit, integrated pressure sensor, voltage transmitter and PI controller. The test platform is built as shown in figure 2.

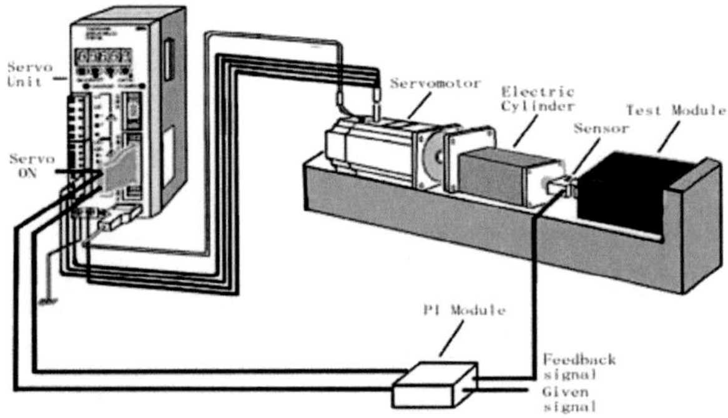

Fig. 2. Experimental system diagram

2.3 The Test Methods

The system reference signal shown in figure 3 consists of a bias voltage signal with 1.5V and a sinusoidal voltage signal with 1V.

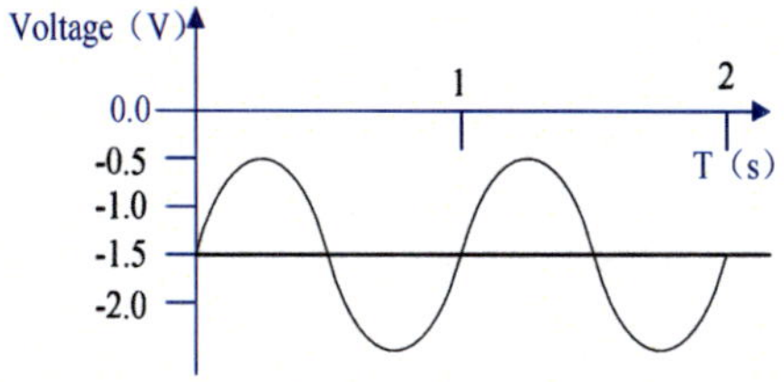

Fig. 3. The given waveform

24V voltage signal is used to start the servo unit. The torque command shown in figure 4 is set as 5V for the rated torque. The parameter is set by the Pn400.

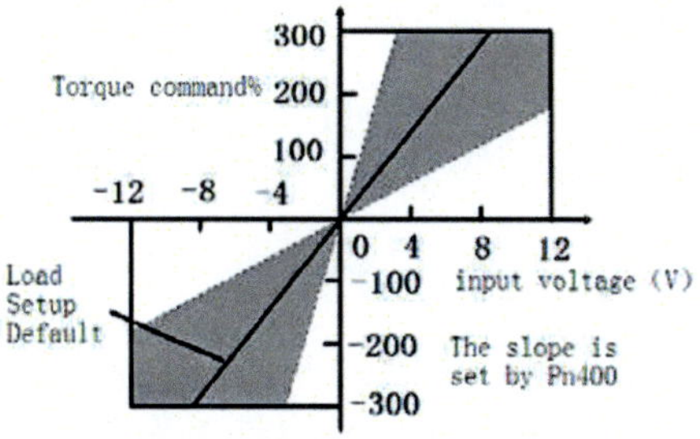

Fig. 4. The setting of rated torque

When the reference signal is not added in the experiment, the motor is still moving in a very low speed. The reason is that when the mode is turned to torque control (analog command), although the directive voltage has been set to 0V, higher plant or the exterior instruction make the voltage have a small amount of "offset(command offset)" which cause the actual output of the command signal is not equal to 0V. After the "offset (= command offset)" is modified to "0V", the motor can work normally. Bias adjustment is shown in Figure 5.

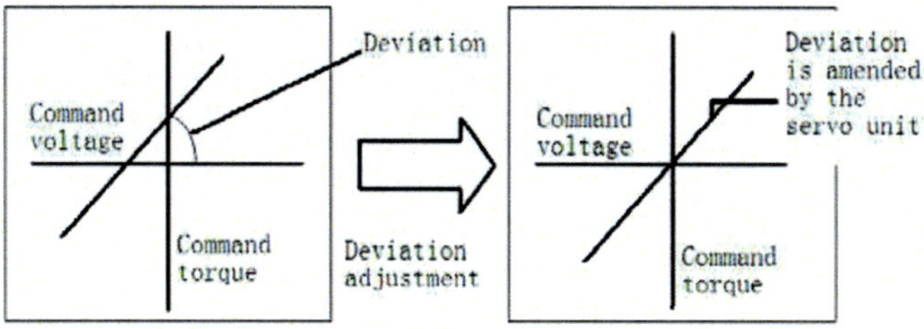

Fig. 5. The adjustment of bias

3 The Design of PI Control Circuit

The proportion integral function is realized by PI control circuit. The PI control circuit is the key of the paper and the core of the control system.

3.1 The DC Bias Circuit

The requirement of practical work is met by DC bias circuit as shown in figure 6. DC bias circuit is constituted by the differential operational amplifier circuit. The output voltage signal is from +5 V to -5V.

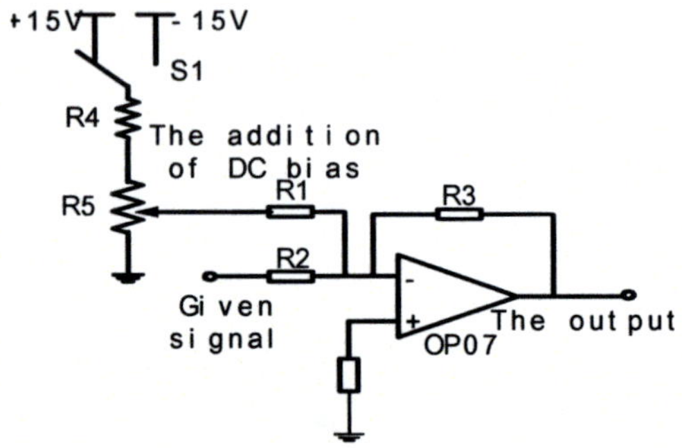

Fig. 6. DC bias circuit

The bias voltage signal can be converted between positive and negative voltage by switch S1. The positive voltage and negative voltage correspond of tension or pressure working conditions respectively in the system. The voltage is divided by R4 and R5. According to the servo unit setting, the DC bias voltage must between +5V and -5V, so R5/R1 is 1/2. The DC bias signal and the sine wave signal are inputted into a differential circuit constituted by OP07. The addition operation for signal is realized in this circuit. The sinusoidal voltage single is outputted with the DC bias voltage. The operation of system is more convenient and more suitable for practical application because of the addition of DC bias circuit. The additive process of DC bias single and given waveform is shown in figure 7.

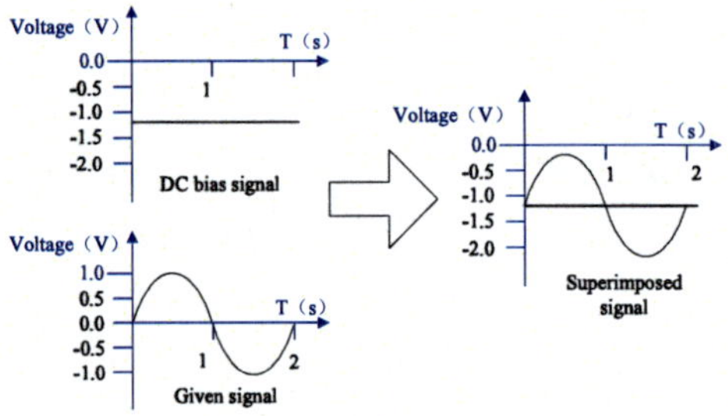

Fig. 7. The signal output of DC bias circuit

3.2 The Proportional Integral Circuit

The PI regulation control is a traditional method [2], and it is suitable for a variety of control site. PI control module is designed to achieve isochronous control in this paper [6-8]. The system principle is shown in Figure 8.

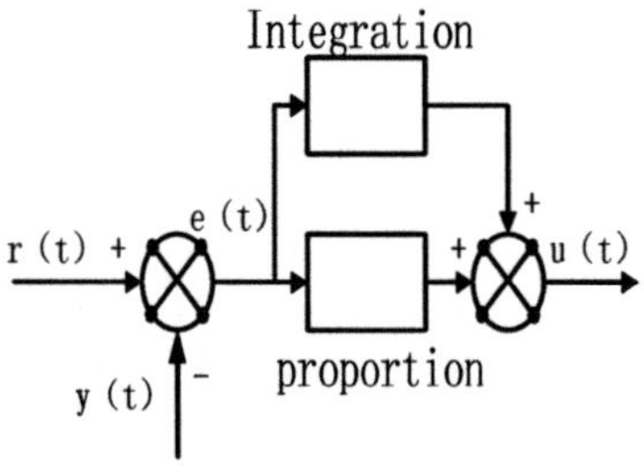

Fig. 8. PI control system diagram

The quick action is realized by proportional element. If the proportionality factor P is too small, the dynamic response will be slow and the ability to eliminate errors will be bad. As the actual system is inertial, the system will be oscillating with the proportionality factor is oversized. The isochronous control is realized by integral element. Integral element must be compatible with system inertia. For the systems with larger inertia the integral action should be weaker. When the errors of system can be eliminated rapidly without the oscillation, the best integral factor is achieved [6-9].

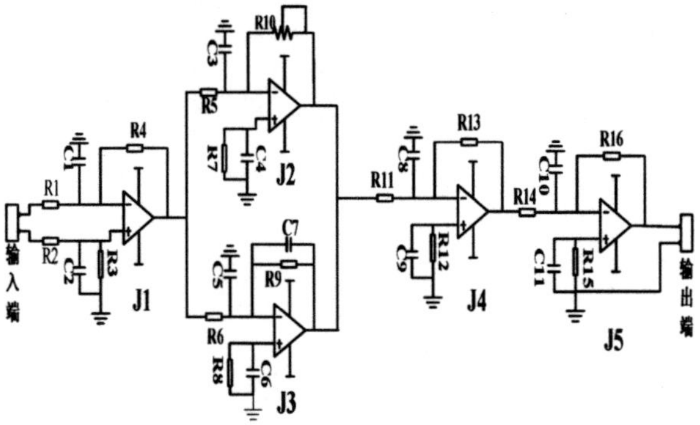

Fig. 9. Proportional integral control circuit

In figure 9, the calculus of differences J1 is used to filter the interfering signal. To facilitate the debugging of the individual parts of system, the proportional-integral element is separated into two independent parts which are proportional element J2 and integral element J3. Through J4, J5 two reverse processes, the output becomes more stable. R10 and C7 are main adjustable parameters in the circuit. Proportionality factor can be changed by adjusting R10. The integral factor can be changed by adjusting C7. The high resistance R9 paralleled in the integral capacity is used to stabilize the output in the experiment. Because of the addition of R9, the Bias current, offset voltage, offset current as well as temperature drift will be eliminated, then integral error brought by voltage drift will be removed[3]. The stability of the system can meet the requirements. Since the whole system is running with a larger interference signal, a simple RC filter circuit is added at each level. Because the signal is 0.1Hz ~ 1Hz, the design of RC filter circuit is low-pass. The entire circuit board is supplied by ± 15V.

4 Experiment Result

Rubber model and spring model is tested from 0.1Hz ~ 1Hz with amplitude of 0.1V ~ 5V, so different waves force are simulated. It is known from the experimental results, hard rubber model is easy to control by changing the PI parameters the movement of motor can be better improved [9]. The coil spring is flexible, the desired result of system can not be achieved by adjusting the parameter. Through experiments it is found that the maximum speed of the motor must be coordinated with the frequency of signal. If the maximum speed of the motor is too high, the phenomenon of motility stasis will appear after the sending of the commend signal. If the maximum speed of the motor is too low, the motor's movements can not keep up with the commend signal. Therefore, the frequency of output signal and a maximum speed of the motor have an optimal operating point. After using the experience parameter the system can be achieved with good control. Waveform results are shown in Figure 10, Figure 11, they are the waveform output and feedback results under 0.1Hz and 1Hz respectively.

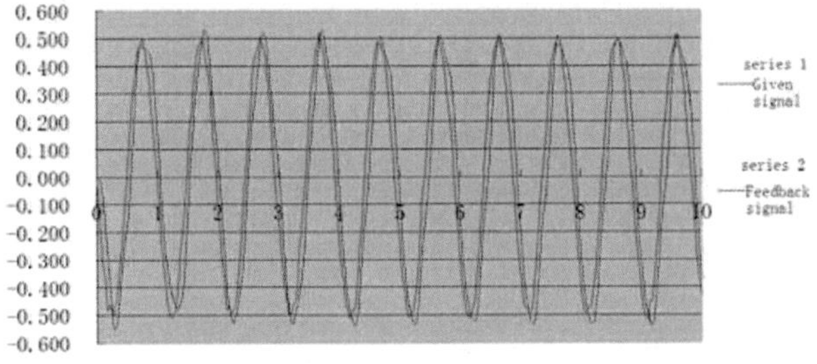

Fig. 10. Detection of 1Hz wave output waveform

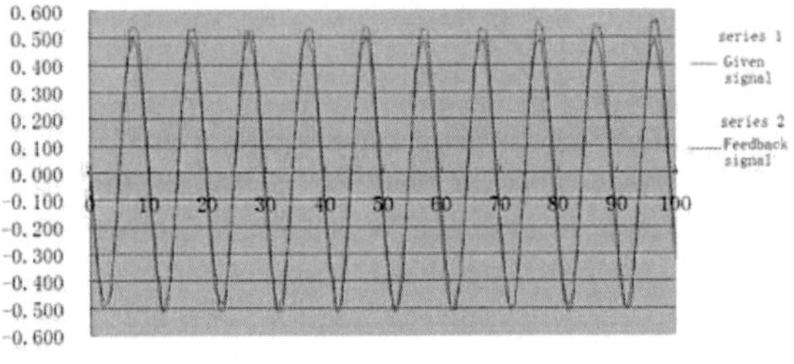

Fig. 11. Detection of 0.1Hz wave output waveform

It can be seen that the feedback voltage signals are synchronous with the given voltage signals from the results. The PI control module designed in this paper makes the torque of motor tracks the given voltage signal well, and recovers the lack that the servo unit can not control cyclic loading torque. The PI module can control the torque of the motor well, and achieves the cyclic loading output of pressure and rally.

5 Conclusions

The PI control module designed in this paper has a strong general performance, reliable operation, and the easily set parameters. The PI control module can solve the problem that servo units can not control and can make the output of the servo motor pulls a better simulation of tidal and wave power. The control module is proved that it has good control effects and is easy to implement simulation requirements by the tests. The control module has a strong ability in engineering control field, and has extensive application values.

References

1. Ao, R., Yuan, K.: Servo system. Aviation Industry Press (2006)
2. Sun, J.: Electrical Machinery & Towage. Mechanical Industry Press (2007)
3. Liu, R.: Analog Electronic Technology. China University of Petroleum Press (2007)
4. Zhu, W., Jia, L., Ding, X.: The Intelligent Control System Based on PI Control. Mechatronics (2003)
5. Wang, M.: Permanent Magnet Synchronous Motor Speed Control Strategy of PI. Electrical Engineering (2008)
6. Wang, S., Liu, X.: Position adaptive PI control in the servo system. Machinery Design & Manufacture (2010)
7. Chen, Y., Wu, G.: The improvement of boiler level measurement and PI parameter setting. Automation in Petro-Chemical Industry (2009)
8. Li, J., Shi, J.: Parameter Regulation of PI Controller for Refrigeration and Air Conditioning System. Refrigeration Journal (1994)
9. Ye, L., Li, S.: PI Controller Auto-tuning for Plant-wide Systems Based on Frequency Domain Performance. Journal of Shanghai Jiaotong University (2007)
10. Gao, M., Wang, Z.: Research on Control System of AC Servo Motor Based on DSP. Mechanical Engineering & Automation (2009)

Design and Development of Intelligent Meter Reading System Based on M-Bus

Xiao-yuan Wang, Peng Gao, and Zhe Yang

School of Electrical Engineering &Automation, Tianjin University, Tianjin, 300072, China
xywang62@tju.edu.cn

Abstract. An intelligent meter reading system based on meter bus(M-Bus) standard is designed in this paper with analyzing the feature of intelligent building system. The intelligent M-Bus system's hardware and software are introduced. The application shows that the system is reliable and stable. The reading data from heat meter are precision. The desired purpose is achieved.

Keywords: M-Bus, C-Sharp, heat meter, level translator.

1 Introduction

With the development of technology and the request of information, the intelligent building will be popular in our country. The main features of intelligent building are faster, high efficiency, more safety and convenient. The intelligent meter reading system is one of the most important embodiments in the intelligent building control system. The intelligent meter reading system is the perfect combination of computer technology, digital communication technology and instrument metering technology. The metering, data acquisition and data processing are made together in the system. It is an synthesis system which can deal with supplying of water, gas and power. The worker labor intensity in different companies can be radically reduced with the reading system. The efficiency of the management department is higher with convenient payment system. The broad areas for development of M-Bus are provided with the development of the intelligent building industry in China.

2 Meter Bus

2.1 The Introduction of Meter Bus

M-Bus is a special bus system. It is originally designed for the use in household meters' industry in Europe. The intelligent meter reading system base on M-Bus is used for measuring facilities such as water meter, electric power meter, gas meter, heat meter. There are many merits with higher reliability, low cost, easy cabling and so on. The two non-polar cables are used in M-Bus system to achieve remote data transmission, remote control, diagnosis, power supply and valuation in different plans.

D. Yang (Ed.): Informatics in Control, Automation and Robotics, Volume 2, LNEE 133, pp. 469–475.
springerlink.com

The standard for data communication was drafted earlier in Europe, especially for heat meter. There are the data communication types of RS-232、RS-485、M-Bus etc. in China. RS-485 is one of the most widely used types in data transferring. RS-485 has the disadvantage of low capacity, low rate comparing with M-Bus. It cannot provide power supply for access equipments. So RS-485 will be substituted by M-Bus in metering system in the intelligent buildings. Currently, China's ministry of construction is popularizing "three meters(water, gas, heat) to be one". The industry of cold-thermal metering is rapidly expanding. M-Bus with unique features of free topological and remote supply will be used widely.

2.2 The Principle of M-Bus

M-Bus is a hierarchical communication system controlled by master with half duplex mode. The system consist of master, slave and cable. The communication mode of calling / response is used in the system. When the host sends command the slave transmit data to it. The exchange of data between slaves is forbidden. The level features of M-Bus are unique because the slave's power can be provided by the master [1].

The process of M-Bus data transmission is divided into two parts: (1) The information in the direction master to slave is transmitted with bus voltage modulation. The correspondence of logic level "1" and logic level "0" is shown in Fig.1(a). (2) The information in the direction slave to master is transmitted with bus current modulation. The correspondence of logic level "1" and logic level "0" is shown in Fig.1(b).

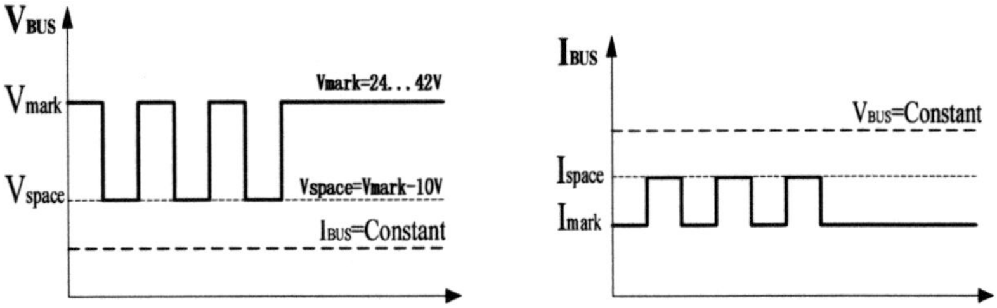

(a) (b)

Fig. 1. Definition of transmission bit

3 The Hardware of the Intelligent M-Bus System

3.1 General Introduction of M-Bus System

The intelligent meter reading system designed in this paper is used to collect information from heat meters. According to the number of heat meters in practical application, the network types can be divided into two kinds. There are single -stage network mode and two-stage network mode. The maximum of meters connected in the former system is 250. The structure is shown in Fig.2 (a). The level translator is connected to PC through RS232 port and the Bus through M-Bus ports. The network topological is achieved in another system. The structure is shown in Fig.2 (b). The first class level translator is used as level conversion. The second class level translator is used as repeater to achieve free topological function.

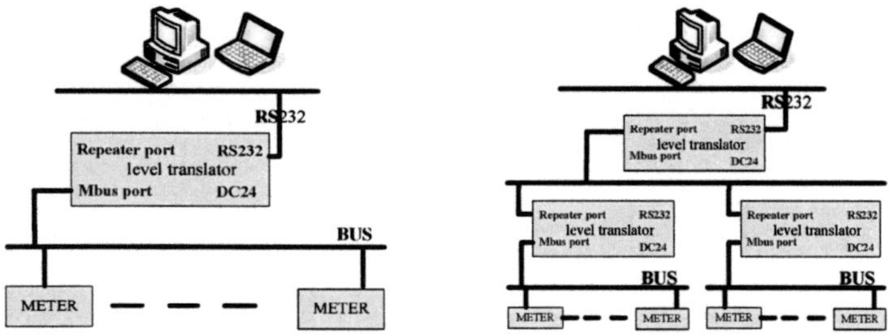

(a) (b)

Fig. 2. The Structure of the intelligent M-Bus system

3.2 Intelligent Meter

The intelligent meter selected in the system is Kamstrup heat meter with Multical401 module. Every Kamstrup heat meter will be set an 8 digit ID code before it is dispatched from the factory. The ID code can be used as registered address code. The last 3 digit is used as first address code. There are two communication modes between master and heat meters, first address mode and full address mode.

The first address mode: the last 3 digit of registered address code is used as communicating address. If the last 3 digit of the registered address code exceeds 250 the last two digit will be used as the first address code. The instruction format as follows:

10	**C**	**A**	**csum**	**16**

"10": start character. "C": command instruction,"08h", initialization of slave,"5Bh", collect data. "A": first address code, a hexadecimal number. "csum": check sum. "16": stop character.

Furthermore, there are four special addresses with the following functions:

Address 000: Ordinary primary address, reserved for non-configured M-Bus Slaves. Address 253: Used for secondary addressing. Only the M-Bus Slave that accepts the address replies. Address 254: All M-Bus Slaves will reply to this address. The address must be used in systems with one connected M-Bus Slave only e.g. for test. Address 255: No M-Bus Slaves will reply to this address but they will all receive the message. This address makes it possible to change the baud rate of a whole system at a time sending one telegram.

The full address mode: In real application, we can take the full address code as communicating address to avoid repetition of first address code. The instruction format as follows:

Long frame identifier	68
Length	0B,0B
Long frame identifier	68
"C"	53 or 73
"A"	FD
"CI"	52
SerialNumber	S_{low}...S_{hign}
Manufacture Code	Man_{low}...Man_{hign}
Meter Version	Ver
Medium Code	Med
Checksum	csum
Stop Byte	16

"Long frame identifier": start character. "Length": the length of. "C": command instruction. "A" and "AI": fixed number. "SerialNumber": full address mode, a hexadecimal number. "Meter Version": version number of the meter. "csum": check sum. "16": stop character.

If the instruction as above format is transmitted successfully, it will be answered"E5h"[2].

3.3 M-Bus Level Translator

When the M-Bus level translator is used as network repeater, RM+ and RM- ports (no polarity) will be connected to the host ports Mbus+ and Mbus- . The M-Bus level translator connected to PC can monitor the answer signal of the master and slave . The whole meter reading system network will become huger, achieving free topology.

4 The Software of Intelligent M-Bus System

4.1 Introduction of System Development Software

C Sharp (C#) language is designed by Microsoft Corporation for NET Framework, which is an object-oriented programming language. C# supports three qualities: encapsulation, inheritance, and polymorphism. As a part of Net frame, C# language is a simple and powerful tool like Visual Basic and C/C++ to application developers [3,4]. Given the above merits, the software of the intelligent M-Bus system is developed in C#.

4.2 The Program Flow Graph for the Software of Intelligent M-Bus System

The program encompasses three major parts: the settings for the port, connecting to the database, and acquainting of heat meter data. The program flow graph for the intelligent M-Bus system software is shown in Fig.3.

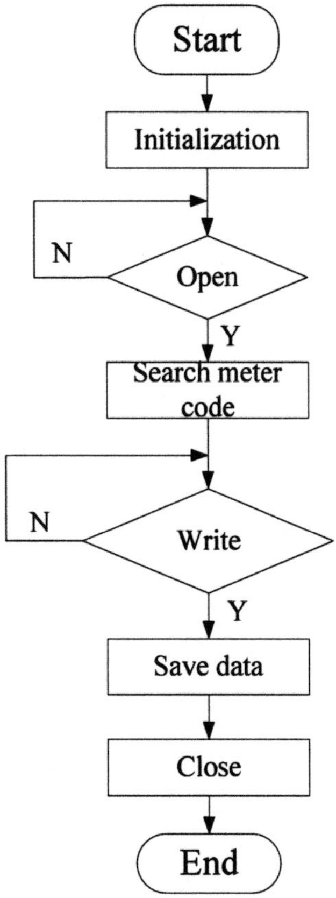

Fig. 3. The program flow graph of the software

4.3 The Application of C# Serial Ports

Serial port is one part of system resource under windows. The application program must apply application (open serial ports) to system before using the serial ports. The source must be released (close serial ports) after completing the communication [3-5]. In the C#, we can find Serialport category. The Serialport category can supply communicating function between PC and level translator. There are baud rate, parity, data bits and stop bits should be set to match with the communication requirements.

4.4 Analysis of key Codes

4.4.1 Data Send

The full address numbers of heat meter are saved in the database. The master need to acquire the full address numbers before collecting heat meter data, making conversion between number systems. The key codes as follows:

{... MyDS_Grid =GetDataSet("Select * from tb_Basic where(subdistrict ='" + comboBox4.Text + "')and(name='" + textBox1.Text + "')and(number ='" +

textBox6.Text + "')and(state='in use')", "tb_Basic");// Acquiring the full address number of meter and other message based on the select conditions.

...arrStr[0] = Convert.ToString (dataGridVie w1[6, 0]. Value).Trim();//Achieving the binding between the data in database and DataGridView control, then searching DataGridView table to acquire the heat meter registered address code and saved in the array arrStr...}.

{...t[0] = Convert.ToByte(Num.ToString("X"), 16);// Converting the registered address code into hex and saved in the array t. The serialPort category is used of controlling file resource of the serial port. The category supplys the synchronous I/O driver and the event I/O driver. It can also allow visiting the port, the interrupt state and the driver of serial port [5] }.

{...serialPort1.Write(t, 0, 5);//Writing the 5 bytes in the array t into the output buffer...}.

4.4.2 Data Reception

Data reception is one of the most important parts. It is also the most important function in the intelligent meter reading system. The key codes as follows:

{...Thread threadReceiveSub = new Thread (new ParameterizedThreadStart(AsyReceiveData)); //Using to receive data asynchronously.

threadReceiveSub.Start(serialPort);//Opening the thread.

int intlength = serialPort.BytesToRead;// Delimiting the length of data in the input buffer.

serialPort.Read(combuffer, 0, intlength); //Reading intlength bytes data, and saved in array combuffer...}.

4.5 Features of Application Software

In this article, an efficient, secure, intelligent M-Bus software is designed and implemented, according to analysis the protocol of serial port. The main characteristics of the M-Bus application software are shown as follows: (1) authority set: The authority functional module is designed and layer of management is set to achieve graded management. Fig.4 shows the login interface. (2) management of the heat meter user : the managers of software can add, delete, and modify the information of users. It can bring great convenience for managers in the practical application process. (3) automatic reading : the software can collect data in one key type from the meter which is effectively connected to the Bus, and make a seismic acquisition to the meter whose data is loss. Meanwhile, the manager can observe the progress of reading meter. (4) data inquiry: inquiring operation of history data, valid data. The result of inquiry can be printed. (5) data maintenance : The data can be imported and exported; it's convenient to maintain the database to avoid data loss according to incorrect manipulation and unexpected problems. (6) report generation: the report of history data can be generated. Fig.5 shows the main interface graph of the system software.

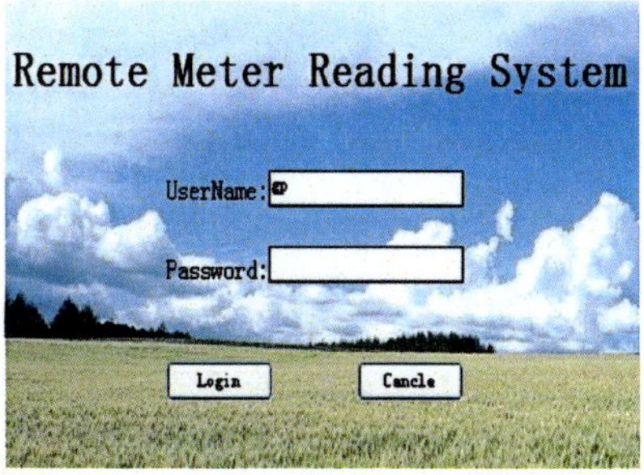

Fig. 4. Login interface

Fig. 5. Main interface graph

5 Conclusion

In this paper, an efficient, intelligent meter reading system is designed and developed, according to analysis of the meter bus protocol and the actual conditions. Practical application shows that the system has the advantages of simple operation, complete function and accurate data acquisition. The system achieves the desired purpose. From a certain extent, the system meets the new requirements of modern people in the intelligent building system.

References

1. The M-Bus: An Overview, http://www.M-Bus.com/infombus.php
2. Technical Description-bus alave for MULTICAL401
3. Wang, X.: The Development Toolkit of C#. Tsinghua University Press (2010)
4. Albahari, B.: The Essence of C#. Beijing Institute of Technology Press (2001)
5. Gong, X.: Realization and Application of Serial Communication in VS2008. Computer & Telecommunication (2011)
6. Xue, H.: Design and Implementation of Kamstrup Heat Meter Communication Module. Science Mosaic (2007)

Fuzzy Path Planning of Two-Wheeled Robot Optimized by Gold Mean

Peng Shu-hua, Hao Cui, and Li Deng-hua

Automation School of Beijing Information Science and Technology University,
Beijing 100101, China

Abstract. Aiming at the problem of path planning of two-wheeled robot, based on the principle that the path curve should be smooth and the total length of the path should be short, a fuzzy control method to the whole velocity and the real-time rotational angle was proposed. The membership functions of the fuzzy controller were optimized by golden mean principle, and which simplified the design of the fuzzy membership function. The experimental results show that, compared with traditional fuzzy controller, the fuzzy controller optimized by golden mean has better control effect. Robot has better real-timing when avoiding obstacles and the global path is shorter.

Keywords: Fuzzy theory, two-wheeled robot, path plan, golden mean.

1 Introduction

In these years, the fuzzy control has developed rapidly in automatic control field. The computers can judge and process some fuzzy information with the fuzzy theory. The fuzzy theory is used widely. In reference [1], the fuzzy control with variable domain was applied to the maximum power tracking system. In reference [2], the application of adaptive fuzzy control used in the electric furnace was introduced and the optimal fuzzy control was applied to automatic robot car in reference [3].

About the problem of robot path planning, at home and abroad, many scholars have explored a lot and do many experiments. Many methods and theories have been proposed. The most basic methods are: artificial potential field method, grid algorithm, genetic algorithm and fuzzy control method. Recently, there are many basic algorithms based on the derivatives of these algorithms. Reference [4] used particle swarm optimization algorithm to accomplish the path planning of soccer robots. Reference [5] applied immune evolutionary algorithm to plan the path of soccer robots. In reference [6] and [7], the path planning algorithms based on optimal cot arc and Bezier curve were proposed, respectively. The genetic algorithm was used to optimize the Bezier curve. In reference [8], the combined method with artificial potential field and genetic algorithm was used in the robot path planning and the effect was very good. According to the research results of real ants, the robot path planning algorithm based on the collaboration of scouts and forager was proposed in reference [9]. In reference [10] and [11], ant colony

D. Yang (Ed.): Informatics in Control, Automation and Robotics, Volume 2, LNEE 133, pp. 477–484.
springerlink.com

immune network method based on dynamic artificial potential field and potential field method were used in the robot route planning, respectively.

In this paper, the optimization method in [12] is used in the path planning problem of two-wheeled robot successfully. The golden segment optimization method was applied to design the fuzzy membership functions and the fuzzy controller based on golden segment optimization is proposed to plan the route. The design process is given and finally, the experiments and simulation results are given.

2 Problem Description

Consider a kind of two-wheeled independent mobile robots with no-sideslip and pure rolling as shown in figure 1. The movement of the robots mainly includes two parts: the translation of center and the rotation around the centroid. Both the two movements are accomplished by controlling the wheel speed of the left and right wheels.

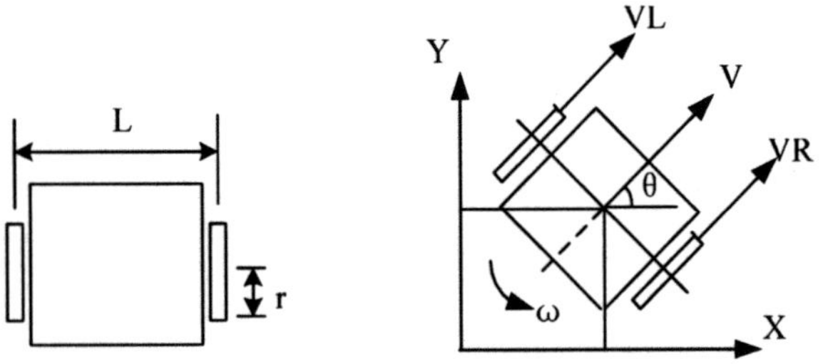

Fig. 1. schematic diagram of two-wheeled robot

The centroid motion equation of the robot is as below.

$$\begin{bmatrix} \dot{x} \\ \dot{y} \\ \dot{\theta} \end{bmatrix} = \begin{bmatrix} \cos\theta & 0 \\ \sin\theta & 0 \\ 0 & 1 \end{bmatrix} \begin{bmatrix} v \\ w \end{bmatrix} \tag{1}$$

Robots meet the rigid motion rule, then:

$$\begin{cases} v = \dfrac{v_L + v_R}{2} \\ w = \dfrac{v_L - v_R}{2} \end{cases} \tag{2}$$

Where, v_L、 v_R are the line speed of the left and right wheels; v is the centroid line speed of robots; w is the centroid angular velocity; L is interval of the two wheels.

Robot dynamic obstacle-avoidance refers to these kinds of problems: In the real-time practical environment, the robots can arrive the target point G from starting point S without any collisions. In order to solve these problems, firstly, the system demands to get the real-time information of the realistic environment, such as the attitude information of the robot, the range information between the robot and the obstacles, the direction information between the robot and target point, etc. Secondly, the impacts of the robot movements caused by obstacles and target point should be determined. As can be known form the analysis of the actual situation, the impacts of obstacles at different directions and different distances are different to the robots. Generally speaking, the impacts of obstacles in front of the robot are great to the robot, but the impacts of obstacles in the side of the robot are smaller; the impacts of obstacles that close to the robot are great to the robot, but the impacts of obstacles which are far away from the robot are little. The relative position among the target point, obstacles and robot decides the direction and side of the turning angle of the robot, for example, 'left big', 'right small', etc. These descriptive languages are the fuzzy language that fuzzy logic system can express and identify well.

The position coordinates of the starting point S and target point G are known. Assume that the position and attitude information (x_R,y_R,θ) of the robot and the position of the obstacles can be got in real-time. In the local coordinate system as shown in figure 2, it can be known from calculating: (1) the distance between the obstacles and robot (d); (2) the distance between the robot and target point G (L); (3) the angle difference is $\theta=\beta-\alpha$, where, α, β are the angular separations of d、L and horizontal axis X of local coordinate system. According to figure 1, the achievement of the obstacle avoidance of soccer robots is that when the robot are constantly close to the target point G, it is determined to turn left or turn right and the size of rotation angle according to angle difference θ. If the distance d of the obstacles is 'far', the problems of obstacle avoidance are not needed to consider of. The rotation direction is changed according to the angle β, the size of speed v is determined by the distance L. If the distance d is 'small', it is going to consider the obstacle avoidance and at this time, when the instantaneous speed v should be kept, the rotation direction should be changed according to the angle θ.

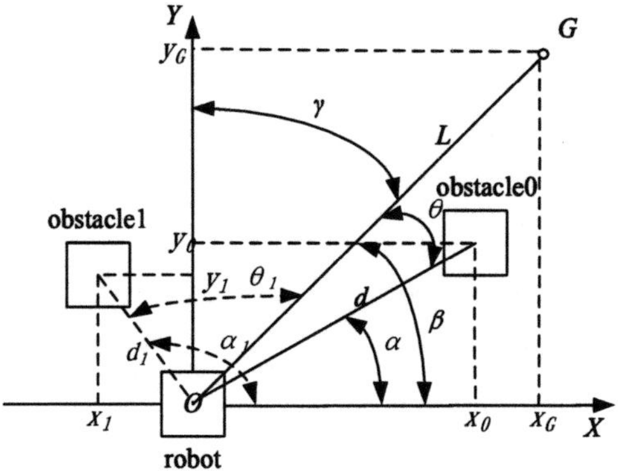

Fig. 2. Local coordinate system schematic diagram

In the current cycle, the position of the obstacle is (x(0),y(0)), then after k time cycles, the position of the obstacle is as below:

$$x(k) = x(0) + \sum_{j=1}^{k} (v(0) + k\Delta v)\cos(\theta(0)+k\Delta\theta)\Delta T$$
$$y(k) = y(0) + \sum_{j=1}^{k} (v(0) + k\Delta v)\sin(\theta(0) + k\Delta\theta)\Delta T \quad (3)$$

3 Fuzzy Route Planning

How to solve the dynamic obstacle-avoidance problem mainly depends on the positions of the starting point, the target point and the position of the obstacle and other factors. Therefore, in the fuzzy controller, the distance d of the obstacle, the angle difference θ and the distance to the target point L are the input variables of the fuzzy system; the speed and the rotation angle are the output variables. From these, the whole structure of the fuzzy controller is shown in figure 3. It can be accomplished by two steps:

(1) Build the fuzzy inference unit of speed;
(2) Build the fuzzy inference unit of angle.

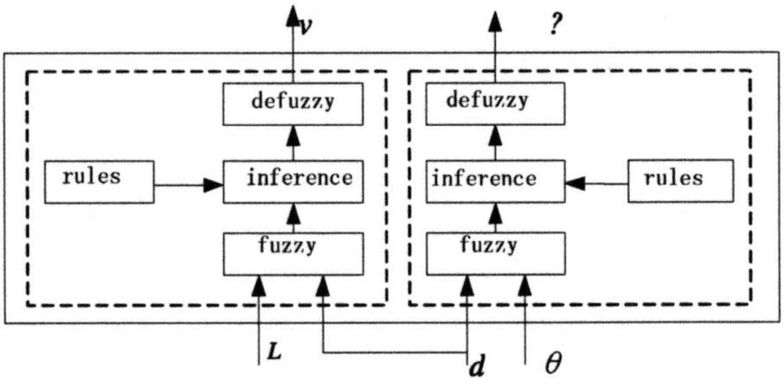

Fig. 3. The structure figure of the fuzzy controller

The conditions of the fuzzy rules in the speed fuzzy controller include the distance d, the distance to the target point L, it needs to consider and analyse the multi-input synthetically using the fuzzy synthetic computation. The general fuzzy synthetic method can be achieved by using 'and' and 'or'. In this paper, the 'and' is used to operate it. There are two 'and' operation methods, which are the min and prod methods. Prod method is applied in this paper.

Before the fuzzy inference, the weight problem of different fuzzy rules must be considered, because for multi rules system, the important degree of each rule is possibly different. Set the weight 1 in this paper, and then the fuzzy implication can be calculated. The input of the fuzzy implication is got from the calculation of the synthetic computation of fuzzy input set. The output is the conclusion fuzzy set of the fuzzy rules derivation. Similar with the fuzzy synthesis method 'and', the fuzzy implication also have two methods: minimum (min) and product method (prod).This paper adopts minimum contain calculation method.

In the fuzzy inference system, the generation of decision depends on all the fuzzy rules, so the above step calculated fuzzy output must pass a certain way synthetic computation result namely output synthesis. In the fuzzy logic system, the results of all the rules are inferred at first, and then all the results are calculated synthetically. The synthetic methods are: maximum method, the probability method, sum method. In this paper, maximum method is adopted to complete the calculation of many rule interference result. In deblurring, the most often deblurring methods: area center method (centroid) is adopted in this paper. Based on the analysis of the fuzzy controller of the input fuzzification, fuzzy inference, and the output, the fuzzy controller model of the speed can be built. The fuzzy rules between input and output are shown in table 1.

Table 1. Fuzzy Rules

d \ v / L	NE	M	FAR
NE	S	S	S
M	S	M	M
FAR	S	M	B

The conditions of fuzzy rules of angle fuzzy controller are multiple-input including distance d, angle θ. In the conditional fuzzy synthesis, the 'prod' method is applied using 'and' adopting operation method. The fuzzy controller of angle can be built and the fuzzy rules between input and output are shown in figure 2.

Table 2. Fuzzy Rules

q / p / d	NL	NM	NS	ZE	PS	PM	PB
FAR	TZ	TZ	TZ	TZ	TZ	TZ	TZ
M	TRS	TRS	TRM	TRM	TRM	TLS	TLS
NE	TRM	TRM	TRB	TRB	TZ	TZ	TZ

According to the reference [12], the optimization of the input and output is designed as below.

Set the domain of the input variable of the fuzzy controller is [-L,L], and it is longitudinal axisymmetric, let the fuzzy subset number is N=2J+1≥3, where, J subsets are negative, one is zero, and J subsets are positive. Every fuzzy subset applies the triangle membership function of golden section ratio. Golden section law was applied, and the distribution of centre point αi of every fuzzy subsets can be expressed as: αi=0.618αi+1 (i=-J, -J+1,…,-1, 0, 1, …,J-1, J), where, the centre point αJ of the Jth fuzzy subset at the boundary L of the domain and α0 is at central shaft. The membership function of golden section optimization is shown as below:

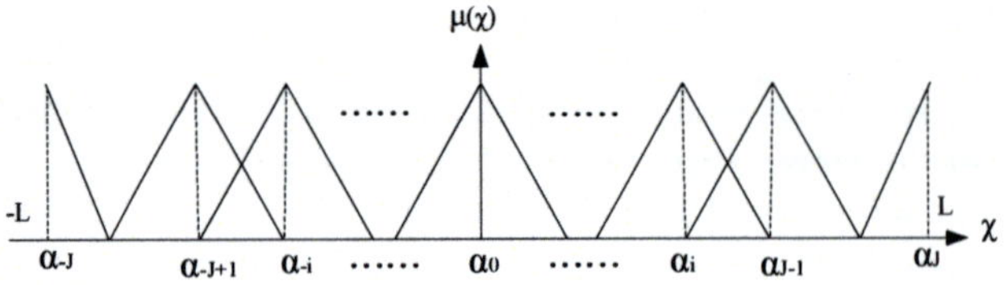

Fig. 4. The membership function of the golden section optimization

4 Experiment Results

The initial point coordinate of the experiment is (0,0), and the target point coordinate is (60,100), the coordinates of the obstacles are (40,60), (28,66), (-10,43), respectively. The experiment results of the soccer robot route planning based on the general fuzzy membership function and golden section ratio theory are shown in figure 5 and figure 6.

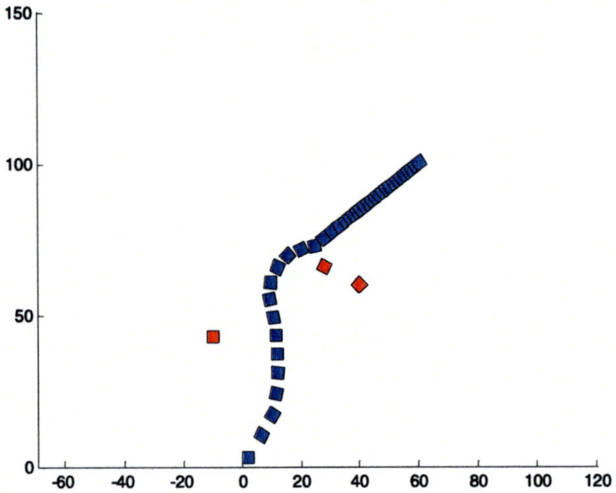

Fig. 5. The route planning result of the soccer robots based on the fuzzy membership function

From the result figures of the two control methods, the fuzzy controller with golden section optimization has better control effect. Robots have better real-time performance while avoiding the obstacles and the overall path is shorter, costing less time to the destination.

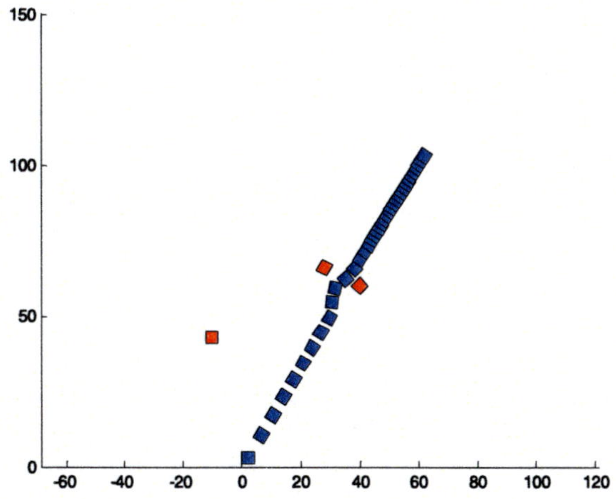

Fig. 6. The route planning result of the soccer robots based on golden section ratio theory

5 Conclusion

The optimization of the membership function in the fuzzy controller design plays a very important role. In the fuzzy control, if the shape and distribution of the membership function can be tot according to the practical experience only, there are many subjectivities and the results are not ideal. The golden section and its subordinate number-theoretic methods have been successfully used in many research and actual work areas, but in control field, the applications are relatively few. In this paper, the golden section law is used to optimize the triangle membership function and is applied to the route planning of the soccer robots. The experiment results show that this method has obtained satisfactory effect on the route planning of the soccer robots. It is simple, clear, easy achievement. It follows the general design rules of the membership function design, and it also has the reliable theory foundation. It can be as the optimization design of the fuzzy control parameters and provide a better process for the industrial process control.

References

1. Yang, Z.-J., Chen, Y.-R., Hu, J.-M.: The research on application of variable universe fuzzy control to maximum power point tracking system. In: 3rd International Conference on Power Electronics Systems and Applications 2009, May 20-22, pp. 1–4 (2009)
2. Teng, F., Li, H.: Adaptive fuzzy control for the electric furnace. In: IEEE International Conference on Intelligent Computing and Intelligent Systems 2009, November 20-22, vol. 2, pp. 439–443 (2009)

3. Farhi, O., Chervenkov, Y.: Optimal fuzzy control of autonomous robot car. In: 4th International IEEE Conference on Intelligent Systems IS 2008, September 6-8, vol. 1, pp. 4-62–4-65 (2008)
4. Wang, L., Liu, Y., Deng, H., Xu, Y.: Obstacle-avoidance path planning for soccer robots using particle swarm optimization. In: 2006 IEEE International Conference on Robotics and Biomimetics, pp. 1233–1238 (2006)
5. Xu, X., Yan, G., Xie, K., Lin, T.Y.: Immune evolution algorithm based dynamic path planning approach for the soccer robot. In: Proceedings - 2007 IEEE International Conference on Granular Computing, pp. 477–480 (2007)
6. Liu, Z., Wang, J.: Soccer robot path planning based on Arc-cotangent optimization algorithm. In: 2008 IEEE International Conference on Robotics, Automation and Mechatronics, pp. 1062–1065 (2008)
7. Yang, L., Luo, Z., Tang, Z., Lv, W.: Path planning algorithm for mobile robot obstacle avoidance adopting Bezier curve based on genetic algorithm. In: Chinese Control and Decision Conference, pp. 3286–3289 (2008)
8. Xu, X., Xie, J., Xie, K.: Path planning and obstacle-avoidance for soccer robot based on artificial potential field and genetic algorithm. In: Proceedings of the World Congress on Intelligent Control and Automation, vol. 1, pp. 3494–3498 (2006)
9. Zhu, Q.-B., Ma, W.: A robot path planning algorithm based on scout ants in collaboration with foraging ants. Control and Decision 24(4), 601–605 (2009)
10. Yin, L., Yin, Y.-X.: Simulation Research on Path Planning Based on Dynamic Artificial Potential Field. Journal of System Simulation 21(11), 3325–3341 (2009)
11. Yuan, M.-X., Wang, S.-A., Li, K.-P., Chen, N.-J.: Ant Colony and Immune Network Algorithm Based on Optimization of Potential Field Method for Path Planning. Journal of System Simulation 21(15), 4686–4690 (2009)
12. Fang, Q.: An optimization design for membership function of fuzzy control. Journal of Electronic Measurement and Instrument 173, 42–45 (2003)

Research on HVDC Model in Transient Voltage Stability Analysis of AC/DC Transmission Systems

Zhichang Yuan[1], Licheng Li[2], Yongjun Liu[3], and Shukai Xu[3]

[1] Department of Electrical Engineering, Tsinghua University, Beijing 100084, China
[2] China Southern Power Grid Company, Guangzhou 510080, China
[3] Electric Power Research Institute of China Southern Power Grid, Guangzhou 510080, China

Abstract. Fast Simplified models of HVDC are widely used in large-scale AC/DC power system transient voltage stability analysis. An important concern of different models is that to what extent the analysis results are impacted by using simplified models. Under voltage reactive power characteristic of HVDC is analyzed based on CIGRE HVDC test benchmark. It is observed that reactive power absorbed by HVDC is proportional to 1.6-th power of AC bus voltage, while the reactive power supplied by fixed capacitors of converter station is proportional to square of AC bus voltage. The above feature makes HVDC harmful to transient voltage stability. Among different simplified models of HVDC, constant power and constant impedance model are quite inconsistent with actual HVDC model, while Quasi-Steady-State (QSS) model is fairly consistent with actual HVDC model. So QSS model is suitable for large-scale AC/DC power system transient voltage stability analysis. Main conclusions are validated through simulations on Chinese Southern Power Grid.

Keywords: AC/DC power system, Transient voltage stability, Simplified models of HVDC, HVDC test benchmark.

1 Introduction

In recent years, several blackout cases caused by transient voltage instability have occurred abroad, therefore transient voltage instability is considered as a major threat of power outage in load centers, transient voltage stability attracts more and more research interests.

Studies suggest that AC-DC parallel transmission is a factor that may cause transient voltage instability, which contains two voltage instability scenarios: Firstly, after outage of HVDC transmission system failure caused by unipolar or bipolar block, power flow may shift to parallel AC lines, which increase reactive loss of AC lines and decrease bus voltage of load side. If reactive power capability of load side is not enough, it may lead to transient voltage instability. Secondly, when AC system faults occur near HVDC converter station, drop of AC bus voltage will cause change of reactive power in HVDC transmission system, which may deteriorate reactive power imbalance and lead to transient voltage instability.

D. Yang (Ed.): Informatics in Control, Automation and Robotics, Volume 2, LNEE 133, pp. 485–493.
springerlink.com

Currently, analysis on transient voltage instability mechanism of AC-DC system focused on the first scenario, that is, the impact on AC system by the shift of power flow caused by DC blocking. However the second scenario is not analyzed thoroughly [5-10].

In the field of AC-DC power system simulation, HVDC models are commonly used in three ways: constant power model, the quasi-steady-state model and the electromagnetic transient model [11]. Constant power model and the quasi-steady-state model have been widely used for their simple structure and less computational burden. However, whether the constant power model and the quasi-steady-state model can reflect the feature of influence between AC and DC system during AC faults has not been reported [12].

Based on electromagnetic transient model of AC-DC system, this paper analyze reactive power characteristic of HVDC under AC system fault and voltage drop, depicts the principle of transient voltage instability caused by reactive power imbalance of AC-DC system. Then, power versus voltage curve of constant power model and quasi-steady-state model are calculated which are compared with electromagnetic transient model, through which the feasibility of simplified models applied to transient voltage stability analysis is discussed. The conclusions are verified through simulation results on Chinese Southern Power Grid.

2 Reactive Power Characteristic of HVDC under Low Voltage

In 1991, CIGRE issued HVDC transmission system benchmark [13], which is widely used in comparative study on different control devices and strategies. Structure and main parameters of the benchmark are listed in [13].

This section describes reactive power characteristic of HVDC system under AC voltage drop, which are conducted on CIGRE HVDC benchmark and EMTDC-PSCAD simulation platform. Electromagnetic transient model of HVDC is applied.

As the load centers are prone to transient voltage instability, in order to test the impact of AC voltage drop on HVDC reactive power, AC voltage of inverter is gradually reduced while voltage of converter remains constant. Different reactive power of HVDC system with AC voltage varied are recorded after AC-DC system get settled. Fig. 1 shows the simulation results in which AC voltage of inverter decreased from 1.0pu to 0.6pu in 10 steps, duration of each voltage remains 0.5 seconds so as to ensure that the oscillating component decay and HVDC power get settled.

Least-squares regression analysis is applied to data in Fig. 1, which shows that voltage-dependent reactive power characteristic of HVDC converter can be roughly expressed as:

$$Q_{HVDC}=Q_{HVDC0}\times(V/V_0)^{1.6} \quad (1)$$

Where V represents AC bus voltage of converter station and V_0 represents rated voltage, Q_{HVDC0} is reactive power absorption by HVDC converter under rated voltage, Q_{HVDC} is reactive power absorption by HVDC converter when voltage drops to V. Curve of reactive power of HVDC converter station with AC bus voltage drops is shown in Fig. 2. For comparison, Fig. 2 also give constant reactive power characteristics, constant current characteristic and constant impedance characteristics, and characteristic of formula (1). It can be see that HVDC reactive power characteristics is roughly between constant current and constant impedance, when AC voltage is at [200kV 227kV], voltage-dependent reactive power characteristics is close to constant current characteristics, when AC voltage drops further, it would be closer to constant impedance characteristics. Overall, HVDC reactive power characteristic can be approximate by formula (1), which shows reactive power of HVDC is proportional to 1.6-th power of AC voltage of converter station.

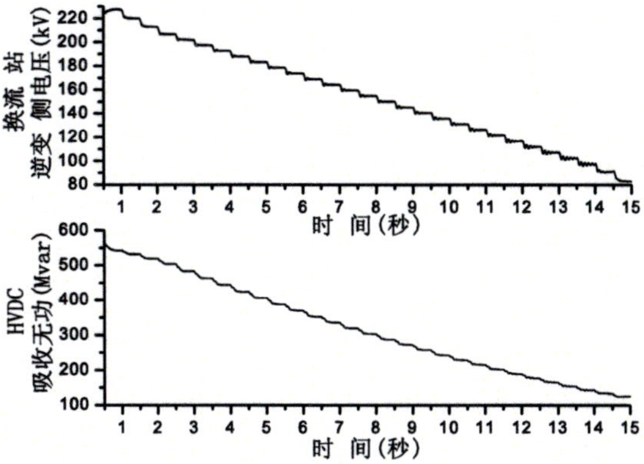

Fig. 1. Simulation results of CIGRE HVDC test bench

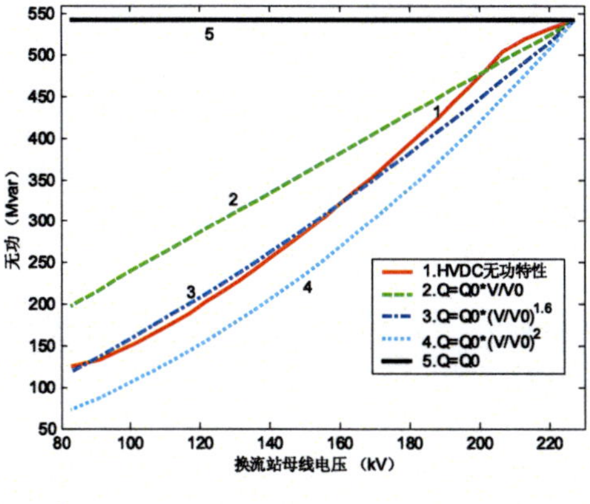

Fig. 2. Under voltage reactive power curve of HVDC

Sufficient shunt capacitor banks usually are installed in converter stations in order to remain balance of reactive power, under rated operating conditions, capacitor banks provide reactive power absorbed by HVDC converter station and reactive power exchange between AC and DC system is close to zero. Under the condition of AC system fault and voltage drop, one hand, reactive power absorb by HVDC converters decrease by the 1.6-th power of AC voltage (as shown in curve 1, Fig. 2), at the other hand, reactive power provided by shunt capacitor banks decreases by the square of AC voltage(as shown in curve 5, Fig. 2). The difference of reactive power dropping rate between HVDC converters and capacitor banks make imbalance of HVDC system reactive power, and the gap must be compensated by AC system nearby. Fig. 3 shows net reactive power exchange between HVDC and AC system. It can be drawn that with AC bus voltage dropping, HVDC converter need absorb more and more reactive power. When voltage is 206kV (0.93pu), reactive power absorbed by HVDC converter reaches peak (55Mvar), after that, reactive power exchange remain between 45Mvar and 55Mvar.

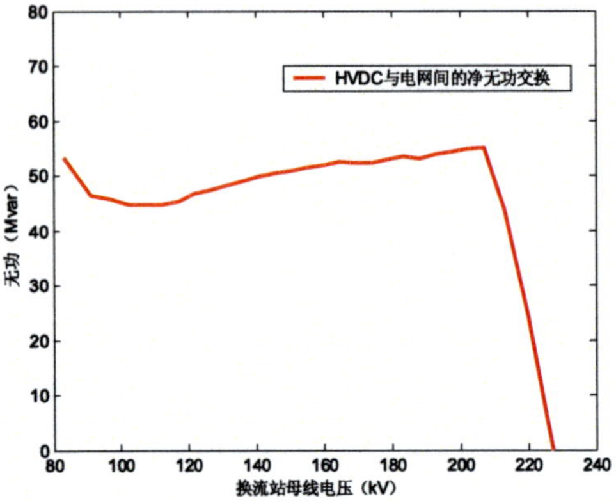

Fig. 3. Net reactive power exchange between HVDC and AC system

As can be seen from the above analysis, under the condition of AC voltage drop, HVDC system can not maintain reactive power balance and need to absorb reactive power from AC system nearby, which is an unfavorable factor to transient voltage stability of AC transmission system. Under the worst case, if AC fault occurs close to converter station, motor load may install under low voltage which make shortage of reactive power, AC system may running at the edge of transient voltage collapse. At the same time HVDC converters will absorb large amount of reactive power which enlarges reactive power gap. This scenario may end up with transient voltage collapse of AC system.

3 Reactive Power Characteristic of Simplified HVDC Models

Quasi-steady-state model of HVDC converter is widely adopted in power system electromechanical transient simulation program, whose basic circuit structure is shown in Fig. 4. Using this model, AC bus voltage is taken as commutation voltage and all controlled angle of converter are defined on the base of commutation voltage, transformer leakage reactance of converter is taken as the commutation reactance. Thus, there are not interactions between bridges in multi-bridge inverter, and all variables can be calculated independently. The quasi-steady-state model of single bridge inverter can be defined as the following formula.

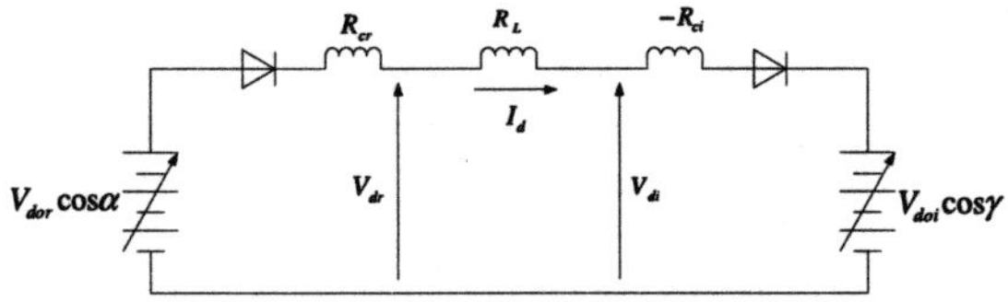

Fig. 4. Stable Model of HVDC

$$V_{d0}=4.242BTE_{ac}/\pi \tag{2}$$

$$V_d=V_{d0}\cos\alpha-3X_cI_dB/\pi \tag{3}$$

$$\cos\varphi\approx V_d/V_{d0} \tag{4}$$

$$P=V_dI_d=P_{ac},\ Q=P\tan\varphi \tag{5}$$

$$I_{ac}=6BTI_d/\pi \tag{6}$$

$$0.707E_{ac}[\cos\alpha-\cos(\alpha+\gamma)]/X_c=I_d \tag{7}$$

Where V_{d0} is ideal DC voltage with no load, E_{ac} is AC bus commutation voltage, B is number of serial converters, T is turns ratio of converter transformer, V_d, I_d, P_d, Q_d are DC voltage, current, active power and reactive power respectively, $\cos\varphi$ is power factor of converter, α and γ are firing angle and commutation angle of rectifier, X_c is leakage reactance of converter transformer converted to the valve side.

During operation, DC voltage and current are controlled through regulating thyristors firing angle of rectifier side and inverter side. Usually the rectifier side is controlled as constant current mode, and inverter side as constant voltage mode. With AC bus voltage of converter station dropping, HVDC may deviate from normal state, when the firing angle of rectifier is less than 5°, minimum firing angle mode will be triggered, and when the extinction angle of inverter is less than 15°, constant-extinction angle mode will be triggered.

According to the above control law, the paper calculated reactive power absorbed by HVDC converters when AC voltage decreased gradually, the results based on quasi-steady-state model are compared with those base on CIGRE benchmark. As shown in Fig. 5, Curves 1 is reactive power characteristic curve of CIGRE benchmark, curve 2 is curve of quasi-steady-state model, curve 3 is curve of constant power model and curve 4 is curve of constant reactance model.

Comparing the four curves in Fig. 5, it can be seen that constant power model is nothing like actual reactive power characteristics of HVDC. While in constant impedance model, HVDC is modeled as constant reactance which is same as capacitor banks, so reactive power absorbed by HVDC is always compensated by capacitor banks, and AC system is not affected by HVDC at all. Both constant power and constant reactance models are inconsistent with the actual characteristic because they can not accurately reflect the influence of HVDC system to transient voltage stability of AC system. Quasi-steady-state model is the most close to CIGRE benchmark in respect of reactive power characteristic, and in can replace the complex standard model in electromechanical transient simulation. It should be noted that the reactive power absorbed by quasi-steady-state model is slightly higher than the standard model, so the transient voltage stability analysis results with quasi-steady-state HVDC model will be slightly conservative.

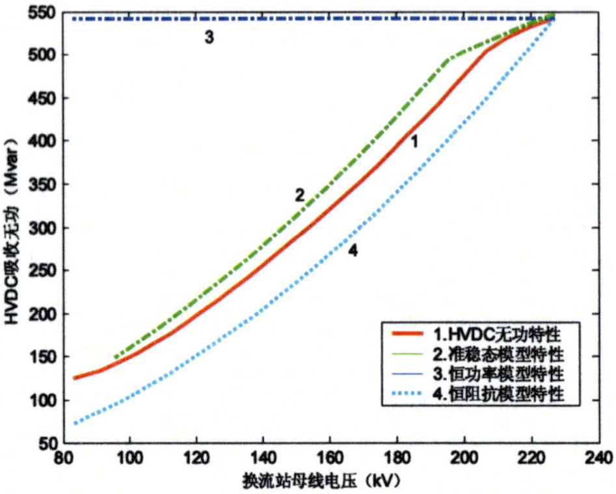

Fig. 5. Under voltage reactive power curve of different HVDC model

4 Simulation

In order to verify the influence of HVDC system under low voltage to transient voltage stability of AC system, the hybrid AC-DC transmission system of Chinese Southern Power Grid simulation is selected as simulation example. West-east corridor in Southern Power Grid is shown in Fig. 6, Electric power from Guangxi, Guizhou and Yunnan province is transferred to Guangdong Province through 8 AC transmission lines and 3 500kV HVDC transmission lines. Xijiang, Luodong, Beijiao, Huadu and Zengcheng 500kV substation are receiving side of HVDC transmission line, and they are also load center. Above factors make this area prone to transient voltage instability.

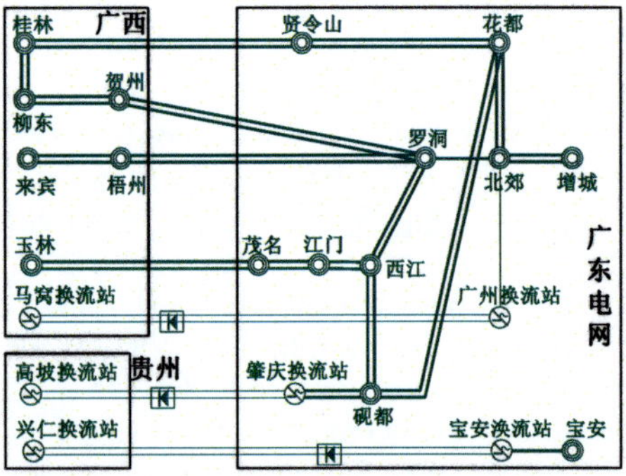

Fig. 6. Power transmission network diagram of Southern Power Grid

Fig. 7-Fig. 8 shows simulation results under three phases to ground fault in Luodong-Xijiang 500kV line, and the fault is cleared by relays after 0.1s. Load model are made up of 60% motor and 40% constant impedance load.

Fig. 7 depicts unipolar reactive power of Gaopo-Tianshengqiao and Zhaoqing-Guangzhou HVDC lines. At the 3rd second, AC fault occurred at 500kV Luodong substation, AC voltage dropped at Zaoqing and Guangzhou converter station and reactive power absorbed by those converters decreased. Reactive power absorbed by converters decreased by the 1.6th power of AC voltage, while reactive power provided by shunt capacitor banks decreased by the square of AC voltage. In order to compensate the reactive power gap due to different dropping ratio of converters and capacitors, reactive power should be compensated by AC system nearby. Reactive power flows from Guangzhou converter station to Beijiao substation is 20 Mvar before fault, while after fault reactive power reverse and the amount becomes 200Mvar. It will deteriorate reactive power imbalance of Beijiao substation and consequently affat nearby Zengcheng and Hengli heavy load area.

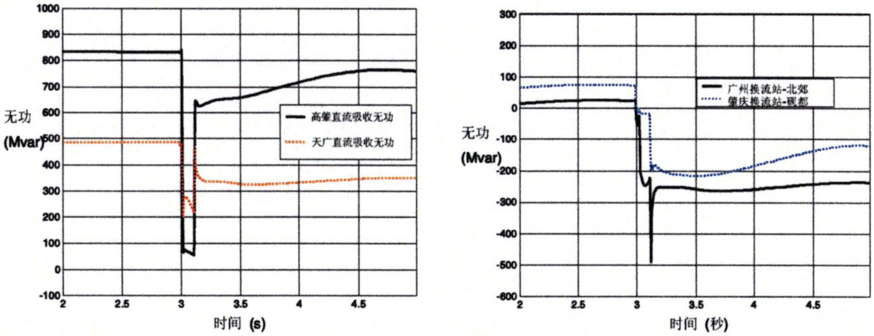

Fig. 7. Reactive power flow after AC fault

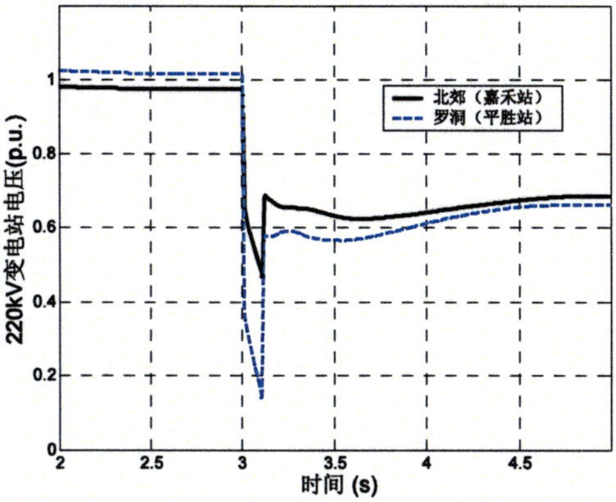

Fig. 8. voltage of AC bus near HVDC

Fig. 8 depicts AC bus voltage of Jiahe and Pingsheng 220kV substations located in Beijiao load area and Luodong load area respectively. After fault, voltage decrease to below 0.7pu and never recover, indicating transient voltage collapses in this load area. It should be noted that transient voltage instability occurs in this area mainly because of heavy load and high proportion of electrical motor load, while reactive power characteristic of HVDC deteriorate reactive power imbalance of this area and make situation even worse.

5 Conclusions

When AC bus voltage drops in HVDC converter stations due to AC disturbances, reactive power absorbed by HVDC system is roughly 1.6th power of bus voltage, while reactive power provided by capacitor banks is square of bus voltage. As a consequence, HVDC will absorb extra reactive power from AC system nearby which may cause or worsen transient voltage instability of AC system. This characteristic of HVDC should be take into consideration when analyzing transient voltage stability of bulk power system. Among commonly used simplified models of HVDC, constant power model and constant impedance model severely deviate real HVDC characteristic, while quasi-steady-state model is good approximation of real characteristic and suitable for transient voltage stability analysis.

References

1. Pal, M.K.: Voltage Stability Conditions Considering Load Characteristics. IEEE Transactions on Power Systems 7(1), 243–249 (1992)
2. Cutsem, T.V.: Voltage stability of electric power systems. Kluwer academic publishers (2001)
3. Taylor, C.W.: Power System Voltage Stability. McGraw-Hill, New York (1994)
4. Lu, C., Cui, W., Li, H., Hu, A.: Problems and solutions of transient voltage stability in load centers of china. Automation of Electric Power Systems 28(11), 9–12 (2004)
5. Bai, Y., Chen, H., Wang, Z.: Discussion on strategies to enhance the system transient voltage stability under the DC Bi-polar block fault. Automation of Electric Power System 30(1), 93–96 (2006)
6. Zhou, S., Lv, J., Zhang, Y.: Influence of HVDC on voltage stability. Journal of Tsinghua University (Science and Technology) 39(3), 4–7 (1999)
7. Liu, M., Cheng, J., Cheng, Y.: Dynamic voltage stability analysis of parallel AC/DC power systems. Automation of Electric Power Systems 23(16), 27–30 (1999)
8. Mo, Q., Zhang, Y., Wu, Z.-G., Guo, L.: Simulation and Analysis on Transient Voltage Stability of AC/DC Hybrid System. In: Proceedings of the CSU2EPSA, vol. 18(6), pp. 87–90 (2006)
9. Xu, M., Li, X., Bai, J., et al.: Analysis on Voltage Stability at Commutation Buses of Parallel AC/DC Power Systems. Automation of Electric Power Systems 33(7), 6–10 (2009)

10. Hu, L., Li, X., Chen, X.: Study on measures to avoid voltage collapse in AC/DC power systems. Automation of Electric Power Systems 23(16), 23–26 (1999)
11. Yang, F., Chen, C., Jin, X., Wu, X., et al.: Comparison between constant PQ Model and Quasi-steady-state Model of HVDC in low frequency oscillation analysis. Automation of Electric Power Systems 32(6), 9–13 (2008)
12. Zhou, C., Xu, Z.: Simulation validity test of the HVDC quasi-steady-state model. In: Proceedings of the CSEE, vol. 23(12), pp. 33–36 (2003)
13. Szechtman, M., Wess, T., Thio, C.V.: A benchmark model for HVDC systems studies. In: International Conference on AC and DC Power Transmission, pp. 374–378 (1991)

A Low-Complexity Tag Number Estimate in EFSA Protocol for RFID Tag Anti-collision*

Zeng Yu, Wu Haifeng**, Tan Yuan, and Liu Jing

School of Electrical and Information Technology, Yunnan University of Nationalities, 650500 Kunming, People's Republic of China
yv.zeng@gmail.com, whf5469@gmail.com,
pllpty@163.com, ningning87109@163.com

Abstract. In many RFID applications, readers need repeatedly identify tags that have been identified in a previous cycle of identification. Some anti-collision protocols, such as efficient framed slotted aloha (EFSA) protocol can retain information obtained from a previous cycle of tag identification, and hence can skip many collisions to quickly re-identify the tags in the current cycle. In EFSA, however, the distribution of tags is not random due to some moving tags. Thus, many exiting estimates will not be suitable. In this paper, we propose a two-dimension binary search (2D-BS) estimate for EFSA. The estimate utilizes a binary search method and nests a 1D search into a final 2D search. The advantage of the proposed estimate is to have lower computational complexity. Simulation results show that the proposed method can reduce the computational complexity and have less error than the existing estimates.

Keywords: RFID, anti-collision, tag number estimate, tag identification.

1 Introduction

One of radio frequency identification (RFID) advantages over traditional bar code is that RFID can support simultaneous multi-tag identification, and hence offer higher identification efficiency. The multi-tag identification is actually a multi-access communication system, where an RFID reader recognizes multiple tags through communication over a shared wireless channel [1]. When the multiple tags simultaneously transmit their signals to the reader, collisions will happen and disturb the reader's identification process. Now, many conventional anti-collision algorithms for multi-access systems have been applied to the RFID tag anti-collision system [1].

In many RFID applications, readers may repeatedly identify the tags that have been identified in a previous process of identification, i.e. a previous cycle. For example, in supply chain operation, tagged objects on a pallet or a conveyor belt will be repeatedly identified at different phases, such as checking in, checking out and taking

* This work is supported by Innovative Research Team in Yunan University of Nationalities and the Scientific Research Foundation of Yunnan Provincial Department of Education (2010Y423).

** Corresponding author.

D. Yang (Ed.): Informatics in Control, Automation and Robotics, Volume 2, LNEE 133, pp. 495–502.
springerlink.com

inventory [2]. For object tracking and locating, tagged objects will also be repeatedly identified since the information of the objects need to be frequently read by monitors [2]. In this scenario, if an anti-collision protocol can retain information obtained from a previous cycle of tag identification, the reader can skip many collisions to quickly re-identify the tags in the current cycle. Myung et al. propose an anti-collision protocol: adaptive binary splitting (ABS) [3], which can successfully reduce the collisions in re-identifying the tag. In ABS protocol, if there is a collision slot, colliding tags in the collision slot will be resolved by binary splitting, while the other unidentified tags is in a waiting state until the colliding tags are successfully resolved. Therefore, the colliding tags in the collision slot would not collide with the other unidentified tags in the other slots. For this reason, ABS's throughput is higher than pure binary splitting protocol [4], pure aloha protocol [5-9]. However, when the number of colliding tags is much greater than 2, ABS's binary splitting may not be an optimal method.

An efficient framed slotted aloha (EFSA) protocol proposed in [10] adopts Q-ary splitting to resolve tag collision where Q is equal to the estimated number of colliding tags. Compared with ABS protocol, EFSA have a better throughput than ABS when tags need to be re-identified. Since Q is set to the number of the colliding tags, estimate of the number should be considered because the number is generally unknown to a reader. In EFSA, distribution of tags is not random due to some moving tags [3, 10]. Thus, many exiting estimates, such as Vogt [6] and MAP estimate [7] will not be suitable. For this reason, EFSA also adopts a tag estimate, which consider the moving tags and has less estimate error than the exiting estimate. However, the estimate in EFSA is a two-dimension (2D) estimate which needs to search a minimum value in two different ranges of tags. Therefore, when the ranges are very wide, the computational complexity of the estimate will also be very high.

This paper proposes a tag number estimate for EFSA. The estimate utilizes a binary search method and nests a 1D search into a final 2D search (2D-BS). Simulation results show that the proposed method can reduce the computational complexity of the estimate and have less error than the existing estimates.

2 2D-BS Estimate

2.1 Description of Problem

In EFSA protocol [10], a reader configures a read cycle with some frames. If a reader realizes that collisions occurred, it starts Q-ary splitting, i.e. a new reading frame for each slot where there was a collision. Colliding tags in the collision slot will be identified in the new frame. The reader needs to adjust the new frames length to reduce the idle slot and collision slot. Compared with tag quantity, a smaller frame length may produce excessive collisions, while a much larger frame length would lead to underutilization of channels. These both result in long tag identification delay. Therefore, the frame length adjustment should be according to the tag quantity. However, the tag quantity is usually unknown to a reader. In general, we can utilize read results collected at the reader, such as idle slot, collision slot and successful slot quantity in a frame to estimate the tag quantity.

Among conventional tag estimates, the simpler ones are lower bound estimate [6, 7], Schoute estimate [8] and idle slot estimate[9]. These three estimates results are not accurate. Vogt estimate [6] and MAP estimate [7] have less error than those three estimates. Vogt estimates tag quantity by searching the minimal distance between the real results of idle slot, successful slot and collision slot quantity and the expected ones. MAP estimates tag quantity also by searching maximum a posteriori probability. If we apply Vogt estimate and MAP estimate to EFSA protocol, the error of the two estimates will increase with the number of staying tags. The reason why this happens is that, the two estimates are based on assumption that the allocation of tags to slots within a frame is random. In the first frame of a cycle of EFSA protocol, however, there are some staying tags which are orderly allocated to slots by a reader in a previous cycle. These tags do not obey the random distribution. Therefore, Vogt estimate and MAP estimate can not be extended to this condition. Certainly, if the first frame has colliding tags, the colliding tags will continue to be identified in some new frames and the allocation of the tags to slots within the frame will be random. Thus, Vogt and MAP estimate can be suitable. Next, we will mainly analyze tag quantity estimate in the first frame.

2.2 Estimate Model

Let A_i, $i>1$ be the set of tags which are inside a reader's range in the ith read cycle of EFSA protocol. To consider the tag's mobility, we defined staying tags set as $B = A_{i+1} \cap A_i$, leaving tags set as $C = A_i - B$ and arriving tags as $D = A_{i+1} - B$. We suppose an EFSA system with α arriving tags and β leaving tags at the beginning of the $i+1$th cycle, where $\alpha>0$, $L>\beta>0$ and L is the first frame length. If the arriving tags and leaving tags' *SSC* is a random number from 0 to L, the probability of finding r arriving tags in a slot of the first frame will be given by

$$p'_r = \binom{\alpha}{r}\left(\frac{1}{L}\right)^r\left(1-\frac{1}{L}\right)^{\alpha-r}, r=0,1,... \quad (1)$$

and the probability of finding 0 and 1 staying tags in a slot will be given by and the probability of finding 0 and 1 staying tags in a slot will be given by

$$p''_0 = \frac{\beta}{L} \quad (2)$$

$$p''_1 = \frac{L-\beta}{L} \quad (3)$$

respectively. Therefore, the probability of finding 0, 1 and κ, $\kappa>1$ tags in a slot are given by $p_0 = p'_0 p''_0$, $p_1 = p'_1 p''_0 + p'_0 p''_1$ and $p_\kappa = p'_\kappa + p'_1 p''_1$, respectively. Substituting (1)-(3) into the three formulas, we have

$$p_0 = \frac{(L-1)^{\alpha}\beta}{L^{\alpha+1}} \tag{4-a}$$

$$p_1 = \frac{(L-1)^{\alpha-1}\alpha\beta + (L-1)^{\alpha}(L-\beta)}{L^{\alpha+1}} \tag{4-b}$$

$$p_{\kappa} = 1 - \frac{(L-1)^{\alpha-1}\alpha\beta + (L-1)^{\alpha}L}{L^{\alpha+1}} \tag{4-c}$$

And the expected number of idle slots, collision slots and readable slots in a frame can be given by

$$a_0(\alpha,\beta) = \frac{(L-1)^{\alpha}\beta}{L^{\alpha}} \tag{5-a}$$

$$a_1(\alpha,\beta) = \frac{(L-1)^{\alpha-1}\alpha\beta + (L-1)^{\alpha}(L-\beta)}{L^{\alpha}} \tag{5-b}$$

$$a_{\kappa}(\alpha,\beta) = L - \frac{(L-1)^{\alpha-1}\alpha\beta + (L-1)^{\alpha}L}{L^{\alpha}} \tag{5-c}$$

Thus, when the first frame with L time slots has c_0 idle slots, c_1 readable slots and c_{κ} collision slots, our estimate can be denoted by

$$(\hat{\alpha},\hat{\beta}) = \arg\min_{\alpha\in\Omega,\beta\in\Xi} \left\|\mathbf{A}(\alpha,\beta) - \mathbf{C}\right\|^2 \tag{6}$$

$$\hat{n} = L + \hat{\alpha} - \hat{\beta} \tag{7}$$

where $\|\bullet\|$ is Euclidean norm and

$$\mathbf{C} = [c_0, c_1, c_{\kappa}]^{\mathrm{T}} \tag{8}$$

$$\mathbf{A}(\alpha,\beta)=\left[a_0(\alpha,\beta),a_1(\alpha,\beta),a_\kappa(\alpha,\beta)\right]^{\mathrm{T}} \tag{9}$$

Since there is at least one arriving tag in a collision slot and one leaving tag in an idle slot, lower bound of α and β can be determined. Thus, the search tag range set, Ω and Ξ can be expressed as

$$\Omega=\{\alpha \mid c_\kappa \le \alpha \le \mathrm{A}\} \tag{10}$$

$$\Xi=\{\beta \mid c_0 \le \beta \le L\} \tag{11}$$

where we suppose that A is a maximum number of arriving tags that the RFID system can read.

2.3 Search Method

To find a minimum, the estimate in (6) needs to search in the range of α, Ω and the range of β, Ξ. If we adopt a brute-force method, the search times will be $(A-c_\kappa+1)\times(L-c_0+1)$, which is very high. To reduce search times, we let

$$\delta(\alpha,\beta)=\left\|\mathbf{A}(\alpha,\beta)-\mathbf{C}\right\|^2 \tag{12}$$

From many simulations of computing $\delta(\alpha,\beta)$ in (12), we have the following results. For a given α, $\delta(\alpha,\beta)$ has a unique minimum at β_m and it will be monotonically decreasing for $\beta<\beta_m$ and monotonically increasing for $\beta>\beta_m$. Therefore, $\delta(\beta)$ is actually a V-shape curve with respect to β. Likewise, for a given β, $\delta(\alpha,\beta)$ is also a V-shape curve with respect to α. Based on the results, we can adopt a two-dimension (2D) binary search method to reduce the search times and hence lower the estimate computational complexity. The 2D binary search can be considered as a nest of two 1D binary searches. We firstly define a 1D function, which searches a minimum of $\delta(\alpha)$ for a given β. Then, we nest the 1D function in the final 2D function. Here, we take an example of searching a minimum of $\delta(\alpha)$ to explain our binary search method. Initially, we let $\alpha left = c_\kappa$, $\alpha right = \mathrm{A}$. Then, $\alpha mid = (\alpha left + \alpha right)/2$. If $\delta(\alpha)$ is monotonically decreasing on some neighborhood of αmid, then $\alpha left = \alpha mid$. Otherwise, $\alpha right = \alpha mid$. If $\delta(\alpha mid-1) > \delta(\alpha mid) < \delta(\alpha mid+1)$, $\delta(\alpha mid)$ will be the minimum of $\delta(\alpha)$. To find a minimum of $\delta(\alpha)$, the times of computing $\delta(\alpha mid)$ is $\log_2(\mathrm{A}-c_\kappa+1)$. Besides, we need compare $\delta(\alpha mid)$ with $\delta(\alpha mid-1)$ and $\delta(\alpha mid+1)$ to determine whether $\delta(\alpha mid)$ is a minimum or not. Thus, the 1D binary search's maximum estimate complexity is

$O(3\log_2(\mathrm{A}-c_\kappa+1))$, and the final 2D binary search's maximum is $O(3\log_2(\mathrm{A}-c_\kappa+1)\times\ 3\log_2(L-c_0+1))$. Table 1 shows estimate complexity of the brute force search and the 2D binary search, respectively.

Table 1. Simulation results for comparison of estimation complexity in the first frame of EFSA

Search Method	Computer Complexity
Brute force search	$O((\mathrm{A}-c_k+1)\times(L-c_0+1))$
2D-BS	$O(3\log_2(\mathrm{A}-c_k+1)\times3\log_2(L-c_0+1))$

3 Performance Evaluation

We evaluate the performance of 2D-BS estimate by computer simulations. We individually perform each simulation 500 times, and average 500 simulation results into the final result. Since EFSA's tag estimate is relate to staying tags number, we define a ratio of staying tags V_y as

$$V_y = \frac{Q(Y_{i+1} \cap Y_i)}{Q(Y_i)} \times 100\% \tag{13}$$

where Y_i is a set of tags which are inside a reader's range in the i th read cycle, and $Q(Y_i) = Q(Y_{i+1})$ =512. Fig. 1 gives estimate error in the first frame of the i th cycle, for lower bound estimate [6, 7], Schoute estimate [8], idle slot estimate [9], Vogt estimate [6], MAP estimate [7] and the proposed estimate in this paper, where the estimate error ε is defined as

$$\varepsilon = \frac{\hat{n}-n}{n} \times 100\% \tag{14}$$

frame length L = 512, tag number n = 512, maximum number of tags that the RFID system can read N =600 and maximum number of arriving tags that the RFID system can read A =600. From Fig. 1, when 40%< V_y <80%, the proposed estimate error is less than the other estimates; when 80< V_y <100%, lower bound estimate error is close to the proposed estimate, while the other estimates errors are larger than the proposed estimate; when V_y <40%, MAP, Vogt and Schoute estimate errors are close to the proposed estimate. These results signify that the proposed estimate does not have a worse performance than the conventional estimates under any value of V_y .

In Table 2, we evaluate computational complexity of several estimates in EFSA protocol. Table 2 gives complexity of brute force search and 2D-BS estimate in the first frame of a cycle. From it, we can see that 2D-BS have much fewer search times than the brute search.

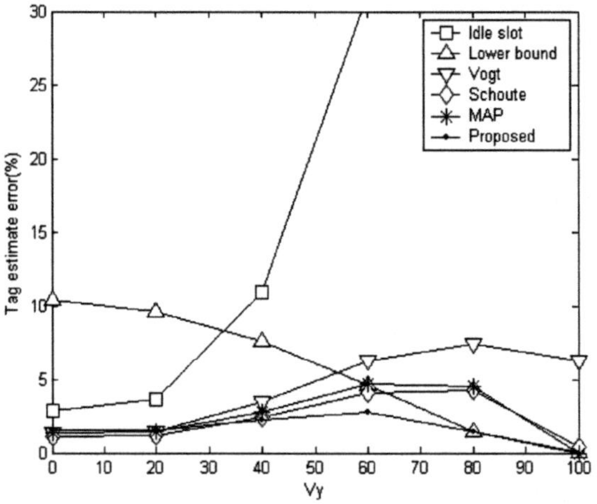

Fig. 1. Simulation result of estimate error against *Vy*

Table 2. Simulation results for comparison of estimation complexity in the first frame of EFSA

V	0	20	40	60	80	100
n	512	512	512	512	512	512
L	512	512	512	512	512	512
A	600	600	600	600	600	600
Brute Search	3.08E5	2.26E5	1.83E5	1.63E5	1.54E5	1.52E5
2D-BS	49	608	420	376	379	374

4 Conclusion

In many RFID applications where readers may repeatedly identify tags, such as supply chain operation, object tracking and locating, some protocols, such as EFSA can reduce time of re-identifying tags. For EFSA protocol, this paper proposes a 2D-BS tag quantity estimate, which introduces a search method and have low computational complexity. Simulation results show that the proposed tag quantity estimate is very suitable for EFSA, and the search method can reduce computational complexity.

References

1. Shin, D.H., Sun, P.L., Yen, D.C., Huang, S.M.: Taxonomy and survey of RFID anti-collision protocols. Computer Communications 29, 2150–2166 (2006)
2. Angeles, R.: RFID technologies: Supply-chain applications and implementation issues. Information Systems Management 22, 51–65 (2005)

3. Myung, J., Lee, W., Srivastava, J., Shih, T.K.: Tag-Splitting: Adaptive collision arbitration protocols for RFID tag identification. IEEE Trans. Parallel and Distributed Systems 18, 763–775 (2007)
4. Hush, D.R., Wood, C.: Analysis of tree algorithm for RFID arbitration. In: IEEE International Symposium on Information Theory, p. 107. IEEE Press, Cambridge (1998)
5. EPC radio-frequency identification protocols class-1 generation-2 UHF RFID protocol for communications at 860MHz-960MHz Version 1.1.0 Draft1. EPCglobal Inc. (2005)
6. Vogt, H.: Efficient Object Identification with Passive RFID Tags. In: Mattern, F., Naghshineh, M. (eds.) PERVASIVE 2002. LNCS, vol. 2414, pp. 98–113. Springer, Heidelberg (2002)
7. Chen, W.T.: An Accurate Tag Estimate Method for Improving the Performance of an RFID Anticollision Algorithm Based on Dynamic Frame Length ALOHA. IEEE Trans. Automation Science and Engineering 6, 9–15 (2009)
8. Schoute, F.C.: Dynamic frame length aloha. IEEE Trans. Comm. Com-31, 565–568 (1983)
9. Khandelwal, G., Yener, A., Lee, K., Serbetli, S.: ASAP: A MAC Protocol for Dense and Time Constrained RFID Systems. In: IEEE International Conference on Communications, pp. 4028–4033. IEEE Press, Istanbul (2006)
10. Wu, H., Zeng, Y.: Efficient Framed Slotted Aloha Protocol RFID Tag Anti-collision. IEEE Transactions on Automation Science and Engineering 8, 581–588 (2011)

Collaboration and Collision Functions for Plan-Based and Event-Driven Mission Control

Wolfgang Meyer and Arne von Drathen

Institute of Automation, Hamburg University of Technology,
21071 Hamburg, Germany
w.meyer@tu-harburg.de, von.drathen@tu-harburg.de

Abstract. Collaboration among operations of several space-time processes is formalized by an analytical expression. For factory scheduling and mission control, time based and event driven approaches are compared based on a unified Petri net representation and implemented as a tool for industrial use.

Keywords: Manufacturing, Mission, Scheduling, Coordination, Petri net.

1 Problem Statement

Consider the following application. Given: a number of services or operations to be provided by several stationary or mobile resources at different locations in a fixed sequence, with service times fixed as well, or roughly specified within time windows. Examples for such processes are: 1. Three stationary robots which cooperate in a manufacturing cell (Fig. 1a). The assembly plan affords some combined handling and assembly operations of several robots inside the hatched area at certain times, whereas at other times, only one robot is allowed to enter the collision area. Question: What about collisions, if a second and third part or process is handled by the cell in parallel? 2. One mobile robot serves several locations in a fixed sequence including loops, according to a predefined process plan (Fig. 1b). Now several other robots enter the field which obey the same process plan, or different ones. Question: When should these robots be released to the field and how to adjust their velocities or routes, to avoid collisions or to achieve collaboration with the already persistent processes?

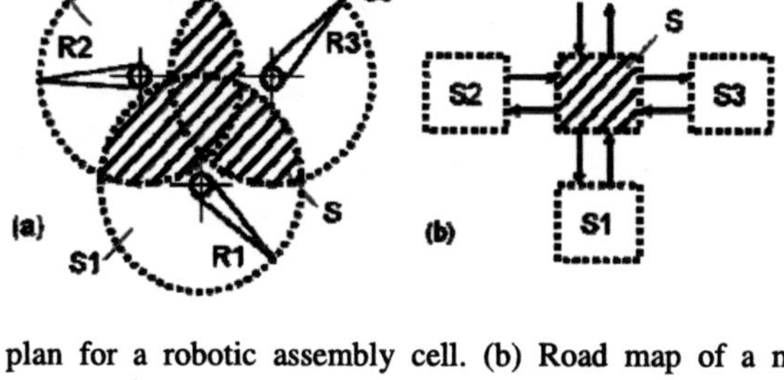

Fig. 1. (a) Floor plan for a robotic assembly cell. (b) Road map of a mission. (*R robot resources, S space resources*).

D. Yang (Ed.): Informatics in Control, Automation and Robotics, Volume 2, LNEE 133, pp. 503–510.
springerlink.com

Both scenarios describe time and spatially distributed *processes* as a sequence of *operations* to be performed by *resources* which are limited. Fig.2 shows the respective resource Gantt chart, with an overlap of bars for *collaboration* of robots in the hatched areas of Fig.1, and collision avoidance (or noninterference of robots, or *exclusion*) in the nonhatched areas.

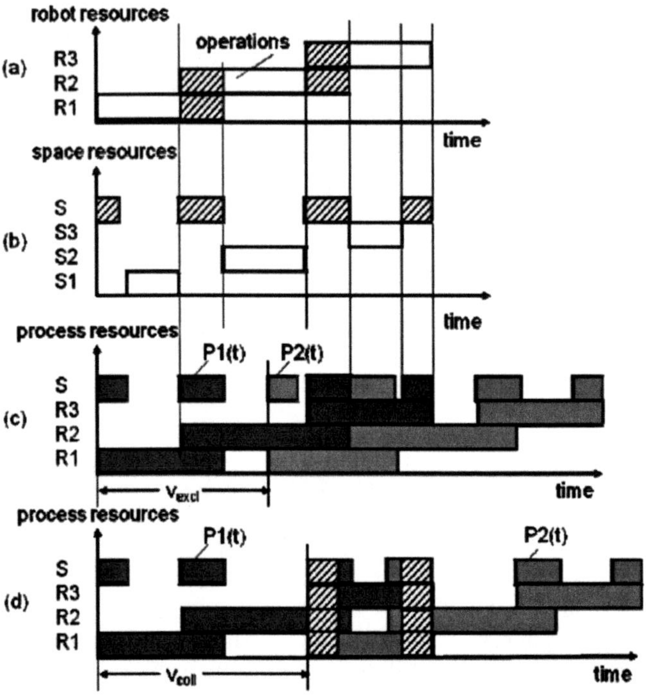

Fig. 2. Gantt charts for floor plans of Fig.1. (a) Robot resource Gantt chart. (b) Space resource Gantt chart. (c) Collision avoidance between two processes. (d) Collaboration among three robots of two processes in the hatched area.

Each bar in Fig.2 may stand for a set of machining, assembly, transport, search or rescue operations to be performed by each robot, and for complex trajectories in the collaborative and exclusive parts of the floor plan. A *mission*, in this context, is the realization of several processes in parallel, under resource constraints. Fig. 2c shows a mission performed by two identical processes P1(t) and P2(t) which avoid collisions among themselves whereas in Fig 2d, collaboration among all three robots is enforced in the hatched space area S for certain periods of the mission flow time. We tackle these two mission problems in this paper: For a number of processes P1(t,v1), P2(t,v2), ..., to be sent to the field with identical or different process plans including loops and varying cooperation strategies, for (a) *fixed* service times: find the optimal schedule v1, v2, ... for each process which minimizes the mission flow time, and (b): for service times prescribed only by time *windows*, find the optimal service times and schedule for each operation per process which minimize the mission flow time or some other time interval. Problem (a) focuses at proactive and central planning, (b) relies on local and event-driven priority rules. Problem (a) is mainly a state centered decision problem, (b) a process oriented optimization problem. Problem (a) is best solved by a discrete space, discrete time model, (b) by a discrete space, continuous time model. From the great range of time representations for scheduling models [1], we choose Petri nets for a unified approach. The idea is to represent the Gantt chart as a logical Petri net and to model the scheduling procedure by a timed dynamic net. The paper builds on our earlier work on cooperative missions and factory control [2] [3].

2 Process-Process Interactions

Fig. 3 expresses parts of the Gantt chart of Fig.2 as a Petri net. It illustrates the simple truth that processes can only operate if the resources needed are available, *time* also being considered as a limited resource. The net makes frequent use of the synchronization-and-splitting-structure of ordinary Petri net theory [4] (Fig.4): The transition (bar) will only fire, i.e. transfer the tokens (black dots) from the k input places (circles) to the n output places, if every pre place is activated, i.e. contains one token. By applying this switching rule sequentially, the process P1 passes through different states each triggered by a switching event of exactly one transition. See, for instance, Fig. 3c which models the collaboration of robots R1 and R2 at the (dashed) space area S over a time interval dS.

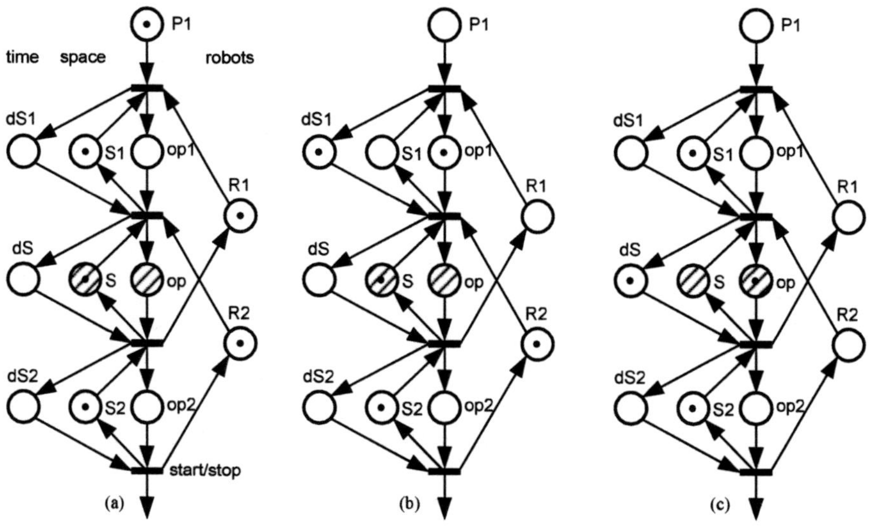

Fig. 3. Petri net representation of the Gant charts from Fig.2c,d. (a) System state 0: Process initialization. (b) System state 1: Operation op1 being performed by robot R1 exclusively in space area S1 over time interval dS1. (c) System state 3: Two robots R1, R2 collaborating through operation op in space area S over time interval dS.

Please note that P1 (the process plan or *A-path* [5]) as a sequence of operations op1, op, op2, … , for instance move operations in special areas of the mission space, is a timeless, purely logical net. The same holds for the availabilities of physical resources like space and robots. Time is then added to the system by the left row of places with the functionality of a clock, meaning that tokens in timed places must stay in place for time intervals dS1, dS, dS2, … until they are allowed to switch their post transitions. - There is a fundamental difference between physical and time resources which is obvious from the direction of arrows in the net and the first three simulations of the model in Fig.3: Physical resources are renewable, time is not - once passed it is gone. Time only comes into existence when operations start. Without activities, there is no time (no token).

However, physical resources are not unlimited, of course. They have capacities of 1, serving one process at a time (mutual *exclusion*), or greater than 1 thus enabling parallel processing (*collaboration*). The respective Petri nets for exclusion and inclusion are given in Fig.5. R stands for any kind of physical resources. In Fig.5a, the exclusive use of R affords sequencing of P1 and P2 as the availability token in R can only be given to P1 or P2, according to some priority rule or scheduling algorithm. In Fig.5b, both processes can collaborate at the same resource S only after P1 and P2 have synchronized. By building networks from these elemental structures, complex collaboration strategies for distributed missions can be designed.

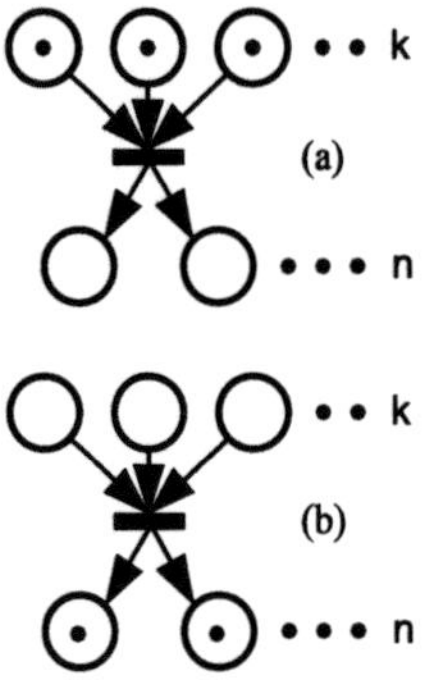

Fig. 4. Flow of tokens across the synchronization and splitting structure (a) before, (b) after switching

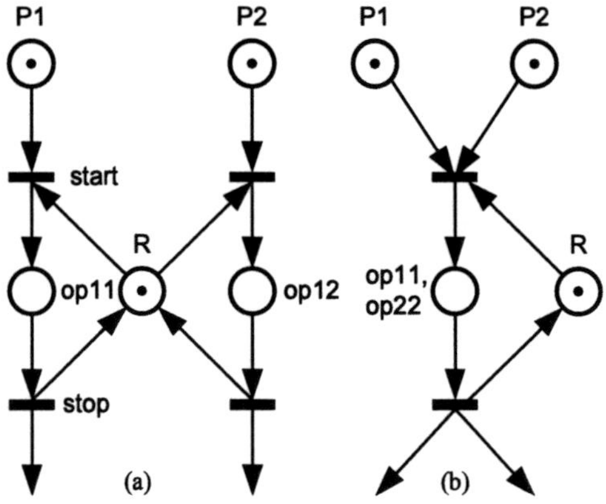

Fig. 5. Basic Petri net structures for (a) mutual exclusion of processes, or *collision avoidance*; and (b) parallel synchronous operation of processes, or *collaboration*

3 Collaboration and Collision Function

The preceding chapter dealt with mission modeling, but for mission control, a quantitative measure for collaboration is needed. In Fig.2, by proper choice of the process release times v_{excl} and v_{coll} , collisions are excluded or collaboration is obtained among the two (identical) processes of this mission example. The shift of the second process along the time axis by a release time v is described by P2(t-v), see Fig.6a. Here, the bars of the Gantt chart, e.g. the uppermost line S in Fig.2b, are encoded as 1/0-pulse trains P1(t), P2(t-v).

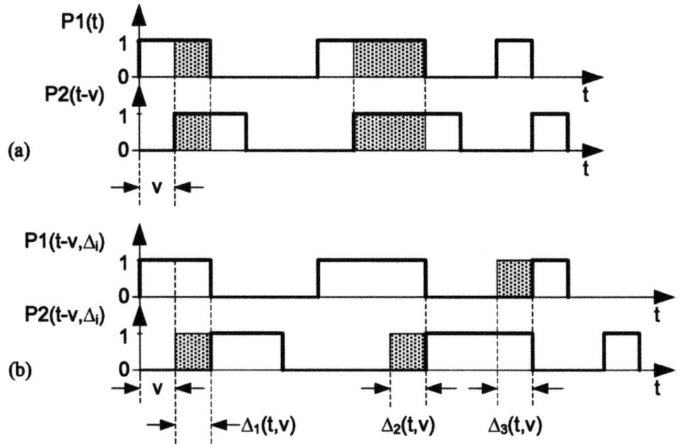

Fig. 6. Two scheduling approaches. (a) Time driven: measures the overlap among fixed processes. (b) Event driven: changes processes and measures the time delay of operations.

The degree of overlap among the bars is calculated by the collision and collaboration function CO(v):

$$CO(v) = \int P1(t) \cdot P2(t-v)\, dt \qquad (1)$$

For CO(v)=0, the two processes do not collide (and do not collaborate) whereas CO(v)>0 means collaboration at varying degrees. For identical processes P1(t)=P2(t), CO(v) as defined by Eqn.1 is the well-known autocorrelation integral for deterministic signals. Eqn.1 has been extended to n identical [6] and n different processes [7]. Eqn.1 is a continuous analytical solution for scheduling problems as visualized by Gantt charts. It delivers a *schedule* as a set of admissible release times $V=\{0\ ,\ v_2\ ,\ v_3,\ \ldots\ ,\ v_n\}$ for two or more processes with fixed service times and predefined collaboration strategies (Problem (a) of Chapter 1). Eqn.1 can be adjusted to cover variable service times or *time windows* (Problem (b). However, event driven models better suit this problem class, where each operation of the mission is scheduled and optimized individually (Fig.6b).

4 Collaboration and Collision Petri Nets

Discrete Time, Time Driven Net. The overlap of bars in Gantt charts and the intersection of sets in set theory bear similarities. In fact, exclusion and inclusion, resp. collision avoidance and collaboration, are most basically modeled by set operators like union and intersection. These can be represented as colored predicate Petri nets including inhibitor and test arcs [4] and have been elaborated by us in [8]. They are combined in Fig.7 which implements Eqn.1. The places P1 and P2 each contain sets of tokens which encode the processes as binary values along the equidistant time axis (sampled values). For each v, starting with v=0, P1(p1) and P2(p2+v) are synchronized and compared. If both values are identical, the transition fires and a 'collision' token is collected at place CO. Place S4 and the connecting arcs increment the release time v by one and integrate. This circuit counts the number of collisions (the overlap) among P1 and P2 as dependent on release time v and stores the result in place CO. For the process plan P1(t) of Fig.8a and identical processes P1=P2, CO(v) is sketched in Fig.8b. A schedule v is collision free if CO(v)=0, otherwise P1 and P2 collaborate.

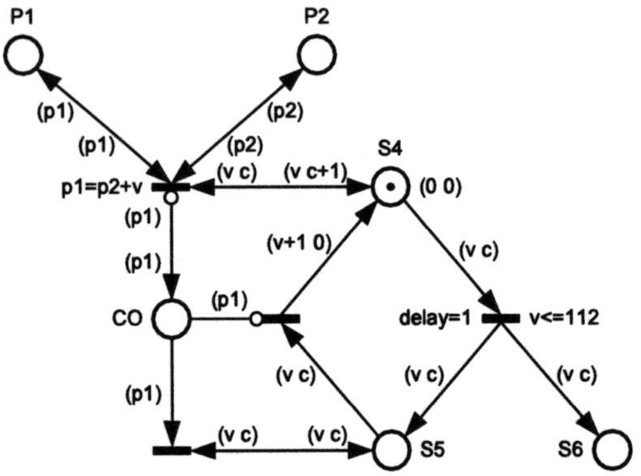

Fig. 7. Time driven Petri net implementation of Eqn.1: $CO(v) = \int P1(t) \cdot P2(t\text{-}v)\,dt$

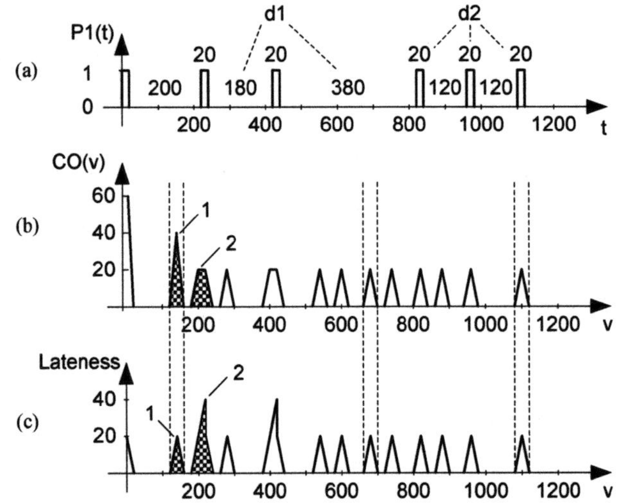

Fig. 8. Collaboration and collisions among two identical processes as dependent on the process release time v. (a) Process example. (b) Collaboration and collision function CO(v) of Eqn.1, as enumerated by the Petri net of Fig.7. (c) Process elongations $\sum\Delta_i(t,v)$ from Fig.6b as obtained from the Petri net of Fig.9.

Continuous Time, Event Driven Petri net. At the core of the network scheduler in Fig.9 we find the collaboration and collision structures of Fig.5. The tokens for P1 and P2 in the black nets of Fig.5 are created and put into operation by place 'S5 Router' [9] [10]. S5 acts as a filter for the process token flow in the outer ring of the net. The process plan (respectively the filter characteristics) is laid down in the token attributes in place S5. Please note: The net of Fig.9 is a timed Petri net as contrasted to Fig.7. Transitions 'delay=d1' and 'delay=d2' are timed transitions which turn the logical process- (or routing-) plan into a time based simulation. The 'generate-and-test cycle' of the combined simulator and optimizer is as follows:

1. Two process tokens P1, P2 are released to the net at times 0 and v.
2. For each operation of P1 and P2, process tokens are attributed according to the process plans by a logical router S5.
3. The token flow is time instantiated by timed delay transitions.
4. Conflicts among process tokens which ask for resource R in parallel, are detected ('tested') by the 'Choice' transition and resolved by simple priority rules.
5. The waiting time for the non-prioritized process tokens is recorded as 'Lateness' in place S6 and stored in place CM.

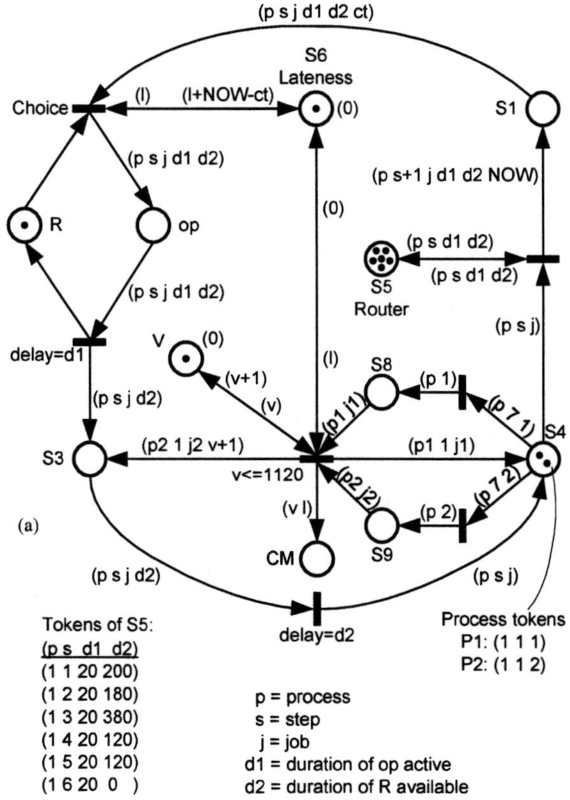

Fig. 9. Event driven implementation of the collaboration structures from Fig.5

The content of place CM is shown in Fig.8c. For collision free schedules v, Figs.8b and 8c coincide. For crowded Gantt charts or small release times, respectively, deviations occur in the collision regions (the shaded areas 1 and 2, for instance). This is due to the completely different conflict resolution mechanisms of the underlying models and is not very astonishing. What is really interesting is the fact that the two diagrams of Fig.8b,c are so similar though such dissimilar quantities like collisions and lateness are on display. Therewith, with an integrated view, the Petri nets of Figs.7 and 9 bridge the gap between time driven and event driven scheduling and control. We applied them to a variety of transport collision problems of industrial size, e.g. the HSP (hoist scheduling problem) [2], load balancing [7] and assembly problems [11], and extend its application range to collaboration problems now. The models are implemented in SMALLTALK and use the Petri net tool PACE [12].

References

1. Leung, J. (ed.): Handbook of Scheduling. Chapman and Hall, Boca Raton (2004)
2. Fiedler, C., Meyer, W., von Drathen, A.: Designing Transitory Routing Controllers for Hybrid Plants. In: IEEE Conf. Control Applications, pp. 934–939. IEEE Press, Toronto (2005)
3. Meyer, W., Fiedler, C.: Mission Control by Coordinating shared Resources. In: IEEE Conf. System of Systems Engineering, pp. 71–76. IEEE Press, Los Angeles (2006)
4. David, R., Alla, H.: Discrete, Continuous and Hybrid Petri Nets. Springer, Berlin (2005)
5. Zhou, M., DiCesare, F.: Petri Net Synthesis for Discrete Event Control of Manufacturing Systems. Kluwer Academic Publishers, London (1993)
6. Meyer, W., Fiedler, C.: Robotic Cycle Shop Control based on deterministic Correlation Maps. In: IEEE Int. Conf. Robotics and Automation. IEEE Press, Rome (2007)
7. Meyer, W., Fiedler, C.: Dynamic Load Balancing of Cycle Shops based on Correlation Functions. In: IANG Int. Conf. Control and Automation. IANG Press, Hongkong (2007)
8. von Drathen, A.: Representing Composition Operators of Automata by Petri Nets. In: IEEE Intl. Symposium Mathematical Theory of Networks and Systems, pp. 2190–2196. IEEE Press, Kyoto (2006)
9. von Drathen, A.: Modeling and Optimization of Hierarchical Routing based on Petri Nets. In: IEEE Intl. Symposium on Assembly and Manufacturing, pp. 88–93. IEEE Press, Ann Arbor (2006)
10. von Drathen, A.: Compact Modeling of Manufacturing Systems by Petri Nets. In: IEEE Intl. Conf. Systems, Man and Cybernetics, pp. 3487–3492. IEEE Press, Montreal (2007)
11. Fiedler, C., Meyer, W.: Transitory Assembly Scheduling based on deterministic Correlation Functions. In: IEEE Intl. Symposium Assembly and Manufacturing. IEEE Press, Ann Arbor (2006)
12. `http://ibe.internet-appliance.net/de/support/download.html`

Quality-First Pattern-Based Clustering Approach with Fuzzed Thresholds*

Qian Ma and Jingfeng Guo

The collage of Information Science and Engineering, Yanshan University, QinHuangdao, Hebei 066004, China
jm8281@163.com, Jfguo@gmail.com.cn

Abstract. Pattern-based clustering has broad applications such as in DNA micro array data analysis, customer segmentation, e-business data analysis, etc. However, pattern-based clustering often returns a large number of highly-overlapping clusters, which makes it hard for users to identify interesting patterns from the mining results. One of the most important factors which cause highly-overlapping is the error thresholds defined. This paper makes error analysis in pattern-based clustering. With a fuzzed thresholds strategy to deduce overlapping which caused by error range, a quality-first clustering algorithm is proposed to mining pclusters of higher quality. Both theoretical and experimental analysis shows that our method is effective and efficient.

Keywords: Pattern-based Clustering, biCluster, subspace Clustering, fuzzed thresholds.

1 Introduction

1.1 Pattern-Based Clustering

Clustering, a method which makes similar objects gather in the same clusters has been an active research area of great practical importance for recent years [1]. An important advance in this area is the introduction of the subspace clustering. This is particularly useful in clustering high dimensional data since not every dimension may be relevant to a cluster [2].

In clustering approaches, most commonly, distance functions, such as the Manhattan and Euclidian distance functions, are used to determine similarity. In other words, similar objects are required to have close values on at least a set of dimensions. However, distance functions are not always adequate in capturing correlations among the objects [3, 4].

For this reason, a number of algorithms that perform simultaneous clustering on the row and column dimensions of the gene expression matrix have been proposed to date. This simultaneous clustering, usually designated by biclustering, seeks to find

* This paper is supported by a grant from the Hebei provincial Applied Basic Research Program, China, (Grant No.10963527D).

sub-matrices, which is subgroups of genes and subgroups of columns, where the genes exhibit highly correlated activities for every condition [5].

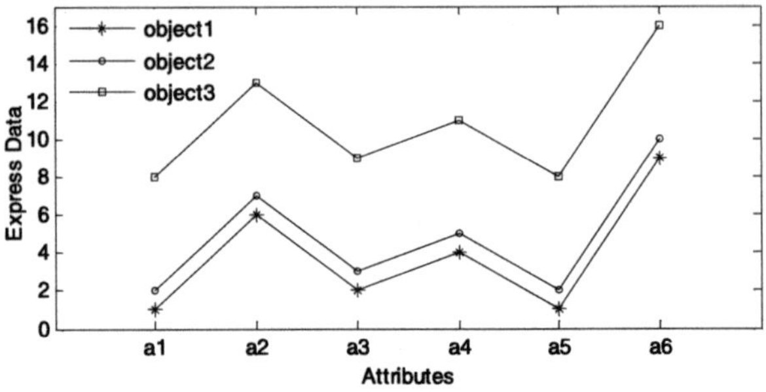

Fig. 1. Three objects follow similar coherence pattern among a set of experiment condition, those pattern similarity which we are interested, are hardly to capture by using traditional distance functions

Pattern-based clustering is an extension work of biclustering. Unlike traditional clustering algorithms, pattern-based clustering algorithms identify groups of objects that show similar activity patterns under a specific subset of the columns, can also used in customer segmentation, e-business data analysis, etc. In this model, similarity between objects is not based on distance but measured by coherence [6].

Definition 1. Pattern

Set D as dataset for a collection of objects, each object has been described by a set of attributes A. (O, T) is a matrix, $O \subset D$ and $T \subset A$. If within error threshold δ, each object of O in T shows a trend similar to the consistency of R, then this consistency of the performance is a "pattern". If the number of element in (O, T) is more than user defined, then (O,T) is a pCluster [7].

1.2 Challenges: Highly-Overlapping Caused by Error Thresholds

In pattern-based clustering, a single object may participate in multiple pathways that may or not be co-active under all attributes. It means that the pattern clusters should not be exclusive and/or exhaustive: an object or attribute should be able to belong to more than one cluster or to no cluster at all and be grouped using a subset of attributes or objects, respectively. So the model of pattern-based clustering determines that most algorithms will obtain numerous highly-overlapping clusters [8]. There are many Factors which cause highly-overlapping, but one of the most important is the error thresholds δ user defined.

All The initial candidate clusters which are mining in original datasets are based on threshold δ. This leads the result that different clusters obtained when taking different central point, however, those clusters are highly-overlapping.

Table 1. Set O as expression data set, δ=0.3. Using Pattern-based Clustering approach, we can get two clusters $C_1\{a_1,a_2,a_3,a_4\}$ and $C_2\{a_2,a_3,a_4,a_5\}$. Because of error δ, C_1 and C_2 are highly overlapped.

		Object1	Object2	Object3	
C_1	a_1	1	1.8	2.8	
C_1	a_2	1	1.9	2.9	C_2
C_1	a_3	1	2	3	C_2
C_1	a_4	1	2.1	3.1	C_2
	a_5	1	2.2	3.2	C_2

In the case above, we do not need to keep a strict error range. Fuzz the error thresholds and combine more significative, less overlapped clusters can be obtain better quality results.

To address this problem, how to increase error thresholds properly, correctly evaluate the quality of clusters, make determine if a cluster should be combined and which should be combined with, has been a key problem.

1.3 Our Contribution

This paper theoretically makes error analysis of pattern-based clustering model; quantify the influence which error threshold δ acts on the quality of Clusters. And then we propose a fuzzed thresholds strategy to implement dynamic error. Based on this a combining policy is proposed to deal with clusters in different error range and center point. We also calculate the error variance as a important parameter in quality score function to drive clustering process. Based on this, a quality-first pattern-based clustering algorithm with fuzzed thresholds is proposed.

2 Error Analysis of Pattern-Based Clustering

For model of pattern-based clustering, error is inevitable. The raw datasets of pattern-based clustering is observed from real application. Either observation error or quantization error is inevitable. Pattern-based Clustering algorithm focus on mining objects that have similar patterns. The definition of "pattern similarity" is in an allowable error range. Start with a given reference point, we can evaluate the overall error excursion of cluster. Different references will obtain different evaluation. In this paper we take shifting pattern for discuss.

Definition 2. Central point

For a given matrix(O,T), O as a set of objects $\{o_1,o_2,\ldots,o_m\}$, T as a set of attributes $(a_1,a_2,\ldots,a_n)$, d_{ij} as the express data of object o_i in attribute a_j. If (O, T) is a pcluster within error threshold δ, there must exist at least one central point (i, j), i∈O and j∈T, depends on the definition of pScore, for each point(k, h)∈(O,T):

$$|\delta_{kh}| \le |\delta|, \quad \delta_{kh} = (di_j + d_{kh}) - (d_{ih} + d_{kj}) \tag{1}$$

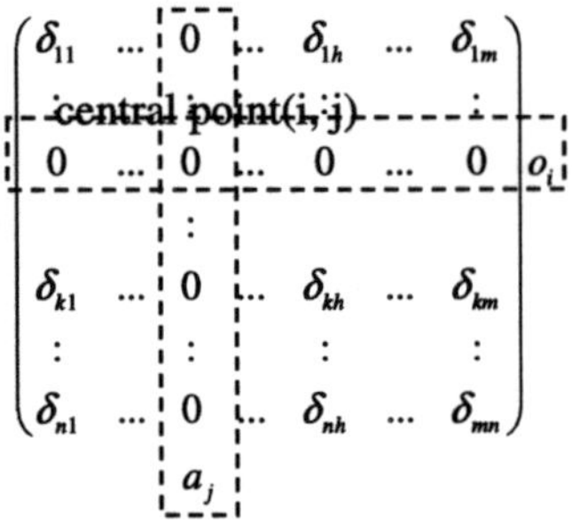

Fig. 2. Error matrix of (O, T) reference to central point (i, j)

To evaluate the overall error distribution of cluster (O, T), we can calculate the average error variance:

$$\overline{V}(O,T) = \frac{1}{|O||T|}\sum_{i\in O,\, j\in T} \delta_{ij} \tag{2}$$

The average error variance can actually reflect the overall error deviations of clusters. Smaller variance means smaller error, more similar patterns objects express, moreover, means higher quality of cluster.

Different central points chosen, different error matrix obtained, it also means different error evaluation score of clusters. There may be more than one point can been chosen to be the central. The best choice should have the smallest average error variance. However, in practical algorithms it will spend too much time to calculate. So we can define mean value of cluster as the central point instead.

For a given Cluster(O,T), O as a set of objects $\{o_1,o_2,\ldots,o_m\}$, T as a set of attributes $(a_1,a_2,\ldots,a_n)$, d_{xy} as the express data of object x in attribute y. For any point (i, j) in (O,T), the relative error is:

$$\begin{aligned} \delta_{ij} &= (M_{OT} + d_{ij}) - (M_{iT} + M_{Oj}) \\ M_{OT} &= \frac{1}{|O||T|}\sum_{x\in O,\, y\in T} d_{xy} \\ M_{iT} &= \frac{1}{|T|}\sum_{y\in T} d_{iy}, \quad M_{Oj} = \frac{1}{|O|}\sum_{x\in O} d_{xj} \end{aligned} \tag{3}$$

Since using mean value as central point can obtain min average error variance of cluster, the average error variance has the most ability to reflect the overall deviations of clusters.

Lemma 1. Set $C_1(O_1, T_1)$, $C_2(O_2, T_2)$ as two given clusters, M_1, M_2 as the central points respectively. C_1 and C_2 can be combined into one cluster $C_3(O_3,T_3)$ if and only if $(T_1 \cap T_2) \neq \emptyset$ and $(O_1=O_2)$, (or $(O_1 \cap O_2) \neq \emptyset$ and $T_1=T_2$). And $O_3=O_1 \cup O_2$, $T_3=T_1 \cup T_2$. The new central point of C3 is:

$$M_3 = \frac{|O_1||T_1|M_1 + |O_2||T_2|M_2}{|O_3||T_3|} \tag{4}$$

For any point d(i, j)∈ C_3,the error distance between d(i, j) and M_3 should be:

$$\delta_{ij} = (M_3 + d_{ij}) - (M_{3j} + M_i) \tag{5}$$

$$M_{3j} = \frac{|O_1|M_{1j} + |O_2|M_{2j}}{|O3|}$$

From the error matrix we can get the error vector of o_i∈O, relative to the central point M:

$$E(M, o_i) = \{0, \delta_{i1}, \delta_{i2}, \delta_{i3}, ..., \delta_{in}\} \tag{6}$$

The error distance between o_i and the central point M is defined as the length of error vectors E(M,o_i):

$$|E(M, o_i)| = \sqrt{\delta_{i1}^{\ 2} + \delta_{i2}^{\ 2} + \delta_{i3}^{\ 2} + ... + \delta_{in}^{\ 2}} \tag{7}$$

During the combine process of candidate clusters, the size of cluster will grow which benefits to the quality, but the error will increases meanwhile. How to keep a balance on size and error becomes a key problem.

Some operating skills can be used to lower the increase of error. Combining clusters which would obtain result clusters of smallest error variance can minimize the increase of error. Since it will spend too much time on calculating error variance, we use a greedy algorithm of taking precedence on combining clusters which have the closest central points. In other words, the smaller error distance between central points are, the smaller error variance of combined result might be.

Set $C_1(O_1,T_1)$,$C_2(O_2,T_2)$ as two given clusters, M_1, M_2 as the central points respectively. Error distance between M_1 and M_2 is:

$$|E(M_1, M_2)| = \sqrt{\delta_1^{\ 2} + \delta_2^{\ 2} + \delta_3^{\ 2} + ... + \delta_n^{\ 2}} \tag{8}$$

$$\delta_i = (M_1 + M_{2i}) - (M_2 + M_{1i})$$

Instead of unchanged error thresholds, we use average error variance as a more flexible score to evaluate similarity of objects in clusters. Average error variance which fuzzes the threshold is inversely proportional to the quality of cluster.

Definition 3. eScore

For a given Cluster (O, T), O as a set of objects {o_1,o_2,…,o_m}, T as a set of attributes (a_1,a_2,…,a_n), eScore of (O,T) defined as:

$$eScore(O,T) = c \times \frac{|O||T|}{V(O,T)} \tag{9}$$

c is a constant, as a regulation parameter in the function.

Lemma 2. Given a set of clusters ($C_1,C_2,\ldots,C_n$). Do combine operator of Ci and Cj if they satisfy conditions below:

1. C_i and C_j are combinable based on Lemma 1;
2. For Cluster C_{ij} is combining result of C_i and C_j, eScore(C_{ij})>eScore(C_i), eScore(C_{ij})>eScore(C_j);
3. For any other cluster C_k, eScore(C_{ij})>eScore(C_{ik}), and eScore(C_{ij})>eScore(C_{jk}).

Based on the fuzzed thresholds strategy of Lemma 2, continuously combining clusters which have better eScore quality, fuzz thresholds to make error range flexible, can effectively eliminate overlapping which error thresholds cause.

3 Quality-First Algorithm

3.1 Approximate eScore

Our goal is to direct the process of the clusters' combination by using eScore, preference for getting clusters with higher eScore. However, In this process, we need a good deal of computing in calculate eScore, especially in calculate error variance. It will take so much time in unnecessary amount of work if the eScore is not good enough. In section 3 we see that combining of clusters which have closer central points would obtain result of smaller error variance. So calculate the error distance between central points of clusters as a approximate eScore, if the approximate score is large enough, do the combination, otherwise do nothing.

3.2 Algorithm Process

Input : D: expression matrix with m rows and n columns; δ: initial error threshold; nc: minimum of columns; nr: minimum of rows; c: regulation parameter of eScore.

Output: pClusters of D;

```
Let C= Ø; Result= Ø ;

For each point x in D, do

        Let Ω=Ø ; // save all points coherence to d(i,j)

        If (mark(i,j)==0)   continue;    End if

        For each point y in D, do

          If   (|δ(x,y )|<|δ|)

               Insert y to Ω;
```

```
            End if
          End for
          Prune(Ω);   //prune points which not satisfy nc or nr condition
          For each points x in Ω, do
            If δx is not the min or max
                mark(x)=0;
            End if
          End for
          S=Clus(Ω); //Generate clusters fromΩ
          Insert S to C
End for
For each pair of clusters (c1,c2) in C,do
          If   aeScore(c1,c2)   is the max
            c3=Combine(c1,c2);
            If eScore(c3)>eScore(c1) && eScore(c3)>eScore(c2)
              insert c3 to C;
            End if
            let aeScore(c1,c2)=0;
          End if
End for
Sort(C); // sort clusters in C by eScore;
For i=1 to |C|
          If ci is not include in any clusters in Result
Insert ci to Result;
          End if
          i++;
End for
Output(Result);
```

4 Experimental Result

We tested algorithm on synthetic data sets with different scales. The system is implemented on windows 2003 machine with 2.11 GHZ cpu and 2.00G main memory. We generate synthetic data sets in matrix forms. Intially, the matrix is filled with random values ranged from 0–500, and then we embed a fixed number of pClusters in the raw data within random noise ranged form (-1, 1). Fix number of columns C=30, nc=7, nr=5%R.

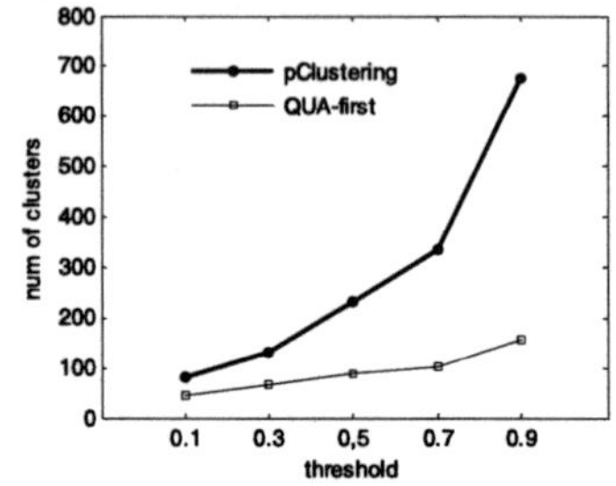

(a). Number of Clusters

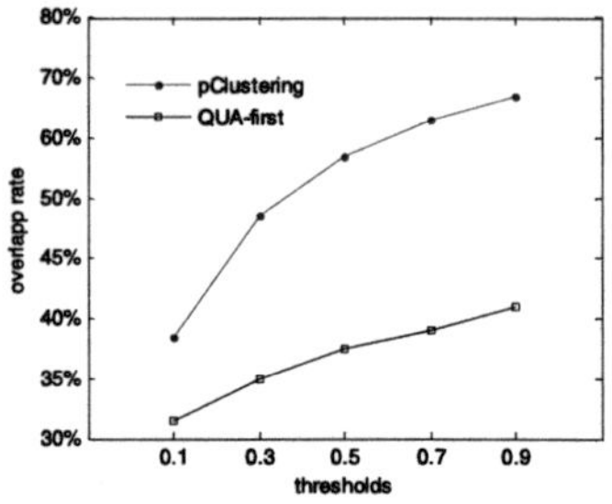

(b). Average overlapping rate

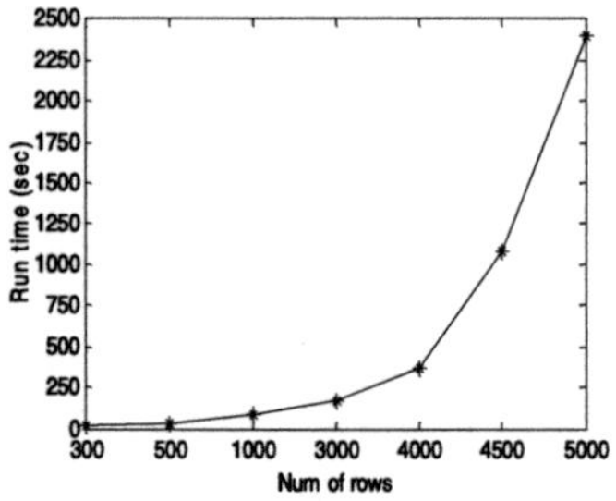

(c). Runtime of algorithm

Fig. 3. Experiment performance of quality-first algorithm compared with pclustering

Since our algorithm use fuzzed error threshold strategy, the number of clusters and overlapping rate perform much more stable.

Conclusion

In this paper, we make detail analysis about how error produce an effect on pattern-based clustering, present a fuzzed error thresholds strategy to eliminate overlapping caused by error threshold. Based on fuzzed error threshold we define eScore as a quality score to evaluate clusters. Finally, a quality-first approach is proposed to mining quality pattern-based clusters. The experimental results show that our method is effective.

Several interesting and important problems still remain open, such as how to identify other important factors influence quality of clusters, and how to use quality score to more complex coherence patterns.

References

1. Aggarwal, C.C., Yu, P.S.: Finding generalized projected clusters in high dimensional spaces. In: SIGMOD, pp. 70–81 (2000)
2. Agrawal, R., Gehrke, J., Gunopulos, D., et al.: Automatic Subspace Clustering of High Dimensional Data for Data Mining Applications. In: Proceedings of the ACM SIGMOD Conference, pp. 94–105 (1998)
3. Cheng, C.H., Fu, A.W., Zhang, Y.: Entropy-based subspace clustering for mining numerical data. In: Proceedings of SIGKDD Conference, pp. 84–93 (1999)
4. Wang, H., Wang, W., Yang, J., Yu, P.S: Clustering by pattern similarity in large data sets. In: Proc. of the 2002 ACM SIGMOD, pp. 394–405 (2002)
5. Cheng, Y., Church, G.M.: Biclustering of expression data. In: Proc. of ISMB 2000, pp. 93–103. AAAI Press (2000)
6. Yang, J., Wang, W., Wang, H., Yu, P.S.: δ-clusters: Capturing subspace correlation in a large data set. In: ICDE, pp. 517–528 (2002)
7. Guo, J., Ma, Q., Liu, H.: 0-SM:A fast algorithm for mining Candidate Clusters in Pattern-based Clustering. In: Proceedings of the 2007 IEEE Symposium on Computational Intelligence and Data Mining, pp. 127–132 (2007)
8. Jiang, D., Peii, J., Zhang, A.: A General Approach to Mining Quality Pattern-based Clusters from Microarray Data. In: Zhou, L.-z., Ooi, B.-C., Meng, X. (eds.) DASFAA 2005. LNCS, vol. 3453, pp. 188–200. Springer, Heidelberg (2005)

Decoupling Algorithm Design of Flight Control System for Attack UAV

Xue-ping Zhu, Jun Yang, and Tao Xu

College of Astronautics, Northwestern Polytechnical University,
710072 Xi'an, China
{zhuxueping,yangjun8}@nwpu.edu.cn,
wddxt@126.com

Abstract. A new decoupling algorithm of flight control system for attack UAV is presented. First the coupling mechanism is analyzed and the decoupling strategy is presented for the serious problems of aerodynamic coupling and rudder operation for plane-symmetry attack UAV; then the decoupling algorithm is designed based on decoupling theory of multivariable control system. Simulation results show that the decoupling algorithm presented in this paper has perfect performance, and could solve the problems of rolling channel coupling on the aerodynamic force and rudder operation.

Keywords: Attack UAV, Flight Control System, Decoupling, Multivariable Control Theory.

1 Introduction

UAV generally uses the plane-symmetry layout, In order to effectively improve the lift-drag ratio for attack UAV, increase the range and endurance. Inclined blowing torque and aileron effect of rudder are strong for plane-symmetry UAV, namely there are serious aerodynamic coupling and manipulation of coupling. Therefore, dealing with the plane-symmetry UAV flight control system decoupling problem is needed.

According to the decoupling problem for aircrafts, engineering and theoretical fields begun the study in depth, and formed a new decoupling method based on the conventional and modern theory of decoupling control[1~3]。Duan Guang-ren proposed a feature structure configuration method with the disturbance decoupling function [4] ; Zhang Wei-guo and so on set forth the flight control system decoupling approximate method based on the output and the partial state feedback, pole placement and the singular perturbation method[5] ; Reference[6] using variable structure control method designed the output decoupling control for BTT missile.

Aircraft decoupling algorithm should both have the completeness in theory and feasibility in engineering. In this paper, aiming at aerodynamic coupling and manipulation coupling problems for plane-symmetry attack UAV, we propose a decoupling algorithm of control system based on the decoupling theory of multivariable control system, and simulation results show that the decoupling algorithm designed has good performance and can be achieved readily.

D. Yang (Ed.): Informatics in Control, Automation and Robotics, Volume 2, LNEE 133, pp. 521–528.
springerlink.com

2 Mathematical Models and Analysis of Coupling Mechanism

2.1 Mathematical Models

Mathematical models for attack UAV :

$$\begin{aligned}
\dot{\alpha} &= \omega_z - \omega_x\beta - a_4\alpha - a_5\delta_z \\
\dot{\beta} &= \omega_y + \omega_x\alpha - b_4\beta - b_5\delta_y \\
\dot{\omega}_x &= -c_1\omega_x - c_3\delta_x - c_{12}\beta - c_{13}\delta_y \\
\dot{\omega}_y &= -b_{12}\omega_x\omega_z - b_1\omega_y - b_2\beta - b_3\delta_y \\
\dot{\omega}_z &= -a_{12}\omega_x\omega_y - a_1\omega_z - a_2\alpha - a_3\delta_z
\end{aligned} \tag{1}$$

where, $a_1 \sim a_5, b_1 \sim b_5, c_1, c_3, c_{12}, c_{13}, a_{12}, b_{12}$ are dynamics coefficient of UAV.

2.2 Analysis of Coupling Mechanism

From the mathematical models for attack UAV mentioned above, BTT control UAV control consists of inertial coupling ($b_{12}\omega_x\omega_z$, $a_{12}\omega_x\omega_y$), movement coupling($\omega_x\beta$, $\omega_x\alpha$), pneumatic coupling($c_{12}\beta$)and manipulation coupling($c_{13}\delta_y$).

Analyzing the modeling accuracy of coupling factors and these factors' influence to UAV flight control system, given the basic characteristics of the coupling factors, see table 1.

Table 1. Basic Characteristics of Coupling Factors

Coupling Factors	Influence Degree	Modeling Accuracy
Kinematics sympathetic	Relatively strong	high
Inertial cross-term	Relatively weak	high
Control surfaces manipulated coupling	Relatively strong	Relatively high
Induced roll	strong	Relatively high

3 Decoupling Theory for Multivariable Control System

3.1 Control Objects of P Standard and V Standard

For an n-inputs $U_1, U_2, \ldots, U_n$ and n-outputs $Y_1, Y_2, \ldots, Y_n$ of the coupling system, in essence, there are two different coupling phenomena. One is the coupling between input and outputs of other channels; the other is the coupling between output and inputs of other channels.

Based on this characteristic, control objects can be divided into two different types of standard forms, that is, control objects of P standard and V standard.

P Standard[7]: The so-called P standard means that any of the output Y_i in this system is subject to the impact of all inputs U_i. See fig 1 (take a two-input and two-output system as an example).

V Standard[7]: The so-called V standard means that every output Y_i in this system is not only influenced by the input U_i of each channel (main channel), but also by the outputs of other channels after the i th channel. See fig 2 (take a two-input and two-output system as an example).

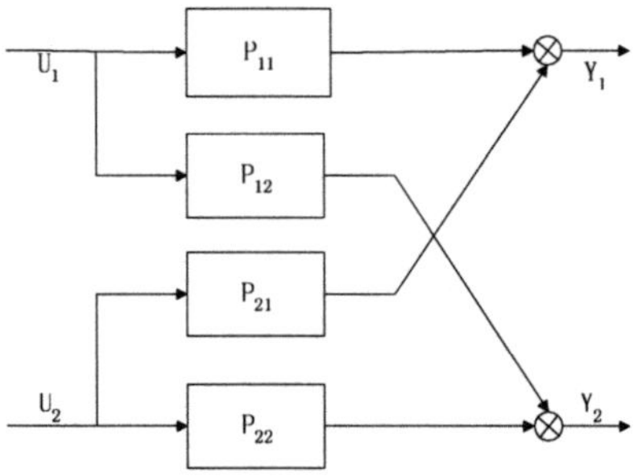

Fig. 1. P Standard Coupling System

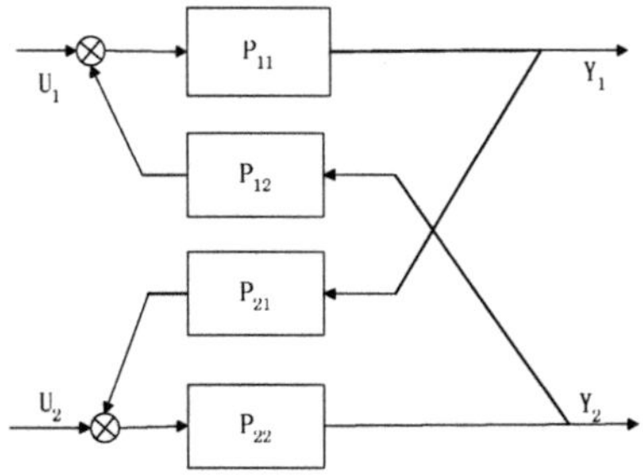

Fig. 2. V Standard Coupling System

For a multi-variable system, in order to eliminate the coupling effect between variables in this system, a necessary decoupling design is needed.

Corresponding with the control objects of P standard and V standard, according to the form in which decoupling part is linked into the system, the forms can be divided into two different types of decoupling forms, that is, P standard and V standard.

3.2 Optimal Decoupling Structure of Dual Variable P Standard in Coupling System [7]

In the engineering, a coupling system is often created as P standard form, the requirements of the optimal decoupling structure of P standard in a coupling system are:

(1) Decoupling part is the most simple one than others, and engineering realization is easy;

(2) Decoupling part is linked only with the objects, nothing to do with the controller parameters;

(3) The determinants of controller parameters are only related with parameters of objects of the main channels, and have nothing to do with the coupling term;
(4) Decoupling system has nothing to do with parameters of coupling terms.

Decoupling parts and controllers are designed respectively, which greatly reduces the design difficulty. Studies have shown that the V standard of decoupling parts is the optimal decoupling structure of objects of P standard, see fig 3.

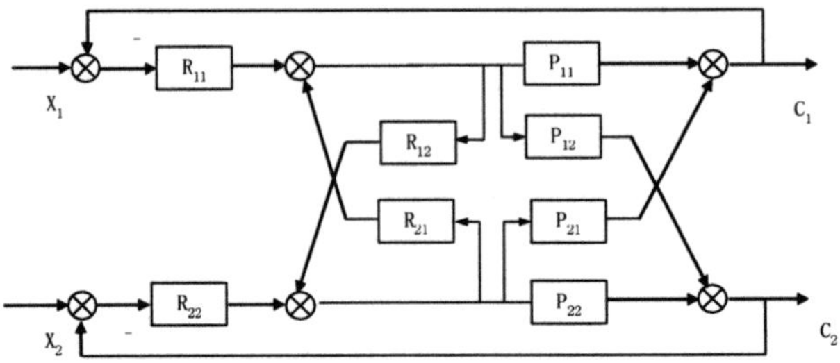

Fig. 3. Optimal Decoupling Structure of P Standard Objects

Decoupling conditions are:

$$R_{12} = -\frac{P_{12}}{P_{22}}$$
$$R_{21} = -\frac{P_{21}}{P_{11}} \tag{2}$$

Characteristic of system after decoupling is:

$$G(s) = \begin{bmatrix} \frac{R_{11}(s)P_{11}(s)}{1+R_{11}(s)P_{11}(s)} & 0 \\ 0 & \frac{R_{22}(s)P_{22}(s)}{1+R_{22}(s)P_{22}(s)} \end{bmatrix} \tag{3}$$

3.3 Full Decoupling of Control Objects of Dual- Variable V Standard[7]

For the control objects of V standard, the full decoupling of input and disturbance can be achieved using feedback decoupling method. The design of dual-variable V standard is shown in Figure 4.

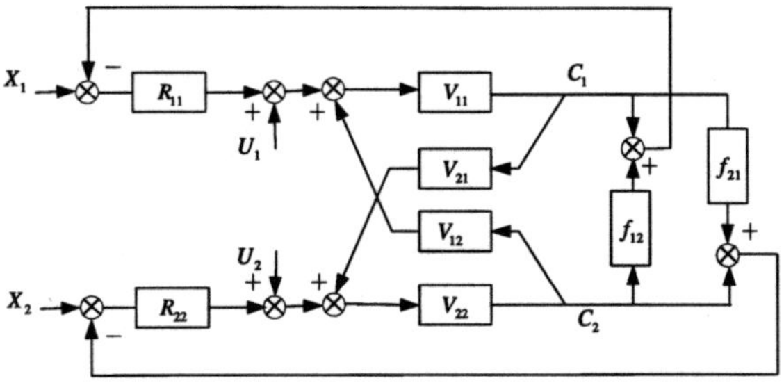

Fig. 4. Full Decoupling of Control Objects of Dual- Variable V standard

where, $f_{21} = V_{21}/R_{22} = V_{21}V_{22}/W_{22}$; $f_{12} = V_{12}/R_{11} = V_{12}V_{11}/W_{11}$;

$W_{11} = V_{11}R_{11}$; $W_{22} = V_{22}R_{22}$ 。

To be :

$$\begin{aligned} C_1 &= (V_{11}U_1 + W_{11}X_1)/(1+W_{11}) \\ C_2 &= (V_{22}U_2 + W_{22}X_2)/(1+W_{22}) \end{aligned} \tag{4}$$

It can be seen in formula (4) :

(1) The system achieves full decoupling of input and disturbance corresponding with output ;

(2) The system after decoupling is fully equivalent to disconnect all coupling links, namely, getting two main- channel control systems after disconnecting $V_{12}, V_{21} \cdot f_{12}, f_{21}$;

(3) Decoupling conditions of f_{12}, f_{21} are only related with main-channel characteristics and coupling-channel characteristics influencing the native channel, while not related with another channel's characteristics.

4 Decoupling Algorithm Design

Transform aerodynamic coupling (induced roll) and manipulation coupling to make them be in an equivalent form with disturbance rudder deflection angle:

$$\begin{aligned} \dot{\omega}_x &= -c_1\omega_x - c_3\delta_x - c_{12}\beta - c_{13}\delta_y \\ &= -c_1\omega_x - c_3\delta_x - c_3\left(\delta_x + \frac{c_{12}\beta}{c_3} + \frac{c_{13}\delta_y}{c_3}\right) \\ &= -c_1\omega_x - c_3\delta_x - c_3\left(\delta_x + \Delta\delta_{x1} + \Delta\delta_{x2}\right) \end{aligned} \tag{5}$$

4.1 Decoupling Design of Induced Roll

According to formula (5), induced roll coupling path is from sideslip angle coupling to aileron angle position, and it is can be described by control objects of dual variable V standard, and also:

$$V_{12} = c_{12}/c_3 \tag{6}$$

For induced roll, using the dual variable V standard control coupling system, the optimal decoupling structure is:

$$f_{12} = \frac{V_{12}}{R_{11}} = \frac{c_{12}/c_3}{R_{11}} \tag{7}$$

Neglecting lateral force contribution of UAV rudder, forms can be expressed with lateral overload information for decoupling:

$$f_{12}' = \frac{c_{12}g}{c_3 b_4 V R_{11}} \tag{8}$$

where, V for UAV flight velocity ; g for acceleration of gravity.

4.2 Decoupling Design of Rudder Manipulation Coupling

According to formula(5), manipulation coupling path is from rudder deflection angle coupling to aileron deflection angle position, and it is can be described by control objects of dual variable P standard, and also:

$$P_{21} = P_{11}c_{13}/c_3 \tag{9}$$

For manipulation coupling, using the dual variable P standard control coupling system, the optimal decoupling structure is:

$$R_{21} = -\frac{P_{21}}{P_{11}} = -\frac{c_{13}}{c_3} \tag{10}$$

5 Simulation Analysis of Decoupling Algorithm

Simulation conditions: under given flight conditions , Longitudinal overload instruction is 0g, lateral overload instruction is 2.5g, roll angle instruction is 0°. Make simulation analysis according to the five degree of freedom model with nonlinear coupling in formula (1). Fig 5 and fig 6 are respectively the curve of sideslip angle and rudder deflection angle; fig 7 shows the curves of roll angle with and without decoupling algorithm under the impact of aerodynamic coupling and manipulation coupling condition.

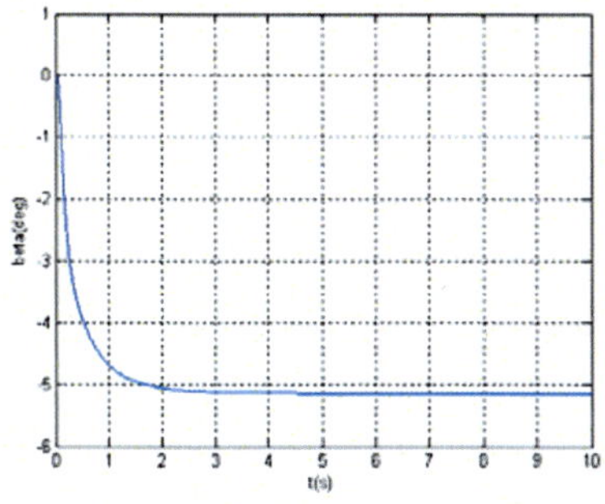

Fig. 5. Curve of Sideslip Angle

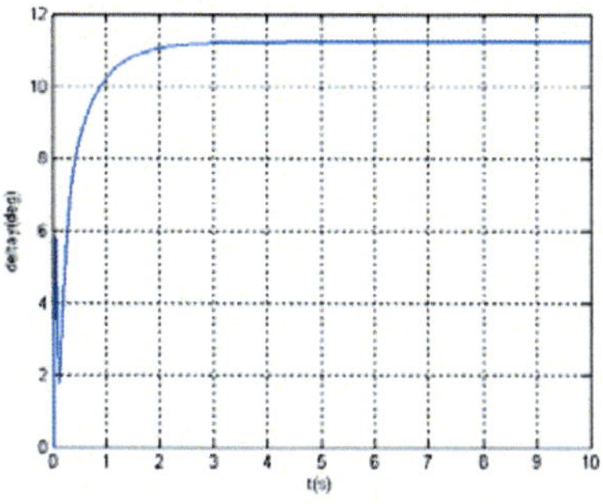

Fig. 6. Curve of Rudder Deflection Angle

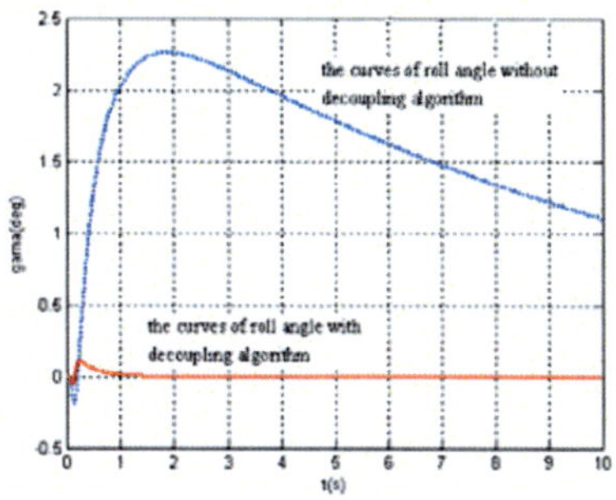

Fig. 7. Decoupling Curves of Aerodynamic Coupling and Manipulation Coupling

According to the simulation results, the decoupling parts of aerodynamic coupling and manipulation coupling have good performance which makes the roll channel has a strong suppression on disturbing torque caused by sideslip angle and rudder deflection.

6 Conclusion

In this paper, coupling mechanism and basic characteristics of coupling factors are analyzed according to the layout of plane-symmetry attack UAV and decoupling strategy is determined; multi-variable control system decoupling theory completed the flight control system decoupling algorithm design, including the main conclusion:

(1) Induced roll coupling can be described by the control objects of dual-variable V standard, and decoupling algorithm can be designed by the optimal decoupling structure of dual-variable V standard control decoupling system;

(2) Rudder manipulation coupling can be described by the control objects of dual-variable P standard, and decoupling algorithm can be designed by the optimal decoupling structure of dual-variable P standard control decoupling system;

(3) Simulation results show that the designed decoupling algorithm has good performance, and achieve effective decoupling of aerodynamic coupling and manipulation coupling about roll channel for plane-symmetry attack UAV.

References

1. Yang, Y.S., Wang, Q.G., Wang, L.P.: Decoupling control design via linear matrix inequalities. In: IEE Proc.-Control Theory Appl., vol. 152(4), pp. 357–362 (2005)
2. Jae, W.C., Ho, C.L.: PD-Eigenstructure Assignment Control for Decoupling and Tracking of Linear Time-Varying Systems. In: Proceedings of the 3th Conference on Decision & Control, Phoenix, Arizona USA (1999)
3. Bajpai, G., Chang, B.C.: Decoupling of Failed Actuators in Flight Control Systems. In: Proceedings of the American Control Conference Arlington, VA, June 25-27 (2001)
4. Duan, G.R.: Assignment of Eigenstructures with Function of Disturbance Decoupling. Control and Decision 5, 13–18 (1990)
5. Duan, W.G., Xiao, S.D.: The Application of an Approximate Decoupling Method to the Flight Control Systems. Aviation Journal 15, 205–210 (1994)
6. Tong, C.X., Wang, Z.J., Zhang, T.Q.: Control System Design of Variable Structure Decoupling for BTT Missile. Aviation Journal 27, 27–30 (2006)
7. Liu, C.H.: Decoupling Theory of Control System for Multi-Variable Process. Water Conservation Electric Power Press (1984)

Linear Precoder for MIMO Wireless Channels Based on SVD

Shan Zhao, Naiqian Zhang, Biaoli Jin, and Dong Li

Communication University of China, Information Engineering School
100024 Beijing, China
{zhaos,ddznq,jlb,lidong}@cuc.edu.cn

Abstract. According to the characteristics of the MIMO channel matrix decomposition, linear precoding techniques can either based on eigenvalue decomposition (EVD) or based on singular value decomposition (SVD). This paper proposes a new MOMO precoding method that based on SVD of the channel matrices. Correlation for transmit antennas and receive antennas will weaken system performance. To optimize the performance effectively, the proposed method add a decorrelation matrix to the conventional system models, which can remove the correlation of transmit antennas and receive antennas. Finally, we have done BER performance simulation for proposed precoding system model, it is shown by means of simulations that the proposed MIMO precoding technique offers significant performance and greatly reduces the BER compared to its conventional counterparts.

Keywords: MIMO, linear precoding, SVD, decorrelation.

1 Introduction

In recent year, precoding for MIMO system has attracted much attention because it is one of the most promising techniques to reduce BER. Many of the earlier designs focused on perfect CSIT, often jointly optimizing both a linear precoder and a linear decoder based on the MSE, the SNR, or the BER [1] [2], more recent work considered partial CSIT. Precoding with mean CSIT was designed to maximize the received SNR [3], minimize the signal error rate (SER) [4], the MSE [5], or the PEP [6] [7]. Precoding with transmit covariance CSIT was similarly developed to minimize the PEP [8], the SER [9], or the MSE [10]. Precoding for both mean and transmit covariance CSIT has been developed to minimize the PEP.

Precoding based on channel decomposition is a hot research, precoding based SVD in this paper is one of the more representative precoding scheme. For MIMO system, correlation always exists between transmit and receive antennas, it will reduce the channel capacity and increases BER, in this case, we proposes a new precoding method for MIMO, which can greatly reducing BER and increase the channel capacity.

D. Yang (Ed.): Informatics in Control, Automation and Robotics, Volume 2, LNEE 133, pp. 529–536.
springerlink.com

2 Theoretical Basis

Precoding can be understood as system pre-processing data in transimitter according to the channel changes. Precoding processing only performance as left multiply send vector by precoding When linear pre-coding system using matrix form. Linear precoding structure is showed in Fig.1.

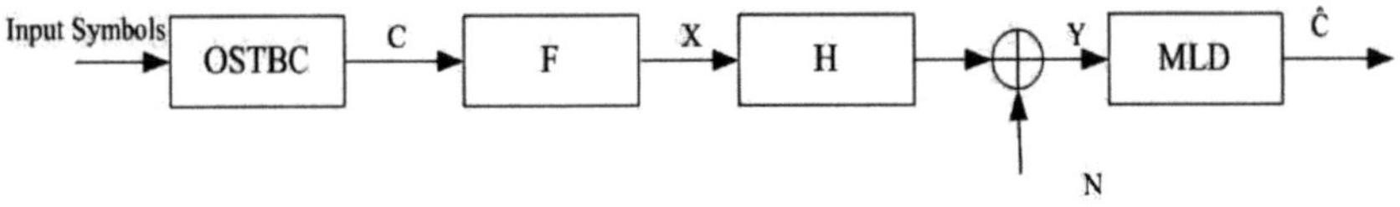

Fig. 1. Linear precoding structure

From the linear pre-code structure picture we can see, receive vector through precoding processing is:

$$Y = HFX + n \tag{1}$$

Where, C is $M \times N_r$ Symbol matrix, M is symbol vector length, F is $N_t \times M$ processing matrix, H is $N_R \times N_T$ channel matrix, Y is $N_r \times M$ receive signal, n is $N_r \times M$ noise matrix.

3 Solution Scheme

3.1 Minimum PEP Precoding

At the receiver, with ML detection the upper bound of PEP can be expressed as:

$$P(C \to \hat{C}) \leq \left(-\frac{\left\| HF(C \to \hat{C}) \right\|_F^2}{4\sigma^2} \right) \tag{2}$$

It can be seen from the above equation, PEP depends on the codeword pair $\left(C, \widehat{C}\right)$, the Chernoff bound of above equation can be written as:

$$f(H, A, F) \leq \exp\left(-\frac{\gamma}{4} tr\left(HFAF^H H^H \right) \right) \tag{3}$$

If the codeword distance A has been selected, then the goal of minimize $E_H\left[f(H, A, F) \right]$ is to choose the best precoder F. According to probability density function of channel H:

$$g(H)=\frac{1}{\pi^{N_tN_r}\det(R_t)^{N_r}}\cdot\exp\left(-tr\left[(H-H_m)^H R^{-1}(H-H_m)\right]\right) \quad (4)$$

Upper bound of average PEP is:

$$\overline{P_e}\leq\frac{\exp\left[tr\left(H_mW^{-1}H_m^H\right)\right]}{\det(W)^{N^r}}\det(R_t)^{N_r}\cdot\exp\left[-tr\left(H_mW^{-1}H_m^H\right)\right] \quad (5)$$

Where, $W=\frac{\gamma}{4}R_tFAF^HR_t+R_t$

In order to obtaining minimum, take the logarithm on both sides of (5) and ignore the constant terms, obtain the following objective function:

$$J=tr\left(H_mW^{-1}H_m^H\right)-N_r\log\det(W) \quad (6)$$

Combining (6) with precoding power constraint condition, target constraint function of minimum PEP is:

$$\begin{cases}\min J=tr\left(H_mW^{-1}H_m^H\right)-N_r\log\det(W)\\ W=\frac{\gamma}{4}R_tFAF^HR_t+R_t\\ tr\left(FF^H\right)=1\end{cases} \quad (7)$$

Considering precoding with an orthogonal STBC, codeword covariance can be showed as:

$$A=\frac{1}{P}(C-\hat{C})(C-\hat{C})^H=\mu I \quad (8)$$

Where, μ represents codeword distance.

μ_0 is defined as code word distance value of matrix A when precoder is best, it can either be the minimum, or the average distance for all codeword pairs. Then (7) can be expressed as:

$$\begin{cases}\min J=tr\left(H_mW^{-1}H_m^H\right)-N_r\log\det(W)\\ W=\frac{\mu_0\gamma}{4}R_tFF^HR_t+R_t\\ tr\left(FF^H\right)=1\end{cases} \quad (9)$$

Let $\xi = \frac{\mu_0 \gamma}{4}$, then (9) can be show as:

$$\begin{aligned} W &= \xi R_t F F^H R_t + R_t \\ \xi &= tr(R_t^{-1} W R_t^{-1} - R_t^{-1}) \end{aligned} \tag{10}$$

Let Φ be the following function of W

$$\Phi(W) = R_t^{-1} W R_t^{-1} - R_t^{-1} \tag{11}$$

Then the constraint in (10) can be rewritten as:

$$tr(\Phi) = \xi \tag{12}$$

Using the Lagrangian function:

$$L(W, v) = tr\left(H_m W^{-1} H_m^H\right) - N_r \log\det(W) + v\left[tr(\Phi) - \xi\right] \tag{13}$$

Where, v is Lagrange multiplier, let $\frac{\partial L(W, v)}{\partial W} = 0$, then:

$$v W R_t^{-2} W - N_r W - H_m^H H_m = 0 \tag{14}$$

Above equation can be regard as a quadratic matrix equation about W. Solving the equation as follow:

$$W = \frac{1}{2v} R_t \left(N_r I \pm \Delta^{1/2}\right) R_t \tag{15}$$

$$\Delta = N_r^2 I + 4v R_t^{-1} H_m^H H_m R_t^{-1} \tag{16}$$

Then the matrix Φ can be written as:

$$\Phi = \frac{1}{2v}\left(N_r I + \Delta^{1/2}\right) - R_t^{-1} = \xi F F^H \tag{17}$$

Let the eigenvalue decomposition (EVD) of Φ:

$$\Phi = U_\Phi \Lambda_\Phi U_\Phi^H \tag{18}$$

Optimal precoding matrix can be obtained according to singular value decomposition (SVD):

$$F = \frac{1}{\sqrt{\xi}} U_{\Phi} \Lambda_{\Phi}^{1/2} V^{H} \tag{19}$$

The left singular vectors and singular values of an optimal precoder are the eigenvectors and square root of the eigenvalues of Φ, while the right singular vector V can be any unitary matrix.

3.2 Improved System Model

When the distance between transmitter or receiver antenna in MIMO system is not large enough, or the angle diffusion is not sufficient, correlation will exist between the transmitter or receiver antenna, that will have impact on system performance, so we consider add a decorrelation matrix D to the conventional model, so as to remove correlation, improved system model structure, showed as Figure 2.

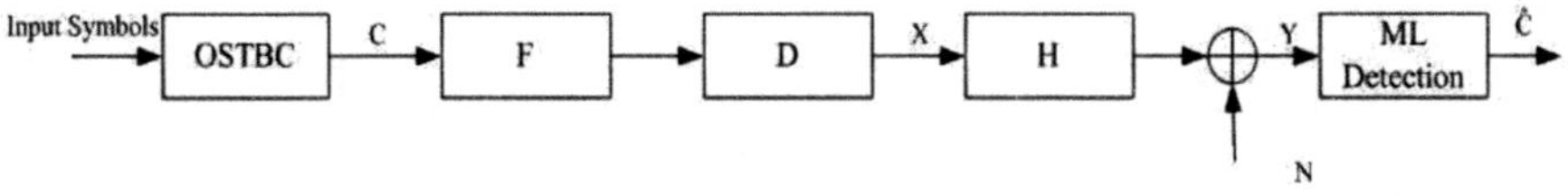

Fig. 2. Improved linear precoding structure

To simplify the analysis of the problem, consider a flat fading MIMO systems, channel matrix H is an Rayleigh fading model with independent distribution, the noise n is Gaussian white noise, and $R_{nn} = \sigma^2 I$.

Rayleigh flat fading channel can be expressed as:

$$H = R_r^{1/2} H_w R_t^{1/2} \tag{20}$$

Where, H_w is independent distribution non-correlation matrix, it is the Gaussian distribution with zero mean and variance is 1. R_t is transmit correlation matrix, R_r is receive correlation matrix, $R_r^{1/2}$ and $R_t^{1/2}$ can be respectively obtained using SVD :

$$R_t^{1/2} = U_t \Lambda_t^{1/2} V_t^{H} \tag{21}$$

$$R_r^{1/2} = U_r \Lambda_r^{1/2} V_r^{H} \tag{22}$$

Where, Λ_t and Λ_r are diagonal matrix, U_t and U_r are unitary matrix, and $U_t U_t^H = I_{N_t}, U_r U_r^H = I_{N_r}$. We assume that only transmit correlation, H can be shown as:

$$H = H_w R_t^{1/2} = H_w U_t \Lambda_t^{1/2} V_t^H \tag{23}$$

Decorrelation matrix is D, the equivalent channel matrix can be expressed as:

$$\tilde{H} = HD = H_w U_t \Lambda_t^{1/2} V_t^H D \tag{24}$$

Correlation of equivalent channel is:

$$\tilde{R} = E\{\tilde{H}^H \tilde{H}\} = D^H (V_t^H)^H (\Lambda_t^{1/2})^H U_t^H E\{H_w^H H_w\} U_t \Lambda_t^{1/2} V_t^H D \tag{25}$$

As the channel H_w is independent distribution non-correlation matrix, so:

$$\tilde{R} = D^H (V_t^H)^H (\Lambda_t^{1/2})^H U_t^H U_t \Lambda_t^{1/2} V_t^H D = D^H V_t \Lambda_t V_t^H D \tag{26}$$

Let decorrelation matrix $D = V_t$, (26) can be express as:

$$\tilde{R} = \Lambda_t \tag{27}$$

As can be seen from above that the correlation is eliminated when equivalent channel correlation matrix convert into a diagonal matrix.

4 Simulation and Discussion

In this section, we present results from extensive simulations to evaluate of the proposed scheme.

Form figure 3 we can see the relationship between SNR and BER under different correlation coefficient. BER of system is 10^{-3} required SNR is about 14dB when the correlation coefficient is 0.5, while required SNR required to 12 dB as the correlation coefficient is 0.2. That is, increased correlation increasing SNR and affecting the system error performance.

The simulation results are shown in figure 4 under 2×2 MIMO system, correlation coefficient is 0.7. As can be seen from the figure, when BER of system is 10^{-4} required SNR is about 18dB if there is no decorrelation, while it reduced to 16dB if adding to decorrelation. Namely, decorrelation makes the system obtain 2dB gain.

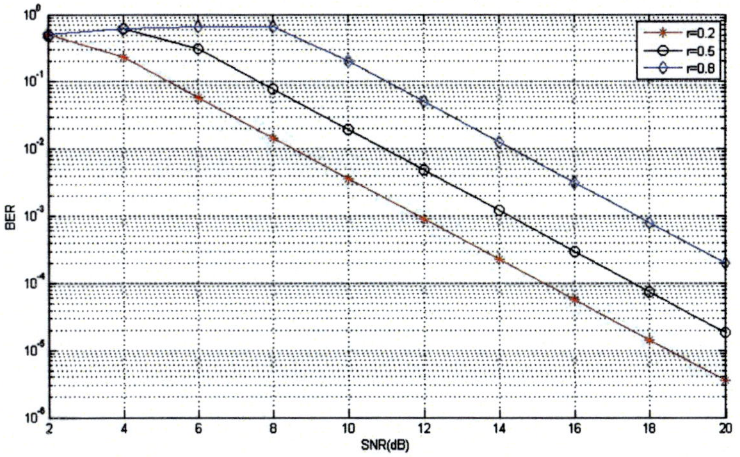

Fig. 3. Precoding gain for space-time coded system with different correlation coefficient

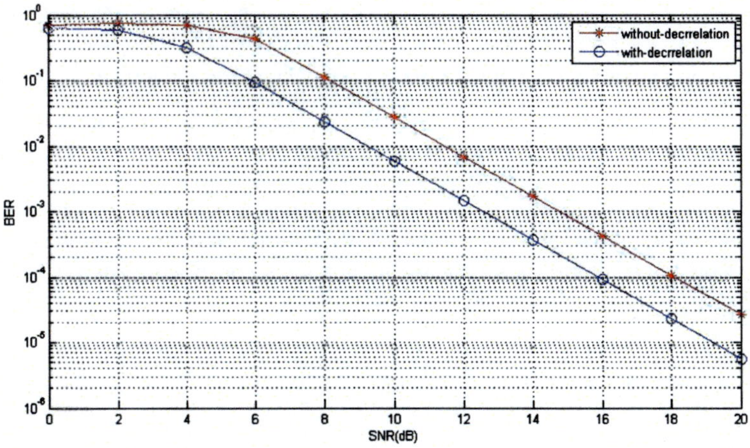

Fig. 4. Precoding gain for space-time coded system with $r = 0.7$

5 Conclusion

This paper has proposed a linear precoder base on SVD for a MIMO wireless system. An improved system model structure has been developed to remove correlation. Computer simulations demonstrated that the proposed precoding method for achieves a good performance, it is also shown that the BER improvement of the proposed scheme comes from decorrelation matrix. Finally, the simulation clarified that the reduction of correlation also reduces SNR that is the system required.

Acknowledgment. This work is Sponsor by Important National Science & Technology Specific Projects (2010ZX03005-001).M.

References

1. Palomar, D., Cioffi, J., Lagunas, M.: Joint Tx-Rx beamforming design for multicarrier MIMO channels: A unified framework for convex optimization. IEEE Trans. Signal Processing 51(9), 2381–2401 (2003)
2. Sampath, H., Stoica, P., Paulraj, A.: Generalized linear precoder and decoder design for MIMO channels using the weighted MMSE criterion. IEEE Trans. Commun. 49(12), 2198–2206 (2001)
3. Narula, A., Lopez, M., Trott, M., Wornell, G.: Efficient use of side information in multiple antenna data transmission over fading channels. IEEE J. Select. Areas Commun. 16(8), 1423–1436 (1998)
4. Zhou, S., Giannakis, G.: Optimal transmitter eigen-beamforming and space-time block coding based on channel mean feedback. IEEE Trans. Signal Processing 50(10), 2599–2613 (2002)
5. Jorswieck, E., Sezgin, A., Boche, H., Costa, E.: Optimal transmit strategiesin MIMO Ricean channels with MMSE receiver. In: Proc. Veh. Tech. Conf., vol. 5, pp. 3787–3791 (September 2004)
6. Jöngren, G., Skoglund, M., Ottersten, B.: Combining beamforming and orthogonal space-time block coding. IEEE Trans. Inform. Theory 48(3), 611–627 (2002)
7. Liu, L., Jafarkhani, H.: Application of quasi-orthogonal space-time block codes in beamforming. IEEE Trans. Signal Processing 53(1), 54–63 (2005)
8. Sampath, H., Paulraj, A.: Linear precoding for space-time coded systems with known fading. IEEE Commun. Lett. 6(6), 239–241 (2002)
9. Zhou, S., Giannakis, G.: Optimal transmitter eigen-beamforming and space-time block coding based on channel correlations. IEEE Trans. Inform. Theory 49(7), 1673–1690 (2003)
10. Haustein, T., Boche, H.: Optimal power allocation for MSE and bit-loading in MIMO systems and the impact of correlation. In: Proc. IEEE Int. Conf. Acoustics, Speech, and Signal Processing, vol. 4, pp. 405–408 (April 2003)

6-Axis Robot Design and Simulation Based on SimulationX

Luyan Bi[1] and Lisheng Liu[2]

[1] Control Science and Engineering Institute
Shenzhen Graduate School of Harbin Industry University, Shenzhen, China
bly8155@163.com
[2] Robot Lab, Cooperate Engineering
Johnson Electric Company, Hongkong, Shenzhen, China
lisheng.liu@johnsonelectric.com

Abstract. This paper presents research and simulation analysis on kinematics and dynamics problem based on the JE 6-axis robot researched and developed independently by Johnsonelectric Company. Firstly, the robot kinematics model is established and robot workspace is obtained based on the robot kinematics analysis by means of Denavit-Hartenberg method. Furthermore, robot system model is founded in software SimulationX and mechanical system and control system are analyzed through simulation. Based on simulation result of SimulationX which includes every property of components in the model of SimulationX, the performance and feasibility of the 6-axis robot model can be clearly evaluated and verified.

Keywords: 6-axis robot, kinematics solution, Denavit-Hartenberg Method, SimulationX, mechanical system.

1 Introduction

Robots are used in a wide range of industrial applications and robotics technology is likely to become the high technology field of the 21st century [1, 2]. Robots were initially applied to jobs that were hot, heavy and hazardous such as die casting, forging, and spot welding. 6-axis robots can be widely used in carrying, assembly, welding, spraying, winding, motor assembly and other industrial sites which alleviate the current shortage of industrial workers and can dramatically reduce the quality of workers because of differences in the impact on product quality [3].

SimulationX is a software for valuation of the interaction of all components of technical systems [4]. It is the universal CAE tool for modeling, simulation and analyzing of physical effects – with ready-to-use model libraries for 1D mechanics, 3D multibody systems, power transmission, hydraulics, pneumatics, thermodynamics, electrics, electrical drives, magnetic as well as controls – postprocessing included[4,5].

Robot system is a multi-disciplinary system which includes control system, mechanical system and electrics system. The objective of applying SimulationX in 6-axis robot design is to provide accurate simulation result for decision making and configuration optimization in the design process.

D. Yang (Ed.): Informatics in Control, Automation and Robotics, Volume 2, LNEE 133, pp. 537–544.
springerlink.com

2 6-Axis Robot Structure

Fig. 1 is the 3D model of the 6-axis robot. The 6-axis robot is composed of control box, robot body and other accessories. The drive equipment of the 6-axis robot is AC servo motor and harmonic gear reducer.

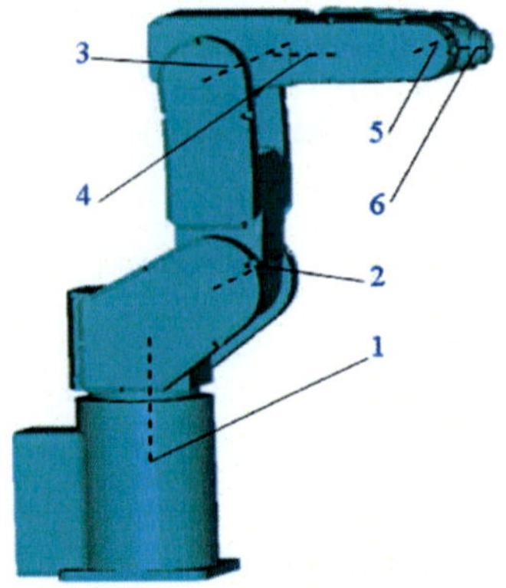

Fig. 1. 3D model of the 6-axis robot

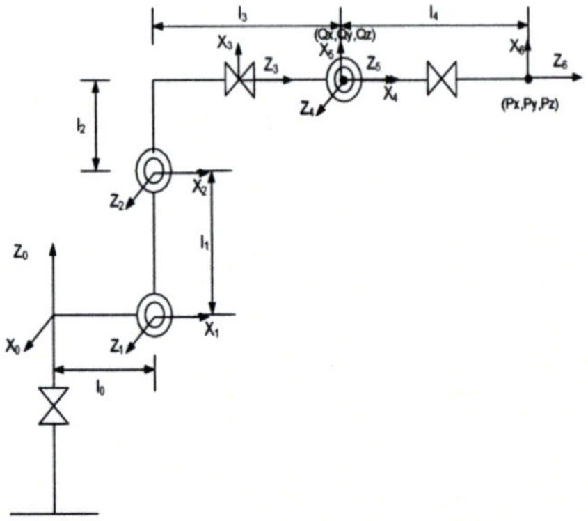

Fig. 2. The joints of 6-axis robot

The robot has six rotate joints. The first axis of robot is waist joint and the rotation axis of it is vertical; the second axis is shoulder joint and the rotation axis of it is horizontal. The third axis is elbow joint and the rotation axis of it is horizontal. The forth, fifth and sixth axis are forearm and wrist joints. Fig 2 is the joints of the 6-axis robot.

3 Kinematic Resolution

The 6-axis robot model is analyzed by means of Denavit-Hartenberg method[6,7]. As shown in Fig. 2, θ_i is rotate angle of every joint, l_i is the every length of the arm, ${}^{i}A_{i+1}$ is transformation matrix from axis i to axis (i+1). Table 1 is the D-H parameter table.

Table 1. D-H parameter

#	θ	d	a	α
1	θ_1	0	l_0	90°
2	θ_2	0	l_1	0
3	θ_3	0	l_2	90°
4	θ_4	l_3	0	-90°
5	θ_5	0	0	90°
6	θ_6	l_4	0	0

Note: $S_i = \sin(\theta_i)$, $S_{ij} = \sin(\theta_i + \theta_j)$, $C_i = \cos(\theta_i)$, $C_{ij} = \cos(\theta_i + \theta_j)$.

Based on table 1, the transformation matrix of every joint is expressed as

$$ {}^{0}A_1 = \begin{bmatrix} C_1 & 0 & S_1 & l_0C_1 \\ S_1 & 0 & -C_1 & l_0S_1 \\ 0 & 1 & 0 & 0 \\ 0 & 0 & 0 & 1 \end{bmatrix} {}^{1}A_2 = \begin{bmatrix} C_2 & -S_2 & 0 & l_1C_2 \\ S_2 & C_2 & 0 & l_1S_2 \\ 0 & 0 & 1 & 0 \\ 0 & 0 & 0 & 1 \end{bmatrix} {}^{2}A_3 = \begin{bmatrix} C_3 & 0 & S_3 & l_2C_3 \\ S_3 & 0 & -C_3 & l_2S_3 \\ 0 & 1 & 0 & 0 \\ 0 & 0 & 0 & 1 \end{bmatrix} $$

$$ {}^{3}A_4 = \begin{bmatrix} C_4 & 0 & -S_4 & 0 \\ S_4 & 0 & C_4 & 0 \\ 0 & -1 & 0 & l_3 \\ 0 & 0 & 0 & 1 \end{bmatrix} {}^{4}A_5 = \begin{bmatrix} C_5 & 0 & -S_5 & 0 \\ S_5 & 0 & C_5 & 0 \\ 0 & 1 & 0 & 0 \\ 0 & 0 & 0 & 1 \end{bmatrix} {}^{5}A_6 = \begin{bmatrix} C_6 & -S_6 & 0 & 0 \\ S_6 & C_6 & 0 & 0 \\ 0 & 0 & 0 & l_4 \\ 0 & 0 & 0 & 1 \end{bmatrix} $$

The transformation matrix from robot base to robot actuator ${}^{0}A_6$ is expressed as.

$$ {}^{0}A_6 = {}^{0}A_1 {}^{1}A_2 {}^{2}A_3 {}^{3}A_4 {}^{4}A_5 {}^{5}A_6 = \begin{bmatrix} n_x & o_x & a_x & p_x \\ n_y & o_y & a_y & P_y \\ n_z & o_z & a_z & p_z \\ 0 & 0 & 0 & 1 \end{bmatrix} \quad (1) $$

$$ \begin{aligned} n_x &= C_1[C_{23}(C_4C_5C_6 - S_4S_6) - S_{23}S_5C_6] + S_1(S_4C_5C_6 + C_4S_6) \\ n_y &= S_1[C_{23}(C_4C_5C_6 - S_4S_6) - S_{23}S_5C_6] - C_1(S_4C_5C_6 + C_4S_6) \\ n_z &= S_{23}(C_4C_5C_6 - S_4S_6) + C_{23}S_5C_6 \\ o_x &= C_1[-C_{23}(C_4C_5S_6 + S_4C_6) + S_{23}S_5S_6] + S_1(-S_4C_5S_6 + C_4C_6) \\ o_y &= S_1[-C_{23}(C_4C_5S_6 + S_4C_6) + S_{23}S_5S_6] - C_1(-S_4C_5S_6 + C_4C_6) \\ o_z &= -S_{23}(C_4C_5S_6 + S_4C_6) - C_{23}S_5S_6 \\ a_x &= C_1(C_{23}C_4S_5 + S_{23}C_5) + S_1S_4S_5 \\ a_y &= S_1(C_{23}C_4S_5 + S_{23}C_5) - C_1S_4S_5 \\ a_z &= S_{23}C_4S_5 - C_{23}C_5 \\ p_x &= C_1[l_0 + l_1C_2 + l_2C_{23} + l_3S_{23} + l_4(C_{23}C_4S_5 + S_{23}C_5)] + l_4S_1S_4S_5 \\ p_y &= S_1[l_0 + l_1C_2 + l_2C_{23} + l_3S_{23} + l_4(C_{23}C_4S_5 + S_{23}C_5)] - l_4C_1S_4S_5 \\ p_z &= l_1S_2 + l_2S_{23} - l_3C_{23} + l_4(S_{23}C_4S_5 - C_{23}C_5). \end{aligned} $$

4 Inverse Kinematics Resolution

The joints of 6-axis robot as shown in Fig.2, the position of sixth arm terminal is (P_x, P_y, P_z) and the position of fifth joint is (Q_x, Q_y, Q_z).

$$ {}^0A_4 = {}^0A_1\,{}^1A_2\,{}^2A_3\,{}^3A_4 = \begin{bmatrix} n'_x & o'_x & a'_x & Q_x \\ n'_y & o'_y & a'_y & Q_y \\ n'_z & o'_z & a'_z & Q_z \\ 0 & 0 & 0 & 1 \end{bmatrix} \tag{2} $$

Based on equation (1) and (2), the relationship between (P_x, P_y, P_z) and (Q_x, Q_y, Q_z) can be expressed as: $\begin{cases} Q_x = P_x - l_4 a_x \\ Q_y = P_y - l_4 a_y \\ Q_z = P_z - l_4 a_z \end{cases}$

1). θ_1

$$ {}^0A^{-1}{}_1 \begin{bmatrix} n'_x & o'_x & a'_x & Q_x \\ n'_y & o'_y & a'_y & Q_y \\ n'_z & o'_z & a'_z & Q_z \\ 0 & 0 & 0 & 1 \end{bmatrix} = {}^1A_2\,{}^2A_3\,{}^3A_4 $$

Based on element (3,4), θ_1 can be expressed as below.

$$ \theta_1 = \arctan(\frac{Q_y}{Q_x}) = \arctan(\frac{P_y - l_4 a_y}{P_x - l_4 a_x}) \text{ or } \theta_1 = \pi + \arctan(\frac{Q_y}{Q_x}) = \pi + \arctan(\frac{P_y - l_4 a_y}{P_x - l_4 a_x}) $$

2). θ_3

$$ {}^1A^{-1}{}_2\,{}^0A^{-1}{}_1 \begin{bmatrix} n'_x & o'_x & a'_x & Q_x \\ n'_y & o'_y & a'_y & Q_y \\ n'_z & o'_z & a'_z & Q_z \\ 0 & 0 & 0 & 1 \end{bmatrix} = {}^2A_3\,{}^3A_4 $$

Based on element (1,4) and (2,4), the following equations are obtained.

$$ \begin{cases} C_2 C_1 Q_x + C_2 S_1 Q_y + S_2 Q_z - C_2 l_0 = l_3 S_3 + l_2 C_3 + l_1 \\ -S_2 C_1 Q_x - S_2 S_1 Q_y + C_2 Q_z + S_2 l_0 = -l_3 C_3 + l_2 S_3 \end{cases} \tag{3} $$

So, θ_3 can be expressed as $\theta_3 = 2\tan^{-1}(\frac{k_1 \pm \sqrt{k_1^2 + k_2^2 - k_3^2}}{k_2 + k_3})$

$$ \begin{cases} k_1 = 2l_1 l_3 \\ k_2 = 2l_1 l_2 \\ k_3 = Q_x^2 + Q_y^2 + Q_z^2 - 2Q_x l_0 C_1 - 2Q_y l_0 S_1 + l_0^2 - l_1^2 - l_2^2 - l_3^2 \end{cases} $$

3). Based on equations (3), θ_2 can be expressed as below.

$$\theta_2 = a\tan 2(S_2, C_2) \begin{cases} C_2\mu_1 + S_2\nu_1 = \gamma_1 \\ C_2\mu_2 + S_2\nu_2 = \gamma_2 \end{cases}$$

$$\begin{cases} \mu_1 = C_1Q_x + S_1Q_y - l_0 \\ \nu_1 = Q_z \\ \gamma_1 = l_3S_3 + l_2C_3 + l_1 \\ \mu_2 = Q_z \\ \nu_2 = -C_1Q_x - S_1Q_y + l_0 \\ \gamma_2 = -l_3C_3 + l_2S_3 \end{cases}$$

4). θ_5

$$ {}^2A_3^{-1}\,{}^1A_2^{-1}\,{}^0A_1^{-1} \begin{bmatrix} n_x & o_x & a_x & p_x \\ n_y & o_y & a_y & P_y \\ n_z & o_z & a_z & p_z \\ 0 & 0 & 0 & 1 \end{bmatrix} = {}^3A_4\,{}^4A_5\,{}^5A_6 = \begin{bmatrix} \cdots & \cdots & C_4C_5 & \cdots \\ \cdots & \cdots & S_4S_5 & \cdots \\ \cdots & \cdots & C_5 & \cdots \\ \cdots & \cdots & 0 & 1 \end{bmatrix} \tag{4}$$

Based on element (3,3), θ_5 can be expressed as $\theta_5 = \arccos(a_xC_1S_{23} + a_yS_1S_{23} - a_zC_{23})$

5). θ_4

Base on element(1,3) and (2,3) of equation (4), θ_4 can be expressed as below.

$$\theta_4 = a\tan 2(S_4, C_4)$$

$$C_4 = \frac{a_xC_1C_{23} + a_yS_1C_{23} + a_zS_{23}}{S_5}$$

$$S_4 = \frac{a_xS_1 - a_yC_1}{S_5}$$

6). θ_6

$$ {}^3A^{-1}{}_4\,{}^2A_3^{-1}\,{}^1A_2^{-1}\,{}^0A_1^{-1} \begin{bmatrix} n_x & o_x & a_x & p_x \\ n_y & o_y & a_y & P_y \\ n_z & o_z & a_z & p_z \\ 0 & 0 & 0 & 1 \end{bmatrix} = {}^4A_5\,{}^5A_6 = \begin{bmatrix} C_5C_6 & -C_5S_6 & S_5 & l_4S_5 \\ S_5C_6 & -S_5S_6 & -C_5 & -l_4C_5 \\ S_6 & C_6 & 0 & 0 \\ 0 & 0 & 0 & 1 \end{bmatrix} \tag{5}$$

Base on element(2,1) and (2,2) of equation(5), θ_6 can be expressed as below.

$$\theta_6 = a\tan 2(S_6, C_6)$$

$$C_6 = -\frac{n_xC_1C_{23} + n_yS_1S_{23} - n_zC_{23}}{S_5}$$

$$S_6 = \frac{o_xC_1S_{23} + o_yS_1S_{23} - o_zC_{23}}{S_5}$$

5 Model the 6-Axis Robot in SimulationX

Base on the 6-axis robot system introduced in section 2, this section illustrates model the robot in SimulationX.

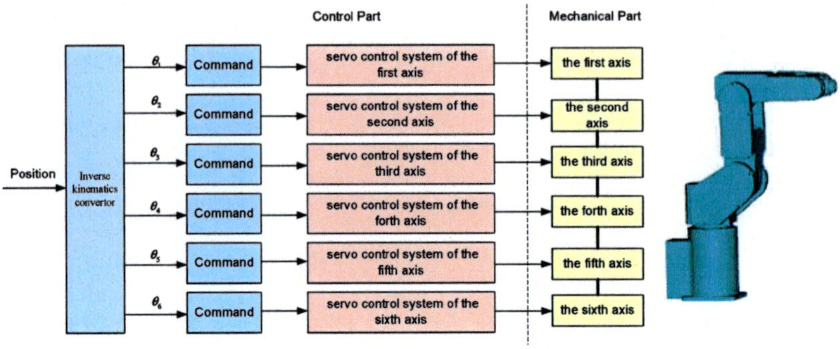

Fig. 3. The 6-axis robot model in SimulationX

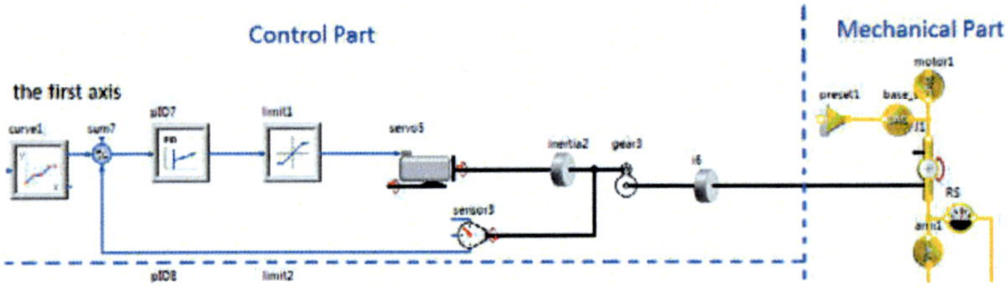

Fig. 4. The detail of the robot first axis in SimulationX

Fig. 3 is the robot model in SimulationX which is composed of control part and mechanical part. Left is control part of the model and right is mechanical part of the model. Control part of the model can be divided into six parts which includes the first, second, third and forth axis part. Fig. 4 is the detail of the first axis in SimualtionX.

6 Simulation Case Study

Based on 6-axis simulation model introduced in section 5, simulation results are obtained and performances are compared through the SimulationX. SimulationX provides an efficiency analysis for robot designers.

Table 2 is input parameters of the model in SimulationX.

Table 2. Input parameters of the model

Axis Num	1#	2#	3#	4#	5#	6#
Motor Power(W)	400	400	150	50	50	50
Rated Torque(Nm)	1.27	1.27	0.448	0.149	0.149	0.149
Rated Speed(r/min)	3000	3000	3000	3000	3000	3000
Transmission Ratio	80:1	80:1	70:1	65:1	65:1	50:1

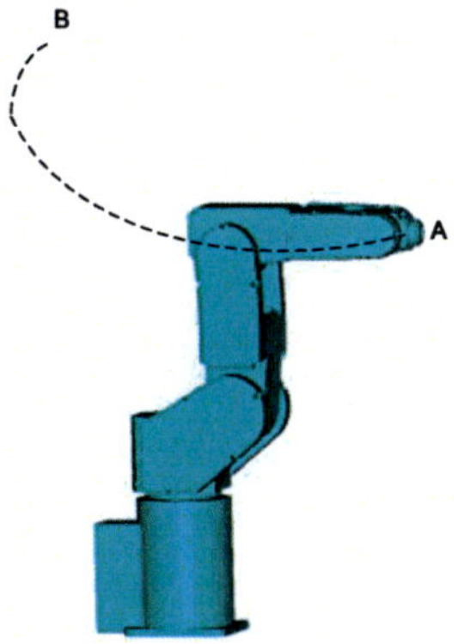

Fig. 5. Moving trace

In this simulation case, robot move from position A to position B as shown in Fig.5. $A = (450,0,360)$ $B = (15,0,645)$ The lengths of robot arm is $l_0 = 75, l_1 = 270, l_2 = 90, l_3 = 295, l_4 = 80$ .According the inverse kinematics resolution discussed in section 3, the joint angles of position A and B are $(0,\frac{\pi}{2},0,0,0,0)$ and $(\pi,\frac{\pi}{2},\frac{\pi}{2},0,0,0)$. According to Transmission Ratio in table 2, the motor of the first axis moves an angle of 80π and the motor of the third axis moves an angle of 35π. Every joint move through S-Curve speed motion. The other joints do not move so that only simulation data of first axis and third axis are provided. Fig.6 is comparison of position command and feedback of first axis and third axis. Fig.7 and Fig.8 is servo motor torque curve and speed curve of the first axis and third axis.

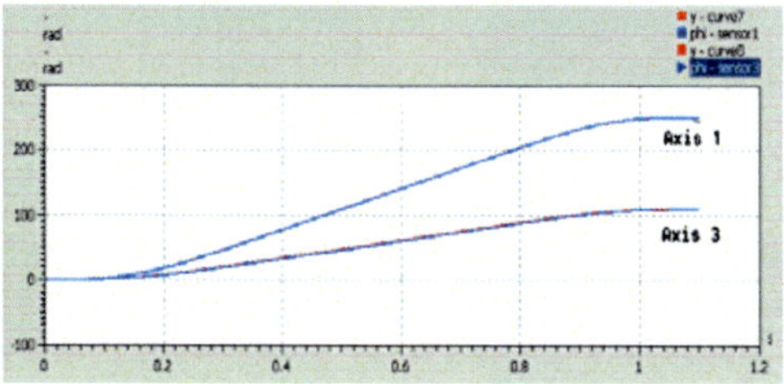

Fig. 6. Comparison of position command and feedback of first axis and the third axis

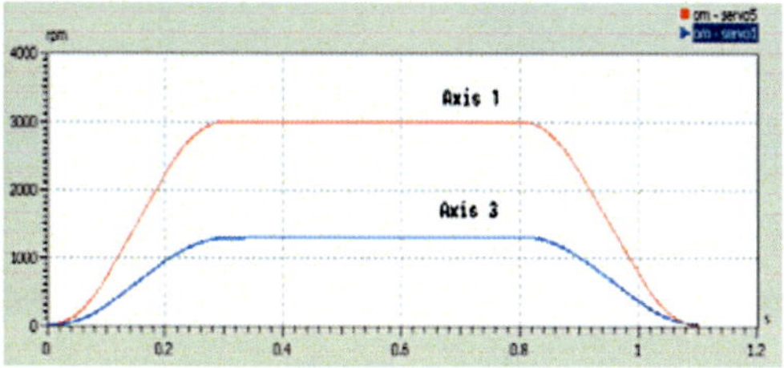

Fig. 7. Servo motor speed curve of the first axis and the third axis

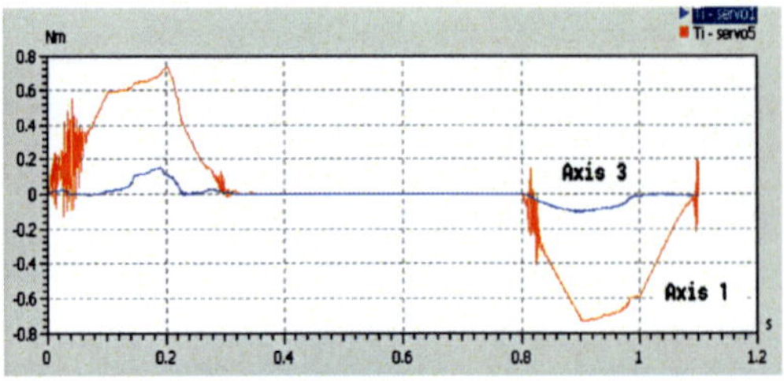

Fig. 8. Servo motor torque curve of the first axis and the third axis

7 Conclusions

This paper presents a 6-axis robot system model and applies SimulationX into robot simulation. Firstly, the mechanical structure and kinematics model is established. Furthermore, robot system model is founded in software SimulationX which includes mechanical system and control system. Based on 6-axis simulation model, the different commands, simulation results are obtained and performances are compared through the SimulationX. SimulationX provides an efficiency analysis for robot designers. The performance and feasibility of the robot system is clearly evaluated and verified based on every property of components in the model of SimulationX.

References

1. Xiong, Y.: Robotics, pp. 21–26. Mechanical and Engineering, Beijing (1998)
2. Cheng, Y.: Research and Simulation Analysis on kinematics of QJ-1 Welding Robot, pp. 1–9. Zhejiang University (2008)
3. Sun, L.: Control System Design and Kinematics Simulation of Six Degrees of Freedom Manipulator, pp. 1–16. Beijing industry University (2009)
4. http://www.robots.com/applications.php
5. http://www.itisim.com
6. Liu, Y.: SimulationX by Example, pp. 1–8. China Mechine Press (2010)
7. Denavit, J., Hartenberg, R.S.: A kinematic Notation for Lower-Pair Mechanisms Based on Matrices. ASME Journal of Applied Mechanics, 215–221 (June 1955)
8. Paul, R.P.: Robot Manipulators, Mathematics, Programming, and Control. MIT Press, Cambridge (1981)

Rov Magnetized Magnetic Field Measurement and Magnetic Dipole Model*

Weiming Cheng**, Chong Kang***, Hui Li, and Jinku Lv

College of Science, Harbin Engineering University, Harbin, China
kangchong@hrbeu.edu.cn

Abstract. Through to the position of the main interference source Rov, I put forward the twelve magnetic dipole model, and then deduced and the model of magnetizing field size of the twelve magnetic dipole near field; and use mathematica to simulate. And in the wild, we established a 3 d coordinate system, and selected certain areas to measure different heightment dates of Rov, through software "surfer" to make magnetized magnetic field figure of Rov, to verify the validity of the magnetic dipole model, this method can calculate the magnetized magnetic field of Rov better.

Keywords: Rov(underwater robot), magnetizing field, magnetic dipole, magnetic-field measurement.

1 Introduction

Scientists are still in the detection of the submarine volcano enrichment mining area and surveying and mapping them. They put the high technology sensors placed in cables, from the ship to the bottom to identify each volcano regions of the mineral species. But compare with the underwater vehicle carrying geomagnetic instrument, the traditional "down type" detection methods has lot of errors for the location of the ocean mineral, and Rov will not be restricted of objective factors, it can be free in the water. But Rov own magnetized magnetic field on the magnetic field measurement will have a great effect, in order to get the high accuracy of geomagnetic data from sensors, we must make up for the measurement of geomagnetic data for the interference of the underwater off carrier magnetic field[1]. This paper puts forward the twelve magnetic dipole near field model, it can calculate the size of the magnetic field of Rov better, the study for the magnetic field of the magnetization Rov has a reference significance.

---

* Fund project: National Natural Science Foundation of China(61174192).

** Author introduction: Weiming Cheng (1986-),male, master graduate student, E-mail: chwm200501012@126.com

*** The author mentor: Chong Kang (1970-),male, professor, doctoral supervisor, E-mail: kangchong@hrbeu.edu.cn

D. Yang (Ed.): Informatics in Control, Automation and Robotics, Volume 2, LNEE 133, pp. 545–551.
springerlink.com

2 Magnetic Dipole Model and Simulation

2.1 Magnetic Dipole

First introduces solve of the magnetic dipole near field model, and then deduces twelve magnetic dipole near field magnetic field intensity[2].

As shown in figure 1, it is the right hand space right Angle coordinate system. The centre of the magnetic dipole circle is Located in the origin of the right hand space right Angle coordinate system. M is a point of any space, and whose spherical coordinate is M (r , ϕ_0, θ_0). R is the distance of centre to M. ϕ_0 is zenith angle. θ_0 is the angle of M in the heliospheric face to X axis in the heliospheric face.

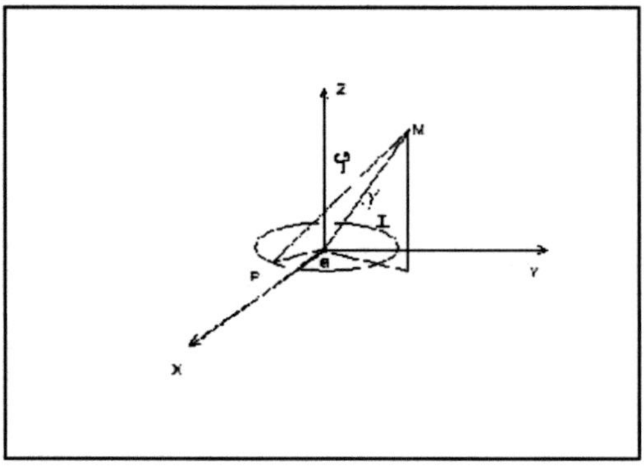

Fig. 1. Magnetic dipole and coordinate system

On the basis of Biot Savart law, the magneticvector of this current element of M in the face of any space[2] is:

$$d\vec{B} = \frac{\mu}{4\pi}\frac{Id\vec{l}\times\vec{a}}{a^3}\overrightarrow{PM} \qquad |a| \neq 0 \tag{1}$$

Supposing the magnetic induction vector[5] is $\vec{B} = B_x\vec{i} + B_y\vec{j} + B_z\vec{k}$, like that:

$$\begin{cases} B_x = \dfrac{3\mu}{8}\dfrac{IR^2}{\left(\sqrt{R^2+r^2}\right)^3}\dfrac{r^2\sin 2\phi_0\cos\theta_0}{R^2+r^2} \\ B_y = \dfrac{3\mu}{8}\dfrac{IR^2}{\left(\sqrt{R^2+r^2}\right)^3}\dfrac{r^2\sin 2\phi_0\sin\theta_0}{R^2+r^2} \\ B_z = \dfrac{\mu}{2}\dfrac{IR^2}{\left(\sqrt{R^2+r^2}\right)^3}\left(1-\dfrac{3}{2}\dfrac{r^2\sin^2\phi_0}{R^2+r^2}\right) \end{cases} \tag{2}$$

This is a point M's three component expression of magnetic induction strength in the face.

If you consider that the concept of the magnetic moment[3] $\overrightarrow{P_m} = I\cdot S\cdot\vec{n}$, and translated into space right Angle coordinate system, formula (2) can variable into:

$$\begin{cases} B_x = \dfrac{3\mu}{4} \dfrac{P_m xz}{\left(\sqrt{R^2+x^2+y^2+z^2}\right)^5} \\ B_y = \dfrac{3\mu}{8} \dfrac{P_m yz}{\left(\sqrt{R^2+x^2+y^2+z^2}\right)^5} \\ B_z = \dfrac{\mu}{4} \dfrac{P_m(2R^2+2z^2-x^2-y^2)}{\left(\sqrt{R^2+x^2+y^2+z^2}\right)^5} \end{cases} \quad (3)$$

2.2 The Near Field Model of Twelve Magnetic Dipoles

As shown in figure 2: According to the main magnetic material's position on Rov (Rov attached pictures last), set up twelve magnetic dipole model.

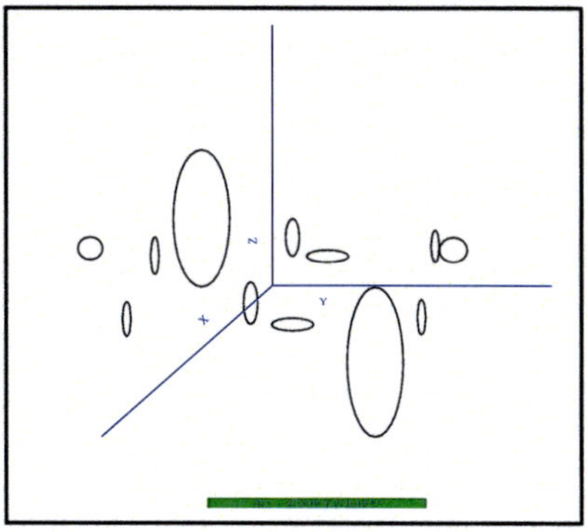

Fig. 2. Twelve magnetic dipole near field model

It can deduce that total magnetic field strength for three components of twelve magnetic dipoles in the near field any point M in space(Please see next page):

2.3 The Simulation of Twelve Magnetic Dipoles Near Field Model

Among them, the horizontal axis is the X axis, and the vertical axis is the Y axis. Measurement range is $\{x \in (-2,2), y \in (-2,2)\}$. Parameters[5] are set to $P_{m1}=0.2$, $P_{m2}=0.3$, $P_{m3}=20$, $P_{m4}=0.6$, $R_1=0.3$, $R_2=0.1$, $\mu=1$, $Z_0=-0.7$ 与 $Z_0=0.7$ (Please see next page):

3 Experimental Data Processing

We use no Rov experimental data are minus the Rov experimental data, so get Rov of magnetized magnetic field.

$$
\left\{
\begin{aligned}
&B_x = B_{x1}+B_{x2}+\ldots+B_{x12} = \frac{3\mu}{4}\frac{P_{m1}x(y-y_1)}{\left(\sqrt{\left(R_1^2+x^2+(y-y_1)^2+z^2\right)}\right)^5}+\frac{3\mu}{4}\frac{P_{m1}x(y+y_1)}{\left(\sqrt{\left(R_1^2+x^2+(y+y_1)^2+z^2\right)}\right)^5}\\
&+\frac{3\mu}{4}\frac{P_{m2}(x-x_2)(y-y_2)}{\left(\sqrt{R^2+(x-x_2)^2+(y-y_2)^2+z^2}\right)^5}+\frac{3\mu}{4}\frac{P_{m2}(x-x_2)(y+y_2)}{\left(\sqrt{R^2+(x-x_2)^2+(y+y_2)^2+z^2}\right)^5}\\
&+\frac{3\mu}{4}\frac{P_{m2}(x+x_2)(y-y_2)}{\left(\sqrt{R^2+(x+x_2)^2+(y-y_2)^2+z^2}\right)^5}+\frac{3\mu}{4}\frac{P_{m2}(x+x_2)(y+y_2)}{\left(\sqrt{R^2+(x+x_2)^2+(y+y_2)^2+z^2}\right)^5}\\
&+\frac{3\mu}{4}\frac{P_{m3}(x-x_2)y}{\left(\sqrt{R^2+(x-x_2)^2+y^2+z^2}\right)^5}+\frac{3\mu}{4}\frac{P_{m3}(x+x_2)y}{\left(\sqrt{R^2+(x+x_2)^2+y^2+z^2}\right)^5}\\
&+\frac{3\mu}{4}\frac{P_{m3}(x-x_2)y}{\left(\sqrt{R^2+(x-x_2)^2+y^2+z^2}\right)^5}+\frac{3\mu}{4}\frac{P_{m3}(x+x_2)y}{\left(\sqrt{R^2+(x+x_2)^2+y^2+z^2}\right)^5}\\
&+\frac{\mu}{4}\frac{P_{m4}\left[2R^2+2x^2-(y-y_4)^2-(z-z_4)^2\right]}{\left(\sqrt{R^2+x^2+(y-y_4)^2+(z-z_4)^2}\right)^5}+\frac{\mu}{4}\frac{P_{m4}\left[2R^2+2x^2-(y+y_4)^2-(z-z_4)^2\right]}{\left(\sqrt{R^2+x^2+(y+y_4)^2+(z-z_4)^2}\right)^5}\\
&B_y = B_{y1}+B_{y2}+\ldots+B_{y12} = \frac{\mu}{4}\frac{P_{m1}\left[2R_1^2+2x^2-(y-y_1)^2-(z+z_1)^2\right]}{\left(\sqrt{R_1^2+x^2+(y-y_1)^2+(z+z_1)^2}\right)^5}+\frac{\mu}{4}\frac{P_{m1}\left[2R_1^2+2x^2-(y+y_1)^2-(z-z_1)^2\right]}{\left(\sqrt{R_1^2+x^2+(y+y_1)^2+(z-z_1)^2}\right)^5}\\
&+\frac{\mu}{4}\frac{P_{m2}\left[2R^2+2(x-x_2)^2-(y-y_2)^2-z^2\right]}{\left(\sqrt{R^2+(x-x_2)^2+(y-y_2)^2+z^2}\right)^5}+\frac{\mu}{4}\frac{P_{m2}\left[2R^2+2(x-x_2)^2-(y+y_2)^2-z^2\right]}{\left(\sqrt{R^2+(x-x_2)^2+(y+y_2)^2+z^2}\right)^5}\\
&+\frac{\mu}{4}\frac{P_{m2}\left[2R^2+2(x+x_2)^2-(y-y_2)^2-z^2\right]}{\left(\sqrt{R^2+(x+x_2)^2+(y-y_2)^2+z^2}\right)^5}+\frac{\mu}{4}\frac{P_{m2}\left[2R^2+2(x+x_2)^2-(y+y_2)^2-z^2\right]}{\left(\sqrt{R^2+(x+x_2)^2+(y+y_2)^2+z^2}\right)^5}\\
&+\frac{\mu}{4}\frac{P_{m3}\left[2R^2+2(x-x_2)^2-y^2-z^2\right]}{\left(\sqrt{R^2+(x-x_2)^2+y^2+z^2}\right)^5}+\frac{\mu}{4}\frac{P_{m3}\left[2R^2+2(x+x_2)^2-y^2-z^2\right]}{\left(\sqrt{R^2+(x+x_2)^2+y^2+z^2}\right)^5}\\
&+\frac{3\mu}{4}\frac{P_{m3}(y-y_3)z}{\left(\sqrt{R^2+(x-x_3)^2+(y-y_3)^2+z^2}\right)^5}+\frac{3\mu}{4}\frac{P_{m3}(y-y_3)z}{\left(\sqrt{R^2+(x+x_3)^2+(y-y_3)^2+z^2}\right)^5}\\
&+\frac{3\mu}{4}\frac{P_{m4}x(y-y_4)}{\left(\sqrt{R^2+x^2+(y-y_4)^2+(z-z_4)^2}\right)^5}+\frac{3\mu}{4}\frac{P_{m4}x(y+y_4)}{\left(\sqrt{R^2+x^2+(y+y_4)^2+(z-z_4)^2}\right)^5}\\
&B_z = B_{z1}+B_{z2}+\ldots+B_{z12} = \frac{3\mu}{4}\frac{P_{m1}(y-y_1)(z+z_1)}{\left(\sqrt{R_1^2+x^2+(y-y_1)^2+(z+z_1)^2}\right)^5}+\frac{3\mu}{4}\frac{P_{m1}(y+y_1)(z-z_1)}{\left(\sqrt{R_1^2+x^2+(y+y_1)^2+(z-z_1)^2}\right)^5}\\
&+\frac{3\mu}{4}\frac{P_{m2}(y-y_2)z}{\left(\sqrt{R^2+(x-x_2)^2+(y-y_2)^2+z^2}\right)^5}+\frac{3\mu}{4}\frac{P_{m2}(y+y_2)z}{\left(\sqrt{R^2+(x-x_2)^2+(y+y_2)^2+z^2}\right)^5}\\
&+\frac{3\mu}{4}\frac{P_{m2}(y-y_2)z}{\left(\sqrt{R^2+(x+x_2)^2+(y-y_2)^2+z^2}\right)^5}+\frac{3\mu}{4}\frac{P_{m2}(y+y_2)z}{\left(\sqrt{R^2+(x+x_2)^2+(y+y_2)^2+z^2}\right)^5}\\
&+\frac{3\mu}{4}\frac{P_{m3}yz}{\left(\sqrt{R^2+(x-x_2)^2+y^2+z^2}\right)^5}+\frac{3\mu}{4}\frac{P_{m3}yz}{\left(\sqrt{R^2+(x+x_2)^2+y^2+z^2}\right)^5}\\
&+\frac{\mu}{4}\frac{P_{m3}\left[2R^2+2z^2-(y-y_3)^2-(x-x_3)^2\right]}{\left(\sqrt{R^2+(x-x_3)^2+(y-y_3)^2+z^2}\right)^5}+\frac{\mu}{4}\frac{P_{m3}\left[2R^2+2z^2-(y-y_3)^2-(x+x_3)^2\right]}{\left(\sqrt{R^2+(x+x_3)^2+(y-y_3)^2+z^2}\right)^5}\\
&+\frac{3\mu}{4}\frac{P_{m4}x(z-z_4)}{\left(\sqrt{R^2+x^2+(y-y_4)^2+(z-z_4)^2}\right)^5}+\frac{3\mu}{4}\frac{P_{m4}x(z-z_4)}{\left(\sqrt{R^2+x^2+(y+y_4)^2+(z-z_4)^2}\right)^5}
\end{aligned}
\right. \quad (4)
$$

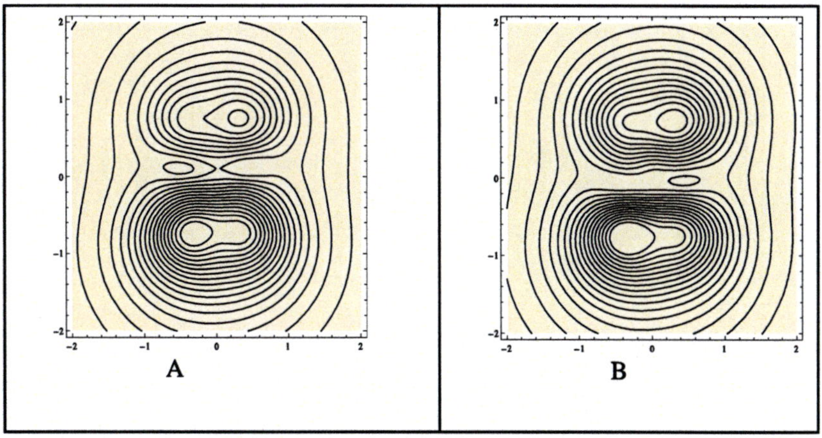

Among them, A figure is $Z_0 = -0.7$, and B figure is $Z_0 = 0.7$.

3.1 Experimental Program

As shown in figure 3, we draw a square area. Its four sides parallel the southeast and northwest direction.

(1) When completed the above experiment, we went on to measure the magnetic field with Rov. Rov general is a rectangle. Field and the experiment used exactly is the same site. As shown in figure 4:

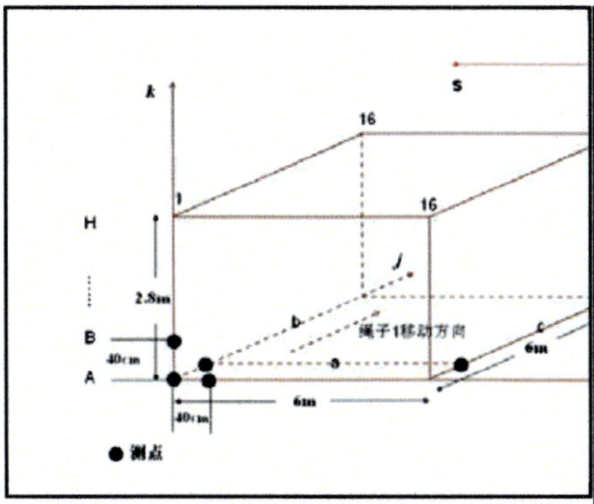

Fig. 3. No Rov measure scheme figure

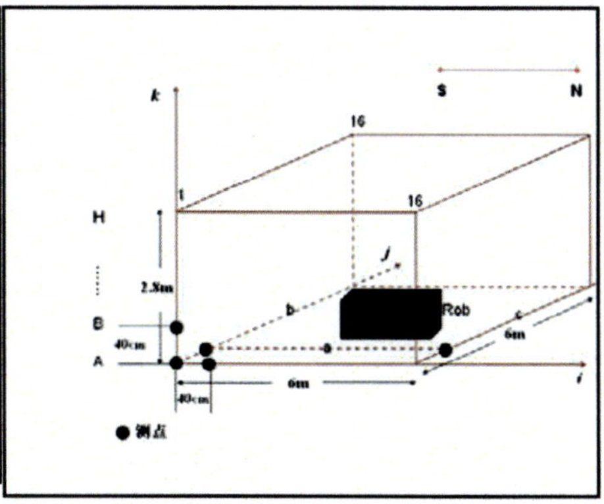

Fig. 4. Have Rov measure scheme figure

The disturbing magnetic field of Rov is the difference of geomagnetic data of the above two experiments. Use surfer software to work out the interference magnetic field figure of Rov, as shown in figure 5 shows. Isoline interval for 20 nT.

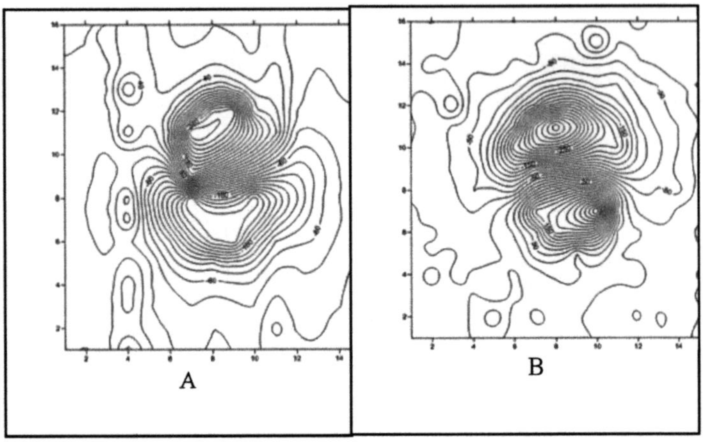

By the calculation process and the experiment indicated that:

1) Geomagnetic chart of Rov made by surfer software are very similar to theory simulation. In the two height of the space with Rov, that has obvious magnetic field changes. The changes present a closed form isoline. The intense change of the magnetic field is corresponding to the Rov magnetic material, chains, screw, etc. In the Rov around, with the increase of around distance, the isoline interval gradually reduced. The influence of magnetic field is more and more small.
2) Due to the influence of magnetizing field about the ground below the surface of the gravel and rock, the theory simulation graphics may be difference from the actual graphics.
3) The magnetic field is a weak magnetic field. Along with the change of the weak magnetic fields, the magnetic Susceptibility of ferromagnetic materials is also change. But in the simulation, we set the magnetic Susceptibility of ferromagnetic materials is fixed[4]. This is one reason for the errors.
4) Because the number of magnetic dipole limit, also can produce errors. The more magnetic dipole number, the closer between theoretical derivation results and the figure made by surfer software. But as the number of magnetic dipole increased, magnetic dipole model will be more and more complex, and computing time will increase. This paper is only for each interference source using a magnetic dipole indicated.
5) Precision of proton magnetometer and the distance of manual method and the distance of manual method, as well as the drift and interference of measurement environment magnetic field. From in figure 5 can see, there are many small circle of the position, and Rov disruption field shape slightly change is because these reasons.

4 Conclusions

Rov on the impact of the geomagnetic field can use twelve magnetic dipoles to indicate. This paper firstly established a fundamental magnetic dipole model, infered its size of the magnetizing field, and then infered the model of magnetizing field size of the twelve magnetic dipoles near field; and use mathematica to simulate. Then use the proton magnetic apparatus to measure the small scope, different height, point of dense geomagnetic field. Through the Kriging interpolation method, we made the high accuracy of local geomagnetic figure with the certain reference significance. Simulation results show that, this method can calculate the magnetized magnetic field better, be used to Rov separation of magnetic field, and be used as an important basis of a Rov's degaussing system design. The related result can also be necessary preparations of magnetic measures for Rov.

References

1. Zhang, Q., Pan, M.-C., Wu, M.-P.: Magnetization field of submarine for underwater geomagnetic navigation. Journal of Chinese Inertial Technology 17, 293–296 (2009)
2. Ren, L.-P., Zhao, J.-S., Hou, S.-X.: Magnetic dipole field space distribution patterns. Marine Surveying and Mapping 22(2), 18–21 (2002)
3. Ye, X.-P., Gong, S.-G.: Ship Physical Fields, pp. 86–99. Weapons Industry Press (1992)
4. Zhou, G.-H., Xiao, C.-H., Yan, H., Gao, J.-J.: A method to calculate the induced magnetic field of ferromagnetic objects in a weak magnetic field. Journal of Harbin Engineering University 30(1), 91–95 (2009)
5. Xie, C.-F., Rao, K.-Q.: Electromagnetic Field and Wave (Version 4), pp. 114–116. Higher Education Press (2006)

Adaptive Modulation and Coding in Cognitive Radio Networks

Yuling Zhang

School of Electrical and Electronics Engineering, Ludong University
264025, Yantai, P.R. China
zhang-yuling@hotmail.com

Abstract. In this paper, we evaluate the performance of a spectral pooling system with adaptive modulation and coding. The whole communication spectrum is divided into N sub-bands, the cognitive users listen to the wireless channel and access to the unused spectrum. At the physical layer, adaptive modulation and coding (AMC) based on rate-compatible low-density parity-check codes (RC-LDPC) is adopted in multiple-input multiple-output (MIMO) fading channels. Numerical results show that the application of AMC based on RC-LDPC codes can achieve considerable spectral efficiency gain in cognitive radio networks.

Keywords: Cognitive Radio, Adaptive modulation and coding, LDPC.

1 Introduction

Cognitive radio technology is a promising key technology that can increase efficiency of wireless communications, which was introduced by Joseph Mitola in 1999[1]. There are two different ways of allowing unlicensed users to access the licensed spectrum. One scheme works by looking for spectrum holes for the use of secondary users, and another allows co-existence of secondary users and primary users (licensed users) in the same frequency band. Motivated by the desire for an effective and practical scheme, in this paper, we focus on the first scheme. The basic idea under the proposed scheme is based on the notion of spectrum pooling [2], which is quite simple: a cognitive radio will listen to the channel and, if sensed idle, will transmit during the voids.

A kind of rate and power adaptation strategy is proposed to optimize data transmission over fading channels in a spectrum sharing system operating under average received-interference constraint at the licensed user (primary user)[3], but they both didn't consider the advanced channel coding scheme, based on these observation, we proposed adaptive modulation and coding framework in spectrum pooling system, in which RC-LDPC codes are considered.

The rest of this paper is organized as follows. In Section 2, we present the system model of the cognitive radio network based on spectrum pooling, while the principle of the adaptive modulation and coding in MIMO fading channels is also illustrated. Simulation Results are given in Section 3, some conclusions are drawn in Section 4.

D. Yang (Ed.): Informatics in Control, Automation and Robotics, Volume 2, LNEE 133, pp. 553–556.
springerlink.com

2 System Model

According to the notion of spectrum pooling, spectral ranges from different spectrum owners are merged into a common pool [2]. The whole spectrum are divided into N sub-bands, at first, there is only one user can simultaneously transmit over the same sub-band under AMC framework. Then the cognitive users listen to the wireless channel and determine which part of the spectrum is unused. Then, they successively adapt their signal to fill detected voids in the spectrum domain.

We adopt the system model in [2], an asynchronous TDD communication system is considered, in which the primary and cognitive users wish to communicate to different receivers, subject to mutual interference in a heterogeneous network where devices operates in a wideband context. The channel between transmitter and receiver for each user is Rayleigh distributed and stays constant over each block fading length, the receiver is able to track the channel variations, and the fed back channel is errorless and in time.

The spectral efficiency for a system with L users can be calculated by the following equation [2].

$$\Phi = \frac{1-\Delta_\infty^L}{1-\Delta_\infty} C_{1,\infty} \tag{1}$$

Where Δ_∞ is the band factor gain, which is defined as the band sensed void from user l to user l+1 over the total bandwidth, and can be expressed as

$$\Delta_\infty = 1-\exp\left(-\frac{\gamma_0}{\gamma}\right) \tag{2}$$

And γ_0 is the cut-off level under which no information will be transmitted. $C_{1,\infty}$ in equation (1) is the spectral efficiency when there is only one user in the system, so firstly, we should discuss the AMC scheme for one user. Figure 1 gives the system model of our AMC system. Assuming that there are N_T transmit antennas and N_R receive antennas. When space time block coding (STBC) is employed, MIMO channel can be equivalent to a SISO channel, the probability density function (PDF) of γ can be calculated as follows [4].

$$\gamma = \frac{P_s}{\sigma^2}\|\mathbf{H}\|_F^2 = \frac{P_T}{\sigma^2 N_T R_c}\|\mathbf{H}\|_F^2 = \frac{\bar{\gamma}}{N_T R_c}\|\mathbf{H}\|_F^2 \tag{3}$$

At the physical layer, there are multiple modulation and coding schemes (MCSs) available. The CSI is estimated at the receiver and then sent back through a feedback channel to the AMC controller, which chooses the appropriate MCS in the next transmission according to the target PER P_{target}.

Suppose that there are N MCSs at the physical layer with increasing rates R_n (n=1, 2, …, N) in terms of information bits per symbol. We will consider the modulation method with the MQAM signal constellation, where M denotes the number of points

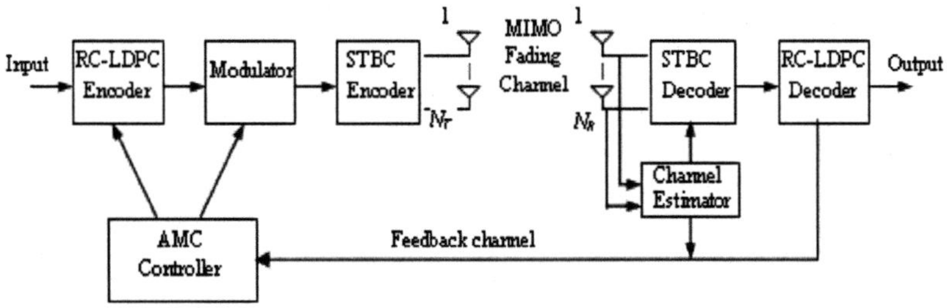

Fig. 1. System Model

in each signal constellation. If the coding rate of a MCS is R_L, we have $R_n = R_L \cdot (\log_2 M)$. The whole SNR range is divided into N+1 intervals based on N thresholds γ_n, n = 1, 2, …, N. When $\gamma_n \le \hat{\gamma} < \gamma_{n+1}$, MCS n with the rate R_n will be chosen for the next transmission. The calculation of thresholds γ_n can be referred to [4].

According to the AMC rule, each MCS n will be chosen with the following probability

$$p_n = \int_{\gamma_n}^{\gamma_{n+1}} p_{\hat{\gamma}}(\hat{\gamma}) d\hat{\gamma} = \frac{\Gamma(K, \lambda\gamma_n) - \Gamma(K, \lambda\gamma_{n+1})}{\Gamma(K)} \tag{4}$$

Consequently, the average spectral efficiency of the whole system is given by

$$C_{1,\infty} = \sum_{n=1}^{N} R_n p_n \tag{5}$$

3 Simulation Results

This section described the numerical results through computer simulation. Assuming that there is two transmit antennas and one receive antenna. At the physical layer, the MCSs were chosen from Table 1, and detailed parameters can be referred to [4].

Table 1. Parameters of MCSs at the physical layer

	MCS1	MCS2	MCS3	MCS4	MCS5	MCS6
Mod.	BPSK	QPSK	QPSK	16QAM	16QAM	64QAM
Coding rate	1/2	1/2	3/4	9/16	3/4	3/4
Rn (b/s)	0.50	1.00	1.50	2.25	3.00	4.50

The performance constraint is assumed to be P_{target} = 0.01. Figure 2 demonstrates the spectral efficiency for different number of users, L=1 means that there is only one user in the whole bandwidth, in those cases that L larger than one, means that there is one primary user and L-1 cognitive users, it is clear that the spectral efficiency of the system with two users is larger than that with only one user.

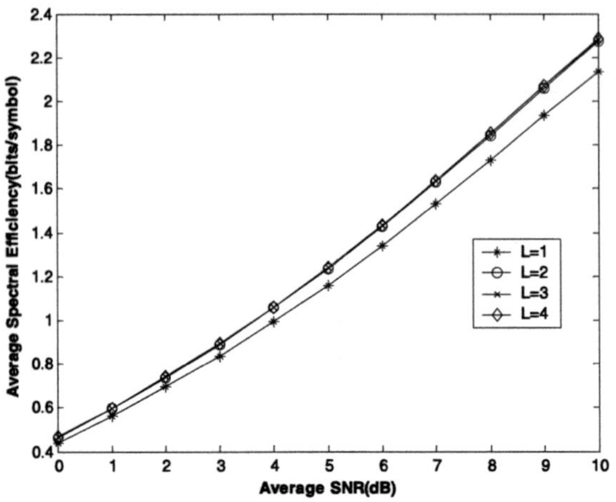

Fig. 2. ASE versus SNR for different L

4 Conclusions

In this paper, we applied AMC scheme in cognitive radio network under the idea of spectrum polling. The relevant MCS is chosen based on the SNR thresholds calculated according to the LDPC PER-SNR relationship. Numerical results show that in cognitive radio context, the AMC yields a considerable increase of the system's spectral efficiency.

Acknowledgments. The authors would like to acknowledge the support of Doctoral Foundation of Shandong Province (BS2009DX022).

References

1. Mitola III, J., Maguire Jr., G.: Cognitive radio: making software radios more personal. IEEE Personal Communications 6, 13–18 (1999)
2. Haddad, M., Hayar, A., Debbah, M.: Spectral efficiency of spectrum pooling systems. IET Communications 2, 733–741 (2008)
3. Asghari, V., Aissa, S.: Adaptive Rate and Power Transmission in Spectrum-Sharing Systems. IEEE Transactions on Wireless Communications 9, 3272–3280 (2010)
4. Zhang, Y., Yuan, D., Wang, C.-X.: Cross-layer design based on RC-LDPC codes in MIMO channels with estimation errors. AEU - International Journal of Electronics and Communications 62, 659–665 (2008)

Design of a FPGA-Based Multiple Motors Control System

Yue Ma, Rong Xiong, and Fan Zhang

Institute of Cyber-Systems and Control, Zhejiang University, Hangzhou, 310027
mayue.zju@gmail.com, rxiong@iipc.zju.edu.cn,
Isaac33zhang@gmail.com

Abstract. In this paper, a FPGA-based motor control system is designed to fulfill the requirement of controlling multiple motors accurately and simultaneously. With the implementation of hardware logic modules with hardware description language, each control module works independently in a parallel mode, so the control frequency of motors is elevated. By modifying the signal sampling method for the motor encoder, the speed measurement accuracy is significantly improved. The proposed control system is successfully applied on the platform of RoboCup small size soccer robot of Zhejiang University, ZJUNlict, in which the improvements on response speed and control accuracy of the robots are verified.

Keywords: Robot, FPGA, BLDC motor, small, parallel.

1 Introduction

Frequently, a motor controller is used for only one motor in robot system, while the synchronous control of the motors is realized through the controller clock synchronization [1-5] . Due to the limits of weight and volume, it is desirable to control multiple motors simultaneously with single motor controller. In this way, the accurate synchronous control of multiple motors can be ensured by the shared controller's clock while the motors is required to work in parallel to guarantee the control frequency.

In general, the DSP or FPGA are both used as motor control chips. The DSP controller calculates fast and works well at floating-point operation. However, each piece of DSP chip usually has only two way outputs. Once there are more than two motors to be controlled, extra DSP chips are thus needed [6-8], which extraordinarily increases the volume and cost of control circuit. Due to its flexible port configuration, FPGA with embedded CPU IP core is widely used in motor control system. Though it can realize multiple motors' closed-loop control, the function of operation is a little bit weak [9-12]. When multiple motors are controlled, each motor's control cycle will be evidently extended, which results in large overshoot and low steady precision of the speed feedback.

In this paper, a parallel multi-motor control system based on FPGA and by the way of logic programming is realized, which uses logic hardware to implement the closed-loop speed control, so that each motor has an independent logic controlling module.

D. Yang (Ed.): Informatics in Control, Automation and Robotics, Volume 2, LNEE 133, pp. 557–567.
springerlink.com

Since multi-motor closed-loop control logic modules works in parallel without any interaction, the motor control frequency will not be deteriorated by the amount of the motors. This logic hardware implementation can minimize the external logic circuit and reduce the volume and cost extremely. Meanwhile, the motor's encoder signal sampling method in the system is improved so that the speed measuring accuracy is enhanced. The multi-motor control system works successfully in Robocup small size robot soccer on-board subsystem of ZJUNlict teams of Zhejiang University. The experimental tests demonstrate that the response speed and control accuracy of the robot are improved by the proposed method.

2 System Architecture

The control system based on FPGA is shown in Fig 1, which is composed of a IP core embedded controller and some motor control modules. IP core controller is the core of the whole system. It is responsible for initializing the robot system, collecting the robot's sensors information, calculating the set points for each motor according to the requirements, and assigning values to the motor control modules through the motor control peripheral interfaces. Each motor has a motor control module, which is responsible for the motor velocity closed-loop control, including filtering the input digital signal, measuring the motor velocity, operating PI, modulating the PWM wave, commutating the brushless motor, protecting the whole circuit against over-current and so on.

IP core embedded controller and motor control modules work in parallel. All of the logic functions about the motor control can be integrated in one FPGA chip, and there is no need of other logical control chips such as chip MC33035 used to modulate BLDC motors. The integrated design makes the control system lighter, smaller, more stable and with lower power consumption. For the mobile robot, the shape will be more elegant if the size of control circuit is smaller. Also, lighter weight and lower power consumption means that the battery capacity can be reduced, so that the battery weight and size will be cut. The system's CPU focuses on the response of external signals, while the traditional method spends a lot of time in calculating the each motor control objects' PI regulators with a serial multi-motor control algorithm. Consequently, the real-time performance of the system is enhanced. Because each motor control module works in parallel, when the system needs one more motors, the real-time performance and control frequency of system will not be hurt.

The multi-motor parallel control system in parallel is only limited by the FPGA's pin number and capacity. With the popularity of the BGA package and increasing capacity of FPGA, these two constraints will not be problems in future. In our system, a single motor control module requires 11 clock control cycles, so the frequency of the motor control modules is limited by the FPGA clock frequency; controlling a brushless motor with hall sensors requires a total of 12 pins: 2 encoder pins, 3 hall signal pins, 1 over-current signal pin, 6 motor drive pins; controlling a brush motor requires 2 encoder pins and 4 motor drive pins. Since each motor control module occupies 25,000 gates, the system should select the FPGA which has an appropriate size and pin number. According to the conditions above, Altera's FPGA EP3C25Q240C8 is selected.

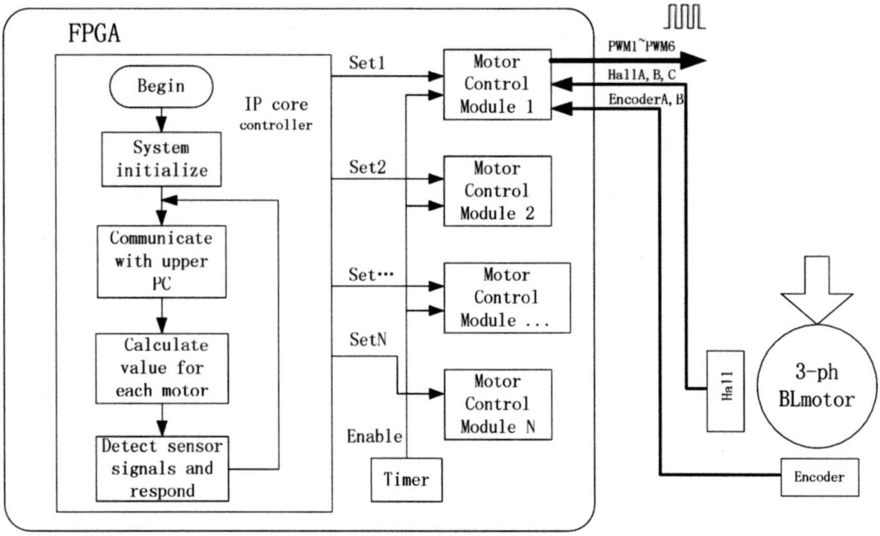

Fig. 1. Multi-motor control system structure

3 Design of Motor Control Module

The motor control module is realized by hardware logic, which consists of signal filtering sub-module、high-precision velocity measuring sub-module, PI controller operating sub-module, Pulse Width Modulation sub-module, over-current protecting sub-module, BLDC commutating sub-module and so on. Sub-module is written by Verilog HDL language and can be encapsulated into block diagram which can also be invoked into upper block diagram. The top-level map of motor control module is shown in Fig 2.

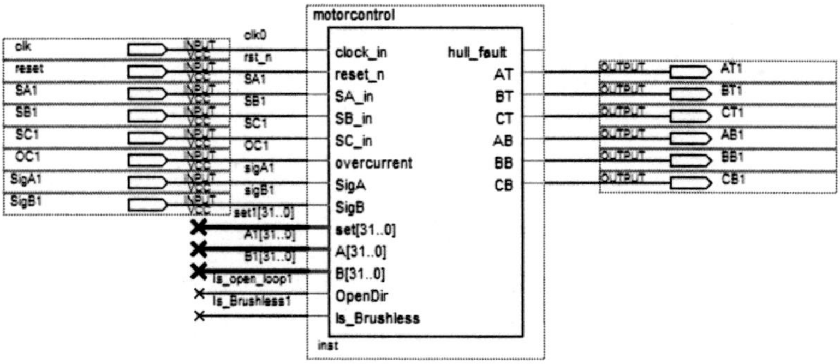

Fig. 2. Top-level map of motor control module

The main difficulty of using hardware logic to realize the motor control module is the correctness of the value involved to calculate. Complex hardware logic function consists of some simple circuit logic units which are all working independently, such as register, trigger, adder, multiplier and so on. PI closed-loop calculating function is executed each control cycle, in which the result of the controller got in the last cycle is also needed. In this paper, the closed-loop function is divided into many sub-modules. There is an enable pin for each sub-module and the number of the clock cycles between its input moment and its output moment has been restricted. There is a state machine sub-module in this system which triggers every sub-module in order of the data flow to make sure that the value in registers of sub-modules will not be replaced by mistake. There is also an enable pin for state machine sub-module which is triggered by controller timer. Every 2ms, the controller timer triggers the state machine and then the state machine triggers the other sub-module to complete the PI closed-loop calculation. What's more, hardware logic development is different from software program development. Designers have to deal with the positive and negative value、the implementing of the base circuit、the temporal logic and the combination logic. Logic simulation is a kind of indispensible development tools, by which designers can check the value of every register and find out the logic mistakes more easily, compared with the hardware debugging.

Fig 3 shows the function structure of the BLDC motor control sub-module. All the modules of this system are triggered by the same synchronous clock signal. A and B are orthogonal input signals of the encoder; SA, SB, SC are the hall commutating signal of the BLDC motor; I is the over-current signal of the motor; AT, BT, CT, AB, BB, CB are signals that trigger the converter of the BLDC motor; SET is from software core controller. The cycle period how long the system executes PI calculation can be set by users, while here it is set to 2ms. The state machine is used to control enable signal involved in order because they are working in parallel while the PID calculating process is executed serially. There is a timer in this system, which triggers the state machine to calculate the PI value once every cycle.

Now the input signal filtering sub-module、high-precision velocity measuring sub-module、PI operating sub-module、BLDC commutating sub-module are to be introduced in detail as follows.

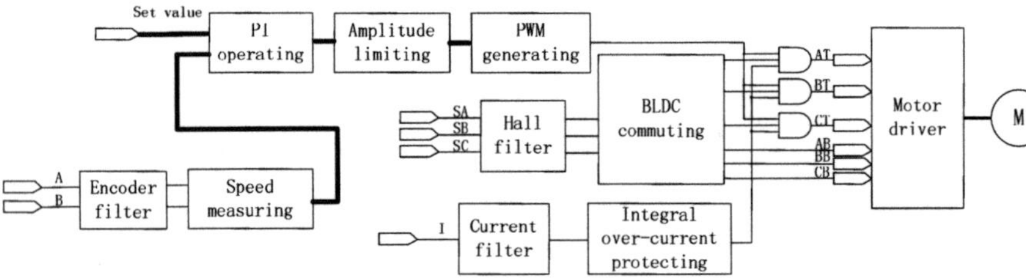

Fig. 3. Functional map of motor control module

3.1 Input Signal Filtering Sub-module

Encoder signals, hall signals, over-current signals and other signals of the motor are all connected to the control circuit boards by wires and connectors. These signals sent to the decision module as judging basis often have electronic interference, which need to be removed via input signal filtering sub-module.

Input signal filtering sub-module uses a signal sequence check method. Samples are taken from the input signals once every clock cycle and compared with those got in the last clock cycle. If they are equivalent, the sampling counter will increase by 1. Otherwise sampling counter will be clear. If the sampling counter is accumulated to number N, which means that the signal has been held on for N times, then the sampling signal is considered as an effective signal and then sent to the next unit to proceed.

3.2 Velocity Measuring Sub-module

Velocity measuring sub-module is designed to detect the direction of the motor's rotation by comparing the two-way pulse signal captured by motor light code wheel and to precisely calculate the speed of the motor rotation through improved-M algorithm. This sub-module consists of two sub slave modules. One of them accounts for the measurement of signal A and B. It produces a pulse defined as "count" once signal A or B changes. The other is applied to record the status of the pulse output from the former module in unit sampling time and then calculates the current speed of the motor.

The traditional M algorithm obtains the angle of the motor rotation in unit sampling time "t" by multiplying by the angle that the motor rolls between two pulses and the number of the pulses "M" in the same period. The error of this algorithm comes from the method that regards the number of the pulses as the real angle of the motor rotation. In this way, the angle between the start of the sampling and the first pulse will be omitted and so does the angle between the final pulse and the end of the sampling. The precision of the sampling can be improved by increasing the number of the code wheel line which, however, has no meaning for eliminating the error. To solve this problem, an improved-M algorithm is proposed herein.

The precision of the sampling in traditional M algorithm is:

$$\Delta = \frac{2\pi}{4*L} = \frac{360^\circ}{4*L} . \tag{1}$$

$$\Delta v = \frac{1}{4*L*t} . \tag{2}$$

in which Δ stands for the angle of the motor rotation between two pulses of the code wheel; Δv is the precision of the velocity indicated by the code wheel sampling; and L is the number of the code wheel line.

Improved-M algorithm: based on the M algorithm, in a given time "t", the pulse count that between start of the sampling and the first pulse, so does the pulse count between the final pulse and the end of the sampling will be add to M as the decimal part which is got by dividing the interval mentioned above by their whole period between two pulse.

The time series of the improved-M algorithm is shown as Fig 4. A and B are the orthogonal input signals of the code wheel; t is the unit sampling time; T1 is the starting time of t; T2 is the ending time of t; t1 is the time from T1 to the first change of signal A or B; t2 is the time from the final change of signal A or B to T2; t3 is the time between two changes of signal A, B before T1; t4 is the time between two changes of signal A, B before T2; t5 is the time between two changes of signal A, B which T1 is contained; t6 is the time between two changes of signal A, B which T2 is contained. Therefore the sampling value of the code wheel calculated through the improved-M algorithm is:

$$MM = M - 1 + \frac{t1}{t3} + \frac{t2}{t4} \quad . \tag{3}$$

in which M is the velocity of the motor rotation measured by the improved-M algorithm; M-1 is the angle that the motor rolls during M pulses; t1/t3 is the angle that the motor rolls from the start of the sampling time to the arrival of the first pulse; t2/t4 is the angle that the motor rolls from the arrival of the final pulse to the end of the sampling time.

Improved-M algorithm improves the precision of the sampling of the signal from the code wheel in unit sampling time so that the value calculated is much closer to the true value of the current velocity of the motor rotation. In this way, the algorithm creates a high-precision feedback which could contribute to the high-precision motor control system.

The error of the improved-M algorithm comes from the assumption that the interval between two pulse are const, i.e., t3=t5, t4=t6. In this assumption, the error will increase when the speed of the motor is at a low level. The measurement of t1-t4 is based on the system clock. The error can be ignored because of the high frequency of the system clock.

Definition:

$$t5 = \Delta 1 + t3 \, . \tag{4}$$

$$t6 = \Delta 2 + t4 \, . \tag{5}$$

The precision of the sampling in improved-M algorithm is:

$$\Delta MM = \frac{t1}{t3 + \Delta 1} + \frac{t2}{t4 + \Delta 2} - \frac{t1}{t3} - \frac{t2}{t4} \quad . \tag{6}$$

in which the value of Δ1 and Δ2 is affected by the acceleration of the motor.

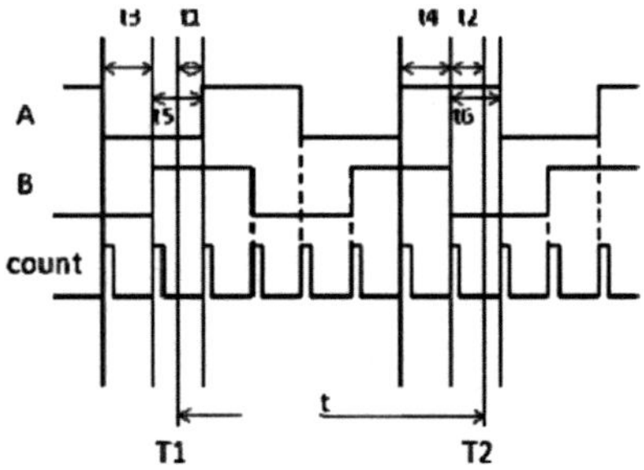

Fig. 4. Timing diagram of improved-M algorithm

3.3 PI Operating Sub-module

The PI operating sub-module is applied to execute PI operation with the set from CPU and the velocity from the velocity measuring sub-module. The result will be the duty value of PWM modulation module after amplitude limitation and parameter compensation.

The PI operating sub-module uses the incremental PI algorithm which calculates the control incremental value Δ based on difference between the setting value of the motor control module and the current rotational velocity measured by improved-M algorithm. With the purpose of reducing the procedure of the operation, the parameter Kp and Ki of the PI formula is combined together. When the program sets the parameters of the motor control module, the setting parameter is also of A and B. It is aimed to ensure that Δ calculation in one cycle.

PI computing formula is:

$$\Delta = A*error - B*pre\_error . \tag{7}$$

$$A = Kp + Ki . \tag{8}$$

$$B = Ki . \tag{9}$$

$$U(k) = U(k-1) + \Delta . \tag{10}$$

3.4 BLDC Motor Commutating Sub-module

The BLDC motor commutating sub-module is used to decode the input hall signal of the motor, then drive the three-phase BLDC motor by measuring the position of the motor. In addition, it can drive the BLDC motor when the FPGA chip sets the hall signal in a fixed value.

The BLDC motor commutation apply the trapezoidal wave commutating mode, which estimate the rotor's position by the output signal of hall sensors, and turn on the MOS-FETs in corresponding phase sequence to drive the rotor. The MOS-FET of upper bridge of the three-phase full controlled bridge is put in sequence of 1, 3, 5, while the MOS-FETs of lower bridge is put in sequence of 4, 6, 2. The hall signal SA, SB, SC is in sequence of 100, 110, 010, 011, 001 and 101. So, when the motor rotates clockwise, the sequence of turning on the MOS-FETs is: 12, 23, 34, 45, 56, 61, while when the motor rotates anti-clockwise, the sequence of turning on the MOS-FETs is: 45, 56, 61, 12, 23 and 34.

4 Test Results

Robocup small size soccer robot is required to control four wheel-drive motors and one dribble-control motor simultaneously. The robot with the limit that its diameter cannot exceed 180 millimeters and its height cannot exceed 150 millimeters is based on typical mobile robot motion control system which strictly limits the volume of control circuit. Therefore, the ZJUNlict small size soccer robot control system is designed using the technological method above and conduct the performance test and comparison.

The wheel-drive motor is disc type motor whose rated power is 45 watts. Its rated rotational speed is 5000 rpm and grid number of its photoelectric encoder is 512. The dribble-control motor is hollow cup motor whose rated power is 15 watts and rated rotational speed is 40000 rpm. Five motors are all blushless motors. The maximum acceleration of robot is $4m^2/s$, corresponding $198rad^2/s$ for motor. The circuit system we design is showed in Fig 5. Circuit boards use laminated structure, which provide sufficient protection for each part of circuit including core board, motherboard, driver board and power board from top to bottom. Since small size soccer robots may encounter strong collision with opponents during competition and circuit boards are easily damaged or burned, adding protection can make circuit boards more stable and easier to maintain.

A Labview-based data collection platform is developed to observe conveniently the dynamic response process of motor. The platform can set up incentive mode, incentive parameter, open or closed loop control mode, PI operating parameter and upload data type. After the robot receives computer instruction, it will move according to the computer specified way and upload real time data at the rate of packet/6ms. Data collection platform can display received data to waveform frame and store data in a text file.

Fig 6 shows comparing data measured by M and improved-M algorithm under the motor no-load running condition. Data measured by M algorithm are scattered in the three integer points, 27 to 29, while data measured by improved-M algorithm are distributed in the decimal point from 27.2 to 28.8. According to the formula 1, the sampling error is 0.17° when M algorithm is used. While improved-M algorithm is adopted with the acceleration set to the maximum, in terms of formula 4, the sampling error gets smaller with the speed increasing. This principle is supported by the data as follows; the sampling error grows from 0.2 ° to $2.76*10^{-5}$ ° while the given motor speed increases from 0 to 1 grids/2ms. According to this principle, we choose M algorithm instead of improved-M algorithm when the speed of the motor is set to 0. Improved-M algorithm is more accurate from theoretical error analysis and experiments. In order to avoid floating point operations, the value of improved-M is enlarged to 128 times.

Fig 7 shows the motor's steady state effect chart motor at control frequency of 500HZ motor homeostatic control diagram in control frequency of 500HZ, and Fig 8 shows the control diagram at control frequency of 250HZ. When the setting speed value of the motor control module is 6 grids/2ms in control frequency of 500HZ, actual speed of the motor fluctuates between 5 grids and 7 grids, the mean value is 6.0031 and the variance is 0.1941; at control frequency of 250HZ, actual speed of the motor fluctuates between 4.5 grids and 7.5 grids, the mean value is 6.0587 and the variance is 0.4885. Experiments prove that the control frequency increase can improve the motor's control accuracy. Feedback value of the motor participates directly in PI closed loop operation without filter processing, thus fluctuation range of motor speed is still large.

Fig. 5. Small size soccer robot circuits

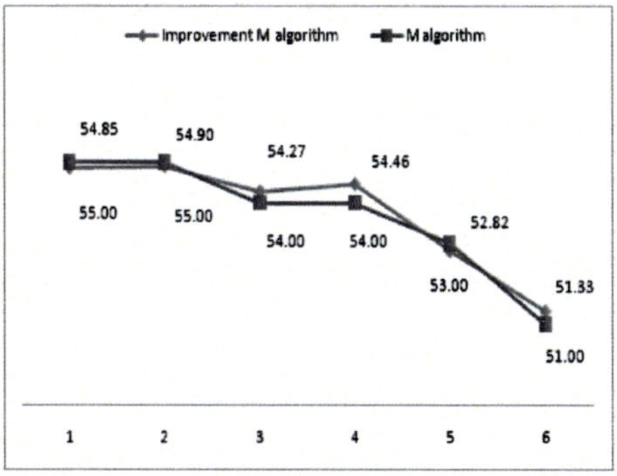

Fig. 6. Chart of M and improved-M algorithm

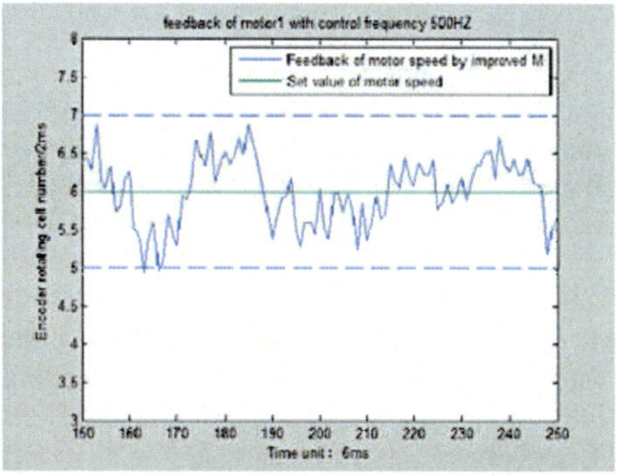

Fig. 7. Steady state effect with motor control frequency 500HZ

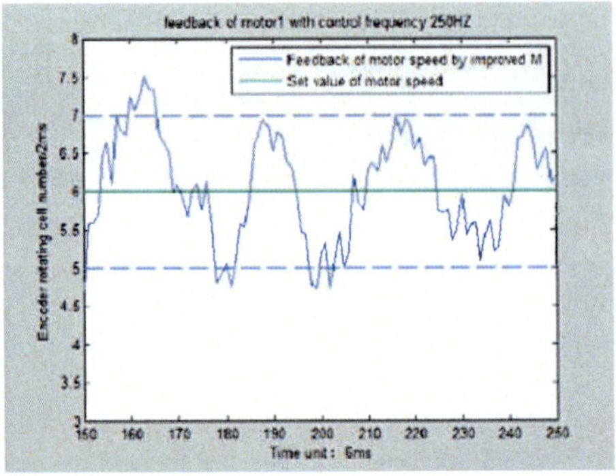

Fig. 8. Steady state effect with motor control frequency 250HZ

5 Conclusion

In this paper, a parallel design is proposed for multi-motor system, and it is successfully used in the Robocup small size soccer robot system of Zhejiang University. Closed loop control for BLDC motor is implemented by hardware description language, which improves real time response and control accuracy of system. Embedded CPU and motor control module can be downloaded into one FPGA chip to compress the volume of circuits. Improved-M algorithm is given in order to increase sampling accuracy under the condition of low line number encoder and reduce hardware cost.

For further development, the effect of motor control can be improved by some advanced algorithm such as fuzzy control algorithm, instead of PI algorithm used at present which leads to larger speed overshoot.

References

1. Joo, J., et al.: Design of a DSP controller and driver for the power-by-wire driving system with BLDC servo motor pump. In: 8th Electrical Engineering/ Electronics, Computer, Telecommunications and Information Technology (ECTI) Association of Thailand - Conference 2011, ECTI-CON 2011, May 17-19. IEEE Computer Society, Khon Kaen (2011)
2. Sivaramakrishna, M., et al.: Development of pid controller algorithm over FPGA for motor control in failed fuel location module in Indian fast reactors. In: 3rd International Conference on Electronics Computer Technology, ICECT 2011, April 8-10. IEEE Computer Society, Kanyakumari (2011)
3. Li, G., et al.: Study on speed and acceleration characteristics of 100 nm scale motion platform driven by linear servo motors. In: 4th Conference on Application of Diamond and Related Materials in China, CADRM 2010 and the 1st International Symposium on Advances in Brazed Superabrasive Tools, ISABS 2010, August 19- 23. Trans. Tech. Publications Ltd., Xiamen (2010)

4. Castano, L.F., Osorio, G.A.: Design of a FPGA based position PI servo controller for a DC motor with dry friction. In: 2011 7th Southern Conference on Programmable Logic, SPL 2011, April 13-15. IEEE Computer Society, Cordoba (2011)
5. Chen, Y., et al.: DSP-based real-time implementation of a neural network observer and hybrid H adaptive controller for servo-motor drives. In: 27th Chinese Control Conference, CCC, July 16-18. Inst. of Elec. and Elec. Eng. Computer Society, Kunming (2008)
6. Sun, Z., et al.: Design of Control System Based on Soccer Robot. In: International Conference on Multimedia Technology, ICMT (2010)
7. Jae-Hun, J., et al.: Design of a DSP controller and driver for the power-by-wire driving system with BLDC servo motor pump. In: 2011 8th International Conference on Electrical Engineering/Electronics, Computer, Telecommunications and Information Technology, ECTI-CON (2011)
8. Niasar, A.H., Moghbeli, H., Vahedi, A.: Implementation of Four-Switch Brushless DC Motor Drive based on TMS320LF2407 DSP. In: IEEE International Conference on Signal Processing and Communications, ICSPC 2007 (2007)
9. Wolf, D.F., et al.: An FPGA-Based Mobile Robot Controller. In: 3rd Southern Conference on Programmable Logic, SPL 2007 (2007)
10. Charitkhuan, C., Bhuripanyo, J., Choomuang, R.: FPGA Implementation of Closed-loop Control System for Small-sized RoboCup. In: IEEE Conference on Robotics, Automation and Mechatronics (2006)
11. Wei, Z., et al.: FPGA implementation of closed-loop control system for small-scale robot. In: Proceedings of the 12th International Conference on Advanced Robotics, ICAR 2005 (2005)
12. Ce, L., et al.: A Multiprocessor System for a Small Size Soccer Robot Control System. In: 4th IEEE International Symposium on Electronic Design, Test and Applications, DELTA 2008 (2008)

Research for All-Around Battlefield Situation Assessment

Jing-Xue Liu and Yun-Yao Yi

National Defence Information Academy, Wuhan, China
ljx_62@sohu.com

Abstract. Aiming at the demand of enhancing the reliability of battlefield situation assessment and easily producing the integrative analysis report of battlefield situation, this paper investigates the problems on the all-around battlefield situation assessment. First, the recognition of supposed battlefield situations is made by using non-monotone reasoning model on the basis of group analysis. Then, the method that collectively estimates the probability of suppositional battlefield situation by means of evidence reasoning is put forward. Last , how to produce the report of analyzing battlefield situations is discussed.

Keywords: Battlefield situation assessment, Non-monotone reasoning, Judgment information, Evidence theory.

1 Introduction

In general, battlefield situation assessment (BSA) bases on military knowledge and experience, adaptively apperceives battlefield situation with rapid dynamic changes, automatically makes plentiful analyses, inferences and judgments on real time data from multi-sources, rationally explains current battlefield scene according to the thinking fashion and experience of military experts, and offers relative integrated and exact analysis report of current battlefield situation for commanders. As a result, BSA is an important measure that heightens the capacity of apperceiving battlefield situation, it plays an important role in enhancing the efficiency of commanding decision-making. At present, there are many literatures involving in the investigation on BSA from diverse views[1-4], but these literatures only emphasize on judging battlefield situations according to the acquired battlefield intelligence. Because battlefield intelligence maybe includes some cheat information released by adversary, there is definite risk in BSA, the result of which is also not all-around. In fact, It is hard that produces exact analysis report of current battlefield situation only from the acquired battlefield intelligence. Thereupon, aiming at keeping away information cheat and easily producing the analysis report of BSA for commanders, this paper bases on the manpower and information in the commanding information system, makes a tentative research of all-round BSA. First, the recognition of each supposed battlefield situation (SBS) is investigated by using non-monotony reasoning. Then, the assessment of battlefield situation is discussed by using evidence theory and situational logic.

D. Yang (Ed.): Informatics in Control, Automation and Robotics, Volume 2, LNEE 133, pp. 569–575.
springerlink.com

2 The Recognition of Each SBS

In order to realize BSA, the set of SBS should be determined first. Then, each SBS should be recognized. A SBS usually derives from elementary analysis of acquired intelligence in the commanding information system. In order to reflect possible changes of battlefield situations, the commanders and the brainmen can extend the set of SBS according to their intuitive judgments.

2.1 The Reasoning Model of Recognizing Each SBS

Under the condition of existing some cheat information, recognizing each SBS should use non-monotony reasoning model. Non-monotonic reasoning is a tolerant error reasoning, it allows conflicts and emphasizes that results are revised by knowledge's change, offsets disadvantages brought by insufficient knowledge under the condition of incomplete information, and widens application scope.

Default reasoning and modal logic are two main methods of non-monotonic reasoning. A general frame based on rules and consequence relations is given in ref.[5]. This frame fuses default reasoning and modal logic, with the form:

$$a : b \parallel\!\!- A \tag{1}$$

where a and b are finite sets of propositions, A is a conclusion. The meaning of expression (1) is that if all propositions from a are assumed (or believed) to hold and no proposition from b is assumed to hold, then infer A. In expression (1), a is called positive premise, b is called negative premise. In order to recognize a SBS, the evidences of supporting and rejecting it should be considered. Consequently, the reasoning mode with the form of expression (1) is quite suited to recognize each SBS.

2.2 The Method of Recognizing Each SBS

In order to fairly gain the result of recognizing SBS and avoid the influence of the mindset, some intelligence experts with various knowledge backdrops should be organized to recognize SBS. Each intelligence expert can independently give the judgment information according to his military knowledge, experience and the acquired intelligence.

2.2.1 The Integration of Fuzzy Lingual Judgment Information

Let $H^0=\{H_1, H_2, ..., H_k\}$ denote the determined set of SBS. Aiming at $H_q \in H^0$, the intelligence experts determined the evidence set of supporting or rejecting H_q is E_{qs} or E_{qr} respectively. The evidences in E_{qs} or E_{qr} must be known facts, military common sense or intelligence that has been affirmed to be true, and must be unattached each other so as to easily form evidence chain. In order to recognize H_q, the intelligence experts need give the judgment information on the integrality of E_{qs} and E_{qr}, the strength of E_{qs} supporting H_q, and the strength of E_{qr} disproving H_q.

Suppose that L_I ={"extremely integrated", "very integrated" , "integrated" , "comparatively integrated" , "moderate", "comparatively frail", "frail", "very frail", "extremely frail"} is the ordered fuzzy lingual term set used for measuring the integrated degree of evidence chain. Similarly, the ordered fuzzy lingual term set that describes the achievable degrees of each evidence chain supporting or disproving the SBS can be set as L_F ={"extremely strong", "very strong", "strong", "comparatively strong" , "moderate" , "comparatively weak" , "weak" , "very weak" , "extremely weak"}. The judgment information in L_I or L_F can be synthesized by using the weighted median operator[6], which is good at dealing with ordinal information. Given that m intelligence experts take part in recognizing H_q, and w_{qj} ($j=1,2,\ldots,m$) is the weight of the jth intelligence expert recognizing H_q. Let $I^s_{qj} \in L_I$ stands for the judgment value that the jth intelligence expert estimates the integrality of E_{qs} ($q=1, 2,\ldots, k$; $j=1, 2 ,\ldots, m$). Thus, the evaluative information that the intelligence experts estimate the integrality of E_{qs} can be denoted as:

$$D_{qs} = \{ (w_{q1}, I^s_{q1}), (w_{q2}, I^s_{q2}), \ldots, (w_{qm}, I^s_{qm}) \}$$

From big to small, ranking I^s_{q1}, I^s_{q2} , ... , I^s_{qm} obtains b^s_{q1} , b^s_{q2} , ..., b^s_{qm} . At the same time , the corresponding weights are turned into u_{q1} , u_{q2} , ..., u_{qm}. The ranked evaluative information may be denoted as:

$$\overline{D}_{qs} = \{ (u_{q1}, b^s_{q1}), (u_{q2}, b^s_{q2}), \ldots, (u_{qm}, b^s_{qm}) \}$$

If p^* makes $\sum_{j=1}^{p^*-1} u_j < 0.5 \le \sum_{j=1}^{p^*} u_j$, then the weighted median of D_{qs} is $b^s_{qp^*} \in L_I$, marked as Weight_Med(D_{qs}) $= b^s_{qp^*}$. Consequently, the collective evaluative result of the integrality of E_{qs} is $b^s_{qp^*}$. Similarly, the collective evaluative result of the integrality of E_{qr} is $b^r_{qd^*}$, the collective evaluative result of the degree of E_{qs} supporting H_q is $b^{ss}_{ql^*}$, the collective evaluative result of the degree of E_{qr} disproving H_q is $b^{rr}_{qt^*}$.

2.2.2 Determining the Acceptable Set of SBS

After the collective evaluative results aiming at each SBS have been acquired, It is necessary to determine the set of acceptable SBS. Therefore, the collective evaluative results should be quantitated so as to determine the synthetical estimating value, which is used to judge whether the corresponding SBS is acceptable. Consequently, let $\mu : L_I \to [0,1]$, $\mu(x)=1-0.1\times f(x)$; where $f(x)=i$, if x is the ith element in L_I . Similarly, let $h : L_F \to [0,1]$, $h(y)=1-0.1\times f(y)$; where $f(y)=j$, if y is the ith element in L_F. If $E(H_q)$ stands for the synthetical evaluative value of H_q, then $E(H_q)$ can be determined by formula (2):

$$E(H_q) = \mu(b^{s}_{qp^*}) \times h(b^{ss}_{ql^*}) - \mu(b^{r}_{qd^*}) \times h(b^{rr}_{qt^*}) \tag{2}$$

Obviously, the bigger $E(H_q)$ is, the higher the degree of the intelligence experts collectively ratifying H_q is. It is assumed that v is the threshold set by the commander, then the acceptable set of SBS can be determined as follows:

$$H^1 = \{H_i | E(H_i) > v, H_i \in H^0\} \tag{3}$$

Let $H^2=H^0-H^1$, then each SBS in H^2 is unacceptable.

2.2.3 Determining the Tentative Likeliest SBS

Given that u is another threshold determined by the commander, moreover $u > v$. If $H_t \in H^0$ satisfies:

$$E(H_t) = \max\{E(H_i) | E(H_i)| > u\} \tag{4}$$

then H_t is tentatively taken for the likeliest SBS. If there are no SBS to satisfy formula (4), then there are no SBS to be tentatively accepted as the likeliest one according to present evidences, it is necessary that further collects evidences to recognize each BAS again.

3 The Assessment of Battlefield Situation

After determining the likeliest SBS, the assessment of battlefield situation requires to estimate the probability that each acceptable SBS occurs, which plays an important role in producing the report of analyzing battlefield situation.

3.1 Estimating the Probability of Each Acceptable SBS

Generally speaking, the probability that each acceptable SBS occurs should be determined collectively by some battle brainmen. D-S evidence theory is an effective tool of uncertain reasoning, It can easily integrate the judgment information given by the battle brainmen to estimate the probability of each acceptable SBS. Suppose that there are L battle brainmen to take part in estimating the probability of each SBS in H^1. The brainmen usually have definite confidence to judge the true SBS to be contained in some subsets of H^1 by analyzing the pertinent information of each SBS. Without losing generality, let $H^1=\{H_1,H_2,...,H_d\}$. In following discussion, this paper actualizes evidence reasoning by using H^1 as the cognitive frame and the judgment information given by the brainmen as the evidences.

Suppose that the judgment information given by the ith battle brainman is (A_{i1},r_{i1}), (A_{i2},r_{i2}),…, (A_{ig},r_{ig}); where A_{ij} is a proper subset of H^1, which contains the true SBS, r_{ij} represents the belief degree that the ith battle brainman bestows on A_{ij}, $0 \le r_{ij} \le 1$ $(j=1,...,g)$. Thus the *basic probability assignment* (*bpa*) given by the ith battle brainman is as follows:

$$m_i(A_{ij}) = r_{ij} \Big/ \sum_{j=1}^{k_i} r_{ij} \quad (j=1,\dots,g; i=1,\dots,L) \tag{5}$$

$$m_i(B) = 0, B \neq A_{ij} \quad (j=1,\dots,g; i=1,\dots,L) \tag{6}$$

After L *bpas* are determined by formula (5) and formula (6), it is demanded that use Dempster's rule (DR) to integrate them. DR can equally treat each *bpa*, and is defined as follows[7]:

Let m_1 and m_2 are two basic probability assignment; $A_1,\dots,A_p$ ($m_1(A_i)>0$, $i=1,\dots,p$) and $B_1,\dots,B_q$ ($m_1(B_j)>0$, $j=1,\dots,q$) are the focus elements associated with m_1 and m_2 respectively. If $\sum_{A_i \cap B_j = \phi} m_1(A_i)m_2(B_j) < 1$, then the rule of combining m_1 and m_2 determined by formula (7)

$$m(A) = \frac{\sum_{A_i \cap B_j = A} m_1(A_i)m_2(B_j)}{1 - \sum_{A_i \cap B_j = \phi} m_1(A_i)m_2(B_j)} \tag{7}$$

Where m is also a *bpa* , and is denoted as $m=m_1 \oplus m_2$.

Let $K = \sum_{A_i \cap B_j = \phi} m_1(A_i)m_2(B_j)$, The bigger K is , the intenser the conflict between m_1 and m_2 is. That sets a threshold $b>0$, if $K \le b$, then m_1 and m_2 are integrated by DR, or else the battle brainmen must revise their judgment information so as to attain the requirement of integration, which insures that $m=m_1 \oplus m_2$ is reasonable .

In order to estimate the probability of each SBS after the evidences given by the battle brainmen being integrated, the degree of the battle brainmen collectively ratifying each SBS should be first determined. $\forall H_j \in H^1$, using

$$pl(\{H_j\}) = \sum_{H_j \in A} m(A) \tag{8}$$

calculates the plausibility of H_j ,which stands for the degree that m don't deny H_j. That standardizes $pl(\{H_j\})$ will educe the subjective probability P_j , which represents the likelihood that the battle brainmen collectively ratify H_j to occur, namely

$$p_j = pl(\{H_j\}) \Big/ \sum_{t=1}^{d} pl(\{H_t\}) \quad (j=1,\dots,d) \tag{9}$$

Let $p = \max_{1 \le j \le d} p_j$, if $p_k = p$, then H_k is the likeliest SBS ratified by the battle brainmen.

3.2 A Calculative Example

Given that $H^1 = \{H_1, H_2, ,H_3 ,H_4\}$ is the acceptable set of SBS according to the recognizing result, four battle brainmen are required to select the likeliest SBS from H^1. After the related intelligence being analyzed ,the judgment information given by the brainmen is as follows: i $(\{H_2, H_3\}, 0.8)$, $(\{H_1, H_4\}, 0.2)$; ii $(\{H_1, H_2\}, 0.3)$, $(\{H_2, H_3\}, 0.6)$, $(\{H_4\}, 0.1)$; iii $(\{H_2, H_3, H_4\}, 0.55)$, $(\{H_1\}, 0.1)$, $(\{H_3, H_4\}, 0.35)$; iv $(\{H_2, H_3,\}, 0.5)$, $(\{H_1, H_3, H_4\}, 0.5)$. $b=0.50$ is the determined threshold, it is required to determine the subjective probability that each SBS in H^1 occurs .

By using formula (5) and (6),the *bpas* given by the four battle brainmen are as follows : $m_1(\{H_2, H_3\})=0.8$, $m_1(\{H_1, H_4\})=0.2$; $m_2(\{H_1, H_2\})=0.3$, $m_2(\{H_2, H_3\})=0.6$, $m_2(\{H_4\})=0.1$; $m_3(\{H_2, ,H_3 ,H_4\})=0.55$, $m_3(\{H_1\})=0.1$, $m_3(\{H_3, H_4\})=0.35$; $m_4(\{H_2,H_3\})=0.5$, $m_4(\{H_1,H_3,H_4\})=0.5$. $\because K=1-(0.8\times0.3\times0.55\times0.5 + 0.8\times0.6\times0.55\times0.5 + 0.8\times0.6\times0.55\times0.5 + 0.8\times0.6\times0.35\times0.5 + 0.8\times0.6\times0.35\times0.5 + 0.2\times0.3\times0.1\times0.5 + 0.2\times0.1\times0.55\times0.5 + 0.2\times0.1\times0.35\times0.5)=0.49<b$, $\therefore$ from formula (7), the combined result is as follows: $m(\{H_1\})\approx0.0059$, $m(\{H_2\})\approx0.1294$, $m(\{H_3\})\approx0.5882$, $m(\{H_4\})\approx 0.01765$, $m(\{H_2, H_3\})\approx0.2588$. According to formula (8), the plausibility of each SBS can be obtained as follows : $pl(\{H_1\})=0.0059$, $pl(\{H_2\})=0.1294+0.2588=0.3882$, $pl(\{H_3\})=0.5882+0.2588=0.8470$, $pl(\{H_4\})=0.01765$. By virtue of formula (9), the subjective probability of each SBS can be determined as follows : $p_1=0.0047$, $p_2=0.3084$, $p_3=0.6729$, $p_4=0.014$. $\because p_3$ is the biggest $\therefore H_3$ is the likeliest SBS. Viewing the judgment information given by the four battle brainmen, this result is reasonable.

3.3 Producing the Report of Analyzing Battlefield Situations

If the likeliest SBS determined by the battle brainmen accords with the identifying result made by the intelligence experts, and one integrated evidence chain can be found to explain this SBS; then it can be confirmed. This evidence chain regards the evidence set that supports the SBS as backbone, and includes other information that supports the SBS. The battle brainmen combine the imaginary conclusions derived from situational logic and the inferable conclusions derived from the evidence set into one integrated "story ". Only this story can effectively explain the likeliest SBS and illuminate the falsehood of each other SBS, can the likeliest SBS be accepted to be true. If one SBS has been accepted to be true, then the report of analyzing battlefield situation should base all intelligence that supports the SBS. In this report, the strategic purpose, center of gravity, main menace, critical vulnerability and main endeavor direction of both army should be described emphatically so as to reflect integrative status of battlefield situation as detailed as possible.

4 Conclusions

Aiming at the demand of enhancing the reliability of BSA and easily producing the integrative analysis report of battlefield situation, this paper investigates the problem of all-around battlefield situation assessment by means of group analysis. That makes use of the judgment information given by the intelligence experts to recognize each SBS by means of non-monotony reasoning model can effectively keep away information cheat actualized by adversary. That bases the judgment information given by the battle brainmen to infer the subjective probability that each SBS occurs by means of evidence theory can validate the result of recognizing each SBS. The consistency of the recognition result made by the intelligence experts with the consequence inferred by the battle brainmen is indispensable for producing the report of analyzing battle situations. The calculative example shows that the proposed method of estimating subjective probability that each SBS occurs has the characteristics of good flexibility and practicality. It is worthy of being noted that a mature commanding information system can easily acquire and integrate the judgment information needed for BSA. So, the research of this paper has definite maneuverability.

Acknowledgments. This paper is supported by the National Defence Information Academy and the Information Engineering Research Institute and Electrical Engineering (Springer).

References

1. Chen, Z.-G., Barkef, G.H.: Situation Assessment of Naval Battlefield. Command Control & Simulation 28(4), 17–20 (2006)
2. Lei, Y.-J., Wang, B.-S., Wang, Y.: Technique s for Battlefield Situation Assessment Based on intuitive Fuzzy Decision. Acta Electronica Sinica 34(12), 2175–2179 (2006)
3. Zhang, X.-M., Gao, X.-G.: Application of Bayesian Networks to Battle Situation Assessment. Fire Control and Command Control 33(10), 30–33 (2008)
4. Hu, Z.-B., Xia, P.-L.: Research on Situation Assessment of Submarine Based on Max-Similarity. Ship Electronic Engineering. 30(12), 35–37 (2010)
5. Alexander, B.: On the Relation between Default and Modal Non-monotonic Reasoning. Artificial Intelligence 101, 1–34 (1998)
6. Yager, R.R., Rybalov, A.: Understanding the Median as A Fusion Operator. International Journal of General Systems 26, 239–263 (1997)
7. Yang, F.-B., Wang, X.-X.: The Synthetical Method of Evidence Theory Dealing with Conflict Evidence, pp. 140–147. National Defence Industry Press, Beijing (2010)

Efficient Spatial Decision-Making for Maritime Search and Rescue Using 3D GIS

Qinghua Qiao[1,*], Tong Zhang[2], and Ming Yang[3]

[1] Key Laboratory of Geo-Informatics of State Bureau of Surveying and Mapping Chinese Academy of Surveying and Mapping, Beijing, China
[2] State Key Laboratory for Information Engineering in Surveying, Mapping and Remote Sensing, Wuhan University, Wuhan, China 430079
[3] Shenzhen Research & Development Centre, State Key Laboratory for Information Engineering in Surveying, Mapping and Remote Sensing, Shenzhen, 518057 China

Abstract. The rate of major accidents at sea tended to increase. Efficient utilization of dynamic information for massive ships in the maritime search and rescue shows more and more importance. The paper imports GIS to emergency data management and discusses many GIS technologies that can be used for maritime search and rescue. We report our experience on the development of a prototype system based on three-dimension GIS. Experiments show that our methodology is applicable and efficient.

Keywords: Decision making, 3D GIS, rescue, Maritime.

1 Introduction

From 2001 to 2005, China's maritime search and rescue system organized emergency rescue operations 8788 times, rescued 71,168 people and 6,877 ships, and saved about 1,71 billion RMB. The statistics show the rate of major accidents at sea tended to increase. Efficient utilization of dynamic information for massive ships in the search and rescue shows more and more importance.

Now dynamic information of ships is provided through AIS(Automotive Identification System), LRIT(Long-Range Identification and Tracking of Ships) , VTS (Vessel Traffic Service) and etc. This information is most important for maritime search and rescue in harsh environments, especially when we need more people and equipments to work together. AIS and LRIT messages are send by different ships from different countries, VTS message are broadcasted by nearby management centers. The explosive growth of this information, is mass-level and even Galaxy-class. For example, single-day peak amount of data could be up to 10 million, three-month total more than 100 million messages, and other ship-related information is also up to totally 10 million or more. Now this information has less relation, especially for space. How to rapidly extract necessary and relevant information for search and rescue work

* Corresponding author. Key Laboratory of Geo-Informatics of State Bureau of Surveying and Mapping Chinese Academy of Surveying and Mapping, Beijing, China. Email: qiaoqh@casm.ac.cn

D. Yang (Ed.): Informatics in Control, Automation and Robotics, Volume 2, LNEE 133, pp. 577–582.
springerlink.com

from mass of dynamic ship information is the problem we must face in the process of preparing.

When an accident occurred, timely information can reduce casualty and property loss effectively. Industry experts need first-hand information to perform analytical work as soon as possible, such as event location and ships nearby. Decision-makers usually need to accurately acquire information of the disaster in the first time, especially for geographic information, which is the basis for scientific decision-making. It means we need not only to quickly get basic geographic information, but also to efficiently transfer to decision makers (ZHOU Xi-feng, 2008). So it is necessary to analyze the characteristics of ship dynamic information storage and possible visualization methods based on the Internat. GISystems have powerful ability for massive data management, and it is the most useful basic platform for multi-source and temporal data integration, while it can provide a graphical display interface for visualizing chaotic data (Bian Fu-ling, 1996).

This article will discuss efficient applications of GIS data management and visualization technology in search and rescue at sea, and finally implemented a prototype system to verify our methodology.

2 Data Management and Related Technology

2.1 Multi-source Data Integration

China Maritime Safety Administration launched a maritime emergency auxiliary command system pilot project, and carried out standard database design of emergency data. In this specification, ship basic information, basic information on the crew, ship dynamic information, shipping information and other business information contained in passenger and freight was carefully combined. And under each category, several subclasses for various types of information are recorded in detail. This business information is very important, but the search and rescue work needs more information such as the surrounding environments of the incident. Usually decision makers pay attention to weather conditions at that time and future development, and the supplies and equipment closer to the incident. This information comes from different channels, and with less relation to each other. It is not conducive to key information extraction and decision analysis. But there are some direct or indirect associations between multiple types of information. At a specific time, ships and ports have fixed spatial locations, and we can access the weather conditions of the designated area. At the same time we can easily find out all the ships and harbours closed to ship accidents. With further analysis, we can get all available rescue personnel, supplies and equipment information.

In practice, all the spatial related information is converted to a unified coordinate system, such as WGS84. And with special index optimization, the capability of information retrieval will be greatly improved. Multi-source information can be well integrated together with the processes mentioned above.

Sailing ship is a typical dynamic target, and the storage of its tracks involved spatio-temporal data model. Research has been conducted on moving object databases for many years (L. Forlizzi, 2000; R. Guting and M. Schneider, 2005). Now researchers have developed a series of spatio-temporal models, query language and physical storage structures (B. Huang and C. Claramunt, 2002; D. Pfoser, 2002). This paper uses the MOST (Move Object Spatial-temporal) model to manage the track of sailing ships.

2.2 Emergency Applications Based on the Internet

Emergency search and rescue is a complex task, which often involves several departments and with varieties of information sources, so network-based decision-making is required. The main objective of web service is to provide cross-platform interoperability, and can be easily integrated by other applications. Now applications of geographic information based on the web are mature, services of many industries are provided by this way. We choose web services as key technology to implement the server software. With series of data services, user can easily access the relevant information. With the descriptive information of each web service, users can integrate this service to special applications according to their own requirement conveniently.

2.3 Visualization with GIS

Overall, the information for search and rescue are usually chaotic, it is critical to efficiently transmit massive information to decision-makers throughout the entire work. With the technology mentioned above, we can quickly access direct information from massive database, but because this information is large amounts of data, user cannot easily understand it in a timely manner.

However, in this case, GIS can often bring out an intuitive way to help users grasp the overall situation of the event quickly. Firstly, different elements in map are display by different intuitive map symbols. Secondly, maps are classified by importance and then determined visible/invisible by map scales, and all the information are shown to users clearly on the map viewport. With topographic maps, we can find out available harbours with a glance. With maritime map, we can easily access the topography of the incident area, and then decided how to rescue. With these light background maps, rescue ships, port facilities and materials distribution are focus marked out, detail real-time information(such as various monitoring data, photos, video and etc) are shown by tables and graphs. As the amount of information is too large, there is no way to show all the information in the screen clearly, so interactive display is necessary. In the interactive environment, users can first find the concerned vessels or ports, and then get their detail information that can be used to rescue. It provides an efficient way to display massive data in limited screen region, which is useful for emergency responses.

3 Architecture of Prototype System

Analyzing the characteristics of emergency applications, taking into account the requirement of search and rescue data service, this paper proposes a four-tier architecture system, which is presented in Figure 1. The objective of this framework is to realize massive emergency data management and access for multi-user through the Internet.

According to the actual data flow of maritime search and rescue, four tiers of this architecture are the data tier, physical tier, services tier, and presentation tier. The data tier is the foundation of whole architecture. Datasets in the data tier are various types which can be accessed through the internet, such as emergency dataset and auxiliary data include vector, DEM and image etc. With the auxiliary GIS data, there-dimension visualization can be closer to the real world. The physical tier provides fundamental data access environment with ODBC and JDBC, which can support system running on heterogeneous environments to adapt to a variety of applications. Services Tier provides available geographical data services to multi-user based on Physical Tier and Data Tier. In the paper we developed emergency data service, image service, DEM service, vector service. Presentation Tier provides human-computer interaction interface to do with various type of emergency data.

4 Experiments

Based on the prototype system implemented above, experiments were performed on DELL OptiPlex 330 with Intel 1.6GHz CPU and 2GB RAM. Client and server are both developed in C#. Experimental data are global SRTM and imagery from NASA, AIS, VTS, LRIT from Maritime sector. SRTM data are about 12G with 90 meters grid size, image are about 500M.

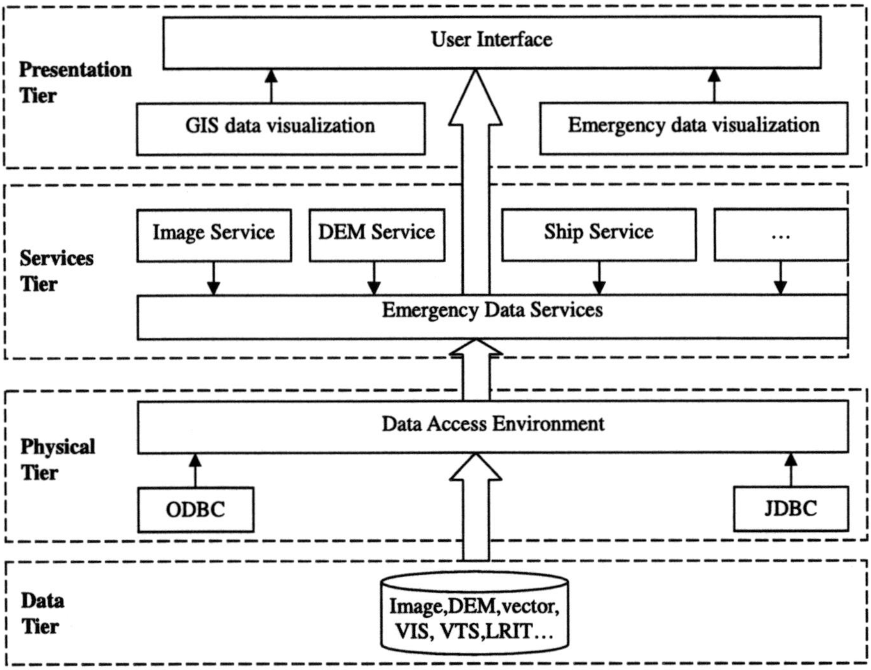

Fig. 1. Architecture of search and rescue system based on GIS

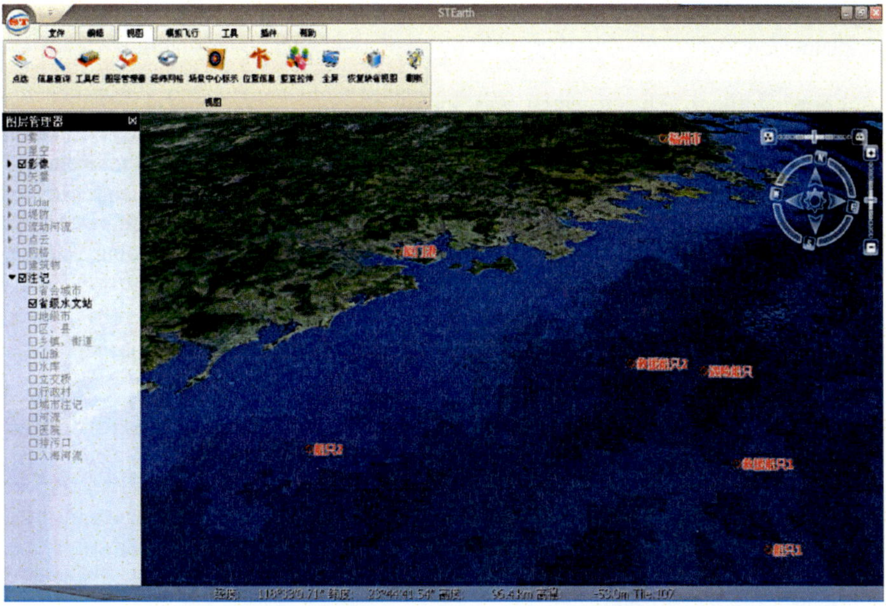

Fig. 2. Maritime search and rescue system

Firstly, geographical data is pre-processed and stored in the server machine. The data access services are provided in web service. Multi-user can access all data through emergency data service, image service, and DEM service on the Internet.

Secondly, client gets various types of data according to user needs and buffers them in local disk. Prototype system automatically loads the necessary data as little as possible when users zoom or pan the scene. If requested data already exists in client disk, system loads them directly, otherwise client sends data requests to server.

Finally, emergency and other geographical data are visualized in map viewport, shown as Figure 2.

Experiments show that with prototype system users can navigate three-dimensional scene in target area smoothly. As users issue different instructions, data are loaded and visualized progressively.

5 Conclusion

Rapid emergency response is an important area that we are promoting. Various emergency data have important practical significance in search and rescue. Development of emergency system based on the Internet is an effective approach to deliver great values for industry applications. In this paper, we analyzed GIS applications in maritime search and rescue, discussed the available technologies, proposed concrete solutions, finally proposed an architecture and implemented a prototype system. Experiments show that the scheme proposed in this paper can lead to good results. Emergency data service based on the Internet can be effectively implemented.

Acknowledgment. This research was jointly supported by National Science and technology support program (2009BAG18B01), National Natural Science Foundation of China (No. 40901190), the Scientific Research Foundation for the Returned Overseas Chinese Scholars, the Open Research Fund of State Key Laboratory of Information Engineering in Surveying, Mapping and Remote Sensing, Wuhan University (No.09Key03, 10I12), and State Key Lab of Subtropical Building Science, South China University Of Technology (2010KB16).

References

1. Huang, B., Claramunt, C.: STOQL: An ODMG-based spatio-temporal object model and query language. In: Richardson, D., Oosterom, P. (eds.) Proc. of the 9th Spatial Data Handling Symposium, Ottawa, Canada, pp. 225–237. Springer (2002)
2. Bian, F.-L.: Principles and Methods of Geographical Information Systems. Beijing Mapping Publishing Company (1996) (in Chinese)
3. Pfoser, D.: Indexing the trajectories of moving objects. IEEE Data Engineering Bulletin 25(2), 3–9 (2002)
4. Forlizzi, L., Güting, R., Nardelli, E., Schneider, M.: A data model and data structures for moving objects databases. In: Proceedings of the 2000 ACM SIGMOD international Conference on Management of Data, Dallas, TX, USA, pp. 319–330 (2000)
5. Guting, R., Schneider, M.: Moving Objects Databases. Morgan Kaufmann, San Mateo (2005)
6. Zhou, X.-F.: Maritime Rescue Plan Based on Cases and Description Logics. Systems Engineering (August 2008)

Development and Application of the Neuro-stimulation Device

Ruituo Huai, Junqing Yang, and Hui Wang

Shandong University of Science and Technology, Qingdao 266510, China
huairuituo@163.com

Abstract. Neuro-stimulation therapy have widely used in the brain stimulation spinal cord stimulation, or muscle incitement to relief the pain, depression, and so on. Neuro-stimulation device mainly contains: implantable neuro-stimulator, non-implantable neuro-stimulator and magnetic stimulator. Implantable neuro-stimulator used in deeply brain stimulation and robo-animal; Non-implantable neuro-stimulator used in neural prostheses; Magnetic stimulator used in the brain and relief depression. This paper review development and application of the neuro-stimulation device and existent problems.

Keywords: implantable neuro-stimulator, non-implantable neuro-stimulator, magnetic Stimulator.

1 Introduction

Neuro-stimulation is the method that stimulate certain nerve through electric current pulse and control or restore the function of the brain, nerve or muscle. With the development of the related technology, neuro-stimulation have widely applied some domains such as deeply brain stimulation, robo-animal, neural prostheses and so on.

Currently, neuro-stimulation device mainly contains: Implantable neuro-stimulator, Non-implantable neuro-stimulator and magnetic stimulator.

2 Implantable Neuro-stimulator in Deeply Brain Stimulation (DBS)

Implantable neuro-stimulator mainly used for DBS, spinal cord stimulation, muscle incitement, etc. DBS is the most invasive therapy option for treatment-resistant depression, pain relief.

2.1 Description of Method

DBS was first developed in the 1950s[1]and was initially used for treatment of chronic pain, it has been used for cure only recently. When using DBS, electrodes connected to implantable pulse generators are implanted in specific brain areas. For individuals suffering from depression, electrodes may be implanted into the following regions of the brain: subthalamic nucleus, internal globus pallidus, ventral internal

D. Yang (Ed.): Informatics in Control, Automation and Robotics, Volume 2, LNEE 133, pp. 583–588.
springerlink.com

capsule, ventral striatum, or the subgenual cingulate. while for individuals suffering from pain, electrodes may be implanted into the following regions of the brain: septal region nuclei, sensory thalamic nuclei, somatosensory thalamic stimulation.

DBS electrodes are implanted using stereotactic guidance. While frameless approaches have been reported, the majority of surgeons use frame-based stereotaxy for electrode implantation. After the frame is applied, patients undergo high resolution stereotactic MRI. Surgery is performed under local anesthesia supplemented as needed by intravenous sedation. A parasagittal frontal burr hole is created through a small incision. Intraoperative physiological stimulation is required to define the exact target for stimulation, which is only approximated by the stereotactic MRI. Thus, the stereotactic coordinates represent only starting points for localization of the physiological targets. Microelectrode recording, microstimulation, and macrostimulation can all be used in the process of localization. Microelectrode recording can help locate targets based on their particular electrophysiologic activity. Once the physiologic targets have been defined with stimulation, permanent electrodes are introduced to those sites, and the leads are externalized through a separate stab wound in the scalp for trial stimulation.

Typically, a postoperative CT scan or MRI is obtained to confirm electrode placement and to assess possible intracerebral hemorrhage. After postoperative recovery and resolution of any peri-electrode edema, test stimulation is initiated to evaluate the patient's pain relief. All possible stimulation combinations are explored during a trial stimulation period that generally lasts 5 to 9 days. If satisfactory pain relief is obtained, the patient is returned to the operating room and the electrodes are connected to an implantable pulse generator. Parameters of stimulation needed to be adjusted individually in order to provide beneficial results.

2.2 The Problem of DBS

The potential complications of DBS have been well elucidated [2-4]. Intracranial hemorrhage is the most significant complication of DBS. It can occur at the time of the insertion or removal of the electrode. The reported incidence of hemorrhage ranges between 1.9% and 4.1%. Using contemporary coaxial DBS electrodes, there has been a significant decrease in the incidence of intracranial hemorrhage.

Permanent neurological deficits have occurred in 14 of the 649 reported patients, with the incidence of such complications ranging from 2.0% to 3.4%. The deficits most commonly resulted from intracranial hemorrhage.

The incidence of infectious complications from DBS ranges between 3.3% and 13.3%. These complications included meningitis, encephalitis, and infection of the scalp. No correlation was found between the time that the electrode was externalized and the occurrence of infection. The majority of cases required wound debridement and removal of all hardware in addition to systemic antibiotics for successful resolution of the infection.

In summary, even historical series of DBS patients have shown acceptable complication rates. Mortality from DBS is rare. Technical advances may already have reduced the morbidity and mortality of DBS.

3 Implantable Neuro-stimulator in Robo-Animal

3.1 Description of Method

Robo-animal have gained some achievements in the recent years[5-7] "Robo-animals" are also called as "cyborgs." The principle of them is "based on the neural science principle of animal of motor behavior, select correlative brain regions and/or nerves in live animals that could generate expected motor behaviors, use adequate artificial electrical signals and adequate methods to micro-stimulate the selected brain regions and/or adequate neural sites, therefore generating certain animal motor behavior under our control and guidance so that the animals move along a predicted path, which is, controlling the neural activity of live animals through electronic signals and turning them into controllable animals.

Based on the concept of population encoding, each chosen nucleus cluster or brain region is planted with several pairs of micro-electrodes, and each relevant neural site is stimulated by the suitable encoded electric signal. Due to numerous constraints, each nucleus cluster is planted with only one pair of stimulating electrodes under most situations, and this pair of electrodes is applied with encoded electric signals in temporal models. The distance between two micro-electrodes is about 1.2-0.5 mm. Although it stimulates and influences several neurons, the stimulating signal is the encoded electric signal of the same tempo. Based on this situation, the encoding model of artificial neuroinformation code is temporal.

Important characteristics of the signal include pulse amplitude, pulse frequency, pulse width, number of pulses in each pulse sequence, and frequency of pulse series. We call this group of data related to such characteristics as the parameter of artificially encoded electric signals or "parameters of stimulating signals." The model of artificially encoded signals is temporal as decided by the group parameters.

3.2 The Problem of the Robo-Animal

Humans have not been able to decode the neuroinformation code, and artificial information codes cannot be recognized or understood by the animal neural system. Thus, it only makes animals understand the rough or even fuzzy connotation expressed by the code. If humans could decode the neuroinformation code, electrical signals will be accurately understood by animals and such phenomena will be reduced greatly. However, we could imagine that it will not be eliminated, since the sensory organs of animals have their own characteristics, and the conflict between two kinds of information will still exist. Only the hesitation caused by information conflict will be reduced.

4 Non-implantable Neuro-stimulator in Neural Prostheses

4.1 Description of Method

Neural prostheses that can restore or augment human functions are now appearing as the result of rapid engineering and biomedical advances in the emerging field of neurotechnology. Devices to restore hearing already are available, while those to reinstate sight and movement are advancing rapidly. While sensory devices have as their goal to inject signals into nervous system (typically the brain), motor prostheses seek to extract signals from the central or peripheral nervous system and deliver them to control devices. It is often forgotten that all voluntary output of the nervous system, whether it is cognitive or a low level reflex, must be produced by a signal from the central nervous system (CNS) to the muscles. This includes a wide range of functions as speech, walking, emotive expression, as well as bowel, bladder and sexual function. Thus loss these actions when the pathways are damaged could be reversed by extracting control signals from the CNS and using them to drive output devices, including physical systems such as computers or robotic devices, other parts of the nervous system, or the muscles themselves.

In a fundamental sense, paralysis restricts the ability of the individual to interact with their environment. In many motor disorders, such as ALS, muscular dystrophy, or spinal cord injury, the individual can be cognitively normal and fully able to generate detailed movement plans using higher motor control structures. Neuromotor prostheses (NMPs) may either recreate the actual lost function or to provide a useful surrogate action to return the ability for the individual to interact with their environment.

4.2 Problem in the Neural Prostheses

Although Neural prostheses have gained some challenging tasks[8-10], the level of the control is considerably less than that of an able-bodied person using a manually controlled object. A number of factors might affect control, including: (1) the small set of randomly selected neurons or EEG recorded by these array, compared to the very large number usually engaged; (2) the influence of spinal cord injury mechanisms or duration since injury; (3) approach to filter building; (4) attention and motivational state of the subject during filter building or control; (5) the user interface. Changes in the recorded population across days may also contribute to both variability and instability of control. Shifting ensembles may result from small motions of the array or through other poorly understoodmechanisms. Despite these variables, it is important to note that useful filters could be created daily from that neural population, and that advances in knowledge and technology are likely to improve recording and decoding.

5 Transcranial Magnetic Stimulation

Transcranial magnetic stimulation (TMS) is another non-implantable neuro-stimulator which involves inducing an electrical current within the brain using pulsating magnetic fields that are generated outside the brain near the scalp. The essential feature is using electricity to generate a rapidly changing magnetic field, which in turn produces electrical impulses in the brain. A typical TMS device produces a fairly powerful magnetic field (about 1.5–3T), but only very briefly (milliseconds). TMS is not simply applying a static or constant magnetic field to the brain. For most TMS applications, it likely is the electricity induced from the pulsating magnet, and not the magnetic field itself, which produces neurobiological effects.

Early TMS devices only emitted a single, brief pulse. Modern devices can generate a rapid succession of pulses, called repetitive TMS (rTMS). These devices are used for behavioral research or clinical treatments and can discharge on and off for several minutes. For example, the typical treatment for depression is a 20–40 min session, 5 days a week for 4–6 weeks. To keep the patient still and the device correctly placed, the patient reclines in a chair and the device is held securely against their head while they are awake and alert without needing anesthesia.

The TMS coil generates a magnetic field impulse that can only reach the outer layers of the cortex [11]. When the TMS device produces a pulse over the motor cortex, descending fibers are activated and volleys of electrical impulses descend through connected fibers into the spinal cord and out to the peripheral nerve where it can ultimately cause a muscle to twitch. The minimum amount of energy needed to produce contraction of the thumb (abductor pollicis brevis) is called the motor threshold (MT) [12-13]. As this is so easy to generate, and varies widely across individuals, the MT is used as a measure of general cortical excitability and most TMS studies (both research and clinical) report the TMS intensity as a function of individual MT (and not as an absolute physical value) [14].

Although this convention has helped in making TMS safer, it is severely insufficient, in that it is referenced only to each machine, and thus is not a universal number. Future work is focusing on more universal, constant, measures of the magnetic field delivered.

6 Conclusion

Neuro-stimualtor have also include vagus nerve stimulation, motor cortex stimualtion, transcranial direct current stimulation. Which application domains is very widely, including stoke , spinal injure, Parkinson's disease, pain, depress .etc. In some disease, several methods are all used, further research need to confirm which is more effective.

Acknowledgement. This study is financially supported by National Natural Science Foundation of China(60875068), International Cooperation Projects of the Ministry of Science and Technology, China(2008DFR00560) and Educational Commission of Shandong Province(J10LG17).

References

1. Heath, R.: Studies in Schizophrenia: A Multidisciplinary Approach to Mind-Brain Relationships. Harvard University Press, Cambridge (1954)
2. Conca, A., Di, P.J., Hinterhuber, H., Kapfhammer, H.P.: Deep brain stimulation: a review on current research. Neuropsychiatr. 25, 1–8 (2011)
3. Mian, M.K., Campos, M., Sheth, S.A., Eskandar, E.N.: Deep brain stimulation for obsessive-compulsive disorder: past, present, and future. Neurosurg. Focus 29, E10 (2010)
4. Zyss, T., Zieba, A., Hese, R.T., Dudek, D., Grabski, B.: Deep brain stimulation -the newest physical method of treatment of depression 44, 301–317 (2010)
5. Talwar, S.K., Xu, S., Hawley, E.S., Weiss, S.A., Moxon, K.A., Chapin, J.K.: Rat navigation guided by remote control. Nature 417, 37–38 (2002)

6. Hirotaka, S., Christopher, W., Berry, B.E., Michel, M.: A cyborg beetle: insect flight control through an implantable tetherless microsystem. In: MEMS (2008)
7. Huai,R.T., Yang, J.Q., Wang, H., Su, X.C.: A new robo-animals navigation method guided by the remote control. In: BMEI (2009)
8. Gopal, S., Stephen, I.R., Byron, M.Y.: A high-performance brain-computer interface 442, 195–198 (2006)
9. Berger, T.W., Hampson, R.E., Song, D., Marmarelis, V.Z., Deadwyler, S.A.: A cortical neural prosthesis for restoring and enhancing memory. J. Neural Eng. 8, 1741–1752 (2011)
10. Velliste, M., Perel, S., Spalding, M.C., Whitford, A.S., Schwartz, A.B.: Cortical control of a prosthetic arm for self-feeding. Nature 453, 1098–1101 (2008)
11. Davey, K.R., Epstein, C.M., George, M.S., Bohning, D.E.: Modeling the effects of electrical conductivity of the head on the induced electrical field in the brain during magnetic stimulation. Clin. Neurophysiol. 114, 2204–2209 (2004)
12. Fox, P.T., Narayana, S., Tandon, N., Fox, S.P., Sandoval, H., Kochunov, P.: Intensity modulation of TMS-induced cortical excitation: primary motor cortex. Hum. Brain Map 27, 478–487 (2006)
13. Sacco, P., Turner, D., Rothwell, J., Thickbroom, G.: Corticomotor response stotriple-pulse transcranial magnetic stimulation:effects ofinter stimulus interval and stimulus intensity. Brain Stimulat. 2, 36–40 (2009)
14. DiLazzaro, V., Pilato, F., Saturno, E., Oliviero, A., Dileone, M., Mazzone, P.: Theta-burst repetitive transcranial magnetic stimulation suppresses Specific excitatory circuits in the human motor cortex. J. Physiol. 565(Pt3), 945–950 (2005)

Hierarchy Modeling and Co-simulation of a Dynamically Coarse-Grained Reconfigurable Architecture*

Ren Chen, Lu Ma, Du Yue, Wen Wen, and Zhi Qi

National ASIC System Engineer Research Center, Southeast University, Nanjing 210096, People's Republic of China
{chenren,malu_cnasic,duyue032,vincentwen}@seu.edu.cn

Abstract. This paper presents a SystemC cycle-accurate simulator, called REmulator, for a dynamically coarse-grained reconfigurable architecture template. By hierarchy modeling, REmulator can either be used at transaction-level, which benefits great for fast simulation of high-level hardware and embedded software, or at register transfer level, which can support more detailed evaluation outcomes on performance and power. In REmulator, the reconfigurable architecture template is described as a parameterized SystemC model so that it can be represented as different architecture schemes, and its architecture parameters would serve as input of the compiler back end, which allows rapid application mapping with architecture modification. A case study on design space exploration of routing network and configuration compression in reconfigurable array demonstrates how exploration is used to adjust and optimize a better architecture template.

Keywords: co-simulation, hierarchy modeling, dynamically coarse-grained reconfigurable architecture, SystemC, SimpleScalar.

1 Introduction

Reconfigurable computing can fill the gap between hardware and software, by combining high performance with flexibility [1][2]. Over the last two decades a few number of coarse grained reconfigurable architectures have been proposed [3].

Nowadays, well-known coarse-grained reconfigurable arrays (CGRAs) diverge in aspects such as interconnect topology, granularity, memory system, and programming model, as well as in how resources for processing, memory and communication are balanced. Many attempts to find systematic approaches to evaluate and explore architectures before a chip is fabricated prove to be valuable while not comprehensive. Zippy [4] presented a multi-context reconfigurable simulator which comprises a reconfigurable unit closely coupled to a CPU core. But reconfigurable array of Zippy is implemented by VHDL, which results in limited reconfigurable array parameters and instances. In WPPA [5], a weakly programmable processor array which supports parameterizable architecture description is described, while it is

* The work was sponsored by the National Scientific Foundation of China (Grant No. 61006029) and Jiangsu Scientific Foundation (Grant No. BK2010165).

D. Yang (Ed.): Informatics in Control, Automation and Robotics, Volume 2, LNEE 133, pp. 589–598.
springerlink.com

customized for specific application domains. PReProS [6] presented a reconfigurable simulator, and derived XPP [7] architecture. However, power information could not be profiled, and no compiler and debugger techniques were reported. In order to enhance simulation performance for parallel processor architectures, an event-driven simulation approach is proposed in [8]. But this approach mainly concerns on simulation speed and omits aspects such as architecture parameterized capability and that components of models can be developed and extended.

Our work presented here mainly focus on two research fields: Organization of CGRAs and methodologies to enable efficient design space exploration of these. REmulator consists of two main SystemC modules. One of that is a SimpleScalar's CPU Core model packaged by SystemC. Another is developed as a parameterizable SystemC architecture model. In order to achieve trade-off between accurate evaluation and fast modeling, hierarchy modeling method is adopted and a compiler back end which takes architecture parameters as input is added into REmulaotr.

In the high-level modeling field, currently the most popular modeling language is the hardware description language SystemC [9], which is based on C++. Experiments presented in [10] show that SystemC can provide a significant speed up for simulations as compared to RTL simulations. And SimpleScalar is an open source processor simulator developed by C language, which is very popular in academia, and widely used in SoC design [11]. In REmulator, SimpleScalar is packaged by SystemC language, since enable the SimpleScalar module driven by the clock. Moreover, by using the transaction level modeling concept in SystemC language, we implemented the behavioral model of the bus and memory with interface method call, which makes the capability of the simulator system expansion has been greatly improved. On this basis, this paper achieves other extensions on the reconfigurable processor simulator, and added debugger and initializer modules. So it provides a friendly interface for users on debugging simulator.

Experimental results show that the simulator framework proposed by this paper can simulate reconfigurable processor system on instruction-level or cycle-level, and it has high value in use in the research of particle size and interconnect of reconfigurable array, memory design and hardware-software partitioning. In addition, the use of SystemC as the interface language, making the external module can be easily added to the simulator framework, and then realize more complex and extensive system simulation and design.

2 REmulator Architecture

2.1 Modeling Methodology

2.1.1 Hierarchy Modeling

For system-level modeling, there are main four abstract layers on communication mechanism, which is shown in Fig. 1.

At message level, sharing resource and timing information are healed. While at transaction layer timing information is retained. Details of clock, register and signal can be implemented at RTL layer. By making full use of this kind of hierarchy layer structure, detailed message of array simulation can be obtained and meanwhile useless module details can be described on a higher level instead.

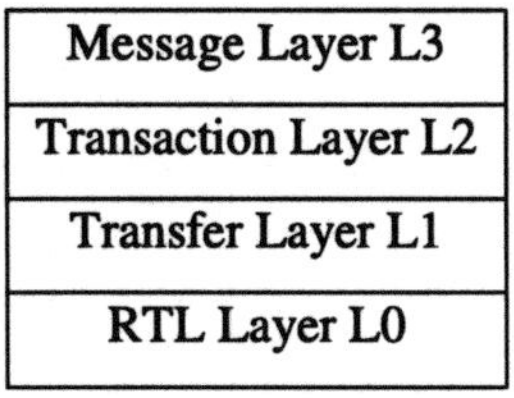

Fig. 1. Communication Abstract Layers on System-level Modeling

In REmulator, cells in reconfigurable array are almost completely described on the RTL-level so that more detailed profile information can be obtained. Moreover, by using the transaction level modeling concept in SystemC language, we implemented the behavioral model of the bus and memory with interface method call, which makes the capability of the simulator system expansion has been greatly improved. Besides, in order to enable the SimpleScalar's CPU core driven by the SystemC clock, we package and extend SimpleScalar by SystemC, and consequently it improves the SimpleScalar simulation performance.

2.1.2 Parametric Modeling

In order to enable fast architecture development and exploration, architecture modules are written as template class and registered at program start-up. These templates are user objects that define how to construct the system in terms of modules and connections from a set of parameters. In REmulator, many important issues such as granularity, routing network, datapath width and size of array etc. can be easily evaluated. The object-oriented concepts as inheritance and polymorphism are also adopted to make simulator easy to maintain and update. Besides, to allow fast application mapping, a set of software tools chain which takes application code, architecture description and control-data flow graph as input are proposed. Details about them are introduced in section 4.

2.2 REmulator Framework

As is shown in Fig. 2, REmulator consists of four main parts, which are system kernel model, system architecture initializer, simulation profiler, and user debugger. System kernel model will be initialized by initializer based on architecture parameters input file so that the whole model in REmulator can be parameterizable. Simulation profiler is used to record customized information such as current cycles, tracing variables, debug signals, etc. User interaction is enabled through user debugger interface, which supports run-time variable access, real-time simulation control and a scripting environment.

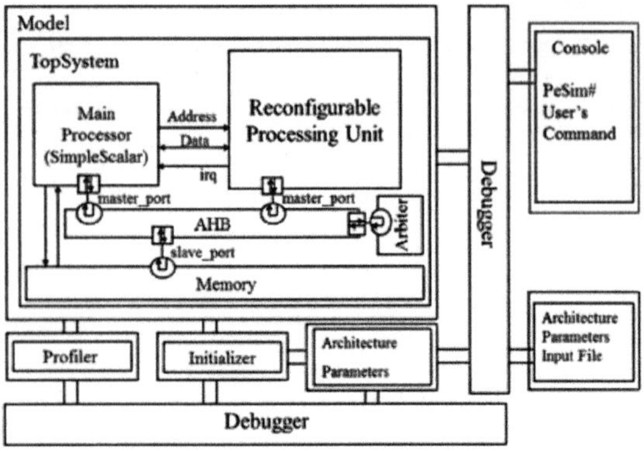

Fig. 2. Framework of REmulator

Besides, for the sake of enabling automatically compiling, we write a compiler back end based on GNU flex&bison, which are very popular lexical and syntax parser tools. And these tools take reconfigurable array architecture parameters as input, then they shall generate configuration words based on architecture parameters input. These tools will be introduced in section 4.

3 Hardware Architecture

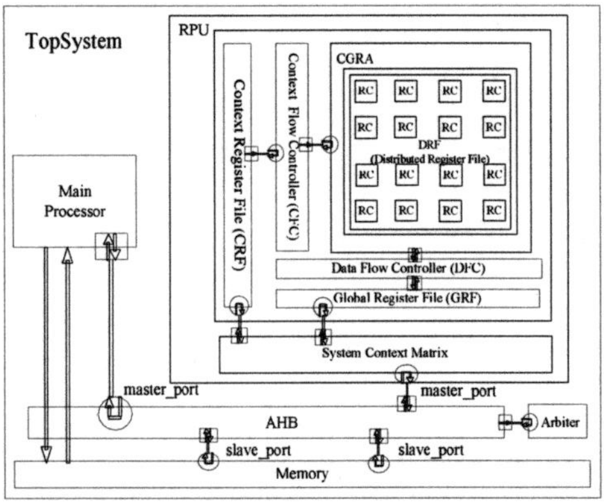

Fig. 3. Dynamically Reconfigurable System Architecture Template

In this paper, REmulator presents a dynamic reconfigurable system architecture template, as is shown in Fig. 3. From the top level, this emulation system consists of one main processor, one reconfigurable processing unit (RPU), corresponding interconnection between them such as AHB bus, and the standard memory. The main processor is a 32-bit RISC processor which is implemented by SimpleScalar in our model. And by using transaction-level modeling method an AHB bus arbitrator protocol is added in the bus model. Memory module in SimpleScalar is transplanted into our top level reconfigurable system and gets packaged by SystemC.

3.1 CPU Architecture Model

In REmulator, we used the SimpleScalar processor simulator as system main controller, which mainly focus on control-intensive computing. SimpleScalar is a 32-bit RISC instruction-set simulation environment for processor architecture research, which includes instruction-set simulator, cross compiler, etc. The CPU core as well as the two-level cache hierarchy is so widely configurable that the CPU model can be parameterized from small low-end to powerful high-end CPUs. The major characteristics of the SimpleScalar's CPU core are

- instruction-set MIPSLike can be extended,
- can be run on most UNIX or NT operation platforms,
- support modification of lots of simulator parameters which describe execution details; such as number of computation units, the sizes of the instruction fetch queue, the branch prediction mode, etc.
- includes multi simulators that are implemented on different abstract layers.

To allow to do exploration on hardware communication between main controller and RPU, we consider two major communication mechanism based on different coupling ways: communication by FIFO or buffer (loosely coupled), set main controller as a heterogeneous part of reconfigurable array (tensely coupled). The former mechanism is adopted by Zippy [4], and the latter one is adopted by ADRES [12].

To attach the RPU to the CPU in REmulator, currently, we have packaged SimpleScalar's CPU model as a System Model and its CPU cycle is now driven by SystemC kernel. Then SimpleScalar's CPU module, reconfigurable processing unit module, interrupt controller module, and debug module are packaged into a top system module which is driven by SystemC clock.

3.2 Reconfigurable Processing Unit (RPU)

As is shown in Fig. 3, the RPU module consists basically of a system connect matrix, a global data register file (GRF), a context register file (CRF), the data flow controller (DFC), the context flow controller (CFC), and CGRA.

The whole RPU running procedure is made up with data flow and context flow. CFC is in charge of controlling flow, which consists of context switching, context preparation, context update and synchronization. DFC mainly takes charge of data flow, which consists of data transfer, data preparation and data organization. Besides, the system connect matrix is responsible for data and context transferring from external memory to register files in RPU.

In CGRA, computing resources are multi heterogeneous reconfigurable cells (RCs), and routing resources include kinds of routing networks such as routing network between RC and DRF or between RC and RC. In addition, memory resources consist of distributed register file (DRF), local register file (LRF), and buffer or FIFO can be configured to be available. Moreover, the CGRA model incorporates a set of parameterizable characteristics. The size of RC is not fixed but can be changed. And the datapath width, the number of operands, and the size of register file in CGRA model can all be parameterized to represent different architecture schemes.

Reconfigurable Cell (RC)

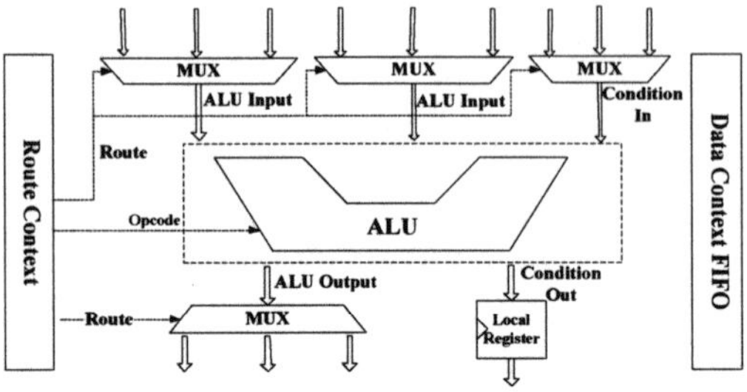

Fig. 4. Reconfigurable Cell

The structure of RC is described in Fig. 4, and each RC contains a common ALU. All of these RCs can be configured as different kinds of heterogeneous cells. Context definition of each RC is divided into two parts: route context and data context. Route context is stored in a context register, and data context is stored in a context FIFO. Data transfer process of each RC is completed by two steps. First of all, data source will be chosen by route context and they can be other RC connected with current RC, DRF or GRF. Secondly, input data will be routing in by mux or by RF read port based on data context.

Routing Network

For compiler-friendly architecture, it is expected that the clock speed is predetermined and the compiler does not need to do timing analysis like in FPGA design. In ADRES [12], most routing is done by direct point-to-point interconnections, consisting of wires and multiplexors, thus timing can be statically analyzed at design time. In MorphoSys [13], routing networks allow for appropriate distribution of computations between RC's so that the un-necessary data movement cycles can be minimized.

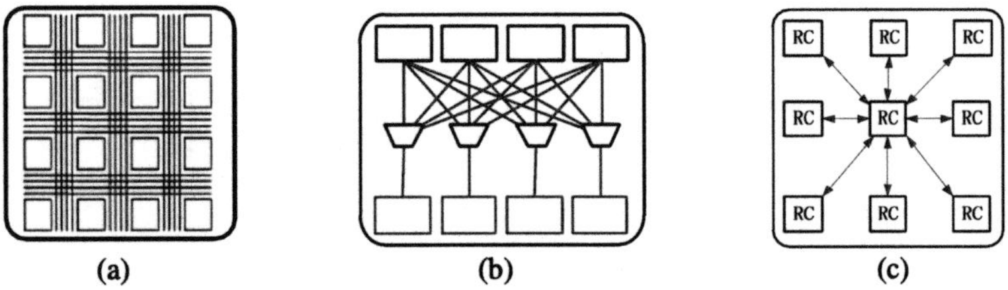

Fig. 5. Different Routing Network among RCs: a) Bus b) Cross-Bar c) 8-N-N

In REmulaotr, routing network between RC and RC can be modified into Cross-Bar or 8-N-N as demanded.

4 Software Tool Chain

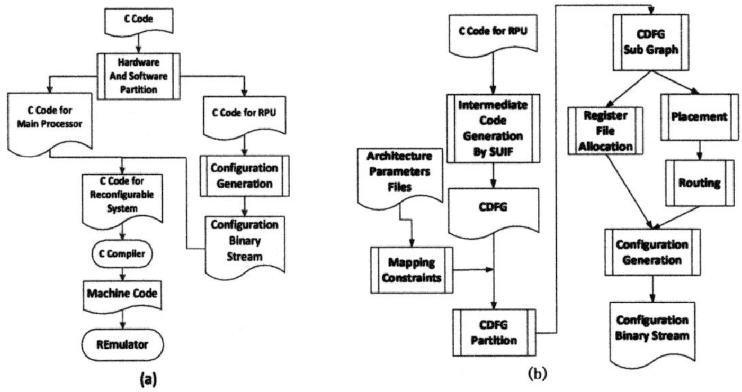

Fig. 6. (a) Compiling flow for Reconfigurable System, (b) The configuration bitstreams generation flow

Fig. 6 shows that the whole compiling process of software tool chain in a reconfigurable system. As is shown in Fig. 6 (a), the general mapping on a hybrid reconfigurable system starts with hardware/software (HW/SW) partitioning stage, e.g. like the ones in [14]. So far, this partitioning step in our simulator is performed manually. The subsequent steps of compiling the application source code and generating the configuration bitstreams are automated.

Fig. 6 (b) gives an overview of the configuration bitstreams generation flow. The configuration bitstreams generation flow is implemented by a set of software tool chain, which consists of the compiler front end and the compiler back end. The compiler front-end takes the application source code as input and it outputs intermediate code such as CDFG or DFG. In this paper, this work is still under way.

The compiler back-end is a placement and routing tool which is described in C and based on lexical and syntax parse tools—GNU flex&bison. It requires two input files: the CDFG input file and the architecture parameters input file. At first, the architecture parameters file will generate mapping constraints, based on which the big DFG of a complete algorithm shall be split into multi sub DFG by the compiler back end. Then, the subsequent steps including register file allocation, placement and routing of sub DFGs are executed to generate configuration bitstreams. Up to now, we've adopted traditional ALSP and ASAP placement and routing algorithms in the compiler back end. And more mapping algorithm optimization work is needed to do to reduce context number and increase reconfigurable array utilization area.

5 Case Study

In this section, some well-known algorithms are chosen as the mapping applications, which are IDCT, FIR, Tone Equalization and Edge Detection. FIR is well-known algorithm kernels widely used in communication and digital processing. IDCT is a fourier-related transformation, often used for signal and image processing applications, especially for lossy data compression. It has two components reflecting similar traits, namely IDCT-row operations and IDCT-column operations. For the experimentation described next, IDCT-row function is taken. Tone Equalization and Edge Detection are often used in graph and image application algorithms.

Table 1. SimpleScalar's CPU model parameter

SimpleScalar's CPU parameter	Setup
Execution units	2 int. ALU; 2int. multiplier; 1FP ALU; 1 FP multiplier
Caches	32-way 16 K L1 I –cache; 32-way 16K L1 D-cache; no L2 cache
Memory interface	32-bit memory bus, 1 memory port
Queue sizes	IFQ:2; RUU:2; LSQ:2
Bandwidth	Decode width:2; issue width: 2; commit width:2
Instruction issuing	Out-of-order
Branch prediction	BPredTaken

In Table 1, by setting SimpleScalar's CPU as a superscalar CPU core, architecture design options, which include routing approach and configuration compression, are evaluated on REmulator. And configuration bits, relative speedup vs. SimpleScalar and easy realizability of software pipelining mapping are taken into account.

1) Routing Approach Exploration

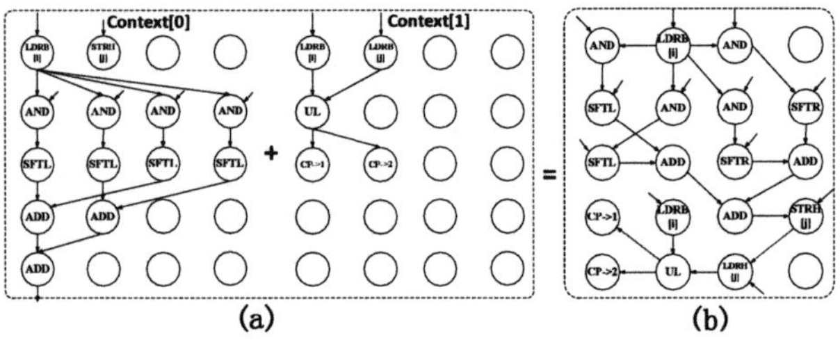

Fig. 7. Application mapping on reconfigurable array: a) mapping instance in a cross-bar array b) mapping instance in 8-N-N array

Fig. 7 is an application mapping result comparison under two kinds of different routing networks between RC and RC: cross-bar routing network, which is shown in Fig. 5 (b), and 8-N-N routing network, which is shown in Fig. 5 (c). This mapping procedure is implemented by compiler back end. Although routing network get changed, this compiler back end need not to be revised and is self-adaptive to these changed. This consideration speeds up application mapping procedure during the process of co-simulation. In order to choose a better routing solution, the size of array is set as 5x4, and condition bit is not adopted as this designation is too specified. Hence, in this situation the multiplexer area of these two approaches are almost equal. In Table 2, we can see that by employing 8-N-N routing networks between RC and RC, configuration bits can be reduced by about 0%-25%, and this reduction will increase with the number of contexts in a algorithm. Meanwhile, with reduction of the number of contexts, simulation cycle of each application is improved.

Table 2. Applications Simulation Performance using Different Routing Network

Application	Configuration bits		Simulation Cycle		Speedup vs. SimpleScalar CPU	
	8-N-N	Cross-Bar	8-N-N	Cross-Bar	8-N-N	Cross-Bar
IDCT(8x8) Row Trans.	6x1240 bits	8x1240 bits	247	342	8.90	6.42
FIR	2x1240 bits	2x1240 bits	372	372	6.88	6.88
Edge Detectection	6x1240 bits	7x1240 bits	7,149	8,253	20.73	17.96
Tone Equalization	5x1240 bits	6x1240 bits	1,188,753	1,276,712	5.04	4.69

2) Configuration Compression Exploration

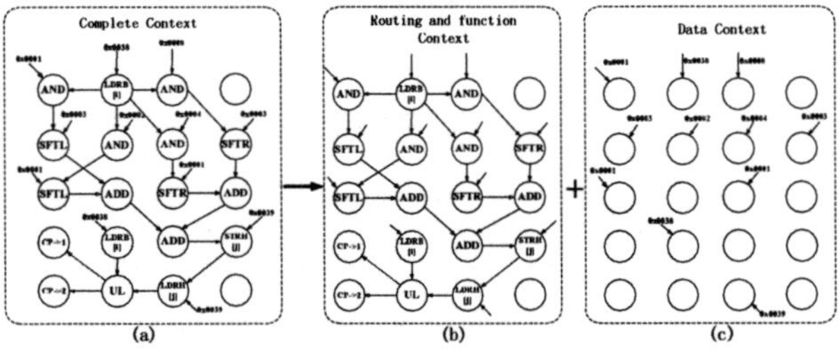

Fig. 8. application mapping on reconfigurable array: a) a complete context b) route and function context after context detachment c) data context after context detachment

Fig. 8 shows that an application mapping result comparison between different context definitions: integrated context definition and separated context definition. Separated context definition means that a complete context is divided into one routing and function context and multi data contexts. Routing and function context is mainly used to describe routing configuration and operation of RCs. And data address is described by data context. As is shown in Table 3, using configuration separation method can reduce configuration bits as many as from four to twenty-six percent, and meanwhile simulation cycle of applications is improved. Moreover, this context separation approach can be used to solve the problem of data organization when the input data of algorithms are mainly 2-dimension data.

In fact, this kind of exploration on context definition requires much flexibility of simulator and compiler. While as compiler back end is tensely coupled with hardware architecture in REmulator, fast application mapping with architecture modification becomes realizable.

Table 3. Reduced Percent of Configuration Bits by using Context Separation Approach

Application	IDCT(8x8) Row Trans.	FIR	Edge Detectection	Tone Equalization
Configuration Bits Reduced(%)	18%	15%	26%	4%

6 Conclusion and Future Works

In this paper, we have presented a cycle-accurate simulator for a reconfigurable system, called REmulator. REmulator supports high-level representations of model and cycle-accurate simulation of the system. Many important issues of reconfigurable system, such like granularity, heterogeneous reconfigurable cells, and different routing networks can be easily evaluated. Compiler back end and debugger are also developed, making up of a part of tool chain. Different algorithms are mapped on REmulator for design space exploration. More trade-off options should be evaluated in this simulator as a further work. And research work on compiler front end need to be done in the near future.

References

1. Compton, K., Hauck, S.: Reconfigurable computing: a survey of systems and software. ACM Comput. Surv., 171–210 (2002)
2. Henkel, J.: A Low Power Hardware/Software Partitioning Approach for Core-Based Embedded Systems. In: DAC, pp. 122–127 (1999)
3. Todman, T., Constantinides, G., Wilton, S., Mencer, O., Luk, W., Cheung, P.: Reconfigurable computing: architectures and design methods. In: IEEE Proceedings – Computers and Digital Techniques, pp. 193–207 (2005)
4. Enzler, R., Plessl, C., Platzner, M.: Co-Simulation of a Hybrid Multi-Context Architecture. Engineering of Reconfigurable Systems and Algorithms, 174–180 (2003)
5. Kissler, D., Kupriyanov, A., Hannig, F.: Teich. J.: A highly parameterizable parallel processor array architecture. In: Proc. of Field Programmable Technology, pp. 105–112 (2006)
6. Brito, A.V.D., Kühnle, M., Melcher, E.U.K., Becker, J.: A General Purpose Partially Reconfigurable Processor Simulator (PReProS). In: IPDPS, pp. 1–7 (2007)
7. Ganesan, M.K.A., Singh, S., May, F., Becker, J.: H.264 Decoder at HD Resolution on a Coarse Grain Dynamically Reconfigurable Architecture. In: FPL, pp. 467–471 (2007)
8. Kupriyanov, A., Kissler, D., Hannig, F., Teich, J.: Efficient event-driven simulation of parallel processor architectures. In: SCOPES, pp. 71–80 (2007)
9. Open SystemC Initiative (OSCI), SystemC 2.0 User's Guide (2002), http://www.systemc.org/
10. Rissa, T., Donlin, A., Luk, W.: Evaluation of SystemC Modelling of Reconfigurable Embedded Systems. In: CoRR (2007)
11. Austin, T.M., Larson, E., Ernst, D.: SimpleScalar: An Infrastructure for Computer System Modeling. IEEE Computer, 59–67 (2002)
12. Mei, B., Vernalde, S., Verkest, D., Man, H.D., Lauwereins, R.: ADRES: An Architecture with Tightly Coupled VLIW Processor and Coarse-Grained Reconfigurable Matrix. In: Y. K. Cheung, P., Constantinides, G.A. (eds.) FPL 2003. LNCS, vol. 2778, pp. 61–70. Springer, Heidelberg (2003)
13. Lee, M., Singh, H., Lu, G., Bagherzadeh, N., Kurdahi, F.J., Filho, E.M.C., Alves, V.C.: Design and Implementation of the MorphoSys Reconfigurable Computing Processor. VLSI Signal Processing, 147–164 (2000)
14. Gajski, D.D., Vahid, F., Narayan, S., Gong, J.: SpecSyn: an environment supporting the specify-explore-refine paradigm for hardware/software system design. IEEE Trans. VLSI Syst., 84–100 (1998)

Battery Energy Storage System Based Power Quality Management of Distribution Network*

Zhigang Lu[1], Guannan Bao[2], Hanchen Xu[2], Xuzhu Dong[1], Zhichang Yuan[2], and Chao Lu[2]

[1] Electric Power Research Institute of China Southern Power Grid, Guangzhou, 510080
[2] Department of Electrical Engineering, Tsinghua University, Beijing, 100084

Abstract. Power quality problems of distribution network include voltage drop, dynamic voltage increases, harmonic pollution et cetera. Battery energy storage system (BESS) can output active and reactive power at the same time and have the four-quadrant operation ability thus can play an important role in the power quality management of distribution network. This paper proposes two control structures of voltage drop governance with BESS, as well as the control structure of harmonic control, and gives the simulation results. Simulation results show that BESS can play a good role in power quality management of distribution network.

Keywords: power quality, battery energy storage system, voltage drop governance, harmonic control.

1 Introduction

Power quality is more and more important with the economic development. Normally the concept of power quality includes voltage quality, current quality, the quality of power supply and the quality of power consumption. In distribution network, widely used power electronic devices deteriorate power quality. Power quality management is in need in distribution network.

The main devices of power quality management in distribution network include uninterrupted power supply (UPS), active power filter (APF), dynamic voltage regulator (DVR), distribution static synchronous compensator (DSTATCOM) etc. UPS can be as a backup power supply of sensitive load in case of power interruption; APF can compensate harmonics; DVR compensates voltage drop through series compensation; DSTATCOM realizes shunt compensation by injecting current to network.

BESS can output active and reactive power at the same time and have the four-quadrant operation ability thus can play an important role in the power quality management of distribution network. Using shunt compensation helps to adopt different strategies for different problems, to achieve different goals.

---

* This work was supported by National Natural Science Foundation of China (51037002) and the Major State Basic Research Development Program of China (2012CB215206).

D. Yang (Ed.): Informatics in Control, Automation and Robotics, Volume 2, LNEE 133, pp. 599–606.
springerlink.com

In this article, Section 2 introduces the structure of BESS for power quality management. Section 3 presents the control strategies for voltage drop governance and harmonic control. Section 4 builds the simulation models in PSCAD and gives the simulation results.

2 The Structure of BESS for Power Quality Management

BESS consists of battery arrays, power conversion system (PCS), and supervisory control system, as shown in figure1. Battery Arrays are energy storage units, controlled by battery management system (BMS). PCS converts the DC power of battery arrays to AC power, and connects to the grid through a step-up transformer. The supervisory control system collects not only battery information from BMS but also grid information such as PMU data and instructions from the dispatch control center. PCS outputs real and reactive power controlled by the supervisory control system to manage power quality.

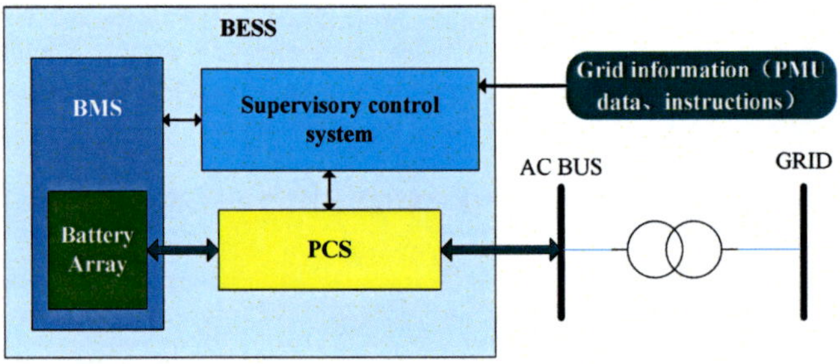

Fig. 1. The structure of BESS

The control structure of power quality management is shown in figure2. External level control chooses the function in need according to the specific problem. Middle level control makes the output to dynamically follow the tracks of the reference value given by external level control, including PI control. Internal level control achieves the pulse modulation of PWM wave.

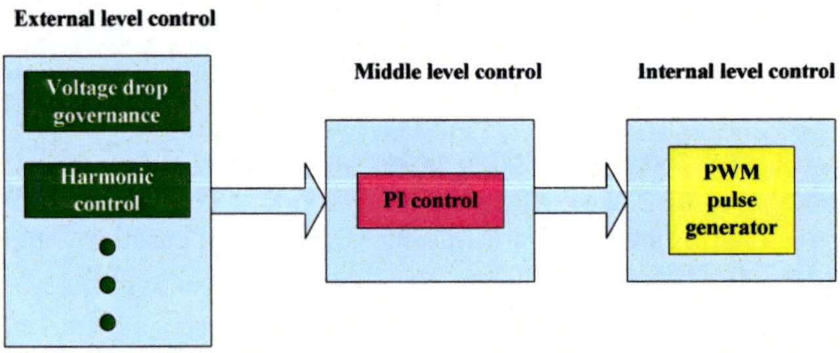

Fig. 2. The control structure for power quality management

3 The Strategies of BESS for Power Quality Management

BESS connects to the grid in parallel, mitigating voltage drop and harmonic by injecting current to the grid. In distribution network, the three-phase current is often unbalanced. The current and voltage values in dq coordinate system fluctuate violently because of the negative sequence component, which reduces the control performance. This paper presents instantaneous control strategies, applied to every phase separately, thus the current can follow the reference values whether the three phases are balanced or not, and the response speed is fast.

3.1 Voltage Drop Governance Strategies

3.1.1 Double-Loop Control Strategy

The first voltage drop governance strategy is double-loop control strategy, in which the outer loop is voltage control loop and the inner loop is instantaneous current tracking control loop, as shown in figure 3.

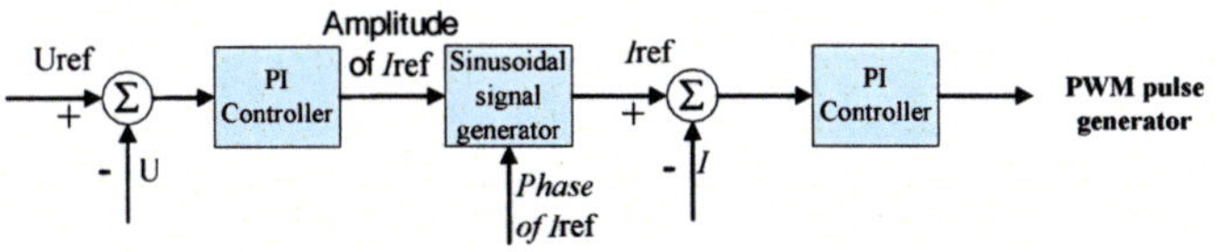

Fig. 3. The structure of double-loop control strategy

The difference of voltage reference value and voltage effective value is used to generate the amplitude of current reference value through the PI regulator. The sinusoidal signal generator generates the current reference value according to the amplitude and the phase of current reference value. The difference of current reference value and instantaneous current is regulated by a PI controller and then sent to the PWM pulse generator. The phase of current reference value is selected from a feasible range in terms of system parameters. If the phase of current reference value is 90°lag behind or ahead of the bus voltage, the BESS injects only reactive power to the grid like a STATCOM.

3.1.2 Instantaneous Current Tracking Control Strategy

The second voltage drop governance strategy is instantaneous current tracking control strategy. Normally some large impact loads lead to voltage drop. If the capacity of BESS is large enough to compensate the large impact loads, the bus voltage will stay in the normal range. The control structure is shown in figure 4.

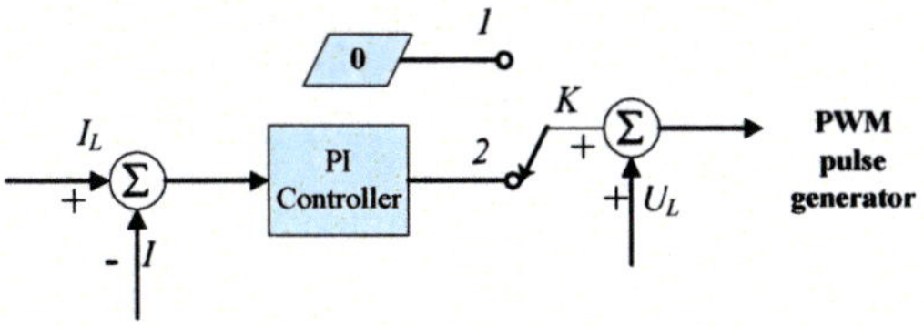

Fig. 4. The structure of instantaneous current tracking control strategy

If the bus voltage is in normal range, switch K is connected to contact 1; if voltage drop occurs, switch K is connected to contact 2, compensating the impact loads. Whether the voltage drop occurs is decided by whether the bus voltage is less than the threshold value. To avoid oscillating, switch K is controlled by a hysteresis loop. The difference of current reference value and instantaneous current is regulated firstly by a PI controller, and secondly by a bus voltage feedforward control before finally sent to the PWM pulse generator.

3.2 Harmonic Control Strategy

If BESS absorbs the harmonic generated by the nonlinear load, the current injected to the grid is sine wave. The FFT to load current is implemented to obtain the fundamental wave of load current. Load current minus its fundamental wave is the harmonic wave. Let the harmonic wave be the reference current and let the BESS current follow the tracks of harmonic wave. The difference of harmonic wave and BESS current is regulated firstly by a PI controller, and secondly by a bus voltage feedforward control before finally sent to the PWM pulse generator, as shown in figure 5.

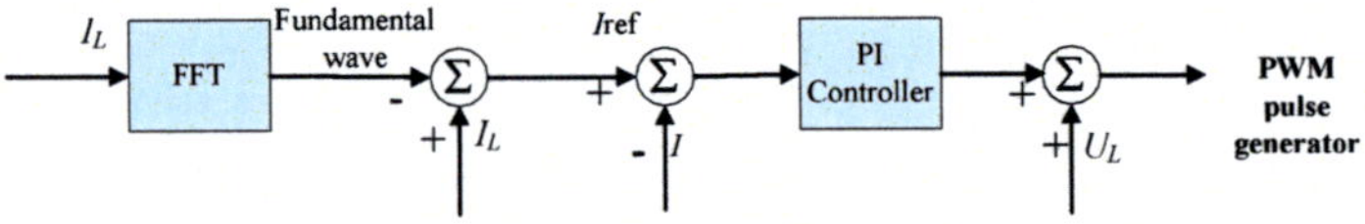

Fig. 5. The structure of harmonic control strategy

4 Digital Simulation

4.1 Testing System

In our testing systems, battery arrays are represented by a DC source, PCS is represented by a three-phase bridge voltage source inverter, forming the single stage DC/AC structure, as shown in figure 6.

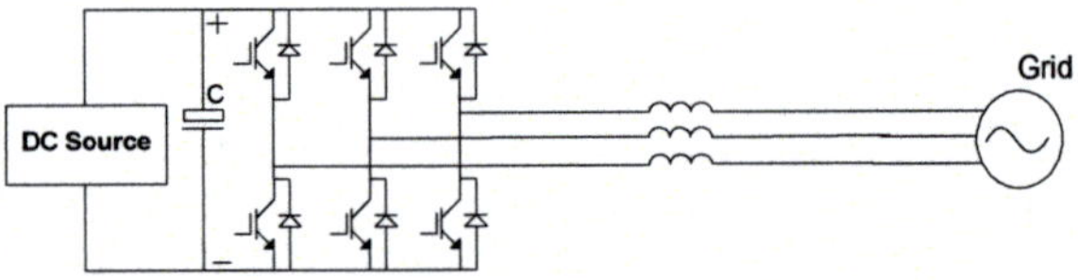

Fig. 6. BESS model in the testing system

4.1.1 Testing System for Voltage Drop Governance

The testing system for voltage drop governance is shown in figure 7.

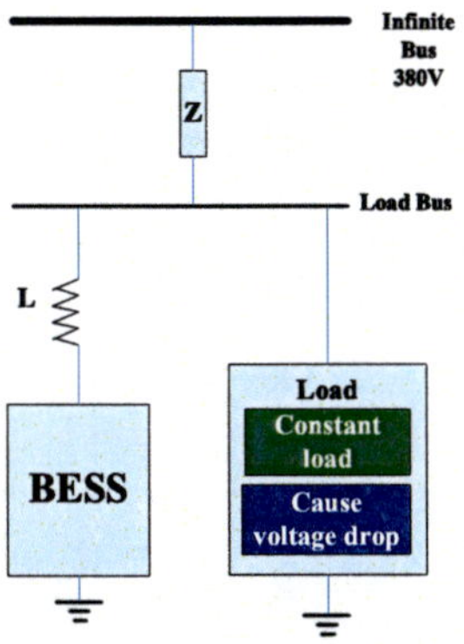

Fig. 7. Testing system for voltage drop governance

In the system, the infinite bus voltage is 380V. The load represents a large factory, in which some facilities like arc furnace may cause voltage drop. Suppose that the constant impedance loads and the loads may cause voltage drop connect to the load bus in parallel. Voltage drop is simulated by three phase impedance grounding.

4.1.2 Testing System for Harmonic Control

If BESS connected to the load bus in parallel can absorb all the harmonic sent by the loads, the current injected to the grid is only fundamental wave, thus the goal of harmonic control is achieved.

If the loads use delta connection scheme or star connection scheme with the neutral point unearthed, the loads would not generate zero sequence harmonic, BESS could use three-phase three-wire system. If the loads use star connection scheme with the neutral point earthed and could generate zero sequence harmonic, BESS must use three-phase four-wire system and its ground point must be the same as the ground point of the loads. The wiring diagram of BESS and loads is shown in figure 8.

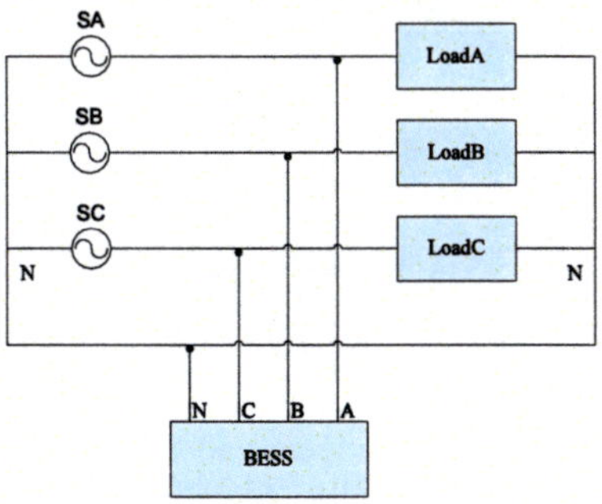

Fig. 8. The wiring diagram of BESS and loads

4.2 Simulation Results

4.2.1 Simulation Results of Voltage Drop Governance

(1) Double-loop control strategy

If the capacity of BESS is large enough and the phase of current reference value is reasonable, BESS can compensate the voltage drop thoroughly, as shown in figure 9.

Fig. 9. The voltage drop is compensated thoroughly

If the capacity of BESS is limited or the phase of current reference value is irrational, the voltage drop may not be compensated thoroughly. Figure 10 shows the voltage curve before and after BESS controlling when the capacity of BESS is not large enough to compensate voltage drop.

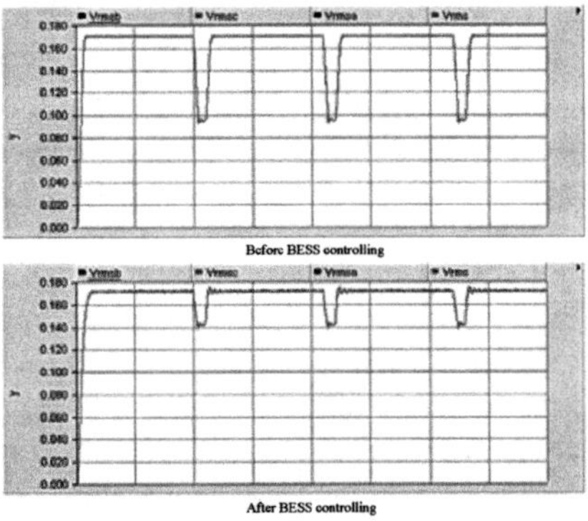

Fig. 10. The voltage drop is not compensated thoroughly

Although the voltage drop is not compensated thoroughly, the voltage drop and the current rise are suppressed. The less the degree of voltage drop is, the better the control effect is.

(2) Instantaneous current tracking control strategy

The effectiveness of instantaneous current tracking control strategy is shown in figure 11.

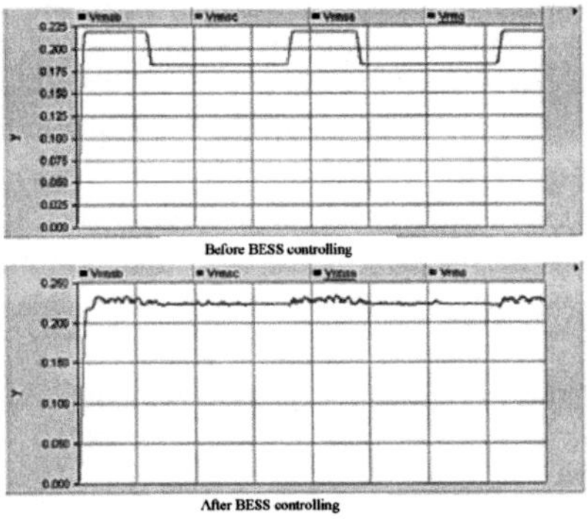

Fig. 11. The effectiveness of instantaneous current tracking control strategy

4.2.2 Simulation Results of Harmonic Control

Figure 12 shows the load curve before BESS controlling which deviates sine wave seriously, the current injected to the grid after BESS controlling, and the current generated by BESS. If the capacity of BESS is large enough, it can absorb all the harmonic and the current injected to the grid is sine wave.

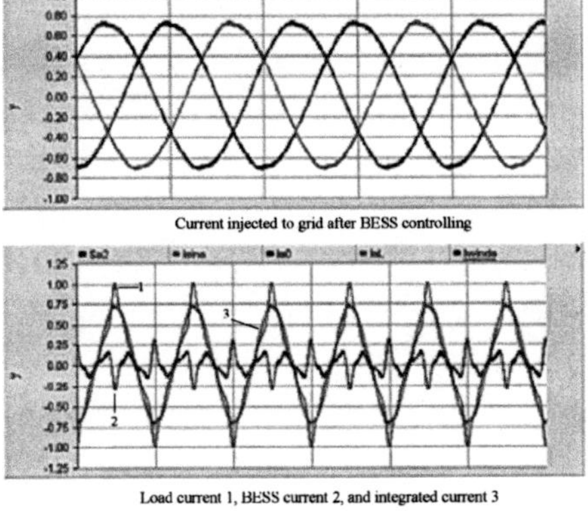

Fig. 12. The current curve before and after harmonic controlling

5 Conclusion

BESS connects to the grid in parallel, managing power quality by injecting current to the grid. Because BESS has four-quadrant operation ability, the current it generates can follow the tracks of current with any phase angle if its capacity is large enough.

BESS can manage voltage drop effectively if its capacity is large enough. Both double-loop control strategy and instantaneous current tracking control strategy are effective to manage voltage drop. In double-loop control strategy, the phase of current reference value should be reasonable. In instantaneous current tracking control strategy, BESS is not used when the voltage is in normal range and a hysteresis loop is used to control switch K. The effectiveness of voltage management is related to the system parameters and the degree of voltage drop.

BESS can control harmonic effectively by absorbing every frequency harmonic wave if its capacity is large enough, thus the current injected to the grid is sine wave.

References

1. Xie, X., Jiang, Q.: Flexible AC Transmission Systems: Principles and Applications. Tsinghua University Press (2006)
2. Lin, H.: Main Problems of Modern Power Quality. Power System Technology 25(10), 5–12 (2001)
3. Kong, L., Liao, L., Zhang, H., Zhao, J.: Application of Battery Energy Storage System in Power System. Electric Switchgear (5), 61–63 (2008)
4. Jin, Y., Song, Q., Chen, J., Zhang, Y., He, W., Sun, K., Liu, W.: Power Conversion System of Large Scaled Battery Energy Storage. Electric Power 43(2), 16–20 (2010)
5. Zhang, B., Zeng, J., Mao, C., Jin, Y., Wang, Y.: Improvement of Power Quality and Stability of Wind Farms Connected to Power Grid by Battery Energy Storage System. Power System Technology 30(15), 54–58 (2006)
6. Yan, J., Zhao, L.: Application of Energy Storage Technology to Distributed Generation. Anhui Electric Power 23(3), 55–69 (2006)
7. Cheng, H., Xu, Z.: Energy Storage for Use with Distribution Power Generation. High Voltage Apparatus 39(3), 53–56 (2003)

Efficient Support Vector Regression with Weighted Constraints

Ling Wang[1,2] and Dong Mei Fu[1,2]

[1] School of Automation & Electrical Engineering, University of Science and Technology Beijing, Beijing 100083, China
[2] Key Laboratory of Advanced Control of Iron and Steel Process (Ministry of Education), Beijing 100083, China
linda_gh@sina.com

Abstract. This paper proposes an efficient training approach for support vector regression, called weighted constraints SVR (WCSVR), which will improve the ability to learn from the noisy samples and overcome the overfitting problem. In the WCSVR learning process, the object function is modified by introducing a weight for each new training sample and the corresponding regularization term. The weighted constraints value to each input sample is confirmed according to its kernel Mahalanobis distance to the center of the dataset. These parameters help us to apply the importance of samples for eliminating the effect of the outliers in the dataset. The overall experiments show prominence of out proposed method in comparison with the standard SVR.

Keywords: support vector regression, weighted constraint, kernel Mahalanobis distance.

1 Introduction

Recently, support vector machine as a novel type of neural networks has received increasing attention in areas ranging from its original application of pattern recognition [1-3] to the extended application of regression estimation [4-5], due to its remarkable generalization performance. For regression problems, Support Vector Regression (SVR) is applied. In SVR, the solution to the problem is represented by sparse data points called support vectors. That is, they are the data points lying on or outside the ε -bound of the decision function. In stand SVR training, the regularization constant C has equal weight to all the ε -insensitive errors between the predicted and actual values. All training samples are treated equally and this may lead to poor performance on indisputable the existence of noisy samples in the training set, which would obtain unusable support vectors. In [6], a fuzzy membership was applied to each input point of SVM and reformulate SVM into fuzzy SVM (FSVM) such that different input points can make different contributions to the learning of decision surface. In[7], C -ascending support vector machines was proposed to model financial time series by taking into account the non-stationarity of financial time

D. Yang (Ed.): Informatics in Control, Automation and Robotics, Volume 2, LNEE 133, pp. 607–615.
springerlink.com

series, which is desirable to place more weights on the recent training data points than the distant training data points. In this paper, we modify the regularized risk function in the SVR method to tolerate outliers in the training set. We define a weighted constraint for the regularized risk function to reach this goal. The weighted constraint introduces a new concept for each sample which is related to position of them. Now we can assign proper value to the concept for eliminating the effect of outliers to obtain useful support vectors.

This paper is organized as follows. In the section 2, our proposed training approach for SVR is explained and the method of assigning the weight factors is discussed. Section 3 presents experimental results and analysis of the proposed approach, and compares it with other approaches. Section 4 gives the conclusion and directions for further work.

2 Weighted Constraints SVR(WCSVR)

We propose a WCSVR approach that uses an adaptive regularization factor C to tolerate the existence of outliers in the training set. First, we apply kernel Mahalanobis distance computation on the original data set to obtain an elliptical area; Second, the regularization factor C of each sample is given with different weight according to the position in the elliptical area. The weight of each sample is defined with γ_i. The value of γ_i can be defined according to the kernel Mahalanobis distance of each sample to the center of training set. So we can assume smaller weights for the outliers to eliminate the effect of them in the training set.

2.1 WCSVR

Consider the SVR training problem that involves a training set D of M samples, $D=\{x_i, y_i\}, x_i \in R^n, y_i \in R, i=1,2,\cdots,M$. SVR makes use of a type of loss function called ε -loss functions defines a tube with radius ε around the hypothetical regression function. If the data is positioned in this tube ε, the loss function can be regard as 0. By introducing the positive slack variables ξ_i and ξ_i^*, the SVR can be formulated to the following constrained optimization problem:

$$\min_{w,b,\xi_i,\xi_i^*} J = \frac{1}{2}\|w\|^2 + c\sum_{1}^{M}(\xi_i + \xi_i^*)$$

$$subject\ \ to \begin{cases} y_i - w^T\phi(x_i) - b \le \varepsilon + \xi_i \\ w^T\phi(x_i) + b - y_i \le \varepsilon + \xi_i^* \\ \xi_i, \xi_i^* \ge 0, i=1,\cdots,M, \varepsilon > 0 \end{cases} \quad (1)$$

Where $\phi(x)$ is the feature mapping in the high-dimensional feature space. w and b are the slope and offset of the regression line. $C > 0$ called regularization parameter, is a positive constant. The positive slack variables ξ_i and ξ_i^* are to measure the deviation of training samples outside tube ε zone.

In standard SVR training, samples are given equal weighting. However, we consider the case when the Training samples tolerate the existence of outliers and have unequal weights. They are tuned by system without any information about importance of samples. So, we focus on deriving a training algorithm for SVR that take into account the sample weights.

To derive the training algorithm, the weighted constraints are incorporated into the regularization function. SVR training can be formulated as

$$\min_{w,b,\xi_i,\xi_i^*} J = \frac{1}{2}\|w\|^2 + c\sum_1^M \gamma_i(\xi_i + \xi_i^*)$$

$$\text{subject to} \begin{cases} y_i - w\phi(x_i) - b \le \varepsilon + \xi_i \\ w\phi(x_i) - b - y_i \le \varepsilon + \xi_i^* \\ i = 1, 2, \cdots, M, \quad \varepsilon \ge 0 \end{cases} \tag{2}$$

Here, the weight of each sample is defined by γ_i which will be discussed in Section 2.2.

In order to solve this problem the constraints should be incorporated into the error function by introducing Lagrange multipliers $\alpha_i > 0, \alpha_i^* > 0$ and constructing the Lagrangian function.

$$\begin{aligned} \min_{\alpha,\alpha^*} w(\alpha,\alpha^*) = \min_{\alpha,\alpha^*} \frac{1}{2}\sum_{i=1}^M\sum_{j=1}^M (\alpha_i^* - \alpha_i)(\alpha_j^* - \alpha_j)K(x_i, x_j) \\ - \sum_{i=1}^M (\alpha_i^* - \alpha_i)y_i + \varepsilon_i \sum_{i=1}^M (\alpha_i^* - \alpha_i) \end{aligned} \tag{3}$$

$$\text{subject to} \begin{cases} \sum_{i=1}^M (\alpha_i^* - \alpha_i) = 0 \\ \alpha, \alpha^* \in [0, \gamma_i c] \end{cases}$$

The Karush Kuhn-Tucker's (KKT) conditions of weighted SVR become

$$\alpha_i(\varepsilon_i + \xi_i - y_i + w \cdot x_i + b) = 0 \tag{4}$$

$$\alpha_i^*(\varepsilon_i^* + \xi_i^* - y_i + w \cdot x_i + b) = 0 \tag{5}$$

$$(\gamma_i C - \alpha_i)\xi_i = 0 \tag{6}$$

$$(\gamma_i C - \alpha_i^*)\xi_i^* = 0 \tag{7}$$

Based on Karush Kuhn-Tucker's (KKT) conditions, the function is given by:

$$f(x)=\sum_{i=1}^{M}\sum_{j=1}^{M}(\alpha_i-\alpha_i^*)K(x_i,x_j)+b. \tag{8}$$

Where $K(x_i,x_j)=\phi(x_i)\phi(x_j)$ denotes the kernel function, which satisfies the Mercer's conditions. This paper shows examples of SVR using radial basis function (RBF) kernel, which is defined as $\exp\left(-\|x_i-x_j\|^2/2\sigma^2\right)$, the parameter σ represents the bandwidth of RBF kernel.

2.2 Defining the Sample Weight

Mahalanobis distance represents the covariance distance of data samples, which can be used to calculate degree of similarity between two samples. Unlike the Euclidean distance, Mahalanobis distance eliminates the different effects on similarity measure which come from the unit of measurement, different attributes and different distribution of samples. In [8], the Mahalanobis distance can be applied to elliptical clustering. Thus, we can make use of the Mahalanobis distance to define an elliptical area to identify the samples which could be important sample or the support vector samples. This technique can be applied in our approach.

The square of Mahalanobs distance between the samples and the center of the ellipse (mean vector) can be defined as follows:

$$D^2(x_i,\mu)=(x_i-\mu)^T\Sigma^{-1}(x_i-\mu) \tag{9}$$

Where each sample x_i is in the n -dimensional space, $\Sigma^{-1}=\frac{1}{M}\sum_{i=1}^{M}(x_i-\mu)(x_i-\mu)^T$ denotes a $M\times n$ covariance matrix, while $\mu=X^T\vec{1}$ is the mean vector, the vector $\vec{1}$ represent an n -dimensional vector and all elements equal to one, X is a $M\times n$ matrix. Moreover, if the covariance matrix Σ^{-1} is singular which can't be inverted, the pseudo-inverse of matrix should be calculated. Based on the theorem[9], the pseudo-inverse of matrix is represented as: $\Sigma^+=pX^T\beta^T\Omega^{-2}\beta X$. Then, the square of Mahalanobs distance can be formulated as

$$D^2(x_i,\mu)=(x_iX-\mu X)^T p\beta^T\Omega^{-2}\beta(x_iX-\mu X). \tag{10}$$

For nonlinear problems, methods of defining non-linear shape kernels are used, called 'kernel trick'. The inner product of the mapping value of the original data in feature space is replaced by a kernel function. With the above notation, the Mahalanobis distance in the kernel-induced feature space can be formulated purely in terms of kernel evaluations as we derive in the following.

$$D^2(x_i,\mu)=(K(x_i,X)-K(\mu,X))^T p\beta^T\Omega^2\beta(K(x_i,X)-K(\mu,X)). \tag{11}$$

According to the kernel Mahalanobis distance, we can define the elliptical area. To eliminate the effect of outliers we propose to introduce a weight for the regularization parameter of each sample which are related to position of the elliptical area with the kernel Mahalanobis distance and assign proper weighted values to the weight for each training sample.

Here, the weight γ_i is defined as:

$$\gamma_i=\begin{cases}(1-\dfrac{D(x_i,\mu)-D_{\min}}{D_{\max}-D_{\min}})^2+\delta, & \varepsilon<D(x_i,\mu)\le D_{\max} \\ 1, & D(x_i)<\varepsilon\end{cases} \quad i=1,\cdots,M \tag{12}$$

Where

$$D_{\max}=\max_{j\neq i,j=1,\cdots,M} D(x_j,\mu)\,,\, D_{\min}=\min_{j\neq i,j=1,\cdots,M} D(x_j,\mu)\,,$$

they are respectively the maximal and minimal kernel Mahalanobis distance square of each sample to the center of the elliptical area. δ represents the small enough positive real number. In this paper, δ =0.003. With the increase of $D(x_i,\mu)$, the

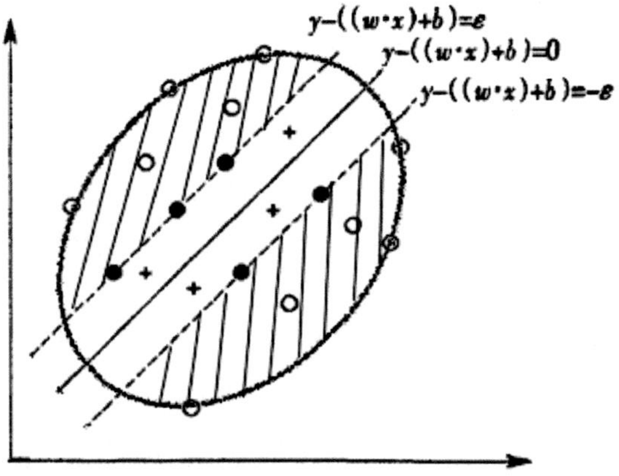

Fig. 1. The support vectors distribution based on kernel Mahalanobis distance

weight γ_i decrease quickly. When $D(x_i, \mu)$ closees to $D_{\max}$, the weight γ_i would approximately equal to δ, which may reduce the effect of these samples on the regression function. From the view of geometry in Fig.1, the support vectors are generally around the boundary of the elliptical area. So we assume the outliers are out of the elliptical area, which are given with smaller weights to eliminate the effect of them in the training set.

If $\gamma_i = 1$ or all the samples have the same weights, the result of the proposed method is the same as the standard SVR method.

3 Experiment and Analysis

We evaluate the performance of the proposed SVR approach on two problems: (1) the first function is single variable function defined as $f^{(1)}(x) = x^{2/3}$ with $x \in [-2,2]$. (2) the second function is $\sin c(x)$ function defined as $f^{(2)}(x) = \begin{cases} \frac{\sin(x)}{x} & if\ x \neq 0 \\ 1 & if\ x = 0 \end{cases}$ with $x \in [-10,10]$.

3.1 The Single Variable Function Approximation

In this case, the training set has 50 points that are generated by randomly selecting, where $x \in [-2,2]$. The outliers are added artificially by moving some points away from designated locations. The testing set has 200 points which is randomly generated by the same function in the same range. Note that the test data do not contain any outliers. The root mean square error (RMSE) of the testing data is used to measure the performance of the proposed SVR method. The RMSE is defined as

$$RMSE = \sqrt{\sum_{i=1}^{l} (y_i - \hat{y}_i)/l} \ . \tag{13}$$

Where y_i is the desired output, $\hat{y}_i$ is the model output and l is the number of the used testing data.

In this investigation, the Gaussian function is used as the kernel function of the SVR. The optimal values of σ, C and ε in the standard SVR are respectively chosen as 0.5, 5 and 0.1 as these values produce the smallest RMSE on the test set in the standard SVR. The same values of the parameters are used in our proposed method for compare. The RMSEs obtained by the model constructed from SVR and from the WCSVR are summarized and listed in Table 1, which indicates that our proposed method could produce a smaller RMSE than the standard SVR and improve the generalization performance of SVR. The learned results from the standard SVR and the WCSVR are respectively illustrated in Fig.2(a) and (b). From Fig.2(a), the approximated results appear oscillation around outliers. Such phenomena can be interpreted as the overfitting phenomenon. Nevertheless, form Fig.2(b), it can be found that the proposed method can detect outliers and reduce the overfitting phenomenon.

Table 1. Comparison with standard SVR and WCSVR with testing samples

Function	SVR		WCSVR	
	RMSE	Number of SVs	RMSE	Number of SVs
$f^{(1)}(x)$	0.07346	50	0.04538	23
$f^{(2}(x)$	0.00763	40	0.00366	17

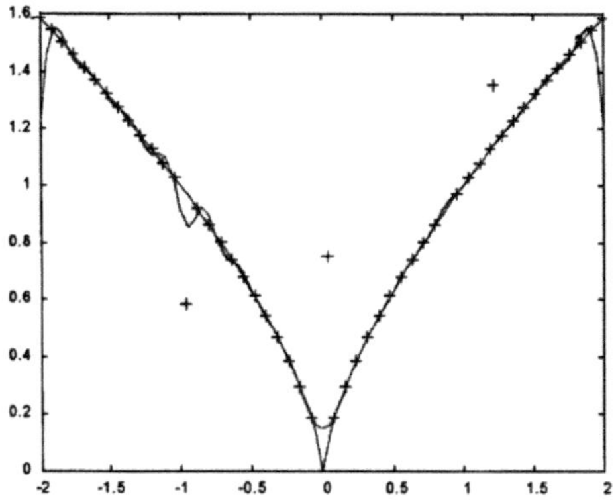

(a)

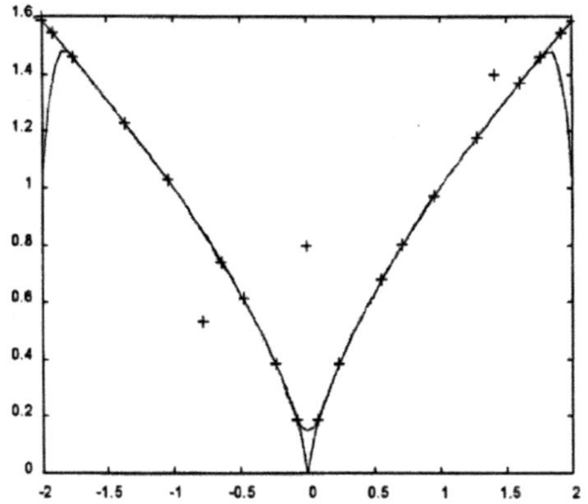

(b)

Fig. 2. (a) the result of SVR for function $f^{(1)}(x)$ with training samples. (b) the result of WCSVR for function $f^{(1)}(x)$ with training samples

3.2 Sinc Function Approximation

In this case, the training set has 50 points that are generated from the function. Among those data, three artificial outliers whose deviation values are 0.3, 0.4 and 0.5, respectively, are created. The testing set has 200 points which is randomly generated by the same function in the same range. Note that the test data do not contain any outliers. Here we also use the Gaussian kernel function. The optimal values of $\sigma = 1.77, C = 50, \varepsilon = 0.02$ are determined from the cross-validation in the standard SVR. After cross-validation the best model was evaluated on the independent test set.

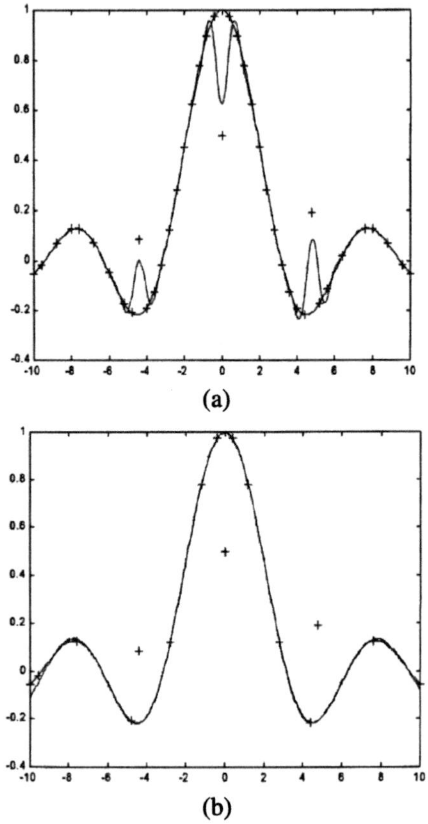

Fig. 3. (a) the result of SVR for function $f^{(2}(x)$ with training samples. (b) the result of WCSVR for function $f^{(2)}(x)$ with training samples

The same values of the parameters are used in our proposed method for compare. The performance comparison between the standard SVR and WCSVR are also made in Table 1. The learned results are shown in Fig. 3(a) and (b) for illustration. These results indicate that the WCSVR maintain the function approximation accuracy and uses less support vectors as compared to the standard SVR. In addition, according to

Table 1 the testing results of WCSVR exhibit better generalization ability and use much smaller number of support vectors. It demonstrates the proposed method have better robustness compared to conventional SVR.

4 Conclusion

In this paper, we proposed an efficient support vector regression with weighted constraints so as to tolerate the existence of outliers in the training set and identify the useful support vectors. For this purpose we modified the regularized factor C with weighted constraints to specify the importance of each sample according to the position in the elliptical area. The weight of each sample is defined by γ_i. These parameters help us to apply the importance of samples for eliminating the effect of the outliers in the dataset. We plan to extend the weighted constraints SVR to the complex industry process and explore new methods of defining the sample weight.

Acknowledgments. This work has been supported by the Fundamental Research Funds for the Central Universities(No. FRF-AS-09-006B).Beijing Key Discipline Development Program (No. XK100080537).

References

1. Susukh, J., Premrudeepreechacharn, S., Kasirawat, T.: Power quality problem classification using support vector machine. In: 2009 6th International Conference on Electrical Engineering/Electronics, Computer, Telecommunications and Information Technology, pp. 58–61 (2009)
2. Kaewarsa, S.: Detection of Power Quality Events Using DOST-Based Support Vector Machines. In: International Symposium on Computer Science and its Applications, pp. 68–71 (2008)
3. Sonkamble, B.A., Doye, D.D.: An overview of speech recognition system based on the support vector machines. In: International Conference on Computer and Communication Engineering, pp. 768–771 (2008)
4. Man, G.N., Won, S.P., Dong, H.L.: Detection and Diagnostics of Loss of Coolant Accidents Using Support Vector Machines. IEEE Transactions on Nuclear Science, 628–636 (2008)
5. Yan, Y.: Application of support vector regression in regional logistics requirement forecasting. In: 2010 3rd IEEE International Conference on Computer Science and Information Technology (ICCSIT), pp. 321–324 (2010)
6. Lin, C.-F., Wang, S.-D.: Fuzzy support vector machines. IEEE Trans. on Neural Networks 13(2), 464–471 (2002)
7. Tay, F.E.H., Cao, L.J.: Modified support vector machines in financial time series forecasting. Neurocomputing 48, 847–861 (2002)
8. Mao, J.C., Jain, A.K.: A self-organizing network for hyper ellipsoidal clustering. IEEE Trans. Neural Networks 7(1), 16–29 (1996)
9. Ruiz, A.: Nonlinear kernel-based statistical pattern analysis. IEEE Trans. Neural Networks 12(1), 16–32 (2001)

Torque Analysis of Spherical Permanent Magnetic Motor with Magnetic Equivalent Circuit and Maxwell Stress Tensor

Bin Li, Chao Liu, Hongfeng Li, and Guidan Li

School of Electrical Engineering and Automation, Tianjin University, Tianjin 300072, China
elib@tju.edu.cn

Abstract. The paper discusses the torque characteristics of spherical permanent magnetic (PM)motor adopting three-dimensional (3D)magnetic equivalent circuit(MEC). Taking into account the unsaturation of the magnetic field, the rotor and the stator are meshed in spherical coordinates and cylindrical coordinates respectively, and then the expressions of the reluctances and magnetomotive force(MMF) have been deduced. According to the relative position between the coordinate systems, the magnetic field in the air gap has been obtained by transforming the coordinate systems. Based on this, the Maxwell stress tensor(MST)method is applied to analyze the electromagnetic force and torque, the corresponding torque matrix is given, and the relationship between the motor torque and winding location is obtained. Compared with the finiteelement method(FEM), this method is proved to have a high accuracy and helpful to the optimization design of the PM spherical motor.

Keywords: PM spherical motor, Electromagnetic torque, Maxwell Stress Tensor, 3D Magnetic Equivalent Circuit.

1 Introduction

Spherical permanent magnetic motors, which can take the place of traditional multi-degree of freedom motion system with the advantages of less complicated mechanical structure, smaller volume, higher precision, and faster response speed, have a wide application prospect in the field of robots, mechanical arms, manipulators and panoramic camera system.

For the PM spherical motor with the stator windings generally using multi-layer hollow cylindrical coils, the magnetic force can be analyzed by Lorentz force law, and then the motor torque characteristics can be obtained[1],[2]. However, this needs computing the volume integral of the vector product of the current density and magnetic flux density in the whole body of the stator windings. Compared to the Lorentz force law, the MST method needs only area integral, which has a smaller computation and a more accurate result[3],[4]. In order to obtain the Maxwell stress tensor, magnetic fields should be analyzed firstly. The analytical method obtains the magnetic fields by solving the Laplace equation under certain boundary conditions [5],[6]. The FEM can accurately describe the nonlinear material properties and

D. Yang (Ed.): Informatics in Control, Automation and Robotics, Volume 2, LNEE 133, pp. 617–628.
springerlink.com

irregular geometry, and can obtain very accurate results [7],[8]. Recently, the MEC method developed gradually, extended from 2-D to 3-D, becoming a powerful tool to analyze and design motors. The MEC method can describe the magnetic circuit geometry, nonlinear material parameters, thus has a higher precision result than analytical method. Compared to the FEM method, the computation time is much shorter in MEC method because the system matrix is sparse[9],[10].

Taking into account the unsaturation of the magnetic field, the paper analyzes the rotor and the stator magnetic fields respectively by meshing them in local coordinate systems with the MEC method, gives the elements equivalent reluctances and magnetic potential sources expressions, and then obtains the magnetic fields in the air gap by 3D coordinate transformations. Based on this, the electromagnetic forces and electromagnetic torques of the stator coilsare analyzed by MST method. Comparing with the FEM method proves that this method has a high accuracy.

2 Principle of 3DMagnetic Equivalent Circuit Method

Similar to the FEM, the MEC method needs to mesh the object and each split unit is equivalent to a reluctance network of the same structure, as shown in Fig.1

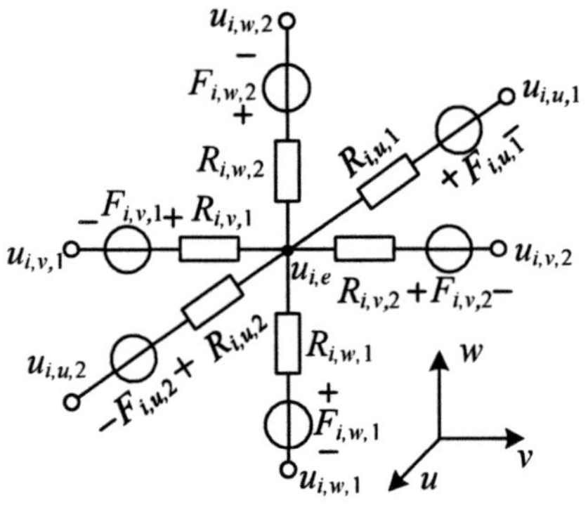

Fig. 1. Analysis model

In Fig.1, three branches of the 3D coordinate system correspond to three components of the magnetic flux respectively, *u* being the magnetic scalar potential, each node labeled by a subscript, where *e* denotes the center node and *R* and *F* denote reluctance and MMF respectively, whose subscripts denote the corresponding branches. The elements are connected to the adjacent elements through the corresponding nodes and constitute the entire MEC network.

In Fig.2, the flux in each branch can be expressed by the node scalar magnetic potential, the MMF and the branch reluctance as

$$\Phi_{i,j,k} = \frac{u_{i,e} - u_{i,j,k} - F_{i,j,k}}{R_{i,j,k}} = G_{i,j,k}(u_{i,e} - u_{i,j,k} - F_{i,j,k})\, j=u,\ v,w k=1,\ 2 \tag{1}$$

where$G$ is the permeance.

According to the conditions of continuous flux, the fluxflowing into the center node is equal to the flux flowing out of the node, as is given by (2)

$$\sum_{j=u,v,w} \sum_{k=1}^{2} \Phi_{i,j,k} = 0 \tag{2}$$

For all the nodes, we can yield from (1) and (2) the following system matrix

$$[G]\{U\} = \{F\} \tag{3}$$

where $[G]$ is the permeance matrix, $\{U\}$ is the scalar potential matrix, and $\{F\}$ is the forcing matrix.

The node scalar magnetic potential can be obtained from the above equation, and then the flux of each branch is available. Assuming that the magnetic field distribution in the element is uniform, it will be possible to obtain the magnetic fluxdensity through the magnetic energy of the element[11].

3 Magnetic Field Analysis of PM Spherical Motor

Fig.2 shows the structure of a PM spherical motor. With alternative variation of the N pole and S pole, permanent magnets are pasted evenly on the spherical rotor both sides along the equator. The rotor yoke is inside the PM. The rotor magnetic pole has the shape of a dihedral cone. Stator windings, each of which can be independently controlled, are fixed on the stator along the latitude direction in three layers. The rotor ball will rotate along the axis under the action of the electromagnetic torque when coils in the same latitude are energized, and pitch and yaw when coils in the same longitude energized. Therefore, the motor can achieve 3 DOF movement by controlling the currents in the coils.

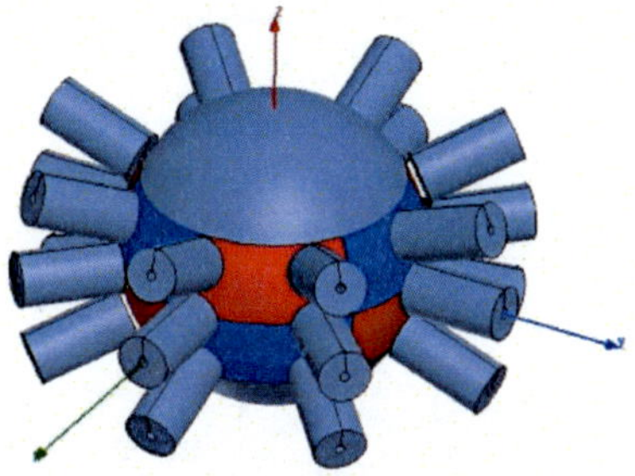

Fig. 2. Structure of the PM spherical motor

Since the stator is hollow coil, flux in the motor being unsaturated, the total magnetic field in the air gap can be obtained by adding the magnetic fields of the stator and rotor, which are computed by MEC respectively.

3.1 Element Parameters of the Stator MEC Model

The motor stator winding with the structure of hollow cylinder, whose section is shown in Fig.3, can be equivalent to cylinder permanent magnet, and the magnetization can be expressed as

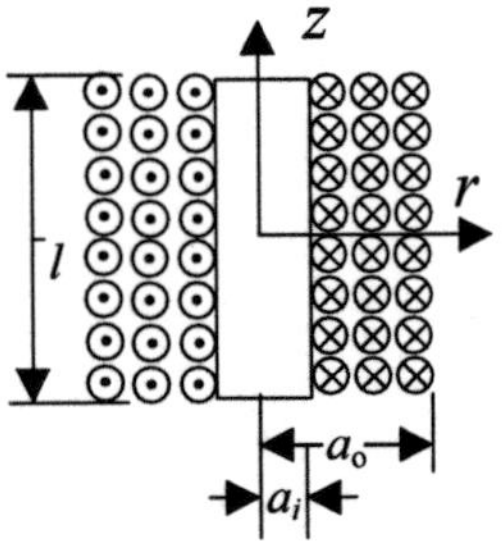

Fig. 3. Structure of the Stator winding

$$M_{r,z}(r) = \begin{cases} \mu_r J(a_o - a_i) & 0 \le r < a_i \\ \mu_r J(a_o - r) & a_i \le r \le a_o \end{cases} \tag{4}$$

where a_i and a_o are the inner and outer radius of the coil, l is the height and J is the Current density of the coil.

Since the magnetic field of the stator windings is symmetric in space and its tangential component is zero, the magnetic field in the whole space can be expanded from the magnetic field in a slice, which is through the center axis of the permanent magnet. Accordingly, the 3D magnetic field is simplified to a two-dimensional (2D) magnetic field. The slice, with tangential angle α_e, and the corresponding air field are meshed along z direction and r direction in cylindrical coordinate. And then the dimension and spatial location of each magnetic reluctance element can be expressed by $(r_{i,in}, r_{i,out})$ and $(z_{i,1}, z_{i,2})$, as shown in Fig.4.

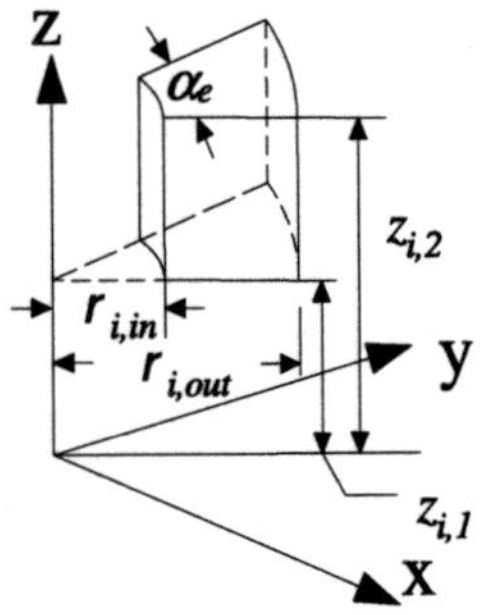

Fig. 4. Space location of the Reluctance element in cylindrical coordinate

The parameters of the magnetic reluctance depend on the geometry, permeability and the magnetic field direction of the element [11]. The element reluctances in cylindrical coordinate can be expressed as

$$R_{i,r} = \frac{1}{\mu\alpha_e l_i}\ln\left(\frac{r_{i,out}}{r_{i,in}}\right) \quad R_{r,z} = \frac{2l_i}{\mu\alpha_e(r_{i,out}^2 - r_{i,in}^2)} \tag{5}$$

where subscript i denotes the i-th element, μ represents the element permeability, which equals to μ_0 in the air domain and $\mu=\mu\mu_m$ in the PM, and μ_m represents the relative permeability, l_i represents the height of the reluctance in z direction, which equals to $z_{i,2}$-$z_{i,1}$.

As a passive network in the air domain, the MMF in each branch of the reluctance is equal to zero, but for the element in permanent magnet, the MMF depends on the magnetization and the element calculated length, which can be expressed by

$$F_{i,z} = M_{r,z} l_i \tag{6}$$

3.2 Element Parameters of the Rotor MEC Model

In order to analyze the magnetic field generated by the rotor permanent magnet, a spherical coordinate, with sphere center of the rotor as the coordinate origin, is established. The PM with the air domain is meshed, under spherical coordinate, into reluctance elements with the shape of a dihedral cone, and the dimension and spatial location of each element are given by ($r_{i,1}$, $r_{i,2}$),($\varphi_{i,1}$, $\varphi_{i,2}$), and ($\theta_{i,1}$, $\theta_{i,2}$), as shown in Fig.5.

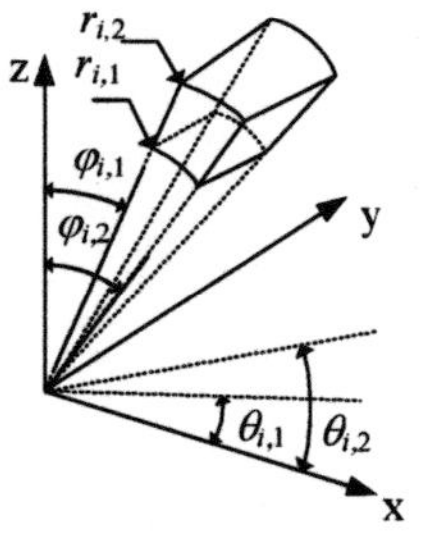

Fig. 5. Dimension and spatial location of reluctance element

The r,θ,φ.-component reluctances are expressed in spherical coordinate as

$$\begin{aligned} R_{i,r} &= \frac{1}{\mu(\theta_{i,2}-\theta_{i,1})(\cos\varphi_{i,1}-\cos\varphi_{i,2})}\frac{r_{i,2}-r_{i,1}}{r_{i,1}r_{i,2}} \\ R_{i,\theta} &= \frac{(\theta_2-\theta_1)}{\mu(r_2-r_1)\ln(\sec\theta+\tan\theta)|_{\theta_1}^{\theta_2}} \\ R_{i,\varphi} &= \frac{[\ln(\tan\frac{\varphi_{i,2}}{2})-\ln(\tan\frac{\varphi_{i,1}}{2})]}{\mu(\theta_{i,2}-\theta_{i,1})\Delta r} \end{aligned} \tag{7}$$

For each element of the PM, the MMF of the equivalentreluctance network can be written as

$$
\begin{aligned}
F_{i,r} &= M_{i,r}(r_{i,2}-r_{i,1}) \\
F_{i,\theta} &= M_{i,\theta}\frac{r_{i,2}+r_{i,1}}{2}\sin(\frac{\varphi_{i,2}+\varphi_{i,1}}{2})(\theta_{i,2}-\theta_{i,1}) \\
F_{i,\varphi} &= M_{i,\varphi}\frac{r_{i,2}+r_{i,1}}{2}(\varphi_{i,2}-\varphi_{i,1})
\end{aligned}
\tag{8}
$$

where $M_{i,r}$, $M_{i,\theta}$ and $M_{i,\varphi}$represent the three components of magnetization respectively[1].

3.3 Magnetic Field Computation in Air Gap

When establishing the MEC models above mentioned, the rotor and the stator are meshed in each local coordinate to facilitate analysis. Fig.6 shows the relative position between the stator cylindrical coordinate and the rotor spherical coordinate. The *z*-axis of the stator cylindrical coordinate points to the rotor sphere center and the coordinate of the stator cylindrical coordinate origin is (r_i, φ_i, θ_i). The rotor Cartesian coordinate and the stator local Cartesiancoordinate are also shown in Fig.6.

In the following analysis, with the rotor coordinate defined as the global coordinate system, the change of the relative position between the stator and the rotor can be expressed by the coordinate change of the stator local coordinate in the rotor coordinate system when the motor runs.

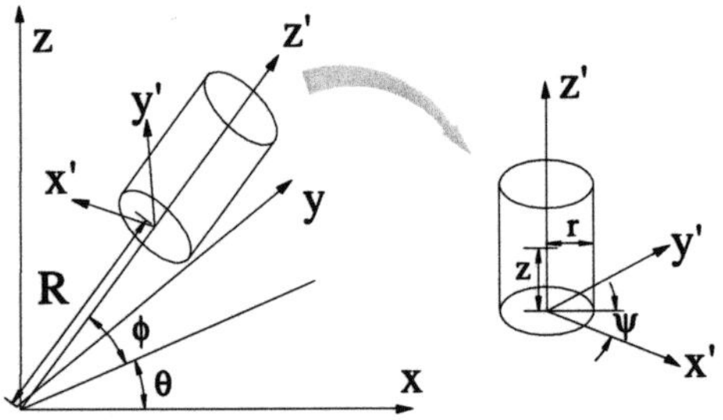

Fig. 6. Relative position in different coordinates

According to the 3D coordinate transformation, the stator flux density in the global Cartesian coordinate system can be expressed as

$$
\begin{bmatrix} B_{x1} & B_{y1} & B_{z1} \end{bmatrix}^T = C_1C_2C_3\begin{bmatrix} B_r & B_\psi & B_z \end{bmatrix}^T \tag{9}
$$

whereC_1,$C_2$and C_3denote the coordinate rotation matrixes, depending on the position of the cylindrical coordinate origin in the rotor spherical coordinate.

Rotor flux density in the global Cartesian coordinate system is written as

$$
\begin{bmatrix} B_{x2} & B_{y2} & B_{z2} \end{bmatrix}^T = C\begin{bmatrix} B_R & B_\theta & B_\phi \end{bmatrix}^T \tag{10}
$$

where C denotes the coordinate transformation matrix between the sphere coordinate and the Cartesian coordinate.

The total flux density in air domain is the linear sum of the rotor flux density and the stator flux density, that is

$$B_i = B_{i1} + B_{i2} \qquad \text{i=x,y,z} \tag{11}$$

4 Force and Torque Computation with the MST Method

Choosing a closed surface surrounding the stator coil as the integration path S, the electromagnetic force of the coil can be expressed as

$$F = \oiint (T \cdot n) dS \tag{12}$$

where n denotes the unit normal vector of the integration surface S, T denotes the Maxwell stress tensor.

$$T = \frac{1}{\mu_0} \begin{pmatrix} B_x^2 - \frac{1}{2}B^2 & B_x B_y & B_x B_z \\ B_y B_x & B_y^2 - \frac{1}{2}B^2 & B_y B_z \\ B_z B_x & B_z B_y & B_z^2 - \frac{1}{2}B^2 \end{pmatrix} \tag{13}$$

where B is the flux density in the closed surface, subscript indicates the corresponding component.

In MST method, different choice of the closed surface denotes different analysis accuracy. In the MEC model of the PM spherical motor, the closed surface is composited by the cylindrical surface S_1, top S_2 and bottom S_3 surrounding the coil, as shown in Fig.7.

Meshing the closed surface evenly in local cylindrical coordinate system, the integrate surface is divided into small pieces. And then (12) can be converted to

$$F = \sum_k f_i = \sum_k T_i n_i s_i \tag{14}$$

where T_i represents the Maxwell stress tensor of the i-th mesh element, n_i the normal vector, and s_i the i-th element area.

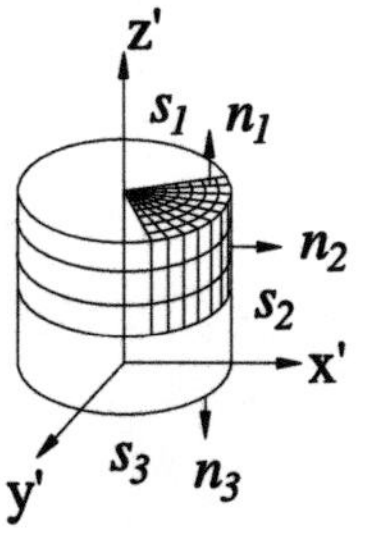

Fig. 7. Maxwell Stress Tensor method

In global Cartesian coordinate, normal vector of the coil side face, bottom and top can be expressed respectively as

$$n_1 = C_1C_2\left[\cos(\psi) \quad \sin(\psi) \quad 0\right]^T \; n_2 = C_1C_2\left[0 \quad 0 \quad -1\right]^T \; n_3 = C_1C_2\left[0 \quad 0 \quad 1\right]^T \tag{15}$$

And then the torque applied on the *i*-th element can be expressed as

$$M_i = r_i \times f_i \tag{16}$$

where M_i denotes the torque of the reluctance element, r_I denotes the space coordinate vector.

The torque applied on the coil is the sum of all element torques.

$$M = \sum M_i \tag{17}$$

The torque of the rotor is equal and opposite of *M*.

5 Simulation

The force and torque of the coil have been computed by MEC and MST methods. In order to verify the solution, a comparison between the FEM and MEC is applied. The structure parameters are shown in Table.1.

Table 1. Parameters of the PM spherical motor

Symbol	Meaning	Value
R_i	Innerradius	20mm
R_o	Outerradius	25mm
α	PMlongitudeangle	30°
β	PMlatitudeangle	45°
μ_0	Airpermeability	4πx10-7
h	Coil height	20mm
r_i	Coil inner radius	5mm
r_o	Coil outer radius	10mm
NI	Ampere turns	1000

Fig.8 shows the *Bx* comparison chart between MEC method and FEM method when coil lies at the center of the interface of N pole and S pole, θ=0°,φ=67.5°,1.75mm far away from the permanence magnet.

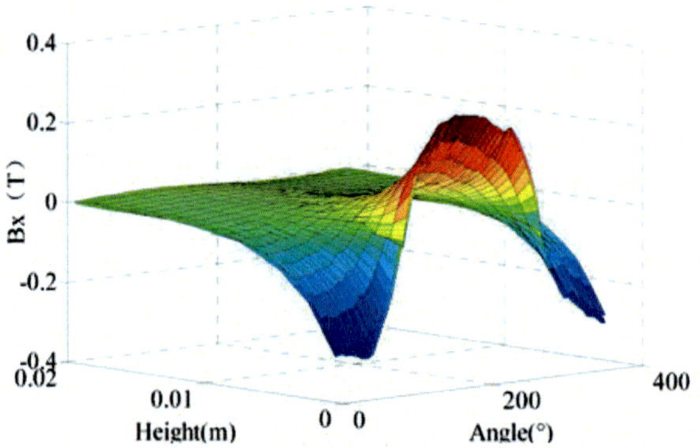

Fig. 8. The total magnetic field B_x

As can be seen from the above figure, flux density is much bigger when the coil is near the rotor, and similar to 0 when far away from the rotor. The magnetic field of the MEC method and the FEM method meets well.

Fig.9 shows the comparison of force and torque when coil lied at the interface of N pole and S pole moves along latitude angle, θ=0°, φ=67.5°~90°, 1.75mm far away from the permanence magnet.

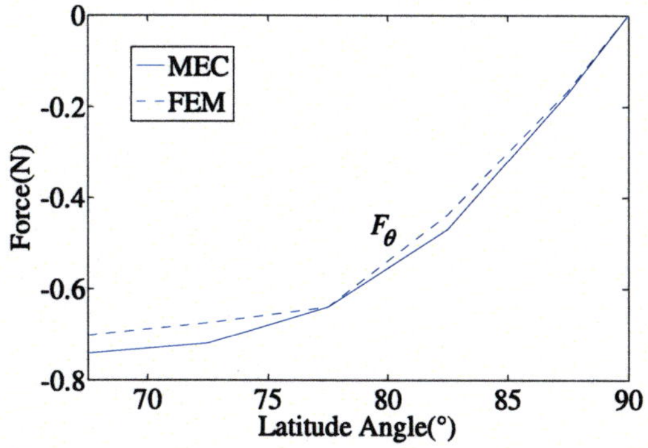

(a)

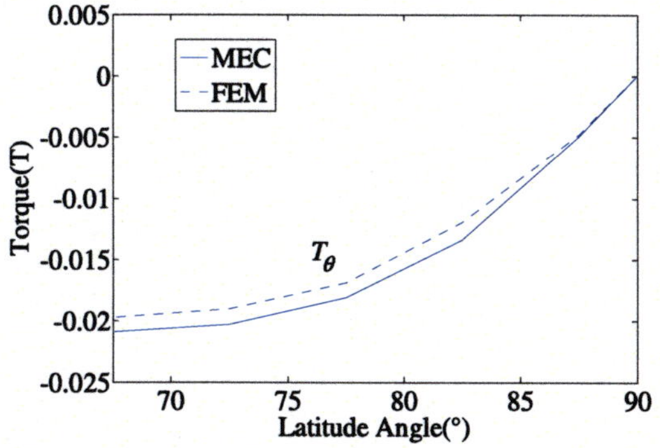

(b)

Fig. 9. (a) Comparison of coil force along latitude angle. (b) Comparison of coil torque along latitude angle.

As shown in Fig.9a, when the coil lies at the interface of the NS pole, only tangential force along longitude direction is applied. Fig.9b shows only tangential torque is applied, which is proportional to the force. When coil is above the equator and at the junction of four magnetic poles, both force and torque are equal to zero.

Fig.10a and Fig.10b show the comparison of force and torque when coil moves along longitude direction, φ=67.5°, θ=-15°~0°.

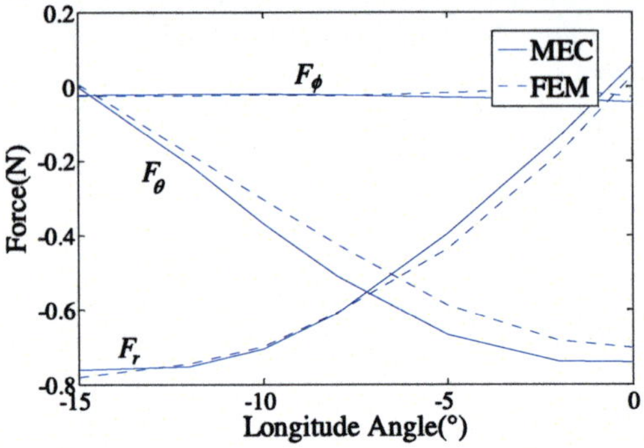

(a)

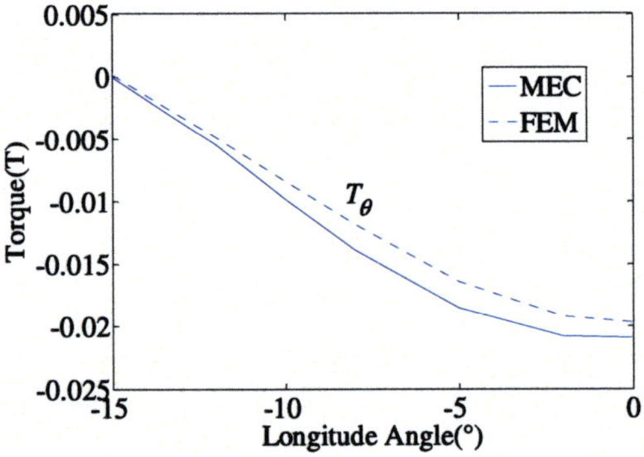

(b)

Fig. 10. (a) Comparison of coil force along latitude angle. (b) Comparison of coil torque along latitude angle.

As can be seen from the above figure, when the coil is on a pole, the force is along *r* direction, and the torque is equal to zero. Furthermore the torque takes the maximum when coil lies at the interface of NS pole.

Fig.11a and Fig.11b show the comparison of force and torque at a common location. θ=12°, φ=0°~22.5°. The preceding analysis shows that there is a good agreement between the results of MEC and those of FEM. A further mesh density can be applied to enhance the accuracy.

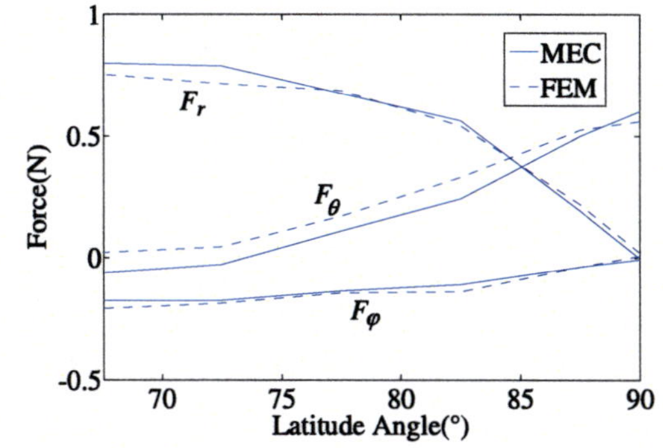

(a)

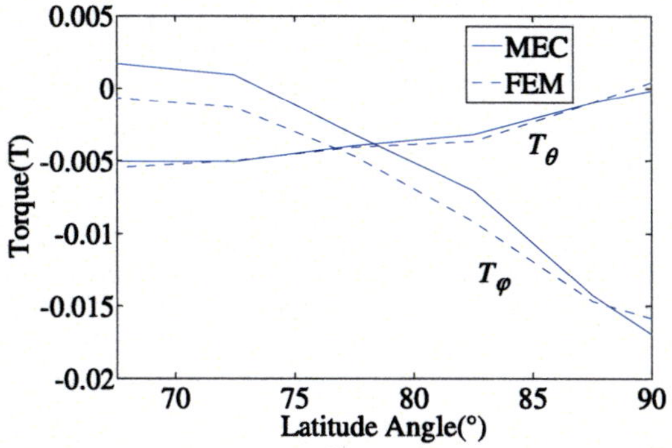

(b)

Fig. 11. (a) Comparison of coil force along latitude angle. (b) Comparison of coil torque along latitude angle.

6 Conclusion

In this paper, the distribution of flux density and coil torque in 3DOF PM spherical motor are analyzed by 3D magnetic equivalent circuit and Maxwell stress tensor. Compared with the FEM, this method with a short solving time has a great advantage when used into the motor optimization design.

Acknowledgments. This work was supported by National Natural Science Foundation of China under Grant 50907043 and 51007061, Tianjin Research Program of Application Foundation and Advanced Technology under Grant 10JCYBJC07900.

References

1. Yan, L., Chen, I.M., Yang, G., Lee, K.-M.: Analytical and Experimental Investigation on the Magnetic Field and Torque of a Permanent Magnet Spherical Actuator. IEEE/ASME Transactions on Mechatronics 11(4), 409–418 (2006)
2. Changliang, X., Hongfeng, L., Tingna, S.: 3-D Magnetic Field and Torque Analysis of aNovel Halbach Array Permanent-Magnet Spherical Motor. IEEE Transactions on Magnetics 44(8), 2016–2020 (2008)
3. Amrhein, M., Krein, P.T.: Force Calculation in 3-D Magnetic Equivalent Circuit Networks With a Maxwell Stress Tensor. IEEE Transactions on Energy Conversion 24(3), 587–593 (2009)
4. Du, Y.M., Xia, P.C., Xiao, L.Y.: Calculating Magnetic Force of Permanent Magnet Using Maxwell Stress Method. IEEE Transactions on Applied Superconductivity 10(1), 1392–1394 (2000)
5. Hongfeng, L., Changliang, X., Tingna, S.: Spherical Harmonic Analysis of a Novel Halbach Array PM Spherical Motor. In: IEEE International Conference on Robotics and Biomimetics, ROBIO 2007, pp. 2085–2089 (2007)
6. Dehez, B., Galary, G., Grenier, D., Raucent, B.: Development of a Spherical Induction Motor with Two Degrees of Freedom. IEEE Transactions on Magnetics 42(8), 2077–2089 (2006)
7. Changliang, X., Peng, S., Hongfeng, L., Bin, L., Tingna, S.: Research on Torque Calculation Method of Permanent-Magnet Spherical Motor Based on the Finite-Element Method. IEEE Transactions on Magnetics 45, 2015–2022 (2009)
8. Sosseh, R.A., Kok-Meng, L.: Finite Element Torque Modeling for the Design of a Spherical Motor, in Control. In: 7th International Conference on Automation, Robotics and Vision, ICARCV 2002, vol. 1, pp. 390–395 (2002)
9. Amrhein, M., Krein, P.T.: 3-D Magnetic Equivalent Circuit Framework for Modeling Electromechanical Devices. IEEE Transactions on Energy Conversion 24, 397–405 (2009)
10. Chun, Y.D., Wakao, S., Kim, T.H., Jang, K.B., Lee, J.: Multiobjective Design Optimization of Brushless Permanent Magnet Motor Using 3D Equivalent Magnetic Circuit Network Method. IEEE Transactions on Applied Superconductivity 14, 1910–1913 (2004)
11. Ostovic, V.: Dynamics of Saturated Electric Machines. Springer, New York (1989)

Levitation Force Analysis for Spherical Permanent Magnet Bearingless Motor

Bin Li, Junjie Shi, Guidan Li, and Hongfeng Li

School of Electrical Engineering and Automation, Tianjin University,
Tianjin 300072, China
elib@tju.edu.cn

Abstract. This paper presents a synthesis method of magnetic levitation force in spherical permanent magnet (PM) motor, which overcomes the adverse impact of friction torque because there is no direct contact between the stator and rotor. Using the theorem of force translation, Lorentz force of a single coil is equivalent to a force and a moment of couple exerted on the rotor center, so the independent control on levitation force can be achieved. This paper analyzed the position change of stator coils in spherical coordinates when the eccentric rotor moves in three degrees-of-freedom (3-DOF), and then the control equation of levitation force is established. The required magnetic levitation force can be synthesized by controlling the stator currents. Simulation model in MATLAB is created, furthermore the relationships between coil currents and magnetic levitation force, eccentric deviation are studied. The results show that the synthesis method of levitation force is feasible and can provide a basis for control system design.

Keywords: Spherical motor, bearingless, magnetic levitation, permanent magnet.

1 Introduction

The traditional multi degree-of-freedom (DOF) motion system consisting of multiple motors has some disadvantages as large volume, low transmission accuracy and control difficulty. The spherical motor can achieve 3-DOF of movement, thus having a wide application prospect in the high-precision control system as robots. The PM spherical motor attracts the attention of domestic and foreign scholars by the advantages as small volume, low price and high energy density.

Ref.[1,2] have proposed that PM spherical motor can achieve two degrees of freedom and three degrees of freedom. Ref.[3,4] have studied the stator commutation strategy for different torque vectors. The PM spherical motor with single and multi-layers rotor poles was proposed, and the torque model was derived based on Laplace equation and the Lorentz force law [5,6,7]. The torque model and position detection of PM spherical stepper motor were studied by Wang [8,9]. Tianjin University has proposed the Halbach array PM spherical motor, and has researched magnetic field distribution and control strategy [10,11].

D. Yang (Ed.): Informatics in Control, Automation and Robotics, Volume 2, LNEE 133, pp. 629–638.
springerlink.com

The spherical rotor supporting mechanism is different from the ordinary motor because of special structure and movements. In the direct contact way, the rotor is placed on the stator wall or on the slip ring directly, friction torque will affect the movement of the rotor although adopting a low friction coefficient coating [1,2]. The transfer bearing has been adopted instead of sliding friction, but the rotor surface may be damaged by the large pressure between the bearings and the rotor [7]. The spherical bearing can achieve a good 3-DOF movement, but it is more suitable for a single-layer spherical motor [12]. The air bearing can levitate the rotor and decrease the friction significantly, but it requires additional air source [13].

In recent years, the bearingless rotation motors have been studied widely and a variety of structure and control methods have been proposed [14,15]. Based on this, the paper has analyzed the synthesis method of magnetic levitation force for the PM spherical motor. The levitation force is produced by the Lorentz force between stator and rotor and is controlled by the stator currents.

2 The Principle of Magnetic Levitation PM Spherical Motor

2.1 The Structure of Motor

Fig.1 shows the structure of a magnetic levitation PM spherical motor, which mainly includes stator coils and spherical rotor. The parameters are given in Table 1. Two layers PM poles are distributed uniformly around the rotor equator. There are six poles with alternate polarity in each layer. Each rotor magnetic pole has the shape of a dihedral cone. The air-core stator coil is a cylindrical structure. The stator coils of three layers, eight coils per layer, are distributed uniformly around the equator of the spherical stator wall, which are made of non-magnetic material (not shown in the figure).

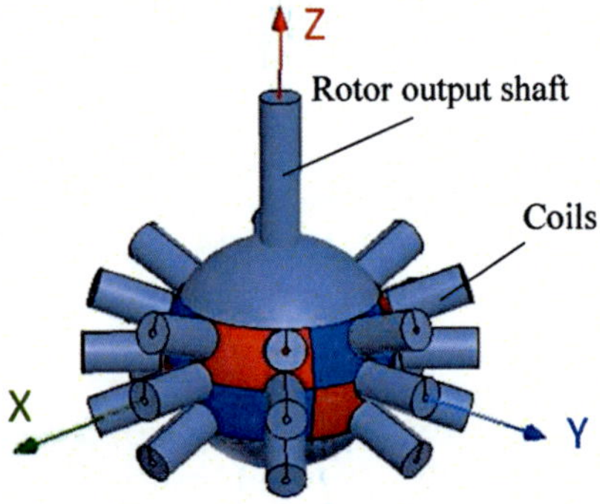

Fig. 1. The motor structure

Table 1. Parameters of structure and material

Symbol	Name	Value
R_{in}	Rotor radius	45mm
R_i	Inner radius of the stator windings	1.5mm
R_o	Outer radius of the stator windings	8mm
H	Height of stator windings	30mm
N	Turns	1000
φ_r	Latitude angle of PM pole	30°
θ_r	Longitude angle angle of PM pole	60°
d_r	Thickness of permanent magnet	5mm
H_c	Magnetic coercivity	900 KA/m
μ_r	Relative permeability of PM	1.15

2.2 The Synthesis Principle of Levitation Force

The initial position and current position of the rotor center are O and O_1 respectively, and the deviation OO_1 is denoted by vector $\Delta \boldsymbol{r}$. The position of stator coil will change relative to the rotor, as shown in Fig.2.

In Fig.2, the distance between a stator coil and the rotor center changes from R to R_1, and the corresponding air gap length changes from δ to δ_1. The stator coil also deviates from the rotor center O_1 by the angle of $\Delta\alpha$(As the deviation is very small, it can be neglected in the following analysis).

Assuming the Lorentz force acting on the i-th stator coil is denoted as F_i, and its reactive force is denoted as F_i'. Without loss of generality, the F_i' is supposed to act on the point A of the rotor surface, as shown in Fig.3.

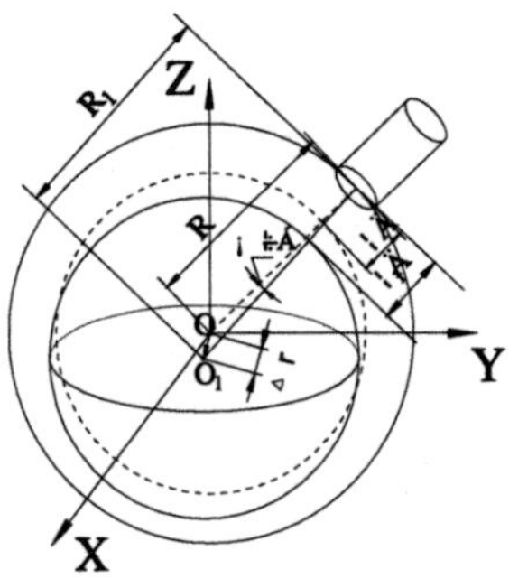

Fig. 2. Schematic diagram of the rotor center deviation

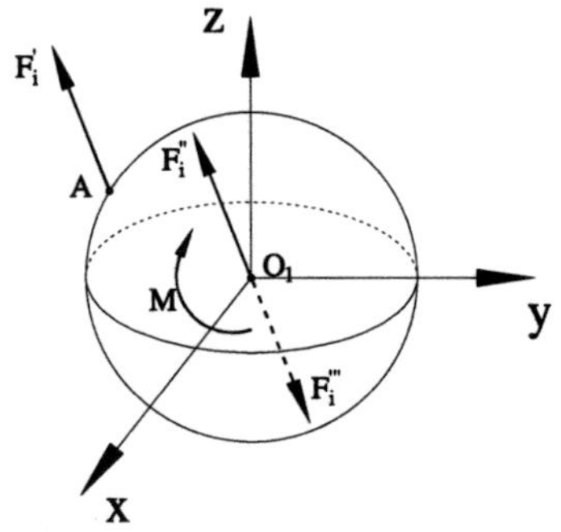

Fig. 3. The equivalent force translation

Based on the theorem of force translation, the F_i' can be translated from the point A to the rotor center O_1. In order to maintain the same effect on the rotor, a pair of forces $F_i''$and F_i''', which are equal and parallel with F_i', are added at the point O_1. The force system (F_i', F_i'', F_i''') is equivalent to the original force F_i'. Since F_i' and F_i''' form a torque M_i, F_i' is equivalent to a force and a moment of couple exerted on the rotor center, and it is written as

$$F_i" = F_i' \qquad M_i = r \times F_i' \tag{1}$$

Where, r indicates the radius vector from point O_1 to point A.

Therefore, all the Lorentz forces generated by the stator coils are translated to the sphere center, the resultant force and torque are available by the vector operation. The resultant force plays the supportive role acting on the rotor center, while the resultant torque can drive the load by the rotor shaft.

When the motor is running, the sensor detect the rotor deviation, and the controller adjusts each coil current to produce the levitation force, which pushes the rotor back to the initial position, then the motor can realize bearingless mode.

3 Current Control of the Levitation Force

3.1 Lorentz Force Generated by the Coil

The Lorentz force generated by a stator coil is related to the current and the air gap magnetic field. When the stator coil current is 1A and air gap δ is 1.5mm, the relationships between the Lorentz force and the coil position are shown in Fig.4, the θ and φ depict the coil position in the rotor spherical coordinates.

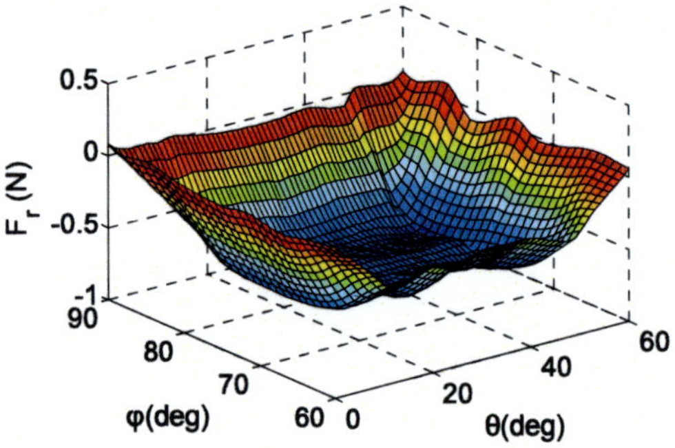

(a)

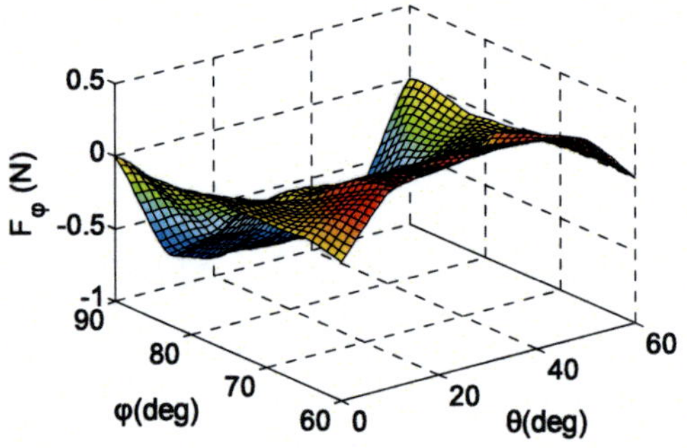

(b)

Fig. 4. Electromagnetic force in $\sum r\theta\varphi$ when δ=1.5mm (a) F_r component (b) F_φ component (c) F_θ component

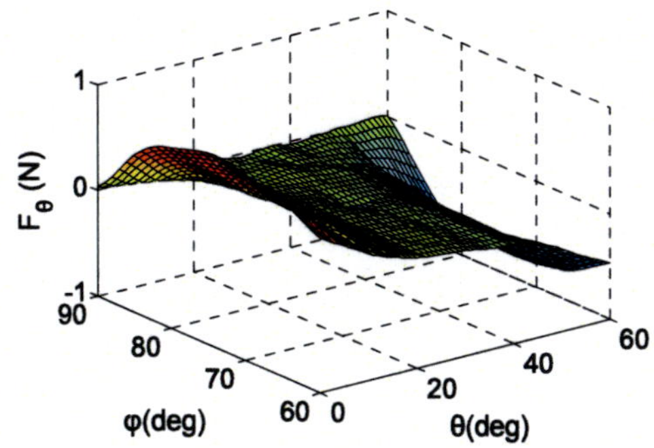

(c)

Fig. 4. (*continued*)

When the rotor deviates from the initial position, the air gap length between a particular stator coil and the rotor would be different from another, so the Lorentz forces would be different even the same current is passed. The forces can be calculated according to the deviation and the stator coils coordinate.

3.2 Translation of Lorentz Force

When the rotor movement, the rotor position can be represented by three rotations and one translation. The Euler angles α, β and γ are used to describe three rotations, and the translation vector $\Delta r=[\Delta x\ \Delta y\ \Delta z]$ denotes the rotor center deviation from O to O_1, as shown in Fig.5. The coordinates $\sum x_i y_i z_i$ (i=1,2,3) denote the rotor Cartesian coordinates.

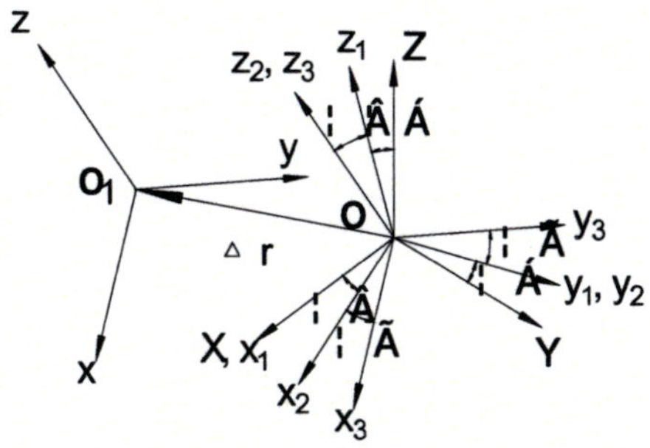

Fig. 5. Coordinates rotation and translation

Given the coordinate (X_i, Y_i, Z_i) of the i-th stator coil in the stator coordinate, the Euler angles (α, β, γ) and the deviation vector Δr, the coil position in the rotor Cartesian coordinates can be obtained as (x_i, y_i, z_i).

$$(x_i, y_i, z_i, 1) = (x_{3i}, y_{3i}, z_{3i}, 1) * L^{-1} \tag{2}$$

$$(x_{3i}, y_{3i}, z_{3i}) = (X_i, Y_i, Z_i) * R(\alpha, \beta, \lambda) \quad i=1\ldots24 \tag{3}$$

where $R(\alpha, \beta, \gamma)$ denotes the rotation transformation matrix between stator and rotor Cartesian coordinates, $L(\Delta x, \Delta y, \Delta z)$ denotes the translation matrix, and (x_{3i}, y_{3i}, z_{3i}) corresponds to (x_3, y_3, z_3) in Fig.5.

Considering that the magnetic field and torque of PM spherical motor are usually analyzed in the spherical coordinates, the coil position can be expressed as

$$\begin{aligned} \varphi_i &= \arccos \frac{z_i}{\sqrt{x_i^2 + y_i^2 + z_i^2}} \\ \theta_i &= \begin{cases} \arccos \dfrac{x_i}{\sqrt{x_i^2 + y_i^2}} & y_i > 0 \\ 2\pi - \arccos \dfrac{x_i}{\sqrt{x_i^2 + y_i^2}} & y_i < 0 \end{cases} \\ r_i &= \sqrt{(X_i - \Delta x)^2 + (Y_i - \Delta y)^2 + (Z_i - \Delta z)^2} \end{aligned} \tag{4}$$

From (4) and the PM spherical model, the Lorentz force could be obtained [6]. In the rotor Cartesian coordinate $\sum xyz$, the components of the force can be expressed as

$$\left| F_x \; F_y \; F_z \right|^T = T \cdot \left| F_r \; F_\varphi \; F_\theta \right|^T \tag{5}$$

The transformation matrix T is

$$T = \begin{bmatrix} \sin\varphi_i \cos\theta_i & \cos\varphi_i \cos\theta_i & -\sin\theta_i \\ \sin\varphi_i \sin\theta_i & \cos\varphi_i \sin\theta_i & \cos\theta_i \\ \cos\varphi_i & -\sin\varphi_i & 0 \end{bmatrix}$$

Finally, the Lorentz force of the i-th coil with a current of 1A in the stator Cartesian coordinate can be written as

$$F_i(X_i, Y_i, Z_i) = F_i(r_i, \theta_i, \varphi_i) * T^T * R(\alpha, \beta, \lambda) \tag{6}$$

3.3 Current Control

The magnetic circuit in the stator coils is not saturated for the air-core coil, so the Lorentz force is linear with the current. The force and torque of each coil are obtained in stator coordinates and let **F** denote the force matrix composed by all the 24 coils, $\mathbf{I_F}$ denote the current vector for levitation. Let the resultant torque is 0 and given the levitation force $[F_x \, F_y \, F_z]$, the relationship between the matrix **F** and the vector $\mathbf{I_F}$ can be expressed as

$$\begin{aligned}
&[F_X\ 0\ 0]^T = \mathbf{F}(X,Y,Z) \bullet \mathbf{I}_{FX} \\
&[0\ F_Y\ 0]^T = \mathbf{F}(X,Y,Z) \bullet \mathbf{I}_{FY} \\
&[0\ 0\ F_Z]^T = \mathbf{F}(X,Y,Z) \bullet \mathbf{I}_{FZ} \\
&\mathbf{I}_F = \mathbf{I}_{FX} + \mathbf{I}_{FY} + \mathbf{I}_{FZ}
\end{aligned} \tag{7}$$

It is known from (7) that the control currents of the stator coils can be obtained by solving the generalized inverse of the force matrix.

4 Simulation Results

A simulation model is built in MATLAB and the block diagram is shown in Fig.6. The inputs are the Euler angles and the deviation of the rotor, while the outputs are the control currents of the stator coils.

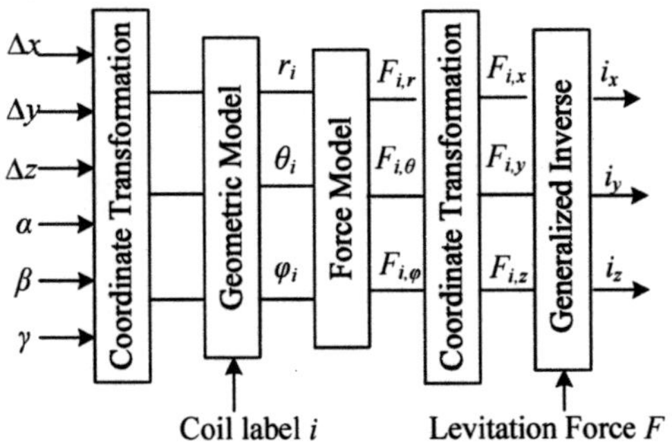

Fig. 6. Simulation block diagram

The rotor coordinates coincide with the stator coordinates in the initial position. The coil numbers of the middle layer are shown in Figure 7 (a), for instance 2-1, in which 2 expresses layer, and 1 expresses the number of coil. Fig. 7 (b) shows the number rules of three layers viewed from Y axis.

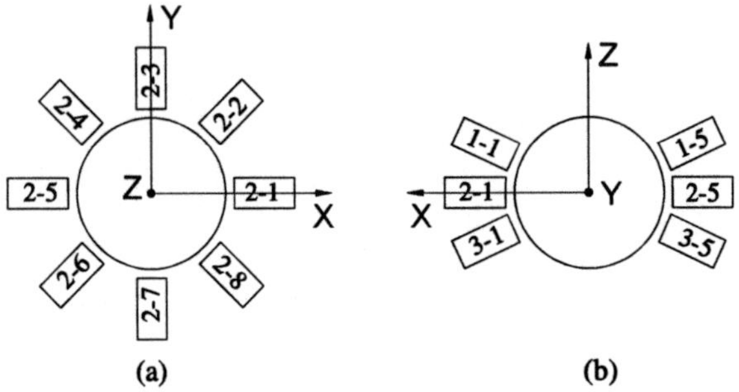

Fig. 7. Coil number rule (a) Top view (b) Side view

When α=0°, β=15°, γ=30° and Δr=[-0.5 0 0], the relationships between force F_X and the currents $I_{2\text{-}1}$ and $I_{2\text{-}2}$ are shown in Fig. 8. Fig. 8 shows that the levitation force is proportional to the coil current when the rotor position is constant.

When α=0°, β=15°, γ=30°, the coils 2-1 and 2-5 are in magnetic neutral line of a pole. The levitation force F_Z is given as 10N, pointing towards the X-axis negative direction. The relationships between deviation Δr and the currents $I_{2\text{-}1}$, $I_{2\text{-}3}$, $I_{2\text{-}5}$ and $I_{2\text{-}7}$ are shown in Fig.9.

Fig. 9 shows that the relationship between the currents and the rotor deviation is complicated when the rotor rotates, which depends on the Euler angles and deviation. However, some coils may be symmetrical about the rotor poles at a particular rotor orientation, and their currents would be changed in same way, as shown in Fig.9.

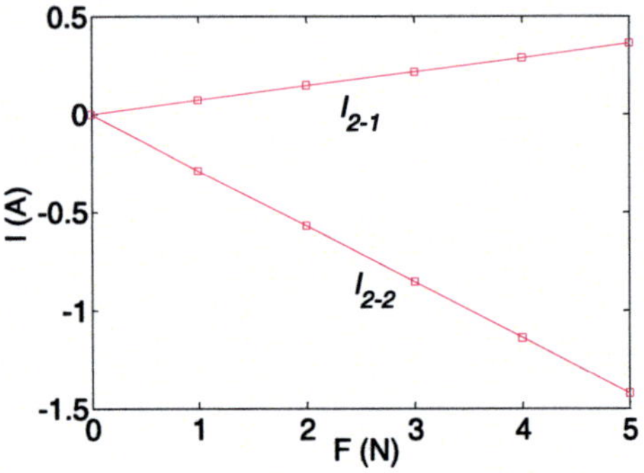

Fig. 8. Currents change with F_x

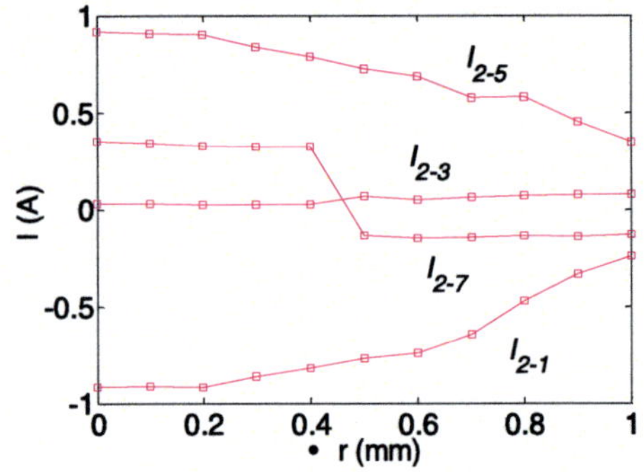

Fig. 9. Currents change with rotor deviation

5 Conclusion

This paper presented a synthesis method of magnetic levitation force in PM spherical motor, which uses Lorentz force between the stator and the rotor to produce magnetic levitation force and torque. The rotor pole structure and the stator coil distribution were given, the current equation of the magnetic force was established, and then the required magnetic force can be synthesized by controlling stator coils currents. To prove the feasibility of the magnetic levitation, a simulation model was built in MATLAB, and the relationships between coil current and levitation force, eccentric deviation were studied. The results show that the levitation force is proportional to the coil current when the rotor position is constant and it is nonlinear function with Euler angle and the deviation when the rotor is eccentric.

Acknowledgments. This work was supported by National Natural Science Foundation of China under Grant 50907043 and 51007061, Tianjin Research Program of Application Foundation and Advanced Technology under Grant 10JCYBJC07900.

References

1. Wang, W., Wang, J., Jewell, G.W., et al.: Design and Control of a Novel Spherical Permanent Magnet Actuator with Three Degrees of Freedom. IEEE/ASME Trans. on Mechatronics 8(4), 457–468 (2003)
2. Wang, J., Wang, W., Jewell, G.W., et al.: A Novel Spherical Permanent Actuator with Three Degrees of Freedom. IEEE Trans. on Magnetics 34(4), 2078–2080 (1998)
3. Kahlen, K.: Control of Multi-dimensional Drives with Variable Pole Pitch. In: The 2002 IAS IEEE Conference on Industry Application, Pittsburgh, PA, pp. 2366–2370 (2002)
4. Kahlen, K., Voss, I., Priebe, C., et al.: Torque Control of a Spherical Machine with Variable Pole Pitch. IEEE Trans. on Power Electronics 19(6), 1628–1634 (2004)
5. Yan, L., Chen, I.M., Lim, C.K., et al.: Torque Modeling of Spherical Actuators with Double-layer Poles. In: IEEE International Conference on Robots and Systems, Beijing, China, pp. 5447–5452 (2006)
6. Yan, L., Chen, I.M., Yang, G.: Analytical and Experimental Investigation on the Magnetic Field and Torque of a Permanent Magnet Spherical Actuator. IEEE/ASME Trans. on Mechatronics 11(4), 409–419 (2006)
7. Liang, Y., Chen, I.M., Chee Kian, L., Guilin, Y., Wei, L., Kok-Meng, L.: Design and Analysis of a Permanent Magnet Spherical Actuator. IEEE/ASME Transactions on Mechatronics 13(2), 239–248 (2008)
8. Wu, L., Wang, Q., Du, S., et al.: Integral Equation Method for Simulation of the Magnetic Field of Permanent Magnet Spherical Stepper. Proceedings of the CSEE 24(9), 192–197 (2004)
9. Wang, Q., Li, Z., Xia, K., et al.: Calculation and Analysis on Configuration Parameters and Torque Characteristics of a Novel Spherical Stepper Motor. Proceedings of the CSEE 26(10), 158–165 (2006)
10. Changliang, X., Hongfeng, L., Tingna, S.: 3-D Magnetic Field and Torque Analysis of a Novel Halbach Array Permanent-magnet Spherical Motor. IEEE Transactions on Magnetics 44(8), 2016–2020 (2008)
11. Chen, G., Changliang, X.: A Dynamic Decoupling Control Algorithm for Halbach Array Permanent Magnet Spherical Motor based on Computed Torque Method. In: IEEE International Conference on Robotics and Biomimetics, ROBIO 2007, pp. 2090–2094 (2094)

12. Chee Kian, L., Chen, I.M.: A novel 3-DOF Sensing Methodology for Spherical Actuator. In: Proceedings of the 2007 IEEE/ASME International Conference on Advanced Intelligent Mechatronics (AIM 2007), Zurich, Switzerland, pp. 1–6 (2007)
13. Kok-Meng Lee, D.E.E., He, T.: Design and Control of a Spherical Air-bearing System for Multi-d.o.f. ball-joint-like Actuators. Mechatronics (13), 175–194 (2003)
14. Stephens, L.S., Dae-Gon, K.: Force and Torque Characteristics for a Slotless Lorentz Self-bearing Servomotor. IEEE Transactions on Magnetics 38(4), 1764–1773 (2002)
15. Nguyen, Q.D., Ueno, S.: Analysis and Control of Non-salient Permanent Magnet Axial Gap Self-Bearing Motor. IEEE Transactions on Industrial Electronics (99), 1 (2010)

The Study of a Sound Category Neural Model on Perceptual Magnet Effect*

Shaobai Zhang and Lili Wang

Computer College, Nanjing University Of Posts & Telecommunications,
Nanjing 210003, China

Abstract. Speech perception is an important part of our perception system. The category learning during infancy leads to the phenomena such as categorical perception and the perceptual magnet effect , which conflicts with the classical formulation of a self-organizing feature map. This paper proposes a model to explain this phenomena and the simulation of model proves the feasibility and correctness of this model. The model further predicts that the magnet effect arises because prototypical vowels have smaller auditory cortical representations than non-prototypical vowels, which can be supported by a functional magnetic resonance imaging (fMRI) experiment.

Keywords: sound category learning, perceptual magnet effect, self-organizing feature map.

1 Introduction

Our ability to sort stimuli into behaviorally relevant categories is a central aspect of human experience. The perception of a continuously varying speech signal as a discrete set of phonemes, syllables, and words is a vivid example of this phenomenon. When a baby was born, just like a "Citizen of the world", she was prepared for discriminating different natural languages. But as the growth of the age and the spread of language learning process, the natural state will be changed by the input of outside language information.

The vowels of the world's languages can be largely differentiated based on the first three formant frequencies of the acoustic signal. These formant frequencies can be viewed as a three-dimensional acoustic space, and different languages break up this space into different vowel systems. In English, for example, approximately twelve "pure" vowel categories occupy the vocalic region of formant space, whereas the Spanish and Japanese languages have only five vowels occupying this same space.[1]

2 Perceptual Magnet Effect

Our ability to discriminate sounds of different parts is not uniform throughout acoustic space. For example, in the process of the sound category learning, two

* This work is supported by the National Natural Science Foundation of China (No.61073115).

D. Yang (Ed.): Informatics in Control, Automation and Robotics, Volume 2, LNEE 133, pp. 639–645.
springerlink.com

sounds that fall into the same phonetic category are difficult to discriminate, whereas two sounds that activate different phonetic categories are esay to discriminate. This phenomena is called categorical perception. And perceptual magnet effect indicates that two prototypical example of a vowel from an individual's native language are harder to discriminate from each other than two non-prototypical examples that are near a category boundary, though prototypical vowels are presumably much more commonly experienced by a listener than non-prototypical vowel-like sounds[2].

In the standard formulation of a self-organizing feature map, increased exposure to a set of stimuli leads to a larger cortical representation for those stimuli. Furthermore, it is widely believed that, all else equal, larger sensory cortical representations lead to better discriminability of the represented stimuli[3]. Although the listeners are more frequently exposed to prototype vowels, but the perceptual magnet effect pointes out that prototype vowels are more difficult to discriminate. This is obviously incompatibe with the classic expression of self-organizing feature map, because in the classical self-organizing feature map, prototype vowels with larger cortical representations, comparing with non-prototypical vowels with smaller cortical representations, should be easier to discriminate. As shown in the next section, we have developed a neural model based on the self-organizing feature map architecture that accounts for the perceptual magnet effect.

3 A Neural Model of Category Learning

In order to explain to the perceptual magnet effect, this paper proposes a model based on self-organizing feature map structure. The model is based on the idea that category learning during infancy changes the distribution of the firing preferences of neurons in auditory cortical maps and thus changes the discriminability of sounds from different parts of acoustic space.

As shown in figure 1, the model consists of two layers: a formant representation layer and an auditory map layer. They are connected by adaptive weights. It is assumed that the early stages of the auditory system process the incoming speech signal and extract important auditory information such as formant frequencies, which are represented in the first layer of the model. Each formant layer cell projects to each cell in the auditory map layer. Cells in the auditory map are tuned to a particular point in the formant frequency space, and thus fire maximally for that particular input. The neural representation of the formants for a sound was defined as follows:

$$x_i^+ = \frac{F_i - F_{iMIN}}{\sqrt{(F_i - F_{iMIN})^2 + (F_{iMAX} - F_i)^2}}$$
$$x_i^- = \frac{F_{iMAX} - F_i}{\sqrt{(F_i - F_{iMIN})^2 + (F_{iMAX} - F_i)^2}} \qquad (1)$$

where the index i indicates the formant number, x_i^+ and x_i^- are the activities of the antagonistically paired cells coding the ith formant, F_i is the value of the ith formant in Mels or Hertz, F_{iMIN} is the minimum value of the ith formant in the simulations, and F_{iMAX} is the maximum value of the ith formant. The constants F_{iMIN} and F_{iMAX} are parameters that insure that the cell activities x_i^+ and x_i^- vary between 0 and 1.

$$m_j = \frac{\sum_i x_i^+ z_{ij}^+ + x_i^- z_{ij}^-}{M} \tag{2}$$

where m_j is the activity of the jth map cell, z_{ij}^+ and z_{ij}^- are modifiable synaptic weights projecting from the ith antagonistic cell pair of the formant representation to the jth map cell, and M is the number of cells in the map.

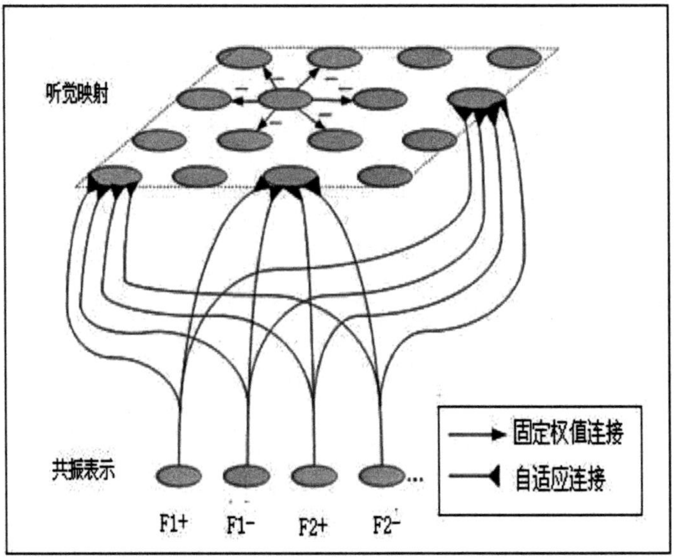

Fig. 1. The basic architecture of model

In the model simulations below, learning of the auditory map depents on a modification of Kohonen's self-organizing feature map algorithm. In Kohonen's formulation, the learning rate η is a scalar that decays exponentially in time and modulates the amount of synaptic weight change that can occur in a given time step. In our model, η is also modulated by the Euclidean distance between the current input and the nearest category center. when the input coincides with a category center the learning rate will be minimal and When the input near the category border, learning rate presents maximum[4].

The model's explanation of the perceptual magnet effect is simple: prototypical examples of a category are more difficult to discriminate from each other than non-prototypical examples because they have a smaller representation in auditory cortical maps. The model further posits that this reduced cortical representation results from phoneme category learning during infancy. In particular, learning to treat sounds from a particular region of acoustic space as members of the same category leads to a decrease in the size of the auditory cortical representation of sounds near the center of that region.

4 Simulation Experiment

4.1 Simulation of the Model

Simulating the model for the region of formant frequency space corresponding to English syllables /r/ and /l/ and Japanese syllable /r/. The left half of figure 2 shows the simulation results of English syllables and the right shows Japanese syllabe.

We can see from the result that the cells representing prototype vowels are less, but the vowels near a category boundary have larger cortical representation. Comparing the two figures,it's not difficult to find there is a dip in the category center of Japanese /r/ corresponding to boundary of English /r/and/l/. This can explain why Japanese are difficult to discriminate English /r/ and/l/, they repeated contact Japanese syllable, so that they have less auditory cortex cells corresponding to formant space region of the boundary.

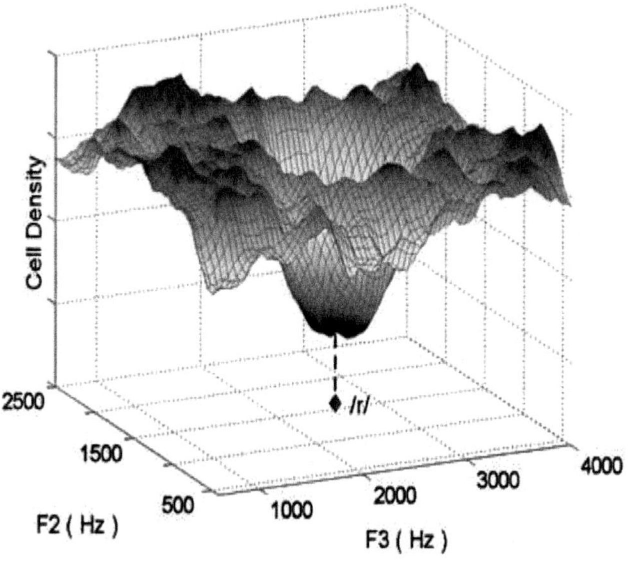

Fig. 2. Simulation results of English phonemes /r/ and /l/ (left) and Japanese phoneme /r/(right)

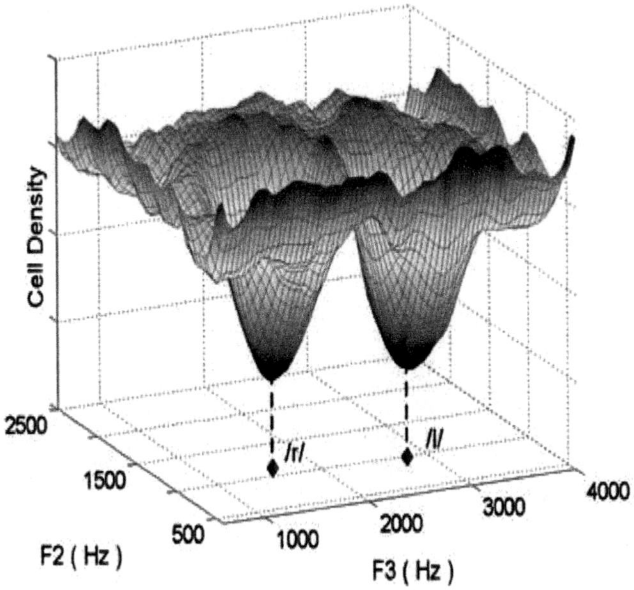

Fig. 2. *(continued)*

4.2 Testing Model Predictions with Auditory Psychophysical Experiments

Guenther et al. had performed a psychophysical experiment in 1999, in this experiment, subjects respectively received categorization training and discrimination training, and then the ability of each subject to discriminate these sounds before and after training was estimated by a d' measure. The result indicated it is the nature of the training task, not just the distribution of the training stimuli, that leads to the perceptual magnet effect[5].

The model is simulated and compared with experimental data of psychophysical experiment above. In the process of auditory map learning, different stimuli are presented at various distances from the prototype (represented in units of the just noticeable difference (JND)).

As same as the psychophysical experiment , pairs of stimuli were presented to the model in a testing phase following training. Gaussian noise was added as described by Zohary . Then calculating the noisy population responses for both stimuli, and if they were within some sensitivity measure, ε, the stimuli were judged to be the same. Otherwise, the stimuli were judged different.[6] Resulting d'scores were calculated and compared with scores obtained using a map with a uniform distribution of cells in the frequency space (corresponding to the pre-training d'test in the psychophysical experiment). The change in d'was then fit to experimental results using a linear least squares fit (see Figure 3). The resulting r^2 value was 0.9061. Therefore the model accounted for over 90% of the variance in the experimental results. This not only proves the feasibility and correctness of this model, but also illuminates this model has neurophysiology significance and practical application value.

We can also further perform a functional magnetic resonance imaging (fMRI) experiment to check the model. The results of experiment show that less activation is seen for the prototypical vowel than non-prototypical vowel throughout most of the auditory cortical areas.

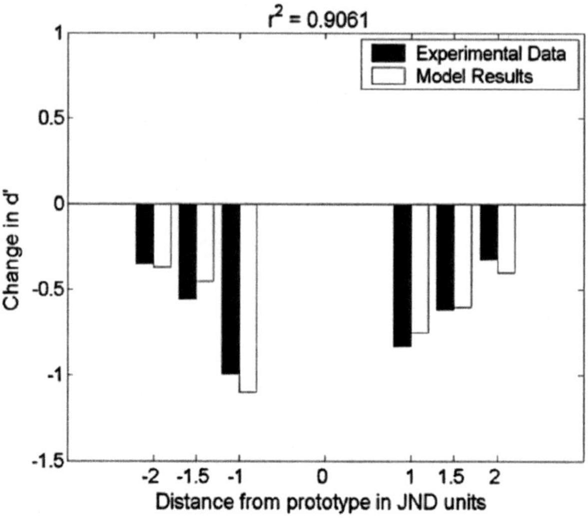

Fig. 3. Comparing of simulation results and experimental data

5 Conclusions

This paper proposes a neural model based on sound category learning which explains the perceptual magnet effec: phoneme category learning in infancy leads to the perceptual magnet effect because it causes a reduction in the size of the auditory cortical representation of prototypical examples of a vowel category. And the simulation experiments of the model demonstration the model comform to related datas of auditory psychophysics and cortical neurophysiology, illuminating this model has neurophysiology significance and practical application value.

This model also can be envisioned as a component of a larger computational modeling framework of speech development, perception, and production called DIVA[7]. Placing the current work within the context of DIVA suggests a scenario in which the sharpening of perceptual representations of phonemic categories. These issues will be investigated further in future work that will incorporate the current model into the DIVA framework.

References

1. Guenther, F.H., Gjaja, M.N.: The perceptual magnet effect as an emergent property of neural map formation. Journal of the Acoustical Society of America 100, 1111–1121 (1996)
2. Guenther, F.H., Nieto-Castanon, A., Ghosh, S.S., Tourville, J.A.: Representation of sound categories in auditory cortical maps. Journal of Speech, Language, and Hearing Research 47(1), 46–57 (2004)
3. Zhang, S., Ruan, X., Cheng, X.: A new constructing method of cerebellum model applying to DIVA model. In: Control and Decision Conference, CCDC 2009, Guilin,Chinese (June 2009)
4. He, J., Hu, G., Huang, H.: Research on the speech recognition system based on self-organizing neural networks. Audio Engineering (07), 56–59 (2006)
5. Guenther, F.H., Husain, F.T., Cohen, M.A., Shinn-Cunningham, B.G.: Effects of categorization and discrimination training on auditory perceptual space. Journal of the Acoustical Society of America 106, 2900–2912 (1999)
6. Zohary, E.: Population coding of visual stimuli by cortical neurons tuned to more than one dimension. Biological Cybernetics 66, 265–272 (1992)
7. Guenther, F.H., Perkell, J.S.: A neural model of speech production and its application to studies of the role of auditory feedback in speech. In: Maassen, B., Kent, R., Peters, H., Van Lieshout, P., Hulstijn, W. (eds.) Speech Motor Control in Normal and Disordered Speech, pp. 29–49. Oxford University Press (2004)

Research on Speaking Rate Effects to the Production of Speech Sounds*

Shaobai Zhang and Yanyan Liu

Computer College, Nanjing University Of Posts & Telecommunications,
Nanjing 210003, China

Abstract. This artical mainly study the speaking rate effects to the production of speech sounds. Based on the DIVA (Directions Into Velocities of Artculators) model, we have discussed and investigated these effects by modify the ODV(Orosensory Direction Vector) and AVV(Articulator Velocity Vector) equations. The simulation results show that increasing rate can increase the velocities of movements corresponding to consonantal gestures, but it increases, or even decreases, the velocities of movements corresponding to vowel gestures. The final conclusion is: despite different speaking rate have differential effects for the movements correspondng to vowels and consonants, the increase in ratios of maximum velocity to movement distance for the vowel and consonant gestures are almost same to vowels and consonants.

Keywords: DIVA model, speech production, convex region, speaking rate.

1 Introduction

Much research in the past twenty years has investigated how different speaking rate affect the production of speech sounds. Several researchers have noted that different speakers tend to use different strategies to increase speaking rate and the two strategies were called velocity strategy and amplitude strategy, respectively[1].

The velocity strategy is already applied to DIVA as a consequence of the GO signal that gates movement commands. Increase the GO signal is directly to increases articulator velocities and speaking rate. The amplitude strategy can be carried out in DIVA by changing the size of the convex region target. When produce a phoneme at slower rates, the convex region will be smaller, and this will lead to clearer, more precise speech.

A common result from their studies is that different speaking rate have different effects to speech production movements, but we still don't know what the effects exactly are. This article studies the effects of changing speaking rate to vowels and consonants. By modifying the equations that govern the ODV and AVV activities.

* This work is supported by the National Natural Science Foundation of China (No.61073115).

D. Yang (Ed.): Informatics in Control, Automation and Robotics, Volume 2, LNEE 133, pp. 647–651.
springerlink.com

2 Speaking Rate Effects

2.1 Speaking Rate Effects to Articulatory Effort

Articulatory effort has been widely discussed with various speech production models. One aspect views speech production as a compromise between ease of articulatory effort and communication accuracy. A good way to understand the mechanism of speech production and related issues is to examine articulatory movement during different speech rates. So Chen-huei Wu and Chilin Shih have done some experiments to study articulatory effort at different speaking rates by examining articulatory trajectory using the Electromagnetic Articulograph AG500[2]. And the results suggest that the articulator undershoots a target and the valley of the target might go deeper while the velocity increases.

Now we already know that different speaking rate have different effects to articulatory effort, then we can study the speaking rate effects to vowels and consonants production.

2.2 Speaking Rate Effects to Vowels and Consonants Production

Shrinking of the convex region target for a sound can be carried out easily in the DIVA neural network: simply add a small positive input to all ODV cells[3]. Since the same input is added to all ODV cells, this input will be referred to as non-specific. Only positive ODV activities can drive movement, the size of the convex region target along that dimension corresponds to the range of values of orosensory feedback that result in no positive activity of either ODV cell in the antagonistic pair. If the same non-specific input is added to all ODV cells, the convex region target shrinks toward the center along all dimensions[3]. Because a larger tonic activity results in a smaller, more precise target, the size of this input should be inversely related to movement speed. To achieve this, we can modify Equations which govern ODV cell activity, as follows:

$$d_{j+} = \sum_i s_i z_{ij+} - f_{j+} + R(1 - G) \tag{1}$$

$$d_{j-} = \sum_i s_i z_{ij-} - f_{j-} + R(1 - G) \tag{2}$$

where G is the value of the GO signal, and R is a parameter that corresponds to the degree to which a particular speaker prefers the amplitude strategy to the velocity strategy. The nonspecific input to the ODV cells is thus R(1-G). Adding a positive input to both channels in an ODV antagonistic pair can have an undesireable side-effect: it can result in positive activities at both ODV cells in the pair[4]. This problem is easily avoided by changing Equations which govern AVV activity during performance, as follows:

$$a_{k+} = \sum_j [[d_{j+}]^+ - [d_{j-}]^+]^+ w_{j+k+} + \sum_j [[d_{j-}]^+ - [d_{j+}]^+]^+ w_{j-k+} \tag{3}$$

$$a_{k-} = \sum_j [[d_{j+}]^+ - [d_{j-}]^+]^+ w_{j+k-} + \sum_j [[d_{j-}]^+ - [d_{j+}]^+]^+ w_{j-k-} \tag{4}$$

In the DIVA simulations, the velocity and amplitude strategies are used simultaneously to increase speaking rate. This is accomplished in the model by changing the parameter R. When R close to 0.0 simulates a speaker who relies more on the velocity strategy than the amplitude strategy, while a larger value of R simulates a speaker who relies more on the amplitude strategy.

Table 1 shows simulation results: With R = 0.0, maximum velocities of movements toward both vowels and consonants increase. When the model use the amplitude strategy with R = 0.2, although the model uses the same strategy to produce vowels and consonants, vowel movement velocities decrease with increased speaking rates, while consonant velocities increase, and this is the result used as evidence for different control structures for vowels and consonants.

The experiment shows another conclusion: No matter what stategy will be taken, the ratios of maximum velocity to movement distance for the vowel and consonant gestures will increase with increased speaking rate. When produce vowel under the amplitude strategy; even though maximum velocity decreases, the ratio of maximum velocity to movement distance increases.

2.3 The Relationship between Speaking Rate and Maximum Velocity /Movement Distance

In DIVA, this results from the multiplicative interaction between the GO signal and movement distance, described by Equation 5. Rearranging this equation yields the Equation 6:

$$v_k = G \times [a_{k+} - a_{k-}] \quad (5)$$

$$\frac{v_k}{[a_{k+} - a_{k-}]} = G \quad (6)$$

where v_k is the velocity along the kth articulatory degree of freedom, G is the value of the GO signal, and a_{k+} and a_{k-} are the antagonistically paired AVV activities corresponding to the ith articulatory degree of freedom. Equation 6 indicates that the velocity/distance ratio scales with the GO signal. It is clear that the ratio of velocity to distance will increase at faster speaking rates. In the case of vowel movements using an amplitude strategy, the decrease in articulator velocity at faster speaking rates occurs because of an even larger decrease in movement distance that outweighs the increase in G of Equation 5[5]; thus, the ratio of peak velocity to distance increases despite a decrease in peak velocity.

Furthermore, the displacements of jaw movements for both vowels and consonants decrease with increased speaking rate in DIVA simulations using the amplitude strategy. Table 2 illustrates this for vowel and consonant gestures produced by the model in the phrase /apapapa/.

3 Simulation Experiments

We experimented with MATLAB, and the results are as follows:

Experiment 1. Effects of Speaking Rate on Vowel and Consonant Movement Kinematics During the Utterance /pap/

Table 1 shows simulation results of the model producing the phrase /pap/ at two different speeds and using two different values of the R parameter. With R = 0.0, the model uses the velocity strategy. We can see that maximum velocities of movements toward both vowels and consonants increase.

But when R = 0.2, the model preferentially uses the amplitude strategy, despite the model uses the same strategy to produce vowels and consonants, vowel movement velocities decrease when speaking rates increases, while consonant velocities increase, and this result prove that vowels and consonants have different control structures.

We also can see that no matter what stategy will be taken, the ratios of maximum velocity to movement distance for the vowel and consonant gestures will increase when speaking rate increase. When produce vowel under the amplitude strategy; even though maximum velocity decreases, the ratio of maximum velocity to movement distance increases.

Table 1. Simulation Results Showing Effects of Speaking Rate on Vowel and Consonant Movement Kinematics During the Utterance /pap/

Velocity Stratedy (R=0.00)			
Quantity measured	Slow rate(G=0.5)	Fast rate(G=1.0)	Percent change
Vowel maximum velocity	0.010	0.019	+90%
consonant maximum velocity	0.088	0.176	+100%
Vowle max.velocity/distance	0.011	0.020	+81%
consonant max.velocity/distance	0.020	0.040	+100%
Amplitude Stratedy(R=0.20)			
Vowel maximum velocity	0.034	0.019	-44%
consonant maximum velocity	0.144	0.176	+22%
Vowle max.velocity/distance	0.010	0.020	+100%
consonant max.velocity/distance	0.025	0.040	+60%

Experiment 2. Jaw Displacement During the Utterance /apapapa/ Using the Amplitude Strategy.

The simulation results shows that the displacements of jaw movements for both vowels and consonants decrease when speaking rate increase in DIVA simulations using the amplitude strategy. But this phenomenon disappears when the parameter R is decreased with the movement displacements being equal at fast and slow rates when R is set to 0.0. This is interesting because some subjects have shown little effect of speaking rate on jaw displacement. The simulation results reported here suggest that these subjects may have used the velocity strategy, whereas the Gay et al.(1974) subjects were known to have used the amplitude strategy[6].

Table 2. Simulation Results Showing Jaw Displacement During the Utterance /apapapa/ Using the Amplitude Strategy

Average Jaw Displacement		
Gesture type	Slow rate(G=0.5)	Fast rate(G=1.0)
vowel	48.7	37.8
consonant	43.4	34.1

4 Concluding Remarks

This artical studies the speaking rate effects to the production of speech sounds. The experiment results shows that:Even use the same process when produce vowels and consonants, increase speaking rate will make different effects to the two type sound: increasing rate causes an increase in the velocities of movements corresponding to consonantal gestures, but it causes less of an increase, or even a decrease, in the velocities of movements corresponding to vowel gestures. Despite different speaking rate have different effects for the movements correspondng to vowels and consonants, the increase in ratios of maximum velocity to movement distance for the vowel and consonant gestures are almost the same to vowels and consonants.

References

1. Zhang, S., Ruan, X., Cheng, X.: A new constructing method of cerebellum model applying to DIVA model. In: Control and Decision Conference, CCDC 2009, Guilin, Chinese (June 2009)
2. Lacerda, F., Klintfors, E., Gustavsson, L.: Multisensory information as an improvement for communication systems' efficiency. To appear in Proceedings from Fonetik 2005, Gothenburg, Sweden (2005)
3. Guenther, F.H., Ghosh, S.S.: A model of cortical and cerebellar function in speech. In: Proceedings of the XVth International Congress of Phonetic Sciences, Barcelona, pp. 169–173 (2003)
4. Guenther, F.H., Perkell, J.S.: A neural model of speech production and its application to studies of the role of auditory feedback in speech. In: Maassen, B., Kent, R., Peters, H., Van Lieshout, P., Hulstijn, W. (eds.) Speech Motor Control in Normal and Disordered Speech, pp. 29–49. Oxford University Press (2004)
5. Guenther, F.H., Hampson, M., Johnson, D.: A theoretical investigation of reference frames for the planning of speech movements. Psychol. Rev. 105, 611–633 (1998)
6. Maeda, S.: In: Hardcastle, W.J., Marchal, A. (eds.) Speech production and speech modelling, pp. 131–149. Kluwer Academic Publishers, Boston (1990)

Multi-pass Dispatching Algorithm Based on Data Mining*

Sun Zijin** and Li Li

School of Electronics and Information, Tongji University,
201804, Shanghai, China
szj_27@sina.com

Abstract. Complex manufacturing systems require appropriate dynamic real-time dispatching approaches to obtain optimal operational performance. Multi-pass dispatching algorithms obtain increasing attentions due to their superiority. At first, the definition and research background of multi-pass dispatching are introduced, and then, the applications of multi-pass dispatching based on data mining (DM) are described. Finally, further research problems and possible research directions are pointed out.

Keywords: data mining, multi-pass dispatching, scheduling, simulation.

1 Introduction

Dynamic real-time dispatching is one of the effective approaches to achieve satisfactory operational performance of enterprises. "Real-time" means instantaneous response to any events. In a manufacturing process, an event could be an arrival of a part, a machine breakdown, or completion of a process. The idea of real-time scheduling is to make a decision fast enough so as not to disrupt the operation of the system. In other words, dynamic real-time dispatching decides how to accomplish the particular tasks dynamically in a certain period and how to allocate the shared resources reasonably in order to obtain optimal performances.

There are two main kinds of dynamic real-time dispatching methods, i.e., single-pass dispatching rules (SPDR) and multi-pass dispatching rules (MPDR). In a SPDR, a set of candidate dispatching rules is simulated and the one with the best long-run performance is selected and used during the whole planning horizon. On the other hand, MPDRs evaluate all the candidate dispatching rules in each relatively short scheduling period and select the best performer to be used in every production area, even in every machine at that interval. Thus,a number of literatures have proved that switching to different rules (multi-pass) yields better performance than using one rule (single-pass) for the entire horizon[1-3].

* This work was supported in part by National Natural Science Foundation of China (No.50905129, No. 61034004) and Program for Scientific Innovation of Shanghai (No.09DZ1120600, 10DZ1120100).

** Corresponding author.

D. Yang (Ed.): Informatics in Control, Automation and Robotics, Volume 2, LNEE 133, pp. 653–660.
springerlink.com

However, application of a MPDR to a large-scale production line covers large enumeration. DM can extract useful knowledge from huge amount of data and be used effectively in this area. The remainder of this paper is organized as follows. In Section 2, the relevant development of multi-pass algorithms is reviewed. Section 3 analyzes several common designs of MPDRs with DM. Finally, Section 4 concludes the paper with a summary and suggests some directions for future research.

2 Multi-Pass Dispatching Rule (MPDR)

2.1 Definition of MPDR

MPDR can be tracing back to the job of Wu and Wysk (1988), in which they combined a learning system with simulation and created a "Multi-pass Framework" for simulation based control. The manufacturing control system learned from its own historical performance [4]. Wu and Wysk (1989) later presented a MPDR using a mechanism controller and a flexible simulator [5]. MPDR is defined as a dispatching algorithm that deals with the scheduling problem of selecting the best dispatching rule, among rules in an alternative space (shown in Fig.1).

2.2 Research Background of MPDR

In 1990s, the research about MPDR mainly focused on dispatching rule selection. And in the 21 century, more researches started to emphasize on other applications (e.g. optimal design, configuration and attribute selection, etc.).

For instance, Ishii and Talavage (1991) proposed a transient-based real-time scheduling algorithm which selected a dispatching rule dynamically for a next short time period in order to respond to changes of system state[6]. In order to solve short-term dynamic scheduling in flexible manufacturing systems (FMSs), Ishii and Talavage (1994) later developed a search algorithm to select an appropriate mixed dispatching rule by using predictions based on discrete event simulation. The results indicated that the mixed dispatching rule approach performed up to 15.9% better than the conventional approach, and was 4% better on average [7].

Jeong and Kim (1998) presented a dynamic real-time scheduling methodology which used simulation and dispatching rules for FMSs. The strategies were formed by combining two factors that might influence the performance of the mechanism: type of simulation model which was used in the mechanism and points of time when new dispatching rules are selected [8].

Li and Olafsson (2005) introduced generating scheduling rules using a data-driven approach. However, unexpected knowledge can be obtained by generating decision tree (DT) model, in a manner that would be impossible if an explicit model of the system or the basic scheduling rules had to be obtained beforehand [9].

In order to avoid the "black box" effect of many optimization methods, Huyet et al. (2006) proposed a method which is based on the synergy between evolutionist optimization and an induction graph learning method. However this method lacks some flexibility if the context has changed [10].

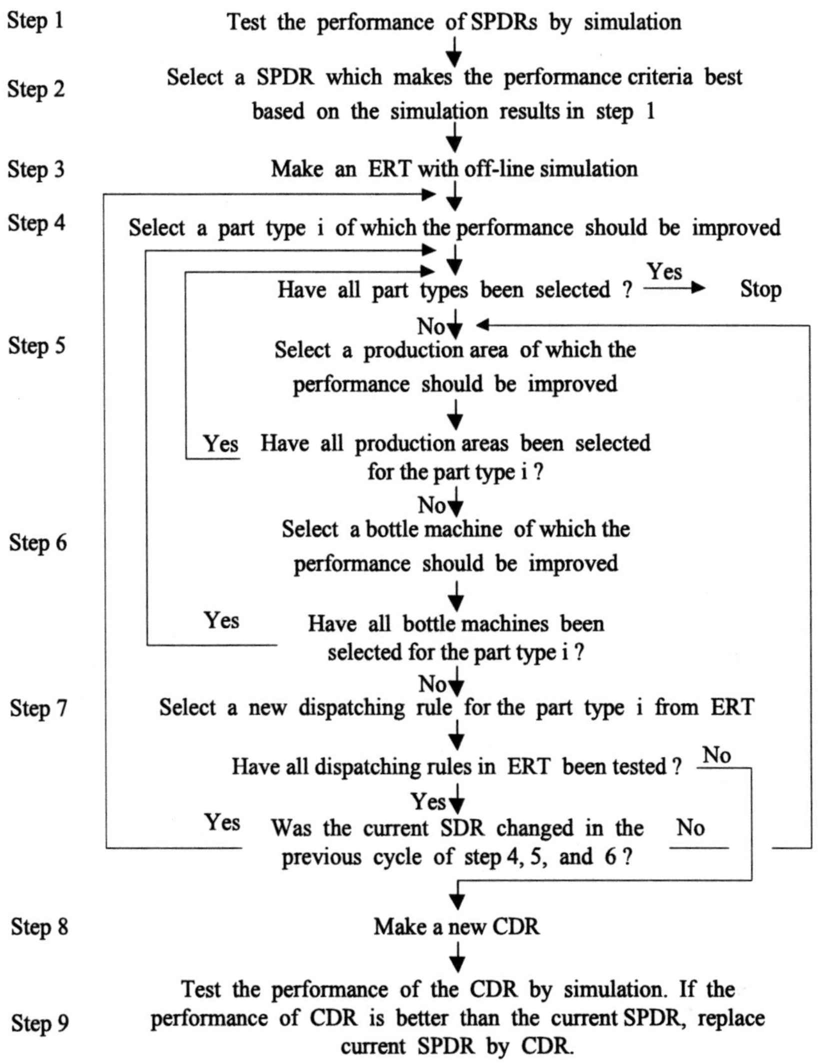

Fig. 1. The MPDR, ERT stands for effective rule table and CDR stands for current dispatching rule

Yildirim and Cakar (2006) designed flexible manufacturing systems based on simulation and neural networks. One of the fundamental differences is that they considered the problem from an operational perspective by continuously selecting a new dispatching rule for each scheduling period [11].

To solve the problem proposed above, Metan and Sabuncuoglu (2011) utilised the statistical process control (SPC) charts to monitor the performance of the DT and dynamically updates this DT whenever the manufacturing conditions change. This gives the proposed system the ability toadapt itself to changes in the manufacturing environment and improve the quality of its decisions [12].

3 Application of MPDR Based on DM

The classical machine learning approach builds up a knowledge base (KB) of dynamic real-time scheduling through MPDR. Its main disadvantage is that the classes (scheduling decision variables) to which training examples are assigned must be provided at the start. Kim and Min (1998) indicated this process becomes intolerably time-consuming because the rules must be determined for each period [13].

DM as a new and cross-disciplinary tool for analyzing data, covers machine learning, statistic, pattern recognition, has broad prospects in the field of study[14]. DM extracts useful knowledge from large-scale database or data warehouse, in a dynamic manufacturing, the form of the extracted knowledge can be mainly expressed as rules. The general idea of the latest research is selecting optimal scheduling strategy under different circumstances based on DM from the best heuristic dispatching rule sets responding to the current system status.

3.1 Decision Tree (DT)

DT is composed of a decision diagram and the possible results, and each decision or event (i.e., the state of nature) is likely to lead to two or more events and produce different results. In DM, a DT describes data but not decisions; rather the resulting classification tree can be an input for decision making.

For dynamic manufacturing systems, Park S. C. and Raman (1997) developed an adaptive and state-dependent scheduling policy, which tailored the dispatching rule to be used at a given point in time to the prevailing state of the system [15]. The inductive learning methodology provides an understanding of the relative importance of the various system parameters in determining the appropriate dispatching rule. Experimental studies indicated the superiority of the suggested approach over the alternative approach involving the repeated application of a single dispatching rule for randomly generated test problems and a real system, particularly, perform well under common sources of disruptions.

Arzi and Iaroslavitz (2000) tested the performance of a DT-based adaptive production control systems (PCSs) operating in a Flexible Manufacturing Cell (FMC) in two different operational environments, distinct by the variability of their demand. A DT-based PCS, which periodically chooses, on the basis of the current state of the system, the most appropriate scheduling rule, out of several predetermined ones, is presented. The DT-based PCS has performed rather well in the produce-to-order environment. However, it does not outperform the alternative scheduling rules in a highly random produce-to-order environment [16].

Kwak and Yih (2004) present a data-mining-based production control approach for the testing and rework cell in a dynamic computer-integrated manufacturing system. A DT-based module generates classification rules on each partitioned data that are suitable for interpretation and verification by users and stores the rules in the competitive decision selector (CDS) knowledge bases. Experimental results show that the CDS dynamic control is better than other common control rules with respect to the number of tardy jobs [17].

Some shortcomings existed in multi-pass scheduling with DT can be induced as following: 1. the training data must choose optimal parameters of scheduling at the decision point; 2. the evaluation of performance on scheduling is the result of classification of knowledge, instead prediction and precision fitting.

3.2 Neural Network (NN)

Rumelhart and Hinto (1986) showed the suitability of NN for automatic learning, association, generalization, and pattern recognition; the NN provides nonlinear transformations for modeling highly complex functions. Owing to the interconnection between dynamic considerations in the manufacturing cell, it is necessary to investigate the possibility of using NN for learning and performing dynamic real-time scheduling [18].

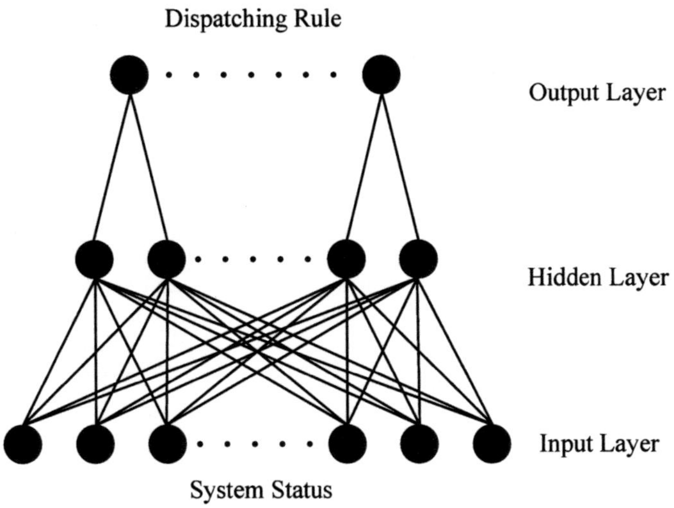

Fig. 2. Multi-pass dispatching with NN

Cho and Wysk (1993, 1995) described three hierarchical control levels and focus on a robust adaptive scheduler to support an intelligent workstation controller (IWC). Firstly, they developed a NN model that generates several part dispatching strategies based on workstation status; secondly, they developed a multi-pass simulator that evaluates the generated strategies and select the best strategy to maximize system efficiency; lastly, they compared the efficiency of the scheduling function with other single-pass strategies with respect to several performance criteria [19-20].

Semiconductor wafer fabrication involves possibly one of the most complex manufacturing processes ever used. Min and Yih (2003) developed a scheduler based on a simulation and a competitive NN for the selection of decision rules for deciding variables in order to obtain the desired performance measures given by a user at the end of a certain production interval. A competitive NN was applied to obtain the scheduling knowledge from the collected data containing the relationship between the change of decision rule set and current system status and the performance measures in the dynamic nature of semiconductor manufacturing fabrication with simulation. The results of the study indicated that applying this methodology to obtain a control strategy is an effective method considering the complexity of semiconductor wafer fabrication systems [21].

Priore and Fuente (2001, 2006) presented a scheduling approach that employs machine learning by analyzing the earlier performance of the system, 'scheduling knowledge' is obtained whereby the right dispatching rule at each particular moment can be determined. Simulation results indicate that the proposed approach produces significant performance improvements over existing dispatching rules [22-23].

Some characters existed in multi-pass scheduling with NN can be induced as following: developing a set of training samples, each of which comprises a vector of system attribute values and the corresponding dispatching rules. A trained NN can provide a preference indicator of candidate dispatching rules for a given system status. As shown in Fig.2, system status act as input of the NN and trained dispatching rules as output.

3.3 Mixture Algorithm

In recent years, many researches find optimal system performance can be obtained with DM at real-time decisions in Shop Floor Control System(SFCS).

Su and Shiue (2005) developed an intelligent scheduling controller (ISC) to support a SFCS to make real-time decisions, robust to various production requirements. The proposed approach integrated genetic algorithms (GAs) and DTs learning to evolve a combinatorial optimal subset of features from possible shop floor information concerning a DT-based ISC knowledge classifier. A GA was employed to search the space of all possible subsets of a large set of candidate features. For a given feature subset, a DT algorithm was invoked to generate a DT. Simulation results indicated that the GA/DT-based ISC improves system performance in the long run over that obtained with classical DT-based ISC and the heuristic individual dispatching rule, according to various performance criteria[24].

A SVM was developed by Vapnik (1999). A special property is that they minimize the empirical classification error and maximize the geometric margin simultaneously; hence, they are also known as maximum margin classifiers. An earlier study by Shiue (2009) indicates that the use of an integrated GA and the SVM approach to build an RTS system KB can yield good system performance compared with classical SVM-based RTS and heuristic individual dispatching rules under various performance criteria over a long period [25].

The combination of DM with computational intelligent methods can be used for real and historical data from real scheduling and simulation. At first, they extracted important scheduling information which improves the operation performance using feature analysis and DM, and then build scheduling model which decides the parameters in the process of production. The intelligent algorithm is used for evaluating the target function and DM used for looking for optimal solution, the results presents faster and better.

4 Conclusions

Multi-pass dispatching algorithm using the combination of intelligent algorithm and DM can get a better performance. Howver, intelligent algorithm is not a panacea, for example, NN needs too much training timeand lack the ablity of interpretation. We conclude main research content as following: 1. using meta-heuristic dispatching rules dynamically; 2. intelligent algorithm used for evaluation and DM used for obtaining optimal solution.

Dynamic real-time is the key to complex manufacturing scheduling. With the development of the technology, dynamic real-time scheduling will have a broad application in all kinds of fields. Owing to the local optimization of intelligent algorithm, most scheduling is still in the stage of study, real application will be emphasized.

References

1. Dar-El, E.M., Wysk, R.A.: Job Shop Scheduling - A Systematic Approach. Journal of Manufacturing Systems 1, 77–88 (1982)
2. Drake, G.R., Smith, J.S., Peters, B.A.: Simulation as a planning and scheduling tool for flexible manufacturing systems. In: Proceedings of the 1995 Winter Simulation Conference, pp. 805–812 (1995)
3. Su, C.T., Shiue, Y.R.: Intelligent scheduling controller for shop floor control systems: a hybrid genetic algorithm/decision tree learning approach. International Journal of Production Research 41(12), 2619–2641 (2003)
4. Wu, S.D., Wysk, R.A.: Multi-pass expert control system - a control/scheduling structure for flexible manufacturing cells. Journal of Manufacturing Systems 7(2), 107–120 (1988)
5. Wu, S.D., Wysk, R.A.: An application of discrete-event simulation to on-line control and scheduling in flexible manufacturing. International Journal of Production Research 27(9), 1603–1623 (1989)
6. Ishii, N., Talavage, J.J.: A transient-based real-time scheduling algorithm in FMS. International Journal of Production Research 29(12), 2501–2520 (1991)
7. Ishii, N., Talavage, J.J.: A mixed dispatching rule approach in FMS scheduling. International Journal of Flexible Manufacturing Systems 6(1), 69–87 (1994)
8. Jeong, K.-C., Kim, Y.-D.: A real-time scheduling mechanism for a flexible manufacturing system: using simulation and dispatching rules. International Journal of Production Research 36(9), 2609–2626 (1998)
9. Li, X., Olafsson, S.: Discovering dispatching rules using data mining. Journal of Scheduling 8(6), 515–527 (2005)
10. Huyet, A.L.: Optimization and analysis aid via data-mining for simulated production systems. European Journal of Operational Research 173(3), 827–838 (2006)
11. Yildirim, M.B., Cakar, T., Doguc, U.: Machine number, priority rule, and due date determination in flexible manufacturing systems using artificial neural networks. Computers & Industrial Engineering 50(1), 185–194 (2006)
12. Meidan, Y., Lerner, B., Rabinowitz, G., Hassoun, M.: Cycle-Time Key Factor Identification and Prediction in Semiconductor Manufacturing Using Machine Learning and Data Mining. IEEE Transactions on Semiconductor Manufacuring 24(2), 237–248 (2011)

13. Kim, C.O., Min, H.S., Yih, Y.: Integration of inductive learning and neural networks for multi-objective FMS scheduling. International Journal of Production Research 36(9), 2497–2509 (1998)
14. Fayyad, U.M., Piatetsky-Shapiro, G., Smyth, P.: The KDD process for extracting useful knowledge from volumes of data. Communications of the ACM 39(11), 27–34 (1996)
15. Park, S.C., Raman, N., Shaw, M.J.: Adaptive scheduling in dynamic flexible manufacturing systems: a dynamic rule selection approach. IEEE Transactions on Robotics and Automation 13(4), 486–502 (1997)
16. Arzi, Y., Iaroslavitz, L.: Operating an FMC by a decision-tree-based adaptive production control system. International Journal of Production Research 38(3), 675–697 (2000)
17. Kwak, C., Yih, Y.: Data mining approach to production control in the compute integrated testing cell. IEEE Transactions on Robotics and Automation 20(1), 107–116 (2004)
18. Rumelhart, D.E., Hinton, G.E., Williams, R.J.: Learning representations by back-propagating errors. Nature 323(9), 533–536 (1986)
19. Cho, H., Wysk, R.A.: A robust adaptive scheduler for an intelligent workstation controller. International Journal of Production Research 31(4), 771–789 (1993)
20. Cho, H., Wysk, R.A.: Intelligent workstation controller for computer-integrated manufacturing: problems and models. Journal of Manufacturing Systems 14(4), 252–263 (1995)
21. Min, H.S., Yih, Y.: Development of a real-time multi-objective scheduler for a semiconductor fabrication system. International Journal of Production Research 41(16), 2345–2364 (2003)
22. Priore, P., De La Fuente, D., Gomez, A., Puente, J.: A review of machine learning in dynamic scheduling of flexible manufacturing systems. Artificial Intelligence for Engineering Design, Analysis and Manufacturing 15, 251–264 (2001)
23. Priore, P., Fuente, D., Puente, J., Parreño, J.: A comparison of machine-learning algorithms for dynamic scheduling of flexible manufacturing systems. Engineering Application of Artificial Intelligence 19(3), 247–255 (2006)
24. Shiue, Y.-R., Guh, R.-S.: Learning-based multi-pass adaptive scheduling for a dynamic manufacturing cell environment. Robotics and Computer-Integrated Manufacturing 22(3), 203–216 (2005)
25. Shiue, Y.R.: Data mining-based dynamic dispatching rule selection mechanism for shop floor control systems using support vector machine approach. International Journal of Production Research 47(13), 3669–3690 (2009)

General TSA Protocol for Coping with the Capture Effect in RFID Systems*

Yuan Tan, Haifeng Wu**, Jihua Feng, Jing Liu, and Zhongting Deng

School of Electrical and Information Technology, Yunnan University of Nationalities, 650500 Kunming, People's Republic of China
pllpty@163.com, whf5469@gmail.com, fengjihua@21cn.com, ningning87109@163.com, seamild@163.com

Abstract. Tag anti-collision is an important issue in radio frequency identification (RFID) systems. In the traditional anti-collision protocol, when two or more tags transmit their signals in a frame, there will be a collision. However, in recent years RFID systems encountered capture effect, where a reader gets only one tag ID exactly even when multiple tags transmit their signals simultaneously. This paper proposes general TSA (Tree Slot Aloha) protocol based on the TSA to cope with capture effect. In this protocol, unrecognized tags hidden by the capture effect in the TSA cycle will be identified in the subsequent TSA cycles, and the general TSA protocol uses FKSS (Fast k-Slot Selection) algorithm which adjusts the initial frame length in dynamic of every cycle. The advantage of general TSA protocol is all tags can be identified under the capture environment and the system can get a high efficiency.

Keywords: RFID, anti-collision, capture effect, FKSS, general TSA.

1 Introduction

RFID is a key technology for IoT (internet of things), and tag anti-collision [1] is the most important cycle in RFID systems. There are currently two traditional types of anti-collision methods: aloha-based protocols [3] and tree-based protocols [4]. Among them, the dynamic frame slot Aloha (DFSA) protocol is being studied more. However, both of the pure tree type protocol and DFSA protocol, their max throughput were lower. In order to enhance the system efficiency, some higher throughput mixed protocols are proposed: Tree Slotted Aloha (TSA) which combined the tree type protocol into the Aloha protocol, and Dynamic Tree Slotted Aloha (DyTSA). Their highest system efficiency can reach about 0.43. Previous research all assumes that a successful slot, i.e., a slot where the reader can decode a tag ID, occurs only when one tag transmits its signal. However, this assumption completely neglects the capture effect, where a reader decodes a tag ID even when multiple tags simultaneously transmit their signals, because the signal of one tag is significantly

* This work is supported by Innovative Research Team in Yunan University of Nationalities.

** Corresponding author.

D. Yang (Ed.): Informatics in Control, Automation and Robotics, Volume 2, LNEE 133, pp. 661–668.
springerlink.com

stronger than that of the others. Some realistic experiments demonstrated that the capture effect actually appears in RFID systems. So, previous anti-collision algorithms cannot recognize all tags, because they do not expand a successful node, causing that some tags hidden by the capture effect are not recognized.

This work proposes general TSA protocol based on the TSA to solve the problem of the capture effect. In this protocol, all of the unrecognized tags hidden by the capture effect will be identified in the subsequent cycles. If the capture effect occurs in the first TSA cycle, we give several more TSA cycles to identify unrecognized tags hidden by the capture effect in TSA cycles until all of the slots in a cycle are idle slots, then we can say that all of tags are identified. However, this kind of solution wastes too many slots, because the frame length of the following cycles is always the original frame length, so the system efficiency is lower. For the sake of enhancing the system efficiency, we use FKSS algorithm to determine the frame length of every cycles dynamically to make the frame length and the tag number remain in a reasonable level, thereby ensuring high system throughput.

The rest of this paper is organized as follows. In Section 2, introduces the basic principle of TSA protocol and TSA under the capture environment, even points out the existing problems. In Section 3, we proposes general TSA protocol using FKSS algorithm into TSA protocol to deal with capture effect, and provides specific mathematical method, basic processes and pseudo-code. In Section 4, we present the simulation results and analyses process. In Section 5, draws conclusions.

2 The Previous Research and the Existing Problems

2.1 The TSA Protocol

For the type of TSA protocol, it contains the basic TSA protocol and DyTSA protocol. The basic thought of TSA protocol is that distributing all of the tags to several slots randomly, if conflict occurs in a slot, only decomposing the conflict tags within the slot, and do nothing with conflict tags in other slots. When all of the tags are assigned to several slots in a frame, and if the conflict occur in the i-th (i>0) layer of the frame and the j-th (j>0) slot, then collision tags will redistribute down in next layer with the slot-length of L_{i+1}^{j} (the slot length in the $i+1$-th layer and the j-th slot) for the frame to identify, and the value of L_{i+1}^{j} should be equal to the number of collision tags $Ncoll_i^j$ (the collision tags in the i-th layer and the j-th slot).In fact, the reader doesn't know the exact number of collision tags in the j-th slots of the i-th layer. Therefore in the actual execution, TSA algorithm using the following method to determine the L_{i+1}^{j}: If there is n_j tags have distributed in the i-th layer of the frame, then engendered $c_{1,j}$ success slots, and $c_{k,j}$ collision slots, so the collision tags of the collision slots in i-th layer of the frame will be distributed to the next layer with the slot length of L_{i+1}^{j} to identify. Among them, when $i>0$, $j>0$, the value of L_{i+1}^{j} will be equal to any j-th collision slot, determined by

$$L_{i+1} = \left\lfloor \frac{n_i - c_{1,i}}{c_{k,i}} \right\rfloor, \; i>0, j\geq 0 \tag{1}$$

In order to minimizing the time of identification process in TSA algorithm to achieve the optimal throughput, besides contented formula (1), we still need to assume the initial frame length L equal to the total number of tags. If meet the above conditions, the TSA protocol can achieve optimal throughput, about 0.43. Fig.6 illustrates the simulation results.

2.2 The Existing Problems in the TSA Protocol under the Capture Environment

Things will be different absolutely if the identification process under the capture environment in the TSA protocol. If we still use pure TSA algorithm to identify tags, a reader can not recognize all tags because a kind of successful slot can not be expanded caused by the capture effect that, a reader decodes a tag ID even when multiple tags simultaneously transmit their signals in a frame, because the signal of one tag is significantly stronger than that of the others. Some realistic experiments demonstrated that the capture effect actually appears in RFID systems and the probability of occurrences depends on the relative attenuation between the tags [4] [5].

3 The General TSA Protocol

3.1 The TSA Protocol under the Capture Environment

Because of capture effect existing in the realistic RFID systems, lots of tags would be not recognized. In this case, we propose TSA_CAP algorithm based on the pure TSA algorithm. The TSA_CAP algorithm is that at the end of the pure TSA protocol cycle, some more cycles will be provided to identify the rest tags, which are hidden by capture effect, until all of the slots in a frame of one cycle (last cycle) are all idle slots that demonstrated all of tags are recognized.

Fig.1 illustrates this TSA_CAP algorithm can identify all of tags under the capture environment and Fig.4 illustrates the system efficiency of this algorithm, but we can see the system throughput is lower because the initial frame length is set unreasonably as same as the frame length of following cycles that all frame length of following cycles equal to the initial frame length. Even the initial frame length is most reasonable, the frame length of flowing cycles are too big for the few unrecognized tags hidden by the capture effect. We propose a new algorithm called FKSS (Fast k-Slot Selection) algorithm that is to choose a reasonable frame length to solve the problem of the unreasonable frame length setting of every cycle.

3.2 The FKSS Algorithm

Previous research demonstrated if we want to get the optimal throughput for TSA protocol, a reader must to have the priori information about tag number, to set the initial frame length equal to initial tag numbers. However, the tag number is difficult to estimated [6]. We cannot receive the observation data until ending the initial frame. So, in order to get the optimal throughput steadily, we propose the FKSS Algorithm to determine the initial frame length to make the frame length and the tag number remain in a reasonable level.

Let k be the threshold of burst collision and burst idle in the first k slots respectively. For example, if the first k slots are all collision slots the initial frame length will be double, and if the first k slots are all idle slots the initial frame length will be reduction of two points. A burst detection mechanism can be designed as:

$$\begin{aligned} L_1 &= 2L_1 \ , \ C_{coll} = k \\ L_1 &= L_1/2 \ , \ C_{idle} = k \end{aligned} \tag{2}$$

Fig.2 illustrates the system efficiency of this algorithm and we can see the system throughput is the highest among the 4 other kind of protocols, TSA, DFSA, Q algorithm, DyTSA.

3.3 The General TSA Protocol

Consider the reality of capture effect in the TSA protocol, we use FKSS algorithm to determine the initial frame length as well as the following frame length of every cycle to make the frame length and the tag number remain in a reasonable level to get high system efficiency, its general TSA protocol.

This part will give specific process of using general TSA protocol to identify tags. Here gives out the reader operation pseudo-code and the tag operation pseudo-code of general TSA protocol respectively.

Pseudo code of General TSA protocol: reader procedure.

```
while Nidleslot~=L
  General TSA (L, k, Pc) (Pc is the probability of capture effect)
End

function General TSA (L, k, Pc)
Broadcast AdjQuery with L
Receive tag responses in the first k slots and detect collision
if k slots are all collision
  L = min(2L, 2^15)
  General TSA (L, k, Pc)
elseif k slots are all idle
  Q = round(L/2)
  General TSA (L, k, Pc)
elseif only a tag response
  level = 1
  s = 0
  TSA(L, level, s, Pc)
end
```

```
  end
end
function TSA(Li, level, s, Pc)
broadcast Query with (Li, level, s, Pc)
for s = 1 to Li do receiveIDs
Li+1 =FrameEstimation(Li)
for s = 1 to Li do
  if (collision[s]= 1)
    identify tag and sendAck;
  elseif (collision[s]> 1)
    level++;
    TSA(Li+1, level, s)
  end
  end
end
end
function tagsEstimation(Li)
compute c0, c1, ck, obtained in frame of length Li;
compute tag number estimation ni
ni = ni - c1
return floor(ni / ck )
```

Pseudo code of General TSA protocol: tag procedure

```
while receive Reader's AdjQuery with L
  do{s=random number form 1 to L
     send myID in slot s
end}
if receive Reader's Query with L
identified = false
  myLevel = 1
  previousValue = 0
  while (not identified)
    do{Receive Query with (Li, level, slot)
       if ((level = myLevel) & (previousValue = slot))
         s = random number form 1 to Li
         myLevel++;
         previousValue = s
         send myID in slot s
         if (receivedAck) identified = true end
       end}
   end
end
```

When a reader start reading tags, the value of initial frame length is L, and then the reader sends a query command with frame length L called AdjQuery. After each tag received orders, they send their ID number at a random slot S between 1 to L. Reader detects the tag response at the first k-slots in the frame. If the first k-slot are all collision, then L=min ($2L$, 2^{15}), reader ends this frame in advance, and send an AdjQuery with a new L, the same that if the first k-slot are all idle, then L=max ($L/2$, 2^{1}), the reader ends this frame in advance, and send an AdjQuery with a new L then detects the next frame tags response. And if the first slot is success, then L changeless, in this case, the algorithm will turn into TSA protocol. And the tag number n_j could be determined by Vogt [2] estimation method. At the end of the TSA cycle, the reader continues to send a query command with frame length L called AdjQuery, and continues detects the first k-slot until find a reasonable L then turn into TSA cycle again. And then repeat this process until all of the slots in a frame are idle slots in one TSA cycle. It is demonstrated that all of tags are identified.

4 Simulation Results

This following part is the simulation results and performance evolution, and all of computational processes are all adopt the Monte Carlo method, we obtained the average results after completing 500 experiments.

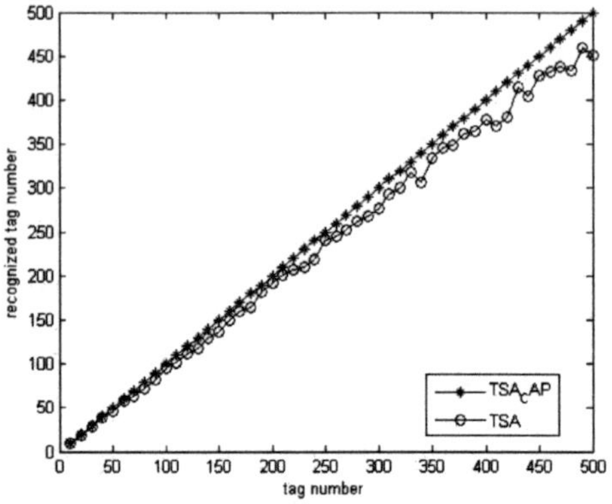

Fig. 1. Illustrates the recognized tag number of TSA_CAP and TSA protocol under the capture environment, where $p_c = 0.3$ (p_c is the probability of the capture effect occurs), here we assuming that the probability of the capture effect is fixed, N from 10 to 500, and the interval is 10, the initial frame length is $L = 128$, and Monte Carlo is 500 times. In this figure, we can see in the TSA_CAP all of tags are recognized by a reader under the capture environment. However, in the pure TSA, a reader can not identify all of tags as capture effect.

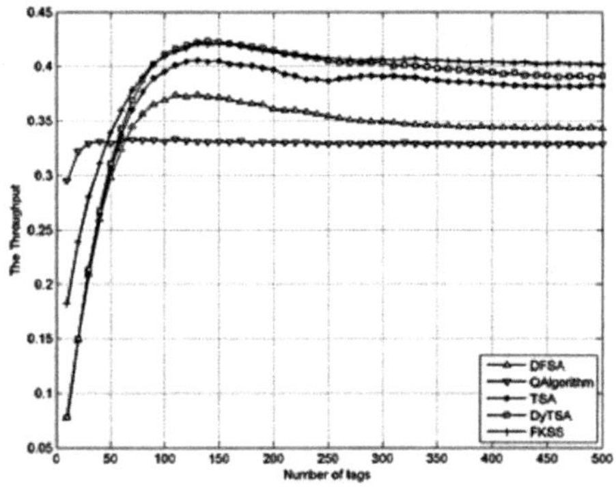

Fig. 2. Illustrates system efficiency of the different algorithms. We set the initial frame length $L = 128$, N from10 to 500, $k = 2$ for FKSS, and Monte Carlo is 500 times. we can see the maximum system efficiency is FKSS protocol which we proposed about 0.43, the same as DyTSA protocol, and no matter tag number increases or decreases, the system efficiency of FKSS will not get significant reduction, will remaining above 0.40, higher than TSA protocol. At the same time, the throughput curve also higher than Q algorithm and DFSA protocol.

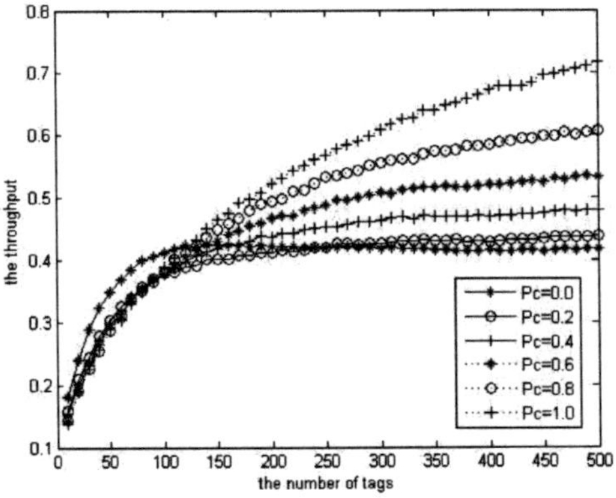

Fig. 3. Illustrates the system efficiency of general TSA in different $p_c = 0.0, 0.2, 0.4, 0.6, 0.8, 1.0$, N from 10 to 500, the initial frame length is $L = 128$, $k = 2$, and the interval is 10, Monte Carlo is 500 times. We can see according to the different p_c , the system efficiency of the general TSA protocol more and more higher by increasing value of p_c .

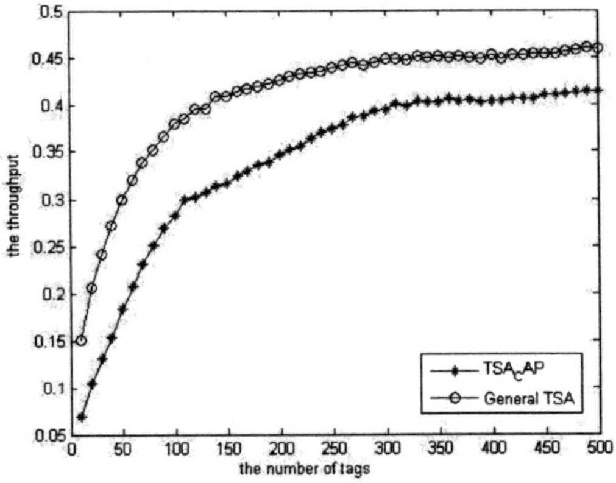

Fig. 4. Illustrates the system efficiency of general TSA protocol and TSA_CAP protocol. We set $p_c = 0.3$, N from 10 to 500, and the interval is 10, the initial frame length is $L = 128$, $k = 2$, and Monte Carlo is 500 times. As we can see the throughput of the general TSA protocol is higher than the TSA_CAP protocol obviously.

5 Conclusions

In this paper, we research tag identification problem under the capture environment. We use several more TSA cycles to cope with the problem caused by the capture effect in the TSA protocol. In addition, we propose general TSA that using FKSS algorithm to determine initial frame length and the following frame length of every TSA cycle, obtaining higher system efficiency than TSA_CAP that only gives several more TSA cycles to cope with the problem capture effect. In section 4, we can see in the Fig 4, the system efficiency of general TSA protocol is higher than TSA_CAP protocol. Therefore, we use FKSS algorithm to determine initial frame length and the following frame length in TSA_CAP protocol is feasible.

References

1. Cha, J.R., Kim, J.H.: Novel anti-collision–algorithms for fast object identification in RFID system. In: Proceedings of the 11th International Conference on Parallel and Distributed System, pp. 63–67. IEEE, Washington D.C., USA (2005)
2. Vgot, H.: Mutiple object identification with passive RFID tags. In: Proceeds of IEEE International Conference on System, Man and Cybernetics, pp. 1–6. IEEE, Hammamet (2002)
3. EPC Radio-Frequency Identity Protocols Class-I Generation-2 UHF RFID Protocol for Communications at 860Mhz-960Mhz version 1.0.9 (2004)
4. Hush, D.R., Wood, C.: Analysis of tree algorithm for RFID abitration. In: Proceeds. of IEEE International Symposium on Information Theory, p. 107. IEEE, Cambridge (1998)
5. Wang, L.-C., Liu, H.-C.: A Novel Anti-Collision Algorithm for EPC Gen2 RFID Systems. In: Proceeds. of IEEE International Conference on System, Man and Cybernetics. IEEE, Hammamet (2006)
6. Wu, H., Zeng, Y.: Bayesian Tag Estimate and Optimal Frame Length for Anti-Collision Aloha RFID System. IEEE Transactions on Automation Science and Engineering, 1–7 (2010)

A Graph-Based Spatial Mapping Algorithm for a Coarse Grained Reconfigurable Architecture Template

Lu Ma, Wei Ge, and Zhi Qi

National ASIC System Engineering Technology Research Center, Southeast University, Nanjing 210096, China
{malu_cnasic,duiker}@seu.edu.cn, zhiqi.cim@gmail.com

Abstract. Coarse Grained Reconfigurable Arrays (CGRAs) have been more and more popular recently due to their high performance with low power consumption. An intelligent compiler is essential to execute applications on CGRAs effectively. However, automatically compilation for CGRAs still faces many challenges though many algorithms have been proposed. In this paper, we present an effective mapping algorithm which is targetable to a parameterized architecture template. The main contributions are an effective priority scheme, a satisfying backtracking algorithm and a fast control data flow graph (CDFG) splitting method. The experimental results demonstrate that our technique gains the performance very close to manual optimization.

Keywords: Coarse Grained Reconfigurable Arrays, Compilation, CDFG, Mapping.

1 Introduction

CGRAs have recently becoming more and more popular for the purpose of either academic research or commercial applications in the past few years. The combination of the flexibility of software and high performance of hardware helps CGRAs gain a better balance among the key indexes, such as performance, power consumption and flexibility.

In the past few years, many CGRAs have been proposed along with various mapping methodologies. In [1], Kinyoung Choi summarized many CGRAs and their application mapping in the recent several years, including some examples of commercial CGRA.

An intelligent compiler is essential to map the compute intensive loops of an application onto the PE array effectively. There are two main classes of existing popular mapping algorithms. One class depends on modulo scheduling. Modulo scheduling is widely used software pipelining technique that exploits loop parallelism by overlapping the execution of successive iterations in a loop. Mei [2] developed a novel modulo scheduling algorithm with the simulating annealing strategy as the main technique to exploit parallelism for the ADRES architecture. An effective abstract architecture representation, modulo routing resource graph (MRRG), is proposed,

D. Yang (Ed.): Informatics in Control, Automation and Robotics, Volume 2, LNEE 133, pp. 669–678.
springerlink.com

to simplify the scheduling problem. In [3], another efficient modulo scheduling algorithm which can target CGRAs with irregular interconnect architectures, is presented using a compact three-dimensional architecture graph and a resource usage aware relocation algorithm. The experiment demonstrates that the proposed algorithm can much improved time-efficiency. Recently, Park[4]proposed the EMS algorithm that outperforms both the conventional modulo scheduling, although it is still architecture and loop structure dependent. However, the performance is heavily depended on the characteristics of the loop structure as well as the CGRA architecture. Even though the above work based on modulo scheduling can achieve satisfying schedule quality, the compiling time is still too long in practice.

The second class schedules a single iteration of a loop each time. Unlike modulo scheduling technique, approaches in this class focus on exploiting inter-iteration parallelism of the loop body. A graph mapping based approach (SPKM, Split-Push Kernel Mapping) is developed in [5]. SPKM is a heuristic approach and is based on the split & push algorithm. The mapping approach focuses on mapping loops to CGRAs with the objective to minimize the required resource. In paper [6], the mapping problem is divided into covering, partition and laying-out. And the three sub-problems are attacked separately. The algorithm first groups the nodes of a kernel into a set of configurations. Then the nodes in a configuration are divided into different clusters. Finally they schedule each node onto the array and build the interconnection network. The performance is demonstrated comparable with hand optimization. The mapping is divided into three levels in [7] while taking the memory operation sharing into consideration. The algorithm in [8] takes into account the spatial locality or connections between the PEs and the temporal locality between operands.

Our work proposes a technique belonging to the second class. Since the target CGRA is parametric, it can be applied to a series of various architectures. The contributions are as following: it proposes a) an effective priority scheme for processing elements (PEs) and operations; b) a backtracking algorithm which improves the performance significantly; c) a fast heuristic for splitting CDFG into sub-CDFGs. The proposed mapping methodology has been applied in our in-house CGRA. The experimental results show that the proposed priority scheme works effectively and the heuristics make a great contribution to gain a better quality of the mapping result.

The rest of this paper is organized as follows. Section 2 describes the background of this work, including the overview of the target CGRA and a brief presentation of compilation flow for CGRAs. Section 3 gives a detailed discussion of the proposed mapping methodology. Section 4 shows the experimental results. Finally, section 5 concludes the paper.

2 Background

2.1 Target Architecture Overview

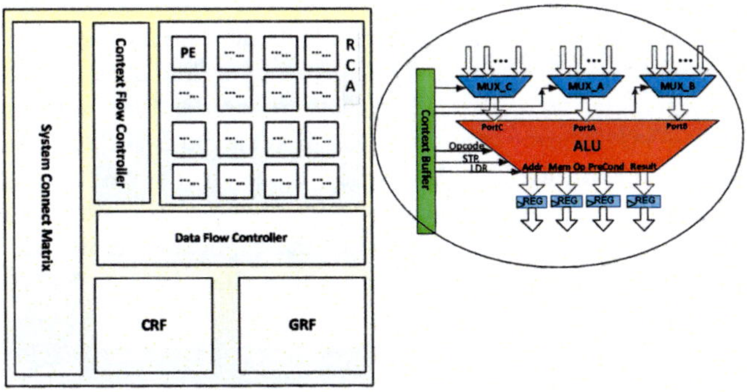

Fig. 1. a) The target CGRA template **b)**The inner structure of PEs

The target CGRA template our work depends on is shown in Fig.1.a). As a coprocessor used for accelerating the running of loops of applications, the CGRA is generally connected to a main processor through a system connect matrix. The GRF (global register file) is used for storing both input data that transported from the external memory to reconfigurable computing array(RCA) and output data generated by the RCA to external memory. The CRF (context register file) is used for storing several contexts. Limited to the size of this memory, when it's not enough for all the contexts, extra time has to be cost to load other contexts from the external configuration memory.

The RCA consists of an array of PEs and several DRFs (distributed register files). The DRFs are used for storing some temporal or intermediate data during the pipeline. As a template, the number and topology are parametric the same as the size and the connection of the array. Fig.1.b) describes the inner structure of PEs. All the operations in the instruction list can be supported by all the PEs. However, some *expensive* operations, like load/store, multiplication, division, etc. may only be performed by certain PEs. Hence the PE array can be heterogeneous. The context buffer stores a set of configuration contexts locally, helping reconfigure PEs quickly. The multiplexors are used for routing data from different sources (other PEs, GRF or DRFs). The results are stored in the output registers. It should be emphasized that the PEs supports a1-bit condition operation, which contributes a lot to both removing the control flow inside loops and guaranteeing data dependence.

2.2 Compilation Flow for CGRA

The compiler design framework is shown in Fig.2. The input is c language description of the application which consists of numerous computation-intensive kernels, typically loops. The HW/SW step chooses the candidate loops that will be mapped to the reconfigurable array, and leaves the remaining code for mapping to the main processor. The generation of intermediate representation (IR) is implemented with the help of an existing compiler which is used as the front end of our compiler. It parses the c code and does some analysis, optimization and transformation. The IR (CDFG), which is used as the input of the compiler back end, is derived after the analysis, optimization and transformation. The CDFG is generally a directed acyclic graph, containing the information concerning the operation nodes and the data & control dependence among them. In the next step, our compiler back end realizes the mapping of the CDFG to the reconfigurable array. The mapping flow consists of several steps, such as mapping initial and intermediate data to register files, assignment to PEs, splitting the CDFG if the array is not big enough to handle the original one, and so on. The configurable code for the loops'execution on the array is emitted by the compiler back end. Then the configurable code and the remaining c code after the HW/SW step are merged. Finally, the merged new code of the application is run on a c compiler and the machine code is generated for the reconfigurable platform.

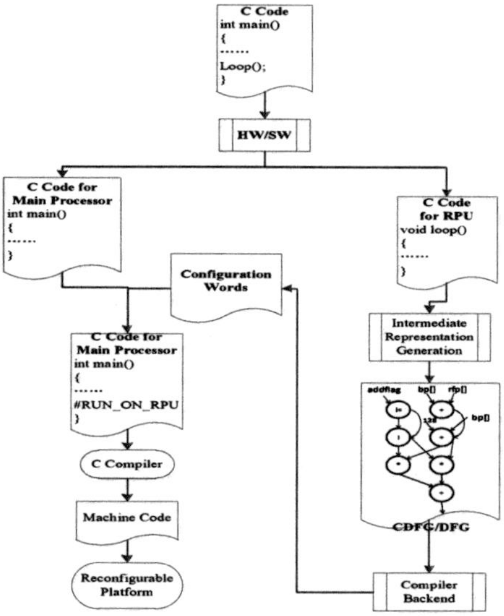

Fig. 2. Compiler design framework for CGRA

Our work in this paper focuses on the part of compiler back end in Fig.2. As mentioned above, this part contains a series of steps. We will describe it in detail in the following sections.

3 Mapping Algorithm

Mapping loops represented by CDFG in the application to CGRAs is a combination of two sub-problems: *placement* and *routing*. Placement assigns the operations to specific PEs for execution while routing connects the operations according to their data dependences. However, unlike traditional FPGA P&R algorithm, the two sub-problems are greatly associated and can't be separated independently. To solve this problem, we take an approach that assigns PEs for operations taking the routability into account simultaneously. The detailed mapping flow is realized as Fig.3.

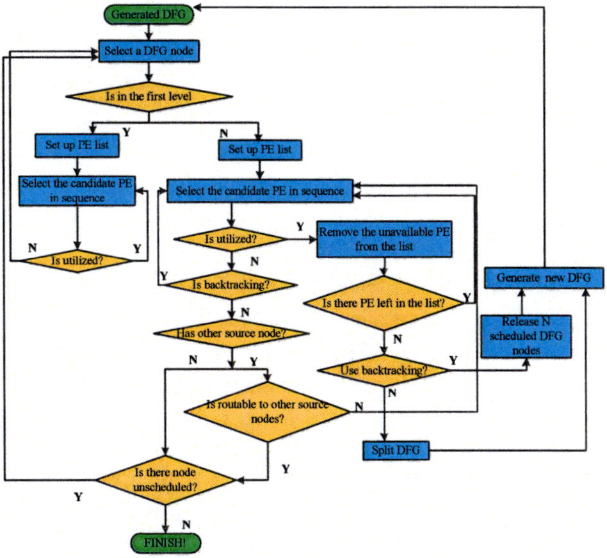

Fig. 3. The detailed mapping flow

The flow starts with the CDFG of the loop body. Next the nodes in the CDFG are mapped on the CGRA array following the mapping algorithm with some effective priority schemes. When failure occurs to find an available PE a backtracking algorithm is performed to have another try. If it is still unable to get an appropriate mapping, the CDFG will be split, leaving with a new CDFG composed of the remaining unscheduled operations. And then the new CDFG is handled in the same way with the steps described above. The iteration stops until all the CDFG nodes are mapped successfully.

3.1 Priority Scheme

■ Priority of operations

The operations are mapped to the reconfigurable array in the order based on the priority mainly considering two sides. First the height-based priority, a popular priority in acyclic list scheduling, is applied. Moreover, instead of only calculating the priority only with the height, we also take the number of successors into account. For the nodes with the same height, the ones who have more successors are given higher priority for the reason that the successors will have more opportunities to get an available PE. Take a sample CDFG shown in Fig.4 as an example. When mapping the sample CDFG onto a CGRA, the candidate operation is queued in the list {OP0, OP2, OP1, OP4, OP3, OP5, OP6 } considering the height-based and successor-based priority.

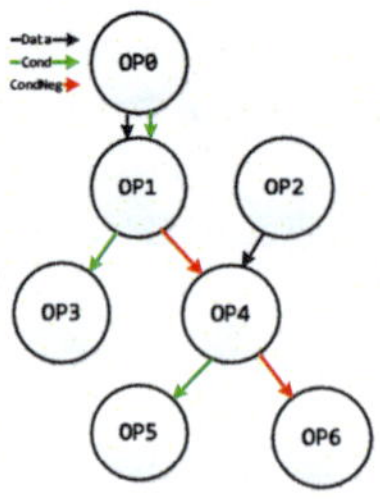

Fig. 4. A sample CDFG

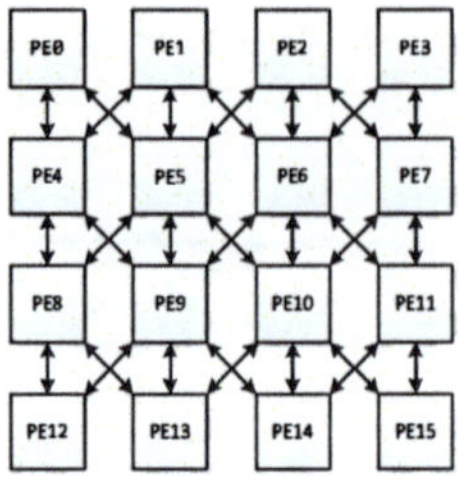

Fig. 5. A sample PE array

- Priority of PEs

As is shown in Fig.5, PEs in the center of the array have more neighbors than the ones in the edge of the array. In other words, PE_5、 PE_6、 PE_9 and PE_{10} are the most routability locations, with PE_1、 PE_2、 PE_4、 PE_7 、 PE_8、 PE_{11}、 PE_{13} and PE_{14} following , and PE_0、 PE_3、 PE_{12} and PE_{15} are the least. This means that if a node mapped to the PE with higher priority, it is more likely for its successors to find an available PE meeting with the demand for routing. When mapping for a selected operation, a list of the candidate PEs ordered in the sequence according to the priority of them. The more connected neighbors, the higher priority. Next the candidates in the list are chosen one by one to check if it is *free*. Once the first satisfying PE is found, a mapping decision can be made and it is unnecessary to go on with the remaining PEs in the list. An example is shown in Fig.6. With OP0-OP4 mapped, three available PEs are left for OP5. And then with the routability taken into consideration, OP5 is assigned to PE10 who has the maximum number of neighbors.

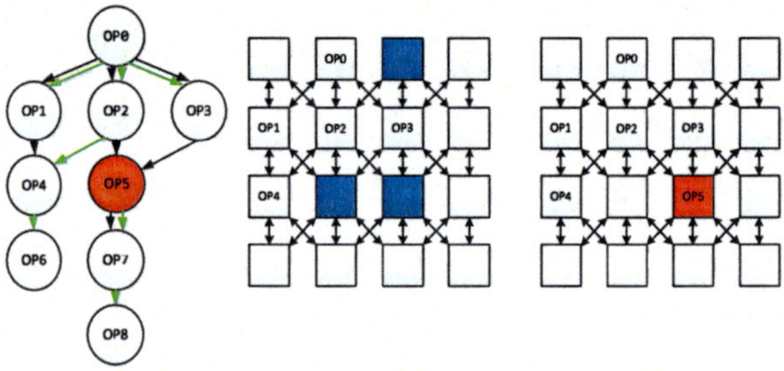

Fig. 6. A mapping example

3.2 Backtracking

Having been proven to be an NP problem [9], mapping application onto a CGRA can hardly be formulated to obtain the optimal solution. Instead, heuristic approaches are used to get the sub-optimal solution that as close as possible to the optimal one. However, heuristic solution can easily bring in the negative consequence of falling into local minima. Therefore, we use a backtracking algorithm to address this problem. When a specific operation fails to find a feasible PE to execute due to the limited hardware resources, several scheduled operations will be removed from the scheduled list and their states are reset unscheduled. Next, these operations will be attempted to be mapped to other available PEs. If the result turns out to be better, or more operations can be scheduled, the new mapping result will be accepted and replace the old one. In order to avoid the algorithm from not converging, the number of the back tracking steps is limited in a certain threshold. If back tracking steps have overcome the set threshold, the algorithm is stopped and a decision is made whether the new mapping result should be adopted. Moreover, if there still exits unscheduled nodes in the CDFG, the CDFG will be split and the mapping will be restarted with the new generated CDFG formed of the remaining unscheduled nodes.

Consider mapping the CDFG shown in Fig. 7a) to a 4×4 array shown in Fig. 7b), illustrating how the back tracking algorithm improves the mapping performance. The result is in Fig. c-d). It can be seen from Fig.7c) that without back tracking there's no more place for OP13 after mapping OP12. Fortunately, with the help of backtracking algorithm, a better solution can be acquired by moving OP11 to another PE. As a consequence, OP12, OP13and OP14 can all be able to find the available schedule. Therefore, the flexibility as well as the performance is upgraded.

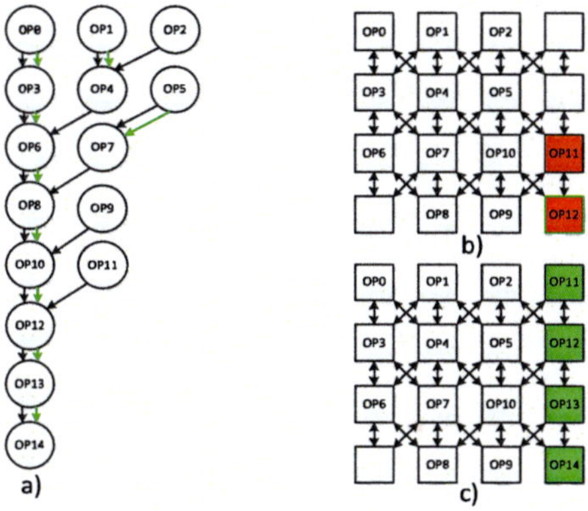

Fig. 7. a) the sample CDFG; b) mapping result without back tracking algorithm; c) mapping result with back tracking algorithm

3.3 Splitting

Due to the limited interconnections among PEs, the size of the reconfigurable array and the high complexity of memory organization, in many cases the original CDFG generated issued by the compiler front-end can hardly be mapped onto a CGRA in one time. When the situation comes up, the CDFG has to be split into several sub-CDFGs to fit the array. A series of CDFG partition algorithms for FPGA-based reconfigurable computing machines have been presented n [10], [11], [12]. However, CDFG splitting for CGRAs is much more complex and difficult. Specific say, the constraints and objective are hard to be extracted, leading it difficult to formulate the problem. The splitting algorithm we proposed in this paper can be divided into three steps. First, the operation which fails being mapped is selected. And then it will be checked to identify whether the PEs that its scheduled predecessors mapped are routable to other free neighbors. If so, a feasible cut is found, otherwise the same process will be repeated for the nodes in the last level. Finally, the corresponding output and input ports are added to the scheduled part and the remaining unscheduled part respectively. A new CDFG is created with the unscheduled nodes and several added input ports. Then the mapping flow is restarted with the new CDFG and the recursive procedure will come to an end until all the CDFG nodes are mapped successfully. The splitting method is proven to be acceptable by the experimental results.

4 Experimental Results

4.1 Experimental Setup

In order to test the proposed mapping algorithm and prove its effectiveness, the following experiments are performed. a) Compare the performance of the proposed mapping algorithm with manual mapping. b) Compare the performance with and without the proposed priority scheme. c) Compare the performance with and without the back tracking algorithm.

We consider the target CGRA consisting of 16 PEs in 4×4 array. Each PE is directly connected to its nearest 6 neighbors. What's more, the 16 PEs are Isomorphism that means all PEs are accessible to GRF and are able to execute every operation the instruction set.

In addition, some assumptions are made for the experiments: i) the PEs are homogeneous which means they are all available for all the operations; ii) there are sufficient registers and data memory; iii) the data bandwidth is not a bottleneck.

The benchmark applications used are shown in table1.

Table 1. The characters of benchmarks

Algorithm	Line of c code	Loops/array variables	Primary data size	description
Tune Equalization	5	240000/2	600×400	Image tone equalization
Edge detection	43	64/2	64×3	Image Edge Detect
FIR	12	18/1	7	Finite impulse response
IDCT	32	0	1x8	IDCT row transformation

4.2 Results

Table 2 shows the experimental results for the whole set of benchmarks. The *IPC* (instructions per cycle) referring to the average number of instructions executed in parallel in the pipelining. The *context number* is the number of contexts which is equal to the number of sub-CDFGs. The *PEs' max utilization* represents the best case among all the contexts. It can be seen from table 2 that our approach yields the results very close to the manual one.

The comparison results for the effectiveness of the proposed priority scheme and backtracking algorithm is shown in table 2. By replacing the proposed priority scheme with the traditional height-based priority for operations and original sequence for PEs, IPC is reducedas expected. Then by disable the backtracking algorithm, IPC is dropped as well. Therefore, the proposed algorithm is demonstrated to be satisfying and efficient.

Table 2. The experimental results of benchmarks

Algorithm	By compiler			By manual mapping		
	Context number	PEs' max utilization	IPC	Context number	PEs' max utilization	IPC
Tune Equalization	2	8	7	2	7	7
Edge detection	15	15	9.7	13	16	11.2
FIR	5	15	10.7	4	15	13.4
IDCT	16	16	10.8	14	16	12.4

Table 3. Comparison of IPC

Algorithm	IPC		
	with the proposed algorithm	without the proposed priority scheme	without backtracking algorithm
Tune Equalization	7	4.6	7
Edge detection	9.7	9.1	8.6
FIR	10.7	7.6	8.9
IDCT	10.8	8.6	10.4

However, as the architecture instances vary in a large range and the benchmarks different from each other a lot, mapping applications onto CGRAs becomes more challengeable and difficult. We are now continuing with the work on trying more benchmarks with more architecture instances.

5 Conclusion

In this paper, a new approach for mapping the loops is proposed. With a parametric template as the target CGRA, it can be applied to a series of various CGRAs. Our method gives higher priority to PEs having more available neighbors, or more routable. Apart from the traditional height-based priority for operations, the number of successors is taken into consideration as well. The set of the proposed priority scheme brings more flexibility for mapping. What's more, a back tracking approach is applied to get the opportunity to achieve better mapping results, playing a significant role in improving the overall performance. And the heuristic for splitting CDFGis demonstrated to be effective. The experimental results show we can achieve satisfying mapping results very close to the manual one.

References

1. Choi, K.: Coarse Grained Reconfigurable Array: Architecture and Application Mapping. IPSJ Transactions on System LSI Design Methodology 4, 31–46 (2011)
2. Mei, B., Vernalde, S., Verkest, D., Man, H.D., Lauwereins, R.: ADRES: An Architecture with Tightly Coupled VLIW Processor and Coarse-Grained Reconfigurable Matrix. In: Y. K. Cheung, P., Constantinides, G.A. (eds.) FPL 2003. LNCS, vol. 2778, pp. 61–70. Springer, Heidelberg (2003)
3. Hatanaka, A., Bagherzadeh, N.: A Modulo Scheduling Algorithm for a Coarse-Grain Reconfigurable Array Template. In: IPDPS, pp. 1–8 (2007)
4. Park, H., Fan, K., Mahlke, S.A., Oh, T., Kim, H., Kim, H.: Edge-centric modulo scheduling for coarse-grained reconfigurable architectures. In: PACT, pp. 166–176 (2008)
5. Yoon, J.W., Shrivastava, A., Park, S., Ahn, M., Jeyapaul, R., Paek, Y.: SPKM: A novel graph drawing based algorithm for application mapping onto coarse-grained reconfigurable architectures. In: ASP-DAC, pp. 776–782 (2008)
6. Ahn, M., Yoon, J.W., Paek, Y., Kim, Y., Kiemb, M., Choi, K.: A spatial mapping algorithm for heterogeneous coarse-grained reconfigurable architectures. In: DATE, pp. 363–368 (2006)
7. Lee, J., Choi, K., Dutt, N.D.: Compilation Approach for Coarse-Grained Reconfigurable Architectures. IEEE Design & Test of Computers, 26–33 (2003)
8. Bansal, N., Gupta, S., Dutt, N., Nicolau, A., Gupta, R.: Interconnect-Aware Mapping of Applications to Coarse-Grain Reconfigurable Architectures. In: Becker, J., Platzner, M., Vernalde, S. (eds.) FPL 2004. LNCS, vol. 3203, pp. 891–899. Springer, Heidelberg (2004)
9. Oliver, C., Shields, J.: Area efficient layouts of binary trees in grids. PhD thesis. Supervisor-Ivan Hal Sudborough (2001)
10. Jiang, Y., Wang, J.: Temporal Partitioning Data Flow Graphs for Dynamically Reconfigurable Computing. IEEE Trans. VLSI Syst., 1351-1361 (2007)
11. Purna, K.M.G., Bhatia, D.: Temporal Partitioning and Scheduling Data Flow Graphs for Reconfigurable Computers. IEEE Trans. Computers, 579-590 (1999)
12. Cardoso, J.M.P.: On Combining Temporal Partitioning and Sharing of Functional Units in Compilation for Reconfigurable Architectures. IEEE Trans. Computers, 1362-1375 (2003)

Solid Oxide Fuel Cell Performance with Developed FeCr Alloy Interconnect

Deni S. Khaerudini[1,2], Mohd Asharaf Othman[1], Shahruddin Mahzan[1], Daniela Fredrick[3], Tjipto Sujitno[4], Puji Untoro[4], and Darwin Sebayang[1]

[1] Faculty of Mechanical and Manufacturing Engineering, Universiti Tun Hussein Onn Malaysia, 86400 Johor, Malaysia
deni.shidqi@gmail.com, {ashraf,sharudin,darwin}@uthm.edu.my
[2] Research Center for Physics-LIPI, Puspiptek Serpong, 15314 Tangerang, Indonesia
deni008@lipi.go.id
[3] Product Developments, Thermal Technology LLC, Santa Rosa, CA 95403, USA
dfredrick@thermaltechnology.com
[4] National Nuclear Energy Agency, 15314 Tangerang, Indonesia
{sujitno,untoro}@batan.go.id

Abstract. FeCr alloy, characterized with high oxidation temperature and electrical resistance, is a potential material for solid oxide fuel cell (SOFC) interconnects. Preparation processes of FeCr alloy were optimized via high energy ball-milling (HEBM), spark plasma sintering (SPS) and lanthanum ion implantation in order to develop an efficiently specimen with low oxide scale growth and electrical resistivity. The commercially available ferritic steel was chosen as a comparison of another high-Cr ferritic alloy. They were evaluated at 900 °C in the air corresponding to the environment of SOFC. The area specific resistance (ASR) of oxide scales was measured. The results show that the developed FeCr alloy has demonstrated a low ASR value after 100 hours, indicating the long-term stability of oxide layer, compared to the commercially available alloy.

Keywords: FeCr alloy, SOFC, interconnect, oxidation, resistivity.

1 Introduction

Solid oxide fuel cell (SOFC), which has been developed within the last four decades, is a promising fuel cell among other types, such as PEMFC, DMFC, PAFC and MCFC. The SOFC may become the first commercially used fuel cell in the future because of its high-energy efficiency and long-term stability. Previous generations of SOFC technology have implemented cells produced thoroughly of ceramics components. When the SOFC operating temperature was lowered, the use of metallic interconnects proved to be encouraging, since they are less expensive and easy to produce than ceramic interconnect. Introducing ferritic alloy to the cell material system, however, present new challenges that must be overcome. Conventional alloys still cannot bear a high temperature for a long-time [1,2]. The thickness of oxide scale usually increases with increasing time, resulting in high electrical resistance and

D. Yang (Ed.): Informatics in Control, Automation and Robotics, Volume 2, LNEE 133, pp. 679–684.
springerlink.com

inferior cell performance. Therefore, the development of an implanted nanostructure Cr-based alloy is paramount in order to hinder Cr vaporization, reduce oxidation velocity, enhance its conductivity and elude the resulting degradation of cathode performance [3-6]. Current research on SOFC interconnects is focusing on these issues.

2 Experimental

Fe-20wt%Cr alloy powders were prepared via solid-state reaction of Fe and Cr powder (99.99% purity). The processes consisted of a high-energy ball milling (HEBM) stage in a Retsch PM-400 planetary ball mill at 300 rpm milling speed for 60 h under Ar atmosphere. The ball milled powder was densified by using SPS system Model 10-3 (Thermal Technology LLC, USA) at a loading pressure of 120 MPa, followed by a fast sintering process of 800 and 900 °C at a rate of 400 °C/min and maintained for 5 min in a vacuum. Then, the alloys obtained will be designed by an abbreviation; for example, the final FeCr alloy as sintered by SPS at 800 °C will be called SPS800. The commercially available ferritic alloy (containing 14.5 wt% Cr, 1.07 wt% Si, 0.51 wt% Mn as the main alloying elements) was chosen as a comparison for this study. Prior to ion implantation stage, the samples were ground and polished according to the standard metallographic procedure. The samples were then implanted with lanthanum (La) at the nominal dose of 1×10^{17} ions/cm^2 using ion implanter of Cockcroft-Walton type. The ion beam energy was maintained at 100 keV with 10 μA/cm^2 beam current density and 200 kV extraction voltages.

Oxidation test of the un-implanted and implanted samples were carried out in the stagnant air using a box furnace at 900 °C for 100 h, with a furnace temperature ramp rate of 5 °C/min. Samples were pulled of the furnace at the interval of 20 h and were weighed to monitor the mass gain using an analytical balance with an accuracy of 0.01 mg.

After the samples were oxidized in the air, the electrical resistivity of oxide scale was measured using the standard two-point electrical measurement to obtain electrical current-voltage (*I-V*) curves. The resistivity measurements were performed under the static air condition after exposing thermal cycles. The large area metal contacts as the counter electrode were prepared by sputtering deposition (JEOL JFC-1600 auto ranging coater) technique using platinum (Pt) target at 20 mA for 60 sec on the surface of samples to ensure good electrical contact between them. The data of electrical properties which presented in this study is in area specific resistance (ASR) value.

3 Results and Discussion

The primary focus of the work was to study the feasibility of using ferritic based alloy as the material interconnects. Several stages processes, nanostructured alloy (through the combination of HEBM and SPS processes) and covered with ion implantation, were evaluated to study their effect on the scale conductivity and long-term stability. The purpose of the surface treatment (ion implantation) is to achieve a dense oxide scale of uniform thickness. In SOFC application, the oxide scale is part of the electrical circuit, so its conductivity is important. The effectiveness of nanostructure, on crystallite size of specimens which calculated by the Williamson-Hall method [7], on material performance was also evaluated. The size calculation results are 17.94, 23.32 and 77.03 nm for FeCr SPS800, SPS900 and commercial alloy, respectively.

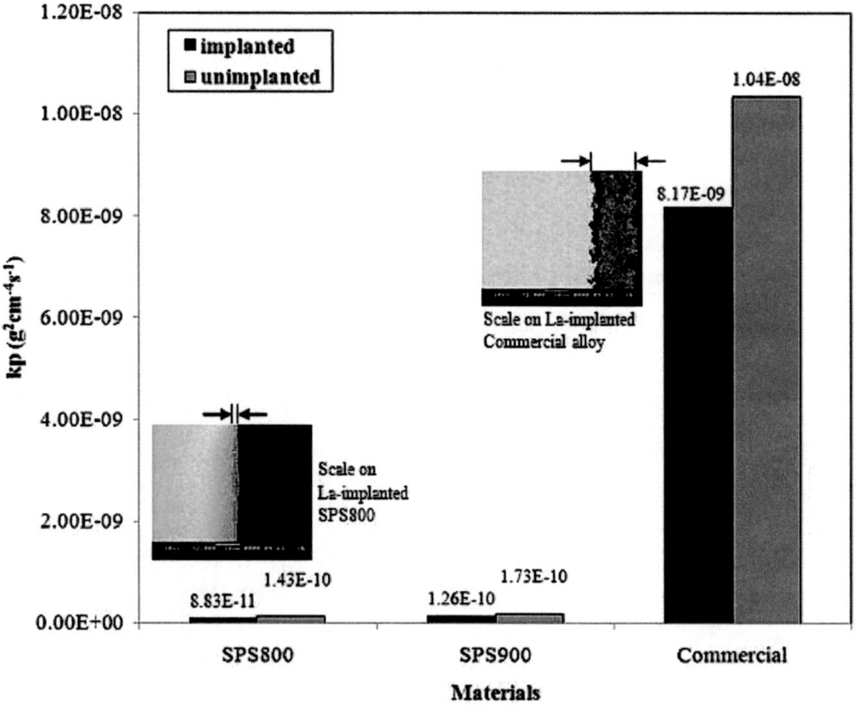

Fig. 1. Values of the parabolic rate constants (k_p), which calculated after plots the mass gain, for un-implanted and implanted samples of SPS800, SPS900 and commercial alloy after 100 h at 900 °C in air under atmospheric pressure (SEM cross section inserted)

The oxidation test of the un-implanted and implanted specimens in the air at 900 °C is shown in Fig. 1. In this study, the resulted oxidation kinetics trend is presented in the parabolic rate constant (k_p) value. The parabolic rate constant deduced from these experiments are indicated in Fig. 1. The samples were also analyzed using scanning electron microscopy (SEM) cross-section morphology for scale thickness and microstructure, presented only for the significantly different of specimens scale thickness, i.e. the SPS800 and commercial specimen. The scale thickness (inserted in Fig. 1) is consistent with parabolic rate constant (mass gain measurement), in that the sample with the thinnest scale also showed the lowest mass gain. The values of the parabolic rate constants of the implanted samples are lower than that of the un-implanted specimens which improved approximately in 30%. Meanwhile, the k_p values of the as developed FeCr alloy (SPS800 and SPS900) are significant lower than that of the commercial alloy, as also evidenced by SEM cross-section for scale thickness between the similar lanthanum ion implanted specimens, SPS800 and commercial alloy. It is indicated that in the case of material performance of this investigation, the nanostructure material has found a superior effect than the ion implantation process. The resulting nanocrystallite size of SPS800 is about 17.94, and the commercial alloy is 77.03 nm. The k_p value of both specimens also shown the appreciably improvement due to the refinement of crystallite size of a specimen which showed the difference of two orders of magnitude. However, in the case of commercial alloy, the effect of lanthanum ion implantation is more clearly visible; the k_p values of the un-implanted and implanted sample are considerably different to each other. From the oxidation studies, the performances of specimens were then further analyzed on the electrical properties.

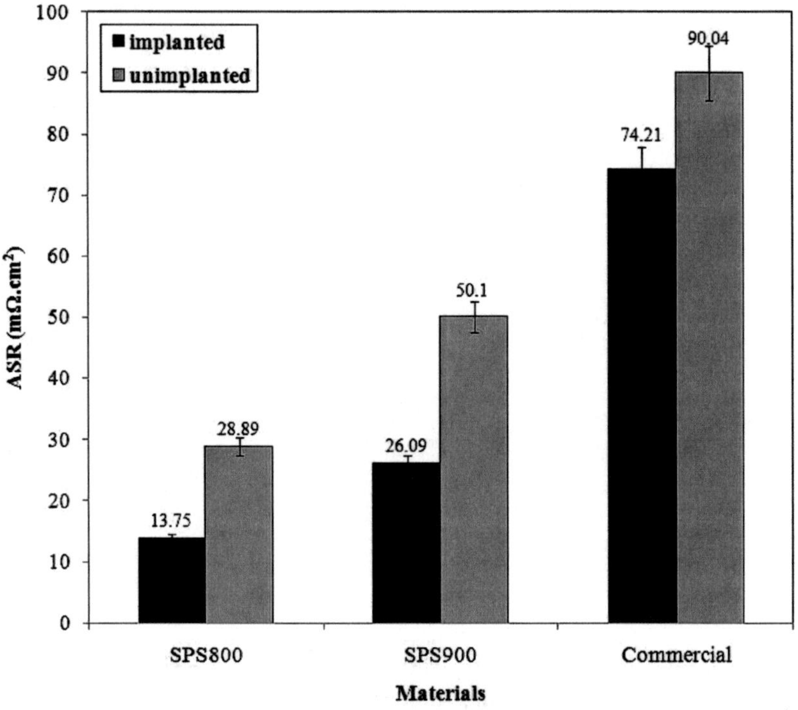

Fig. 2. Values of ASR parameter for un-implanted and implanted samples of SPS800, SPS900 and commercial alloy after 100 h at 900 °C in air under atmospheric pressure

The study of electrical properties (Fig. 2) of the oxide scales formed on the un-implanted and implanted SPS800, SPS900 and commercial alloy showed that they were well adapted for use as interconnect SOFC. After 100 h at 900 °C, the ASR values are still lower than the limit of 100 $m\Omega cm^2$ as generally reported in the literature [8]. An evaluation shows that the ASR parameter is equal to 13.75 $m\Omega cm^2$ for SPS800, 26.09 $m\Omega cm^2$ for SPS900 and 74.21 $m\Omega cm^2$ for commercial alloy; these values are for the La implanted specimens. For the un-implanted specimens, it also showed still below the ASR limit. However, it showed that the presence of the reactive elements modified and improved the electrical conductivity of the oxide scale. Generally, the electrical properties of alloys depend on the metallic substrate (nanostructure material). In this study, it had also shown depends on the reactive element. As expected, in the case SPS800 (and other specimens) the lanthanum implantation decreased the electrical resistivity of the scale compared to the un-implanted alloy. This phenomenon can be explained by the fact that, after 100 h at 900 °C, lanthanum active element reacts with the thermally grown chromia (Cr_2O_3) to form lanthanum chromite ($LaCrO_3$) upon thermal exposure. Clearly the presence of lanthanum chromite, which has a good electrical resistivity [8], explains the decrease in the ASR value which considerably improves the specimen's performance and then expected to be suitable in interconnect SOFC material application.

From these ASR values, the commercial alloy showed the highest ASR value which could be considered due to the increasingly high oxide scale formation, as shown in Fig. 1. Moreover, the decreased ASR value in SPS800 and SPS900 shown to be superior to the commercial alloy. The numbers of studies are being carried out to find the metallic interconnects, which have near or ideal the required ASR value. In this study, the interesting phenomenon due to the nanostructure specimens could be considered and distinguished, which remarkably increased with finer of nanocrystalline structure or increase in a volume fraction of the interface. Due to the large fraction of the interface, many properties of nanocrystalline materials have been found to vary significantly from those of coarse-grained materials [8]. Among these is the electrical conductivity. It is noted that the improved electrical conductivity (low ASR value) in nanostructure alloy arises mainly because of the faster diffusion of elements that can form passive oxide film and/or because of improved adhesion of passive film formed over nanostructure alloy. However, the high or low electrical resistance observed is not completely understood; nevertheless, it can be related to the scale surface. Since, it surface is non-ideally flat, as can be observed by SEM cross-section at the interface scale/alloy in Fig. 1. These observations suggest that the ASR value can, to some extent, be controlled mainly by the alloy structure and also the surface treatment, specifically the lanthanum ion implantation process in this work.

4 Conclusion

ASR value and growth rate of the developed and lanthanum implanted FeCr alloy resulted in better stability than the commercial alloy in the SOFC relevant atmosphere and temperature. The measured ASR values of the developed FeCr alloy of 13.75-50.10 $m\Omega cm^2$ in the air at 900 °C were substantially lower than the target resistance and stable over a limited number of thermal cycles exposure carried out during the project. Additional development work is needed to improve the stable phase content in the scale to enhance the further stability. Successful completion of these tasks will play a major role in enabling the SOFC system as a commercial reality.

Acknowledgment. This work was supported by the Ministry of Higher Education-Malaysia under FRGS Grant Scheme project with contract number VOT.0759.

References

1. Zhu, W.Z., Deevi, S.C.: Development of Interconnect Materials for Solid Oxide Fuel Cells. Mater. Sci. Eng. A 348, 227–243 (2003)
2. Elangovan, S., Balagopal, S., Timper, M., Bay, I., Larsen, D., Hartvigsen, J.: Evaluation of Ferritic Stainless Steel for Use as Metal Interconnects for Solid Oxide Fuel Cells. J. Mater. Eng. Perform. 13, 26–273 (2004)
3. Gannon, P.E., Tripp, C.T., Knospe, A.K., Ramana, C.V., Deibert, M., Smith, R.J., Gorokhovsky, V.I., Shutthanandan, V., Gelles, D.: High-Temperature Oxidation Resistance and Surface Electrical Conductivity of Stainless Steels with Filtrated Arc Cr-Al-N Multilayer and/or Superlattice Coatings. Surf. Coat. Technol., 55–61, 188--189 (2004)
4. Fergus, J.W.: Lanthanum Chromite-based Materials for Solid Oxide Fuel Cell Interconnects. Solid State Ionics 171, 1–15 (2004)
5. Zhu, W.Z., Deevi, S.C.: Opportunity of Metallic Interconnects for Solid Oxide Fuel Cells: A Status on Contact Resistance. Mater. Res. Bull. 38, 957–972 (2003)
6. Chen, X., Hou, P.Y., Jacobson, C.P.: Protective Coating on Stainless Steel Interconnect for SOFCs: Oxidation Kinetics and Electrical Properties. Solid State Ionics 176, 425–433 (2005)
7. Saryanto, H., Khaerudini, D.S., Untoro, P., Saleh, M.H., Sebayang, D.: Determination of Nanocrystalline Fe80Cr20 Powder based Alloys using Williamson-Hall Method. Adv. Mater. Res. 999, 129–131 (2010)
8. Quadakkers, W.J., Abellan, J.P., Shemet, V., Singheiser, L.: Metallic Interconnectors for Solid Oxide Fuel Cells - A Review. Mater. High Temp. 20, 115–127 (2003)

Morphological Edge Detection Method of Multi-structure and Multi-scale Based on Image Fusion in Wavelet Domain

Jing Liu[1,*], Xinze Luan[1], Xiaolin Tian[1], and Yankui Sun[2]

[1] Faculty of Information Technology, Macau University of Science and Technology, Macao, China
[2] Department of Computer Science and Technology, Tsinghua University, Beijing, China
am.liujing@hotmail.com

Abstract. To extract the edge information from images in a better way, a new morphological edge detection method of multi-structure and multi-scale based on image fusion in wavelet domain has been proposed. The new method merges the edge detected by multi-structuring elements and multi-scale elements in wavelet domain via wavelet transformation to these detected edge images. Experimental results have shown that more detail edges and texture information could be detected by the proposed method, comparing with results from most recently proposed similar edge detection algorithm.

Keywords: Edge detection, Wavelet transform, Image fusion, Multi-structure and multi-scale morphology.

1 Introduction

Edge is the reflection of image grayscale discontinuity. It contains a lot of information of image and reflects the nature of the object. In recent years, in order to improve the accuracy of the edge detection, and improve the performance of the edge detection, the application of mathematical morphology has being paid more and more attentions. In mathematical morphology operation, structure elements with different shape are different in aspect of denoising and maintaining the detail of images. Structure elements of different scale also have different denoising results.

Image fusion is a data fusion technology which integrates images from difference sensors for the same scene together. Because there are information redundancy and complementarily among images of the same scene obtained from different sensors, the image fusion technology could obtain more accurate info from these images of the same scene, so the image fusion technology has being now widely used in military, remote sensing, computer vision, medical image processing and other field [1-2].

A new edge detection algorithm based on multi-structure and multi-scale has been proposed in this paper. The new algorithm has mainly improved fusion method used in edge detection, which merges images in wavelet domain by wavelet transformation

* Corresponding author.

D. Yang (Ed.): Informatics in Control, Automation and Robotics, Volume 2, LNEE 133, pp. 685–691.
springerlink.com

for these images obtained by morphological edge detection of different structures and different scales. Experimental results have indicated that the new algorithm could get better denoise results and maintain more detail edge information of the image. The Experimental results also shown that the new algorithm could be used in different types of images, which means the new algorithm has better applicability too.

2 Morphological Edge Detection

2.1 Mathematical Morphology

In image processing, Mathematical Morphology can simplify image data and maintain their basic shape characteristics, it also can remove irrelevant structure of images, which is widely used in edge detection [3].Mathematical Morphology is composed by a set of Morphological algebraic operators, it has four basic operations: dilation, erosion, opening and closing.

The dilation can be obtained by:

$$A \oplus B = \{z \in E \mid (B^s)_z \cap A \neq \varnothing\} \tag{1}$$

Where B^s denotes the symmetric of *B*.

The erosion of the binary image *A* by the structuring element *B* is defined by:

$$A \Theta B = \{z \in E \mid B_z \subseteq A\} \tag{2}$$

Where B_z is the translation of *B* by the vector z .

The opening of *A* by *B* is obtained by the erosion of *A* by *B*, followed by dilation of the resulting image by *B*.

$$A \circ B = (A \Theta B) \oplus B \tag{3}$$

The closing of *A* by *B* is obtained by the dilation of *A* by *B*, followed by erosion of the resulting structure by *B*.

$$A \bullet B = (A \oplus B) \Theta B \tag{4}$$

Gray dilation, erosion, opening and closing operations will produce the following effect[4] :

1) According to the characteristics of the dilation, dilation operation can be used to eliminate dark details of image and to enhance the edge of the light area.
2) According to the characteristics of the erosion, erosion operation can be used to eliminate light details of image and to weaken the edge of the light area.
3) Gray opening operation can eliminate light details of smaller size compared with the structure elements.
4) Gray closing operation can eliminate dark details of smaller size compared with the structure elements.

2.2 The Method of Morphological Edge Detection

The traditional morphological edge detection algorithm is as follows:

Set $E(x, y)$ as edge function, $f(x, y)$ as input gray scale image, $b(x, y)$ as structure element.

Grey scale morphological dilation edge detection operation:

$$E_d(x, y) = f(x, y) \oplus b(x, y) - f(x, y) \tag{5}$$

Grey scale morphological erosion edge detection operation:

$$E_e(x, y) = f(x, y) - f(x, y) \Theta b(x, y) \tag{6}$$

Edge detector of dilation and erosion (also be known as Morphological gradient):

$$G(x, y) = \{f(x, y) \oplus b(x, y)\} - \{f(x, y) \Theta b(x, y)\} \tag{7}$$

An improved morphological edge detection method is proposed in the [5] paper: filter the image with complex mathematical morphological filter. Improve the traditional mathematical morphological filter and increase the signal to noise ratio. Carry out research on the morphological edge detection of anti-noise. Introduce multi structure element method to improve the traditional mathematical morphological edge detector and increase the positioning accuracy so that the edge can be positioned accurately. The method is as follows:

1) At first, use morphological filter for image filtering, the filtering formula is:

$$C = \frac{1}{2}[(A \bullet B) \circ B + (A \circ B) \bullet B] \tag{8}$$

2) Select the appropriate structural elements B_1, B_2, B_3 the size of structuring element selected according to need.
3) Using (multi-structuring elements) anti-noise-type mathematical morphology edge detection algorithm for edge detection, the formula is as follows:

$$D_1 = [(A \Theta B_1) \oplus B_2] \oplus B_3 - [(A \Theta B_1) \oplus B_2] \bullet B_3 \tag{9}$$

$$D_2 = [(A \oplus B_1) \Theta B_2] \circ B_3 - [(A \oplus B_1) \Theta B_2] \Theta B_3 \tag{10}$$

$$D_3 = [(A \Theta B_1) \oplus B_2] \oplus B_3 - [(A \oplus B_1) \Theta B_2] \Theta B_3 \tag{11}$$

The algorithm proposed in the paper [5] has improved the accuracy of edge detection, which can position the edge more accurately. But there are still some details from the edges of image which cannot be well extracted. To better preserve various edge details of images, this paper will propose a new algorithm for edge detection based on the paper [5]. The algorithm is as follow.

3 A New Method of Morphological Edge Detection

3.1 Morphological Edge Detection Method of Multi-structure and Multi-scale

Morphological edge detection algorithm can suppress noise well and maintain details of edge, which can considerably reduce degrees of sensitivity to noise. In order to completely keep the details of image and extract ideal image edge, we hope to use elements with various structures and various scales to extract edges of the image.

Multi-structure and multi-scale edge detection operation:

$$E_i(x,y) = f \oplus b_i - f + 0.5\begin{Bmatrix} \max\{f \oplus b_i - f, f - f\Theta b_i, G_i\} \\ -\min\{f \oplus b_i - f, f\Theta b_i, G_i\} \end{Bmatrix} \tag{12}$$

Where f is the input image, b_i is a structure element.

The selection of different structural elements will affect the results of morphological operations. Here we use a variety of structures with different scales so as to achieve different edge detection for image $E_1, E_2, \cdots\cdots, E_i$.

3.2 Image Fusion in Wavelet Domain

Wavelet transform can decompose the original image into a series of sub-images with different spatial resolution and frequency domain characteristics, which as a result fully reflects partial variation characteristics of the original image. The original image is decomposed into a series of channels, using appropriate integrating rules and integrating operators. Fuse the images by integrating their own characteristics and details within multiple disintegrating layers, channels with various scenes [6]. Extract the structure and details of the information effectively from the source image, which results in a more visually integrated image [7].

According to the paper [8], four rules of wavelet-based fusion is proposed:

Rule 1: Low-frequency coefficients using the largest absolute value, and high-frequency coefficients using the weighted average method.

Rule 2: Low-frequency coefficients using the average, while the high frequency coefficients using the largest absolute value.

Rule 3: Low-frequency coefficients using the average, while the high frequency coefficients using the regional energy method, the threshold value of 0.6.

Rule 4: Low-frequency coefficients using the average, while the high frequency coefficients using the regional energy method, the threshold value of 0.8.

We can use these rules of wavelet image fusion to deal with the images $E_1, E_2, \cdots\cdots, E_i$ to get edge of the images detected.

3.3 A New Morphological Edge Detection Method Based on Image Fusion in Wavelet Domain

According to the analysis above, a new edge detection algorithm, which is based on image fusion in the wavelet domain and multi-scale and multi-structural morphological edge detection, could be described step by step as below.

1) Use the formula (8), morphological filter, to eliminate image noise by filtering the image.
2) Select a number of suitable structure-elements b_i which have different structures and different scales. The selection of these structure-elements can be based on the characteristics of actual images.
3) Get different edge-detection images $E_1, E_2, \cdots\cdots, E_i$ by using multi-scale and multi-structural edge-detection factors (12).
4) Select appropriate rules of wavelet image fusion to deal with the images $E_1, E_2, \cdots\cdots, E_i$ to get edge of the images detected.

4 Experimental Results and Analysis

The new algorithm proposed has been tested in the MATLAB R2009b environment. The experiment has selected structures from four directions of horizontal, vertical, slant upward and slant downward, and three scales as 3*3, 5*5, 9*9 have been used for testing respectively. The testing results of the new algorithm have been shown in (c) from the fig.1-fig.3 below, and the experimental results of the algorithm of the paper [5], which is recently proposed in the area, have been shown too for the comparison purpose.

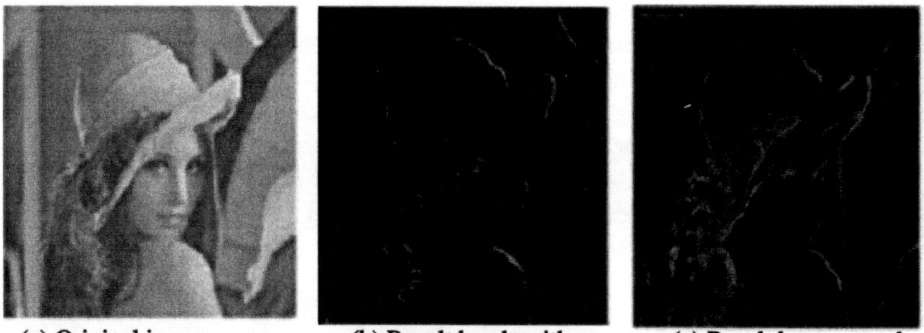

(a) Original image (b) Result by algorithm of the paper [5] (c) Result by proposed new algorithm

Fig. 1. Testing image by different algorithms

Fig.1 (a) is a common testing image in image processing. The Fig.1 (b) is the result of the paper [5]; and we could see that the peak, feather, nose, mouth and other detail edges of the image are not very clear. On the other hand, from the Fig.1 (c) which is the result of the new algorithm, everyone could see that those detail edges are much clearer than what in the Fig.1 (b).

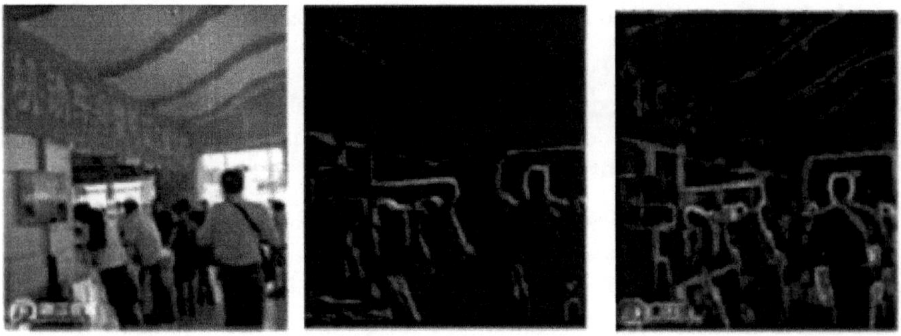

(a) Original image (b) Result by algorithm of the paper [5] (c) Result by proposed new algorithm

Fig. 2. Common image by different algorithms

Fig.2 (a) is a common image in daily life. Fig.2 (b) is the result by using the algorithm of the paper [5]; you cannot see the logo from bottom-left clearly, neither as the words on the banner in the picture. But in Fig.2 (c) which used the algorithm of this paper, all the details can be seen very clearly.

From above two types of testing results, it could be easier to see that the new edge detection algorithm of wavelet-based multi-scale multi-structure morphological edge detection is able to extract details of the edge from images clearly, and the detected edges are continuously smooth, with clearer outline and better effect. Also, the accuracy of the new algorithm is higher. The new algorithm can be applied to general images. Better detected edges indicate that the new algorithm has good versatility, which could be used in various fields.

Optical Coherence Tomography (OCT) is a high-resolution and nondestructive imagining mean with real-time imaging ability. With its special coherence technology and advantages of interferometry, people can observe tissues and structures of organisms under their surface without harming. These technologies have won great development in medical imagining field, especially in ophthalmic diagnosis [9]. In clinical medicine, the quantitative measurement of the central corneal thickness and the measurement of the corneal curvature play a very important role in the pathology diagnosis of cornea. But a preprocessing of anterior chamber OCT image, namely edge detection, is needed before the measurement.

(a) Original image (b) Result by algorithm of the paper [5] (c) Result by proposed new algorithm

Fig. 3. OCT image by different algorithms

Fig.3 (a) is a example of anterior chamber OCT images. Fig.3 (b) and Fig.3 (c) are results from edge detection which use the algorithm of the paper [5] and the one proposed in this paper respectively. It can be seen that the edge of Fig.3 (b) is apparently weaker than that in Fig.3 (c). Morphological edge detection method of multi-structure and multi-scale based on wavelet transformation image fusion can achieve more edge information and lead to more distinct edge details.

Acknowledgments. This work is supported by the Science and Technology Development Fund of Macao Special Administrative Region, China with the project number 063/2010/A. The work is also supported by the National Natural Science Foundation of China, the project number is 60971006. Many thanks to Shenzhen Moptim Imaging Technique Co. Ltd too, for providing the anterior chamber OCT image data, which have been used in testing for the new algorithm.

References

1. Liu, G.-X., Zhao, S.-G.: A multi-resolution image fusion scheme using biorthogonal wavelet transform. Opto-Electronic Engineering, 1003-501X, 04-0050-04 (2004)
2. Chao, R., Zhang, K., Li, Y.: An Image Fusion Algorithm Using Wavelet Transform. Acta Electrica Sinica 0372-2112, 05-0750-04 (2004)
3. Gonzalez, R.C., Woods, R.E.: Digital Image Processing. Electronic Industry Press (2003)
4. Xu, B.-G., Wang, J.: Adaptive algorithm of edge detection based on mathematical morphology. Journal of Computer Applications, 1001-9081, 04-0997-03 (2009)
5. Wang, G.-F., Zhan, G.-L.: Research and application of edge detection operator based on mathematical morphology. Computer Engineering and Applications 45(9), 223–226 (2009)
6. Wang, Z.-J., Zion, D.: Acomparative analysis of image fusior methods. IEEE Transaction on Geoscience and Remote Sensing 43(6), 1391–1402 (2005)
7. Wang, Z., Li, D.: Image fusion with wavelets analysis. Journal of Wuhan Technical University of Surveying and Mapping 25(2), 137–141 (2000)
8. Li, Y.-C., Wu, J.-G.: Analysis on image fusion rules based on wavelet transform. Computer Engineering and Applications 46(8) (2010)
9. Liu, X.-W., Wang, H.-N.: Denoising process of image based on wavelet transform. Acta Photonica Sinica 35(6), 935–938 (2006)

Large Time Behavior of Solutions to a Problem for Conservation Law

Huiping Cui

Basic Department, Guangdong Pharmaceutical University, Guangzhou, 510006, China
cuihuiping.x@163.com

Abstract. The generalized Benjamin-Bona-Mahony-Burgers equation is studied in this paper. When the flux function is non convex, the existence of the global solutions are proved under the different conditions by combining the local existence result with the a priori estimate.

Keywords: BBM-Burgers equation, non-convex flux function, global solution.

1 Introduction

The BBM equation which was advocated by Benjamin-Bona-Mahony[1] is used to describe the unidirectional propagation of long waves in certain nonlinear dispersive systems. As a refinement of the KdV equation, BBM equation has more expedient mathematical properties. Then, studying this kind of equations plays an important role in both theoretical analysis and practical computation. One of the equations that model unidirectional propagation of planar wave with dispersion is $u_t - \gamma u_{xxt} - \nu u_{xx} + \beta u_x + uu_x = 0$. Since the dispersive effect of this equation is the same as the BBM equation, while the dissipative effect is the same as the Burgers equation, this equation is called the BBM-Burgers equation[2].

Since 1972 the BBM equation was proposed, the periodic boundary value problem, the initial value problems and the initial-boundary value problems, for various generalized BBM-Burgers equations have been studied, cf. [3, 4]. On the other hand, many mathematicians have studied the large time behaviors of solutions to the initial value problems for various generalized BBM-Burgers equations, cf. [4, 5, 8, 9]. Under certain assumptions both L^2 and L^∞ rates of decay of the solutions to these problems were established, cf. [6, 7].

When the flux function is non-convex, the problem will become complex and interesting. Thus, this paper is devoted to proving the existence of the global solutions of the generalized BBM-Burgers equation

$$u_t + f(u)_x = u_{xx} + u_{xxt}, x > 0, t > 0 \tag{1}$$

with the following initial data and boundary data

D. Yang (Ed.): Informatics in Control, Automation and Robotics, Volume 2, LNEE 133, pp. 693–696.
springerlink.com

$$\begin{cases} u(x,t)\big|_{x=0} = u_-, t \geq 0 \\ u(x,t)\Big|_{t=0} = u_0(x) = \begin{cases} u_-, x = 0 \\ u_+, x \to \infty, \end{cases} \end{cases} \tag{2}$$

where $u_\pm$ are constant satisfies $u_- < u_+$, assumed $u(0) = u_-$. The flux function f is a general smooth function and satisfies:

$$\begin{cases} f \in C^2 \\ f(0) = f'(0) = 0, \ f''(0) > 0 \\ f(u) > 0, u \in [u_-, 0). \end{cases} \tag{3}$$

2 Main Theorem

In this section, under the conditions that $u_- < u_+ \leq 0$ and $u_- < 0 < u_+$, respectively, we prove that solutions for the initial-boundary value problem (1) exist and converge time-asymptotically to a stationary wave and the superposition of a stationary wave and a rarefaction wave.

By the condition (3) and the properties of continuous function, there exist positive constants r, τ such that

$$f''(u) \geq \tau > 0 \text{ for } |u| \leq r \text{, and } f(u) \geq \tau \text{ for } u \in [u_-, -r]. \tag{4}$$

Theorem 1. Assume $u_- < u_+ \leq 0$, $f(u_+) < f(u)$, for $u \in [u_-, u_+)$ and $v_0 \in H^2$. If there exists a positive constant η such that $\|v_0\|_{H^2}^2 + h(\tilde{u}) \leq \eta$, then the stationary problem corresponding to (1) has a unique global solution $v \in X(0,\infty)$ satisfying

$$\lim_{t\to\infty} \sup_{x \in R^+} \left| \frac{\partial^i v(x,t)}{\partial x^i} \right| = 0 \ (i = 0,1), \tag{5}$$

where $\tilde{u} = u_+ - u_-$, $h = h(\tilde{u})$ satisfying $\lim_{\tilde{u}\to 0} h(\tilde{u}) = 0$.

Let $v(x,t) = u(x,t) - \Phi(x,t) = u(x,t) - \phi(x) - \psi(x,t)$, where ψ is the smooth approximation of the rarefaction wave for (1). Then the problem (1) is reformulated in the form

$$\begin{cases} v_t + (f(\Phi+v) - f(\Phi))_x - v_{xx} - v_{xxt} = F(\phi,\psi),\ x > 0, t > 0 \\ v(0,t) = 0, t \geq 0 \\ v(x,0) = v_0(x) = u_0(x) - \phi(x) - \psi(x,0),\ x > 0. \end{cases} \tag{6}$$

where

$$F(\phi,\psi) = -(f'(\phi+\psi) - f'(\phi))\phi_x - (f'(\phi+\psi) - f'(\psi))\psi_x + \psi_{xx} + \psi_{xxt}.$$

Theorem 2. Assume $u_- < 0 < u_+ < r$ and $v_0 \in H^2$, then there exists a positive constant η such that, if $\|v_0\|_{H^2}^2 + d_0 \leq \eta$, then the problem (6) has a unique global solution $v \in X(0,\infty)$ satisfying

$$\lim_{t\to\infty} \sup_{x\in R^+} \left| \frac{\partial^i v(x,t)}{\partial x^i} \right| = 0 \ (i = 0,1), \tag{7}$$

where $d_0 = \max\{h(-u_-), h_1(u_+)\}$, and h_1 is a function of u_+ satisfying $\lim_{u_+\to 0} h_1(u_+) = 0$. The proof of Theorem 1 and Theorem 2 both combine the local existence result with the a priori estimate via L^2 weighted energy method.

References

1. Benjamin, T.B., Bona, J.L., Mahony, J.J.: Model equations for long waves in nonlinear dispersive system. Phil. Trans. R. Soc. A272, 47–78 (1972)
2. Yin, H., Zhao, H.J., Kim, J.S.: Convergence rates of solutions toward boundary layer solutions for generalized Benjamin-Bona-Mahony-Burgers equations in the half-space. J. Diff. Equ. 245, 3144–3216 (2008)
3. Liu, T.P., Matsumura, A., Nishihara, K.: Behavior of solutions for the Burgers equations with boundary corresponding to rarefaction waves. SIAM J. Math. Anal. 29, 293–308 (1998)
4. Liu, T.P., Nishihara, K.: Asymptotic behaviors for scalar viscous conservation laws with boundary effect. J. Diff. Equ. 133, 296–320 (1997)
5. Mei, M.: Large-time behavior of solution for generalized Benjamin-Bona-Mahony-Burgers equations. Nonlinear Analysis 33, 699–714 (1998)
6. Mei, M.: Lq-decay rates of solutions for Benjamin-Bona-Mahony-Burgers equations. J. Diff. Equ. 158, 314–340 (1999)

7. Zhang, L.H.: Decay of solution of generalized Benjamin-Bona-Mahony-Buegers equations in n-space dimensions. Nonlinear Analysis 25, 1343–1369 (1995)
8. Naumkin, P.I.: Large-time asymptotic of a step for the Benjamin-Bona-Mahony-Buegers equation. Proc. Roy. Soc. Edinburgh Sect. 126, 1–18 (1996)
9. Hashimoto, I., Matsumura, A.: Large-time behavior of solutions to an initial-boundary value problem on the half line for scalar viscous conservation law. Meth. Appl. Anal. 14, 45–60 (2007)

The Analytical Algorithm of Program Quaternion in the Arbitrary Spacecraft Attitude-Adjusting Control

Shaobo Ni[1], Jianwu Wu[1], Jiayuan Shan[1], Lei Liu[2,*], and Mao Su[2]

[1]School of Aerospace Engineering, Beijing Institute of technology, Beijing 100081, China,
Beijing Aerospace Automatic Control Institute, Beijing 100854, China
nsbxx@hotmail.com
[2] Key Laboratory of Education Ministry for Image Processing and Intelligent Control,
Department of Control Science & Engineering,
Huazhong University of Science and Technology, Wuhan 430074, China
lei.liu.chn@gmail.com

Abstract. The analytical algorithm of program quaternion is studied, aiming at the problem of the arbitrary spacecraft attitude-adjusting control. It also provides the analytical constructor method of the program quaternion for any given initial quaternion and angle velocity, terminal quaternion and angle velocity. The advantages of this method are as follows: (1) Good adaptability. Both of the initial and terminal conditions are arbitrary; (2) Simple model. The calculation process only involves the simple algebraic operation; (3) Easy to optimize. By using the program quaternion model, the time consumption in the process of spacecraft attitude adjustment and the maximum angular velocity can be optimized directly. According to the Mathematical simulation results, the program quaternion tracking can be realized by a simple PD control law.

Keywords: Spacecraft; Dynamic Modeling; Attitude Control; Quaternion.

1 Introduction

Aircraft often need attitude-adjusting in flight. To solve this problem, many researchers have carried out extensive researches which combined with quaternion model and proposed many control method based on Lyapunov function[1], Optimal control[2], Adaptive control[3], Feedback control[4,5] and Quaternion method respectively.

Among the researches on attitude-control based on quaternion, literature[6] considered the reorientation of spacecraft, and discusses the quaternion model of rotation from the orthogonality of the shafting. But, this quaternion model given by this paper can't optimize the consumption of time in the process of aircrafts' attitude-adjusting, meanwhile, this method can't estimate and limit the generated max angular velocity, and that, there is a non-solving singular case in this model. According to optimize the consumption of time in the process of attitude-adjusting, literature[7] gave a program quaternion calculation method which based on the bang-bang control strategy if the connection caused by the unequal moments of inertia in each channel were neglected.

* Corresponding author.

D. Yang (Ed.): Informatics in Control, Automation and Robotics, Volume 2, LNEE 133, pp. 697–704.
springerlink.com

Aiming at the situation of study of the attitude control method which based on program quaternion, this paper will study an analytical algorithm of program quaternion in the arbitrary situation. For any given initial quaternion, angular velocity, terminal quaternion and angular velocity, this paper will give an analytical algorithm of program quaternion based on construction in sections and use the quaternion model to optimize the consumption time and the max angular velocity in the process of attitude-adjusting. On this basis, this paper will validate the traceability of quaternion by mathematical simulation.

2 Dynamic Model

In spacecraft coordinate system $O_1x_1y_1z_1$, the kinetic equation which describe the attitude motion of spacecraft is:

$$\frac{d}{dt}(\mathbf{J}\boldsymbol{\omega})+\boldsymbol{\omega}\times(\mathbf{J}\boldsymbol{\omega})=\mathbf{M}, \tag{1}$$

which $\omega=[\omega_{x1}, \omega_{y1}, \omega_{z1}]^T$ is the angular velocity of the spacecraft, J=diag$\{J_{x1},J_{y1},J_{z1}\}$ is the rotational inertia matrix of the spacecraft, M=$[M_{x1}, M_{y1}, M_{z1}]^T$ is the controlling moment which act on the spacecraft, according to using Euler angle describe spacecraft attitude would appear singular phenomenon, this article use unit quaternion to describe the attitude of spacecrafts.

When using unit quaternion $q=q_0+q_1k_1+q_2k_2+q_3k_3$ (k_1,k_2 and k_3 are the three imaginary unit of the quaternion) to describe the attitude of spacecrafts, the kinematical equation of attitude is:

$$\frac{d}{dt}\begin{bmatrix}q_0\\q_1\\q_2\\q_3\end{bmatrix}=\frac{1}{2}\begin{bmatrix}0&-\omega_{x1}&-\omega_{y1}&-\omega_{z1}\\\omega_{x1}&0&\omega_{z1}&-\omega_{y1}\\\omega_{y1}&-\omega_{z1}&0&\omega_{x1}\\\omega_{z1}&\omega_{y1}&-\omega_{x1}&0\end{bmatrix}\begin{bmatrix}q_0\\q_1\\q_2\\q_3\end{bmatrix}, \tag{2}$$

which, the conversion relations between quaternion q and Euler angles as follows:

$$\tan\varphi=\frac{2(q_1q_2+q_0q_3)}{q_0^2+q_1^2-q_2^2-q_3^2},\quad \sin\psi=2(q_0q_2-q_1q_3),\quad \tan\gamma=\frac{2(q_0q_1+q_2q_3)}{q_0^2-q_1^2-q_2^2+q_3^2}. \tag{3}$$

Letbe $p(t)=\tilde{q}\circ q(t)$, which, "∘" is the product sign in quaternion, $\tilde{q}$ is the arbitrary unit quaternion, that , it is proved that $p(t)$ is also unit quaternion, and satisfy that:

$$\frac{d}{dt}\begin{bmatrix}p_0\\p_1\\p_2\\p_3\end{bmatrix}=\frac{1}{2}\begin{bmatrix}0&-\omega_{x1}&-\omega_{y1}&-\omega_{z1}\\\omega_{x1}&0&\omega_{z1}&-\omega_{y1}\\\omega_{y1}&-\omega_{z1}&0&\omega_{x1}\\\omega_{z1}&\omega_{y1}&-\omega_{x1}&0\end{bmatrix}\begin{bmatrix}p_0\\p_1\\p_2\\p_3\end{bmatrix}. \tag{4}$$

According to the spacecraft attitude-control system represented by equation(1) and equation(2), this paper will make analytic solution for program quaternion $q_{cx}(t)$, program angular velocity $\omega_{cx}(t)$ and program control moment $M_{cx}(t)$ and that on this basis, designed a control law to make the spacecraft to move from the initial state $q(0)=q_{\text{initial}}$, $\omega(0)=\omega_{\text{initial}}$ and $\dot{\omega}(0)=0$ to the expected terminal state $q(t_f)=q_{\text{final}}$, $\omega(t_f)=\omega_{\text{final}}$ and $\dot{\omega}(t_f)=0$.

3 Program Quaternion

A. The construction of program quaternion

Combined with the character of quaternion, it is obtained respectively from equation(1) and equation(2) that:

$$\begin{cases}(M_{cx})_{x1}=J_{x1}(\dot{\omega}_{cx})_{x1}-\left(J_{y1}-J_{z1}\right)(\omega_{cx})_{y1}(\omega_{cx})_{z1}\\(M_{cx})_{y1}=J_{y1}(\dot{\omega}_{cx})_{y1}-\left(J_{z1}-J_{x1}\right)(\omega_{cx})_{z1}(\omega_{cx})_{x1},\\(M_{cx})_{z1}=J_{z1}(\dot{\omega}_{cx})_{z1}-\left(J_{x1}-J_{y1}\right)(\omega_{cx})_{x1}(\omega_{cx})_{y1}\end{cases} \tag{5}$$

$$\begin{bmatrix}(\omega_{cx})_{x1}\\(\omega_{cx})_{y1}\\(\omega_{cx})_{z1}\end{bmatrix}=2\begin{bmatrix}-(q_{cx})_1 & (q_{cx})_0 & (q_{cx})_3 & -(q_{cx})_2\\-(q_{cx})_2 & -(q_{cx})_3 & (q_{cx})_0 & (q_{cx})_1\\-(q_{cx})_3 & (q_{cx})_2 & -(q_{cx})_1 & -(q_{cx})_0\end{bmatrix}\begin{bmatrix}(\dot{q}_{cx})_0\\(\dot{q}_{cx})_1\\(\dot{q}_{cx})_2\\(\dot{q}_{cx})_3\end{bmatrix}. \tag{6}$$

Therefore, if the analytical solution of the program quaternion $q_{cx}(t)$ is obtained, the analytical solution of the program angular velocity and control moment can be directly calculated by equation(5) and equation(6).

Then, we analyzed the sufficient conditions which make quaternion $q_{cx}(t)$ meeting the terminal constraints. Because $q_{cx}(t)$ is the unit quaternion, so it can be represented as follows:

$$q_{cx}(t)=\tilde{q}\circ\left[\cos(\Omega(t))+\sin(\Omega(t))\sum_{i=1}^{3}C_i(t)\hat{k}_i\right], \tag{7}$$

which, the constant $\tilde{q}$ is the unit quaternion, $C(t)=[C_1(t),C_2(t),C_3(t)]^T$ meets $C(t)^TC(t)=1$.

Vector $C(t)$ corresponds to the direction of the spacecraft rotational Euler axis. Substituting Equation(7) into Equation(6), we can obtain:

$$\begin{bmatrix}(\omega_{cx})_{x1}\\(\omega_{cx})_{y1}\\(\omega_{cx})_{z1}\end{bmatrix}=\begin{bmatrix}C_1(t)\dot{\Omega}(t) & \dot{C}_1(t) & \dot{C}_2(t)C_3(t)-C_2(t)\dot{C}_3(t)\\C_2(t)\dot{\Omega}(t) & \dot{C}_2(t) & \dot{C}_3(t)C_1(t)-C_3(t)\dot{C}_1(t)\\C_3(t)\dot{\Omega}(t) & \dot{C}_3(t) & \dot{C}_3(t)C_1(t)-C_3(t)\dot{C}_1(t)\end{bmatrix}\begin{bmatrix}1\\\sin(\Omega(t))\cos(\Omega(t))\\\sin^2(\Omega(t))\end{bmatrix} \tag{8}$$

According to equation (7) and equation(8), there is a conclusion as follows:

For arbitrary given q_{initial}, ω_{initial}, q_{final} and ω_{final}:

(i) If $\tilde{q}=q_{\text{initial}}$, $\Omega(0)=2k\pi(k=0,\pm1,\pm2,\ldots)$, $C(0)\dot{\Omega}(0)=\omega_{\text{initial}}$, $C(t_f)\dot{\Omega}(t_f)=0$, $\Omega(t_f)=2l\pi(l=0,\pm1,\ \pm2,\ldots)$, then equation (7) and equation (8) satisfies $q_{cx}(0)=q_x(t_f)=q_{\text{initial}}$, $\omega_{cx}(t)=\omega_{\text{initial}}$, $\omega_{cx}(t_f)=0$;

(ii) If $\tilde{q}=q_{\text{initial}}$, $\Omega(0)=2k\pi(k=0,\pm1,\pm2,\ldots)$, $\cos(\Omega(t_f))=(q_{\text{initial}}^{-1}\circ q_{\text{final}})_0$, $C_i\sin(\Omega(t_f))=(q_{\text{initial}}^{-1}\circ q_{\text{final}})_i$ $(i=1,2,3)$, $,\dot{\Omega}(0)=\dot{\Omega}(t_f)=0$, $\dot{C}(t_f)=0$, then equation (7) and equation(8) satisfies $q_{cx}(0)=q_{\text{initial}}$, $q_x(t_f)=q_{\text{final}}$, $\omega_{cx}(t)=\omega_{cx}(t_f)=0$;

(iii) If $\tilde{q}=q_{\text{initial}}$, $\Omega(0)=2k\pi(k=0,\pm1,\pm2,\ldots)$, $C(0)\dot{\Omega}(0)=0$, $C(t_f)\dot{\Omega}(t_f)=\omega_{\text{final}}$, $\Omega(t_f)=2l\pi(l=0,\pm1,\pm2,\ldots)$, then equation (7) and equation(8) satisfies $q_{cx}(0)=q_x(t_f)=q_{\text{final}}$, $\omega_{cx}(t)=0$, $\omega_{cx}(t_f)=\omega_{\text{final}}$;

(iv) For any situation in(i), (ii) and (iii), if it also satisfies $\dot{\mathrm{C}}(0)=\ddot{\mathrm{C}}(0)=0$, $\dot{\mathrm{C}}(t_f)=\ddot{\mathrm{C}}(t_f)=0$, $\ddot{\Omega}(0)=0$, $\ddot{\Omega}(t_f)=0$, then $\dot{\omega}_{cx}(0)=\dot{\omega}_{cx}(t_f)=0$.

For any situation in(i), (ii) and (iii), if it also satisfies $\dot{\mathrm{C}}(0)=\ddot{\mathrm{C}}(0)=0$, $\dot{\mathrm{C}}(t_f)=\ddot{\mathrm{C}}(t_f)=0$, $\ddot{\Omega}(0)=0$, $\ddot{\Omega}(t_f)=0$, then $\dot{\omega}_{cx}(0)=\dot{\omega}_{cx}(t_f)=0$.

The above conclusions indicate that, the program quaternion can be piecewise constructed, that , make the system represented by equation(1) and equation(2) to move from the initial state $q(0)=q_{\mathrm{initial}}$, $\omega_{cx}(0)=\omega_{\mathrm{initial}}$to the first intermediate state $q_{cx}(t_{m1})=q_{m1}$, $\omega_{cx}(t_{m1})=\omega_{m1}$, where $q_{m1}=q_{\mathrm{initial}}$, $\omega_{cx}(0)=\omega_{\mathrm{initial}}$, this process is called angular velocity zero-adjusting segment (if$\omega_{\mathrm{initial}}=0$, omit this process); then make the system move to the second intermediate state $q_{cx}(t_{m2})=q_{m2}$, $\omega_{cx}(t_{m2})=\omega_{m2}$, where, $q_{m2}=q_{\mathrm{final}}$, $\omega_{m2}=0$, this process is called quaternion-adjusting segment(if $q_{m1}=q_{m2}$, omit this process); finally, make the system to move to the expected target state $q_{cx}(t_f)=q_{\mathrm{final}}$, $\omega_{cx}(t_f)=\omega_{\mathrm{final}}$,this process is called angular velocity adjusting segment(if $\omega_{\mathrm{final}}=0$, omit this process).

Based on the above analysis, the program quaternion $q_{cx}(t)$is constructed as follows:

(1) Angular velocity zero-adjusting segment ($t\in[0,t_{m1}]$, which $t_{m1}=\pi/|a|$)

$$q_{cx}(t)=q_{\mathrm{initial}}\circ\{\cos\left[2n(\sin(at)+at)\right]+\sin\left[2n(\sin(at)+at)\right]\sum_{i=1}^{3}A_i\hat{k}_i\}\tag{9}$$

which , n is an arbitrary nonzero integer,

$$a=\frac{\sqrt{(\omega_{\mathrm{initial}})_{x1}{}^2+(\omega_{\mathrm{initial}})_{y1}{}^2+(\omega_{\mathrm{initial}})_{z1}{}^2}}{8n}\tag{10}$$

$$\begin{bmatrix}A_1\\A_2\\A_3\end{bmatrix}=\frac{1}{\sqrt{(\omega_{\mathrm{initial}})_{x1}{}^2+(\omega_{\mathrm{initial}})_{y1}{}^2+(\omega_{\mathrm{initial}})_{z1}{}^2}}\begin{bmatrix}(\omega_{\mathrm{initial}})_{x1}\\(\omega_{\mathrm{initial}})_{y1}\\(\omega_{\mathrm{initial}})_{z1}\end{bmatrix}\tag{11}$$

(2) Quaternion adjusting segment ($t\in[t_{m1},t_{m2}]$, which $t_{m2}=\pi/|a|+2\pi/|b|$)

$$q_{cx}(t)=q_{\mathrm{m1}}\circ\left\{\cos\left[2m\left(\sin(bt-bt_{m1}-\pi)+(bt-bt_{m1})\right)\right]+\sin[2m(2m\left(\sin(bt-bt_{m1}-\pi)+(bt-bt_{m1})\right))]\sum_{i=1}^{3}B_i\hat{k}_i\right\}\tag{12}$$

which, b is the arbitrary nonzero real number,

$$m=\frac{1}{4\pi}\arccos\left((q_{m1}{}^{-1}\circ q_{m2})_0\right),\tag{13}$$

$$\begin{bmatrix}B_1\\B_2\\B_3\end{bmatrix}=\frac{1}{\sin(4m\pi)}\begin{bmatrix}(q_{\mathrm{m1}}{}^{-1}\circ q_{m2})_1\\(q_{m1}{}^{-1}\circ q_{m2})_2\\(q_{m1}{}^{-1}\circ q_{m2})_3\end{bmatrix}.\tag{14}$$

(3) Angular velocity adjusting segment ($t \in [t_{m2}, t_f]$, which $t_f = \pi/|a| + 2\pi/|b| + \pi/|c|$)

$$q_{cx}(t) = q_{\text{final}} \circ \left\{ \cos\left[2l\left(\sin(ct - ct_{m2} - \pi) + (ct - ct_{m2} - \pi)\right)\right] + \sin\left[2l\left(\sin(ct - ct_{m2} - \pi) + (ct - ct_{m2} - \pi)\right)\right] \sum_{i=1}^{3} C_i \hat{k}_i \right\} \tag{15}$$

which, l is an arbitrary nonzero in integer,

$$c = \frac{\sqrt{(\omega_{\text{final}})_{x1}^2 + (\omega_{\text{final}})_{y1}^2 + (\omega_{\text{final}})_{z1}^2}}{8l}, \tag{16}$$

$$\begin{bmatrix} C_1 \\ C_2 \\ C_3 \end{bmatrix} = \frac{1}{\sqrt{(\omega_{\text{final}})_{x1}^2 + (\omega_{\text{final}})_{y1}^2 + (\omega_{\text{final}})_{z1}^2}} \begin{bmatrix} (\omega_{\text{final}})_{x1} \\ (\omega_{\text{final}})_{y1} \\ (\omega_{\text{final}})_{z1} \end{bmatrix}; \tag{17}$$

substitute equation(9) and equation(12) into equation(6) respectively and combine equation(5), the analytical expressions of $\omega_{cx}(t)$ and $M_{cx}(t)$ are as follows:

(1) If $t \in [0, t_{m1}]$

$$\begin{cases} (\omega_{cx})_{x1} = 4naA_1(\cos(at)+1) \\ (\omega_{cx})_{y1} = 4naA_2(\cos(at)+1), \\ (\omega_{cx})_{z1} = 4naA_3(\cos(at)+1) \end{cases} \tag{18}$$

$$\begin{cases} (M_{cx})_{x1} = -4na^2A_1J_{x1}\sin(at) - 16n^2a^2A_2A_3(J_{y1}-J_{z1})(\cos(at)+1)^2 \\ (M_{cx})_{y1} = -4na^2A_2J_{y1}\sin(at) - 16n^2a^2A_3A_1(J_{z1}-J_{x1})(\cos(at)+1)^2; \\ (M_{cx})_{z1} = -4na^2A_3J_{z1}\sin(at) - 16n^2a^2A_1A_2(J_{x1}-J_{y1})(\cos(at)+1)^2 \end{cases} \tag{19}$$

(2) If $t \in [t_{m1}, t_{m2}]$

$$\begin{cases} (\omega_{cx})_{x1} = 4mbB_1(\cos(bt - bt_{m1} - \pi)+1) \\ (\omega_{cx})_{y1} = 4mbB_2(\cos(bt - bt_{m1} - \pi)+1), \\ (\omega_{cx})_{z1} = 4mbB_3(\cos(bt - bt_{m1} - \pi)+1) \end{cases} \tag{20}$$

$$\begin{cases} (M_{cx})_{x1} = -4mb^2B_1J_{x1}\sin(bt - bt_{m1} - \pi) - 16m^2b^2B_2B_3(J_{y1}-J_{z1})(\cos(bt - bt_{m1} - \pi)+1)^2 \\ (M_{cx})_{y1} = -4mb^2B_2J_{y1}\sin(bt - bt_{m1} - \pi) - 16m^2b^2B_3B_1(J_{z1}-J_{x1})(\cos(bt - bt_{m1} - \pi)+1)^2; \\ (M_{cx})_{z1} = -4mb^2B_3J_{z1}\sin(bt - bt_{m1} - \pi) - 16m^2b^2B_1B_2(J_{x1}-J_{y1})(\cos(bt - bt_{m1} - \pi)+1)^2 \end{cases} \tag{21}$$

(3) If $t \in [t_{m2}, t_f]$

$$\begin{cases} (\omega_{cx})_{x1} = 4lcC_1(\cos(ct - ct_{m2} - \pi)+1) \\ (\omega_{cx})_{y1} = 4lcC_2(\cos(ct - ct_{m2} - \pi)+1), \\ (\omega_{cx})_{z1} = 4lcC_3(\cos(ct - ct_{m2} - \pi)+1) \end{cases} \tag{22}$$

$$\begin{cases} (M_{cx})_{x1} = -4lc^2C_1J_{x1}\sin(ct - ct_{m2} - \pi) - 16l^2c^2C_2C_3(J_{y1}-J_{z1})(\cos(ct - ct_{m2} - \pi)+1)^2 \\ (M_{cx})_{y1} = -4lc^2C_2J_{y1}\sin(ct - ct_{m2} - \pi) - 16l^2c^2C_3C_1(J_{z1}-J_{x1})(\cos(ct - ct_{m2} - \pi)+1)^2. \\ (M_{cx})_{z1} = -4lc^2C_3J_{z1}\sin(ct - ct_{m2} - \pi) - 16l^2c^2C_1C_2(J_{x1}-J_{y1})(\cos(ct - ct_{m2} - \pi)+1)^2 \end{cases} \tag{23}$$

B. Optimization for program quaternion

From equation(10) and equation(16), the consumption time of the angular velocity zero-adjusting segment $t_1 = \pi/|a|$ and the angular velocity adjusting segment $\bar{t}_1 = \pi/|c|$

are monotone increasing with the absolute value increasing of n and l respectively, because n and l are nonzero integer, $t_1+\bar{t}_1$ gets the minimal value if $n^2=l^2=1$. For $\lim_{n\to 0} t_1 = \lim_{m\to 0} \bar{t}_1 = 0$, in order to reduce the consumption time of the angular velocity zero-adjusting segment and the angular velocity adjusting segment, n and l can be decimal fraction according to the allowable range of the amplitude of control moment. But it need to notice that, now, the intermediate quaternion is no longer $q_{m1}=q_{\text{initial}}$ and $q_{m2}=q_{\text{final}}$, need to be modified respectively as follows:

$$q_{m1} = q_{\text{initial}} \circ \left[\cos(2|n|\pi)+\sin(2|n|\pi)\sum_{i=1}^{3} A_i\hat{k}_i\right], \tag{24}$$

and

$$q_{m2} = q_{\text{final}} \circ \left[\cos(2|l|\pi)-\sin(2|l|\pi)\sum_{i=1}^{3} C_i\hat{k}_i\right]. \tag{25}$$

And because the consumption time of the attitude-adjusting segment $t_2= 2\pi/|b|$ only depends on b, so if the control moment is big enough, we can always construct the program quaternion which the consumption time of spacecraft attitude-control.

From equation(18), the angular velocity zero-adjusting segment satisfied $|(\omega_{cx})_j|\leq|(\omega_{\text{initial}})_j|$ $(i=x_1,y_1,z_1)$;From equation(22), the angular velocity-adjusting segment satisfied $|(\omega_{cx})_j|\leq|(\omega_{\text{final}})_j|$ $(i=x_1,y_1,z_1)$. From equation(20), the max value in the quaternion-adjusting segment relates to b, the limit of the max angular velocity in the whole attitude-adjusting process can be realized by adjusting b. It need to pay attention that, with the increasing of, the max value of the angular velocity in the quaternion adjusting segment monotonically increases, but the consumption time monotonically decreases, so in the optimization process, it need to trade off both the indexes.

4 Mathematical Simulation

A. *Simulation conditions*

The deviation of the quaternion in the attitude-adjusting process is

$$\Delta q(t) = q(t)^{-1} \circ q_{cx}(t), \tag{26}$$

the deviation of the angular velocity is:

$$\Delta\omega(t) = \omega_{cx}(t) - \omega(t), \tag{27}$$

the selected PD control law is:

$$\begin{cases} M_{x1}=k_{q1}\Delta q_1+k_{\omega 1}\Delta\omega_{x1} \\ M_{y1}=k_{q2}\Delta q_2+k_{\omega 2}\Delta\omega_{y1}, \\ M_{z1}=k_{q3}\Delta q_3+k_{\omega 3}\Delta\omega_{z1} \end{cases} \tag{28}$$

which, the gain coefficients satisfy $k_{q1},k_{q2},k_{q3},k_{\omega 1}, k_{\omega 2}, k_{\omega 3}>0$.

Consider the moments of inertia of the three axial directions in the spacecraft respectively: $J_{x1}=350Kg{\cdot}m^2$, $J_{y1}=1000Kg{\cdot}m^2$, $J_{z1}=1200Kg{\cdot}m^2$.

Assume the initial values of the attitude are: $q_{initial}$=0.8365-0.1294k_1+0.4830k_2+0.2241k_3, $\omega_{initial}$=$[0.1,-0.3,0.2]^T$ rad·s^{-1} the target values are: q_{final}=0.4830+0.2241k_1+0.8365k_2-0.1294k_3, $\omega_{initial}$=$[-0.2,0.1,0.3]^T$ rad·s^{-1}

The selected parameters in the simulation process are n=l=0.05 and b=0.8, the gain coefficients in feedback control law(28) are selected as: k_{q1}=k_{q2}=k_{q3}=20000 and $k_{\omega1}$=$k_{\omega2}$=$k_{\omega3}$=15000

B. Simulation results

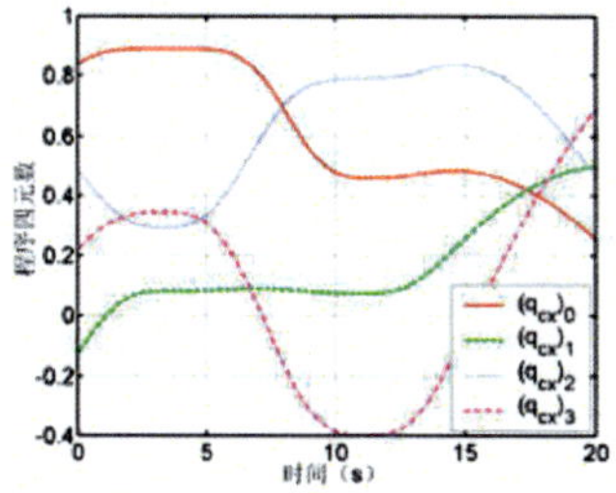

Fig. 1. Program quaternion $q_{cx}(t)$

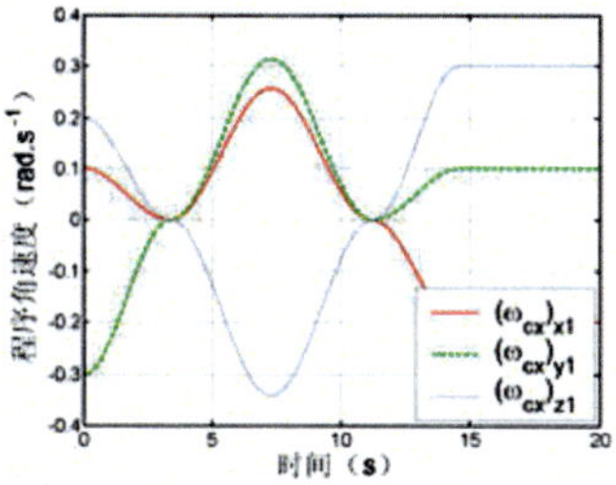

Fig. 2. Program angular velocity $\omega_{cx}(t)$

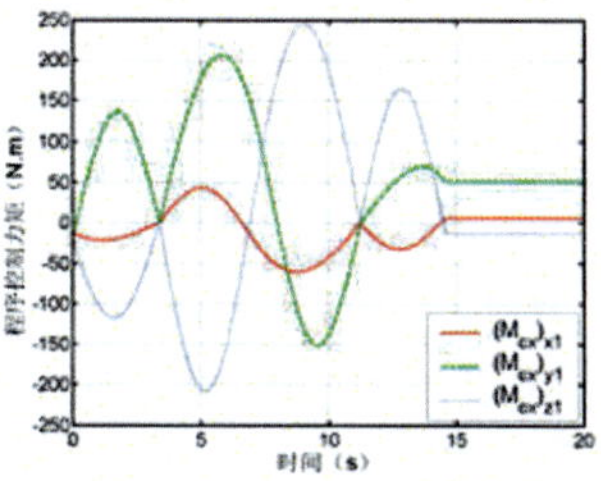

Fig. 3. Program control moment $M_{cx}(t)$

According to the equation(9) to equation(23), the program quaternion, program angular velocity and program control moment, the corresponding variation curve shows in Figure 1 to Figure 3.

The variation curve of the practical quaternion, angular velocity, control moment and attitude angles shows in Figure 4 to Figure 6. Comparing Figure 1, Figure 2, Figure 4 and Figure 5, it shows that : in the attitude-adjusting process, the effect of tracking the program quaternion and the program angular velocity is good. And it reached the attitude control goal at t_f=14.5710.

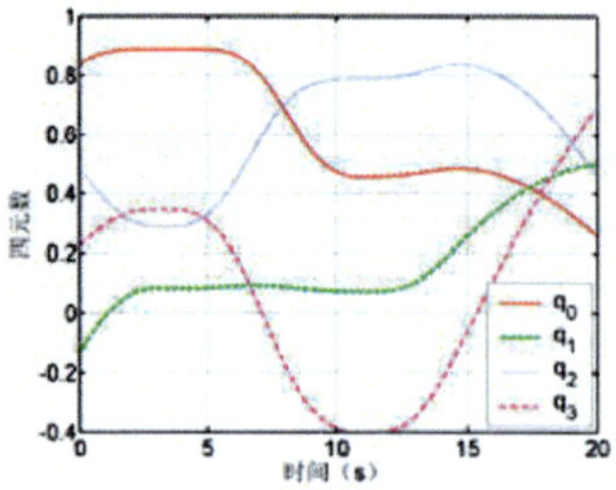

Fig. 4. Quaternion $q(t)$

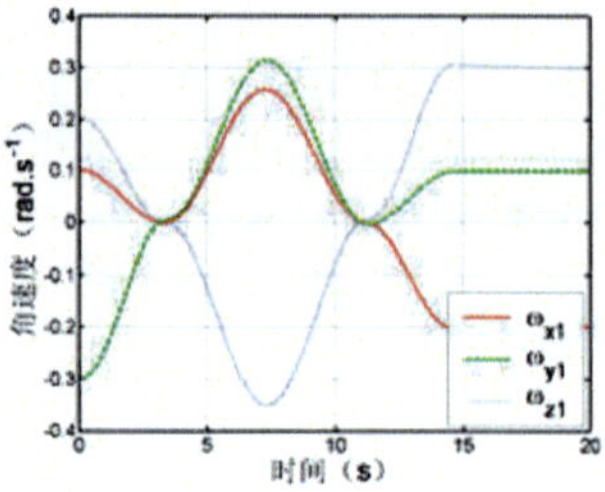

Fig. 5. Angular velocity $\omega(t)$

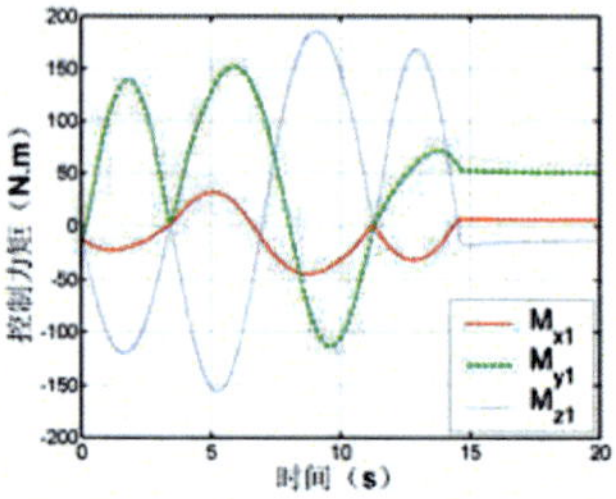

Fig. 6. Control moment $M(t)$

5 Conclusion

Aiming at the arbitrary attitude control of spacecraft , this passage studies the analytical algorithm of program quaternion. For any given initial state $q(0)=q_{initial}$, $\omega(0)=\omega_{initial}$ and $\dot{\omega}(0)=0$,the expected terminal state $q(t_f)=q_{final}$, $\omega(t_f)=\omega_{final}$, $\dot{\omega}(t_f)=0$, it provides the analytical constructor method of the program quaternion and in the basis, the time consumption in the process of spacecraft attitude adjustment and the maximum angle velocity can be optimized directly. According to the Mathematical simulation results, the program quaternion tracking can be realized by a simple PD control law.

References

1. Fragopoulos, D., Innocenti, M.: Stability considerations in quaternion attitude control using discontinuous Lyapunov functions. IEEE Proceeding of Control Theory and Applications 151(3), 253–258 (2004)
2. Chowdhry, R.S., Cliff, E.M., Lutzet, F.H.: Optimal Rigid-Body Motions. Journal of Guidance 14(2), 383–390 (1991)
3. Hong, H., Queiroz, M.S., Kapila, V.: Adaptive tracking control using synthesiz velocity from attitude measure. Automatica 37(6), 947–953 (2001)
4. Jensen, B.H.: Wisniewski R.: Quaternion Feedback Control for Rigid-body Spacecraft. AIAA 2001-4338, pp. 1-6 (2001)
5. Oh, H.S., Yoon, Y.D., Chang, Y.K., et al.: Near-Eigenaxis Rotation Cobtrol Law Design for Moving-to-Rest Maneuver. Journal of Guidance, Control and Dynamics 24, 1288-123 (2001)
6. Liu, H., Usbinski, U.B.: Approach of Orientation of Spacecraft on the Basis of Quaternary Models of Rotation of Rigid Body. Aerospace Control 2, 22–27 (2000)

A Self-adaptive Fault Monitoring Method Based on EMD

Zhao Zhenhao and Sun Yongjin

China Aerospace Science and Technology Corporation,
No.104 YouYi Road, Haidian District, Beijing, China
zhenhao@126.com, sunyongjin_73@126.com

Abstract. The effect of traditional fault monitoring methods was usually affected by the variations of equipment's working condition. In view of this situation, a self-adaptive fault monitoring method by using empirical mode decomposition (EMD) and the energy comparison of different order intrinsic mode functions (IMF) was presented. The method can avoid the negative impact of equipment state changing and human intervention, has strong adaptability, and is proved by simulation.

Keywords: EMD; IMF; Energy Comparison; adaptability.

Background

In the traditional fault monitoring methods, designing the filter of some system signal or setting the threshold of equipment parameters is the common approach. By observing the output of the filter or whether the variation of these feature parameters is beyond the threshold, we can judge the system condition. Normally the methods are effective, especially when the system state is stable. But because rely on the human intervention for presetting the filter's bandwidth or the threshold value, the effect of traditional fault monitoring methods was usually affected by the variations of equipment's working condition. So how to use limited sampling parameters to monitor system failure effectively become the problem which we need to solve [1-3].

In view of the above, a self-adaptive fault monitoring method by using empirical mode decomposition and the energy comparison of different orders intrinsic mode functions was presented. The method can avoid the negative impact of equipment state changing and human intervention, and achieve good effect of failure monitoring.

1 Introduction of EMD

In 1998, to analysis non-stationary and nonlinear data, Intrinsic Mode function (IMF) and Empirical Mode Decomposition (EMD) were presented by N. E. Huang. By the EMD method, we can decompose any signal to the intrinsic mode functions which from the high frequency to the low frequency, and each intrinsic mode function is separated by the signal own characteristic. Without human intervention, EMD method has strong adaptability, and can be applied to analysis the non-stationary signal. Now

D. Yang (Ed.): Informatics in Control, Automation and Robotics, Volume 2, LNEE 133, pp. 705–710.
Springerlink.com

the method has been got effective application in many fields, and get more and more attention [4,5].

Before introducing empirical mode decomposition, the intrinsic mode function should be defined. An intrinsic mode function is a function that satisfies two conditions:

(1) In the whole data set, the number of extrema and the number of zero crossings must either equal or differ at most by one;
(2) At any point, the mean value of the envelope defined by the local maxima and the envelope defined by the local minima is zero.

Obviously, the second condition may not be completely satisfied, but only approximately.

The procedure of EMD is shown in figure 1.

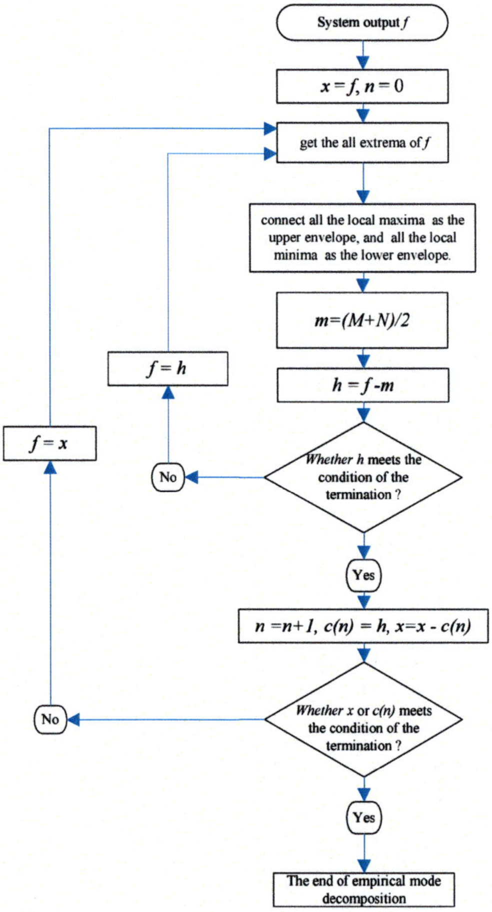

Fig. 1. The flow chart of EMD

The specific process as follows:

The first IMF can be got by equation (1), and $x(t)$ is the original signal which is need to deal with. Then get all the extrema of $x(t)$, connect all the local maxima by a cubic spline line as the upper envelope, and all the local minima by a cubic spline line as the lower envelope. Their mean is designated as m_1, and the mean of h_{i-1}'s envelopes is designated as m_i.

$$\begin{aligned} & x(t) - m_1 = h_1, \\ & h_1 - m_2 = h_2, \\ & \quad \vdots \\ & h_{k-1} - m_k = h_k, \\ & \Rightarrow c_1 = h_k \end{aligned} \tag{1}$$

When h_k satisfies the IMF's definition, the iteration will be stop, and then we can get the first order IMF c_1.

(2) Then designate $r_1 = x(t) - c_1$, repeat the iteration above, and we can get $c_2, c_3 \cdots c_n$. The procedure is illustrated by equation (2).

$$\begin{aligned} & x(t) - c_1 = r_1, \\ & r_1 - c_2 = r_2, \\ & \quad \vdots \\ & r_{n-1} - c_n = r_n \\ & \Rightarrow x(t) = \sum_{j=1}^{n} c_j + r_n \end{aligned} \tag{2}$$

(3) When the r_n has no more than two extrema, the empirical mode decomposition will be stop, and r_n can be defined as a residue. So finally the signal is decomposed into n IMFs and one residue.

2 A Self-adaptive Fault Monitoring Method

Because of the strong adaptive characteristics of EMD, the method can be used in system fault monitoring to avoid the effects of manual intervention and equipment state change. So a self- adaptive fault diagnosis method is presented which using EMD method and energy comparison of different orders intrinsic mode functions, and the procedure is as follow.

First, sample system output such like temperature, current or voltage by fixed length sampling window ,and the window has the same number of data every time by using first-in, first-out method. Namely at each sampling time, the sampling window accepts a current sample value, and eliminates the oldest one. So we can get the sampling sequence $\{x(i), i=1,2,...,N\}$, and N is the length of the sampling sequence.

Second, decompose the sampling sequence by EMD method. To facilitate comparison between different sampling times, we can only calculate S and higher order IMFs in each EMD processing, and treat the lower order IMFs as a part of the residual. Because the boundary effect of empirical mode decomposition will pollute data at the ends of the sample sequence, we take out 10% of the amount of data on both ends of the IMF, keeping the results of only 80% of the data in the middle, and then calculate the sums of every IMFs' absolute value, namely find out the energy of every order IMFs. Now we get S sums of IMFs' absolute value.

Third, record the energy values of n successive sampling windows, and average energy of every order IMFs, and the value of n depends on the computer performance and the requirement of system response time. Then judge the system condition by comparing the value between the every average energy of various order IMFs and their thresholds. If the every average energy is less than their thresholds, the system is in a good condition, otherwise there will be a fault. In practical application, the threshold value can be set based on the experimental data of the system, or autonomous learning of Initial system state.

Finally, repeat these three steps in every sample time, we can realize the real-time failure monitoring.

3 Simulation

In this section, we make a simulation to verify the effectiveness of the presented method in this paper.

Assuming sampling frequency is 4 Hz, and simulation time is 600 seconds, so we will have 2400 sampling points. To Simulate different faults, we add different kinds of failure data in 800-840, 1500-1560, 2100-2200 sampling points respectively, and the system output is shown in figure 2.

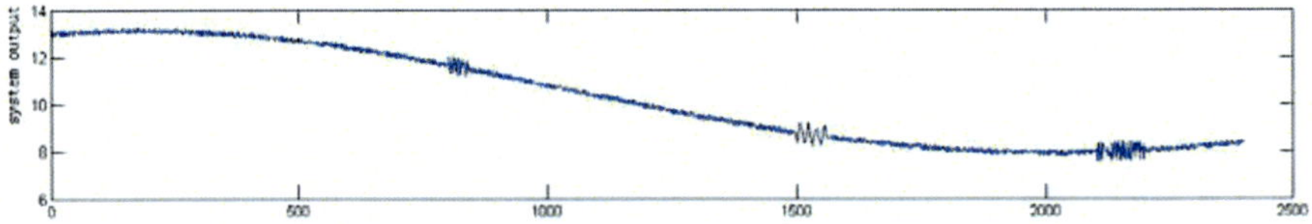

Fig. 2. The curve of system output

Select the length of sample window as 240, record every energy value of first and second order IMFs in 10 successive sampling windows. Then average different order IMFs' energy values, and record all the averages in 10 minutes which are drawn as shown in figure 3, figure 4. In this simulation, based on the autonomous learning of Initial 100 seconds sampling data, the threshold value can be set to average 1.2 times to reduce the probability of misstatement, then judge the system condition by comparing the first and second order IMFs with their thresholds.

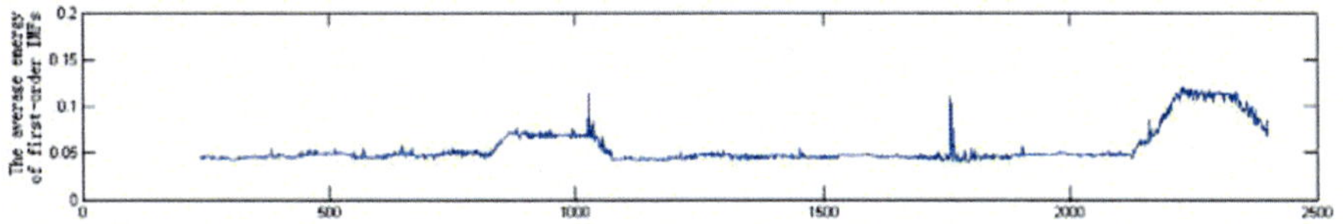

Fig. 3. The average energy of first-order IMFs

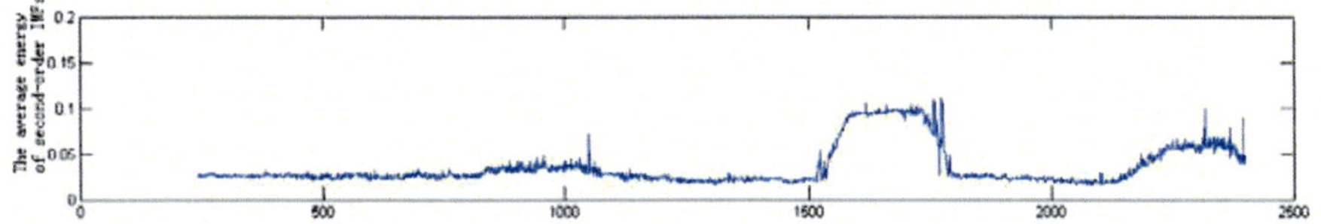

Fig. 4. The average energy of second-order IMFs

In this simulation, to avoid the boundary effect of empirical mode decomposition, we eliminate 10% of the data in every sample window, and use 10 successive sampling windows to average the energy values, so the moment when the fault alarm appears will late than when the fault happens about 30 sampling cycles. Finally there are two fault alarms at the 820th point and 2130th point in the first-order IMF average energy, and another two fault alarms at the 1520th point and 2130th point in the second-order IMF average energy. So the failure monitoring method presented in this paper is accurate and effective.

4 Conclusion

In this paper, a self-adaptive fault monitoring method is presented. By using empirical mode decomposition and the energy comparison of different order intrinsic mode functions, the method can avoid the negative impact of equipment state changing and human intervention, and shows good effect of system failure monitoring in the simulation.

References

1. Zhu, D.-Q., Liu, Y.-A.: Information fusion method for fault diagnosis. Control and Decision 22(12), 1321–1328 (2007)
2. Zhang, Y.-F., et al.: Study on Fault Diagnosis of Power-Shift Steering Transmission Based on Spectrometric Analysis and SVM. Spectroscopy and Spectral Analysis 30(6), 1586–1589 (2010)

3. He, X.-Y., He, Q.-H.: Fault diagnosis of construction machinery hydraulic system based on multi-network model. Journal of Central South University 41(4), 1385–1490 (2010)
4. Huang, N.E., et al.: The empirical mode decomposition and the Hilbert spectrum for nonlinear and non-stationary time series analysis. Proc. R. Soc. Lond. A 454, 903–995 (1998)
5. Huang, D.J., Zhao, J.P., Su, J.L.: Practical implementation of Hilbert-Huang Transform algorithm. Acta Oceanologica Sinica 22(1), 1–14 (2003)

NEAT versus PSO for Evolving Autonomous Multi-agents Coordination on Pursuit-Evasion Problem

Jin Zhao and Gang Peng

Department of Control Science and Engineering,
Huazhong University of Science and Technology, Wuhan, China
Key Laboratory of Image Processing and Intelligent Control of Education Ministry
zhaojinhust@gmail.com, penggang@mail.hust.edu.cn

Abstract. A method based on multi-agents and ANN (Artificial Neural Network) was proposed to solve the pursuit-evasion task in continuous time-varying environment. According to this method, several autonomous agents with 8 circular sector sensors and an ANN controller were used to form a coordinated behavior to capture the evaders. To evolve the controller, NEAT (NeuroEvolution of Augmenting Topologies) and PSO (Particle Swarm Optimization) method were used to optimize the network. The simulation experiments show that both methods can successfully evolve the controller to capture the evaders, while NEAT requires less swarm members and consume less time comparing to PSO method.

Keywords: Pursuit; Multi-agents, Artificial Neural Network (ANN), NeuroEvolution of Augmenting Topologies (NEAT), Particle Swarm Optimization (PSO).

1 Introduction

Pursuit-Evasion task is an interesting problem of Artificial Intelligence field. According to how the strategies of the Agents are generated, the solution of the problem can be separated into two kinds: one is based on manmade strategies and the other is based on machine learning. And according to the environment of the task, the problem can be separated into discrete and continuous. A lot of work has been done in discrete and manmade strategies situation, while the continuous and machine learning aspect of the problem has not been explored thoroughly. Leo.H.Langenhoven and Geoff.S.Nitschke has compared the *Neuro-Evolution (NE)* and *Particle Swarm Optimization* (PSO) for a competitive co-evolution situation [1], while it was in a discrete environment and there are only one predator and one prey.

In this paper, an ANN solution of the problem with multi-pursuers and multi-evaders on a continuous environment was proposed. A swarm of pursuers with sensors and ANN controllers were added to an enclosed stage, and then with optimization algorithm some multi-agent coordinated capture behaviors will emerge. In order to evolve the ANN controller, two ANN optimization methods: PSO and *NeuroEvolution of Augmenting Topologies* (NEAT) were carried on, and the performances of the two methods on evolving the controller were compared.

D. Yang (Ed.): Informatics in Control, Automation and Robotics, Volume 2, LNEE 133, pp. 711–717.
Springerlink.com

2 Method

2.1 Particle Swarm Optimization (PSO)

PSO was first proposed by J. Kennedy and R. Eberhart in 1995[1, 2]. The initial purpose of this algorithm was to simulate the flying of the bird swarm, but the method was found an effective solution on optimization problems, especially when the dimension of the problem was high.

PSO is initialized with a group of random particles; each particle's position represents a solution. In every iteration process, each particle is updated according to two values. The first one is the best position of the particle itself (*pbest*), and the other one is the swarm's best position in history (*gbest*).

The update of the particle abides by the following steps.

(a) Initialize the swarm. Every particle will be assigned a random position and velocity.
(b) Calculate the fitness scores of all the particles.
(c) Update the *pbest* and *gbest* for each particle.
(d) Update the positions and velocities of the particles according to equation (1) and (2).
(e) Check if the end condition is met, if not, goes to (a) and start another loop.

$$v = v + c1 \cdot \mathrm{rand} \cdot (pbest - present) + c2 \cdot rand \cdot (gbest - present) \quad (1)$$

$$present = present + v \quad (2)$$

Where *v* is the velocity of the particle, *present* is the position of the particle. *pbest* and *gbest* are defined as stated before. *rand* is a random number between 0 and 1. *c1*, *c2* are learning factors.

The Inertia Weight. In PSO, the exploration of local and global is difficult to balance, to solve this problem, Yuhui Shi proposed an improvement called the inertia weight [3], and the equation is shown as below.

$$v = \omega \cdot v + c1 \cdot \mathrm{rand} \cdot (pbest - present) + c2 \cdot rand \cdot (gbest - present) \quad (3)$$

ω is the inertia weight. Its value is usually set in the range of [0, 1.4]. When omega is high, the PSO has a relatively good ability on global exploration, and a lower omega will bring better performance on local exploration. In this paper, we set the ω to 0.7, which will give a balance on local and global exploration.

2.2 NeuroEvolution of Augmenting Topologies (NEAT)

NEAT is a method to evolve ANN proposed by Stanley and Miikulainen [4], different from the traditional methods, it does not need a designed ANN topology; instead the algorithm can generate and evolve the topology and weights of an ANN from scratch. This method has been successfully applied in a FPS (First Person Shooter) video game to evolve the bots' behaviors [5], in the experiment autonomous shooting, dodging and hiding behaviors are evolved, which proves the suitability of the method in complex environment.

3 Agents

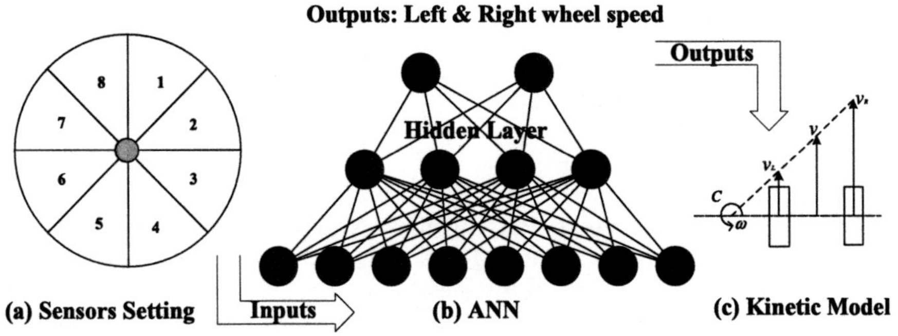

Fig. 1. (a) is the sensory setting of the agents; (b) is the structure of the ANN controller; (c) is the kinetic model of the agents. During the simulation, the sensory information is sent to the ANN and calculated by the ANN, and then the ANN outputs are used to drive the agent.

3.1 Kinetic Model

The movement of the agent is driven by the two wheels; the kinetic model is shown as in figure 1(c). C is the rotation center of the agent, ω is the angular speed, V_L and V_R are the speed of the left and right wheels; x, y are the agent's location; θ represents the direction of the agent, and if we set $\boldsymbol{q} = [x \quad y \quad \theta]^T \in R^3$, $U = [v_R \quad v_L]^T \in R^2$ .The kinetic model can be described as the following equations[6].

$$S(q) = \begin{pmatrix} (\cos\theta)/2 & (\cos\theta)/2 \\ (\sin\theta)/2 & (\sin\theta)/2 \\ (2r_{cp})^{-1} & -(2r_{cp})^{-1} \end{pmatrix} \tag{4}$$

$$\dot{q} = S(q)U \tag{5}$$

3.2 Sensors

The detection space of the agent is a round area, and it was separated into 8 segments, as shown in figure 1(a). Each segment works as an individual sensor. When the evaders were detected, the respond was negative, when the obstacles, wall or other pursuers were detected, the results are positive.

3.3 ANN Controllers

The agent is controlled by an ANN controller, which maps sensory inputs to wheel speeds of the pursuer agents. In the NEAT solution, the algorithm will generate an ANN automatically, while in the PSO situation, a designed ANN is needed (figure 1(b)).

For the PSO situation, there are 8 inputs and 2 outputs in the ANN and the hidden neurons number is 4. The output neurons use logistic sigmoid functions to regulate the output values, in order that they can be used to drive the agents directly.

For both NEAT and PSO situation, the inputs are normalized to [-1, 1], and the output of the network is between 0 and 1.

3.4 Controller Optimization

PSO Solution. In this case, the PSO algorithm is used to optimize the connection weights of the ANN. In our case, the ANN has 8 input neurons, 4 hidden neurons and 2 output neurons. Each neuron was connected to every neuron in its neighboring layer. And considering the bias weight, we have 46 connection weights to optimize, which means the dimension of the particle in the swarm is 46. Because the simulation environment cannot hold too much agents at a time, we only have 8 pursuers to evolve, so the swarm also has 8 particles.

The initial positions of the particles are set by random, and immediately used to generate the first generation of controllers. And after each iteration, the new positions of the particles in the swarm will be assigned to the controllers of the agents, and then the simulation process will run to calculate the fitness score of the controller. After each simulation the fitness score will be given to the PSO algorithm to generate new *gbest* and *pbest* for the particles.

NEAT Solution. Comparing to the PSO method, NEAT doesn't need a human designed ANN controller, only the numbers of input and output neurons are required. In this case, we only have to tell the algorithm that there should be 8 input and 2 output units, and decide the evolving parameters; the remaining work including generating and adjusting an ANN will be handled automatically. The author has successfully applied NEAT to evolve an ANN controller in a former work [8].

3.5 Fitness Calculation

The fitness of the agents is the feedback of the controller on how well it works. It plays a very important role on guiding the evolution of the optimizer. The fitness is calculated according to equation (8).

$$f = w \times 10000 + i + a \times 800 + b \times 5 - (c + d + e) \tag{6}$$

Where i is the frames of the simulation, a is the flag on whether the agent has found any evader; b is the collision times with evaders, c、d、e are the collision times with the obstacles, other pursuers and walls, w shows how many evaders have been captured by the pursuer, and f is the fitness.

3.6 Evader Strategies

The evaders are controlled by a manmade strategy which was called DODGE. According to this strategy, the evader will not move unless they found any pursuers close to them. The dodge direction is calculated according to the distances of the pursuers detected. The sensory range of the evaders is the same as the pursuer, and the moving speed of them are also the same as their enemy. The purpose to use this manmade strategy is to assure a fair competition between PSO and NEAT method.

4 Experiment Settings

4.1 Simulation Environment

The pursuit-evasion simulation process is held on an enclosed continuous 2-D space. The size of the stage is 800×600 pixels, all the pursuers, evaders and obstacles are round object with 10 pixels of radius. The sensory range is 100 pixels. All the agents can only move inside the stage, and the obstacles cannot be moved by the agents.

4.2 Experiment Process

There are 8 pursuers, 4 evaders and 4 obstacles in the stage. Their locations are hand-arranged as shown in figure 2, and they will be reset at the start of every simulation. This is to assure that the controller will face the same environment in each time.

For each simulation process, there are 2000 time steps; the simulation rate is 60 steps/second. The whole experiment will run for 200 simulations for each optimizer (PSO and NEAT). During a simulation, the 8 pursuers will move according to the sensory information and their ANN controller. After each simulation, the fitness of all the pursuers will be calculated, and the optimizer will use them to update the weights of the ANN controller.

To eliminate the influence of the random initial conditions, both trails are run for 4 times.

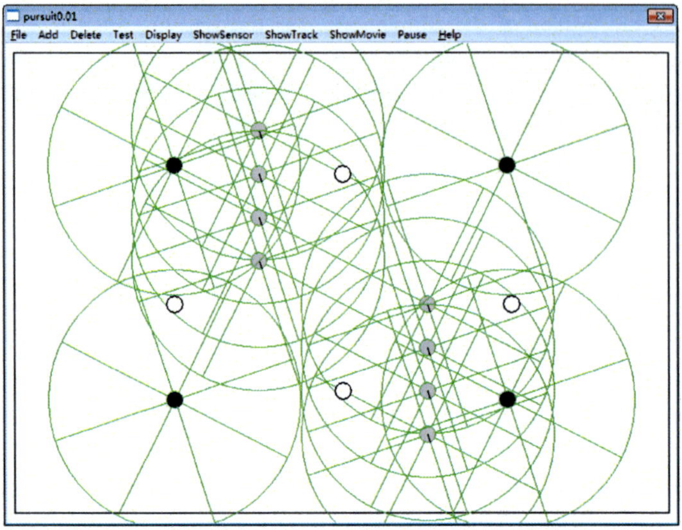

Fig. 2. Locations of the pursuers (black), evaders (grey) and obstacles (white) at the start of simulation. The sectors are the sensors of the agents.

4.3 Result Analysis

For the NEAT method, each time the experiment will get an average fitness that is over 10000 after 150 generation, which means when enough iteration are run, the NEAT can always find a strategy to capture the evaders.

While for the PSO method, only for one time the average score after 150 generation is over 10000, for the other 3 trials, the swarm is trapped in local best value and their average fitness is below 5000.

Figure 3 shows the best NEAT and PSO result.

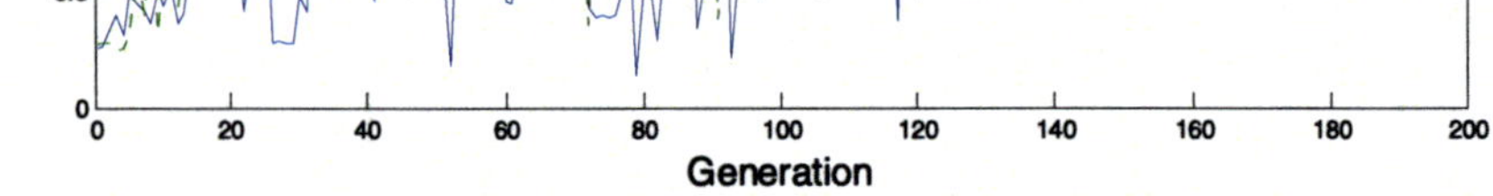

Fig. 3. PSO & NEAT simulation results of 200 generations

From the result we can draw the following conclusions.

1. At the beginning of the trial, the fitness score of the PSO raise faster than NEAT. Which means PSO works faster than NEAT.
2. During the middle generations, the NEAT fluctuate more than PSO method. This is because NEAT will always try to innovate while PSO will explore the space according to a relatively steadier route.
3. At the last 50 generations, for both the methods, the average fitness scores are almost the same, which means the methods' best performances are the same.
4. The NEAT method is more suitable on the small swarm, while PSO needs a relatively large swarm size to suit the high dimension of the controller.

5 Conclusions

In this paper, a new solution based on ANN and optimization methods (NEAT, PSO) are proposed to cope with the pursuit-evasion problem. The experiment proves the effectiveness of this method, and the performances of the two optimization methods are compared.

References

1. Langehoven, L.H., Nitschke, G.S.: Neuro-Evolution versus Particle Swarm Optimization for Competitive Co-evolution of Pursuit-Evasion Behaviors
2. Kennedy, J., Eberhart, R.: Swarm Intelligence. Morgan Kaufmann Publishers, San Francisco (2001)
3. Eberhart, R., Simpson, P., Dobbins, R.: Computational Intelligence PC Tools. Academic Press Professional, Boston (1996)
4. Shi, Y., Eberhart, R.C.: A Modified Particle Swarm Optimizer. In: Proceedings of the IEEE International Conference on Evolutionary Computation, pp. 69–73. IEEE Press, Piscataway (1998)
5. Stanley, K.O., Miikkulainen, R.: Evolving Neural Networks Through Augmenting Topologies. Evolutionary Computation 10(2), 99–127
6. Stanley, K.O., Cornelius, R., Miikkulainen, R., et al.: Real-Time Learning in the NERO Video Game. In: Proceedings of the Artificial Intelligence and Interactive Digital Entertainment Conference (AIIDE 2005), Demo Papers
7. Liu, L., Xiang, P., Wang, Y., et al.: Trajectory tracking of a nonholonomic wheeled mobile robot. Tsinghua Univ. (Sci. & Tech.) 47(S2) (2007)
8. Zhao, J., Peng, G.: A Multi-Robots pursuit-evasion problem based on NEAT solution. Proceeding of Huazhong Univ. of Sci. & Tech. (Natural Science Edition) (2011)

Visual Sensing Method Based on Texture Analysis for Automatic Seam Tracking Systems

Zou Yirong, Du Dong, and Wang Li

Key Laboratory of Ministry of Education for Advanced Materials Processing Technology, Tsinghua University, Beijing, P.R. China

Abstract. Seam tracking is a key technology for automated welding process and post-welding NDT. Manuel monitoring supervised by naked eye or industrial TV is experience-dependant, labor-intensive and risks to be biased. It is necessary to develop low cost yet efficient seam tracking system. In this paper, an approach based on texture analysis is proposed to provide a possible solution for seam tracking. CCD camera is used as the vision sensor. Textural features are extracted through the Grey Level Co-occurrence Matrix (GLCM) generated from the captured image. The region of welded seam was located according to their textural features. Experimental results show that the proposed method can satisfy the requirement of seam tracking.

Keywords: Vision sensing, texture analysis, seam tracking, robotic welding, post-welding NDT.

1 Introduction

Seam tracking is a crucial problem for the automation of welding process and post-welding NDT (Non-Destructive Testing). During the recent years, vision sensors for seam tracking are widely applied in welding robotics and other forms of automatic welding devices. To achieve better productivity and welding quality, it is of great importance to develop reliable and efficient visual method for seam tracking. Low-cost yet efficient and flexible systems are particularly in demand.

Various visual methods have been proposed as solution. Among them, the active vision methods utilize structured light or laser stripes to detect seam geometrics[1-3], and have seen industrial applications[4].

However, when comes to seam tracking during multi-layer/multi-pass welding, the existing methods are no longer helpful. After one layer of weld is finished, the next is to be performed on top of it. The vision sensor should be able to recognize the surface of weld, which helps to indicate the right path for the welding of the next layer. Since there is neither geometric feature for the application of active vision, nor obvious grey-level gradient for passive vision, it is necessary to develop novel visual method.

Similar applications could possibly appear in post welding operations, such as ultrasonic testing, where weld needs to be recognized through its appearance for the navigation of ultrasonic sensors.

D. Yang (Ed.): Informatics in Control, Automation and Robotics, Volume 2, LNEE 133, pp. 719–723.
springerlink.com

In this research, an approach based on texture analysis was proposed, in order to extract the textural features of welding ripple on the surface of weld, through the analysis of the image. Different textural feature descriptors were first defined. Then the analysis of the features was carried out, in order to distinguish the weld seam region and the base metal region in the image. The method was tested for CCD captured image of carbon steel CO_2 weaving welding, and satisfactory results were obtained.

2 Method

2.1 System Configuration for Robotic Welding

A schematic representation of the seam tracking system is shown in Fig.1. The CCD camera is positioned in front of the torch, separated by a certain distance in between. Images of the welded seam are captured then processed with the texture analysis algorithm. Position of the welded seam is provided to the control module, in order to monitor the welding path for the latter layers during a multi-layer welding process.

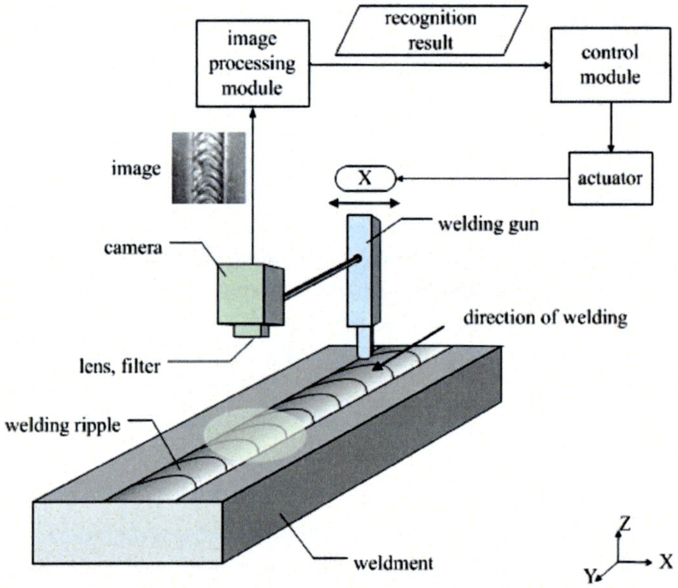

Fig. 1. Configuration of seam tracking system for multi-layer welding

2.2 Texture Analysis Approach

The conventional image segmentation usually bases on grey level differences from different regions. However for the recognition of welded seam, the thresholding according to grey level can not lead to the discrimination of the seam region and the base metal region. (Fig. 3(a))

Instead of the conventional grey-level thresholding, the textural feature which represents the local distribution of grey level can be more efficient in describing the characteristics of the welded seam region. Analysis based on Grey Level Co-occurrence Matrix (GLCM) helps to extract the textural feature of the welding ripple, which reveals the most important visual difference between the seam region and the base metal region. Thus the seam can be recognized. (Fig. 3(b))

2.3 Texture Feature Descriptors Based on GLCM

It is important to define effective descriptors of features, which allow us to extract different texture features, from the grey level image. Generally, the methods of textural feature description include statistical ones and structural ones. The Haralick features, generated from GLCM (Grey-Level Co-occurrence Matrix), can give statistical measures of image texture. Human vision is able to detect second-order features during texture perception; the GLCM method has been proved to achieve similar level of performance [5].

A Grey Level Co-occurrence Matrix (GLCM) is a tabulation of the distribution of co-occurring grey-level values at a given offset. Mathematically, it is defined over an image as shown in equation 1. Each matrix element is an estimate of the probability that two image pixels, separated by the displacement vector (d,θ), have intensities i and j.

$$P(i,j\mid d,\theta)=\frac{\#\{k,l\in D\mid f(k)=i,f(l)=j,\|k-l\|=d,\angle(k-l)=\theta\}}{\#\{m,n\in D\mid \|m-n\|=d,\angle(m-n)=\theta\}} \tag{1}$$

The GLCM is parameterized by an offset (d,θ) between the pair of pixels. Images are generally requantized to 16 or 32 discrete grey levels [12-13]. Feature descriptors shown as in equation 2~6 are the most frequently used ones from GLCM.

$$W_M=\sum_h\sum_k \mathbf{M}_{h,k}^{\;2} \tag{2}$$

$$W_{Ct}=\sum_h\sum_k |h-k|\mathbf{M}_{h,k} \tag{3}$$

$$W_{Cr}=\frac{1}{\sigma_x\sigma_y}\sum_{h=1}^{N}\sum_{k=1}^{N} hkm_{h,k}-\mu_x\mu_y \tag{4}$$

$$W_E=-\sum_h\sum_k \mathbf{M}_{h,k}\log \mathbf{M}_{h,k} \tag{5}$$

$$W_H=\sum_h\sum_k \frac{\mathbf{M}_{h,k}}{1+|h-k|} \tag{6}$$

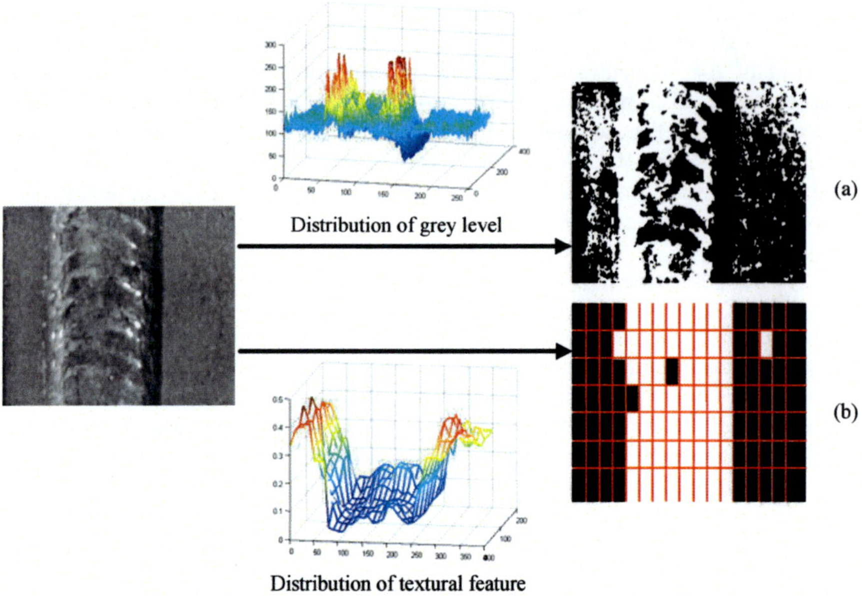

Fig. 2. The texture analysis approach compared with grey-level thresholding

2.4 Analysis on Results of Experiments

Experiments were run in order to test the algorithm described above. The results demonstrated that the texture analysis approach could achieve satisfactory accuracy. Table 2 presents the analysis of the recognition accuracy, in terms of both pixel and actual distance.

Table 1. Accuracy of welded seam recognition using the texture analysis approach

	Error on the left edge of the seam			Error on the right edge of the seam			Error on the central line		
	max	min	mean	max	min	mean	max	min	mean
Distance on the image plane / pixel	6	1	4	3	0	1	3	0	2
Actual distance on the weldment / mm	0.5	0.1	0.3	0.2	0	0.1	0.3	0	0.2

3 Conclusions

A visual method based on texture analysis was developed for seam tracking. According to the experimental result, the following conclusions have been drawn:

a) Visual method based on texture analysis was proposed for seam tracking of robotic welding and post-welding NDT processes.
b) For seam images processed in the study, the optimal parameterization has been determined as following: 24-level grey scale, 90° orientated GLCM with a distance of 3 pixels, 60×10 pixel-sized subimages. Experimental result shows the precision of seam recognition was within ±0.5mm.

References

1. Agapakis, J.E.: Vision-Aided Robotic Welding: An Approach and a Flexible Implementation. International Journal of Robotics Research 9(5), 17–34 (1990)
2. Umeagukwu, C., McCormick, J.: Investigation of an array technique for robotic seam tracking of weld joints. IEEE Transaction on Industrial Electronics 38(3), 223–229 (1991)
3. Kim, J.S.: A Robust Method for Vision-based Seam Tracking in Robotic Arc Welding. In: Proceedings of the 1995 IEEE International Symposium on Intelligent Control (1995)
4. Wilson, M.: The role of seam tracking in robotic welding and bonding. Industrial Robot: an Industrial Journal 29(2), 132–137 (2002)
5. Haralick, R.M., Shanmugam, K., Dinstein, I.: Textural Features for Image Classification. IEEE Transactions on Systems, Man, and Cybernetics SMC-3 (6), 610–621 (1973)
6. Julesz, B., Gilbert, E.N., Shepp, L.A., Frisch, H.L.: Inability of humans to discriminate between visual textures that agree in second-order statistics revisited. Perception 2, 391–405 (1973)
7. Walker, R.F., Jackway, P.T., Longstaff, D.: Genetic algorithm optimization of adaptive multi-scale GLCM features. International Journal of Pattern Recognition and Artificial Intelligence 17(1), 17–39 (2003)

Evaluation on the Categorical DBMS for the Manufacturing Intelligent System

Yuanping Xu, Hongping Shu, and Jun Lu

Chengdu University of Information Technology
ghxpy@hotmail.com

Abstract. This paper starts with a brief examination of current problems on knowledge conversion between knowledge base and database in the field of advanced manufacturing intelligent systems. To solve the problems highlighted, a categorical Database Management System (DBMS) has been developed to bridge the knowledge gap between knowledge base and database. This paper focuses on giving a short evaluation on the categorical DBMS to test its usability and functionality.

Keywords: knowledge base, categorical object model, categorical DBMS.

1 Introduction

During the last a couple of decades, manufacturing industry becomes more and more flexible and global through outsourcing. Geographically dispersed (remote) design and manufacturing practices are rapidly increasing [1]. This move is one of the major contributing factors to the establishment of a number of universal standards and rules to unify the characteristics of machining and measuring workpieces. However, these standards and rules are often too complex, abstract, and theoretical to be applied efficiently by the designers and engineers in manufacturing industry. For example, it is difficult for engineers who are not familiar with Geometrical Product Specification (GPS) matrix system to search for a single parameter among thousands of paper based files and even more difficult to cross-refer and link them to form useful information.

Therefore, in order to overcome current implementation problems of these manufacturing standards and rules, researches relating to manufacturing standards computerizing had been hotly investigated [2-4]. However, the major software systems at present are still weak on functionality and failed to stand for the complexity of the intelligent design and measurement world (e.g. Functional features such as product function translation, reasoning of design properties and the related verification principles, measuring equipments, calibration requirements, as well as management of uncertainty are often largely ignored [5]). One of the major reasons for causing these drawbacks is that the traditional database systems applied by all the current engineering aided design and manufacturing software tools can not efficiently support complex data structures to reflect the complicated relationships among geometrical parts and manufacturing standards, which are essential for comprehensive analysis and data manipulations to solve practical production problems. This research

D. Yang (Ed.): Informatics in Control, Automation and Robotics, Volume 2, LNEE 133, pp. 725–731.
springerlink.com

solves this problem by constructing a categorical DBMS based on the categorical object model to handle large amount of synchronized text-based ISO standard data for guiding product design and manufacturing. Thus, expert information and knowledge can be clearly extracted from ISO manufacturing standards and new generated knowledge can be directly stored into database for supporting further operations and improving the traceability of users' works. The detailed rationales for justifying the research and development of the categorical object model can be found in a separate publication [6]. This paper only focuses on giving a short evolution on the categorical DBMS with current mainstream relational and object-oriented DBMSs. This paper can also be used as a guide for readers on choosing suitable DBMSs for their advanced knowledge systems relating to aerospace, manufacture and biology areas.

2 Evaluation

As a potential solution, the Category Theory has been widely recommended as a mathematical tool to overcome these application problems [7]. Category Theory is a form of constructive mathematics, which is devised to describe various structural concepts from different mathematical areas in a uniform foundation. Category Theory can grasp the essence of the researching targets as it focuses on the properties of mathematical structures instead of on their detail representations. For example, the diagram in Category Theory is similar to the graph in Topologic Theory, which is used to model pairwise relations between objects in a certain domain instead of focusing too much on precise positions of those objects.

This project has researched into this subject area and devised a Category Theory-based object model and a DBMS based on it—named 'categorical DBMS' to handle complex objects and object relationships. In the categorical object model, this research uses the Category Theory to deal with the modeling of specialization and generalization inheritances, view, typing, operations, encapsulation and other user constraints. Through developing a DBMS based on the categorical object model, the knowledge base and database can be unified into single mathematic foundation. Thus, knowledge in manufacturing systems can be directly stored and queried without the requirement to grogram any mapping codes for knowledge conversions between knowledge base and database. As this research uses categorical object model and categorical DBMS to fill the knowledge gap between knowledge base and database, the evaluation part starts with the comparison of the categorical object model with other two data models that are widely used at present: relational data model and Object Database Management Group (ODMG) object model. Nowadays, there are around 40 relational DBMS products developed by various vendors (e.g. Oracle, SQLSever and MySQL), which have dominated the market for last three decades. The most important reason for the success of relational DBMSs is that they have a universal formal basis – relational data model based directly on Set Theory. The current ODMG standard 3.0 for object-oriented DBMSs has been adopted by several mainstream object-oriented DBMSs at present such as Objectivity, Versant and ObjectStore [8]. Table 1 demonstrates a comparison of these three data models in aspects of modelling capability and mathematical support.

Table 1. Comparison of three data models

	Relational data Model	**ODMG Object Model**	**Categorical Object Model**
Modeling Capability			
Formal relationship structure (including n-ary)	YES (Based on the descartes in Set Theory)	NO	YES (Based on the product construct in Category Theory)
Trees/Collections/Arrays	NO	YES	YES
Inheritance	NO	YES	YES
Aggregation	NO	YES	YES
Multi-level mappings	NO	NO	YES
Object nests	NO	YES	YES
Mathematical Support			
Manipulations	YES (Based on set operations, algebra and calculus)	NO	YES(based on arrow mapping, arrow composition and functor composition)
Methods/Dynamic Constraints	NO	YES(Based on object definition Language without mathematical support)	YES(Based on method arrows)
Normalization	YES (Based on functional dependency checking on sets)	NO	YES (Based on arrow composition checking on categories)
Referential Integrity	YES (Based on foreign key definitions)	YES(Based on object identifiers)	YES(Based on initial internal objects of categories)
Membership/cardinality	YES(by labels)	NO	YES(by typing functors)

Based on the Table1, the key feature for relational data model can be highlighted as:

The relational data model has good mathematical foundation which gives users a clear and formal construct ("Table") to model data and a rigor manipulating mechanism based on relational algebra and calculus on sets. However, it is weak in modeling of complex object structures especially for modeling of multi-level constraints/mappings and object nests.

The key feature for ODMG object model can be highlighted as:

The ODMG object model has good capability of modeling complex object structures but it is lack of mathematical foundation. So it is difficult to ensure the integrity and consistency of database schema after any manipulations such as deletion, updating or addition.

Table 1 shows the categorical object model can satisfy both objectives – having good capability of modeling complex object structures especially in handling the multi-level constraints and mappings while offering a rigor mathematical foundation based on Category Theory, like the relational data model based on Set Theory. It provides a uniform way to model both static (attributes) and dynamic (methods) aspects of objects by using different types of morphism. Moreover, it can define a manipulation language based on functor mappings and compositions, so the integrity and consistency of database schema can be ensured through diagram chasing and algebraic deduce.

In this section, a case study will be discussed to evaluate the research outputs. This case study focuses on generating a complete callout symbol (technical drawing) from Geometrical Product Specification (GPS) matrix system. The complete callout extracts from a set of matrices defined in GPS, which is in form of nested and multi-level structures, see Fig.1.

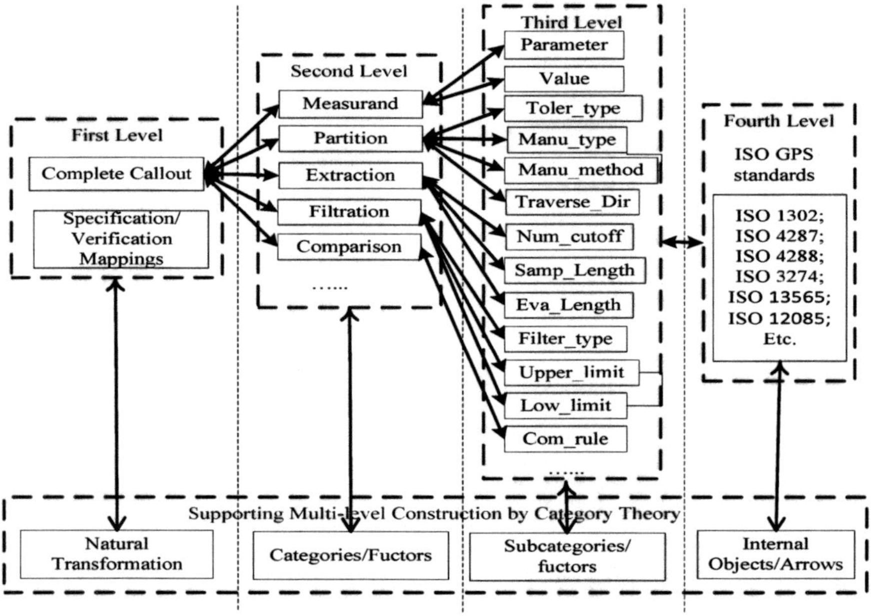

Fig. 1. The multi-level structure of a complete callout

To implement this case in relational DBMS, the main problem is that it is very difficult to separate these complex multi-level mappings and nested matrices into the strict formats of flat tables. In order to conform to normalization rules, database developers need to create a lot of tables for entities and their linkages. This process is very time consuming and it is also difficult for developers who are not familiar with GPS standards to keep all mappings and constraints consistently during persistence.

The main problem for other object-oriented DBMSs to implement this case is that they can not directly support the categorical object model. So different database developers may have different ways to define classes, which is an error-prone process. As objects in database are different from objects in applications, misunderstandings may occur between GPS knowledge base designers and the database developers. Because of the absence of a multi-level mapping constructs in traditional object-oriented DBMSs, the multi-level constraints can be easily lost during persistence. Moreover, in other object-oriented DBMSs, they often directly involve one class into another class to form a relationship between the two classes. This has three problems:

- There is no a class definition to hold the information generated from the relationship link. Therefore, queries on the relationship information are difficult to get. The query closure is also difficult to achieve without a formal relationship structure.

Table 2. Comparison of categorical DBMS with other object-oriented DBMSs

	Categorical DBMS	**Object-oriented DBMSs**
Structures		
Formal Relationship Structure (including n-ary)	YES	NO
Trees/Collections/Arrays	YES	YES
Inheritance/Aggregation	YES	YES
Multi-level Mapping	YES	NO
Rules		
Normalization Support	YES	NO
Referential Integrity	YES	NO
Membership/Cardinality	YES(by typing functors)	YES(by labels)
Manipulation		
Algebra/Calculus	YES(based on functor composition)	NO
Declarative Query	YES	YES
Closure	YES	NO
View	YES	NO
Methods	YES	YES

- It is difficult to check the cardinality and membership for a relationship. This also leads to the complexity of updating or deleting objects involved in a relationship from the database.
- The object-oriented query strategy is difficult to generate.

Table 2 gives a comparison of the categorical DBMS with other object-oriented DBMSs (Objectivity9.0 and Versant7.0 in this paper).

Therefore, comparing with other conventional object-oriented DBMSs, the advantages of the categorical DBMS are highlighted as follows:

- Design and implement a categorical object model. This categorical object model can map complex object structures into mathematical formalizations in Category Theory. Thus, Algebras and calculus defined in Category Theory enable categorical DBMS to manage both dynamic and static aspects.
- The categorical object model is especially good at representing the multi-level architecture, which enable advanced constraint specifications and good extensibility of designed models.
- The algebras and calculus such as arrow composition, arrow mapping, functor composition and functor mapping can be used as basis for implementation of an object-oriented query language with closure.
- Involving integrity checking mechanism in both intra and inter category levels. Thus, the integrity and consistency of the database schema can be ensured in the categorical DBMS.

3 Conclusions and Future Work

The categorical DBMS based on categorical object model provides a feasible way to bridge the knowledge gap between knowledge base and database in three main aspects: Firstly, it can handle various complex knowledge structures with ease (e.g. multi-level and nested features). Secondly, this model formally supports basic object-oriented database constructions such as inheritances, aggregations, and associations through using safety notions in Category Theory such as product, coproduct, pullback and pushout. And finally, different types of morphism (e.g. arrows, functors and natural transformations) provide a powerful and unified style for natural modelling of both dynamic (methods, operations, inferences) and static (attributes, properties, classes) aspects of the knowledge and database schemas. It also facilitates the manipulation of complex data structures, queries, and views.

The categorical DBMS is not intended to support more database concepts than other DBMSs. Rather it aims give a formal mathematical basis for modern object-oriented DBMS. After the initial tests and analyzing the results, it is evident that the categorical object model and the categorical DBMS can facilitate the intelligent system designers and programmers communicating with manufacturing engineers, as well as contribute to advanced knowledge manipulations. Future work of this research focuses on applying Category Theory to control knowledge refinement and fuzzy inference.

References

1. Bennich, P., Nielsen, H.: An Overview of GPS – A Cost Saving Tool. Institute for Geometrical Product Specifications, 1st edn. (2005)
2. Bhaskar, R.N., Raja, J.: A knowledge-based system for selection of surface texture parameters: a preliminary investigation. Surface Topography 1, 445–454 (1988)
3. Bui, S.H., Gopalan, V., Raja, J.: An internet based surface texture information system. International Journal of Machine Tools and Manufacture 41(13-14), 2171–2177 (2001)
4. CATIA 3D, http://www.catia.cz/fileadmin/Pictures_Menu/Catia/Catia./FTA.pdf
5. Bennich, P., Nielsen, H.: An Overview of GPS – A Cost Saving Tool. Institute for Geometrical Product Specifications, 1st edn., Denmark (2005)
6. Xu, Y.: Developing a Knowledge-based system for complex geometrical product specification (GPS) data manipulation. International Journal of Knowledge-Based System 24, 10–22 (2011)
7. Goguen, J.A.: A Categorical Manifesto. Mathematical Structures in Computer Science 1(1), 49–67 (1989)
8. Cattell, R.G.G., et al.: The Object Data Standard:ODMG3.0. ODMG. Morgan Kaufman, San Francisco (2000)

Research of Buoy with Floating Movement

Shiming Wang, Linlin Jiang, Jing He, Xianzhu Ai, and Xin Tang

Shanghai Ocean University, No.999, Huchenghuan Road, Shanghai, P.R. China
linlinjung@163.com

Abstract. This paper was proposed a new kind of method to change buoyancy of buoy. The buoyancy of the buoy was changed through lengthen and shorten hydraulic cylinder. This paper was given the structure diagram of the buoy and introduced the control progress of the buoy. At the second part the thrust of cylinder was adjusted. At the third part, this paper deduced the velocity of buoy's uniform motion by using mechanical analysis. At the last part, the equation of instantaneous acceleration and velocity starting from quiescence to uniform motion of buoy was deduced in two parts. One was the buoy getting provision volume by instance, the other was the buoy getting provision volume gradually.

Keywords: Buoy, cylinder, Intelligent Control Systems, Sensor, mechanical analysis, Computer Technology, instantaneous acceleration, velocity.

1 Introduction

There is a great deal of resources such as minerals, biotechnology, oil, and so on. And human exploration of the sea has never stopped. Ocean profile detection buoy system is the necessary comprehensive equipment for development of the sea and use of ocean resources. And it is one of the most important parts of the all day long stereoscopic observation. Argo buoy is currently widely used in the world. The United States, France and Canada have the production capacity of this submerged buoy, China had also been made into prototype and had a test in the East China Sea [1].

2 The Structure of Profiling Buoy

2.1 Introduction of Foreign Buoy Structure

The type of APEX buoy was developed by company of Webb of America. French Ifremer Institute developed Marvor buoy in 1990, then cooperated with Metocean of Canada company to develop PROVOR profiling buoy. The last two profiling buoys has now been accepted and applied by all the oceanographer around the world [2].

This type of buoy uses piston and bladder to achieve the target of rising and sinking: The oil is injected into the bladder by the pressure of piston pump, thereby to achieve the purpose of rising according to increase the volume of bladder volume; The oil is drown out of the bladder by the piston pump, thereby to achieve the purpose of sinking according to reducing the volume of bladder [3].

D. Yang (Ed.): Informatics in Control, Automation and Robotics, Volume 2, LNEE 133, pp. 733–740.
springerlink.com

2.2 Introduction of the Buoy Structure

The design of the buoy had Intelligent Control Systems, and it was made up by a hydraulic cylinder, hydraulic pump, vacuum pump, battery, circuit boards, Sensor and other components. The profiling buoy used hydraulic piston pump to push hydraulic cylinder elongation and shorten, then it could drive buoy's two shells (outer and inner) to overlap and separation through the length of the cylinder: When the hydraulic cylinder elongation, the hydraulic cylinder pushed the inner and outer shell separated from each other. The buoyancy of the buoy was increased by increased volume, thereby achieved the target of rising; When the hydraulic cylinder shorten, the inner and outer shells were pulled close to each other by inner vacuum and outer pressure. So the size of the buoy was decreased. And the purposed of sinking was achieved by reducing buoyancy.

The design of the buoy structure diagram showed in Figure 1.

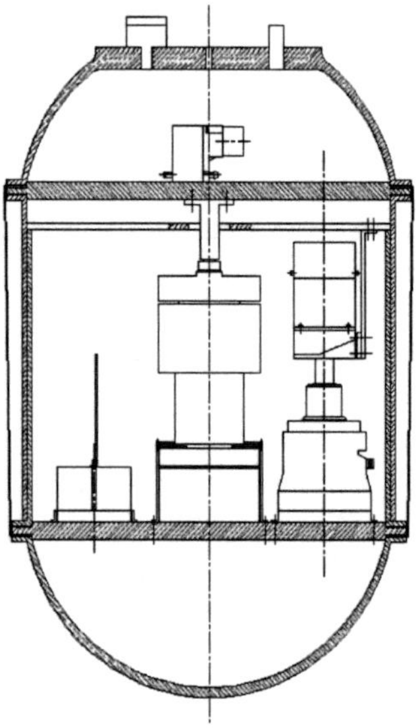

Fig. 1. The structure of buoy

2.3 Buoy Control Process

The manufactured buoy was thrown into the sea after setting up all the parameters. At this time the buoy's shells were separated from each other and the volume was the biggest. Also the buoyancy was the maximum. Then it communicated with the satellite to transfer the initial data. After this it dived into the sea at about 1000 meters for some time. The buoy continued to draw back shells and dived into about 2000 meters. Then the size of the buoy was increased steadily because of separating shells pushed by hydraulic cylinder. So it rose steadily. At the same time, the sensors started to profile survey and store the data. After reaching the sea surface, the shells stretched to the maximum in order to ensure that the buoy could float stable and communicate with satellite easily and conveniently. The buoy went into the next cycles after the data had been transferred completed until the battery ran out.

Buoy control process flow chart shown in Figure 2:

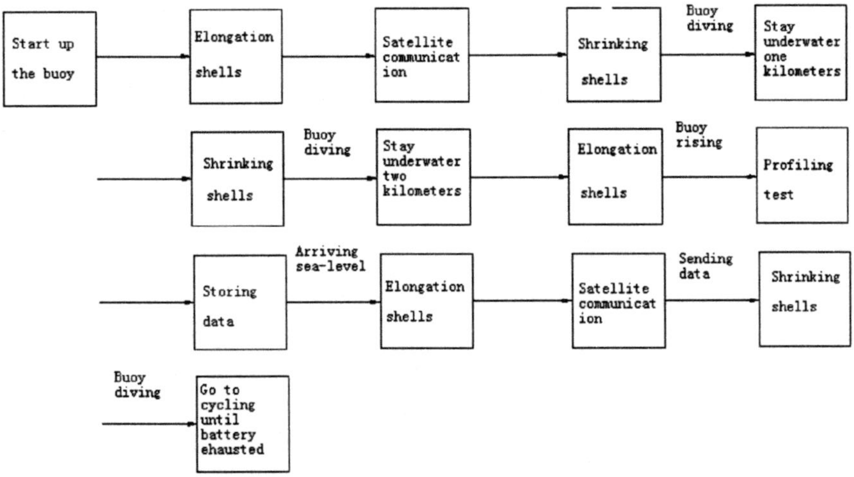

Fig. 2. The flow chart of buoy control process

3 The Push Force Calculation of Hydraulic Cylinder

The buoy needed to overcome the pressure generated by sea-water and the negative pressure generated by the vacuum to extend hydraulic cylinder.

$$F_{push} \geq F_{sea} + F_{vacu} \tag{1}$$

F_{push} stood for the thrust of hydraulic cylinder. F_{sea} stood for the thrust generated by sea- water. F_{vacu} stood for the negative thrust generated by the vacuum.

$$p_{sea} = \bar{\rho}_{sea} g z \tag{2}$$

$$F_{sea} = p_{sea} S_{buoy} = \bar{\rho}_{sea} g z S_{buoy} = \pi r^2 \bar{\rho}_{sea} g z \tag{3}$$

$$F_{vacu} = p_{vacu} S_{buoy} = \pi r^2 p_{vacu} \tag{4}$$

$$F_{push} = (A_1 p_1 - A_2 p_2) = \frac{\pi}{4}[(p_1 - p_2) D^2 + p_2 d^2] \tag{5}$$

ρ_{sea} density of sea-water

g acceleration of gravity

z the distance from sea level to buoy

r radius of the buoy

p_{sea} the pressure of the sea-water

p_{vacu} vacuum pressure

S_{buoy} the project area of buoy meet the flow

A_1 effective area of hydraulic cylinder rodless chamber

A_2 effective area of hydraulic cylinder rod chamber

p_1 the oil inlet pressure of hydraulic cylinder

p_2 the oil outlet pressure of hydraulic cylinder

D piston diameter

d piston rod diameter

The design of the parameters shown in Table 1

Table 1. Parameter value

Para	$\bar{\rho}$	g	z	r	p	p	p_1	p_2	D	d
Valu	1040	9.8	2000	20	20.3	-5	65	-5	25	12
Unit	kg/m^3	N/m	m	cm	MPa	MPa	MPa	MPa	cm	cm

After calculation F_{sea} =2.55 M Newton, F_{vacu} =0.628 M Newton, F_{push} =3.37 M Newton, so $F_{push} \geq F_{sea} + F_{vacu}$. Therefore, the hydraulic cylinder had the enough thrust to push shells separated from each other at the 2000 meters harbour..

4 The Basic Equation of the Buoy Rising Motion

For the text of the profiling process of the buoy, we could use the following assumptions. First, between 0-2000 meters there was no flow in the sea. Second, the destiny of the sea had no change between 0-2000 meters. Third, the buoy rose as the uniform motion.

Assume that the buoy rose as the uniform motion to the sea-level, then we could got the following equation:

$$F_D + mg = \bar{\rho}_{sea} g V_{buoy} \tag{6}$$

Because $F_D = C_D \frac{\bar{\rho}}{2} u_0^2 A$ [5], so the equation could be represented as follows:

$$C_D \frac{\bar{\rho}_{sea}}{2} u_0^2 A + mg = \bar{\rho}_{sea} g V_{buoy} \tag{7}$$

F_D resistance by the flow around the buoy

C_D flow resistance factor

u_0 buoy speed

A the projected area meet the flow of the buoy

m buoy weight

V_{buoy} buoy volume

Based on the equation above, we could got the speed of buoy at uniform motion as below.

$$u_0 = \sqrt{\frac{2(\bar{\rho}_{sea} g V_{buoy} - mg)}{C_D \bar{\rho}_{sea} A}} \tag{8}$$

According to the formula $\mathrm{Re} = \frac{u_0 d}{\gamma}$, we could got the Reynolds' Number .It was 2.1×10^4. The d represented buoy diameter, and the γ represented the water's kinematical Viscosity. When the Reynolds' Number was $\mathrm{Re} = 10^3 \sim 3 \times 10^5$, the flow resistance factor was $C_D \approx 0.55 \sim 0.65$. Then we chose the average value 0.6. The buoy's weight was 35 kilograms. Put the data into equation(8), then we could get the minimum volume. It was 33.7 L.

5 The Study of Buoy Movement

5.1 The Buoyancy Remain the Same during Rising

5.1.1 The Establishment of the Instantaneous Rising Acceleration

From the static status to the uniform motion, the speed of the buoy was a accelerating process. And acceleration became smaller and smaller. Assuming this process remained the same buoyancy buoy, the buoy rising process of dynamic balancing equation can be expressed as:

$$m\frac{du}{dt} = F_{float} + F_{weight} + F_{vm} + F_b \tag{9}$$

F_{vm} represented false mass force, F_b represented Basset force [6]. These two forces were too smaller than buoyancy and gravity, so they could be omitted. The equation of dynamic balance process could be expressed as:

$$m\frac{du}{dt} = \bar{\rho}_{sea} g V_{buoy} - m g - C_D\frac{\bar{\rho}_{sea}}{2} u^2 A \tag{10}$$

Therefore, the instance acceleration of rising process was as following:

$$\frac{du}{dt} = \frac{\bar{\rho}_{sea}}{m} g V_{buoy} - g - C_D\frac{\bar{\rho}_{sea}}{2m} u^2 A \tag{11}$$

5.1.2 The Establishment of Speed Equation during Acceleration Process

According to equation(11), we could conclude that:

$$\frac{m}{\sqrt{2(\bar{\rho}_{sea} g V_{buoy} - m g) C_D \bar{\rho}_{sea} A}} \ln\left|\frac{u - \sqrt{2\left(\frac{\bar{\rho}_{sea} g V_{buoy} - m g}{C_D \bar{\rho}_{sea} A}\right)}}{u + \sqrt{2\left(\frac{\bar{\rho}_{sea} g V_{buoy} - m g}{C_D \bar{\rho}_{sea} A}\right)}}\right| = t + C \tag{12}$$

From the static status at the depth of 2000 meters to sea-level, the buoy began to rise. We put the initial time $t=0$ and $u=0$ to equation (12), then we concluded that $C=0$. So the equation(12) could express as [7]:

$$\frac{m}{\sqrt{2(\bar{\rho}_{sea} g V_{buoy} - m g) C_D \bar{\rho}_{sea} A}} \ln \left| \frac{u - \sqrt{2(\frac{\bar{\rho}_{sea} g V_{buoy} - m g}{C_D \bar{\rho}_{sea} A})}}{u + \sqrt{2(\frac{\bar{\rho}_{sea} g V_{buoy} - m g}{C_D \bar{\rho}_{sea} A})}} \right| = t \quad (13)$$

We could get the time $t = 10s$ when the buoy reached the final velocity.

5.2 The Buoyancy Increased Gradually during Rising

5.2.1 The Establishment of the Instantaneous Rising Acceleration

The buoy began to rise from the initial stationary state a the 2000 meters. In 4.1 we assumed that its buoyancy had remain unchanged. But buoy volume was increased through elongation of the hydraulic cylinder, and this process required a certain amount of time. In this process the buoy had risen by some distance. Therefore, the buoyancy increased gradually until the required volume. Because buoy volume increasing and elongation of the hydraulic cylinder was linear, combined with the computer technology, we might wish to make the following assumptions:

$$V_{buoy} = kt + b \quad (14)$$

We could concluded by calculation that $k = 5.2 \times 10^{-4}$, $b = 0.02$. So the instantaneous rising acceleration of the buoy could express as:

$$\frac{du_p}{dt} = \frac{\bar{\rho}_{sea}}{m} g\ (kt+b) - g - C_D \frac{\bar{\rho}_{sea}}{2m} u^2 A \quad (15)$$

5.2.2 The Establishment of Speed Equation during Acceleration Process

Putting all the parameters into equation(15), we could conclude the following equation:

$$\frac{du_p}{dt} = -1.12 u_p^2 + 0.15t - 3.976 \quad (16)$$

This was an ordinary differential equation, so the speed could be calculated by Matlab that [8], [9].

u_p =-i*(-1875/7*i)^(1/3)*(C1*AiryAi(1,-1/35*(-1988+75*t)*35^(2/3)/(-1875/7*i)^(2/3))+AiryBi(1,-1/35*(-1988+75*t)*35^(2/3)/(-1875/7*i)^(2/3)))*35^(2/3)/(140*C1*AiryAi(-1/35*(-1988+75*t)*35^(2/3)/(-1875/7*i)^(2/3))+140*AiryBi(-1/35*(-1988+75*t)*35^(2/3)/(-1875/7*i)^(2/3)))

(17)

C1 is an arbitrary constant. We put the initial time t=0 and the initial speed u_p =0 into the equation (17), then we could get the equation of velocity as follows:

u_p =5*(-AiryBi(1,-1988/375*7^(1/3)*3^(1/3))/AiryAi(1,-1988/375*7^(1/3)*3^(1/3))*AiryAi(1,1/375*7^(1/3)*3^(1/3)*(-1988+75*t))+AiryBi(1,1/375*7^(1/3)*3^(1/3)*(-1988+75*t)))*7^(1/3)*3^(1/3)/(-28*AiryAi(1/375*7^(1/3)*3^(1/3)*(-1988+75*t))*AiryBi(1,-1988/375*7^(1/3)*3^(1/3))/AiryAi(1,-1988/375*7^(1/3)*3^(1/3))+28*AiryBi(1/375*7^(1/3)*3^(1/3)*(-1988+75*t))) (18)

6 Conclusion

1、This paper proposed a new kind structural approach of floating and sinking for the profiling float. The volume of buoys was changed by pushing the inner shell and outer shell separation and overlap through hydraulic cylinder. So it could ultimately to change the buoyancy of the profiling float.

2、The thrusting of the hydraulic cylinder was calculated check. It was calculated that there had enough thrusting to overcome the pressure by the sea water and buoy vacuum chamber. So the inner and outer shell could be pushed by the hydraulic cylinder.

3、The method of stress analysis was used to derivate the equation of the process of floating buoy motion. The speed formula was obtained and the minimum volume required for the buoy to float was calculated.

4、The equation of the instantaneous acceleration and the equation of the speed from stationary to uniform motion was derived by Newton's theorem.

References

1. Shi, S., Wang, J., Jiang, X.: The study on the influence factors of micro - bubble floating process in the water. Journal of Naval University of Engineering 6(3), 83–87 (2008)
2. Wang, S., Wu, A.: Application of hydraulic technology in ARGO buoy. Fluid Power Transmission and Control, 37-40 (2010)
3. Liu, Y., Guo, Z., Zhu, B.: Underwater tool system driven by seawater hydraulic. Marine Technology 12, 65–69 (2006)
4. Yu, L., Zhang, S., Shang, H.: Design and research of Argo buoys in China. Marine Technology 6, 121–129 (2005)
5. Xia, T.: Engineering fluid mechanics. Shanghai Jiaotong University press, Shanghai (2006)
6. Che, D., Li, H.: Multiphase flow and its application. Xi'an Jiaotong University press, Xi'an (2007)
7. Zhang, Z.: Higher mathematics. Lanzhou University press, Lanzhou (2006)
8. Zhang, D., Ma, Y., Xu, M.: MATLAB higher mathematics experiment. Electronics industry press, Beijing (2008)
9. Fang, Y.: The finite recursive method for solving a class of constant linear coefficients ordinary differential equation. Practice and Cognition of Mathematics 9(17), 246–250 (2009)

A Image Retrieval Method Based on Color Histogram

Wu Junyun

School of Information Engineering, Nanchang University, 330029, Nanchang, China

Abstract. This paper first analyzes the key technologies of image retrieval based on color features: the transfer of RGB, HSI and HSV space, Principles and characteristics of the color histogram, proposes a image retrieval method by feature of color histogram. Finally, the paper verifies the effectiveness of the algorithm.

Keywords: Image retrieval, Color specification, Color histogram, Similarity distance.

1 Introduction

Currently, content-based retrieval is the most popular study at image retrieval, the color is one of the most common and basic content. This paper blocks an image through the study of the color histogram, Compares the color distribution of each small block with the sample images, then finds some small piece of similar picture, arranges the image according to the number of small block images similar and image similarity, the higher similarity image arranged in the front queue of similarity images. It is no longer appeared after when the similarity below a certain value.

2 Color Model

Color is the most significant feature of color image, compared with other features, color features are very stable, and insensitive to rotation, translation and scale changes, showing a strong robustness, because the calculation of the color characteristics is simple, so color-based queries become most basic method of application at content-based image retrieval system.

MPEG-7 international standard officially recommended a color descriptor [1]--color space. It gives the colors can be used to describe the color spaces: RGB, HSI, HSV, etc.

2.1 RGB Model

There are several ways to quantitatively express the color, such as the color of pixel in color digital image. The most direct way is to use red, green and blue brightness value, the size can be limited to a certain extent, such as {0, 1, ...255}, we call this convention known as RGB format. Each pixel (actually any color may be quantified) can be used in the first quadrant of the three-dimensional space to represent a point.

D. Yang (Ed.): Informatics in Control, Automation and Robotics, Volume 2, LNEE 133, pp. 741–746.
springerlink.com

2.2 HSI Model

Another useful program by the color hue, saturation and intensity (or brightness) component of the HIS(Hue, Intensity,Saturation)format. It is Munseu's color system format, often used by artists. This design reflects the way that people observe the color, and conducive to image processing.

In the HSI format, I expresses intensity or brightness. H which determined by wavelength of the visible spectrum, is the basic characteristics of color light. S reflects the color shades, depending on the color content of white light, which is the color purity of light reflected. HSI coordinate S is color space for cylindrical, gray-tone black along the axis to change the bottom to the top of the white. The color with the highest brightness in the maximum saturation lies on the top surface of the cylinder circumference.

2.3 HSV Model

Although the RGB color space is the most common, but RGB space structure is not similar to the color of people's subjective judgments. Therefore, it was proposed a model based on the HSV(Hue, Saturation, Value) space. HSV color model corresponds to a conical subset in a cylindrical coordinate system, cone at the top of the face should be V = 1, which contains RGB model R = 1, G = 1, B = 1 of the three faces, representing the colors bright[4].

Color H by the angle of rotation around the V-axis given. Red corresponds to the angle of 0 degrees, green corresponds to the angle of 120 degrees, blue corresponds to the angle of 240 degrees. In the HSV color model, each color and its complementary color difference of 180 degrees. Saturation S values from 0 to 1, so the top surface of the cone radius is 1. In the cone of the vertex, V = 0, H and S is not defined, represents the black.Cone of the top representatives the white center. The circumference of the cone on the top color, V = 1, S = 1, the color is solid. HSV model is shown in Figure 1:

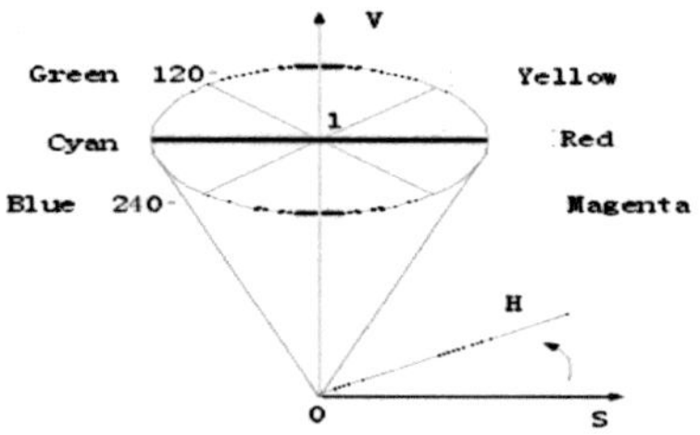

Fig. 1. HSV model

3 Color Feature Extraction and Expression

One of the key technology by using Color feature for image retrieval is the color feature extraction [2]. Image color characteristics can be the proportional distribution of various colors and color spatial distribution, Currently, most color-based retrieval systems use color ratio distribution to obtain the color feature, namely the field of color histogram method [3]. It is relative to the geometry, has the inherent rotation invariance and scale invariance.

Color histogram is a representation of the most common features. The basic idea is statistical probability of each color in the color space. Histogram describes the statistical characteristics of image color, reflecting the statistical distribution of the image colors and basic colors. Specifically, for an image I, its color components from the L (C_1 , C_2 ,..., C_I ,...C_L) level, C_I is the color value of the i-th level. In the whole image, the pixel value with the CI for the hi, then a group of pixel statistics h_1, h_2, ...,h_L is called the color histogram of the image. With H H(h_1,h_2, ...,h_L)representation.

Traditional color histogram are mainly three kinds: Jain proposed three separate one-dimensional histogram; Swain, who proposed three-dimensional histogram; Mehtre and others refer to the color table based on the color histogram. These three methods have advantages and disadvantages, one-dimensional histogram method reduces the number of dimensions, but did not take into account the correlation between color; three-dimensional color histogram method takes into account the correlation, but the reference color table is static, making it only suitable for trademark image database retrieval [5,6].

4 Color Characteristics of the Similarity Measure

For a color feature representation, the distance measurement methods can have a variety, there are eight kinds of commonly used metrics about color histogram: histogram intersection, histogram Euclidean distance, histogram cosine distance, the second set of values, Hammimg distance, histogram quadratic distance metric, and second, the second value sets the distance, Mahalanobis distance histogram. Here we only briefly describe the system used histogram intersection and histogram Euclidean distance [7,8].

4.1 Histogram Intersection Method

Histogram intersection algorithm: according to the color index to retrieve the color histogram of query image, its image library with the histogram of each image to take the intersection, according to the value of the intersection to select the best match image.

$$P(Q,D)=\frac{\sum_{k=0}^{L-1}\min[H_Q(k),H_D(k)]}{\sum_{k=0}^{L-1}H_Q(k)} \tag{1}$$

Where P is the query image histogram, Q is the image library in the histogram of any image, K is the two images match, its value closer to 1, the more similar the two images.

Histogram intersection method can conduct a detailed comparison of two images, however, for many synthetic images, such as trademarks, they have a lot of the same color, three-dimensional histogram is only one domain value is high, and many other changes in the domain of color information is limited. Thus, too small to compare this image is unnecessary.

4.2 Euclidean Distance Method

We can also use the Euclidean distance formula (Euclidean Distance) to calculate the image distance.

$$D=\sqrt{(H_1-H_2)^2+(S_1-S_2)^2+(V_1-V_2)^2} \qquad (2)$$

This distance formula is simple, but it has nothing to do with a specific color distribution, it is easy to calculate.

5 Evaluation Criteria

In this paper, we use the recall rate and precision to evaluate the results.

At present, people often use recall and accuracy of statistical data to curves for evaluation. As a comprehensive search query, description, matching, extraction processing, there is the possibility of success and failure. According to the principle of pattern recognition, as shown in Table 1 obtained the four circumstances.

Table 1. Basic parameters of retrieval system performance

Correlation / Search Results	association	No association
retrieved	A（Correct retrieval)	B（Error retrieval)
Not retrieved	C（Leak check)	D(Correctly rejected)

With the basic parameters of Table 1, defines the recall and precision of the retrieval system used to indicate the performance.

$$recall=\frac{A}{A+C} \qquad precision=\frac{A}{A+B}$$

The method of qualitative evaluation is too simple, individual query results with one-sidedness, the system can not fully explain the effectiveness of retrieval methods.Therefore, we use recall and precision for the retrieval evaluation.

6 Test Results

In this study, all the images are a number of continuous fishing pictures and other miscellaneous pictures. There are 64 pictures in the library, which 10 are about the fishing picture. Figure 2 shows a sample retrieval.

Fig. 2. Sample retrieval

This paper takes characteristics of image the sample and the characteristics of the image database similarity matches of greater than or equal to 50% of the images as search results; meanwhile, the search results of the recall rate and precision can only show the sub-query results. Figure 3 shows the results of search

Fig. 3. Search results

According to Figure 3 search results, we can get conclusion that:

$$recall = \frac{10}{10} = 100\%$$

$$precision = \frac{10}{10+5} = 66.7\%$$

To test the search results and efficiency, this experiment used a total of seven different pictures made seven experiments, the results shown in Table 2:

Table 2. Analysis of search results

Sample retrieval	1.JPG	2.JPG	3.JPG	4.JPG	5.JPG	6.JPG	7.JPG
Associated with the correct result	8	8	8	8	8	8	8
Retrieved all the results	10	13	4	11	8	21	15
recall	100%	100%	50%	100%	100%	100%	100%
precision	80%	61%	100%	73%	100%	38%	53%

7 Conclusion

Image retrieval techniques based on color feature have been studied and discussed in this paper, and analysis of some search algorithm by color feature of image, design and implement a typical image retrieval system through use of the color histogram similarity matching. The system analyzes the color characteristics based on user-supplied sample images, searches image database, and similarity to the query image according to the similarity with the sample image from small to large sort, finally returns the result to the user.

References

1. Sun, J.D., Wu, X.S.: Image retrieval based on main color and shape of image blocks. Opto-Electronic Engineering 33(12), 85–90 (2006)
2. Liu, Z.W., Zhang, Y.J.: A Comparitive and Analysis Study of Ten color Feature-based Image Retrieval Algorithms. Signal Processing 16(1), 79–84 (2000)
3. Hang, Y., Yang, Y.B., Chen, Z.Q.: A Survey of Content-based Image Retrieval. Application Research of Computers (09), 9–14 (2002)
4. Song, M.L., Li, F.: An Image Retrieval Technology Based on HSV Color Space. Computer Knowledge and Technology (01), 200–207 (2007)
5. Gao, L., Xu, C.M.: Image Content Retrieval Based on Color Histogram Similarity. Computer Engineering and Science 25(3), 52–55 (2005)
6. Yin, J.Q., Tian, G.H., Nin, C.L., Lu, F.: A Color-based Image Retrieval Method. Journal of Image and Graphics 8(9), 1033–1036 (2003)
7. Hu, Y.J.: Image Retrieval Based on Sub Regions and Primary Colors. Journal of Hunan Industry Polytechnic 5(3), 19–21 (2005)
8. Huang, Y.Y., He, Y.F.: A Color Based Image Retrieval Method. Journal of Image and Graphics 11(12), 1768–1772 (2006)

The THP Three-Dimensional Model of Modern Enterprise Operation*

Lili Zhang[1] and Jianhua Ren[2,**]

[1] Postgraduate, Teachers College, Shihezi University, China
[2] Dean of Education Department, Teachers College, Shihezi University, China 69121 Shihezi, China
zjcrjy@126.com, jkyrjh@263.net

Abstract. The essence of operation in the modern sense is a kind of purposeful activity of organizations. It's a series of planning and management activities implemented by organizations in order to achieve their fundamental purposes. It has hierarchy and type features. Based on this understanding we defined the concept of modern operation, and proposed the THP model of modern enterprise operation, to provide a new theoretical perspective for modern business operation.

Keywords: Operating, modern operation, operation mode, THP three-dimensional model of modern enterprise operation.

1 Introduction

At present the word of operation is extremely widespread used. In addition to enterprise organizations commonly used operation concept, the frequency of the concept being used by non-business organizations increased year by year. There were something like city operation, school operation, land operation, expert operation, operating life, etc., began to be familiar and used. We believe that the meaning of operation used by non-business organizations should have relationships and differences with that used by business organizations. It is a meaning in modern sense, which enriches the operation meaning in business. This paper defined the connotation of modern operation, by systematically collected the English meaning, the Chinese meaning and enterprise operation three aspects of the operation concept, and further proposed a THP three-dimensional model of modern enterprise operation.

* The article is stage achievement of Theoretical framework and practice model research of school operation (RCSX200903), high-level professionals project of Shihezi University.
** Corresponding author.

D. Yang (Ed.): Informatics in Control, Automation and Robotics, Volume 2, LNEE 133, pp. 747–752.
springerlink.com

2 The Three Sources of Modern Business Meaning

2.1 Chinese Intention of the "Operation"

Operation in "Modern Chinese Dictionary" has two connotations, one is "planning and management (enterprise, etc.)", and the other is "planning and organization". "Ci Hai" interprets operation as "manager's planning and management. It has meanings of exploitation, development and planning future." There are two intentions of operation in "Ci Hai", one is "Longitude to create", comes from the "Poetry • Daya • Lingtai": "The beginning of Lingtai, after the camp." means planning and management, also refers to artistic conception. Another explanation is "still contacts, "which comes from Du Fu's "Dancited": "Chao said that the general Buddhist silk elements, images dismally in." These show that the Chinese intentions of operation are "planning", "management", "conception ", and so on.

2.2 English Intention of the "Operating"

The English words correspond to operating are manage, run, operating and so on. On the level of internal resources allocation, manage and operating are semantic same. The main meaning of run in business operation is included in "operating". Therefore, we use operating to express it. According to Webster's Ninth New Collegiate Dictionary, Operating mainly has four significances. The first is "to perform a function", the second is "to produce an appropriate effect", the third is "to perform an operation or a series of operation, to perform surgery, to carry on a military or naval action or mission", and the last one is "to follow a course of conduct that is often irregular ". Thus, the English intention of operating on is mainly refers to "operation ","work", "implement" and "activities".

2.3 The Meaning of Operation in Enterprise Business

Generally, "operation" refers to "businesses' operating". Business activities fully reflect the meaning of "operation" in various levels. Enterprise operation is constituted by technology, business, finance, security, accounting and management six activities, says Fayol, one of the main representatives of classical organization theories. Chen Jiagui argues that corporate operation is decision-makings for its survival, development and achievement of its strategic objectives and efforts on all aspects in order to implement these decisions. It includes economic forecasts, economic decision-makings, embodying the enterprise's development directions and goals in a variety of programs and work in order to achieve the development goals of the enterprise (such as financial activities, supplying of production, production and sales, marketing development, new product development, production organization and administration reform, the development of collaborative relationships with other companies, etc.). Zeng Zhaoxiang states that operation is an enterprise engages in production, trading (buying and selling) as well as all of these directly related substantive business activities. Liu Jisheng put it this way: Business operation and management is people who have abilities to dominate the enterprise's economic activities under the socialist market operating conditions, facing the needs of the market and users, to balance all internal and external conditions and possibilities of the enterprise, to consciously use the law of value, and to achieve business goals through a series of planning and operating activities. Zheng Guangcai considers enterprise operation is a series of organized activities of the enterprise that makes markets as its objects, commodity production and trading as means, and dynamically balances its business activities such as investment, production and marketing, and its external environment to achieve its business goals.

2.4 The Understanding of the Meaning of Operating by Li Wuwei

Li Wuwei believes that operating has a broad meaning at present stage. it includes both commodity operating and assets operating. It is operating not only on owned assets, but also on all available resources of society. It's operating not only on the material resources, but also on human resources, knowledge and information resources. As per Li Wuwei, the transform of enterprise from commodity operating to assets operating is a leap. Now the enterprise operating faces a new jump, that is, the altering from its own assets operating to social resources operating. Therefore, Li Wuwei gives an innovative definition of the operation concept: operation is using possibly minimal capital investment and mobilizing more social resources, to realize the value-added by a certain innovative activities, such as design, processing, manufacturing, packaging, reorganization, trade, exchange, etc. Broadening the operation concept can greatly enlarging companies' business ideas. Many companies feel stress and are at a loss under the situation of the relative surplus of supply and the under-utilized social production capacity of the current society. However, with new operation point of view, the other side of the relative surplus of production is the relative rich of idle resources or available resources of the society. Thus, it is more likely to mobilize a large social resource with a smaller piece of capital input, obtaining value-added and hence a considerable output efficiency through a certain innovation. This mode of operation requires businesses transforming from commodity operator to assets operator becoming resource integrators, business operation becomes operating enterprise and operating resources.

3 The Meaning and Characteristics of Modern Operation

Through analyzing the Chinese meaning and the English meaning of operation, the evolution of the use of operation in enterprise operation, and the actual use of operation in business operation in modern sense, we have a basic outline about the meaning of modern operation.

3.1 The Essence of Operation

The essence of operation is a kind of purposeful activity of organizations. The purpose of operation is improving resources efficiency and resources operating efficiency, through optimal allocation and integration on their owned and controlled resources, this is the essential property of operation. For enterprise organizations, the purpose of operation is profit. The fundamental way of making profit is to fully enhance the efficiency of resources and operational efficiency, to realize outputs excess inputs to make profit, through optimal allocation and integration on resources. We believe that the basic approach to make profit in enterprise operation is to improve the resource allocation efficiency and benefit. Pursuing higher efficiency and higher benefit is the fundamental purpose of operation activities, and also the new essential implication of modern operation. Business organizations make it the basic way of making profit, whereas non-business organizations regard it as the essential nature of operating activities.

3.2 The Basic Forms of Organization's Operation Activities

Operation is a series of planning and management activities implemented by organizations in order to achieve their fundamental purposes. The basic forms of organization's operation activities are operation environment analysis, the establishment of operation ideas, the determination of operation target, the selection of operation strategy and the determination of business operation mode, etc.

3.3 Operation Activity has Hierarchy Feature

The essence of operation is implementing planning and management activities for its owned resources or controlled resources. Since resources are not only the resources owned by organizations, but also those owned by non-organizations, according to whether they are owned by organizations, there are classified as operation on the organizations' owned resources and operation on external resources. This is the hierarchy feature of operation. For different levels of operation activities, the content and form of its basic operation activities take different modes.

3.4 Operation Activity Has Type Feature

Because operation activity is an organizational behavior, organizations are of different type and organizational form. There are enterprise organizations and non-business organizations, and among enterprise organizations there are organizations of different business areas. So operation activity have type feature. Organizations of different types have different forms of operation activities.

These four aspects mentioned above are the basic characteristics of modern operation. Among those characteristics the optimal allocation of resources which enhances the efficiency and effectiveness of activities is the essential property of organizational and operational activities. The process of operation activity, the hierarchy of operation activity and the type of operation activity are three basic features of modern operation. Accordingly, we define the operation concept in the modern sense as that operation is the organization from the reality of their own behavior characteristics, aims at improving the efficiency and effectiveness of resources, lets the multi-level optimal allocation and integration of resources as basic path, implements a series of planning and management activities including the operating environment analysis, the establishment of operation ideas, the determination of operation targets, the selection of operating strategies and the determination of business operating modes, etc.

4 The THP Three-dimensional Model of Modern Enterprise Operation

The definition of modern operation shows that the modern operation has relatively independent internal regulations of type of operating, hierarchy of operating and process of operating three aspects, constituting three-dimensional states of modern operation activities. It is constituted by the type of operating, hierarchy of operating and process of operating three dimension, we call it THP three-dimensional model of modern operation. Any operating activities, operating behaviors or operating model, are integral parts of operating state.

4.1 Dimension of Operating Type

Operations of organizations of different types have different characteristics. Operating work must proceed from the organization specific positioning. Corresponding to the classification of business organization, there operations of different types. From the angle of business organization forms, it's mainly divided into a single enterprise operating, multiple enterprise operating, enterprise Consortium operating and enterprise group operating, etc. To differentiate from the angle of enterprise property rights, there are mainly state-owned enterprises operating, multiple investor enterprises operating and private enterprises operating, etc. To divide operating according to the attributes of products provided by company, there are mainly operations of enterprises which provides simple public goods, operations of enterprises provides quasi-public goods and operations of enterprises which provides human capital goods, etc. From the business areas view, it is mainly divided into industrial enterprises operating, commercial enterprises operating, service enterprises operating and Information enterprises operating, etc.

4.2 Dimension of Operating Hierarchy

The operation work of modern enterprise has hierarchical feature. For operation work of different hierarchy, the means and ways of its resources allocation have different forms. From the view of enterprise resource allocation, the enterprise operation generally is divided into micro-level of internal resources operation, mid-level of external similar resources operation and macro-level of the operation of all external resources. Enterprise operation of Micro-level was called enterprise production operation, whereas mid-level and macro-level operations were called industry operation. At each level, because of different forms of resources, enterprise operation has resources operation, assets operation and capital operation three forms. The enterprise operation form of micro-level is mainly resources operation, aiming at improving resources efficiency. This is more similar to the content involved in enterprise management. The industry operation of mid-level and macro-level must rely on the means in assets operation and capital operation, or the means of information technology, such as chain operation, monopolized operation and network operation, etc.

4.3 Dimension of Operating Process

Modern enterprise operation is an organic entirety composed by five basic activities of relatively independent significance of the analysis of operation environment, the establishment of operation ideas, the determination of operation goals, the choice of operation strategies and the determination of means operation. The whole operation process contains the five aspects of operation activities.

References

1. Jin, X., Ren, J.: Concerning School Operation. Beijing normal University Journal (Humanities Social Science Edition) (4) (2002)
2. Ren, J.: Operating schools: A New Education Economics Theory Category. Journal Victory Education (3) (2002)
3. Ren, J., Geng, L.: Do school management, Help Midwife Present. China Education Quote (7) (2002)
4. Ren, J., Geng, L.: School Operation theory, the Basic Framework. Journal of Education and Economic (2) (2003)
5. Mao, Y., Ren, J.: The School Types and Characteristics of Education Economics Analysis. Journal of Modern Education Scientific Research (6) (2003)
6. Ren, J.T.: The business activities of 3dModel. Journal of Business Research (8) (2003)
7. Ren, J.-H.: Operating School: a New Theory of Education Economy. The Victory of Education (3) (2002)

Drowsiness Detection from EEG Spectrum Analysis

Yabo Yin, Yunkai Zhu, Shi Xiong, and Jiacai Zhang

College of Information Science and Technology, Beijing Normal University,
No.19, Xinjiekouwai Street, Haidian District, Beijing, P.R. China, 100875
jiacai.zhang@bnu.edu.cn

Abstract. Drowsiness is a transient state between consciousness and sleep. In some situations, the operators' drowsiness raises the risk of major accidents for the lack of agility. Prediction of drowsiness using EEG signals has become the hot topic in the EEG research community. drowsiness states. However, modeling the drowsiness level mathematically from EEG data suffered from the label problems. Drowsiness is subjects' state of feeling lacking motivation and alertness, which is difficult to be measured at the time of recording the EEG signal. In this paper, we put forward an EEG labeling method employing K-means clustering to separate EEG signal recorded in consciousness and drowsiness states. EEG dataset is divided into two categories according to the EEG rhythmss' spectrum pattern, and assigned label of drowsiness or consciousness. Comparative study showed thatαand β wave in EEG correlated with the drowsiness level. We also designed a LDA classifier trained with the labeled EEG data, and used it to classify the EEG data into consciousness and drowsiness states. The high classification accuracy illustrates the method put forward in this paper can distinguish these two states (i.e. drowsiness and consciousness) with a high recognition rate.

Keywords: Drowsiness, EEG, K-means, LDA.

1 Introduction

Facing stronger and stronger competition in the process of industrialization, the increasing of the operators' working strength and working time leads to a situation that they will continue working while sleepy. This will inevitably lead to the decline of the work efficiency, and even cause many serious safety misadventures. Drowsiness is the cause of a large number of severe car accidents and industrial accidents. Preventing such accidents caused by drowsiness has become a serious concern to society but requires techniques for continuously detecting, estimating, and predicting the level of drowsiness of drivers or operators.

There are various signals for drowsiness detection, such as frequent blinking and heavy eyelids, frequent yawning or rubbing eyes, drifting from lane or tailgating, and so on. However, these indexes are not central nervous signals and there exists some defects when using them to describe drowsiness which belongs to the brain' subjective feeling. Eye-activity variables as examples, according to those informed researches, they can reflect the drowsiness of subjects, But calculating the eye-motion frequency

D. Yang (Ed.): Informatics in Control, Automation and Robotics, Volume 2, LNEE 133, pp. 753–759.
springerlink.com

requires a relatively long moving- averaged window aiming to track slow changes in vigilance. What's more, these non-central nervous signals can be fabricated easily. On the other way around, EEG signal shows the brain tissue's activities and its functional state. And the EEG-based method can use a shorter moving-averaged window to track second-to second fluctuations in the subject performance [1].

Research in cogntive science has proved that normal subjests' EEG signal showed enhanced power in high frequency (beta rhythm) around their prefrontal in attention states, at the same time, the power in low frequency(alpha rhythm) decreased[2]. NASA also researched on the prediction of engagement and decided whether the operator performed manually or automatically. NASA measured the engagement with the ratio of the power of high frequency to low frequency[3-4].

At present, there is still no systemic study for EEG features about drowsiness. We put forward a labeling way for drowsiness level in this paper. This method combines the design of experiments for data acquisition with the analysis of K-means clustering, and labels the data with drowsiness or consciousness automatically. By using the clustering algorithm, we divide the EEG data which record the subject's complete state-changing from consciousness to sleepiness and then consciousness again into two categories. This prevents the problem of manually finding the divide point of drowsiness and consciousness. And then, we divided the labeled EEG data into two sets. One is the training set and the other is testing set. The former is used to train the LDA classifier, while the latter is used for testing it. The consequence shows the method can distinguish these two states (i.e. drowsiness and consciousness) with a high recognition rate.

2 Data Acquisition and Analysis

2.1 EEG Dataset

Our subject is a normal college student (age of 22). He has accepted pre-training before the experiment. The Off-line training include manually trained him to rest regularly.after his lunch. For we need a whole state-changing process from wakefulness to drowsiness and then wakefulness again, we recorded his lunch break (about 30min) and manually trained him to rest regularly. Keep training for a few days. We can begin the formal experiment when the subject formed a regular lunch break.

During the experiment time, the subject is seated in a small room with limbs relaxed. He is asked to keep the head still and try not to make big action. As the experiment is conducted during his lunch break, he will soon feel sleepy. Since the system starts to collect data, we whisper to the subject every 3 minutes. If hearing what we say, he is asked to move his index finger to show wakefulness. Otherwise, we will mark that he has stepped into drowsiness at a specific time.

After the experiment, ask the participant whether he has been drowsy. Those data collected in the whole process from wakefulness to drowsiness and then wakefulness again are effective.

We use the Biosemi ActiveTwo EEG acquisition system in our laboratory to attain the EEG data. This system, whose sampling rate is 256Hz, will automatically record the EEG signal which meets needs (i.e. the amplitude upper limit is 100μV and the frequency ranges from 0.16Hz to 100 Hz). The experiment used 64 electrode channels distributed on the scalp according to the 10-20 reference system. The data acquisition lasts 9 minutes, during which the Biosemi system has recorded the subject's EEG signal.

2.2 Feature Selection

EEG signal contained several wave bands(rhythms) according to the frequency, such as δ wave (0.1~3Hz), θ wave (4~7Hz), α wave (8~12Hz), SMR wave (12~15Hz), βl wave (13~20Hz), βh wave (21~30Hz). The powerspetrum of wave bands are correlated with wakefulness states.

Before processing the data, we should do some pretreatment, such as head-moving artifact and power frequency interference removal. Then we use the Level 3 Chebyshev filter to extract six rhythms waves mentioned above. Through observation, we find out that δ, θ, α and βl showed the highest correlation with the subjects' drowsiness level. Therefore, we use these 4 waves to do following processes. We calculate the waves' power with the unit of 4 seconds. So the whole 9 minutes can be divided into 135 parts without overlap.

In addition, in view of the nonstationarity of the EEG signal, we do the feature extraction based on the concept of windowed Fourier transformation [5]. We treat the EEG data as a series of Short-term stationary signal. But no matter how short the time window, it is still nonperiodic. In order to reduce this influence, the time window is moving in the half overlapped way. So we ensure the locality in either time domain or frequency domain and make the result reliable.

2.3 K-Means Clustering

J.B.MacQueen is the one who firstly put forward the theory of K-means [6]. It's a basic clustering method. Using Euclidean distance to measure the distance between sample points, K-means method divides the dataset in m-dimensional space into K subsets. In this study, we expect to decide whether the sample time-points of the subjects' EEG data belong to the cluster of drowsiness or not, which means the exact number of clusters here is 2. Therefore, the K-means clustering method, using iteration to optimize the result, is adequate for our processing.

Defined our data matrix (n elements) as D in n-dimensional space where m_i (i=1,2) represents the initial central of the two subsets and p_j (j=1,…,n) denotes one point in the set D. Meanwhile, d (i,j) represents a measure of the distance between points.

Specific steps as follows.

1. Select the two starting points (subsets' central) randomly from D as m_i.
2. Calculate the distance from the two central to all the other data points in D.

$$d(i,j) = \sqrt{|p_j - m_i|^2} \quad (1)$$

3. Find the minimum value min(d(*i*,*j*)), and then put every p_j into the closet class.
4. Recalculate the centroid of the set separately as the new central of each subset.

$$m_i = 1/n_i \sum_{j_i=1}^{n_i} p_{j_i} \quad i=1,2; \tag{2}$$

5. According to

$$\mathrm{E} = \sum_{i=1}^{K} \sum_{p \in p_i} |\mathrm{p} - \mathrm{m_i}|^2 \tag{3}$$

we get the squared error E(t) of each sample in D. Then compare them to the previous ones E(t-1).

6. If E(t)- E(t-1)<0, turn to step 2.Otherwise, it indicates the convergence of the centroids, thus, comes to the end.

After K-means, the similarity between points in the same class become bigger, while that in different classes become smaller. And the result of this classification is consistent with our record in the experiment.

2.4 Fisher LDA Classifier

Linear Discrimination Analysis, LDA for short, is one of the most popular methods of supervised classification [7]. It has the basic thought projecting the pattern of high dimension into the best identification space, so that we can successfully get the clustering information and the impression of dimension reduction. And Fisher function is just constructed to dig out that projecting orientation. This procedure can be conceptualized as a nonparametric method because it maximizes between-group variability relative to within-group variability [8](i.e. it can make the best separability).

In view of our data sets after K-means, 2 groups have already existed in dimension of 4. Under this 2-groups circumstance, Fisher criterion can define an optimal between-group variability [9].

Define the between-group measure of dispersion as S_b.

$$S_b = \sum_{i=1}^{K} P_i(\mu_i - \mu)(\mu_i - \mu)^T \tag{4}$$

where

K = number of the groups. In the study, it's 2.
K_1,K_2 = the 2 labeled clusters after K-means,
P_i = prior probability,
μ_i = mean of cluster K_i ,
μ = mean of K_1 and K_2.

Define the within-group measure of dispersion as S_w.

$$S_w = \sum_{i=1}^{K} P_i S_i \tag{5}$$

where

$S_i = E[(x-\mu_i)(x-\mu_i)^T \mid x \in K_i]$

J(W) represents the Fisher criterion function.

$$J(W) = |W^T S_b W| / |W^T S_w W| \tag{6}$$

where

W= the optimal projecting orientation for the data.

As we talk about above, our data's time courses' feature has a intuitionistic relation to α、βl、θ and δ wave bands. It means after K-means the two clusters labeled with drowsiness or wakefulness has a dimension of 4. So every sample point will be projected onto a beeline, and the data dimension changes from 4 to 1. LDA's ultimate goal is to make the diversity of the marked 2 clusters as big as possible.

3 Results and Analysis

3.1 Result of K-Means

From the figure, we can find α wave's proportion of the total power of the rhythms waves will grow when the subject is drowsy, while the high frequency βl is on the contrary. These results are accordance with informed researches. Based on the result, K-means clustering method classifies the data efficiently, and the dividing point in the figure of K-means' result corresponds with our record in the experiment.

The figure shows the point where α wave and βl wave have a sudden change is basically in accord with the K-means' result, which means α wave and β wave have a high correlation with the drowsiness level.

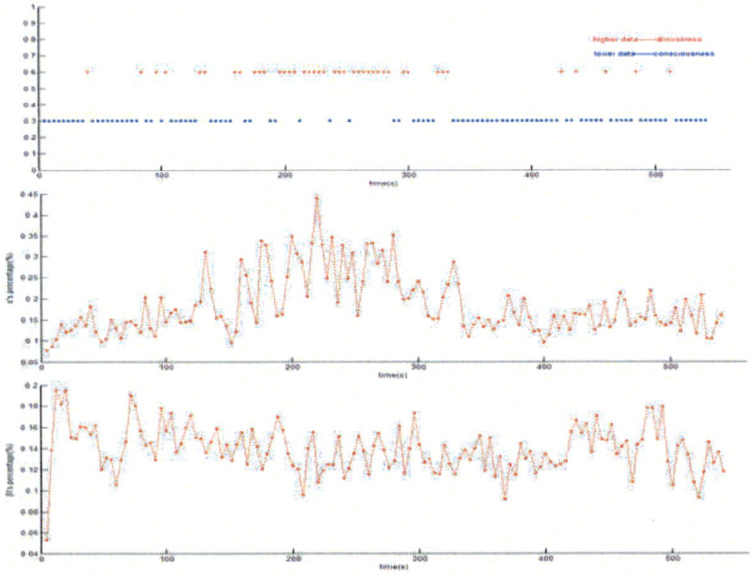

Fig. 1. The result of K-means method, α wave and βl wave

3.2 Result of LDA

This figure shows the data points distributed according to the acquisition time for better observation (calibration on the horizontal axis is only for observation). In the figure, '*' represents the points of drowsiness, while '+' represents the points of consciousness. It is obvious that these two types of data points flank a horizontal line, and their gathering situation agrees with our process of experiment, that is from consciousness to sleepiness and then consciousness again. The value of the vertical axis is the basis of the LDA clustering. According to the regularity of the data's distribution, we choose the mean value of the data in the training set as the threshold of our classifier.

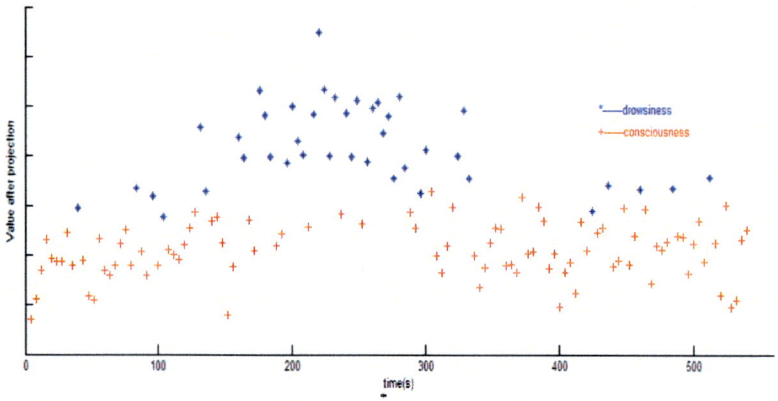

Fig. 2. Results of LDA

Randomly select half of the data respectively from the two sets after K-means, and then project them onto the optimal projecting orientation. According the distribution of the data, we set the mean of them as the boundary threshold. And the rest data are used for testing. Repeat the above procedure 10000 times. Then we get the accuracy of our LDA classifier. The accuracy of drowsiness is 0.9516, and the accuracy of wakefulness is 0.9764. According to this, the LDA classifier has a great distinguishability and sensibility.

4 Discussion and Conclusion

With repeatedly experimenting and analyzing, we have verified that α wave and βl wave have a high correlation with drowsiness level. Considering that the amplitude of α will increase significantly, while that of βl will decrease when the subject gets sleepy, we think that using α/βl as the index to quantize drowsiness level can be a good choice. Comparing α/βl to α or βl, we can discover that the ratio α/βl can better reflect the state of drowsiness. However, the scientificity of α/βl as the index should be further studied.

Besides, we find a phenomenon that is also about α wave. That is the amplitude of α wave will relatively reach a high level when the subjects are sober with their eyes closed, while nearly zero if they keep sober but with eyes open. All these are helpful for further study.

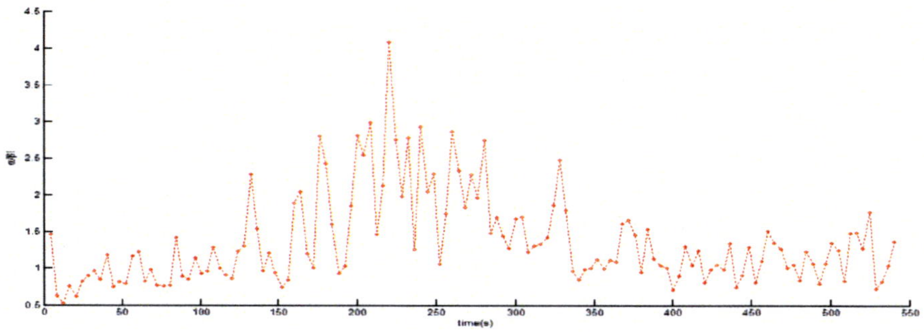

Fig. 3. $\alpha/\beta l$

By repeated experiment, we find that drowsiness and consciousness can be distinguished. Facing the problem of the deficiency of marking information, this paper choose K-means dynamic clustering as a labeling method to precisely label the EEG data with drowsiness or consciousness. And then design a classifier to predicte the drowsiness and consciousness from EEG signals. The classifier is on the basis of the Fisher criterion on the result of K-means. After all these work, we find that the accuracy of the classifying result is quite satisfactory and accuracy.

Acknowledgements. We thank Dr. Li Yao, Faculty and students in Signal Processing Lab for their technical assistance. This work is supported by the Natural Science Foundation of China (NSFC) under Grant Number 61075111.

References

1. Lin, C.-T., Wu, R.-C.: Estimating Driving Performance Based on EEG Spectrum Analysis. EURASIP Journal on Applied Signal Processing (2005)
2. Deiber, M.P., Missonnier, P., Bertrand, O., et al.: Distinction between perceptual and attentional processing in working memory tasks: A study of phase-locked and induced oscillatory brain dynamics. J. Cognitive Neurosci. 19(1), 158–172 (2007)
3. Scerbo, M.W., Freeman, F.G., Mikulka, P.J.: A brain-based system for adaptive automation. Theoretical Issues in Ergonomic Science 4(1-2), 200–219 (2003)
4. Fairclough, S., Ewing, K., Roberts, J.: Measuring Task Engagement as an Input to Physiological Computing. In: Workshop on Affective Brain-Computer Interfaces & IEEE International Workshop on Social Signal Processing ABCI 2009, vol. II, pp. 33–41 (2009)
5. Gabor, D.: Theory of communication. J. Inst. Elec. Eng. (1946)
6. MacQueen, J.: Some Methods for Classification and Analysis of Multivariate Observations. In: Proc. of the 5th Berkeley Symposium on Mathematical Statistics and Probabilit., Berkeley, USA, pp. 281–297 (1967)
7. Rausch Joseph, R., Ken, K.: A comparison of linear and mixture models for discriminant analysis under non-normality. Behavior Research Methods 41, 1 (2009)
8. Fisher, R.A.: A method for simulating non-normal distributions. Psychometrika 43 (1936)
9. Yu, B., Jin, L., Chen, P.: Normalized LDA Method for Face Recognition. Journal of Computer-Aided Design & Computer Graphics (2003)

High Precision BLDCM Servo Control with Nonlinear Identification*

Di Bao and Wei Huo

The Seventh Research Division, Science and Technology on Aircraft Control Laboratory
Beijing University of Astronautics and Aeronautics, Beijing 100191, China
hogo@vip.sina.com

Abstract. Nonlinear factors cause great influence to brushless DC motor (BLDCM) servo control system. One common result is that servo system can not respond rapidly when the direction of reference signal changes which is a serious problem to high speed aircraft. In this paper, a simple nonlinear identification method based on sliding mode variable structure algorithm is proposed to identify those nonlinear factors, and a novel fitting model modified from LuGre model is presented aiming to be utilized in control algorithm to compensate identified nonlinear factors. Simulation results show that the control precision can be improved significantly by the proposed method.

Keywords: BLDCM servo control, nonlinear identifier, sliding mode variable structure.

1 Introduction

Brushless DC motor based electrical servo system has been widely used in many industries. Its performance with high-precision and low velocity is greatly affected by nonlinear friction, torque fluctuation, environment disturbance and unknown plant parameters. Thus, it is critical to propose new schemes to identify those nonlinear factors and compensate them in control algorithm to improve servo performance.

One major problem for BLDCM servo control system is that when tracking low frequency sine signal, the servo position can hardly track the reference signal on wave crest which called “top out” phenomenon. The reasons lie in the Stribeck effect of friction and transmission clearance [1]. In all the nonlinear factors, nonlinear friction is a major cause that often degrades performance of precise servo control systems [2]. To counter its effect, one possible solution is to design an on-line observer to identify the nonlinear factor and managed to eliminate them through control. Since neural network has the advantage to fit any nonlinear functions, genetic algorithm and particle swarm optimization algorithm are self-adjusted, they are commonly used for this purpose, for example in [3], [4], [5]. However, this method needs large amount of calculations and great hardware data processing ability. Another alternative solution is to identify the

* This work was supported by National Science Foundation of China under Grant No.61074010.

D. Yang (Ed.): Informatics in Control, Automation and Robotics, Volume 2, LNEE 133, pp. 761–768.
springerlink.com

nonlinear factors off-line and find a suitable fitting function for compensation. [6] and [7] provide two interesting nonlinear control methods to identify nonlinear factors. [8], [9] and [10] provide several typical friction models to describe the Stribeck effect accurately.

In this paper, an off-line identifier is designed to identify the nonlinear factor with sliding mode variable structure control algorithm; and a simple Gaussian function is applied to fit the identified nonlinear factor. Simulation results show that this nonlinear identifier combined with regular PID controller can improve control precision effectively.

2 Mathematical Model of Servo System

The BLDCM servo control system diagram is shown in figure 1. The servo system contains brushless DC motor, controller, speed reduction gear, PWM power circuits and position indicator. A reference angular position signal θ_r is received by PC. Our target is to design control input U_c such that the servo angular position θ track the reference angular signal θ_r with high precision and satisfied dynamic/static performances.

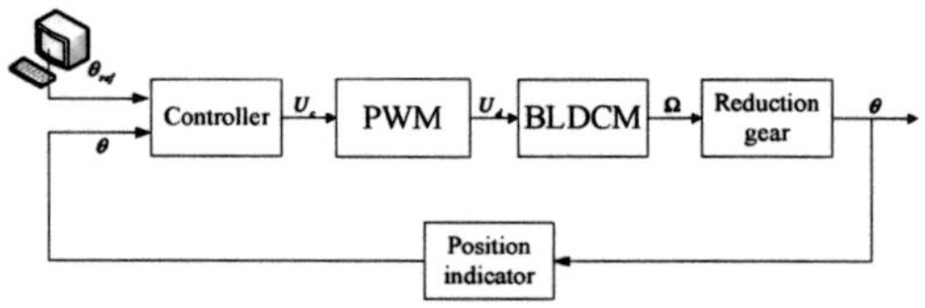

Fig. 1. Servo system structure

The brushless DC motor has three stator windings and permanent magnets on the rotor. Since both the magnet and the stainless retaining sleeves have high resistivity, rotor-induced currents can be neglected and no damper windings are modeled. A block diagram of the BLDCM is shown in figure 2, where U_d is busbar voltage, L_a is armature inductance, r_a is armature resistance, T_{em} is motor output torque, T_L is load torque, T_f is disturbance torque, C_T is electromagnetic torque coefficient, C_e is back electromotive force coefficient, J is electromotor moment of inertia, B_v is motor viscous friction coefficient, Ω is rotor angular speed, I is the phase current, E is the back electromotive force.

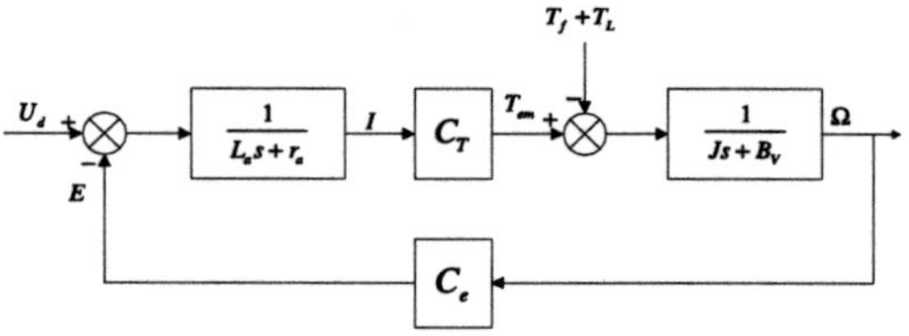

Fig. 2. Transfer function of BLDCM

The transfer function of the BLDC motor can be written as:

$$G_u(s) \triangleq \frac{\Omega(s)}{U_d(s)} = \frac{C_T}{r_a B_v + C_e C_T} \cdot \frac{1}{T_m s + 1} \tag{2.1}$$

$$G_L(s) \triangleq \frac{\Omega(s)}{T_f(s) + T_L(s)} = -\frac{r_a}{r_a B_v + C_e C_T} \cdot \frac{1}{T_m s + 1} \tag{2.2}$$

where $T_m = (r_a J + L_a B_v)/(r_a B_v + C_e C_T)$ is mechanical time constant.

The PWM circuit is a three-phase bridge inverter. Its transfer function

$$U_d = K \cdot U_c \tag{2.3}$$

where K is the equivalent duty ratio of PWM.

The speed reduction gear outputs the servo angular

$$\theta = \frac{i}{s}\Omega \tag{2.4}$$

where i is reduction ratio.

Synthesizing (2.1)-(2.4), we can get servo system model

$$\ddot{\theta} + a\dot{\theta} + F = bU_c \tag{2.5}$$

where $a \triangleq 1/T_m > 0$, $b \triangleq C_T iK/(r_a J + L_a B_v) > 0$, and $F \triangleq r_a i(T_L + T_f)/(r_a J + L_a B_v)$ reflects effects of nonlinear torques T_L and T_f.

Since the exact values of a and b in model (2.5) are always hard to get. Let $a = \hat{a} + \Delta a, b = \hat{b} + \Delta b$ Then the system (2.5) can be written as:

$$\ddot{\theta} + \hat{a}\dot{\theta} + \Delta a\dot{\theta} - \Delta bU_c + F = \hat{b}U_c \tag{2.6}$$

where $\hat{a}$ and $\hat{b}$ are nominal values of a and b, Δa and Δb are parameter uncertainties. These uncertainties and the nonlinear torque F are unknown to control designer, so an identifier is needed to identify them.

3 Nonlinear Identifier

Construct a nominal system as follows:

$$\ddot{\hat{\theta}} + \hat{a}\dot{\hat{\theta}} + G = \hat{b}U_c \tag{3.1}$$

where G is a self-tuning function to be designed. From (3.1) and (2.5) we get

$$\ddot{\tilde{\theta}} + \hat{a}\dot{\tilde{\theta}} + \Delta a\dot{\theta} - \Delta bU_c + F = G \tag{3.2}$$

where $\tilde{\theta} = \theta - \hat{\theta}$ is position error between servo system and nominal system. Clearly, if $\tilde{\theta}$ is controlled to zero, we know

$$G = \Delta a\dot{\theta} - \Delta bU_c + F \tag{3.3}$$

Which means when position error $\tilde{\theta}$ converges to zero, G identifies unknown model uncertainties $\triangle a\dot{\theta}-\triangle bU_c+F$. Then the significant problem is how to design the self-tuning function G such that $\tilde{\theta}$ converges to zero.

Proposition: Suppose that the parameter uncertainties and unknown nonlinear factor F are bounded, i.e. there exist known positive constants K_1, K_2 and K_3 such that

$$|\triangle a|\le K_1,|\triangle b|\le K_2,|F|\le K_3 \tag{3.4}$$

Design
$$G=-K_G s-\varepsilon\text{sign}(s)-(c-\hat{a})\dot{\tilde{\theta}} \tag{3.5}$$

where $K_G>0$; $s=\dot{\tilde{\theta}}+c\tilde{\theta}$, $c>0$; $\varepsilon>K_1|\dot{\theta}|+K_2|U_c|+K_3$; sign() is the sign function; then the error $\tilde{\theta}\to 0$ as $t\to\infty$

Proof: Select sliding surface:

$$s=\dot{\tilde{\theta}}+c\tilde{\theta},c>0 \tag{3.6}$$

From (3.2) and (3.5) we get

$$\begin{aligned}\dot{s}&=\ddot{\tilde{\theta}}+c\dot{\tilde{\theta}}=(c-\hat{a})\dot{\tilde{\theta}}-\triangle a\dot{\theta}+\triangle bU_c-F-K_G s-\varepsilon\text{sign}(s)-(c-\hat{a})\dot{\tilde{\theta}}\\&=-K_G s-\varepsilon\text{sign}(s)-\triangle a\dot{\theta}+\triangle bU_c-F\end{aligned}$$

$$\begin{aligned}s\cdot\dot{s}&=s\cdot\left[-K_G s-\varepsilon\text{sign}(s)-\triangle a\dot{\theta}+\triangle bU_c-F\right]\le -K_G s^2-|s|\left(\varepsilon-\left|\triangle a\dot{\theta}-\triangle bU_c+F\right|\right)\\&\le -K_G s^2-|s|\left(K_1|\dot{\theta}|+K_2|U_c|+K_3-|\triangle a\dot{\theta}|-|\triangle bU_c|-|F|\right)\le -K_G s^2\end{aligned}$$

It implies that $s\cdot\dot{s}<0$ when $s\neq 0$. Based on the sliding model variable structure control theory [13] we know that $\tilde{\theta}\to 0$ as $t\to\infty$ □

Structure of the off-line nonlinear identifier is shown below:

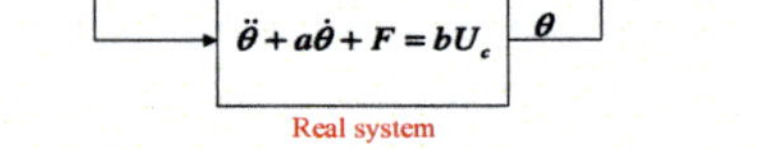

Fig. 3. Diagram of the nonlinear identifier

4 Fitting Function for Nonlinear Factor

The identified G contains friction, backlash and parameter variation. In order to compensate these nonlinear factors in control algorithm, it is necessary to find a suitable function to fit identified G.

When servo system traces a low frequency sine signal, a special phenomenon called "top out" on wave crest often happens. Researchers believe that this phenomenon is caused by static and viscous frictions as well as transmission gear clearance.

The curve in Fig.4 is called Stribeck curve. When servo system begins from low speed, friction is very large in a period, and then it will reduce greatly to mark 2. After that, the friction value is a linear function of speed. The friction from mark 1 to mark 2 results to the "top out" phenomenon, because when the movement direction changes, speed of servo system is very small, so that the friction suddenly increases and keep large for a while. Due to this large friction value, it is difficult for servo system to change its movement quickly. In this paper, the transmission gear clearance can also be treated as a torque at low speed. As a result, the nonlinear factor can be handled as a friction torque. Furthermore, compared with friction, parameter variation contributes less to the "top out" phenomenon, so the identified nonlinear function G can be modeled as a friction function.

The wide-accepted friction model is LuGre friction model. It is a dynamic model to describe the Stribeck curve accurately. The formula of LuGre model is:

$$\begin{cases} F = \sigma_0 z + \sigma_1 \dot{z} + \alpha\dot{\theta} \\ \dot{z} = \dot{\theta} - \sigma_0 \left|\dot{\theta}\right| z \Big/ \left[F_c + (F_s - F_c)\exp\left(-\left(\dot{\theta}/V_s\right)^2\right)\right] \end{cases}$$

where σ_0 and σ_1 are dynamic friction coefficients; F_c, F_s, α and V_s are static friction coefficients. F_c indicates coulomb friction, F_s indicates static friction, α is viscous friction coefficient; V_s denotes the switching velocity between static and dynamic frictions.

As described above, determination of LuGre friction model needs six coefficients, which is difficult for both on-line and off-line identifications. Also, in practical use, the model used for compensation of nonlinear factors in control algorithm does not need extremely precise. So a simpler model is needed and practicable in this situation.

The friction model proposed in this paper is given below and can be called Gaussian friction model:

$$\begin{cases} F = F_s \exp\{-\dot{\theta}^2/(2c_s^2)\}\text{sign}(\dot{\theta}) + \text{sign}(\dot{\theta}), & \text{if } \left|\dot{\theta}\right| < v_s \\ F = \dot{\theta}, & \text{if } \left|\dot{\theta}\right| \ge v_s \end{cases}$$

where c_s is a static coefficient to decide the span of static friction; the definitions of F_s and v_s are the same as that in the LuGre model. Clearly, only 3 parameters (c_s, F_s, v_s) is needed. Its curve is shown in Figure 5.

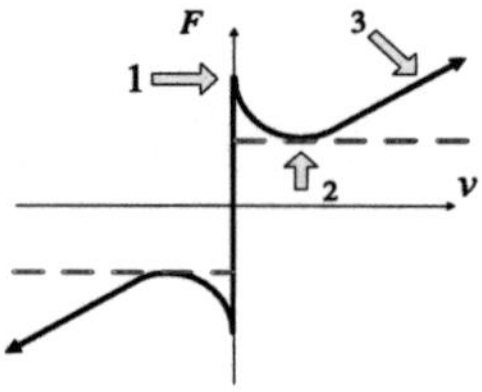

Fig. 4. Typical friction curve

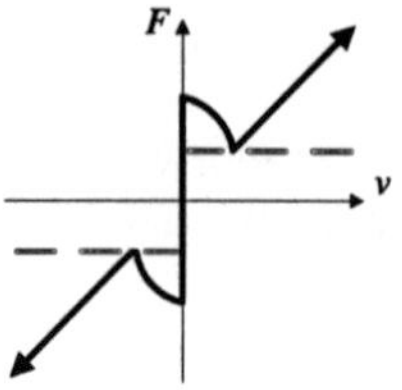

Fig. 5. Proposed Gaussian friction curve

5 Controller Design

Once a suitable Gaussian function $\hat{G}$ is found to fit the identified nonlinear function G, design control

$$U_c = (\hat{G} + U_c')/\hat{b} \tag{4.1}$$

the servo system (2.6) becomes

$$\ddot{\theta} + \hat{a}\dot{\theta} = \hat{G} - F + U_c' \approx U_c' \tag{4.2}$$

This is a linear system, U_c' can be designed with traditional PID approach. Given a reference signal θ_r, define tracking error $e = \theta_r - \theta$, the error equation is obtained from (4.2) :

$$\ddot{e} + \hat{a}\dot{e} - \ddot{\theta}_r - \hat{a}\dot{\theta}_r = U_c' \tag{4.3}$$

Design

$$U_c' = \ddot{\theta}_r + \hat{a}\dot{\theta}_r + K_p e + K_I \int e dt + K_D \dot{e} \tag{4.4}$$

Substituting it to (4.3) yields closed-loop system equation:

$$\ddot{e} + (\hat{a} + K_D)\dot{e} + K_p e + K_I \int e dt = 0$$

Based on the Hurwitz stability criterion, in order to guarantee the asymptotical stability of the error equation, controller parameters K_P, K_I, K_D should satisfy

$$K_D + \hat{a} > 0; (\hat{a} + K_D)K_P > K_I$$

Combining (4.4) and (4.1) get controller

$$U_c = (\hat{G} + \ddot{\theta}_r + \hat{a}\dot{\theta}_r + K_p e + K_I \int e dt + K_D \dot{e})/\hat{b} \tag{4.5}$$

Besides, in progress of off-line identification the following controller can be used:

$$U_c = U_c' = \ddot{\theta}_r + \hat{a}\dot{\theta}_r + K_p e + K_I \int e dt + K_D \dot{e}$$

6 Simulation Results

To illustrate performance of the proposed method in this paper, a simulation is implemented in follows. The parameters of controlled brushless DC motor model (2.5) are listed: $r_a = 0.3\,\Omega$, $B_v = 8\times10^{-5}$ Nm/rad, $J = 63\times10^{-5}$Kgm^2, $L_a = 0.002$H, $K = 0.024$, $i = 0.02$, $C_e = 0.095$ V/radps, $C_T = 0.095$ Nm/A

The real values of *a*, *b* and *F* in (2.5) are:

$$a = \frac{1}{T_m} = \frac{r_a B_v + C_e C_T}{r_a J + L_a B_v} = 47.8, b = \frac{C_T iK}{r_a J + L_a B_v} = 251, F = \frac{r_a i(T_L + T_f)}{r_a J + L_a B_v} = 33.6(T_L + T_f)$$

The load torque T_L is added gradient and LuGre friction model is chosen for T_f. In order to simulate the clearance of transmission gears, a typical dead-zone is added.

The nominal values $\hat{a} = 40$, $\hat{b} = 200$. The boundaries of uncertainties $|\Delta a|, |\Delta b|$ and $|F|$ are: $|\Delta a| \le K_1 = 10$, $|\Delta b| \le K_2 = 60$, $|F| \le K_3 = 50$

The PID controller parameters are $K_P = 80$, $K_I = 0.1$ and $K_D = 0.1$. The parameters of self-tuning function (3.5) are $K_G = 350$, $c = 200$, $\varepsilon = 60$.

Given the reference signal $\theta_r = 5\sin t$, Figure 6 shows the output *G* of nonlinear identifier and the Gaussian fitting function $\hat{G}$. Figure 7 shows the servo effects using controller (4.2) and (4.3). Figure 8 shows magnified effects of wave crest. Figure 9 presents the tracking errors. Obviously, the "top out" phenomenon is eliminated remarkably.

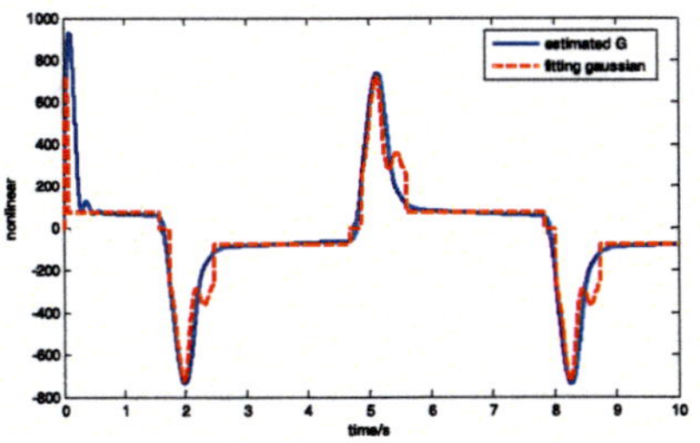

Fig. 6. Identified G and Gaussian $\hat{G}$

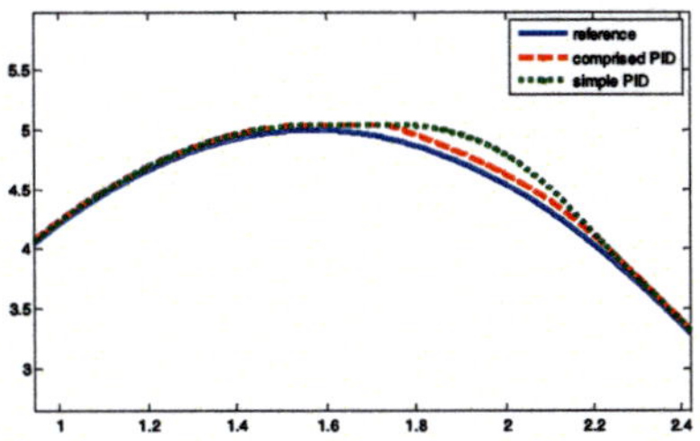

Fig. 8. Magnified effects of wave crest

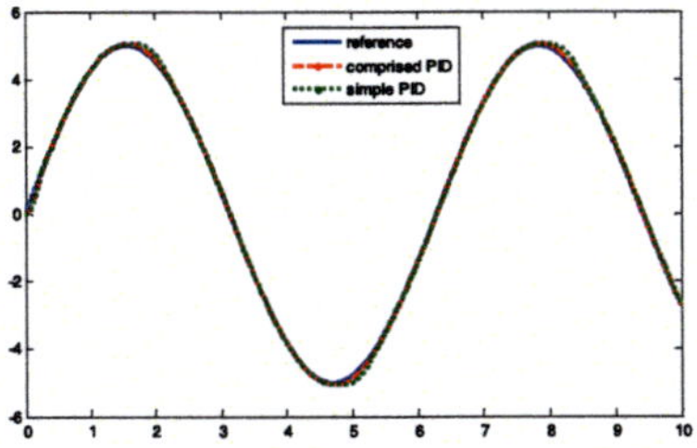

Fig. 7. Servo effects

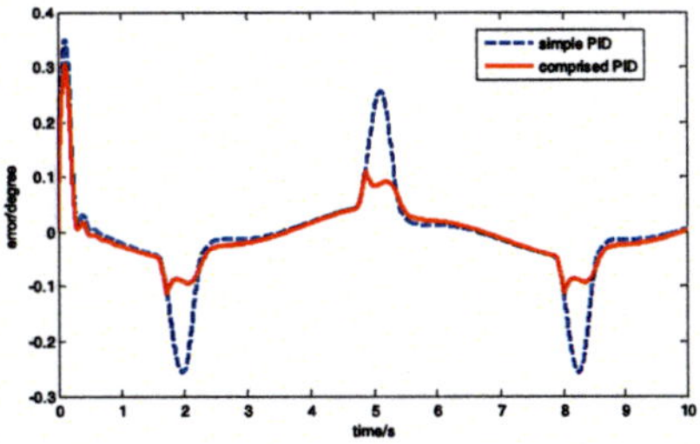

Fig. 9. The error effects

7 Conclusions and Discussions

In this paper, an off-line nonlinear identifier is proposed using sliding mode variable structure method. This identifier is based on the tracking of real system and nominal system. Simulation demonstrates that this method can drive the error between two systems converging to zero quickly and then nonlinear factors can be identified. Secondly, a simple Gaussian friction model is proposed to fit the identified nonlinear function. Analysis is made to find out the reason to the "top out" phenomenon so that the fitting function is confirmed to be suitable for the nonlinear function. The final test proves that this control scheme can reduce the "top out" phenomenon efficiently.

Some improvements are worth to be made later. In this paper, the parameters of the fitting function are fixed. Some self-learning strategy such as neuro network can be proposed to better fit the identified nonlinear function.

References

1. Pillay, P., Krishnan, R.: Modeling, Simulation and Analysis of Permanent-Magnet Motor Drivers, Part II: The Brushless DC Motor Drive. IEEE Trans. on Industry Applications, 274–279 (1989)
2. Armstrong-Hélouvry, B., Dupont, P., Canudas de Wit, C.: A Survey of Models, Analysis Tools and Compensation Methods for the Control of Machines with Friction. Automatica 30(7), 1083–1138 (1994)
3. Seong, I.H., Kwon, L.S.: Recurrent Neural Network Control for Nonlinear Friction with Sliding Mode and Friction Estimator. In: 6th International Conference on Natural Computation, pp. 1387–1392 (2010)
4. Liu, D.: Parameter Identification for LuGre Friction Model Using Genetic Algorithms. In: 5 th International Conference on Machine Learning and Cybernetics, pp. 3419–3422 (2006)
5. Zhang, W.: Parameter Identification of LuGre Friction Model in Servo System Based on Improved Particle Swarm Optimization Algorithm. In: 26th Chinese Control Conference, pp. 135–139 (2007)
6. Xian, B., Dawson, D.M., de Queiroz, M.S., Chen, J.: A Continuous Asymptotic Tracking Control Strategy for Uncertain Multi-input Nonlinear Systems. IEEE Trans. Automatic Control 49(7), 1206 (2004)
7. Makkar, C., Hu, G., Sawyer, W.G., Dixon, W.E.: Lyapunov-Based Tracking Control in the Presence of Uncertain Nonlinear Parameterizable Friction. IEEE Trans. on Automatic Control 52(10), 1988–1993 (2007)
8. Zhang, C., Liu, Q.: Friction Modeling and Compensation of Positioning Stage Driven by Linear Motor. Journal of Beijing University of Aeronautics and Astronautics 34(1), 47–50 (2008) (in Chinese)
9. Canudas de Wit, C., Olsson, H., Astrom, K.J.: A New Model for Control of System with Friction. IEEE Trans. on Automatic Control 40(3), 419–425 (1995)
10. Olsson, H., Astrom, K.J., Canudas de Wit, C., Gafvert, M.P., Lischinsky: Friction Models and Friction Compensation (1997)
11. Olsson, H., Astrom, K.J.: Observer Based Friction Compensation. In: Proceedings of 35th IEEE Conference on Decision and Control, Japan, pp. 4345–4350 (1996)
12. Hu, H., Yue, J., Zhang, P.: A Control Scheme Based on RBF Neural Network for High-Precision Servo System. In: International Conference on Mechatronics and Automation, pp. 1489–1494 (2010)
13. Gao, W., Huang, J.: Variable Structure Control of Nonlinear Systems: a New Approach. IEEE Trans. on Industrial Electronics 40(1), 45–55 (1993)

The Traveling Salesman Problem Base on Triple-Stranded DNA Structure Model*

Jing Yang[1], Zhi-xiang Yin[1], and Kai-feng Huang[2]

[1] Department of Mathematics and Physics, Anhui University of Science and Technology Huainan, Anhui province, China 232001
[2] Department of Information and Electrical Engineering, Huainan Vocational & Technical College Huainan, Anhui Province, China 232001
jyang@aust.edu.cn

Abstract. Traveling salesman problem is the NP problem of graph theory. The superior solution for traveling salesman problem is the Hamilton ring that finds out to have minimum power in the graph. The triple-stranded DNA computing model has a low rate of wrong solutions. Because the pool of data generated, there are the double helix structure of DNA chain, and the stability of the double-stranded DNA than single strand of DNA stability. The use of triple-stranded DNA model, can be solved with plans to optimize the combination of some of the NP problem. In this paper we show that triple-stranded DNA structure model can be used for solving the traveling salesman problem.

Keywords: traveling salesman problem, triple-stranded DNA, DNA computing.

1 Introduction

Since Adleman in 1994 firstly presented the DNA computing model of DNA molecule 7 vertex of a directed graph for solving Hamilton path problem[1], research in theory of DNA computing and DNA computers has been a spectacular biological technology development. Since made a number of scholars after a different DNA computing model, mainly single-stranded, single, double-stranded and single-double-stranded mixed model[2-5]. Recently they proposed triple-stranded DNA computing model. Now triple-stranded DNA computing model has been solving the satisfiability problem, 3-vertex-coloring problem, 0-1 integer programming problems[6], the working operation problem[7], the postman problem[8].Here author will give the triple-stranded structure of DNA computing model for traveling salesman problem.

2 The Working Operation Problem

The two chain of the double-stranded DNA is in the opposite direction,parallel to each other and interaction of the chain winding through Watson-Crick base pairs between the hydrogen bonds linked. As the base of bias on the side of the screw shaft,

* Project supported by CNSF (61170172, 60873144) and fund for outstanding young talent in Anhui Province (2010SQRL052, 2011SQRL198).

D. Yang (Ed.): Informatics in Control, Automation and Robotics, Volume 2, LNEE 133, pp. 769–775.
springerlink.com

the two chains form a large groove and a small groove.For homopruine/homopyrimidine triplex structures recognition of the third strand occurs through the major groove by formation of Hoogsteentype or reversed Hoogsteen-type hydrogen bonds between the bases of the oligopyrimidine or oligopruine the third strand and the purine-rich strand of the duplex target. We can make use of this unique structure to study some possible or feasible computing Models of the graph and combinatorial optimization.

3 The Traveling Salesman Problem

The traveling salesman problem is the NP problem of graph theory. A traveling salesman want to vist some cityies,and return his place of departure. Given travel distance (time) required between the various cities, how should he plan his route so that he can access the every city just once and total distance (time) the shortest? Empowered by the terminology of graph theory is a complete weight graph to find out the smallest weight Hamilton ring. We call for the optimal ring.But there is no efficient algorithm for solving traveling salesman problem. Commonly used algorithms for Brute force attack, the greedy method, Divide and conquer, dynamic programming, simulated annealing, neural networks, etc. We will try triple-stranded DNA computing model to solve traveling salesman problem.

3.1 The Basic Algorithm

Step 1: Search out all the rings graph G corresponding to the traveling salesman problem ;

Step 2: Retain all the rings graph G of starting from a fixed vertex and returning to the fixed vertex;

Step 3: Retain the Hamilton rings passing through all vertices and only at least once of the graph G;

Step 4: Search the Hamilton rings of the smallest weight summation,and this is the optimal solution for the traveling salesman problem.

3.2 Biological Method

1, Structure some single-stranded DNA to represent these vertices and arcs of the graph G.And after the reaction,we obtain these double-stranded DNA on behalf of the G graph of all the directed path through.These double-stranded DNA are as data pool.

2, Mix and incubate the strands that make up the vertex strands and RecA protein in a reaction mixture which contain $ATP_{\gamma}S$ under certain conditions.Thus, the nucleoprotein filament will come into being.

3, Mix the probe matching to the first vertex in date pool, and then library strands encoding truth assignments satisfying the first vertex are attached with these probes. In result, RecA protein mediated triple-stranded DNA structure come into being.

4, Mix the magnetic particles coated with streptavidin and the product form step3 in a proper reaction mixture.Due to the high affinity of biotin-streptavidin interaction, the formed triple-stranded DNA structures will be arrested by the magnetic particles and separated form other library strands. After deproteinizing the products, the desired library strands will be acquired. At this time the data pool, contain the first vertex of the graph G.

5, Repeat steps 3 and 4 for each of the remaining 4 vertices, and the finally acquired library strands have been contained all vertices of the graph G.

6, The use of gel electrophoresis measure the shortest double-stranded DNA. This is a smallest weight Hamilton ring, and also is the smallest time or the smallest distance of order traveling salesman problem.

7, A product of the above-mentioned steps will be sequenced in order to find the process order of the traveling salesman problem.

3.3 Biological Steps

Step 1:Encode the vertices and arcs of the graph G given on automated DNA synthesizer. Mix and incubate the oligonucleotide fragments corresponding to the vertices and arcs. Add a buffer solution and DNA connected Enzyme to make it connect response. So as to generate G graph of all the paths. Here is a mixture of single-double-stranded DNA, which has a strand of DNA instability. Accession to the polymerase, primer, the use of the Watson-Crick principle of complementarity, in the $3'$end constantly amplified DNA molecule, so that the mixture of all single-double-stranded DNA are in the form of double-stranded in order to increase the stability of the DNA strand. In response to add a product of the substrate DNA molecules, the amount of primer (G graph of the vertex corresponding to the oligonucleotide fragments make up the strand), DNA polymerase and buffer for PCR-amplification, so that the polymerase the role may include those vertices of the G graph of the DNA strand f starting from the fixed vertex and returning to the fixed vertex into increasing in the index. These retain the DNA strands. These products are purified, and then the productis purified after the separation. These double-stranded DNA as the original data pool(library).

Step 2: In graph G, said the vertex of each of the oligonucleotide fragments make up the strand of DNA single strand of the $5'$ end are added with poly(A).and then biotinylated. Mix and incubate the strands that make up the vertex strands and RecA protein in a reaction mixture which contain $ATP_{\gamma}S$ under certain conditions. Thus, the nucleoprotein filament will cone into being. Making use of the nucleoprotein filament as a template structure probes.

Step 3: Mix the probes matching to the ith vertex with the date pool, and then library strand encoding truth assignments satisfying the ith vertex are attached with these probes in accorda-nce with the principle of the probe Hoogsteen pairing. In result, RecA protein mediated triple-stranded DNA structure come into being. Mix the magnetic particles coated with streptavidin and the product form step3 in a proper reaction mixture. Due to the high affinity of biotin-streptavidin interaction, the formed triple-stranded DNA structures will be arrested by the magnetic particles and separated form other library strands. After deproteinizing the products, the desired library strands will be acquired. The data pool at this time, there are elements of the ith vertex v_i. Repeat this step will be to contain every vertex of the double-stranded

DNA. This means that we can find that contains all the graph G vertices of the directed Hamilton rings.

Step 4: The third step of the plasma products are agarose gel electrophoresis, to run in the previous paragraph is the shortest of the DNA strands.

Step 5: The above steps will be the product of sequenced (this technology is now mature), the process in order to find a traveling salesman problem processing the order.

3.4 Case Studies

We use a specific example to illustrate the molecular processes operating methods of the problem. Assuming a traveling salesman want to visit 5 cities. He start from the first city(v_1),and only access at least once in every city, and return to the first city. Every city has direct access. We have given from v_i city to v_j city in distance of ω_{ij} . Search the Hamilton rings of the smallest weight summation .We have given here is a simple traveling salesman problem, to do so just for the encoding operation and convenience. We will introduce to the above-mentioned algorithm corresponding to the molecular steps.

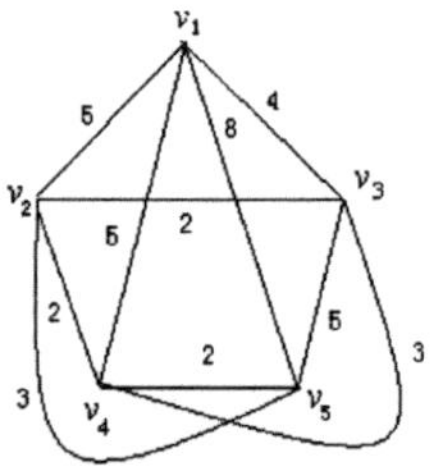

Fig. 1. Process corresponding to the traveling salesman problem

Step 1: Encode the vertex and arc of the graph G,and have the specific encoding as follows: We plans for any of the vertex v_i , with a length of the 20 oligonucleotide fragments that,and recorded as $M_i, i=1,2,\cdots,5$. The arc (v_i, v_j),the use of oligonucleotide fragments M_{ij} that it have three parts:the first part is last 10 base pairs of the M_i make up strand, and the third part is the first 10 base pairs of the M_j make up strand. The second part is weight of arc that can be said that the base $3\times t_{ij}$. Fig.2 specific code is as follows. We will mix the vertex v_i oligonucleotide fragments M_i and arc (v_i, v_j) contain the nucleotide- M_{ij} fragments in a test tube, adding buffer, primer and the polymerase reaction to a certain temperature to generate double-stranded DNA in Fig. 2. After connecting a product that contains all the path of the graph G. These products will be purified,we get these double-stranded form of DNA strands as a path to the data pool.

Step 2: In graph G,the 5′ end of the first vertex v_1 of the oligonucleotide fragments M_1 make up the strand of DNA single strand(*GGATAGCTTTCCCTCCTCCT*) are added with poly(A),and then biotinylated. Mix and incubate the strands that make up the vertex strands and RecA protein in a reaction mixture which contain $ATP_{\gamma}S$ under certain conditions.Thus,the nucleoprotein filament will come into being.Making use of the nucleoprotein filament as a template structure probes in Figure 4.3.

M_1 = 5′ − CCTATCGAAA GGGAGGAGGA − 3′		$\overline{M_1}$ = 5′ − GGATAGCTTT CCCTCCTCCT − 3′	
M_2 = 5′ − AAGAGAAAAG GGATATTCCT − 3′		$\overline{M_2}$ = 5′ − TTCTCTTTTC CCTATAAGGA − 3′	
M_3 = 5′ − CCTCTAAGGG AGGAGTTGAC − 3′		$\overline{M_3}$ = 5′ − GGAGATTCCC TCCTCAACTG − 3′	
M_4 = 5′ − GCCTAGAGGA AAGAAGGAGT − 3′		$\overline{M_4}$ = 5′ − CGGATCTCCT TTCTTCCTCA − 3′	
M_5 = 5′ − AAAGGAGTTC CAGGAAGAGA − 3′		$\overline{M_5}$ = 5′ − TTTCCTCAAG GTCCTTCTCT − 3′	

M_1 M_2
CCTATCGAAAGGGAGGAGGATTAGCTAAGTTGCCAAAGAGAAAAGGGATATTCCT
···CCCTCCTCCTAATCGATTCAACGGTTTCTCTTTTC···
M_{12}

M_2 M_1
AAGAGAAAAGGGATATTCCTTTAGCTAAGTTGCCACCTATCGAAAGGGAGGAGGA
···CCTATAAGGAAATCGATTCAACGGTGGATAGCTTT···
M_{21}

Fig. 2. The DNA coding of the vertex and arc

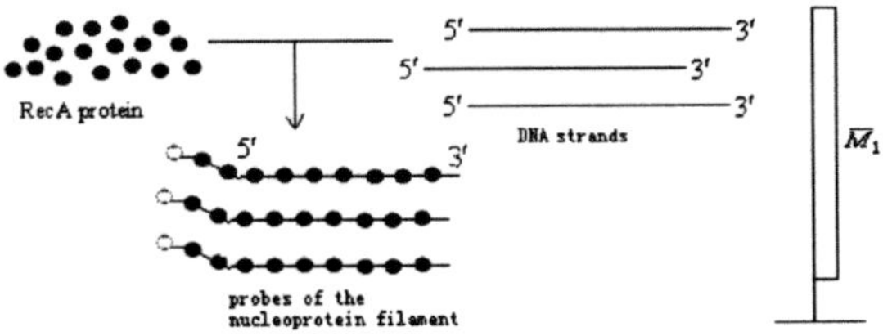

Fig. 3. The group of 5 probes with nucleoprotein filament

Step 3: Mix the $\overline{M_1}$ probes matching to the first vertex with date pool,and then library strand encoding truth assignments satisfying the first vertex are attachedwith these probes in accordance with the principle of the probe Hoogsteen pairing.In result , RecA protein mediated triple-stranded DNA structure come into being. Mix the magnetic particles coated with streptavidin and the product form step3 in a proper reaction mixture.Due to the high affinity of biotinstreptavidin interaction,the formed triple-stranded DNA structures will be arrested by the magnetic particles and separated form other library strands.After deproteinizing the products,the desired library strands will be acquired in Fig.4. The data pool at this time, there are elements of the first vertex v_1. The new data pool will be purified.Repeat this step will be to contain every vertex of the double-stranded DNA. This means that we can find a graph G that contains all the 5 vertices of the Hamilton ring.

Step 4: The third step of the plasma products are agarose gel electrophoresis, to run in the previous paragraph is the shortest of the DNA strands.

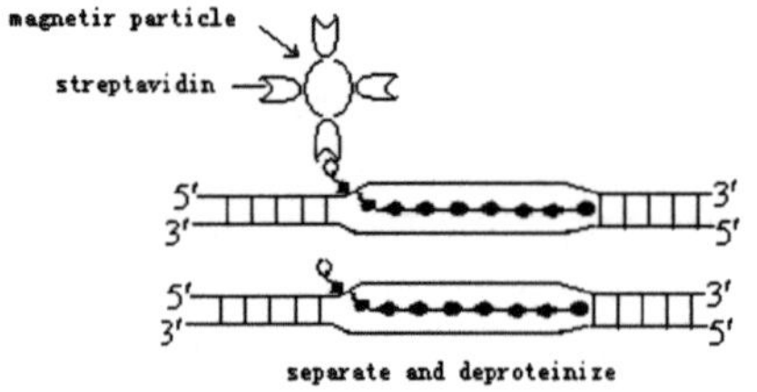

Fig. 4. Separation and deproteinize

Step 5: The above steps will be the product of sequenced (this technology is now mature),the process in order to find a traveling salesman problem processing the order. Cases have been on the order of processing $v_1 \to v_4 \to v_5 \to v_2 \to v_3 \to v_1$, and weight summation is 16.

6 Conclusion

With the development of molecular biology techniques, the biological model of DNA computing operations will be better able to achieve, the model will further enhance the versatility. There are some algorithm for searching the smallest weight Hamilton ring of the traveling salesman problem. But for bio-algorithm time to short, because biochemical reactions with a high degree of parallelism. Compared with other model, there is wrong solution to lower the rate in this model. There are the double-stranded DNA structure in the generated data pool. And the stability of the double-stranded DNA is better than single strand DNA. Compared to other DNA computing methods, interprobe and interlibrary stand hybridization and because of coding problems "hairpin" structures cannot happen. Then the biochemical reactions will be more fully, and more efficient. As for the three chains of the DNA double-stranded DNA separation is relatively easier than the separation. Of course, for solving practical problems, there are a number of biotechnical problems need further study. For the right to a greater or lesser value, the code can easily solve the problem, and the weight difference is too large when the weight of the code should be relatively difficult. However, the vertex increase in the code but not much. Just vertex increase in the code, the weight to the value of the code is then determined. In further we will combine with biological technology to achieve the versatility and effective-ness of the triple-stranded DNA computing model.

References

1. Adlenlan, L.: Molecular computation of solutions to combinatorial problems. Science 266(11), 1021–1024 (1994)
2. Lipton, R.J.: DNA solution of hard computation problem. Science 268(4), 542–545 (1995)
3. Ouyang, Q., Kaplan, P.D., Liu, S.M., et al.: DNA solution of the maximal clique problem. Science 278(17), 446–449 (1997)

4. Head, T., Rozenberg, G., Bladergroen, R.B., et al.: Computing with DNA by operating on plasmids. Biosystems 57(2), 87–93 (2000)
5. Sakamoto, K., Gouzu, H., Komiya, K., et al.: Molecular computation by DNA hairpin formation. Science 288(5), 1223–1226 (2000)
6. Yang, J., Yin, Z.-X.: 0-1Integer Programming Problem Based on RecA-mediated Triple-stranded DNA Structure. Computer Engineering and Applications 44(2), 76–79 (2008)
7. Yang, J., Yin, Z.-X., et al.: The Working Operation problem on Triple-stranded DNA Structure mode. Global Congress on Intelligent Systems, 295–299 (2009)
8. Yang, J., Yin, Z.-X., et al.: The postman problem on Triple-stranded DNA Structure model. In: 2010 First International Conference on Cellular, Molecular Biology, Biophysics and Bioengineering, pp. 335–338 (2010)

Robust Control of Strip Thickness for Cold Rolling Mill

Zhang Xiaofeng and Zhang Qingdong

School of Mechanical Engineering, University of Science and Technology Beijing,
100083 Beijing, China
zxf1024@gmail.com

Abstract. In this paper, we derive a mathematical model for the final stand of the tandem cold rolling mill and design control systems to improve the thickness precision for the strip. On the basis of a linearized model which discussed the uncertainty of stand module, strip stiffness and the time delay of gauge measurement of the rolling mill system, we propose a thickness controller using robust control which effectively rejects the uncertainties and disturbances. The comparison of suboptimal H∞ control and μ Control for strip thickness control is given. It is illustrated by simulations that the robust μ thickness controller yields better performance than H∞ methods when the system uncertainties and a periodic thickness disturbance exist.

Keywords: cold rolling mill, thickness control, robust control.

1 Introduction

This instruction file for Word users (there is a separate instruction file for LaTeX users) may be used as a template. Kindly send the final and checked Word and PDF files of your paper to the Contact Volume Editor. This is usually one of the organizers of the conference. You should make sure that the Word and the PDF files are identical and correct and that only one version of your paper is sent. It is not possible to update files at a later stage. Please note that we do not need the printed paper.

Thickness is one of the most important index of hot or cold strip. In order to meet increasing the precision of strip thickness, various types of Automatic Gauge Control(AGC) methods have been developed for hot or cold rolling processes. Use of these methods has resulted in good strip thickness accuracy in a steady state, but it is very difficult to maintain good thickness accuracy in an unsteady state, such as head and tail strip thickness. The main reasons for this problem are many disturbances and uncertainties in this process which lead to large variations in strip thickness.

Various advanced control methods have been developed for cold or hot rolling mill to meet the requirements for users, The H_2 (Linear Quadratic) control [1], [2], [3] is known as an efficient algorithm since it guarantees the optimality. However, it has the weak robustness with respect to the uncertainties, it may suddenly become unstable. H_∞ control [4], [5], [6] has a good robustness with for the uncertainties, but the constraints on the input are not considered. Moreover the major task of these research is improving the accuracy of strip thickness in steady state and the uncertainties during unsteady-state phase were not been fully considered.

D. Yang (Ed.): Informatics in Control, Automation and Robotics, Volume 2, LNEE 133, pp. 777–785.
springerlink.com

To promote the thickness precision control in the unsteady-state phase of cold rolling, we attempt in this paper to apply the new method in the control for final stand of a tandem cold mill. A linearized model of the rolling mill system which based on the quantitative description of uncertainty for mill and strip is built. Base on this model, the frequency responses of the perturbed open-loop system be computed. A robust controller is designed by using μ-synthesis method for final stand of a tandem cold mill. Comparison with the effect of industrial gaugemeter feedback controller, Simulation results show that the robust controller has better robust and disturbance attenuation performance.

The paper is organized as follows: Modeling considerations is given in Section 1. The robust μ-synthesis method is applied to the cold rolling mill control problem in Section 3. In Section 4 the controller implementation on the simulator is described, and the results from the implementation are discussed. Of this paper, which concludes in Section 5.

2 System Model

2.1 Thickness Control Model

The typical thickness control for final stand is depicted in Figure 1. Thickness feedback control(monitor control) used the actual thickness deviation measured by the strip thickness measuring unit on the exit side of the mill. The thickness deviation signal is weighted by a feedback controller and the output value of thickness feedback control is added to the reference value for subordinated control loops in roll gap control.

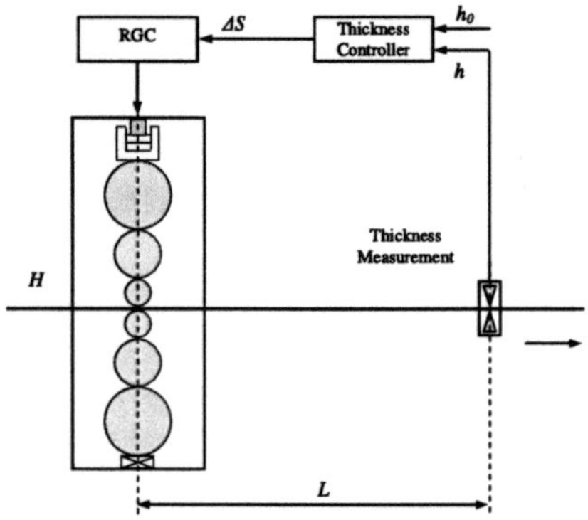

Fig. 1. The structure of the thickness control system in final stand

By the theory of continuous rolling, the thickness deviation Δh and the block diagram of the closed-loop system can be described by the following structure.

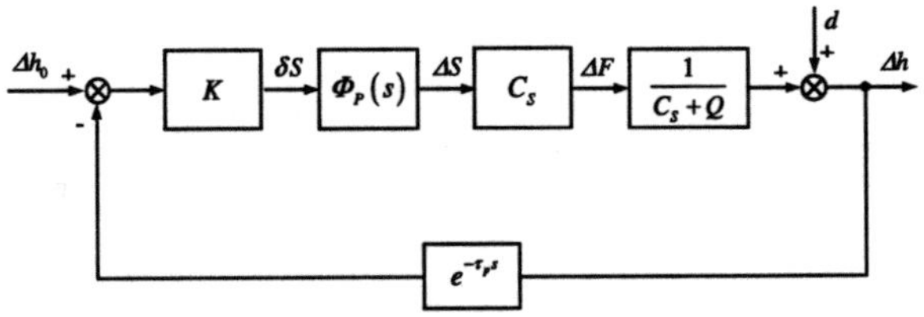

Fig. 2. The block diagram of the closed-loop system

where K is thickness feedback controller ; $\Phi_P(s)$ is the transform function of RGC, $\Phi_P(s)=\frac{\omega_p^2}{s^2+2\xi_p\omega_p s+\omega_p^2}$, the ξ_p and ω_p are normal ; C_S is the stand module of final stand ; Q is strip stiffness. $e^{-\tau_p s}$ is the death time which depends on the distance between roll gap and thick gauge.

2.2 Modelling of Uncertainties

In the realistic system, the there physical parameters C_S, $1/(C_s+Q)$ and $e^{-\tau_p s}$ in Figure 2 are not known exactly. We note that the stand module C_S may be represented as a linear fractional transformation (LFT) in p_{C_S} and δ_{C_S} ($-1\le\delta_{C_S}\le 1$), which represent the possible (relative) perturbations on the parameter.

$$C_S=F_U\left(M_{C_S},\delta_{C_S}\right)$$
$$M_{C_S}=\begin{bmatrix}0 & \bar{C}_S\\ p_{C_S} & \bar{C}_S\end{bmatrix} \tag{1}$$

Where $\bar{C}_S$ is the so-called nominal values of C_S.

Similarly, let $M=1/(C_S+Q)$, the parameter of M may be represented as an upper LFT in δ_M, let $\gamma=\frac{\tau_P}{2}=\bar{\tau}_\gamma(1+p_\gamma\delta_\gamma)=\frac{\bar{\tau}_P}{2}\left(1+\frac{p_{\tau_P}}{2}\frac{\delta_{\tau_P}}{2}\right)$, the time delay $e^{-\tau_p s}$ may be represented as an upper LFT in δ_γ. Tables 1 shows the uncertain parameters of the full stands tension system.

Table 1. Parameters and uncertainties

Parameter	Nominal value	Unit	Uncertainty
C_S	3.8	MN/mm	±10.0%
Q	34	MN/mm	±10.0%
L	1.62	m	0
V	3.3	m/s	±5.0%
ω_p	10	—	0
ξ_p	0.707	—	0

Let G_{S_5} denote the input/output dynamics of the final stand thickness system, which takes into account the uncertainty of parameters.

$$G_{S5}=\left[\begin{array}{c|cc} A & B_1 & B_2 \\ \hline C_1 & D_{11} & D_{12} \\ C_2 & D_{21} & D_{22} \end{array}\right]$$

$$A=\begin{bmatrix} -2.989 & -8.165 & 0 \\ 8.165 & -11.151 & 0 \\ 1.655 & 1.655 & -4.074 \end{bmatrix}$$

$$B_1=\begin{bmatrix} 0 & 0 & 0 \\ 0 & 0 & 0 \\ 0.043 & -1.630 & -0.050 \end{bmatrix}, B_2=\begin{bmatrix} 2.021 \\ -2.021 \\ 0 \end{bmatrix} \tag{2}$$

$$C_1=\begin{bmatrix} 7.678 & 7.678 & 0 \\ 0.203 & 0.203 & 0 \\ 1.655 & 1.655 & -4.074 \end{bmatrix}, C_2=\begin{bmatrix} -0.203 & -0.203 & 1.000 \end{bmatrix}$$

$$D_{11}=\begin{bmatrix} 0 & 0 & 0 \\ 0.005 & -0.200 & 0 \\ 0.043 & -1.630 & -0.050 \end{bmatrix}, D_{12}=\begin{bmatrix} 0 \\ 0 \\ 0 \end{bmatrix}$$

$$D_{21}=\begin{bmatrix} -0.005 & 0.200 & 0 \end{bmatrix}, D_{22}=\begin{bmatrix} 0 \end{bmatrix}$$

The uncertain behavior of the original system can be described by an upper LTT representation

$$h=F_U\left(G_{S5},\Delta\right)h_0 \tag{3}$$

with diagonal uncertainty matrix $\Delta=diag\left(\delta_{C_S},\delta_M,\delta_\gamma\right)$. It should be noted that the unknown matrix Δ has a fixed structutre, such uncertainty is thus called structured uncertainty.

3 Robust μ-Thickness Controller Design

3.1 Design Requirements of Closed-Loop System

The block diagram of the closed-loop system showing the feedback structure and including the elements reflecting model uncertainty and performance requirement, is given in Figures 3.

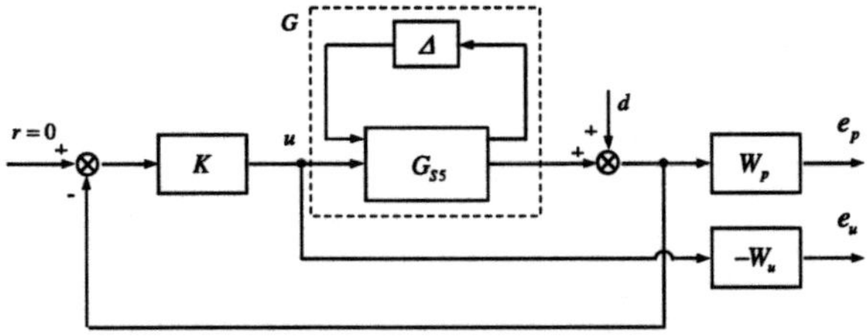

Fig. 3. The block diagram of the closed-loop system

In the given case, the performance weighting function w_p and w_u are chosen as

$$W_p = 0.4\frac{s^2 + 1.8s + 10}{s^2 + 8.0s + 0.01}, \quad W_u = 5\frac{s + 100}{s + 500000} \tag{4}$$

3.2 μ Synthesis

The μ-synthesis is executed by the M-file dkit from the Robust Control Toolbox of Matlab, which automates the procedure by using D-K iterations[7].

After three iteration, the value of μbecome equal to 0.7211, which means that robust performance has been achieved. The μ plot of the closed-loop system with the newly obtained a 22nd-order controller K at Iteration 2 is shown in Figure 4.

$$\left.\begin{aligned} \dot{x}_\mu &= A_\mu x_\mu + B_\mu u \\ y_\mu &= C_\mu x_\mu + D_\mu u \end{aligned}\right\} \tag{5}$$

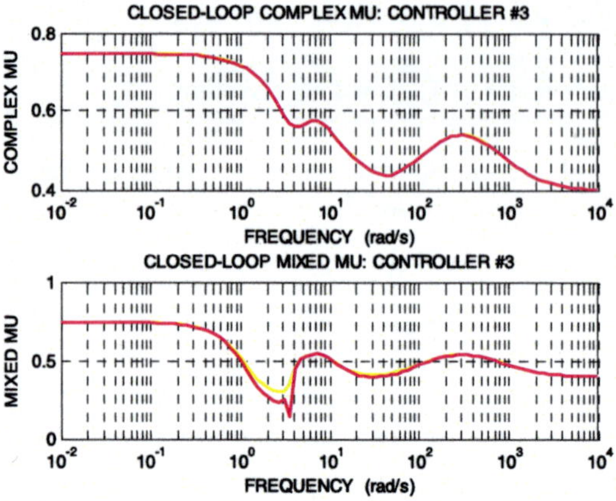

Fig. 4. μ plot after fourth iteration

4 Comparison of Suboptimal H_∞ Controller and μ Controller

4.1 Robust Stability

The frequency responses of μfor two control systems are shown in Figure 5 and Figure 6. It is seen that in this case the robust stability of two control systems is achieved. But the robust μcontrol system is characterized with better robust stability since in this case the destabilizing perturbations have the larger norm.

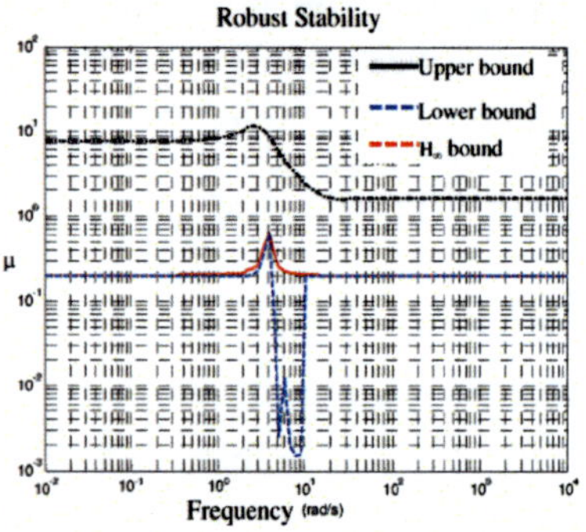

Fig. 5. Robust stability for H_∞ controller

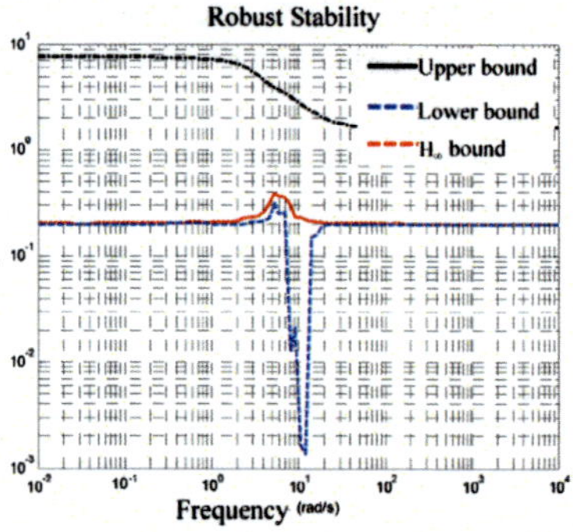

Fig. 6. Robust stability for μ controller

4.2 Robust Performance

The robust performance of the closed-loop system with two controller be tested by means of the μ-analysis. The μ values over the frequency range for two systems are plotted in Figure 7 and Figure 8. We can see clearly that although the two controllers achieve robust performance, the H_∞ controller almost fails to satisfy the robust performance criterion.

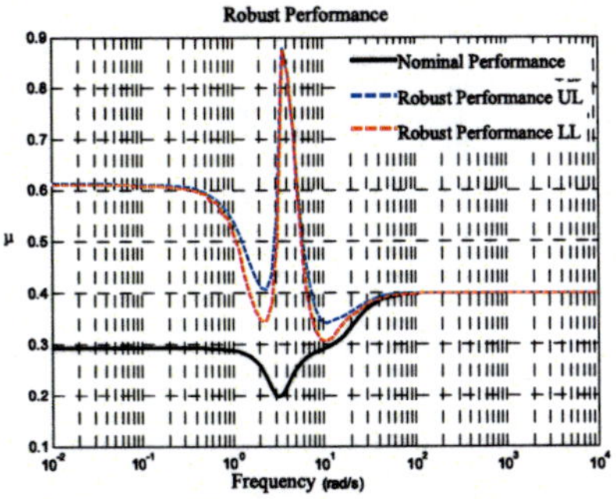

Fig. 7. Robust performance for H_∞ control

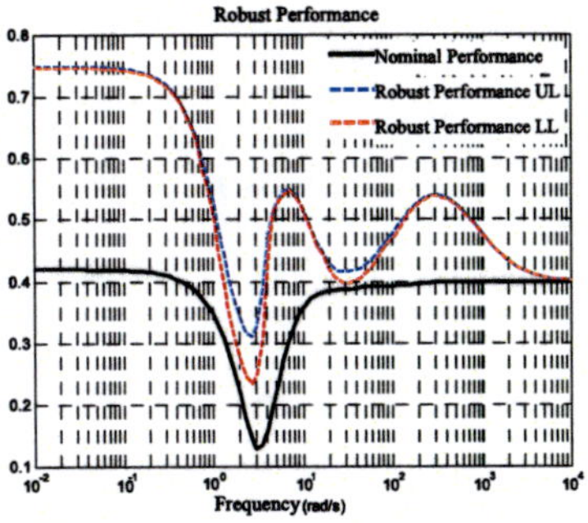

Fig. 8. Robust performance for μ control

4.3 Time Domain Simulation

In Figures 9 and Figure 10 we show the transient responses to disturbance of the thickness control system with two controllers. Comparing with the responses in the case of the H_∞, we see that the time responses of theμcontroller is faster than H_∞ controller, and with smaller overshoot. It means that the possibilities of having large transient variations are decreased with robust μthickness controller.

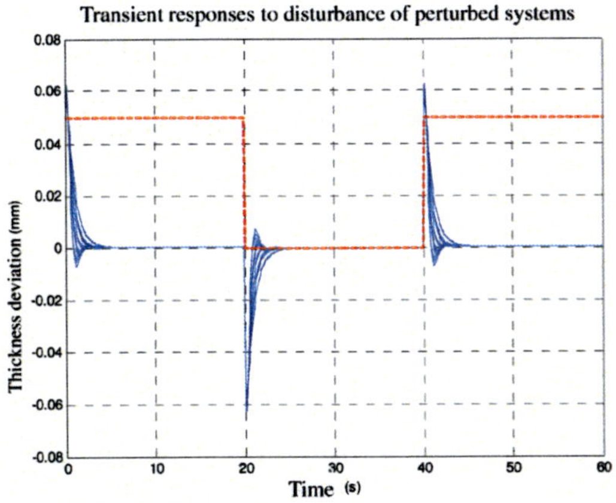

Fig. 9. Transient responses to disturbance for H_∞ controller

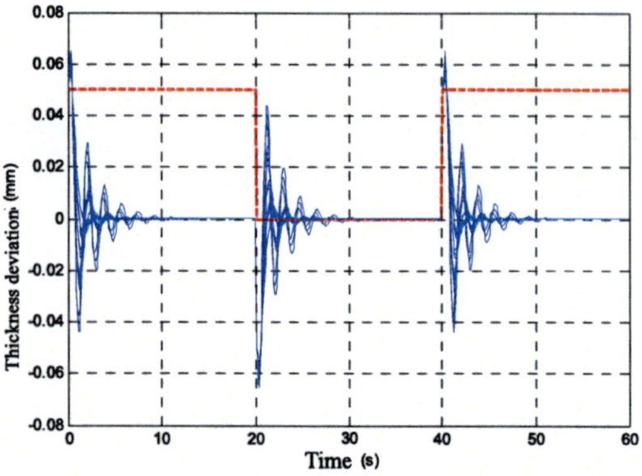

Fig. 10. Transient responses to disturbance for μcontroller

In summary, both the H_∞ controller and robustμcontroller ensure robust stability and performance of the closed-loop system with respect to the parametric perturbations that are included in diagonal uncertainty matrix. But the time domain comparison results indicate that in the present exercise the robust μcontroller might be the better one, it is more satisfactory and less conservative control system.

5 Conclusion

It is difficult to produce the very high standard for quality of cold strip at unsteady state because of the Uncertainties and large disturbances. To solve this problem, Based on the quantitative description of uncertainty of stand module, strip stiffness and the time delay of thickness measurement, this paper proposes an robust μcontrol for thickness feedback control of tandem cold mill. Comparison of suboptimal H_∞ controller and μ controller ,the simulation results show that the robust stability and robust performance of two control systems are achieved, But the robust μthickness control system more satisfactory and less conservative control system than the H_∞ controller.

References

1. Grimble, M.J., Hearns, G.L.: controllers for state-space systems with pure transport delays: Application to hot strip mills. Automatica 34(10), 1169 (1998)
2. Pittner, J., Simaan, M.A.: State-dependent Riccati equation approach for optimal control of a tandem cold metal rolling process. IEEE Transactions on Industry Applications 42(3), 836 (2006)
3. Postlethwaite, I., Geddes, J.: Gauge control in tandem cold rolling mills: A multivariable case study using H∞ optimization. In: Proceedings of IEEE International Conference on Control and Applications, New York, USA, p. 1551 (1994)

4. Hwang, G., Ahn, H.S., Kim, D.H., Yoon, T.W., Oh, S.R., Kim, K.B.: Design of a robust thickness controller for a single-stand cold rolling mill. In: Proceedings of the 1996 IEEE International Conference on Control Applications, Dearborn, MI, pp. 468–473 (1996)
5. Geddes, J., Postlethwaite, I.: Improvements in product quality in tandem cold rolling using robust multivariable control. IEEE Transactions on Control Systems Technology 6(2), 257 (1998)
6. Kim, I.S., Chang, Y.S., Hwang, I.C., Lee, M.H.: Design of LMI-based H∞ servo controller for cold tandem mill. In: Proceedings of the 1999 IEEE International Symposium on Industrial Electronics, Bled, Slovenia, Spain, p. 957 (1999)
7. Gu, D.W., Petkov, P.H., Konstantinov, M.M.: Robust Control Design with MATLAB, pp. 47–52. Springer (2005)

A Digital Entertainment System Based on Augmented Reality

Mandun Zhang[1], Lei Wu[1], Lu Yang[1], and Yangsheng Wang[2]

[1] School of Computer Science and Engineering, Hebei University of Technology, Tianjin 300401
[2] Digital Interactive Media Laboratory, Institute of Automation,
Chinese Academy of Sciences, Beijing, 100190
zhangmandun@126.com, wl83937456@hotmail.com, scseyl@126.com,
Yangsheng.wang@ia.ac.cn

Abstract. Augmented Reality Digital Entertainment System uses Augmented Reality technology in the field of digital entertainment. It merges the virtual scene produced by computer and the reality world scene into a whole perfectly. Then the scene is displayed on the HMD after computer processing. User can see the image clearly through HMD and interact with virtual objects. We choose the Video See-through HMD in experiment and propose a method for the no-markers' situation. The theory of coordinate transformation is analysised in 3D registration and the coming development of Augmented Reality is talked about at last.

Keywords: Augmented Reality, Registration, SIFT features, Feature Recognize.

1 Introduction

Augmented Reality is a new technology developed from basic of virtual reality in early 1990s.In recent years, it has become a hot technology research in the field of digital media. Augmented Reality is the technology that integration of virtual object and the real world perfectly through computer. It makes people to feel living in a real and virtual world .It is a amazing experience for users and there are have good prospects in many fields such as mechanical engineering, medical, digital entertainment industry and so on.

Augmented Reality technology has been actually applied in the field of military, medical, machinery manufacturing and many other areas. It demonstrates the extraordinary prospect. For example, we could make a magic book [6] through Augmented reality technology for education. We demonstrates the some different animation or images above the corresponding markers and they can replace the word perfectly. It not only make education becoming easy and it will improve the passion of student learning. Augmented Reality technology makes our life more convenient and colorful. Naturally, augmented Reality technology will be a hot research in a long time.

This paper describes a head-mounted augmented reality system[2].We mainly introduce some crucial related technology, principles and the improved method of artificial marker recognition.

D. Yang (Ed.): Informatics in Control, Automation and Robotics, Volume 2, LNEE 133, pp. 787–794.
springerlink.com

2 Augmented Reality Related Technology

2.1 Helmet Display

If you want augmented reality technology to offer a wonderful experience for users, you must prepare the technology of head mounted display which is crucial part of all. We need a good device in order to achieve the effect of augmented reality and make users to feel the perfect integration of real world and virtual object. At present, we have two choice: the video see-through and optical see-through[3][5].But there is a optical lag because optical see-through is integration of the virtual object generated by computer and the real world directly. Taking into account these reasons, we choose the video see-through HMD as the experimental tool HMD.

Video see-through HMD captures a picture of real world through the camera in the top of HMD and then puts images into the computer graphic workstation. Graphic workstation will process the image and the virtual object. Then send the image which have been processed to the display screen and users will see it through display sreeen.

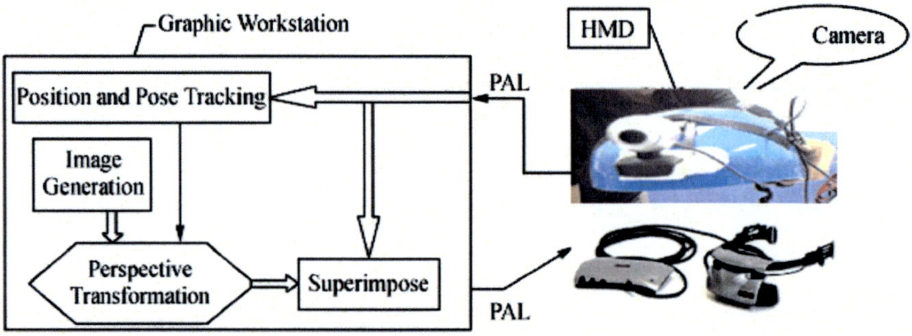

Fig. 1. Helmet structure

2.2 Marker Recognition and the Improve Method

In our experiment, we will use the ARtoolkit[7] which is a second development package based on C language for developing augmented reality system. It mainly uses video tracking capabilities in order to calculate the real camera position and orientation relative to square physical markers in real time. Once the real camera positionis known as a virtual camera can be positioned at the same point and 3D computer graphics models draw exactly overlaid on the real marker. The marker of ARtoolkit is black and white square graphic. This marker is easy for recognition and good enough for real time, but it is artificially producted and unnatural. Further more, some places can not lie on the marker, it is a great disadvantage. So we want to use Natural scenes recognition to replace the artificial marker recognition. At present, SIFT[4] Descriptor is widely used by many people for pattern recognition due to its strong performance of feature recognition and matching. SIFT descriptor can robustly identify objects even among clutter and under partial occlusion, because his SIFT feature descriptor is invariant to uniform scaling, orientation, affine distortion and partially invariant to illumination changes. The following is some advantage about SIFT descriptor:

(1)SIFT feature is a kind of local feature extracted from training images. It is invariant to uniform scaling, orientation, affine distortion and partially invariant to illumination changes.
(2)SIFT feature is unique and informative and it could match fastly and robustly in a large amounte of data.
(3)Even a few objects can also generate a lot of SIFT features.
(4)SIFT feature is qualified for real time system when it is optimized.

Before startint to run program, the first step is to capture a image from real world and take it as the model image. Then we will extract SIFT feature points as the model feature points. The camera captures the real world image as target image continuously and the computer extracts the SIFT feature points automatically from the interesting area of the image. Next, we find candidate matching features based on Euclidean distance of their feature vectors when model image's SIFT vector and target image's vector are generated. This method is to compare a feature point of model image with two feature point of target image which the two feature points' Euclidean distance is shortest about the model image's feature points[4][8]. If Euclidean distance of the shortest feature point divided by second Euclidean distance of feature point is smaller than the threshold, then we will think this pare matching points are correct. If we reduce the value of the threshold, the correct matching points also will reduce. But the result will be robust. We use RANSAC algorithm to remove the wrong matching points when the number of the matching points achieves the specified value and the step of point matching is done. Next, we choose the optical flow method[12] to track the good enough feature points. At last, we uses the POSIT[10][11] algorithm to compute transform matrix about world coordinate transform to camera coordinate.

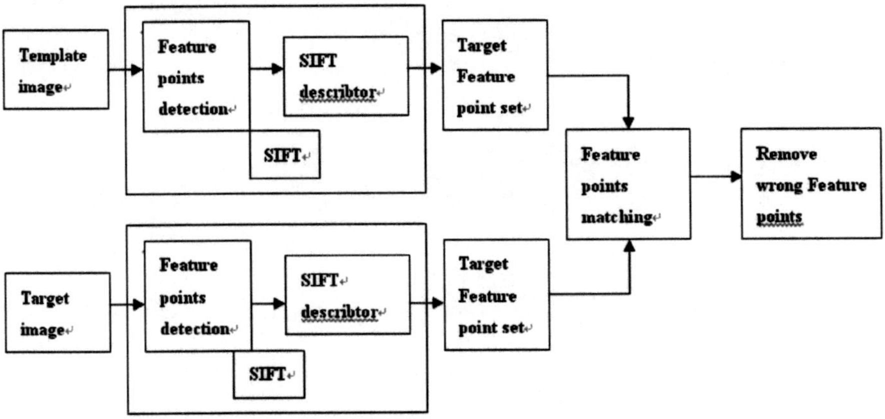

Fig. 2. SIFT feature point matching

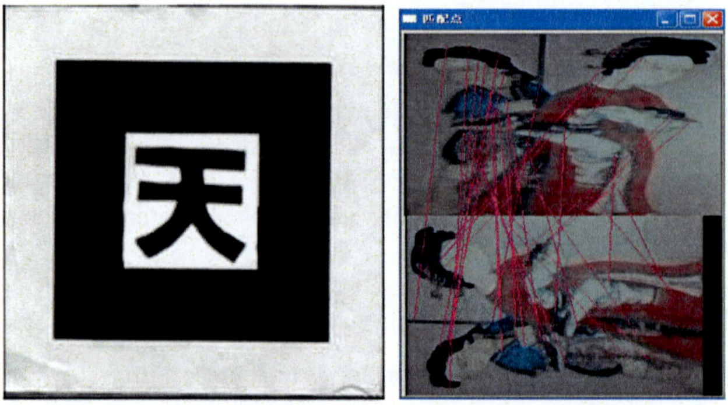

a. artificial marker(the picture from [7]) b. SIFT feature point matching

Fig. 3. ARtoolkit artificial marker and the camera picture

Because the SIFT algorithm is the method of extract the local feature points, we can uses it to solve the problem about part of image occlusion and the problem of illumination changes, orientation, affine distortion and so on. Due to these reasons, SIFT descriptor has a high practical value. In Fig. 4, Fig. a is artificial marker and the marker is easy to be recognised. Fig. b is the effect image of integration of marker and the virtual object. Fig. c is a picture on the wall we capture by camera. Fig. d is the effect image of integration of picture and the virtual object. From Fig. e, we can see clearly, the virtual object and the picture can still combine perfectly even if we occlude the picture.

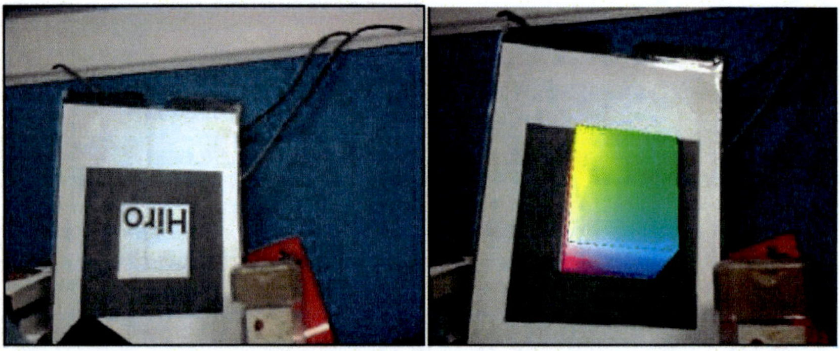

a. artificial marker b. integration of marer and virtual object

Fig. 4. Contrast effect between artifial image and nature screen image

c. a picture on the wall d. integration of image and virtual object

e. some occlusion by hand

Fig. 4. (*continued*)

2.3 Image Registration and Coordinate System

In order to integrate the virtual object into the real world perfectly, the step of 3D image registration is most crucial. The image registration is mainly to compute the transform matrix about the world coordinate, the camera coordinate, the image coordinate and observed screen coordinate.

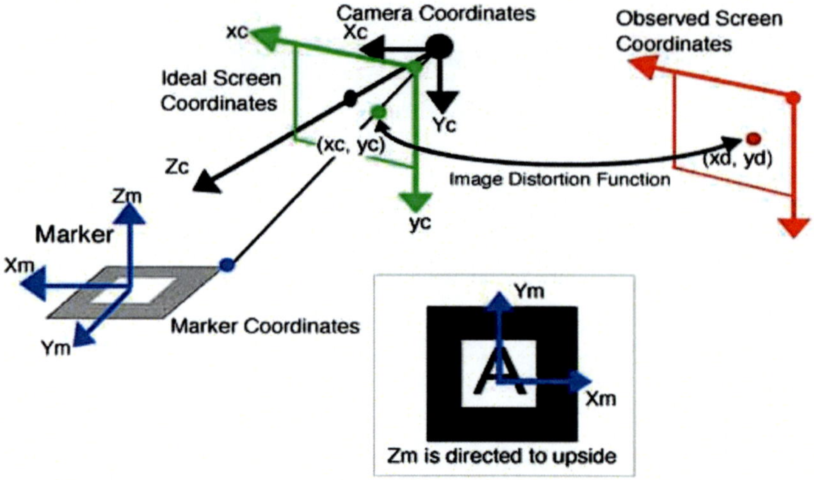

Fig. 5. Augmented reality coordinate system (picture from[7])

The orientation of virtual object in the world coordinate depends on the function you want the system achieved. We have already known the transform relationship about the world coordinate and the marker coordinate. So the geometric description of virtual object in the marker coordinate can transform to the geometric description of virtual object in the world coordinate.

$$[\, x(w) \quad y(w) \quad z(w) \quad 1\,] = [\, x(m) \quad y(m) \quad z(m) \quad 1\,]\, A \tag{1}$$

In formula (1),(x(w), y(w), z(w)) denotes world coordinate, and (x(m) , x(m) , y(m)) denotes marker coordinate. A denotes transform matrix。

Now, we want to compute the transform matrix about the marker coordinate and the camera coordinate, then we could transform the geometric description of virtual object in the marker coordinate to the Geometric description of virtual object in the camera coordinate. Because we have already known the transform matrix about the marker coordinate and the world coordinate, we could compute the transform about the camera coordinate and the marker coordinate if we could compute the transform matrix about the world coordinate and the camera coordinate .

$$[\, x(c) \quad y(c) \quad z(c) \quad 1\,] = [\, x(w) \quad y(w) \quad z(w) \quad 1\,] \quad B \tag{2}$$

$$[\, x(c) \quad y(c) \quad z(c) \quad 1\,] = [\, x(m) \quad y(m) \quad z(m) \quad 1\,]\, A\, B \tag{3}$$

We can project the virtual scene into the screen coordinate system when we get the transform matrix B and the step of image registration is done.

2.4 Interactive Technology

Scientific and technological progress is to better serve humanity. In the field of digital entertainment, the Quality of the game depends on the users interaction with the game. Most people like interacting naturally like feeling temperature, weight and mutual force of virtual object. But it is difficult to achieve these function and these targets are driving people to conquer them.

In our shooting experiment, we choose picking the object method to achieve the function of shooting and take the laser ray gun to replace the mouse. There is a selection mechanism with basic name stack from Opengl to solve the problem of picking. The following is detailed steps:

(1)First, we will enter into the picking mode when clicking the mouse.

(2)In picking mode, We re-draw the virtual scene (re-drawing the virtual scene is automatic and we can not see it).

(3)We need to name these object you want to pick when re-drawing the virtual scene.

(4)We will get the result of which object is selected depending on the model view transform and the projection transform.

(5)At last the image is drew again.

These information of selected object store in a one-dimensional array. These information includes the number of name and the depth value. The smallest depth value is what we want.

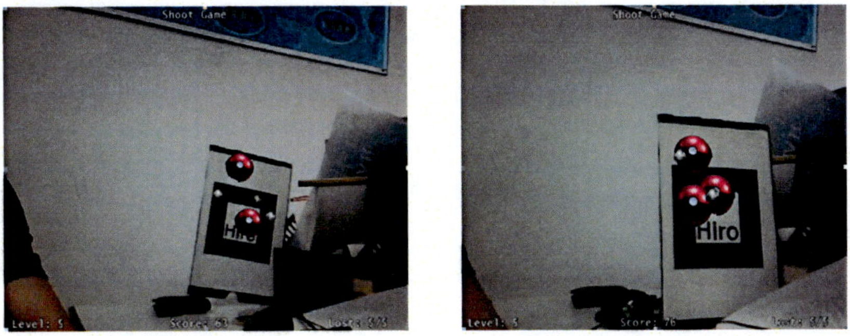

a. further shooting image b. closer shooting image

Fig. 6. Contrast about two defferent distance image

3 Conclusion

This paper introduces the related technology with our experiment and we solve the problem about some place can not lay in marker. We choose the method of extract SIFT feature point from the nature scene image to recognize the nature scene. We are basically satisfied with the results: SIFT descriptor has good matching results in the condition of scaling changes, affine distortion and illumination changes, and it also good enough if the target image is part occlusion. So the results of the experiment show that SIFT has good prospect in the field of Augemented Reality. The next focus point is how to make the users to interact with virtual scene more naturally. It is hard and urgent need in the future.

Acknowledgements. This paper is sponsored by Youth Foundation of Higher Education Scientific Research of Hebei Province 'Research of fast personalized virtual human modeling for human-computer interaction' (Grant No.2010228).

References

1. Bo, K., Peng, R.: Natural texture-based tracking algorithm for augmented reality. Systems Engineering and Electronics 31(10) (2009)
2. Kato, H., Billinghurst, M.: Marker Tracking and HMD Calibration for a video-based Augmented Reality Conferencing System. In: Proceedings of the 2nd International Workshop on Augmented Reality, San Francisco, USA (October 1999)
3. Rolland, J., Holloway, L., Fuchs, H.: A comparison of optical and video see-through head-mounted displays (1994)
4. Lowe, D.G.: Distinctive image features from scale-invariant keypoints. International Journal of Computer Vision 06(02) (2004)
5. Ha, Y.-G., Zhou, Y., Wang, Y.-T., Yan, D.-Y.: Head mounted display for augmented reality visualization. Optical Technique 26(4), 1 (2000)
6. Billinghurst, M., Kato, H., Poupyrev, I.: The MagicBook: A Transitional AR Interface. Computers and Graphics, 745–753 (November 2001)
7. `http://www.hitl.washington.edu/artoolkit/`
8. Chen, J., Wang, Y.-T., Liu, Y., Weng, D.-D.: System initialization algorithm based on SIFT key points for markerless augmented reality applications. Infrared and Laser Engineering 36(6) (2007)
9. Chen, M., Chen, Y.-M., Huang, S.-H., Yao, Z.-W.: Video Object Tracking Algorithm for Augmented Reality. Computer Engineering 36(12) (2010)
10. Oberkampf, DeMenthon, D., Davis, D.F.: Iterative Pose Estimation Using Coplanar Points. Computer Vision and Image Understanding 63(3), 495–511 (1996)
11. DeMenthon, D., Davis, L.S.: Model-Based Object Pose in 25 Lines of Code. International Journal of Computer Vision 15, 123–141 (1995)
12. Baker, S., Matthews, I.: Lucas-kanade 20 years on: A unifying framework. IJCV 56, 221–255 (2004)
13. System initialization algorithm based on SIFT key points for markerless augmented reality applications

The Fluid-Solid Interaction Analysis of WDPSS-8 Based on ANSYS

Shaoni Jiao[1], Yaqing Zheng[2], and Gui Lin[3]

[1,2] Mechanical Engineering and Automation, Huaqiao University, 361021 Xiamen, China
[3] Automatic Weapon and Ammunition Engineering,
Institutes of Technology of Nanjing, 210094 Jiangsu, China
jiaoshaoni1986@126.com, yq_zheng@hqu.edu.cn,
lingui350322@foxmail.com

Abstract. Based on the numerical analysis function of ANSYS, the motion of the standard aircraft driven by WDPSS-8 (Wire-driven Parallel Suspension System with 8 Wires) in the laminar flow is simulated by the fluid-solid interaction function. Firstly, the deformations of the wires simulated by beam4 element without wind pressure agree with those of the theoretical results. Secondly, by solving the solid field in ANSYS and analyzing the solution to forward kinematics of WDPSS-8, it is found that the wires sustain the wind pressure against wind and the aircraft tends to move along the wind. So the wires in simulation bend against wind. It is also demonstrated that by increasing the wire axial tension, the coupled fluid-solid interaction phenomena can be eliminated. Finally, we analyzed the wind field affected by WDPSS-8. On the one hand, by observing the wind vector graphs and the pressure distributed in the wind field, WDPSS-8 is superior to the rigid support in reducing the aerodynamical interference. On the other hand, by analyzing the wind velocity and pressure on and under the wing, it is found that the simulation is fit to the flying state of the aircraft in the air.

Keywords: WDPSS-8, laminar flow, the fluid-solid interaction, against wind, forward kinematics.

1 Introduction

With the development of robotic technology and force control method, the wire-driven parallel technology has a wide field of application, e.g. the crane robot, the cable-cabin coarse tuning system of the large radio telescope [1], the robot with large working space, the wind tunnel support system [2] and machining operations and so on. Wires can only sustain tension and has small lateral stiffness, so under the disturbance of the external load, wires always have the strongly coupled fluid-solid interaction phenomenon which occurs in many fields, e.g. biomechanical engineering [3], the flying aerospace, the bridge accident caused by the wind vibration behavior and the oil pool exploitation effecting the rock [4] and so on. The forcedly oscillatory experiments using wire-driven suspension systems [5] in low-speed wind tunnel can obtain derivatives of the aerodynamic forces and moments. And the wind vibration

D. Yang (Ed.): Informatics in Control, Automation and Robotics, Volume 2, LNEE 133, pp. 795–802.
springerlink.com

behavior which induces the end effector's vibrations or even damage also happens during the tests. So it is important to investigate the coupled fluid-solid interaction phenomenon between the wires and the air or choose the reasonable ropes.

The task of my workgroup is dedicated to doing the forcedly oscillatory experiments of scaled aircraft model suspended by the soft WDPSS-8 (Wire-driven Parallel Suspension System with 8 Wires) in wind condition. Because of the flexibility of wires, it makes sense to simulate the motion of the aircraft driven by the soft wires and the flow field around the aircraft in the laminar flow at first, and finally we can ignore the interaction phenomenon or take measures to reduce the rope vibration and the aerodynamical interference. Then the aerodynamic force can be measured exactly by the internal balance. In fact, the wire deformations in fluid-solid phenomenon may lead to the complex vibration of ropes which influences the single degree of freedom oscillation accuracy of the scaled aircraft in the application of WDPSS-8 system. And it can also be solved by adding dampers [6] or using other motion control methods.

2 Research Method

Based on the development of the computational fluid dynamics and computational structural dynamics, it is affordable to use computational resources to simulate the interaction behavior. By simulating the interaction between the fluid field and the structure field in ANSYS, the explicit deformations of the eight wires under wind pressure can be easily studied. Generally, the complete components of the ANSYS computational strategy for the modeling of fluid-solid interaction are as follows: (1) The fluid field which is controlled by the Navier-Stokes equation set containing momentum and continuity equations; (2) The solid field; (3) The interface between the fluid and solid domains is studied by the arbitrary Lagrangian-Eulerian (ALE) description [7] incorporating the advantages of Lagrangian and Eulerian methods and overcoming their disadvantages. This method is based on a discrete time integration scheme, and one of its advantage is that it can avoid the mesh distortion induced by the element twisted excessively. The computational grid described by ALE can move independently, so it is useful to solve many kinds of questions such as the strong nonlinear free surface flow problems, the frictional contact problems and the coupled fluid-solid interaction problems. And in the process of the whole deformation, the fluid particle can not pass through the grids and the solid surface at the fluid-solid interface. At the same time, the solid grids and the fluid grids should alwasys keep conincidence at the interface.

The whole computational strategy can be regarded as the following three steps. Firstly, assuming the initial structural surfaces as the accurate boundary of the fluid flow, the pressure that acts on the solid can be obtained in the fluid field; Secondly, the displacement of the solid under wind pressure takes place, and then it can repeat calculating the pressure after modifying the fluid mesh according to the deformed solid figuration; Thirdly, repeat executing the above two steps until the wind pressure on the interaction is less than the tolerance settled before.

3 Fluid-Solid Interaction Analysis

3.1 Modeling of WDPSS-8 and Wind Fields

The scaled aircraft model driven by WDPSS-8 can implement 5 degrees-of-freedom motions including roll, pitch, yaw, heave and translation. Due to space limitations, here the model rolling at 0°transient condition is just analyzed. First, design the WDPSS-8 system and the scaled aircraft model in Pro/E and preserve the model as '.igs'. Then, the fluid field which is surrounded by the areas of the aircraft driven by WDPSS-8 and the six rectangles outside is modeled by using Boolean operation in ANSYS. And at last the model of fluid and solid fields is shown in Fig.1. The dimension of the air volume is $4.2m \times 3.1m \times 1m$ and the wind with the arrow represents the wind direction paralleling with the z axis of the global coordinate system. 1~8 mean the eight wires with $\varphi\, 8\, mm$. The lengths of 1-4 are 1.5m and others are 2m. According to the design rules for cranes, wires used here can sustain the minimum breaking force 45000N.

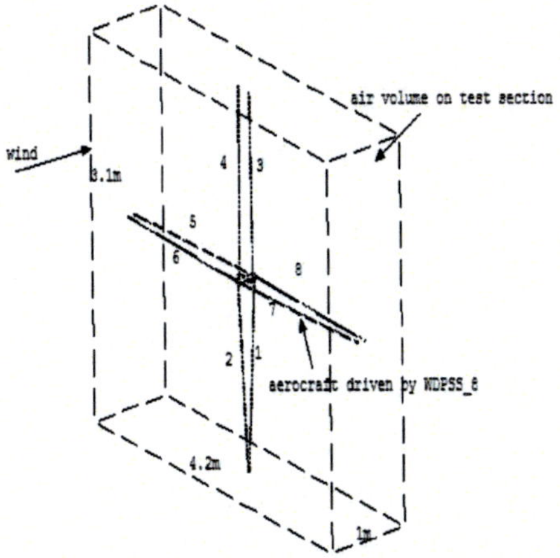

Fig. 1. The model of structural and fluid fields

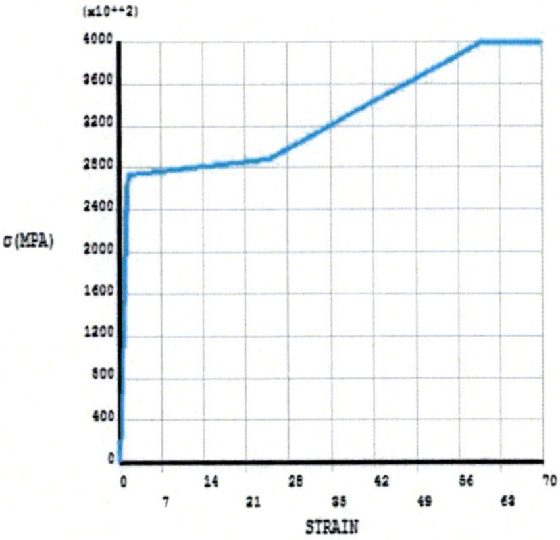

Fig. 2. Strain-stress curve of steel

3.2 Mesh and Define Loads

In this section, the air and wire elements are set to be types of 3D fluid142 and beam4 respectively. Fluid142 element can be used to model transient or steady fluid and thermal systems that involve fluid regions. Beam4 is a uniaxial element with tension, compression, torsion and bending capabilities. Reference [8] points out that beam4 element is good at modeling the dynamic characteristic of wires and can be widely used in engineering application. Considering the air resistance and gravity, the dynamic model of the flexible rope element simulated by Link167 element in ANSYS/LS-DYNA is analyzed in Reference [9]. And considering the viscoelasticity of the suspension line of the parachute, it is assumed as spring model in reference [10] where the stress passing path in the canopy fabric is analyzed and the test results confirms with that simulated by spring element in ADAMS.

The attribution table of wires is as follows: (1) assuming the wire is isotropic, with the elastic modulus 2.06×10^5 Mpa and the Poisson ratio 0.45. The elastic limit of wire is 2.61×10^5 Mpa from Fig.2. (2) The stiffness per meter of wire is

$$K = \frac{EA}{L} \ . \tag{1}$$

So the wires with length of 1.5m and 2m are extended by 1.16mm and 1.5mm respectively in theory under the tension 8000N. Then the attribution table of the fluid field necessitates the following parameters: (1) The flow density is $1.205kg/m^3$ at normal temperature. (2) The dynamic viscosity η is calculated from the equation [2] as given below, where η_0 is the fluid dynamic viscosity at temperature $T_0 = 0$, and n is a constant. (3) The Reynolds number is low here, so it is regarded as laminar flow. (4) The Mach number is so small that the air field can be considered as incompressibility.

$$\eta = \left(\frac{T}{T_0}\right)^n \times \eta_0 \ . \tag{2}$$

Mesh the solid and fluid fields with the defined elements. The smooth mesh grid rarely causing problems with ill-conditioned element is important for analysis. First of all, mesh all the areas with the "mesh-only" element mesh200. Then sweep the wire model and free mesh the fluid field which is irregular. At last, the number of all the elements is 498,609.

3.3 Governing Equations and Solve

The WDPSS-8 system is modeled as the linear elastic structure. And in the elastomeric region, the constitutive equation explaining the relationship between the displacement vector and the Cauchy stress tensor is

$$\sigma_s = \lambda_s (\nabla \cdot x_s) I + 2\mu_S (\nabla x_s + \nabla x_s^T) \ . \tag{3}$$

The structure stiffness matrix is changed for the solid geometry is influenced by the fluid field. So the equilibrium equation should be established on the exact position after deformation.

And then the air volume is modeled as a viscous incompressible Newtonian fluid. By omitting the external body forces, the Navier-Stokes equation described in the ALE coordinate system is:

$$\frac{\partial v_f}{\partial t} + \left[(v_f - v_g) \cdot \nabla \right] v_f + \frac{\nabla p}{\rho_f} = \mu_f \nabla^2 v_f \ . \tag{4}$$

Here v_f is the fluid velocity vector and v_g is the velocity of the moving fluid domain in the ALE frame. The other parameters are the fluid pressure p, the fluid density ρ_f and the kinematic viscosity μ_f. The continuity equation obtained from the conservation of mass is

$$\nabla \cdot v_f = 0 \ . \tag{5}$$

And at the interface the structure velocity $\partial x_s / \partial t$ equals to the fluid velocity v_f. The above equations constitute the closed set of equations of solving the interior of the fluid domain and the structure deformed under the wind pressure.

Then the interaction behavior can be simulated and solved. Firstly, analyze the wire only sustain the tension. Fig.3 shows the wire deformations without wind under F=8000N and Fig.4 shows wire length condition under F=20000N and the initial strain which can simulate the pretension here is 0.001. F represents the axial force on wires; *MX* and *MN* are the maximum and minimum displacements.

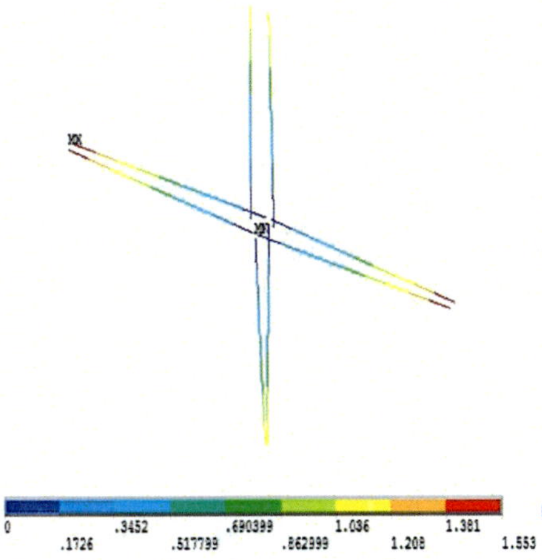

Fig. 3. F=8000N

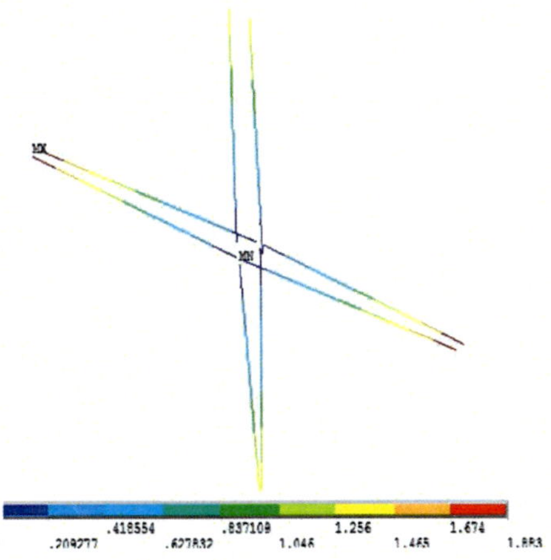

Fig. 4. F=20000N

Secondly, load wind pressure from fluid field on the eight ropes with the initial strain 0.001. Preloads in wires and other structural components often have significant effect on deflections and stresses. Reference[11] points out three methods simulating the initial tension of bolt by finite element method. And the pretension element method is better than the decreasing temperature and penetrative contact methods. Considering the stress stiffness effect, the nodal displacement nephogram results are in Fig.5~6. Fig.7~8 show the deformed shape under different tension.

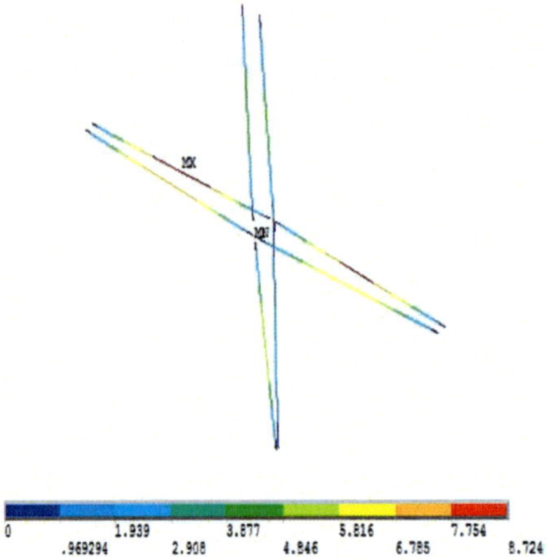

Fig. 5. *F*=8000N

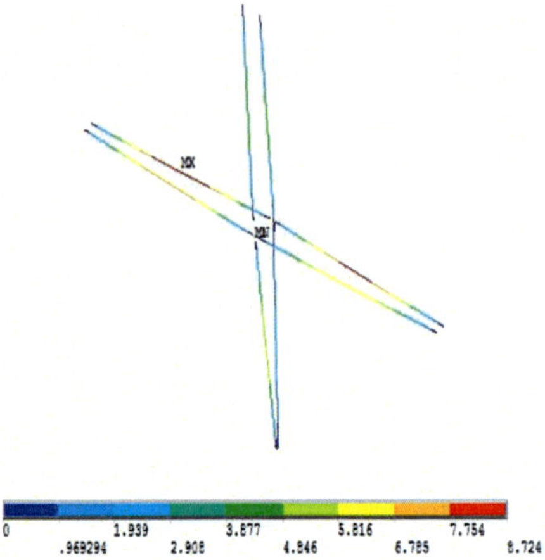

Fig. 6. *F*=20000N

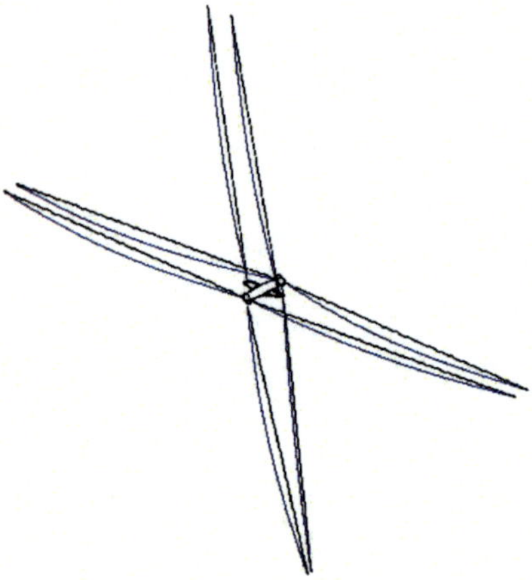

Fig. 7. *F*=8000N

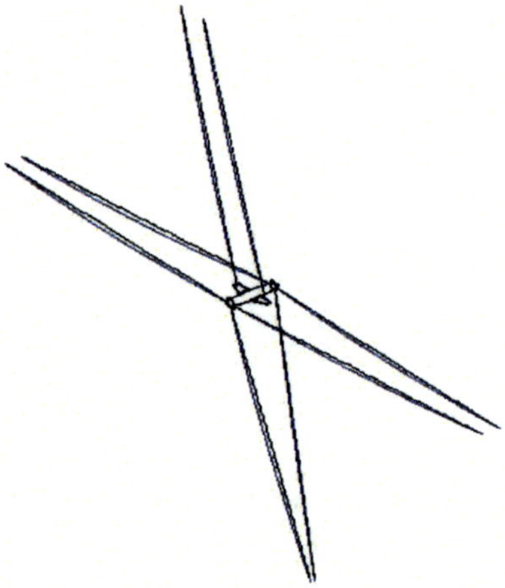

Fig. 8. *F*=20000N

Fig.3 shows the 5th-8th wires are extended by 1.555mm which is 1.55mm in theory. Fig.5~8 show that the lager the axial force is exerted, the less deformation and the bending degree occurs. The wire deformations in Fig.7 and Fig.8 are ten times that of the real displacement. Thirdly, the scaled aircraft can not achieve the expected position and attitude due to the fluid-solid interaction, so it is important to analyze the solution to forward kinematics of WDPSS-8 in MATLAB. Initially it is found that the aircraft tends to move along the wind and the tendency can be cut down by increasing the axial tension.

3.4 The Air Field Infected by the Solid Field

One advantage of the technology of wire-driven parallel robots is that it can reduce the aero-dynamical interference around the model. Here generate the wind vector graphs and analyze the wind pressure distributed throughout the field. The former shows that the velocity is low around the WDPSS-8 and becomes high after the barricade in the air field and it ranges from 35m/s~39.5m/s. The latter shows that the pressure on the first windward side is 0.7~0.8Mpa and nearly 0Mpa on the exit area. At the same time the negative zone is found near the wing. Then make four paths to reflect the velocity and pressure respectively above and below the wings. It is found that the pressure below is higher than that above and just as the opposite the velocity is higher above the wing.

4 Conclusion

The author attempts to investigate the fluid-solid interaction of WDPSS-8.

(1) Wire can be meshed by beam4 element by accessing to some relevant information and analyzing the wire deformations without wind.
(2) It is shown that the wires bend in wind field by simulation function of fluid-solid interaction and the bending deformation can be cut down by enhancing the axial tension on the ropes. Of course the wire vibration can also be improved by adding viscoelastic dampers or using other motion control methods in application.
(3) It is useful to compensate the position and attitude by solving the forward kinematics of WDPSS-8, otherwise the control is so inaccuracy that the aircraft is not up to the destination.
(4) Observing the wind velocity vector graphs and the pressure in the air, it is proven that the application of WDPSS-8 can cut down the air flow field affected by the suspension system.
(5) The four paths reflecting the velocity and pressure below and above the wing show that on the same streamline the more the velocity is, the less the pressure is. This is just right according to Bernoulli equation. And the negative pressure zone exiting around the wing shows that there are wake vortices in the forcedly oscillatory experiments.

References

1. Liu, X., Qiu, Y.Y., Sheng, Y.: Stiffness Enhancement and Motion Control of a 6-DOF Wire-driven Parallel Manipulator with Redundant Actuations for Wind Tunnels. ACTA Aeronautica ET Astronautica Sinica 30, 1156–1165 (2009)
2. Magill, J.C., Wehe, S.D.: Initial Test of a Wire Suspension Mount for Missile Virtual Flight Testing. In: 4th AIAA Aerospace Sciences Meeting and Exhibit, pp. 1–14 (2002)
3. Kazui, T., Takamoto, S.: Simulation Study of Aortic Valve Function Using the Fluid-structure Interaction Finite Element Method. In: Advances in Understanding Aortic Diseases, pp. 53–60. Springer (2009)

4. Xu, X.R., Fan, X.P.: Math Simulation of Coupled Fluid Flow and Geomechanics for Multiple Phases Reservoir. Chinese Journal of Rock Mechanics and Engineering 21, 93–97 (2002) (in Chinese)
5. Zheng, Y.Q., Lin, Q., Liu, X.W.: One Wire-driven Parallel Suspension Systems for Static and Dynamic Derivatives of Aircraft in Low-speed Wind Tunnels. ACTA Aeronautica ET Astronautica Sinica 30, 1449–1554 (2009) (in Chinese)
6. Li, J.Q., Liu, J.B.: Viscoelastic Damper on Wind-induced Response Analysis for Tower. Engineering Mechanics, 534–538 (2002) (in Chinese)
7. Dettmer, W.G., Peric, D.: A Fully Implicit Computational Strategy for Strongly Coupled Fluid-Solid Interaction. Arch. Comput. Methods Eng. 14, 205–247 (2007) (in Chinese)
8. Fang, Z.F., Wu, J.H.: Dynamic Simulation of Steel Cable Structure Modeling. Hoisting and Conveying Machinery 2, 71–75 (2009)
9. Hu, X.N., Zheng, J.B., Zhao, G.W.: Dynamics Simulation of Flexible Rope with Air Resistance. Mechanical Science and Technology for Aerospace Engineering 29, 1334–1337 (2010) (in Chinese)
10. Zhang, H.Y., Tong, M.B., Wang, Y.Q.: Emulation Analysis for Suspension Line Fracturing. Journal of System Simulation 17, 2525–2527 (2005) (in Chinese)
11. Li, H.X., Hu, Y.C., Zhang, J.Z.: Study on Simulating Bolt Pretension by Using ANSYS. Journal of Shandong University of Science and Technology 25, 57–59 (2006) (in Chinese)

Comprehensive Evaluation on the Urban-Rural Integration Process within Harbin Metropolitan Area Based on the Grey Multi-level Evaluation Model

Lihong Han, Guangji Tong, and Lei Shi

College of Economics and Management, Northeast Forest University
Harbin, 150040, China
hithlh@yahoo.com.cn, tonggj63@126.com, eyu@sina.com

Abstract. This paper establishes a comprehensive evaluation index system of urban-rural integration process, and uses the AHP and grey relational analysis methods to evaluate the urban-rural integration process within Harbin metropolitan area. It concludes that GDP per capita and rural industrial added value are best fit the trend of Harbin metropolitan urban-rural integration curve. Therefore, we should develop deep processing of agricultural products and light industry to promote the level of industrialization and the prosperity of the tertiary industry. The rural areas can exploit their favorable geographic location relative to the central city and consolidate the "urban-rural differences", then complete the urban-rural integration finally.

Keywords: Metropolitan area, urban-rural integration, comprehensive evaluation.

1 Introduction

According to the 2005 "Research Report of Harbin Metropolitan Area Planning", Harbin metropolitan area is an economic zone based on the administrative division. It includes urban district of Harbin city, Acheng, Shuangcheng, Wuchang, Shangzhi, Zhaodong and Binxian six satellite towns and their villages (Acheng has become a district of Harbin city in 2006; Zhaodong belongs to the administrative jurisdiction of Suihua).

As a spatial structure, Metropolitan area plays the role of concentration and diffusion, at the same time it also promotes the process of urban-rural integration within the metropolitan area. Promoting the urban-rural integration is bound to need the help of some spatial structure, and the metropolitan area is a good spatial structure which promotes the urban-rural integration process. Therefore, metropolitan area and urban-rural integration can develop and promote mutually.

Knowing the stage of urban-rural integration within the metropolitan area correctly will provide a reference for us in the following process of formulation and implementation of urban-rural integration development strategy. The process of urban-rural integration within Harbin metropolitan area can be evaluated by the grey multi-level evaluation model (combination of grey relational analysis methods and AHP).

D. Yang (Ed.): Informatics in Control, Automation and Robotics, Volume 2, LNEE 133, pp. 803–806.
springerlink.com

2 Comprehensive Evaluation on the Urban-Rural Integration Process within Harbin Metropolitan Area

Urban-rural integration is a dynamic process of constant development, which involves urban-rural natural, social, economic and other aspects relation. Therefore, it needs to establish a comprehensive evaluation index system to evaluate.

2.1 Establishing a Comprehensive Evaluation Index System

According to the connotation of urban-rural integration, the principles of establishing evaluation system and data availability, this paper chooses 17 indexes from the scope of comprehensive economic, coordinating urban-rural industry, urban-rural social differences, the regional environment governance, infrastructure and public services, they are: the per capita GDP (X_{i1}), the proportion of added value of second and third industry to GDP (X_{i2}), the level of urbanization (X_{i3}), rural industrial added value (X_{i4}), the coefficient of urban-rural dual structure (X_{i5}), the ratio of urban-rural per capita consumption expenditure (X_{i6}), the ratio of urban-rural Engel coefficient (X_{i7}), the proportion of reinforced brick construction housing(X_{i8}), the ratio of urban-rural education spending(X_{i9}), the proportion of rural transferring labor force to the total rural labor force(X_{i10}), the standard rate of industrial wastewater(X_{i11}), the rate of comprehensive utilization of industrial solid waste(X_{i12}), the ratio of urban-rural per capita income(X_{i13}), gross agricultural machinery power(X_{i14}), the proportion of employees in second and third industry(X_{i15}), rural per capita electricity consumption(X_{i16}), doctors and health workers per 1000 rural agricultural population(X_{i17}).

We select seven years numerical value of these indexes from 2003 to 2009 to evaluate (as shown in Table 1).

Table 1. Numerical value of the urban-rural integration Indexes

indexes	2003(X_1)	2004(X_2)	2005(X_3)	2006(X_4)	2007(X_5)	2008(X_6)	2009(X_7)
X_{i1}	13832	16175	18061	20635	23936	28141	31456
X_{i2}	83.07	82.32	82.72	84.62	85.08	86.57	86.81
X_{i3}	48.73	51.94	51.09	51.8	53.45	52.75	52.59
X_{i4}	164.13	192.07	207.09	219.41	260.17	313.73	335.42
X_{i5}	2.84	2.358	2.45	2.95	2.77	3.12	3.07
X_{i6}	3.08	3.23	3.48	3.41	3.21	3.16	3.15
X_{i7}	0.79	0.8	0.75	0.77	0.83	0.84	0.84
X_{i8}	71.4	72.5	77.7	78.9	81.3	83.8	85.9
X_{i9}	3.66	4.41	3.72	3.22	3.85	3.83	3.48
X_{i10}	38.03	41.93	47.45	53.28	48	50.8	50
X_{i11}	92	90.8	91.1	76.3	77.9	92	97.4
X_{i12}	73	74.3	74.5	63.8	70.9	74.8	76.3
X_{i13}	2.68	2.47	2.51	2.55	2.52	2.44	2.34
X_{i14}	341.7	373.8	397.5	526.4	584.4	619.7	718.8
X_{i15}	63.3	66.4	65.1	65.1	67.3	67.7	68.2
X_{i16}	465.95	502.93	513.23	553.47	594.67	757.66	806.11
X_{i17}	17.45	17.68	17.23	17.45	19.7	19.3	20.67

2.2 Evaluating the Process of Urban-Rural Integration within Harbin Metropolitan Area

From the data of Table 1, we can determine the optimal index set:

$$[X_{0i}]=[31456, 86.81, 53.45, 335.42, 2.358, 3.08, 0.84, 85.9, 3.22, 53.28, 97.4, 76.3, 2.34, 718.8, 68.2, 806.111, 20.67]. \quad (1)$$

According to the judging matrix given by the experts, we use the software of AHP to calculate the relative weight of the evaluation indexes as follows:

$$P(n)=(0.152,0.159,0.0716,0.1096,0.0443,0.031,0.0244,0.0132,0.031,0.0817,0.0366, 0.0446,0.05,0.0295,0.0544,0.0462,0.0229). \quad (2)$$

After normalizing and calculating the original data set of indexes and optimal index, we can get the correlation coefficient matrix R. Transposing matrix R to matrix R_1, according to the rule of $E = P * R_1$, then the correlation vectors R_{11} of X_1 year multiplies the index weight vector P (n), we get the correlation value of X_1 year, thus we can see the development progress of urban-rural within Harbin metropolitan. Correlation value of each year as follows:

$$E= [0.403003, 0.466706, 0.461223, 0.509101, 0.585476, 0.748629, 0.910113]. \quad (3)$$

3 Results Analysis

According to the correlation value of every year and several key indexes, we draw the curves of urban-rural integration development trend, per capita GDP and rural industrial added value within Harbin metropolitan area, as shown in Figure 1.

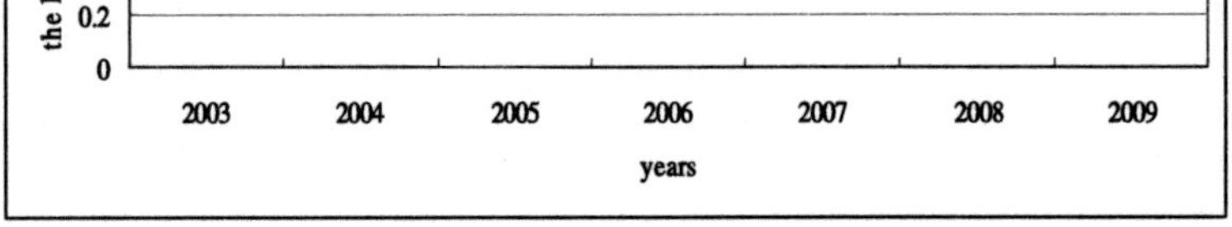

Fig. 1. Urban-rural integration process within Harbin metropolitan area

From the curve of the level of integration, we can see that the highest level of urban-rural integration is the year of 2009 and there are 3 distinct stages:

The first stage: 2003 -2004. The urban-rural integration process within Harbin metropolitan area declines in the process of growth, since the 2003 SARS and the 2004 inflation affected economy to a certain extent. The second stage: 2005-2006.This stage increases slowly. This shows that the national policy of agriculture, rural areas and peasants has begun to play the role of stimulating economic growth. The third stage:2007-2009. This stage increases significantly. This shows that the effect of policy of construction of new rural areas and urbanization has appeared obviously.

In addition, as shown in Figure 1, the curves of GDP per capita and rural industrial added value are best fit the trend of Harbin metropolitan urban-rural integration curve. This shows, under the background of the binary division of the big industry and big agriculture, the surrounding rural areas are in need of industrialization urgently. Therefore, in the follow-up process of urban-rural integration and harmonious development, the central city should further support the surrounding areas.

4 Conclusions

From the scope of strategic position of a commodity-grain producing area, transferring heavy industry within Harbin metropolitan areas is not appropriate. Therefore, we should develop deep processing of agricultural products and light industry to promote the level of industrialization and the prosperity of the tertiary industry. The rural areas can exploit their favorable geographic location relative to the central city and consolidate the "urban-rural differences", then complete the urban-rural integration finally.

Acknowledgements. It is a project supported by Natural Science Foundation of Heilongjiang Province of China (G200907) and the Fundamental Research Funds for the Central Universities (DL11AC34).

References

1. Epstein, S., Jezeph, D.: Development-There is another Way: a Rural-Urban Partnership Development Paradigm. World Development 29(8), 1443–1454 (2001)
2. Du, D., Pang, Q.: Modern Comprehensive Evaluation Methods and Cases. Tsinghua University Press, Beijing (2008) (in Chinese)
3. Ren, P., Zhou, J., Zhang, G.: Evaluation on Urban-Rural Integration Process in Chengdu City. Journal of Sichuan Normal University (Natural Science) 29(6), 747–751 (2006) (in Chinese)
4. Chen, G.: Comprehensive Evaluation for Urbanization Level Based on Gray Relation Analysis. Journal of Anhui Agriculture Science 39(9), 5610–5612 (2011)

Recognition of Combined Arm Motions Using Support Vector Machine

Yanjuan Geng[1], Dandan Tao[1,2], Liang Chen[1,2], and Guanglin Li[1]

[1] Shenzhen Engineering Laboratory for Neurorehabilitation Technology of Limb Functions, Shenzhen Institutes of Advanced Technology, Chinese Academy of Sciences
[2] Harbin Institute of Technology Shenzhen Graduate School
11068, Xueyuan Road, Nanshan District, Shenzhen, 518055, China
{yj.geng,dd.tao,chenliang,gl.li}@siat.ac.cn

Abstract. To investigate the classification performance of combined arm motions only using surface electromyography (EMG) signal, six different feature sets were adopted to match support vector machine (SVM) classifier respectively. Four unilateral transradial amputees participated in multi-channel surface EMG signal collection. The results show that the wavelet features outperforms others with average classification accuracy 98%±2% for intact arm and 89%±6% for amputated arm across all subjects. And the classification performance of intact arm motions was significantly better than that of amputated arm motions.

Keywords: Combined motion, SVM, pattern recognition, surface electromyography.

1 Introduction

Surface electromyography (EMG) is the electrical signals that are generated by the electrical activity of the muscle fibers during contraction and detected over the skin surface. For transradial amputees, when they image and actuate a movement involved in their lost limbs with their "phantom limbs", the motor nerve signals from brain cause residual forearm muscle contracting, thus EMG signals related to some wrist and hand movements can be measured with surface electrodes. Based on this hypothesis, pattern recognition technique is used to identify a variety of intended movements since each specific motion has its own distinguishing characteristics represented by multi-channel EMG signals. Hence, various pattern recognition schemes, consisting of feature extraction and classification, have been proposed and investigated for intuitive control of multifunction prostheses.

Many previous pattern recognition studies in identifying the multiple classes of single arm and hand movement showed a high classification accuracy (>%93) [1-3] that suggested this control method holds a promising in control of multifunction prostheses. Yet, with an output of single movement at a time, the current signal processing paradigms only allows sequential, on/off control of multifunctional prostheses, i.e. only one pattern or class can be active at a time. By contrast, many

D. Yang (Ed.): Informatics in Control, Automation and Robotics, Volume 2, LNEE 133, pp. 807–814.
springerlink.com

natural limb movements in our daily living life involve two or more joints to produce a combined motion simultaneously. This disparity makes the current control schemes rigid and lack of natural in operating a motorized prosthesis, resulting in low clinical acceptance.

Recently, there are a few studies to address this issue. For example, a research group in Denmark focused on proportional and simultaneous control of multifunctional prostheses at multiple degrees of freedom (DOF). In particular, Jiang et al [4] proposed a non-negative matrix factorization (NMF) model, the multi-channel mean square values (MSVs) of EMG can be modeled as a non-linear mixture of the forces produced at multiple DOF of a muscular joint, both the simulation and experimental results showed that NMF algorithm is able to extract the multidimensional control information simultaneously. Nielsen et al [5] also showed that their method satisfactorily works even for estimation of forces exerted by the hand contra-lateral respect to the side where the EMG is recorded during mirrored bilateral movement. These researches all based on a same condition that the kinematics data and EMG were provided by normally limbed subjects. However, the force function target of the amputated limb is no longer available for unilateral upper-limb amputees, and thus this approach is clinically inapplicable.

In this paper, only using surface EMG as the control information, we investigated the classification performance of a single classifier in identifying two types of forearm motions: the single degree of freedom (DOF) movements and combined movements in amputees. Six feature sets extracted from surface EMG recordings were chosen to match with support vector machine (SVM) classifier, which proposed by Vapnik et al. [6] and has been recently used for multifunction prostheses control [7-9].

2 Methods

2.1 Subjects

Four male and one female subjects with unilateral transradial amputation, aged from 21 to 46 years, participated in this study. The protocol of this study was approved by the Shenzhen Institutes of Advanced Technology Institutional Review Board. All subjects gave written informed consent and provided permission for publication of photographs for scientific and educational purposes.

2.2 Experiment and EMG Data Collection

Each participant was asked to participate into the experiment with two trials. In the first trial, the participants performed ten classes of forearm movements using their intact arm, and then in the next trial, they did same movement task using their amputated arm. These ten classes of upper-limb motions were composed of three single DOF motions (finger pinch, hand close, hand open) and six combined motions (wrist flexion with hand close, wrist extension with hand close, wrist pronation with hand close, wrist supination with hand close, wrist flexion with hand open, wrist extension with hand open) plus a "no movement" class, as shown in Figure 1.

A commercial EMG acquisition system (BagnoliTM, DELSYS Inc., USA) was used in this study to collect EMG sugnals. A total of 10 channels of surface EMG signals were collected from forearm muscles with metal bipolar electrodes. The EMG signals were amplified and band-pass filtered (5-450Hz), and then acquired with a sample rate of 1 kHz. Of the 10 bipolar electrodes, 6 were uniformly placed around the proximal portion of the forearm and other 4 electrodes were positioned on the distal end of residual forearm. These electrodes recorded EMG signals from muscles physiologically related to wrist and hand movements. A large circular electrode was placed on the elbow bony area of the test arm as a reference.

During experiment, subjects were instructed to watch a video demonstration of each target movement and to perform the movement in synchrony with the video. In particular, when these subjects performed forearm movements with their amputated arms, they were asked to conduct the same movements with their mirrored intact arm, so that we can see if they can follow the video well. Each experimental trial contained ten repetitions of each of ten motion classes. For each repetition, the subjects were asked to exert a comfortable level of contraction with a medium force and to hold the contraction for approximately 4 seconds, and then to relax for the next 5 seconds. The ten repetitions of each movement totally produced 40-second active EMG recordings. To avoid muscle and mental fatigue, the subjects were allowed to take a rest of 3 to 5 minutes between trials.

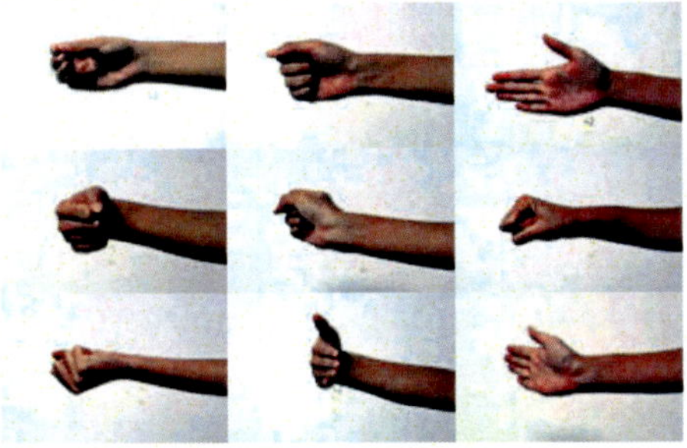

Fig. 1. Nine classes of forearm motions included in this study

2.3 Feature Extraction

EMG feature extraction was performed on analysis windows. For each movement, 40-second active EMG recordings were segmented into a series of analysis windows with a time length of 150 ms and a time increment of 100 ms. Six different feature sets were obtained to represent EMG characteristics, respectively. The feature sets were:

Set 1: Commonly used four time-domain features including mean absolute value, number of zero crossings, waveform length and number of slope sign changes[1-3,10].

Set 2: Six-order auto-regressive (AR) model coefficients via Yule-Walker method were extracted from each analysis window.

Set 3: The combination of six-order auto-regressive model coefficients with two time-domain features (mean absolute value and variance).

Set 4: Three frequency-domain features including mean power frequency, median frequency and peak frequency which were often used in sports medicine [11-12].

Set 5: The wavelet coefficients that describe EMG features in time-frequency domain. Sym5 mother wavelet was used in 3-level wavelet decomposition, The mean absolute value of the reconstructed coefficients with db4 wavelet constructed the feature set.

Set 6: Compared with wavelet decomposition transformation, the wavelet packet analysis decomposes not only low-frequency level but also high-frequency, and selects the frequency band matching the signal spectrum adaptively. For wavelet packet feature set, each EMG signals were decomposed into 3 levels with db4 wavelet packet, and eight mean values of reconstruction coefficients at the 3rd level composed the sixth feature set:

$$S = [\, S_{31}\ S_{32}\ S_{33}\ S_{34}\ S_{35}\ S_{36} \,] \tag{1}$$

2.4 Classification Using SVM

The motion classification problem could be considered as a multi-pattern recognition using SVM. The EMG features from the first half of EMG recordings were used as the training data set to train a SVM classifier for the ten motion classes, and the EMG features from the second half of EMG recordings were used as the testing data set to evaluate the performance of the trained SVM classifier in identifying the motion classes.

Basically, SVM only solves the binary classification problem based on the following principle. Given a set of training data $(x_1, y_1), \ldots, (x_l, y_l)$, where $x_i \in X \subseteq R^n$ is the feature vector, $y_i \in -1,1$ is the class label, and l is the size of data set, SVM is proposed to find the optimal separating hyperplane (OSH), which minimizes the following cost function with constrains:

$$\min_{\omega,b,\xi} J(\omega,\xi) = \frac{1}{2}\|\omega\|^2 + C\sum_{i=1}^{l}\xi_i$$
$$s.t.\ \begin{cases} y_i\,[\omega . x_i + b] \geq 1-\xi_i \\ \xi_i \geq 0 \quad (\mathrm{i}=1,\ldots,l) \end{cases} \tag{2}$$

The cost function $J(\omega,\xi)$ consists of two parts: the regularized term $\frac{1}{2}\|\omega\|^2$, of which the minimum value maximizes the distance between the hyperplane and the closest point of each class, and the penalty term $\sum_{i=1}^{l}\xi_i$, which represents the empirical risk for linear unseparable cases. For a linear separable problem, $\xi_i = 0$. The parameter C is the trade-off factor between regularized term and penalty term, of which a larger value emphasizes more on classification errors.

For the typical quadratic programming problem with linear constrains, Lagrange multiplier method is a conventional way to solve the optimal solution by transforming the original minimization problem into a dual maximization problem, which is given by:

$$\max\ W(\alpha)=\sum_{i=1}^{l}\alpha_i-\frac{1}{2}\sum_{i=1}^{l}\sum_{i=1}^{l}\alpha_i\alpha_j y_i y_j x_i.x_j$$

$$s.t.\ \begin{cases}0\le\alpha_i\le C\ \ (\mathrm{i}=1,...,l)\\ \sum_{i=1}^{l}\alpha_i y_i=0\end{cases}\tag{3}$$

Where $\alpha=(\alpha_i)$ is the vector of Lagrange multipliers.

Given a function $\phi(x)$ which maps the input feature space into a higher-dimensional one, the objective function of can be rewritten as:

$$\begin{aligned}W(\alpha)&=\sum_{i=1}^{l}\alpha_i-\frac{1}{2}\sum_{i=1}^{l}\sum_{i=1}^{l}\alpha_i\alpha_j y_i y_j\phi(x_i).\phi(x_j)\\&=\sum_{i=1}^{l}\alpha_i-\frac{1}{2}\sum_{i=1}^{l}\sum_{i=1}^{l}\alpha_i\alpha_j y_i y_j K(x_i,x_j)\end{aligned}\tag{4}$$

where $K(x_i,x_j)=\phi(x_i).\phi(x_j)$ is the kernel function.

If denoting the obtained optimum vector of Lagrange multipliers, by solving the above optimization problem, as α^*, the corresponding optimum linear weight vector ω^*, the optimum bias b^*, and the OSH are given by following equations respectively:

$$\begin{aligned}&\omega^*=\sum_{i=1}^{l}\alpha_i^* y_i\phi(x_i)\\&\mathrm{b}^*=-\frac{\max_{y_i=-1}(\omega^*.\phi(x_i))+\min_{y_i=1}(\omega^*.\phi(x_i))}{2}\\&\sum_{i=1}^{l}\alpha_i^* y_i K(x_i,x)+\mathrm{b}^*=0\end{aligned}\tag{5}$$

Therefore, the decision function for identifying the class of the input x is given by

$$y(x)=\mathrm{sgn}(\sum_{i=1}^{l}\alpha_i^* y_i K(x_i,x)+\mathrm{b}^*)\tag{6}$$

The above procedures only realize a binary-classification SVM. Two approaches can extend SVM to multi-classification, the one-against-one and one-against-all. The kernel concept of each approach is to convert the multi-classification into several binary ones. In this work, we only adopt the one-against-one method with linear kernel function.

The performance of the trained classifier in identifying a movement was measured by the classification accuracy (CA), which is defined as:

$$CA=\frac{Number\ of\ Accurate\ Classification\ Decisions}{Total\ Number\ of\ Classification\ Decisions}\times 100\%\tag{7}$$

3 Results

For all subjects, each EMG feature set was used to match with SVM classifier, resulting two groups of performance index expressed as percentages, as shown in Table 1. For each subject, the left column is the CA from their intact arm and the right column is the CA from their amputated arm.

Table 1. Classification Accuracy for All Subjects in Two Trials

SVM	TR01		TR02		TR03		TR04		TR05	
	Intact	Amputated	Intact	Amputated	Intact	Amputated	Intact	Amputated	Intact	Amputated
TD	97.16	77.11	95.63	80.15	93.40	89.70	99.15	81.36	94.02	91.95
ARYL	95.27	71.33	95.16	77.12	93.61	85.05	99.32	82.71	97.61	90.85
TDAR	92.7	70.44	94.53	70.00	89.38	85.36	96.44	70.00	94.02	90.73
FD	85.41	74.89	82.81	78.18	92.27	88.87	96.10	87.63	91.41	91.22
WT	99.46	82.67	95.47	85.15	99.07	91.24	99.83	89.49	95.22	97.20
WPT	99.19	80.67	96.41	83.48	96.29	91.34	99.83	89.15	95.11	98.29

3.1 Effects of Each Feature Set

In order to choose the best suited features for SVM classifier, the average CA across all five subjects with respect to all six feature sets was calculated, as shown in Figure 2. It can be clearly seen from that the wavelet features (98%±2% for intact arm and 89%±6% for amputated arm) and wavelet packet features (97%±2% for intact arm and 89%±7% for amputated arm) had better performance than other feature sets. The commonly used TDAR feature combination which has clear advantage matching with other classifiers seemed not suitable to match with SVM classifier, especially for amputated limb.

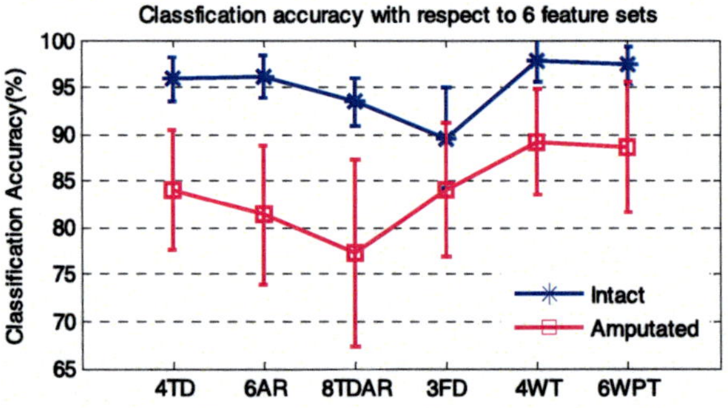

Fig. 2. Comparison of six feature set across all subjects

3.2 Comparison of CA for Intact Arm and Amputated Arm

Previous studies have shown that the CA of EMG from amputated limb is generally lower than that from intact limb due to only limit EMG information could be restored after limb amputation [1-3,10]. This was proved again by the result shown in Figure3, i.e. no matter which feature set applied. However, it is not necessary all the same for all subjects may due to surface electrodes space distribution. Fig.4 shows the average CA using wavelet features and wavelet packet features, the CA for intact limb is much higher except TR05. But as a whole, the result shown in Fig.4 is consistent with previous conclusion, the average CA over all five subjects was 89%±6% for the amputated arm and 98%±2% for the intact arm. The paired t-test result ($p < 0.02$) shows the significance between amputated arms and intact arms.

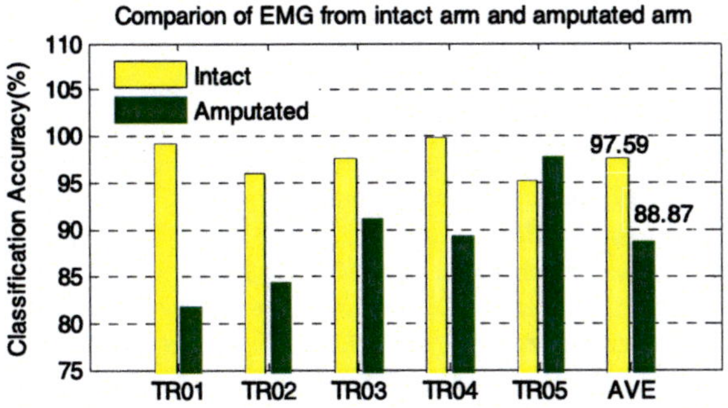

Fig. 3. Difference between intact limb and amputated limb

4 Conclusions

In this study, six feature sets were applied in combination with SVM classifier to investigate the performance of single classifier in identifying 10 combined forearm motions. The results showed when using a SVM classifier, the wavelet decomposition and wavelet packet decomposition outperformed than other feature sets. And the motions activated by amputated arm were harder to be identified by the classifier than that by intact arm, due to loss of muscles and nerves after amputation. The average classification accuracy across all five subjects was about 89% for amputated arms when using the wavelet features and SVM classifier. The result showed a promising application of single classifier in complex motion classification.

Note that only six 2-DOF forearm movements were considered in this study due to anatomical factors. That is, the muscles responsible for flexion/extension are indeed superficial and their activity can be easily recorded from surface electrodes. However, the surface detection of the supinator and pronator teres muscles activity is more challenging because these muscles are deep and partially covered by the bicipital aponeurosis. More combined motion classes and parallel classifier will be investigated in the further study.

Acknowledgments. This work was supported in part by the Shenzhen Governmental Basic Research Grants #JC200903160393A and #JC201005270295A, the National Natural Science Foundation of China under Grant #60971076, Guangdong Innovation Research Team Fund for Low-cost Healthcare Technologies (GIRTF-LCHT), and Shenzhen Key Laboratory of Neuropsychiatric Modulation Grant.

References

1. Kuiken, T.A., Li, G., Lock, B.A., Lipschutz, R.D., Miller, L.A., Stubblefield, K.A., Englehart, K.B.: Targeted muscle reinnervation for real-time myoelectric control of multifunction artificial arms. JAMA 301, 619–628 (2009)
2. Huang, H., Zhou, P., Li, G., Kuiken, T.: A Analysis of EMG Electrode Configuration for Targeted Muscle Reinnervation Based Neural Machine Interface. IEEE Trans. Neural Systems and Rehabilitation Engineering 16(1), 37–45 (2008)
3. Parker, P., Englehart, K., Hudgins, B.: Myoelectric signal processing for control of powered limb prostheses. J. Electromyogr. Kinesiol 16, 541–548 (2006)
4. Jiang, N., Englehart, K.B., Parker, P.A.: Extracting Simultaneous and Proportional Neural Control Information for Multiple-DOF Prostheses From the Surface Electromyographic Signal. IEEE Trans. on Biomedical Engineering 56(4) (April 2009)
5. Nielsen, J.L.G., Holmgaard, S., Jiang, N., Farina, D., Parker, P.: Enhanced EMG Signal Processing for Simultaneous and Proportional Myoelectric Control. In: Proc. 31st IEEE EMBS Annu. International Conf., Minnesota, pp. 4335–4338 (2009)
6. Corts, C., Vapnik, V.: Support vector networks. Mach. Learn. 20, 273–297 (1995)
7. Liu, Y.H., Huang, H.P., Weng, C.H.: Recognition of Electromyographic Signals Using Cascaded Kernel Learning Machine. IEEE/ASME Trans. on Mechatronics 12(3) (2007)
8. Yang, D.P., Zhao, J.D., et al.: An Anthropomorphic Robot Hand Developed Based on Underactuated Mechanisum and Controlled by EMG Signals. Journal of Bionic Engineering 6, 255–263 (2009)
9. Lucas, M.F., Gaufriau, A., Pascual, S., Doncarli, C., Farina, D.: Multi-channel surface EMG classification using support vecter machines and signal-based wavelet optimization. Biomedical Signal Processing and Control 3, 169–174 (2008)
10. Li, G., Schultz, A.E., Kuiken, T.A.: Quantifying Pattern Recognition- Based Myoelectric Control of Multifunctional Transradial Prostheses. IEEE Trans. Neural Systems and Rehabilitation Engineering 18(2), 185–193 (2010)
11. Dieen, J.H.V., Putten, E.P.W., Kingma, I., Looze, M.P.: Low-level activity of the trunk extensor muscles causes electromyographic manifestations of fatigue in absence of decreased oxygenation. Journal of Electromyography and Kinestiology 19, 398–406 (2009)
12. Szeto, G.P.Y., Straker, L.M., Sullivan, P.B.: EMG Median frequency changes in the neck-shoulder stabilizers of symptomatic office workers when challenged by different physical stressors. Journal of Electromyography and Kinestiology 15, 544–555 (2005)

Optimization and Verification for a Robot Control System Based on Learning Petri Net Model

Liangbing Feng, Masanao Obayashi, Takashi Kuremoto, and Kunikazu Kobayashi

Division of Computer Science & Design Engineering, Yamaguchi University, Ube, Japan
{n007we,m.obayasi,wu,koba}@yamaguchi-u.ac.jp

Abstract. The optimization and verification for a robot control system have been studied respectively. In this paper, we propose a robot control model which combines optimization and verification based on Learning Petri Net (LPN). In our former work, learning Petri net model has been constructed based on high-level time Petri net and Reinforcement Learning (RL). The reinforcement learning is applied to optimize the parameters of Petri net. These parameters adjusting may make the properties of learning Petri net change. In this paper, the verification algorithms for learning Petri net are proposed. Using this model, a robot control system with optimization and verification capability is constructed. Finally, for the purpose of certification of the effectiveness of our proposed methods, this robot control system is used to control a robot system – E-puck as an example. The result of the experiment shows the methods are correct and effective.

Keywords: Petri net, reinforcement learning, robot control system, e-puck.

1 Introduction

Petri net combines a well defined mathematical theory with a graphical representation of the dynamic behavior of systems. The theoretic aspect of Petri net allows precise modeling and analysis of system behavior, while the graphical representation of Petri net enable visualization of a modeled system state changes [1]. So, Petri net is widely used to model and verify various dynamic systems. But the traditional Petri net doesn't have the learning capability, and it can't optimize its parameters when modeled system is running. All the parameters which describe the system characteristic need to be set individually and empirically when dynamic system is modeled. Recently, there are some researches for making the Petri net have learning capability and making it optimize itself. In paper [2], the global variables are used to record all state of colored Petri net when it is running. The global variables are optimized and colored Petri net is updated according to these global variables. A learning Petri net model which combines Petri net with neural network is proposed in [3]. And this learning method was applied to nonlinear system control. In our former work [4, 5], a learning Petri net model has been proposed based on reinforcement learning. RL is applied to optimize the parameters of Petri net. And, this learning Petri net model has been applied to robot system control. In these optimization methods,

D. Yang (Ed.): Informatics in Control, Automation and Robotics, Volume 2, LNEE 133, pp. 815–823.
springerlink.com

the parameters adjusting may make the property of learning Petri net change, for example, live and safe properties are broken. In this paper, the verification methods for learning Petri net are proposed. An optimization and verification method for a robot control system based on learning Petri net model is illuminated.

The rest of this paper is organized as follow. Section 2 gives the optimization and verification model based on learning Petri net. Section 3 describes the application of this model to robot control system for optimization and verification. Finally, Section 4 summarizes the paper and discusses some directions for future work.

2 The Optimization and Verification Model

An optimization and verification model based on learning Petri net is constructed. This model combines the learning method and verification method. It is adapted to robot system control.

2.1 Definition of Learning Petri Net

Learning Petri Net (LPN) is an expanded Petri net which introduces reinforcement learning into Petri net [5].

Definition 1. LPN has a 7-tuple structure, $LPN=(NG, C, W, DT, VW, DT, M_0)$ [1].

where: *(i)*. $NG=(P, Tr, F)$ is called net graph. P *is* a finite set of nodes which called Places. Tr *is* a finite set of Transitions, which disjoints from P, $P \cap Tr = \varnothing$; $F \subseteq (P \times Tr) \cup (Tr \times P)$ is a finite set of directional arcs, known as the flow relation;

(ii). C is a finite and non-empty color set for describing different type of data;

(iii). W: $F \rightarrow C$ is a weight on F. If $F \subseteq (P \times Tr)$, the a set of weight W is W_{in}. If $F \subseteq (Tr \times P)$, the weight W is W_{out}.

(iv). DT: $Tr \rightarrow R$ *is a set of* delay *time function of a transition which has a Time delay* for an enable transition fired or the fire of a transition lasting time.

(v). VW (value of weight): $W_{in} \rightarrow R$, is a weight marking on W_{in}. An arc $F \subseteq (P \times Tr)$ has a set of weight W_{in} and each W_{in} has a reward value item $VW \in$ real number.

(vi). VT (value of delay time): $DT \rightarrow R$, is a function marking on DT. A transition has a set of DT and each DT has a reward value item $VT \in$ real number.

(vii). M_0: $P \rightarrow \cup_{p \in P} \mu C(p)$ such *that* $\forall p \in P$, $M_0(p) \in \mu C(p)$ is the initial marking function which associates a multi-set of tokens of correct type with each place.

In LPN, RL algorithm is used to find the optimal policy of the dynamic system through environment state observation and improvement of its behavior through trial and error with the environment. RL agent senses the environment and takes actions. It receives numeric rewards or punishments from some reward functions. The system learns to choose adaptive actions to maximize a long term sum or average of the future rewards it will receive [6].

In Fig. 1, colored tokens C_{1j} are input to p_1. Token C_{1j} fires which transition that is decided by the value $VW_{C1j,1,k}$ (k=1,···, n). Transition which has the biggest $VW_{C1j,1,k}$ will fire or transition which has bigger $VW_{C1j,1,k}$ has bigger fire probability. In weight value learning, $VW_{Cij,i,j}$ is randomly set firstly. When the system runs, formula (1) is used to update $VW_{Cij,i,j}$ and the optimal transition to C_{ij} will get the biggest $VW_{Cij,i,j}$.

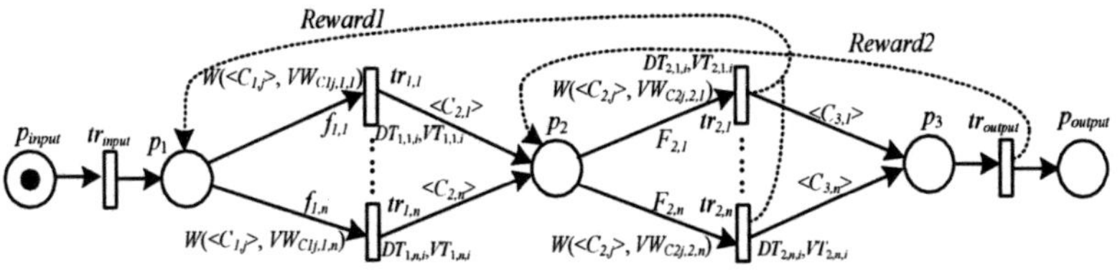

Fig. 1. An example of LPN model

$$VW_{Cij,i,j} = VW_{Cij,i,jj} + \alpha[r + \gamma(\overline{VW'_{cij,i,j}}) - VW_{Cij,i,j}] \quad (1)$$

where, (i). α is the step-size, γ is a discount rate.

(ii). r is reward which $W(<C_{ij}>, VW_{Cij,i,j})$ gets when $Tr_{i,j}$ is fired by $<C_{ij}>$.

(iii). $(\overline{VW'_{cij,i,j}})$ is the feedback value from all next action of c_{ij} fired transition.

In the transition delay time learning algorithm, a transition has a set of different delay times. For example, tr_{11} in Fig. 1 has m different delay times $DT_{11,i}$ and every delay time has a value item $VT_{11,i}$. After tr_{11} fired at delay time $DT_{11,i}$, it gets a reward r. The value of VT is updated by formula 2.

$$VT_{11,i}(p,tr) \leftarrow VT_{11,i}(p, tr) + \alpha[r + \gamma VT_{2j,i}(p', tr') - VT_{11,i}(p, tr)] \quad (2)$$

where, $VT_{11,i}(p, tr)$ is value of transition tr at Petri net state p. $VT_{2j,i}(p',tr')$ is the value of next transition tr' of tr at next state p' of p. α is a step-size, γ is a discount rate.

After renewing of V, the optimal weight or delay time will be selected. For example, in Fig. 1, when $DT_{11,1}...DT_{11,m}$ get value $DV_{11,1}...DV_{11,m}$, the m transitions are looked as m different transitions and are selected by the soft-max method according to a probability of Gibbs distribution which is shown in formula (3). Which transition has a bigger value will obtain a bigger selection probability.

$$\Pr\{tr_t=tr|p_t=p\} = \frac{e^{\beta V(p,tr)}}{\sum_{b\in A} e^{\beta V(p,b)}} \quad (3)$$

where, $\Pr\{tr_t=tr|p_t=p\}$ is a probability selecting transition tr at state p, β is a positive inverse temperature constant and A is a set of available transitions.

2.2 Deadlock Detection and Safe Test in the LPN

If the parameter of learning Petri net changes, the properties of LPN may be changed. So, some properties of leaning Petri net need to be verified.

Before verifying, some definitions are presented firstly. The sets of input and output transitions, color tokens of place p are defined respectively as follows [7]:

$*p = \{tr \in Tr : (tr, p) \in F\}, p* = \{tr \in Tr : (p, tr) \in F\}$,

$*p$(c) is the color tokens set of $*p$ can deal; $p*$(c) is the token set of $p*$ put into p.

S is a nonempty subset of places of a Petri net. If

$$(*S \subseteq S*) \wedge (*S(c) \subseteq S*(c)) \tag{4}$$

then the learning Petri net is in deadlock. If

$$(S* \subseteq *S) \wedge (S*(c) \subseteq *S(c)) \tag{5}$$

then the learning Petri net is trapped.

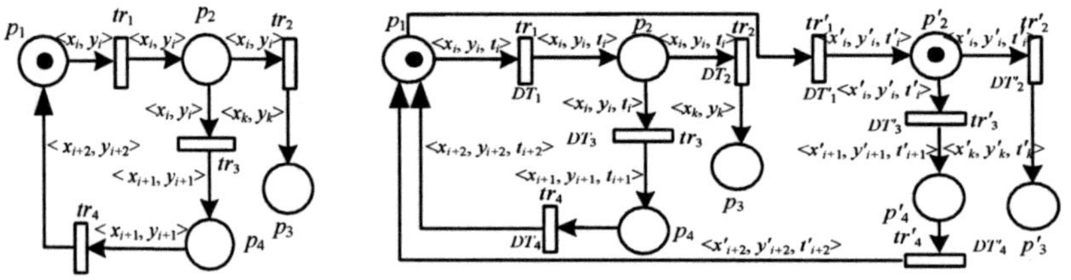

(1) (2)

Fig. 2. Deadlock and 1-safeness in the LPN model

A LPN is shown in Fig. 2 (1). In this model, the transition's weights are learnable. After many times adjusting for weight, it may make the output $<x_{i+j}, y_{i+j}>$ equate to $<x_i, y_i>$. if $S=\{P_1, P_2, P_4\}$, then

$*S=\{tr_1, tr_3, tr_4\} \subseteq S* = \{ tr_1, tr_2, tr_3, tr_4\}$ and $*S(c)=\{<x_i, y_i>, <x_{i+1}, y_{i+1}>, \cdots, <x_{i+j-1}, y_{i+j-1}>\} \subseteq S*(c) = \{<x_i, y_i>, <x_{i+1}, y_{i+1}>, \cdots, <x_{i+j-1}, y_{i+j-1}>, <x_k, y_k>\}$. So, the learning Petri net is in deadlock.

Places of a Petri net are often used to represent information storage areas or product and tool storage areas. A marking of a Petri Net is called *k-safe* if the number of tokens in all places is at most k. A Petri Net is k-safe if the initial marking is *k-safe* and the marking of all traces is k-safe [8]. In high-level Petri net model for robot control system, a marking of a high-level Petri Net is called *k-safe* if the number of same colored tokens in all places is at most k. Thus,

$$M_t(C(p)) \le k \; (\forall p \in P) \tag{6}$$

For example, a LPN is shown in Fig. 2 (2) and this Petri net needs to be a 1-safeness system. If the weight and the delay time are learnable, it may make the output $<x_{i+2}, y_{i+2}, t_{i+2}>$ equate to $<x'_{i+2}, y'_{i+2}, t'_{i+2}>$ when the weight and delay time are adjusted. This makes the place p_i be input two same colored tokens at the same time.

This breaks the rule of 1-safeness system. Because, there is a place p_i at time t_i has two same colored tokens, i.e. $M_{ti}(C(p))\leq 1$ is broken. Thus, its modeling system is not safe. Therefore, the safeness of LPN needs to be tested while the LPN is running.

2.3 The Learning and Verification Algorithm for LPN

Now, the learning and verification algorithm for the LPN is proposed. And it is listed in Fig. 3.

Learning and verification algorithm for LPN:

Step 1. Initialization: Randomly set all V of weight and delay time.

Step 2. Initialize the learning Petri net. i.e. make the Petri net state as M_0.

Repeat i) to v) until system becomes end state.

i) When a place gets a colored token C_{ij}, there is a choice that which transition is fired. The transition firing probability obeys formula (3).

ii) When a transition is selected, there is a choice of transition delay time. According to the value of every delay time, the delay time selection probability obeys formula (3).

iii) The selection of weight and delay time is set to Petri net. The deadlock, trap and safe for Petri net are detected according to formula (4), (5), (6). If violation of these properties, the reward is -1.

iv) If Petri net isn't in deadlock or trap or unsafe, the reward of weight and delay time are gotten from environment.

v) Update weight value and delay time value according formula (1), (2).

Fig. 3. Learning and verification algorithm for LPN

3 Application to Robot Control System for Optimization and Verification

3.1 Experiment Environment Construction

"e-puck" is a mini mobile robot which has a variety of devices equipped with camera, microphone, speaker and motor. The experiment environment is shown in Fig. 4 (left). The e-puck moves in a 1 m × 1 m square field. The goal is a green brand at one corner of this square field. Each of e-pucks has the follows actions:

1. E-puck runs a special distance. Then, it looks for the goal. If it recognizes the goal through the camera equipped in the e-puck, it will directly run towards the goal.

2. When an e-puck reaches the goal, it shouts aloud for telling other e-pucks to finish searching goal and run to the goal directly according to the sound direction.

3. If the e-puck doesn't find the goal, it will listen for an interval. If it hears a voice which comes from other e-puck, it will recognize the voice place and directly run towards. If it hasn't heard any voice, it turns a special angle and runs again.

There are three tasks in the experiment. I). An e-puck control model is constructed using the LPN; II). The e-puck behaviors are controlled and optimized using the LPN model. An e-puck's actions sequence consists of "run", "search" and "listen" when it is looking for the goal. Because, it is not able to execute actions such as look, listen and run at the same time. So, how long does the e-puck run and how long does it listen and look? For example, if the e-puck runs for 5 seconds and then takes a look (the time is very short and omits) and listens for 1 second, it may miss several times of the voice of other e-puck which has found the goal and sounds a time every 2 seconds. III) Correctness of behaviors is verified using verification method of Petri net. For example, if the e-puck turns 90° at 4 times continuously, it will come back to its fore state. If these actions are repeated, the e-puck will run a circle and its action will be in deadlock.

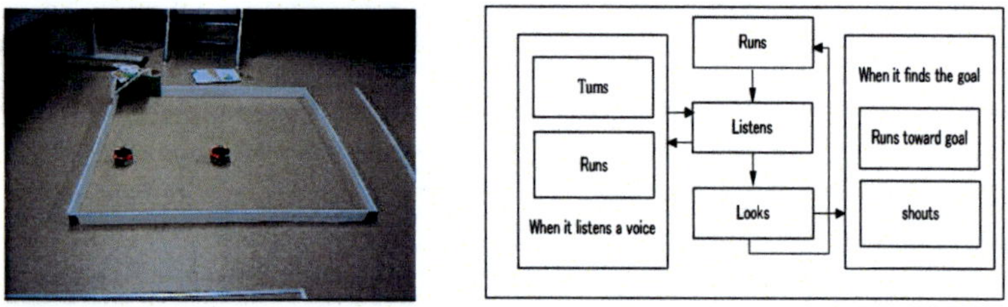

Fig. 4. The experiment environment (left) and the control flow for e-pucks (right)

3.2 LPN Model

The control flow for e-puck is shown in Fig. 4 (right). According to the control flow, the LPN model for the e-puck is constructed. The LPN control flow for two e-pucks is shown in Fig. 5. At the same way, the model for one or three e-pucks could be founded. The meaning of place, transition and token in Fig. 5 is listed Table 1.

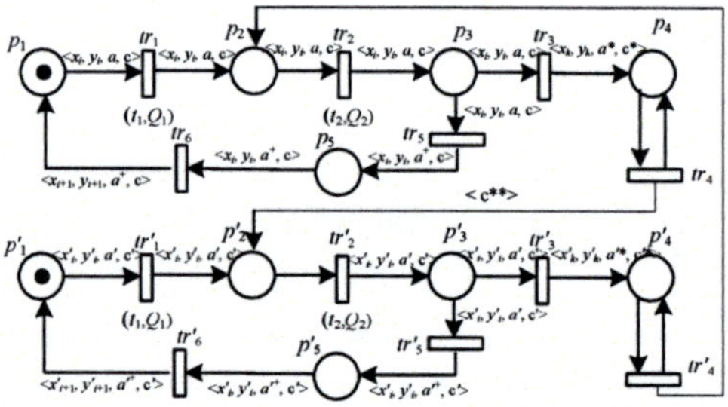

Fig. 5. The LPN control model for two E-pucks

Table 1. Meaning of places, transitions and tokens in Fig. 5

p_1	System starting state	p_2	State of getting search information
p_4	Arrival at goal	p_3	State of getting voice information
tr_1	Look for goal	p_5	State after adjusting angle
tr_2	Listen to other	tr_3	Directly run to goal
tr_4	Shout	tr_5	Turning a special angle
tr_6	Run	p'_1- p'_5	same to p_1- p_5
a, a'	Turning angle of e-puck	x', y'	Place of E-puck2 (coordinates)
x, y	Coordinates of E-puck1	tr'_1- tr'_6	same to tr_1- tr_6
c**	Sound signal	c, c'	Control signals of e-puck

3.3 Result of Simulation and Experiment

Each e-puck in this experiment turns a special angle after moving a certain distance. This turns may make the e-puck run in a circle. Using the computer simulation, one e-puck which turns at different angle is simulated. Here, the velocity of e-puck is 1cm/s; e-puck will turn an angle after every 2 second moving.

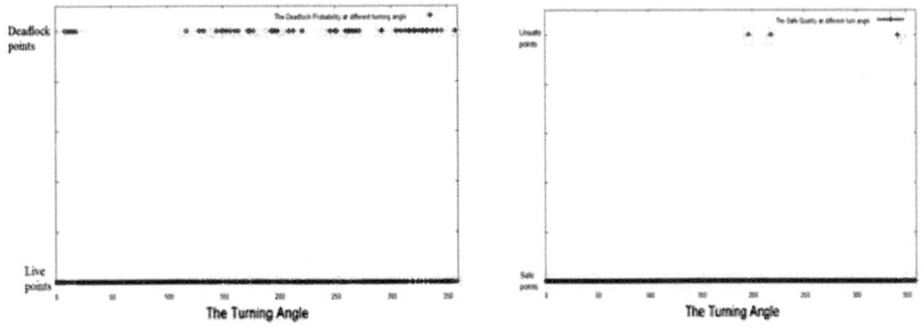

(1) (2)

Fig. 6. Result of deadlock and safe detection in control system. (1) shows deadlock points and live points when one e-puck turning 0°~360° different angles. (2) shows unsafe points and safe points when 2 of e-puck turning 0°~360° different angles.

The result is shown in Fig. 6 (1). The simulation result shows that there are many turning angles making the e-puck in deadlock. After computer simulated learning, the turn angles which do not make the e-puck deadlock are set into e-puck and the experiment is executed with real robots.

If more than one e-puck is used in this experiment, they may run to one point at the same time. Then there is not safe in this case. Two e-pucks which turn at different angles were tested to find the unsafe points. The result is shown in Fig. 6 (2) and it shows that there are three points which are unsafe.

Each e-puck in this experiment has three actions: run, turn and listen. When it finds the goal, it directly runs towards the goal and shouts out. The others hear the sound and run towards the sound. If one e-puck shouts out, the other one hears immediately and stops searching. Then, the efficiency is high. But how long does it run and how long does it listen? This is a problem for e-puck action control. If the listening time is too short, it may miss the calling. If the listening time is too long, it will waste time. Now, it will be learnt in Fig. 5 that delay time of transition tr_2 is what percent to delay time of transition tr_6.

In the case that two e-pucks are used, when two e-pucks arrive the goal, they will get a reward as $r=1/(t_1+t_2)$. Here, t_1 is the time the e-puck-1 used to find the goal. And, t_2 is the time the e-puck-2 used.

In the case that three e-pucks are used, they will get a reward as $r=1/(t_1+t_2+t_3)$. Here, the e-puck-1 finds the goal which used time t_1, the e-puck-2 finds the goal which used time t_2 and the e-puck-3 finds the goal which used time t_3.

In the experiment, the e-puck runs 2 seconds then executes listening. The proportion of running time is used to represent the listen time. For example, the proportion of running time is p and running time is t, the listening time is $p*t$. The total training times are 10,000 seconds. The results are shown in Fig.7. When e-puck listen time is as this proportion of running time, the total time for all e-pucks arriving goal is minimal. Explanation of Fig.7:

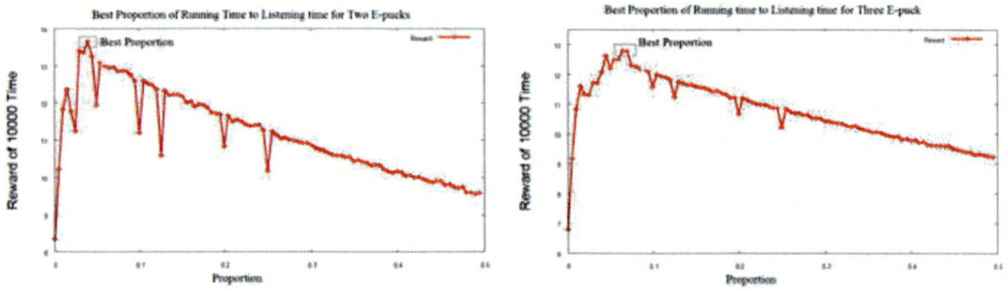

Fig. 7. When e-pucks run for a special time, they listen for a part of running time. The best proportion which compares running time with listen learning is shown for two (left) and three (right) e-pucks cases. When e-puck listen time is as this proportion of running time, the total time for all e-pucks arriving goal is minimal.

4 Conclusion

An optimization and verification model for robot control system was proposed. This model was constructed based on learning Petri net. The reinforcement learning is applied to optimize the parameters of Petri net. These parameters adjusting may make the properties of learning Petri net change. The verification algorithm for learning Petri net was combined with its learning algorithm. Using this model, a robot control system with optimization and verification capability was constructed. This robot control system was used to control a robot system – "e-puck" as an example. The result of the experiment showed the methods are correct and effective. In the future, we will construct other control model for robot system.

References

1. Wang, J.: Petri nets for dynamic event-driven system modeling. In: Fishwick, P.(ed.) Handbook of Dynamic System Modeling, pp. 1–17. CRC Press (2007)
2. Baranaushas, V., Sarkauskas, K.: Colored Petri Nets-Tool for control system Learning. Electronics and Electrical Engineering 4(68), 41–46 (2006)
3. Hirasawa, K., Ohbayashi, M., Sakai, S., Hu, J.: Learning Petri Network and Its Application to Nonlinear System Control. IEEE Transactions on Systems, Man and Cybernetic, Part B: Cybernetics 28(6), 781–789 (1998)
4. Feng, L., Obayashi, M., Kuremoto, T., Kobayashi, K.: A Learning Petri Net Model Based on Reinforcement Learning. In: Proceedings of the 15th International Symposium on Artificial Life and Robotics, AROB 2010, February 4-6, pp. 290–293 (2010)
5. Feng, L., Obayashi, M., Kuremoto, T., Kobayashi, K.: An Intelligent Control System Construction Using High-Level Time Petri Net and Reinforcement Learning. In: Proceedings of International Conference on Control, Automation, and Systems (ICCAS 2010), October 27-30, pp. 535–539 (2010)
6. Sutton, R.S., Barto, A.G.: Reinforcement Learning. The MIT Press (1998)
7. Karatkevich, A., Bieganowski, J.: Detection of deadlocks and traps in petri nets by means of thelen's prime Implicant method. International Journal of Applied Mathematics and Computer Science 14(1), 113–121 (2004)
8. von Bochmann, G., Jourdan, G.-V.: Testing *k*-Safe Petri Nets. In: Núñez, M., Baker, P., Merayo, M.G. (eds.) TESTCOM 2009. LNCS, vol. 5826, pp. 33–48. Springer, Heidelberg (2009)

Reviews on the Cognitive Radio Platform Facing the IOT

Yajuan Tang and Qing Li

Department of Electronic Engineering, Shantou University, Shantou 515063, China

Abstract. The rapid development of the Internet of things (IOT) proposed new challenges to the spectrum sources in great shortage. For Cognitive Radio technology can improve the spectrum efficiency, we review the platform for the Cognitive Radio in the paper and put forward the application prospect of the IOT. This paper focuses on the cognitive radio-based platform. We summarize the hardware platform at home and aboard, make a contrast and classification, to prepare for the platform facing the IOT.

Keywords: IOT, frequency spectrum, Cognitive Radio.

1 Introduction

As the second fusion of information and communication technology, the Internet of things is the integrate product of information sensing, transformation and processing. When the concept is put forward, the governments, enterprises and academic institutes pay high attention to it. It quickly spreads all over the world under the push of demand and research. The rapid development of the IOT needs the support of diverse wireless technology. As a result, it proposed new challenges to the management and utilization of the spectrum resources. The IOT will be used in all kinds of industry and scene in the future. Thus, it will bring mass terminal and form far greater wireless access data than human communication interconnected mobile communication. If not taken seriously, the shortage of spectrum resources will become the bottleneck of the IOT's development in the near future.

Because the Cognitive Radio(CR) technology can automatically detect radio environment, adjust the transmission parameters, multidimensional share the wireless spectrum from space, time, frequency, modulation mode, it dramatically improve spectrum efficiency. Apply CR to the IOT will better meet the increasing demand of frequency. In view of this, we review the realization platform for cognitive radio in order to better use it in the IOT.

The emerging of cognitive radio starts a new situation for solving the frequency resources shortage, realizing the dynamic management and improving the frequency spectrum efficiency. As the U.S. FCC open the TV white space, the mature products based on cognitive radio becoming more and more people's pressing needs. As a result, the researchers at home and abroad and relevant organizations made specific research, developed some hardware platform. The platform supports the practical application of CR, especially the IOT.

D. Yang (Ed.): Informatics in Control, Automation and Robotics, Volume 2, LNEE 133, pp. 825–833.
springerlink.com

2 Current Cognitive Radio Platforms

From the hardware to classification, it can be classified into two categories: one is based on the USRP, the other is based on non-USRP.

Classify the platform according to the USRP is because, using the Universal Software Radio Peripheral as a signal processing software platform has the following several advantages : complete software signal processing has very big flexibility; use general high level language (such as C/C++) for software development, the extendability and portability is strong, the development cycle is short; based on general hardware platform, the cost is low, and can enjoy computer technology progress with all kinds of advantages[1].

2.1 USRP (Universal Software Radio Peripheral) [2]

The Universal Software Radio Peripheral (USRP) is produced by Ettus Research, it enables engineers to rapidly design and implement powerful, flexible software radio systems. A typical USRP product family includes two parts: one with high speed signal processing FPGA motherboard, and one or more daughter boards cover different frequencies range. They together realized from the bit stream to host computer (i.e. antenna receiver), or from the host computer transmit to the antenna (namely send).

In all kinds of daughter boards, the USRP product family cover from DC to 5.9 GHz, including the whole range frequencies from AM radio to far than Wi Fi.

The true value of the USRP is in that it enables engineers and designers to create on a low budget and with a minimum of effort. A large community of developers and users have contributed to a substantial code base and provided many practical applications for the hardware and software. The USRP and GNU Radio, powerful combination of flexible hardware, open-source software and a community of experienced users makes it the ideal platform for software radio development.

GNU Radio is a free software development toolkit that provides the signal processing runtime and processing blocks to implement software radios using readily-available, low-cost external RF hardware and commodity processors [3]. It is a completely open software radio project, initiated by the Eric Blossom, which aims to encourage global technicians to collaborate and innovate in this field. At present, it already has certain influence. It primarily based on Linux operating system, can also be transplanted into other operating system, using c + + combining Python script for programming, and its code completely open [4].

2.2 Cognitive Radio Platforms Based on USRP

The CR platforms based on USRP in the existing publications:

(1) Hydra[5] of UT Austin

Hydra is a flexible wireless network testbed being developed at UT Austin. It focuses on networks that support multiple wireless hops and where the network, especially the MAC, takes advantage of sophisticated PHY techniques, such as OFDM and MIMO. Using the frameworks of the Click modular router and GNU Radio and coding in C++ makes it relatively easy to implement working prototypes of cross-layer designs that require custom MACs and PHYs.

Hydra's RF front-end is the Universal Software Radio Peripheral for use with GNU Radio. PHY is implemented in C++ using the GNU Radio framework and MAC is implemented in C++ using the Click modular router framework [6].

All of the communication protocols for Hydra are implemented in a software framework that runs on a general purpose processor(GPP) instead of FPGA or ASIC. Thus, it has low development time and accessibility, also can reduces cost.

(2) IRIS of Trinity College Dublin[7]

IRIS (Implementing Radio In Software) has been under development at Trinity College Dublin since 1999. It is a highly flexible and highly reconfigurable software radio platform for a GPP running either Windows or Linux.

Focusing on runtime reconfiguration, Iris offers support for all layers of the network stack and provides a platform for the development of not only reconfigurable point-to-point radio links but also complete networks of cognitive radios.

Iris is designed specifically to support maximum reconfigurability while the radio is running. Iris was originally designed and implemented in2004 (v1.0) . In 2008, they redesigned the Iris 2.0 with increased flexibility. The runtime system and components are implemented completely in portable C++[8].

(3) VT-CORNET[9] of VT

CORNET(Cognitive Radio Network Testbed) is a unique heterogeneous wireless communication network testbed based on the cognitive radio build up by Wireless@VT(Virginia Tech). The network includes 48 radio nodes distributed in the 4-layer building. Key goal in cognitive engine design, self-organization network algorithm and network security [10]. USRP2 used as the interface between PC and RF front-end. The frequency of custom RF front-end (Motorola transceiver) ranges from 100MHz to 4GHz [11].

The testbed caters to the needs of the CR research, develops wireless systems on medium scale radio system, emulates wireless systems and integrates new ones, such as distributed CR operation, CR network security, spectrum coordination and dynamic spectrum access tests[12]. The testbed current in operation is version 1.0.

(4) ORBIT[13] of Rutgers University WINLAB (Wireless Information Network Laboratory)

ORBIT (Open Access Research Testbed for Next-Generation Wireless Networks) is a large-scale wireless network testbed with 400nodes[14]· In October 2005, the research team made it public for everyone to use. In 2009, it incorporated 10 USRP boards,50 high-performance USRP2 and large areas of programmable WiMax base station. Each node consists of a small form factor PC and several 802.11 a/b/g network cards. The testbed is laid out in a 20 × 20 grid with each node separated by 1 m. The research on ORBIT is primarily focused on higher-layer network protocols. A valuable feature of

ORBIT is that it is accessible from a web interface and refined for remote experiments.

ORBIT uses a two-tier architecture with a lab-emulator matrix and field-trial network. The emulator matrix is constructed with a large two-dimensional array of static IEEE802.1 la/b/g radio nodes, which can dynamically be interconnected into a specified topology for the targeted wireless network [15-18].

(5) ADROIT of BBN Technologies

The ADROIT(Adaptive Dynamic Radio Open-source Intelligent Team) project is building an open-source software-defined data radio, intended to be controlled by cognitive applications[19]. It desired to create cognitive radio teams and have an open-source system. It incorporated the neural nets to improve throughput in simulation [20].

Table 1. Platforms based on the USRP

Platform	Development environment	Frequency range	Function layer	Main function
Hydra	GNU Radio Click C++	Multiple frequency (400-500MHz, ISM, etc)	PHY MAC	PHY ➔ GNU Radio ➔ C++ MAC ➔ Click ➔ C++
IRIS	IEEE 802.22 C++ OFDM	Multiple frequency (54-862MHz)	PHY	Opportunistically spectrum access Cross-layer optimization
VT-CORNET	GNU Radio OSSIE	Multiple frequency (100MHz-4GHz)	PHY	Medium scale Radio system Emulate wireless system
ORBIT	IEEE 802.11a C++	Single frequency (2.4GHz)	PHY MAC	Two-tier testbed, indoor testbed and outdoor ones
ADROIT	GNU Radio Click C++	Single frequency (2.4GHz)	PHY MAC	create cognitive radio teams open-source system

2.3 Cognitive Radio Platforms Based on Non-USRP

(1) WARP [21]of Rice University

WARP (Wireless Open-Access Research Platform) is an open-access and scalable wireless community platform. It is a complete system containing custom hardware, support package, design tools and application library. It uses Xilinx Virtex-II Pro FPGA to process digital signal optimally. As an open platform, WARP has been adopted by many universities, companies and institutes. Institute for Networked Systems of RWTH Aachen University in Germany used WARP for CR wireless network research, they developed new CR MAC protocol, experimented on all kinds of distributed and centralized CR wireless network resource management algorithm, studied various spectrum sensing problem, such as use machine learning to classify the waveform [22-24].

(2) KUAR[25] of Kansas University

Kansas University Agile Radio (KUAR) is an experimental SDR platform that is intended for research in frequency- agile and cognitive radios. The KUAR composed of three separate printed circuit boards: the digital signal processing board, power supply board, and RF board. On the boards, there are five main subsystems: power supply, control and process host, data board with programmable signal processor and D\A and A\D, RF transceiver, emitting antenna. It has low cost, flexible RF front-end, portable and powerful computation ability. The main digital communication and process operation is accomplished by the FPGA of Xilinx [26].

(3) WiNC2R[27] of WINLAB(Rutgers University)

The WINLAB Network Centric Cognitive Radio Platform (WiNC2R) is started from September of 2004, as a four year teamwork of WINLAB, Lucent Bell-Labs and GA Tech, aimed to design a high-performance CR platform with complete PHY and network layer function. It based on the virtual flow pipeline with increased design flexibility and software definition function. The WINC2R board includes baseband processor (modem module), networking module, RF module. It used the Xilinx Virtex-4 SX series FPGA（e.g.XC4SX35） for physical layer baseband processing, the networking module as a daughter board can be plugged into the modem module ,using Xilinx 4FX12 FPGA[28].

(4) OCRA[29,30] of GEDC (GIT)

On 25th April 2006, Georgia Institute of Technology presented the world's first complete cognitive radio platform. The Broadband and Wireless Networking Laboratory developed the CR testbed OCRA (OFDM-based cognitive radio network) based on OFDM. The testbed contains cognitive user accomplished by Cognitive Transceiver Unit, cognitive base station and various network modules. The function modules of Cognitive Transceiver Unit including modems, RF front-end and cross-layer cognition realized by software and FPGA[31].

(5) BWRC-BEE2[32]

BWRC (Berkeley wireless research center) has set up a testbed to verify sensing algorithms and presented a indoor CR network working prototype. The testbed based on BEE2, has multiple FPGA simulation environment, can connect 18 radio RF front-end, validates various baseband sensing algorithms and cooperative sensing scheme. They selected 2.4GHz for experimentation at first, used the existing devices on the BEE2 and simulated easily controlled main user [33]. They focused on energy and cyclostationary detectors and indoor network cooperation [34].

Berkeley Emulation Engine2 (BEE2)[35] is a generic multipurpose FPGA based, emulation platform for computationally intensive applications[36]. BEE2 consists of 5 Vertex-IIPro70 FPGAs and integrates 500 giga-operations per second. One FPGA is used for control, while the other four are targeted for user applications.

Table 2. Platforms based on non-USRP

platform	Main technology	Frequency range	Main description
WARP	OFDM	Multiple frequency	Open-access scalable Contains custom hardware, support package, design tools and applications library
KUAR	NC-OFDM	Multiple frequency (5-6GHz, 2.4GHz)	Low-cost flexible RF front-end portable Powerful computation
WiNC2R	OFDM QPSK/DSSS	Multiple frequency	Contains baseband processing (modem modules), networking module, RF module
OCRA	OFDM	Multiple frequency (54MHz-6GHz)	Contains cognitive user, cognitive base station and various networking module
BWRC-BEE2	OFDM	Single frequency (2.4GHz)	Based on BEE2 , can connect 18 radio front-end

3 Domestic and Other Research on Realization of Cognitive Radio

In [37], they proposed an architecture that combines the best of both simulators/emulators and real-time test beds for scaling up the experiments while keeping testing as realistic as possible with the harsh wireless environments. At present, it is based on USRP, there are 5 USRP2.0 and the RF board is plugged into the main FPGA board. It currently working in 2.4 GHz ISM band, in the future plans to use 5GHz and other unauthorized frequency band.

In [38], they presented a heterogeneous and reconfigurable cognitive radio platform based on SoC (System on Chip). It contains GPP (general purpose processor), DSRH (domain specific reconfigurable hardware), DSP, FPGA and ASIC. Compared with the single reconfigurable processing platform, this heterogeneous reconfigurable cognitive radio platform has such merits: dynamic configuration, high efficiency and energy saving, etc. But this kind of heterogeneous SoC realization design method is different from the single processor or other similar systems. This test bed mainly realizes the physical function, its realization complexity is relative higher and not include the performance research above the MAC layer.

In thesis [39], the author conducted research and design on the technology difficulty of digital signal processing in CR. He realized the function of digital signal processing on FPGA, designed and implemented Numerical Controlled Oscillator module (NCO), Cascaded Integrator Comb module (CIC), Finite Impulse Response module (FIR) and Fast Fourier Transform module(FFT).

4 The Future and Application Prospects of Cognitive Radio Platform

In conclusion, there are two main methods for realizing CR hardware platform. One is based on the generic hardware platform, like USRP, with low cost, short development cycle, easy to learn. It suits for beginners and general staff, but it's performance and flexibility is limited. Another is based on programmable logic devices, like DSP, FPGA and ARM. We can choose different board according to need to realize all sorts of function. But due to the development language is different, its' compatibility is not strong. Thus researchers need to have professional knowledge in order to better use the devices.

In the future, the two methods will be neck and neck, won't reciprocal. On the one hand, the universal platform also is in developing, with new products and function, like USRP2; On the other hand, the programmable devices are under great development, DSP, FPGA, ARM have developed a series of products with more powerful functions and better performance.

Information is developing rapidly, and the trend of global economic integration is more and more obvious, the high technology fields of IOT will be the commanding heights of the race for all countries. Along with the development of the IOT, it will form a IOT communication framework based on software radio and cognitive radio[40,41]. At present, the study of cognitive radio platform focuses on the TV spectrum and 2.4 GHz ISM band. Future, we will open and use more frequencies in order to meet the needs of the developing IOT.

5 Conclusion

The IOT technologies developing rapidly and the spectrum resources are in great shortage, it all urge us to set up the cognitive radio platform facing the IOT. In this paper, we review the cognitive radio hardware platform at home and abroad, make a classification and contrast, prepare to build cognitive radio platform for IOT. In the future, a high efficiency hardware platform will be build according to the practical need of IOT.

References

1. Cao, H., Wang, W.: GNU Radio: open software platform. J. Telecommunications Information 4, 31–34 (2007)
2. Ettus Research LLC, `http://www.ettus.com/`
3. GNU Radio, `http://gnuradio.microembedded.com/chinese`
4. Blossom, E.: GNU Radio: Tools for Exploring the Radio Frequency Spectrum (EB/OL). Linux Journal (2004)
5. Hydra: A MIMO-OFDM Multihop Testbed, `http://netlab.ece.utexas.edu/hydra/`
6. Mandke, K., Choi, S.-H., Kim, G., Grant, R., Daniels, R.C., Kim, W., Nettles, S.M., Heath Jr., R.W.: Early Results on Hydra: A Flexible MAC/PHY Multihop Testbed. In: Proceedings of the 65th IEEE Vehicular Technology Conference, pp. 1896–1900. IEEE Press (2007)

7. Sutton, P.D., Lotze, J., Lahlou, H., Fahmy, S.A., Nolan, K.E., Ozgul, B., Rondeau, T.W., Noguera, J., Doyle, L.E.: Iris: an architecture for cognitive radio networking testbeds. IEEE J. of Communications Magazine 48(9), 114–122 (2010)
8. Sutton, P.D., Lotze, J., Lahlou, H., Ozgul, B., Fahmy, S.A., Nolan, K.E., Noguera, J., Doyle, L.E.: C.: Multi-platform demonstrations using the Iris architecture for cognitive radio network testbeds. In: 2010 Proceedings of the Fifth International Conference Cognitive Radio Oriented Wireless Networks & Communications (CROWNCOM), pp. 1–5 (2010)
9. VirginiaTech, `http://www.wireless.vt.edu/research/Cognitive_Radios_Networks/`
10. Cognitive Wireless Technology, `http://www.cognitiveradio.wireless.vt.edu/dokuwiki/doku.php?id=home`
11. Integrated SCM & Project Management, `http://cornet.wireless.vt.edu/trac/`
12. Rieser, C.J., Rondeau, T.W., Bostian, C.W.: Cognitive Radio Testbed: Further Details and Testing of a Distributed Genetic Algorithm Based Cognitive Engine for Programmable Radios. In: IEEE MILCOM 2004, October 31-November 4, vol. (3), pp. 1437–1443 (2004)
13. ORBIT, `http://www.orbit-lab.org/`
14. Raychaudhuri, D., Seskar, I., Ott, M., Ganu, S., Ramachandran, K., Kremo, H., Siracusa, R., Liu, H., Singh, M.: Overview of the ORBIT Radio Grid Testbed for Evaluation of Next-Generation Wireless Network Protocols. In: IEEE WCNC 2005 (2005)
15. Singh, M., Ott, M., Seskar, I., Kamat, P.: ORBIT Measurements Framework and Library (OML): Motivations, Design, Implementation, and Features. IEEE Tridentcom (2005)
16. Ott, M., Seskar, I., Siracusa, R., Singh, M.: ORBIT Testbed Software Architecture: Supporting Experiments as a Service. IEEE Tridentcom (2005)
17. Ganu, S., Kremo, H., Howard, R., Seskar, I., Addressing Repeatability, S.: in Wireless Experiments using ORBIT Testbed. IEEE Tridentcom (2005)
18. Mahindra, R., Bhanage, G., Hadjichristofi, G., Seskar, I., Raychaudhuri, D.: Integration of heterogeneous networking testbeds. IEEE Tridentcom (2008)
19. Troxel, G.D., et al.: Adaptive dynamic radio open-source intelligent team (adroit): Cognitively-controlled collaboration among sdr nodes. In: First IEEE Workshop on Networking Technologies for Software Defined Radio (SDR) Networks, pp. 8–17. IEEE (2006)
20. Troxel, G.D., Caro, A., Castineyra, I., Goffee, N., Haigh, K.Z., Hussain, T., Kawadia, V., Rubel, P.G., Wiggins, D.: Cognitive Adaptation for Teams in ADROIT. In: Global Telecommunications Conference, GLOBECOM 2007, pp. 4868–4872. IEEE (2007)
21. WARP, `http://warp.rice.edu/`
22. Amiri, K., Yang, S., Murphy, P., Hunter, C., Cavallaro, J.R., Sabharwal, A.: WARP, a Unified Wireless NetworkTestbed for Education and Research. In: Microelectronic Systems Education, MSE 2007, pp. 53–54. IEEE (2007)
23. Murphy, P., Sabharwal, A., Aazhang, B.: Design of WARP:a Wireless Open-Access Research Platform. In: Proc. EURASIPXIV European Signal Processing Conference (2006)
24. Gupta, S., Murphy, P., Hunter, C., Sabharwal, A.: Cognitive Radio Platforms: Wireless Open-Access Research Platform for Networks
25. `https://agileradio.ittc.ku.edu`
26. Minden, G.J., Evans, J.B., Searl, L., DePardo, D., Petty, V.R., Rajbanshi, R., Newman, T., Chen, Q.: KUAR: A Flexible Software-Defined Radio Development Platform. In: 2nd IEEE Symposium on New Frontiers in Dynamic Spectrum Access Networks (DySPAN), Dublin, Ireland (2007)

27. WiNC2R,
http://www.winlab.rutgers.edu/pub/docs/focus/WiNC2R.html
28. Miljanic, Z., Seskar, I., Le, K., Raychaudhuri, D.: The WINLAB Network Centric Cognitive Radio Platform – WiNC2R. In: Proc. CrownComm. (2007)
29. Cognitive Radio Networks,
http://www.ece.gatech.edu/research/labs/bwn/CR/testbed.html
30. http://www.gedcenter.org/news14.html
31. Akyildiz, I.F., Li, Y.: OCRA: OFDM-based cognitive radio networks, Broadband and Wireless Networking Laboratory Technical Report (2006)
32. Cognitive Radio Research,
http://bwrc.eecs.berkeley.edu/Research/Cognitive/default.htm
33. Mishra, S.M., Cabric, D., Chang, C., et al.: A Real Time Cognitive Radio Testbed for Physical and Link Layer Experiments, 562–567 (2005)
34. Tkachenko, A., Cabric, D., Brodersen, R.W., Asilomar, C.: Cognitive Radio Experiments using Reconfigurable BEE2. In: Conference on Signals, Systems and Computers, pp. 2041–2045 (2006)
35. BEE2, http://bee2.eecs.berkeley.edu/
36. Chang, C., Wawrzynek, J., Brodersen, R.W.: Bee2: A high-end reconfigurable computing system. IEEE Design and Test of Computers 22(2), 114–125 (2005)
37. Budihal, R., Jamadagni, H.S.: Hybrid approach to Cognitive radio test bed. In: Ultra Modern Telecommunications & Workshops, ICUMT 2009, pp. 1–7 (2009)
38. Hu, S., Wu, Y.: A heterogenes reconfigurable cognitive radio platform and its design method. Gui zhou Normal University 26(1), 73–76 (2008)
39. Wu, J.: Design and realization of Cognitive Radio hardware platform. J. Beijing University of Posts and Telecommunications (2010) (in Chinese)
40. Ning, H., Xu, Q.: Research on Global Internet of Thing's Development and It's Lonstruction in China. Acta Electronica Sinica 38, 11 (2010)
41. Gu, J., Chen, S., Zhuang, Y.: Wireless sensor network-based topology structures for the Internet of Things localization. Chinese Journal of Computers 33(9) (2010) (in Chinese)

Author Index